MICROMACHINED TRANSDUCERS

SOURCEBOOK

McGraw-Hill Series in Electrical and Computer Engineering

Senior Consulting Editor
Stephen W. Director, University of Michigan, Ann Arbor

Circuits and Systems
Communications and Signal Processing
Computer Engineering
Control Theory
Electromagnetics
Electronics and VLSI Circuits
Introductory
Power and Energy
Radar and Antennas

Previous Consulting Editors

Ronald N. Bracewell, Colin Cherry, James F. Gibbons, Willis W. Harman, Hubert Heffner, Edward W. Herold, John G. Linvill, Simon Ramo, Ronald A. Rohrer, Anthony E. Siegman, Charles Susskind, Frederick E. Terman, John G. Truxal, Ernst Weber, and John R. Whinnery

Electronics and VLSI Circuits

Senior Consulting Editor
Stephen W. Director, University of Michigan, Ann Arbor

Consulting Editor
Richard C. Jaeger, Auburn University

Colclaser and Diehl-Nagel: *Materials and Devices for Electrical Engineers and Physicists*
DeMicheli: *Synthesis and Optimization of Digital Circuits*
Elliot: *Microlithography: Process Technology for IC Fabrication*
Fabricius: *Introduction to VLSI Design*
Ferendici: *Physical Foundations of Solid-State and Electron Devices*
Fonstad: *Microelectronic Devices and Circuits*
Franco: *Design with Operational Amplifiers and Analog Integrated Circuits*
Geiger, Allen, and Strader: *VLSI Design Techniques for Analog and Digital Circuits*
Grinich and Jackson: *Introduction to Integrated Circuits*
Hodges and Jackson: *Analysis and Design of Digital Integrated Circuits*
Huelsman: *Active and Passive Analog Filter Design: An Introduction*
Ismail and Fiez: *Analog VLSI: Signal and Information Processing*
Kasap: *Principles of Electrical Engineering Materials and Devices*
Kovacs: *Micromachined Transducers Sourcebook*
Laker and Sansen: *Design of Analog Integrated Circuits and Systems*
Long and Butner: *Gallium Arsenide Digital Integrated Circuit Design*
Millman and Grabel: *Microelectronics*
Millman and Halkias: *Integrated Electronics: Analog, Digital Circuits, and Systems*
Millman and Taub: *Pulse, Digital, and Switching Waveforms*
Neamen: *Electronic Circuit Analysis and Design*
Neamen: *Semiconductor Physics and Devices*
Ng: *Complete Guide to Semiconductor Devices*
Offen: *VLSI Image Processing*
Roulston: *Bipolar Semiconductor Devices*
Ruska: *Microelectronic Processing: An Introduction to the Manufacture of Integrated Circuits*
Schilling and Belove: *Electronic Circuits: Discrete and Integrated*
Seraphim: *Principles of Electronic Packaging*
Singh: *Optoelectronics: An Introduction to Materials and Devices*
Singh: *Physics of Semiconductors and Their Heterostructures*
Singh: *Semiconductor Devices: An Introduction*
Singh: *Semiconductor Optoelectronics: Physics and Technology*
Smith: *Modern Communication Circuits*
Sze: *VLSI Technology*
Taub: *Digital Circuits and Microprocessors*
Taub and Schilling: *Digital Integrated Electronics*
Tsividis: *Mixed Analog-Digital VLSI Devices and Technology*
Tsividis: *Operation and Modeling of the MOS Transistor*
Wait, Huelsman, and Korn: *Introduction to Operational and Amplifier Theory Application*
Yang: *Microelectronic Devices*
Zambuto: *Semiconductor Devices*

MICROMACHINED TRANSDUCERS SOURCEBOOK

Gregory T.A. Kovacs

Stanford University

Boston Burr Ridge, IL Dubuque, IA Madison, WI New York San Francisco St. Louis
Bangkok Bogotá Caracas Lisbon London Madrid
Mexico City Milan New Delhi Seoul Singapore Sydney Taipei Toronto

WCB/McGraw-Hill

A Division of The **McGraw-Hill** *Companies*

Micromachined Transducers Sourcebook

4 5 6 7 8 9 0 DOC DOC 3 2

ISBN 0-07-290722-3

Editorial director: Kevin Kane
Publisher: Tom Casson
Sponsoring editor: Lynn Cox
Editorial coordinator: Nina Kreiden
Marketing manager: John Wannemacher
Project manager/production supervisor: Natalie Durbin
Cover designer: Francis Owens
Printer: R. R. Donnelley & Sons Company

Library of Congress Cataloging-in-Publication Data

Kovacs, Gregory T. A.
    Micromachined transducers sourcebook / Gregory T.A. Kovacs.
        p.   cm.
    Includes bibliographical references and index.
    ISBN 0-07-290722-3
    1. Transducers.   2. Microelectromechanical systems.   I. Title
  TK7872.T6K68 1998
  681'.2--dc21
                                    98-4846
                                      CIP

Photo credits
Front Cover: Scanning electron micrograph of a single-crystal silicon leaf-spring that has been fabricated using a deep reactive ion etching technique. The structure is 200 microns tall.
Courtesy Dr. Nadim Maluf, Lucas Novasensor, Fremont, CA.

Back Cover: Top - Scanning electron micrograph of a membrane of silicon dioxide, with an underlying support mesh of boron-doped silicon. Center - Scanning electron micrograph of a silicon substrate with an array of holes fabricated using plasma etching. Bottom - Scanning electron micrograph of a single-crystal silicon accelerometer. The inset shows the silicon beams of the sensing capacitor plates.
Courtesy Dr. Nadim Maluf, Lucas Novasensor, Fremont, CA.

http://www.mhhe.com

This book is dedicated to my parents, George and Klara Kovacs, who raised me with love and inspired me to wonder about the world around me.

ABOUT THE AUTHOR

GREGORY T. A. KOVACS received the B.A.Sc. degree in Electrical Engineering from the University of British Columbia, the M.S. degree in Bioengineering from the University of California, Berkeley, the Ph.D. degree in Electrical Engineering and the M.D. degree from Stanford University.

He is an Associate Professor of Electrical Engineering at Stanford University, where he has been a member of the faculty since 1991. He teaches courses in electronic circuits and micromachined transducers. His present research areas include solid-state sensors and actuators, micromachining technologies, biological and medical applications of fluidic devices, and analog circuits for transducer applications, all with emphasis on solving practical problems. He has authored over eighty technical publications and holds several patents. He held the Robert N. Noyce Family Faculty Scholar Chair from 1992 to 1994, received a National Science Foundation Young Investigator Award in 1993, was appointed a Terman Fellow in 1994, was appointed to the Defense Sciences Research Council in 1995, and was appointed a University Fellow in 1996. He has broad industry experience in the design of circuits and instruments, commercial product design, and intellectual property law consulting. In addition, he has been one of the founders of several technology companies, most recently Cepheid, Inc., in Sunnyvale, CA.

PREFACE

The field of micromachined sensors and actuators, often referred to as "MEMS" (microelectromechanical systems), has been growing at an exciting pace in recent years. Using tools originally developed for the silicon integrated circuit industry, people are now fabricating miniaturized transducers and structures from silicon and other materials. In many (but not all) cases, these new devices offer advantages over their "conventional" counterparts, including great reductions in size, new functions that could not otherwise be realized, the capability to include on-chip signal processing/control circuitry (if the substrate is a semiconductor), reductions in per-unit cost, and the ability to fabricate scaled and multimodal arrays. Of course, new technologies bring with them new problems, such as the need for new packaging and power sources, system integration issues that need to be addressed, the need for CAD and simulation tools, and process compatibility considerations. All of these challenges are gradually being dealt with.

This sourcebook began, as many books do, as the notes for a graduate course. The main goal I set for myself was to provide a fairly complete overview of the field, beginning with micromachining approaches and including all major categories of transduction (including some that may not be familiar to the reader). The approach I have taken is, where possible, to study the way individual devices were fabricated and what the key design issues were. In this way, I hope that the reader will gain an intuitive sense for how to develop his or her own processes and designs.

I also sought to provide sufficient examples of important transducers or structures for the reader to compare the performance obtainable through different approaches. In many cases this was not possible. In studying hundreds of papers in this area, spanning many years, I found the lack of metrics by which to compare different designs quite striking. Even for relatively simple devices, such as pressure sensors, where metrics for their macroscopic counterparts are well-established, reporting of performance in the literature is sparse and inconsistent. One could certainly speculate that reasons for this might be that the authors choose to omit poor characteristics of prototypes, have a need to protect proprietary information, or lack familiarity with existing specifications. In any case, it is likely that this situation will have to improve in order for a significant number of designs to succeed against competitive "conventional" devices.

This book does not cover simulation methods as they relate to micromachining. The tools for these purposes are still evolving and proving themselves useful to designers, and covering them is beyond the scope of this book. Second, circuits specific to transducer applications are not covered — one could write another entire book on that subject!

i

While the book is a fairly comprehensive overview of the field, it is not meant to be a complete catalog. I have carefully selected examples based on a number of criteria. I certainly wanted to focus on those with interesting operating principles, designs, or fabrication processes. I also sought to choose papers wherein, after close inspection, real experimental verification of device operation was provided (I decided not to cite several interesting examples where even basic functionality had not been verified — in this type of paper the word "simulated" appears next to the word "data"). It was quite interesting to follow up, via electronic searching, on many of the papers where a pretty picture was presented with some phrase like "testing is in progress." For a large number of them, there never was a follow-on publication, even many years later. But that is to be expected from a discipline that is young and dynamic. Priorities change, graduate students move on, and some things simply don't work. I have also included a few examples of devices that have not been micromachined but that could be.

Many people have made substantial contributions to this work. Most notably, Bruce Darling, who read and critiqued the entire manuscript twice. Invaluable comments, additions, and general advice on the text also came from Nadim Maluf, Tom Kenny, Kurt Petersen, Henry Baltes, Godfrey Mungal, Paul Yager, Roger Howe, Fred Forster, Dorian Liepmann, Chris Storment, Tony Flannery (who receives special thanks for his ongoing support), Dave Borkholder, Ken Honer, Micki Leder, and Josh Molho. I am also very grateful to my EE312 students for their helpful comments, for "beta testing," and for bringing many interesting papers and patents to my attention, particularly to Peter Ho, for providing many key references on thin-film batteries. Special thanks go to everyone who helped me with several of the figures, including Stephen Ryu, Tamara Ahrens, Sasha Mittelman, Peter Krulevitch, Rosanna Foster, Tony Flannery, Dave Borkholder, and Ken Honer. I am sure I have not remembered everyone, but I extend my most sincere thanks to those above and any I have regrettably forgotten to name. Last but not least, special thanks to my wife, Laurel Joyce, for putting up with my many grumpy or spaced-out, computer-fogged comments while I hacked away at this project, and for her massive assistance with the final set of changes to the manuscript.

Although I have made my best effort to bring this material together into a useful form, I have undoubtedly made errors of fact or omission, and sincerely apologize for them. I welcome hearing from readers about such errors, key concepts, papers that may have been omitted, format suggestions, etc., via electronic mail (kovacs@cis.stanford.edu). Updates, corrections and other information will be available at the McGraw-Hill web site: http://www.mhhe.com.

Here's to your enjoyment of this exciting field!

Gregory T. A. Kovacs
Stanford, CA, January, 1998

PROLOGUE

Kurt Petersen, Ph.D., Cepheid, Inc., Sunnyvale, CA

Several defining events independently contributed to the unambiguous emergence of microelectromechanical systems (MEMS) as a vital technical discipline in the early 1980's. During 1982 and 1983, the first mass-produced MEMS products were released into high volume production, the micromachined automotive MAP (Manifold Absolute Pressure) sensor and the micromachined disposable medical blood pressure sensor. In 1983, the first international conference on Solid State Sensors and Actuators was held at Delft Technical University in the Netherlands. In 1981 and 1982, the first start-up companies were founded that were committed to the commercialization of additional applications (other than pressure sensors) for this new technology, Microsensor Technology, Inc. and Transensory Devices, Inc. In 1982, the first review paper on the field was published in the *Proceedings of the IEEE*, "Silicon as a Mechanical Material" (Petersen (1982)). Pioneered by Professor James B. Angell at Stanford University (who had research projects in the field of micromachining dating from the early 1970's), every major university in the United States had a formally established major research program in MEMS by about 1983.

During these early years, only about five Fortune 500 companies had research programs in MEMS; today over 25 such companies are contributing to the technology. During this time, only one type of MEMS product existed, the micromachined pressure sensor; today accelerometers, micro-valves, projection display chips, biosensors, ink jet nozzle arrays, and other products are all manufactured and shipped in commercial volumes. During this time, fundamental MEMS processes were invented, researched, and defined; today, these processes are being refined in manufacturing environments and are being applied to additional products. During this time, little technical and communication infrastructure existed; today, several MEMS journals are thriving and an entire series of scheduled conferences are held every year. During this time, MEMS experts were few and far between; today, well-established and well-respected university programs graduate dozens of Ph.D. MEMS students every year in the United States alone. During this time, government research funding was sparse, unorganized, and low profile; today, funding from DARPA and other government agencies amounts to over $100M/year and is targeted and organized around specific militarily and commercially strategic applications.

As a result of all this progress, the future of micromachining and MEMS is more than bright — it is dazzling. With a strong infrastructure in place, well-established university programs, many trained and experienced engineers, and a proven diverse series of successful products, a strong momentum has developed which is beginning to change the nature of many large technical disciplines, including biotechnology, storage technology, instrumentation, input/output, and telecommunications.

Because MEMS technology employs many of the same manufacturing techniques as integrated circuits, the business of MEMS is often compared to the IC business. However, this comparison is misleading. The integrated circuit is used only to manipulate electronic signals. MEMS devices are actually the interface between the physical world and the electronic world. As such, they must function in a much wider, more diverse, and much more complicated overall environment, interacting effectively and accurately both in the electronic and in the physical worlds. On one hand, this reality greatly complicates the design, production implementation, and performance qualifications for any single application. On the other hand, a vastly broader and more diverse range of utility for MEMS devices, from optical fiber switching, to fluid control, to micromechanical electrical relays, to biosensors is made possible.

In fact, no single aspect characterizes the field of MEMS more succinctly than its wide diversity. This diversity is expressed both in the multi-disciplinary approach required for design, engineering, and manufacturing as well as in the wide range of markets and application areas that employ MEMS devices. The most important technical disciplines required for MEMS research and development include electrical engineering, mechanical engineering, integrated circuit processing, circuit design, materials science, chemistry, instrumentation, fluidic engineering, optics, and packaging. This diverse range of technical disciplines creates a very complex and challenging background from which devices and systems must be designed, built, and tested. In addition, the important markets and application areas for MEMS devices and systems are also very diverse, including sensors, fluid control and manipulation, optics, displays, printing, electrical switching, chemical analysis, biochemical fluid processing, precise mechanical motion and actuation, and data storage systems. Such a wide diversity of engineering disciplines and varied fields of application creates a challenging situation for anyone trying to understand, summarize, and learn MEMS technology.

Previous efforts to review MEMS technology have been made either very early in the history of the technology (Petersen (1982)), in limited reviews (for example, reviews of micromachined sensors in Ristic (1994)), or in anthologies of important papers (as in Muller, et al. (1991), and Trimmer (1997)). In fact, the very prospect of reviewing the entire field of MEMS, entirely because of the vast breadth and diversity of the technology (especially in recent years), has been viewed as a daunting task to say the least.

With this backdrop, Gregory Kovacs has undertaken the ambitious and formidable task of comprehensively reviewing the current status of micromachined transducers and MEMS technology in a single book. The effort is phenomenally successful. This book is the most thorough review of micromachining technology available. It meticulously covers virtually every proposed application of MEMS. It is unusually detailed and incisive, delving into the physics and theory behind the applications and into the reasons and rationale for miniaturization. It is extremely well-researched and referenced with well over 1,500 references. No published work summarizes and characterizes the diversity of the MEMS field better and more completely than *Micromachined Transducers Sourcebook*.

"Silicon as a Mechanical Material" served as a technical introduction and reference during the nascent years of MEMS development. The comprehensive *Micromachined Transducers Sourcebook* will serve as the primary technical reference for the next ten years. If you will be working in the field of MEMS, or using MEMS devices or products, this book is indispensable.

Kurt Petersen
San Jose, CA, January 1998

PROLOGUE REFERENCES

Muller, R. S., Howe, R. T., Senturia, S. D., Smith, R. L., and White, R. M. [eds.], "Microsensors," IEEE Press, New York, NY, 1991.

Petersen, K. E., "Silicon as a Mechanical Material," Proceedings of the IEEE, vol. 70, no. 5, May 1982, pp. 420 - 457.

Ristic, Lj. [ed.], "Sensor Technology and Devices," Artech House, London, 1994.

Trimmer, W. S., "Micromechanics and MEMS: Classic and Seminar Papers to 1990," IEEE Press, New York, NY, 1997.

TABLE OF CONTENTS

Chapter 1:
INTRODUCTION AND OVERVIEW

1. INTRODUCTION TO MICROMACHINED DEVICES

The use of lithographic and other microfabrication technologies to create miniaturized sensors, actuators, and structures has become increasingly popular in many areas of science and engineering. In order to fabricate such devices, the addition, subtraction, modification, and patterning of materials are typically done using techniques originally developed for the integrated circuit industry. In the late 1960's, researchers began to appreciate the fact that silicon and other semiconductors could be used to fabricate not only discrete and integrated electronic circuits, but also transducers and other devices with new properties due to the materials used and their miniature size. The term *micromachining* broadly refers to the use of lithographic and other precision techniques to carry out such fabrication. (The resulting devices are sometimes referred to as "MEMS," for microelectromechanical systems, although the term is misleading since many micromachined devices are not mechanical in any sense.)

The interdisciplinary nature of both micromachining techniques and their applications can and does lead to exciting synergies. Use of these technologies has resulted in an unprecedented range of devices that can be employed in applications through either displacement of macroscopic competitors or by enabling functions that are otherwise impossible. It is the latter case where the use of micromachining can be most effective in creating new capabilities and products.

In many cases, the use of these miniaturization technologies confers advantages beyond the obvious decreases in physical volume and weight, such as increased performance and reliability, and decreased cost. However, this is certainly not always the case for a number of reasons, discussed below. For the newcomer to micromachining or the seasoned professional, a key aspect of successful design is the *a priori* consideration of the important issues of scaling device properties, the need to integrate circuitry, packaging, testability, and many others. In addition, if performance comparisons are useful, it is critical to choose appropriate metrics, which are not always obvious for some classes of devices.

2. WHAT ARE TRANSDUCERS?

Transducers, by definition, convert one form of energy into another. This term encompasses both sensors and actuators. For practical purposes, a transducer is a conduit for transforming energy between two or more domains (i.e., chemical to electrical). In some cases, an intermediate transformation is required, as in a sensor that converts light to heat and then to electricity. The noun *transducer* and the verb *transduce* are relatively new in the English language, as can be seen from the definitions in *Webster's Ninth New Collegiate Dictionary*:

---

transduce \tran(t)s-'d(y)üs, tranz-\ vt trans•duced; trans•ducing [L transducere to lead across, transfer, fr. trans- + ducere to lead] (1947) 1: to convert (as energy or a message) into another form...

transducer \-d(y)ü-ser\ n (1924): a device that is actuated by power from one system and supplies power usually in another form to a second system (as a telephone receiver that is actuated by electric power and supplies acoustic power to the surrounding air)

In practice, the vast majority of work in the field has been on sensors rather than actuators. Sensors measure something in their environment and (generally) provide an electrical output that relates to the parameter(s) they measure. As discussed in this book, there are a large number of things that can be sensed, including the broad areas of mechanical, optical, (electro)magnetic, ionizing radiation, thermal, and chemical phenomena. Some authors, such as Lion (1969), White (1987), and Ristic (1994) divide sensors into "domains" such as:

Thermal (temperature, heat, and heat flow)

Mechanical (force, pressure, velocity, acceleration, and position)

Chemical (concentration of chemicals, composition, and reaction rate)

Magnetic (magnetic field intensity, flux density, and magnetization)

Radiant (electromagnetic wave intensity, wavelength, polarization, and phase)

Electrical (voltage, current, and charge)

It is inevitable that, in applying such categories, there will be devices that either do not fit well into any category or belong in several. The organization of this book is partly by the type of input/output energy associated with the transducer (e.g., optical), but also by function (e.g., fluidics). The reader will quickly realize that there are considerable overlaps between domains in terms of applications. For example, a mechanical sensor such as a strain gauge can be considered a biomedical sensor if it is implanted in the body. Thus, extraordinary efforts to define a rigid "domain-based" taxonomy for micromachined devices was felt to be unwarranted.

Actuators are still a developing area in micromachining, since very few micromachined actuators have "actuated" objects other than themselves and even fewer have been used in commercial products. Actuators are, however, often incorporated into sensors (an example is the use of electrostatic or thermal expansion actuators to provide "self-test" capability for accelerometers).

Micromachined structures, such as holes, channels, cantilevers, membranes, etc., are also very useful and important as building blocks for more complex micromachined devices. In particular, they are being used in microfluidic systems, which are a relatively new development in the field.

3. COMPONENTS OF TRANSDUCER SYSTEMS

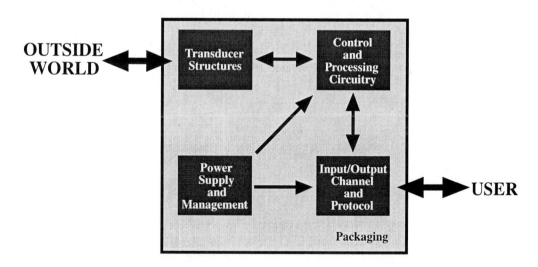

Illustration of the components of a transducer system.

The focus of this book is on the transducers themselves, but it is important to keep in mind the other components of real transducer systems, as illustrated above. Packaging for micromachined transducers, for example, tends to be very case-specific. In some instances, such as for inertial sensors, the packaging can be hermetically sealed. In other cases, such as pressure sensors, an opening, or port, is required, potentially subjecting the sensor to environmental degradation. In addition, testing is a major issue that needs to be considered when trying to build micromachined transducer systems. In some research circles, the bulk of the research efforts focus on the transducers themselves, yet moving them out into the "real world" requires some or all of the other aspects illustrated above. Naturally, these are prime considerations among industrial users.

4. WHAT ARE MICROMACHINED TRANSDUCERS?

As mentioned above, micromachined transducers are (generally) those that are *fabricated using tools and techniques developed for the integrated circuit industry*, such as microlithography, etching, etc., or some of the newer techniques developed specifically by and for the micromachining community. Micromachined devices are generally not at the forefront of lithography, typically requiring no features smaller than 1 μm, and generally much larger. So-called "nanotechnology" is beyond the scope of this book.

As discussed in the following chapter, micromachining can be grossly categorized into "bulk" and "surface." *Bulk micromachining* involves removal of significant regions of the substrate (e.g., etching cavities in a silicon substrate) and can thus be thought of as a *subtractive* process. *Surface micromachining* involves building up and patterning thin-film layers to realize the desired structures on the surface of the wafer (an *additive* process). As mentioned above, not all micromachining is done using tools for integrated circuits, and some "traditional" methods, such as electric discharge machining or injection molding, have been modified down to produce micron-scale features.

5. WHEN DOES IT MAKE SENSE TO "MICROMACHINE"?

When surveying the rapidly growing body of literature related to micromachining, many publications will cause the reader to question the need to micromachine a device in the first place. This is certainly a more common question among industrial, as opposed to academic workers. Sometimes, an "advanced, micromachined transducer" actually offers no advantages in terms of performance, size/weight, reliability, cost, etc., and such devices are generally not worth fabricating unless they are stepping stones to more successful designs. Whether or not a device "should" have been micromachined based on the measured performance, cost, size, path to future micromachined devices, etc., is a healthy question to ask.

Quite often, micromachining a transducer can confer great advantages in one or more areas, and the key is to evaluate the benefits of micromachining *before* undertaking a project. It is a key goal of this book to provide the reader with some of the background to make such judgments. Since there is no simple, generic "algorithm," the only practical way to get to this point is to obtain an overview of the field and then compare proposed micromachined devices to others (both micromachined and "conventional") in terms of performance. A general approach to this is to learn the underlying physical mechanisms of importance to a proposed device, to understand the scaling of desired and undesired (parasitic) effects as the design is miniaturized, and, if things make sense, to plan a realistic fabrication process. While a book of this nature cannot possibly cover the entirety of these areas, it can at least help the reader build intuition.

As stated by Prof. Kurt Petersen during his address at the Transducers '95 conference in Stockholm, Sweden, "MEMS has competition." The point is that one is not always micromachining a device that is unique. Typically, there is another way to accomplish the same result. In cases where "there is no other way," or there are clear advantages, micromachining can be a very powerful technique.

5.1 SCALING AND PERFORMANCE

In terms of performance, while some physical effects scale favorably into the microns-to-millimeters scale of micromachined devices, others do not. Thermal transport properties of microstructures allow for local regions to be highly coupled or isolated, and this can be a problem or a great benefit. Enhanced mass transport properties of electrochemical microelectrodes can be extremely advantageous and are readily harnessed. Increased surface-area-to-volume ratios at the microscale often provide tremendous improvements in separation techniques such as capillary electrophoresis or gas chromatography. However, while fluidic channels can be scaled down, there are dimensions at which bubbles cannot be purged for any reasonable applied pressures due to capillary forces. Actuation mechanisms at the microscale tend to be quite limited, and if substantial forces are required, may involve considerable power dissipation in greatly reduced volumes. There are numerous other examples, and, as mentioned above, one must be keenly aware of the scaling laws pertaining to the project at hand. As with any engineering endeavor, there are multitudes of trade-offs, and skillful design can lead to good, functional compromises.

5.2 COST REDUCTION ISSUES

In terms of cost, an often-touted advantage of micromachining is that devices are fabricated in large batches, in parallel. This is certainly true. Unfortunately, however, for most microsensors, the parallelism does not extend into packaging and testing, which are domains that, when taken together, can account for as much as three-quarters of the final cost of a device. For conventional integrated circuits, testing is not parallel, but it is at least rapid and inexpensive. Testing is generally carried out by rapidly making temporary connections to each chip (still on a silicon wafer) and exercising it electrically. Some optical or magnetic devices can be tested in a similar way, by adding the necessary non-contact stimuli. For the vast majority of micromachined transducers designed to interact with physical phenomena in other domains (e.g., acceleration, chemical interaction, fluidic effects), such rapid testing is not possible. In those cases, usually individual chips are tested and often only after they are packaged. For some types, such as accelerometers, this testing can be fairly rapid, as they still can be tested using fast methods such as mechanical shakers. For devices such as wet chemical sensors, however, there currently seems to be no way to carry out testing rapidly. For at least some chemical sensors, however, this problem may be mitigated through the use of redundant arrays, differential measurement techniques, and calibration at the point-of-use.

As for any product successful in volume production, mass market drivers must be found or created. In the micromachined transducers marketplace, the automotive industry has provided a major market pull for such devices as manifold air pressure sensors and accelerometers for air bag systems (for an overview of current and future automotive sensors needs, see Giachino and Miree (1995)). Unfortunately, while needing high volumes, automotive and other durable goods markets are extremely cost-sensitive, making it difficult for manufacturers to amortize their net development costs. One approach taken by commodity sensor manufacturers to move "up the food chain" has been adding more value through the manufacture of subsystems that perform functions beyond simple measurements (e.g., moving from production of valves to that of flow controller modules). In this way, they can often achieve more realistic profit margins, necessary for long-term survival and growth. For chemical and biomedical applications, disposable devices are likely to increase in importance, since volumes can be orders of magnitude greater than those for comparable items permanently affixed in durable goods. In the case of single-use devices for common applications such as clinical diagnostics, it appears that there is tremendous potential for growth and profitability.

5.3 COMPLEXITY OF MICROMACHINED DEVICES AND SYSTEMS

Surveying the micromachining literature, one observes that the complexity of the sensors, actuators, and subsystems being fabricated is increasing over time. For example, research laboratories have produced a broad range of sensors that include on-chip circuitry to locally process sensor outputs into forms more palatable to computers. In the industrial world, such highly integrated sensors are certainly becoming available, although the bulk of the micromachined sensors being sold today do not contain transistor circuits. As an example of a trend toward integration in another area, microfluidic systems have been realized on an experimental basis where interconnecting channels, reaction chambers, and other functions are combined with micromachined valves and pumps. At present, however, there are few, if any, fully monolithic fluidic systems.

A principal issue with increasing the complexity of micromachined systems is that, as for hybrid assemblies in the electronics industry, the net yield of fully functional devices is usually negatively affected by increased complexity. The often-debated question of whether or not to integrate circuitry with micromachined sensors, appears to have an answer that is domain- and case-specific. For example, integrating circuitry might make sense for an accelerometer where there may be great market pressures for reduction of a subassembly (including a sensor and electronics) to a single chip. On the other hand, for disposable micromachined pressure sensors for medical applications, there may be considerably less pressure to integrate since the installed base of equipment that uses the sensors could not take advantage of such advanced capabilities. This latter market, however, may still be quite profitable and benefit from the high yields obtained with relatively simpler processes. A sensible strategy for developing micromachined devices that may some day have on-chip circuitry is to plan an evolutionary path such that the micromachining technologies used do not preclude the later addition of such circuits.

At the system level, many researchers make statements about building a "chemistry lab on a chip" but do not mention that sample volumes and supporting hardware may not scale down comparably. For example, it has long been possible to fabricate compact mass spectrometers (e.g., quadrupole designs), yet the technology for fabricating the necessary gas pumping systems has lagged far behind. Similarly, for miniaturized systems requiring self-contained electrical power, the power requirements of many crucial microactuators (e.g., valves, pumps) have in many cases remained roughly comparable to their macroscopic counterparts while power densities of batteries have not increased sufficiently to allow overall miniaturization. When miniaturizing clinical diagnostic instruments, in many cases the sample volumes required to obtain statistically valid samples of bacteria or viruses are many times larger than the microstructures themselves (potentially reducing throughput if they must be routed through microfluidic devices). Discussion of these issues by no means implies that miniaturized systems will not be successful. On the contrary, they certainly will have a major impact. It is important, however, that the "big" picture be considered rather than immediately miniaturizing parts of a system *simply because one can*.

Additional systems issues are faced by those seeking to use micromachining technologies in (bio)chemical, clinical, or similar domains. For example, for devices that come into contact with body fluids or are implanted, the issues of *biocompatibility* and *bioresistance* must be considered. Biocompatibility refers to the ability of materials to function in contact with biological media and not adversely affect cells or tissues (e.g., by invoking inflammatory responses or leaching out toxic compounds). Bioresistance refers to the ability of materials to withstand the harsh chemical environments faced in such settings (e.g., corrosive attack by body fluids and cells). The materials commonly available in micromachining research facilities are often not the best for such applications, and considerable work is needed in the area of coating materials and surface modification methods.

There are also several categories of systems where the optimal implementations will be in the mesoscopic ("sugar-cube-to-fist-sized") domain, rather than micro-scopic. These include chemical reactors, synthesizers, and analyzers; heating and cooling units; and combustors and fuel cells. For example, it is possible to revisit chemical reactions that are very unstable at the macroscopic scale, and use them effectively through a massively parallel array of microreactors. The enhanced mass and energy transport properties at this scale make this possible, and a mesoscopic array of microdevices could still produce substantial volumes of product. This parallel approach can, in many other applications, bring the advantages of microscale devices to bear on a larger physical scale. It is worth noting that such mesoscopic systems may include micromachined elements, or be made from them. Thus, the advantages of operating over an expanded range of physical scales can be harnessed.

An area that requires considerable research for all micromachined systems is the development of suitable interconnect and packaging methods. Just as the integration of wiring onto the integrated circuit transformed electronic systems a few decades ago, it is likely that analogous developments for fluidic devices, for example, will have a major impact in our ability to integrate micro- and mesoscopic systems in many domains. It is also probable that such interconnects and packages will not all be fabricated using relatively expensive silicon technologies, but perhaps, instead, using ceramics and injection-molded plastics, which can already provide the necessary precision to interface with micromachined structures. Unfortunately, discussion of packaging for micromachined transducers and structures has not appeared frequently in the literature, and standard integrated circuit packaging is seldom appropriate.

6. ISSUES TO CONSIDER

There are several important issues to consider when studying an existing micromachined device or planning a project to develop one. As discussed above, the most important question to answer is whether or not a device should be fabricated using micromachining technologies. The table below can serve as a starting point for a "checklist" of these issues.

Establish commercial or research need in light of "conventional" competition.
Understand the basic physics and operating principles, including scaling laws.
Understand the important issues in designing macroscopic and micromachined versions.
Survey prior work in micromachined versions, as well as "natural" (biological) analogs.
Consider the potential need to integrate on-chip circuitry (now or in the future).
Design a feasible, not overly complex, and reasonably priced fabrication process. If active circuits are, or may be, required, be sure to allow for that by avoiding incompatible steps.
Consider the issues of packaging. Can existing packages be adapted?
Consider realistic testing methods that suit the market (e.g., 100% test or statistical).
Estimate the final cost of the "ready-to-use" or "ready-to-ship" device (does it make sense?).
Consider the possibilities of evolving the design in the future to improve performance, reduce cost, etc. (this may, for example, feed back into the process design).
Make an overall decision as to feasibility prior to embarking on the research effort.

Table of issues to consider before undertaking a micromachining research and development effort or when studying an existing device. While this table is not intended to be all-inclusive, it should serve as a starting point.

7. WHAT ARE THE MARKETS FOR MICROMACHINED TRANSDUCERS?

In general, market projections for "MEMS" devices tend to be optimistic. Several authors have indicated that the market for all micromachined devices will reach $12 to $13 billion by the year 2000 (only a few years away). In contrast, Ristic, et al. (1994) project that the *total* sensor market (not just micromachined devices) will total $13 billion in the year 2000. The critical issue for those interested in the micromachining area is what fraction of that monetary total those devices will represent. They explain that most of the growth will be due to the further development of "silicon sensor technology."

WORLD SENSOR MARKET BREAKDOWN

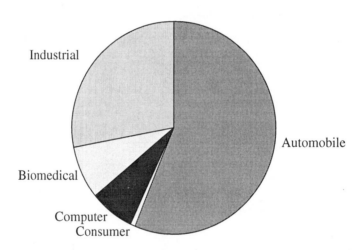

Breakdown of the world sensor market by function. After Ristic, et al. (1994).

A market study by System Planning Corporation (SPC (1994)) predicts that the MEMS market alone will grow to $13.9 billion in the year 2000. Despite the fact that they consider a different population of devices, these two estimates are reasonably close, and both are probably quite optimistic. (Ristic, et al. (1994) consider sensors only, and micromachined sensors are lumped into the total, while SPC (1994) considered all micromachined devices, including actuators.) Others, such as Petersen (1996a) have estimated that the markets will be substantially smaller, but nonetheless very significant (e.g., $2.6 billion in device sales, to enable $21 billion in system sales). Similarly, Bryzek (1996) provides more conservative estimates of $10.5 billion for all micromachined devices in the year 2005. Bryzek also notes that, when making such estimates, there is generally a ten to twenty year time lag between first prototypes and large-scale sales.

In terms of breaking down sensor markets by category, most experts agree that automotive applications will dominate for the near future. The two sources referred to above break the market down quite differently (Ristic, et al. (1994) by market type and SPC (1994) by device type). In any case, it is likely that markets using disposable devices, such as the medical device market, will increase relative to the others since non-disposables mainly represent durable goods.

WORLD MARKET FOR MICROMACHINED DEVICES BY FUNCTION

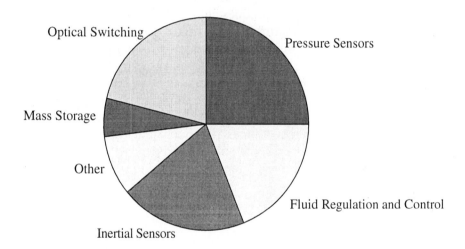

Breakdown of the world market for micromachined devices by function. After SPC (1994).

7.1 MICROMACHINING AS AN "ENABLING TECHNOLOGY"

A key point is that micromachined sensors and actuators constitute *enabling technologies*, wherein their greatest power lies in enabling and adding value to systems. The SPC report cited above estimates (perhaps somewhat optimistically) that by the year 2000, the $13.9 billion annual sales of micromachined devices will enable nearly $100 billion in overall annual sales. However, when looked upon as an important enabling technology, one could make similar statements for injection molding, printed circuit boards, adhesives, etc. Further discussion of the concept of "enabled" sales may well be academic, but it will undoubtedly continue.

7.2 IS THE MARKET "READY TO EXPLODE"?

The market for micromachined transducers is still being developed and does not have the explosive growth rate of, for example, integrated circuits in the 1970's. Some authors like to compare the two, but this does not really make sense, since there is no "dominant technology" in micromachining that is analogous to MOS (metal oxide semiconductor) circuitry (which led, in a large part, to the exponential growth of the digital electronics industry). Perhaps a better comparison would be likening micromachining to analog integrated circuit technology, which is certainly enabling, but with nowhere near the market volume of digital circuits such as memories and microprocessors. However, time will ultimately provide the answers as to how important this new set of technologies are.

At the moment, a great deal of research in this area is focused on "surface" micromachining (fabricating thin-film structures above the substrate, as discussed in detail in the Micromachining Techniques chapter), yet in industry the vast majority of shipping devices are manufactured using much older "bulk" micromachining. While some surface micromachined devices are now being produced in volume, it will take a few more years for this approach to make a large impact in the market. It is unlikely that one approach will dominate, at least in the short run, as new devices using both surface and bulk micromachining (circuits can be integrated in either case) continue to be marketed. An interesting and important aspect of the development of micromachined devices has been that their "incubation times" from conception to volume shipment often exceed a decade, so it is likely to be too early to rely too heavily on technology predictions.

8. INFORMATION RESOURCES

8.1 ON-LINE RESOURCES

Micromachining is a relatively new field, and is growing rapidly at present. At the moment, most of the interesting publications appear first in conference proceedings and then later, if at all, in more easily accessible journals. Thus, computer databases are often an extremely useful tool for tracking new developments. Several excellent on-line databases exist, but even papers as recent as the 1970's may not appear in some databases (particularly if higher-cost access to older records is not subscribed to).

The Information Sciences Institute (ISI) in Marina del Rey, CA, operates a DARPA-sponsored MEMS "bulletin board" (BBS) that serves as a "clearinghouse" for MEMS information. For general information, one can send e-mail to mems@isi.edu, or visit their web site at http://mems.isi.edu/mems.html. Their web pages contain convenient links to some of the major university sites, although in general, web links may not be as frequent as one would hope.

8.2 MICROMACHINED TRANSDUCERS MEETINGS

There are a large number of meetings whose focus is micromachined transducers. A few of the best-known examples are:

• *Solid-State Sensor and Actuator Workshop*, or *"Hilton Head"* (small, North American only, limited attendance meeting held at Hilton Head, SC, on alternate years, e.g., 1992, 1994, 1996, ...).

• *International Conference on Solid-State Sensors and Actuators*, or *"Transducers"* (large, international meeting held in Asia, North America, or Europe on alternate years, e.g., 1991, 1993, 1995, ...).

• *Micro Electro Mechanical Systems Workshop*, or *"MEMS"* (moderate sized, international workshop, with a focus on actuators and mechanical devices, held annually).

• *Micro Total Analysis Systems*, or *"μTAS"* (international, focus on micromachined chemical systems, held alternate years in Europe, e.g., 1994 (first), 1996, ...).

• *Eurosensors* (European, with broad coverage, held annually).

There are many micromachining-specific meetings in North America and abroad. In addition, many long-established "mainstream" conferences in specific disciplines are adding "MEMS" tracks. This seems natural since as the micromachining field matures, the end-users in other disciplines will adopt the useful aspects of the technology.

The proceedings from these meetings are an excellent source of material if one can find them. In some cases, copies of prior meeting proceedings can be obtained by writing to the host organizations. A few examples are given below.

Meeting	Host Organization
Hilton Head	Transducers Research Foundation, Inc., P.O. Box 18195 Cleveland Heights, OH, USA.
Transducers	Variable, depends on host country: 1995 is available from the Royal Swedish Academy of Engineering Sciences, IVA, Box 5073, S-102, 42 Sweden, 1997 is available from the IEEE, at the address given below.
MEMS	IEEE Service Center, 445 Hoes Lane, P.O. Box 1331, Piscataway, NJ, USA.
μTAS	1994 - Kluwer Academic Publishers, P.O. Box 17, 3300 AA Dordrecht, The Netherlands. 1996 - AMI Editorial Office, Münsterplatz 6, Postfach 1955, CH-4001, Base, Switzerland.

Table of organizations from which copies of micromachining-related conference proceedings can be ordered. Note that searching the web or otherwise contacting the host organizations is often critical to obtaining such material.

8.3 TEXTBOOKS AND PAPER COLLECTIONS

There are several relevant books on the market, covering micromachined and/or macroscopic transducers in various proportions. Each book tends to have its own particular set of strengths, and in general, a researcher in this field will want to own several. Some examples, without particular endorsements (*caveat emptor!*) are listed in the General References section at the end of this chapter. In addition, two bound collections of "classic" papers on micromachined devices, Muller, et al. (1991) and Trimmer (1997) are available from IEEE, and are good sources for background information.

8.4 JOURNALS

There are several journals now available that cover the micromachined transducer area. Some examples are:

A) Journals with a primary focus in the MEMS area:

• *Sensors and Actuators A (Physical)*

• *Sensors and Actuators B (Chemical)*

• *Sensors and Actuators C (Materials)*

• *IEEE/ASME Journal of Microelectromechanical Systems (JMEMS)*

• *Journal of Micromechanics and Microengineering*

B) Journals with information of interest to the MEMS community and occasional MEMS papers:

• *IEEE Electron Device Letters*

• *Journal of the Electrochemical Society*

• *Journal of the Vacuum Society*

• *Proceedings of the SPIE — International Society for Optical Engineering*

8.5 THESES

An often-overlooked but rich information resource are M.S. and Ph.D. theses. These can be obtained through University Microfilms, Inc. (300 North Zeeb Road, P.O. Box 1346, Ann Arbor, MI 48106-1346, Phone: (313) 761-4700) or by directly contacting the authors or their advisors. Unfortunately, the major difficulty is knowing about them, since they are not necessarily searchable on-line.

8.6 PATENTS

Another often-overlooked (at least by academics) yet critical information resource is patents. Quite frequently, companies or research groups will file for patents long before publishing, and sometimes companies will *never* publish. Patents can be searched on-line (through patent service providers or on the web, e.g., at IBM's patent server site: http://patent.womplex.ibm.com/) or at patent libraries.

9. CONCLUSION

Micromachining technologies are clearly going to have a major impact on domains including basic science, consumer goods, aerospace, environmental monitoring, and medicine. Their successful application typically requires broad knowledge of physical principles and scaling laws. Equally important are system-level issues of complexity, packaging, testing, and manufacturability. As these technologies mature, and are adopted by more end-users, exciting developments will certainly occur, particularly in interdisciplinary domains. For further information on trends in micromachining, predictions of future growth and reviews of new application areas, the reader is referred to Baltes (1996), Bryzek (1996), Fujita (1996) and Petersen (1996b).

INTRODUCTION AND OVERVIEW REFERENCES

GENERAL REFERENCES

Fraden, J., "Handbook of Modern Sensors," Second Edition, American Institute of Physics Press, Woodbury, NY, 1996.

Gardner, J. W., "Microsensors: Principles and Applications," John Wiley and Sons, Inc., Chichester, West Sussex, UK, 1994.

Hauptmann, P., "Sensors: Principles and Applications," Prentice-Hall International (UK), Hertfordshire, UK, 1991.

Janata, J., "Principles of Chemical Sensors," Plenum Press, New York, NY, 1989.

Khazan, A. D., "Transducers and Their Elements: Design and Application," Prentice-Hall, Inc., Englewood Cliffs, NJ, 1994.

Kress-Rogers, E. [ed.], "Handbook of Biosensors and Electronic Noses: Medicine, Food and the Environment," CRC Press, Inc., Boca Raton, FL, 1997.

Madou, M., "Fundamentals of Microfabrication," CRC Press, Inc., Boca Raton, FL, 1997.

Madou, M. J., and Morrison, S. R., "Chemical Sensing with Solid State Devices," Academic Press, Inc., San Diego, CA, 1989.

Middelhoek, S., and Audet, S. A., "Silicon Sensors," Academic Press, Inc., Boston, MA, 1989.

Muller, R. S., Howe, R. T., Senturia, S. D., Smith, R. L., and White, R. M. [eds.], "Microsensors," IEEE Press, New York, NY, 1991.

Noltingk, B. E. [ed.], "Instrumentation Reference Book," Second Edition, Butterworth-Heinemann, Ltd., Oxford, UK, 1995.

Norton, H. N., "Handbook of Transducers," Prentice-Hall, Inc., Englewood Cliffs, NJ, 1989.

Ristic, Lj. [ed.], "Sensor Technology and Devices," Artech House, London, 1994.

Sze, S. M., "Semiconductor Sensors," John Wiley and Sons, Sommerset, NJ, 1994.

Trimmer, W. S., "Micromechanics and MEMS: Classic and Seminar Papers to 1990," IEEE Press, New York, NY, 1997.

SPECIFIC REFERENCES

Baltes, H., "Future of IC Microtransducers," Sensors and Actuators, vol. A56, nos. 1 - 2, Aug. 1996, pp. 179 - 192.

Bryzek, J., "Impact of MEMS Technology on Society," Sensors and Actuators, vol. A56, nos. 1 - 2, Aug. 1996, pp. 1 - 9.

Fujita, H., "Future of Actuators and Microsystems," Sensors and Actuators, vol. A56, nos. 1 - 2, Aug. 1996, pp. 105 - 111.

Giachino, J. M., and Miree, T. J., "The Challenge of Automotive Sensors," Proceedings of the SPIE Conference on Microlithography and Metrology in Micromachining, Austin, TX, Oct. 23 - 24, 1995, SPIE vol. 2640, pp. 89 - 98.

Lion, K. S., "Transducers: Problems and Prospects," IEEE Transactions on Industrial Electronics and Control Instrumentation, vol. IECI-16, no. 1, July 1969, pp. 2 - 5.

Petersen, K. E., personal communication, 1996.

Petersen, K., "From Microsensors to Microinstruments," Sensors and Actuators, vol. A56, nos. 1 - 2, Aug. 1996, pp. 143 - 149.

Ristic, Lj. [ed], "Sensor Technology and Devices," Artech House, London, 1994

System Planning Corporation, "Microelectromechanical Systems (MEMS): An SPC Market Study," July 1, 1994, available from System Planning Corporation, 1429 North Quincy Street, Arlington, VA 22207, Phone: (703) 351-8461, Fax: (703) 351-8662.

White, R. M., "A Sensor Classification Scheme," IEEE Transactions on Ultrasonics, Ferroelectrics and Frequency Control, vol. UFFC-34, no. 2, Mar. 1987, pp. 124 - 126.

1. CAPABILITIES AND LIMITATIONS OF MICROMACHINING

Micromachining is a very powerful approach for fabricating sensors, actuators, microstructures, and systems. However, it has limitations that are as important to know about as its strengths if these technologies are to be applied effectively. One of the most important goals in learning about micromachining is to realize what can and cannot be fabricated using these methods and to be able to make meaningful comparisons to "conventional" fabrication techniques (e.g., drilling, end milling, injection molding, electric discharge machining, etc.). For example, structures such as those illustrated below would be difficult or impossible to fabricate lithographically, since this involves either stacking two-dimensional (2-D) layers or using processes that yield "extruded" shapes.

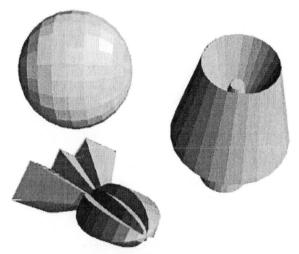

Illustration of some structures that would be difficult to fabricate with sequential lithographic steps.

There are a few true "three-dimensional" (3-D) processes (i.e., capable of yielding arbitrary, overlapping curved surfaces such as spirals), but they are often serial, as opposed to most microfabrication methods, which have the great advantage of parallelism. Essentially, the common micromachining tools available, mainly derived from the mainstream integrated circuit industry, involve adding 2-D layers, etching the 2-D layers (sometimes forming cavities by removing "sacrificial" spacer layers), adding tall "extrusion-like" structures (e.g., plating within template structures that are later removed), bonding on additional substrates (e.g., glass, silicon, etc.),

19

and etching the substrate. Since microfabrication tools are almost applied with successive photolithographic patterning, any 3-D aspects of micromachined devices are due to patterning of the 2-D layers, interaction of stacked and patterned 2-D layers, selective etching of the layers or the substrate, or the use of bending or hinged structures. Getting a sense for the geometric, chemical, and thermal interactions of the large assortment of processes discussed below is key to their successful application. Studying the examples provided in this book and found in the literature should aid in this process.

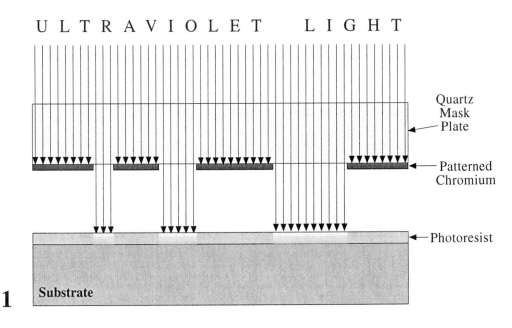

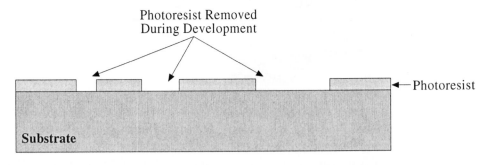

Illustration of the basic photolithographic process employed in the vast majority of micromachining processes. A photoresist layer is first spun-on and baked (not shown) and exposed to ultraviolet (UV) light through a mask (a quartz plate covered with a thin film of chromium, which is selectively removed to allow light through). The exposed photoresist (positive photoresist is shown) then dissolves in a developer solution, leaving only the non-exposed regions behind.

Process Type	Examples
Lithography	photolithography, screen printing, electron-beam lithogaphy, x-ray lithography
Thin-Film Deposition	chemical vapor deposition (CVD), plasma-enhanced chemical vapor deposition (PECVD), sputtering, evaporation, spin-on application, plasma spraying, etc.
Electroplating	blanket and template-delimited electroplating of metals
Directed Deposition	electroplating, stereolithography, laser-driven chemical vapor deposition, screen printing, transfer printing
Etching	plasma etching, reactive-ion enhanced (RIE) etching, deep reactive ion etching (DRIE), wet chemical etching, electrochemical etching, etc.
Directed Etching	laser-assisted chemical etching (LACE)
Machining	drilling, milling, electric discharge machining (EDM), diamond turning, sawing, etc.
Bonding	fusion bonding, anodic bonding, adhesives, etc.
Surface Modification	wet chemical modification, plasma modification
Annealing	thermal annealing, laser annealing

Table of example processes used in micromachining.

The table above shows a variety of example micromachining processes, which are discussed below. It must be noted that this book cannot replace a solid background in standard integrated circuit fabrication techniques, for which the reader is referred to Sze (1988), Runyan and Bean (1990), or the wide variety of other texts on that subject. The discussion below concerns processes of use in micromachining, not necessarily mainstream integrated circuit fabrication. In addition to the material presented herein, the reader is referred to Delapierre (1989), Mastrangelo and Tang (1994), Ristic (1994), Kovacs, et al. (1996), and Madou (1997) among others, for further discussion of micromachining processes.

2. MATERIALS FOR MICROMACHINING

2.1 SUBSTRATES

Substrate materials for micromachining need not be silicon. Micromachining efforts to date have primarily focused on silicon, partly for historical reasons, and partly for practical ones. In general, crystalline semiconductors (Si, Ge, GaAs, etc.) are used to take advantage of their inherent features: 1) well characterized and readily available, 2) multitude of mature processing techniques available (at least for Si), 3) useful crystal plane anisotropy for micromachining, and 4) potential for integration of active circuits. Silicon is certainly the dominant substrate material for transducers, but it is important to pay attention to work in progress with other materials and to consider them for new projects. Aside from the ability to integrate circuitry directly on a transducer substrate (not always useful in the final analysis), a great number of other suitable, non-semiconductor substrates exist. A partial list includes metals, glasses, quartz and other crystalline insulators, ceramics, plastics and polymers, and other organic and inorganic materials. It is also important to note that the functional material(s) of the transducer may be added above the substrate, or be the substrate itself.

2.2 ADDITIVE FILMS AND MATERIALS

The category of additive films and materials to be deposited on the substrate includes conductors, semiconductors, and insulators. A partial list of possible thin-film materials includes silicon (epitaxial [single crystal], polycrystalline, amorphous); silicon compounds (Si_xN_y, SiO_2, SiC, etc.); metals and metallic compounds (e.g., Au, Cu, Ni, Al, ZnO, GaAs, IrO_x, CdS); a wide variety of ceramics (e.g., Al_2O_3 and more complex compounds); and organics (i.e., diamond, polymers, enzymes, antibodies, DNA, RNA, etc.). By adding such materials to semiconductor substrates, the ability to co-fabricate active circuitry is retained. The trade-off, however, is that the additive materials are generally subject to variations in their physical properties that are far wider than those for single-crystal semiconductor substrates, which are remarkably uniform. It is worth noting that active circuitry can also be fabricated using thin-film layers, allowing them to be added to a wide variety of substrates such as glasses and plastics. Despite the apparent attractiveness of such a capability, it has not been used in any significant way in mainstream micromachined transducers, although its use in active-matrix liquid-crystal displays is routine. Thin-film transistors are discussed briefly in Section 6.1.1 below.

3. MICROMACHINING TERMS

People working in the micromachining field use some specialized terms, but the bulk of the vocabulary is drawn directly from that of integrated circuit fabrication. It should be noted that these terms are not perfect, nor are there "official" definitions.

Surface micromachining is processing "above" the substrate (i.e., not involving the substrate in a big way: in other words, mainly using it as a base to build upon). *Bulk micromachining* is processing that removes "bulk" substrate (i.e., large pits, holes to the back side of a wafer, sawing, etc.).

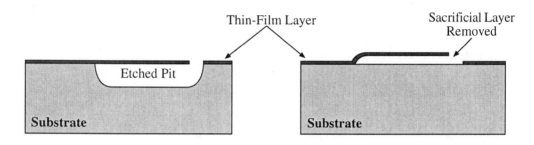

Examples of "subtractive" (left) and "additive" (right) cantilever beams.

The two major types of processing are *subtractive* (etching, laser machining, mechanical milling, etc.) and *additive* (deposition of dielectrics, metals, etc.). When categorizing an overall microstructure, some define "subtractive" structures as any that involve etched pits, voids, holes, etc., and "additive" structures such as those that are only built "up" from the substrate. These definitions for structures tend to lose their meaning for some of the more complex structures (i.e., made from multiple components).

Aspect ratio is a key term referring to the ratio of the depth to the width of an etched hole or a grown structure. It is often used as a metric by which to compare different processes.

Isotropic and *anisotropic* refer to how perpendicular (e.g., to the wafer surface) the sidewalls of an etched void or structure are for a given fabrication process (etching, deposition, etc.). Isotropic processes (e.g., etching) progress with equal rates in all directions, but in practice etch rates are not always entirely omnidirectional. Isotropic processes tend to produce structures with "rounded" features. Anisotropic processes tend to progress in a preferred direction (usually perpendicular to the substrate) and produce features with sharply defined features in the vertical direction.

These terms are used more as relative than absolute measures since there is really a continuum of possible process profiles (e.g., an etch may be described as "fairly anisotropic," but stating the aspect ratio is much more descriptive).

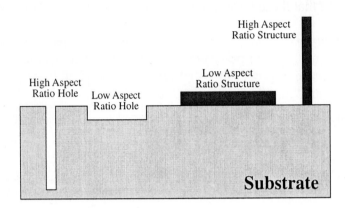

Examples of high and low aspect ratio features.

A wide variety of micromachining processes are available, and these form an important part of the vocabulary. A variety of examples are discussed below.

4. GENERAL PROPERTIES OF COMMON SEMICONDUCTORS

GENERAL PROPERTIES @ 300 K	Si	Ge	GaAs
Atomic Weight	28.09	72.60	144.63
Density (g/cm^3)	2.328	5.3267	5.32
Atomic Density (atoms/cm^3)	5.0×10^{22}	4.42×10^{22}	4.42×10^{22}
Lattice Constant Å	5.43095	5.64613	5.6533
THERMAL PROPERTIES			
Melting Point (°C)	1,415	937	1,238
Specific Heat (J/g•K)	0.7	0.31	0.35
Linear Coeff. of Thermal Expansion ($\alpha = \Delta L/(L\Delta T)$, in K^{-1})	2.6×10^{-6}	5.8×10^{-6}	6.86×10^{-6}
Thermal Conductivity (at 300 K) (W/cm•K)	1.5	0.6	0.46
Thermal Diffusivity (cm^2/s)	0.9	0.36	0.24
ELECTRICAL PROPERTIES			
Energy Gap (eV) at 300 K	1.12	0.66	1.424
Intrinsic Carrier Concentration (cm^{-3})	1.45×10^{10}	2.4×10^{13}	1.79×10^6
Intrinsic Resistivity (Ω•cm)	2.3×10^5	47	10^8
Dielectric Constant (DC only)	11.9	16.0	13.1
* Breakdown Field (V/cm)	$\approx 3 \times 10^5$	$\approx 10^5$	$\approx 4 \times 10^5$
* Minority Carrier Lifetime (s)	2.5×10^{-3}	10^{-3}	$\approx 10^{-8}$
* Electron Mobility (cm^2/V•s)	1,500	3,900	8,500
* Hole Mobility (cm^2/V•s)	450	1,900	400

Table of basic physical, thermal, and electrical properties of Si, Ge, and GaAs. Note that the properties marked with asterisks at the left are for the highest-quality, undoped samples and are never fully achieved in practice. From Sze (1988).

4.1 MECHANICAL PROPERTIES OF SILICON

Single-crystal silicon is a brittle material. It yields via catastrophic failure (like glass), rather than plastic deformation (like most metals). It has a Young's modulus, hardness, and tensile yield strength that approach those of many commonly used metals such as stainless steel. The *mechanical properties of silicon are anisotropic* and hence are orientation-dependent.

There is a tendency for cracks to propagate along certain crystal planes. These planes tend to reveal themselves when wafers are cleaved (or unintentionally broken). The more perfect ("polished") the surface and edges of a piece of silicon, the less likely it will crack under mechanical loading, since imperfections serve as starting points for cracks (as do edges and corners, even if they are perfectly smooth). It is important to note that micromachining processes vary in the degree with which they damage silicon surfaces, thus greatly affecting the mechanical properties of the finished devices.

Material	Yield Strength (10^9 N/m^2)	Knoop Hardness (kg/mm^2)	Young's Modulus (GPa)	Density (g/cm^3)	Thermal Conductivity (W/cm•K)	Thermal Expansion Coefficient (10^6/K)
*Diamond	53	7,000	1,035	3.5	20	1
*SiC	21	2,480	700	3.2	3.5	3.3
*TiC	20	2,470	497	4.9	3.3	6.4
*Al$_2$O$_3$	15.4	2,100	530	4	0.5	5.4
*Si$_3$N$_4$	14	3,486	385	3.1	0.19	0.8
*Iron	12.6	400	196	7.8	0.803	12
SiO$_2$ (fibers)	8.4	820	73	2.5	0.014	0.55
*Si	7	850	190	2.3	1.57	2.33
Steel (max strength)	4.2	1,500	210	7.9	0.97	12
W	4	485	410	19.3	1.78	4.5
Stainless Steel	2.1	660	200	7.9	0.329	17.3
Mo	2.1	275	343	10.3	1.38	5
Al	0.17	130	70	2.7	2.36	25

Table of mechanical properties of silicon and other materials. From Petersen (1982). (Note that the values marked with asterisks at the left are for single-crystal materials and are not necessarily useful for amorphous or polycrystalline forms.)

4.2 NATIVE OXIDES OF SILICON AND OTHER SEMI-CONDUCTORS

It is important to note that Si, unlike Ge and GaAs, has a broadly useful native oxide that can come in handy for many simple micromachining tasks (it was particularly useful in the early days of micromachining, before a large variety of high-quality thin films could so easily be deposited). All semiconductors have native oxides, but of the three common ones listed above, only silicon has an oxide (SiO_2) that is very suitable for photolithographic masking, has excellent thermal stability, is very adherent (since it is grown *in situ*), and has sufficiently low electronic defect densities at the Si/SiO_2 interface to allow MOSFETs (metal oxide semiconductor field-effect transistors) to be successfully realized.

Germanium's native oxide, GeO_2, is water soluble (making it useless for wet processing except where one would like to remove it, as was the case for early point contact diodes and transistors).

Gallium arsenide has over ten different native oxides, the most common being Ga_2O_3, As_2O_3, As_2O_5, GaO, and AsO_2. Each oxide has very different properties and the relative proportions of each type varies with the type of processing (several of the oxides can transmute between stoichiometries during thermal processing). Arsenic oxides tend to be more soluble in acidic solutions, while gallium oxides tend to be more soluble in alkaline solutions, providing for both problems and advantages in GaAs processing.

Oxide growth is very different for GaAs and other III-V compounds from that for Si or Ge. In Si, O_2 diffuses through the existing oxide and growth occurs at the Si/SiO_2 interface. In GaAs, for example, the Ga and As are more volatile and diffuse through the existing oxide, and growth occurs at the oxide/gas interface (resulting in large electronic defect densities at the GaAs/oxide interface in addition to poor adhesion of the oxide to the substrate).

4.3 TYPICAL SILICON WAFER TYPES

At present, many groups doing micromachining research and/or manufacturing are using 100 mm (4 in.) or, in some cases, 150 mm (6 in.) silicon wafers (considered completely obsolete by the "mainstream" semiconductor industry). Typical 100 mm wafers are ≈ 500 μm thick, but wafers of nearly any thickness can be obtained from commercial vendors or fabricated by chemical/mechanical thinning). While the mainstream integrated circuit industry is scaling from 8 in. to 12 in. wafers, the bulk of the micromachining community has not even faced the transition to 6 in. wafers. It is interesting to note that one of the main drivers for the transition in the micromachining field is the diminishing availability of 4 in. silicon wafers.

Wafers are typically marked with "flats," as shown below, to indicate the orientation of certain crystal planes (note that the flats may have different meanings for some wafer sizes, as well as for III-V materials for which international standards differ).

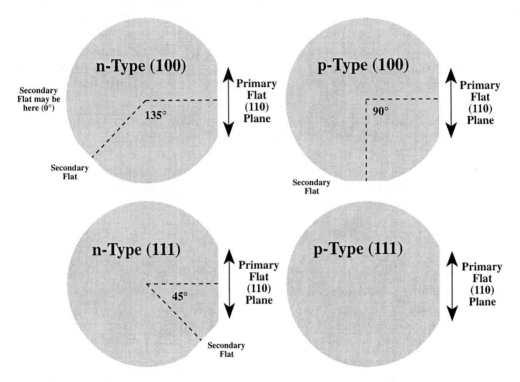

Illustration of coded "flats" as typically used on 4 in. wafers to help identify them (the configuration shown at upper left is the most common). (SEMI standard.)

Notation for Miller indices:

(i j k) ....a specific crystal plane or face
{i j k} ....a family of equivalent planes
[i j k] ....a specific direction of a unit vector
<i j k> ....a family of equivalent directions

---

Table showing standard notation for Miller indices.

The table above shows the standard notation for Miller indices, which are often used to indicate crystal orientations, crystal planes, etc. It is possible to order wafers of nearly any crystal orientation and nearly any thickness (although they may be quite expensive). Wafers of glass, quartz, sapphire, and a variety of ceramics can also be obtained.

5. "BULK" (SUBTRACTIVE) PROCESSES

Many etching methods are available for bulk processing of silicon. Wet and dry etchants are selected on the basis of a large number of characteristics, including selectivity to etch masks; exposed metallization and other materials; availability of etch-stop methods (e.g., doping or electrochemical); etch rate; degree of surface roughness created; safety; disposal difficulty; circuit compatibility; availability in a given fabrication facility; cost; etc. In practice, would-be etch users often make their selection based more on the historical availability of etchants in a given facility (including staff/institutional acceptance and familiarity) rather than starting from scratch. Quite often, the etchants are used for post-processing of prefabricated structures that may contain CMOS (complementary metal oxide semiconductor) circuitry, aluminum metallization, etc., which places significant constraints on the choice of etchants (interesting early papers considering these issues are Parameswaran, et al. (1988, 1989) and Moser, et al. (1990)).

The most common etches used for micromachining are those for silicon, and representative types are presented below in a comparison table (the table is intended for general comparison only, since the etchants themselves are described in detail below). Etches for other materials, including typical thin-film layers used in micro-fabrication are discussed in Section 6.1 below. An excellent summary of etch processes for micromachining applications, suitable for comparison of a large number of approaches, can be found in Williams and Muller (1996). Another useful survey of such techniques can be found in Madou (1997).

Comparison of Example Silicon Etchants							
	HNA (HF+HNO$_3$ +Acetic Acid)	Alkali-OH	EDP (ethylene diamine pyrochat-echol)	TMAH (tetramethyl-ammonium hydroxide)	XeF$_2$	SF$_6$ Plasma	DRIE (Deep Reactive Ion Etch)
Etch Type	wet	wet	wet	wet	dry [1]	dry	dry
Anisotropic?	no	yes	yes	yes	no	varies	yes
Availability	common	common	moderate	moderate	limited	common	limited
Si Etch Rate µm/min	1 to 3	1 to 2	1 to 30	≈ 1	1 to 3	≈ 1	> 1
Si Roughness	low	low	low	variable[2]	high[3]	variable	low
Nitride Etch	low	low	low	1 to 10 nm/min	?	low	low
Oxide Etch	10 to 30 nm/min	1 to 10 nm/min	1 to 80 nm/min	≈ 1 nm/min	low	low	low
Al Selective	no	no	no [4]	yes [5]	yes	yes	yes
Au Selective	likely	yes	yes	yes	yes	yes	yes
p++ Etch Stop?	no (n slows)	yes	yes	yes	no	no (some dopant effects)	no
Electrochemical Stop?	?	yes	yes	yes	no	no	no
CMOS Compatible?[6]	no	no	yes	yes	yes	yes	yes
Cost [7]	low	low	moderate	moderate	moderate	high	high
Disposal	low	easy	difficult	moderate	N/A	N/A	N/A
Safety	moderate	moderate	low	high	moderate?	high	high

1	Sublimation from solid source.
2	Varies with wt% TMAH, can be controlled to yield very low roughness.
3	Addition of Xe to vary stoichiometry in F or Br etch systems can yield optically smooth surfaces.
4	Some formulations do not attack Al, but are not common.
5	With added Si, polysilicic acid or pH control.
6	Defined as 1) allowing wafer to be immersed directly with no special measures and 2) no alkali ions.
7	Includes cost of equipment.

Table comparing the general features of various etchants suitable for micromachining applications. Note that the values presented are approximate and depend on exact formulations and on the chemistry of the material being etched.

5.1 WET ETCHING OF SILICON

Many wet silicon etchants are available, with a wide variety of etching properties, costs, safety levels, degrees of compatibility with prefabricated electronic circuits, etc. By appropriate choice of substrate type, thin-film masking layers, and etchant(s), a wide variety of subtractive structures (holes, trenches, mesas, etc.) can be fabricated. A fundamental distinction between etchants is whether their behavior tends to be isotropic or anisotropic (in typical micromachining, anisotropic etchants are used in the vast majority of cases). As discussed below, for wet etchants, such anisotropy generally arises due to unequal etch rates for different crystal planes of the substrate.

5.1.1 ISOTROPIC WET ETCHING

Isotropic etchants etch in all directions at nearly (and sometimes exactly) the same rate. They can slow down in long, narrow channels due to diffusion limiting of the reaction, and in such cases, agitation of the etchant can control the etch rate and the geometry of the resulting etched structures. Pits and cavities with rounded surfaces (even nearly perfectly hemispherical shapes) can be obtained with good agitation. Chemical mass transport issues generally keep such etchants from exhibiting perfectly isotropic properties, and the function of agitation is to help speed the transport of reactants and products and to keep the transport more uniform.

ISOTROPIC WET ETCHING: AGITATION

SiO_2 Mask

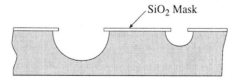

ISOTROPIC WET ETCHING: NO AGITATION

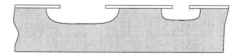

Illustration of isotropic etch cross sections showing the effects of mask geometry and agitation. Note that this type of etchant is an option for silicon and is, in general, the only wet etch option for glasses. After Petersen (1982).

HF/HNO₃/ACETIC ACID ("HNA")

The most common isotropic silicon etch is "HNA," a mixture of hydrofluoric acid (HF), nitric acid (HNO₃), and acetic acid (CH₃COOH). The HNO₃ drives the oxidation of the silicon, while fluoride ions from HF then form the soluble silicon compound H_2SiF_6. The acetic acid, which is much less polar than water (smaller dielectric constant in the liquid state), helps prevent the dissociation of HNO₃ into NO_3^- or NO_2^-, thereby allowing the formation of the species directly responsible for the oxidation of silicon,

$$N_2O_4 \longleftrightarrow 2NO_2$$

Despite this, the same mixture without the acetic acid is found to be nearly as effective (for short etch times, until the NO_2 is depleted). The etching chemistry is complex (due to HNO₃'s autocatalytic ionization), and etch rates depend on chemical mixture and silicon doping. The overall reaction (Williams and Muller (1996)) is,

$$18HF + 4HNO_3 + 3Si \rightarrow 2H_2SiF_6 + 4NO_{(g)} + 8H_2O$$

A potential drawback of this etchant is that it attacks SiO_2 relatively quickly (30 to 70 nm/min). It is also noteworthy that the HNA etch is slowed down in etch rate ≈ 150 times by regions of *light* doping ($<10^{17}$ cm^{-3} n- or p-type) relative to more heavily doped regions. Further information on HNA etchants for silicon can be found in Bogenschütz, et al. (1967). Discussion of the related HF:H₂O₂:CH₃COOH etchant for selectively etching $Si_{1-x}Ge_x$ relative to Si is found in Carns, et al. (1995).

In general, the basic mechanism of wet single-crystal silicon etchants is: 1) injection of holes into the Si to form Si^{2+} or Si^+, 2) attachment of OH⁻ groups to the Si^{2+} to form $Si(OH)_2^{2+}$, 3) reaction of the "hydrated" Si (silica) with a complexing agent in the solution, and 4) dissolution of the reaction products into the solution.

Thus, for such an etch, one needs a source of holes, OH⁻, and a complexing agent. It should be noted that the idea of chemical reactions producing holes is nothing more than moving the electrons used to the other side of the equation. The isotropic HNA system, described above, is used as an example to illustrate this point, as discussed in Petersen (1982),

$$2e^- + HNO_2 + HNO_3 + H_2O \rightarrow 2HNO_2 + 2OH^-$$

which can be written,

$$HNO_2 + HNO_3 + H_2O \rightarrow 2HNO_2 + 2OH^- + 2h^+$$

This is therefore an electrochemical reaction. A source for holes on one side of the equation is equivalent to reduction (addition of electrons) on the other side.

Since the etching is basically a charge-transfer-driven process, it makes sense that the dopant type/concentration and externally applied electrical potential should modulate it. In general, this does occur, and is discussed below.

Etchant (Diluent)	Reagent Quantities	Temp. °C	Etch Rate (μm/min)	(100)/(111) Etch Ratio	Dopant Dependence	Masking Films (etch rate)
HF	10 ml				$\leq10^{17}$ cm^{-3} n or p reduces etch rate $\approx150\times$	
HNO$_3$	30 ml	22	0.7 to 3.0	1:1		SiO$_2$ (30 nm/min)
(water, CH$_3$COOH)	80 ml					
HF	25 ml					
HNO$_3$	50 ml	22	4	1:1	no dependence	Si$_3$N$_4$
(water, CH$_3$COOH)	25 ml					
HF	9 ml					
HNO$_3$	75 ml	22	7	1:1	---	SiO$_2$ (70 nm/min)
(water, CH$_3$COOH)	30 ml					

Table of HNA etchant formulations, assuming standard acid concentrations (note that water or acetic acid can be used) and their properties. From Petersen (1982).

5.1.2 ANISOTROPIC WET ETCHING

Anisotropic (or "orientation-dependent") etchants etch much faster in one direction than in another. Those described here slow down markedly at the (111) planes of silicon, relative to their etch rates for other planes. Classic papers describing these crystal-plane-dependent etches are Bean (1978) and Bassous (1978). Most such etchants can be dopant- or electrochemically modulated but slow down at the (111) planes regardless of the dopant(s). In general, the slowest etching planes are exposed as the etch progresses. It is also important to note that from the top view (illustrated below), etching at "concave" corners on (100) silicon stops at (111) intersections, but "convex" corners are undercut. A considerable amount of effort is often expended in designing mask patterns to take these effects into account.

ANISOTROPIC WET ETCHING: (100) SURFACE

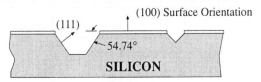

ANISOTROPIC WET ETCHING: (110) SURFACE

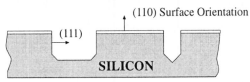

Illustration of anisotropic wet etching of (100) [top] and (110) silicon [bottom]. After Petersen (1982).

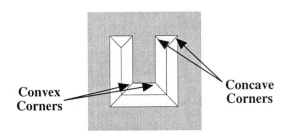

Illustration of convex and concave corners on (100) silicon. The former are rapidly undercut by anisotropic silicon etchants.

While the reaction mechanisms have generally been elucidated, the mechanisms of dopant modulation and anisotropic etching along crystal planes have not been fully explained. The phenomena of crystal cleavage and anisotropic etching stopping on certain crystal planes are commonly thought to result from the plane with the "least surface density" of atoms, but this theory cannot account for all of the behavior seen. As an example, for cubic crystals (zincblende and diamond structures), the surface density of atoms does not vary by more than a few percent over all possible directions, and this cannot possibly account for the >100:1 anisotropies and cleavage preferences seen in practice. Another factor influencing this anisotropy is likely to be "screening" of the surface by attached H_2O molecules, which will be determined by crystal orientation. Typical values for the relative etch rates (for KOH etchants, discussed below) for the three planes of interest are (111) (reference) = 1, (100) = 300 to 400, and (110) = 600. These values are extremely dependent on the chemical composition, concentration, and temperature of the etchant solutions used. In general, however, the etch rate ratios are very large, and account for the tremendous anisotropy seen.

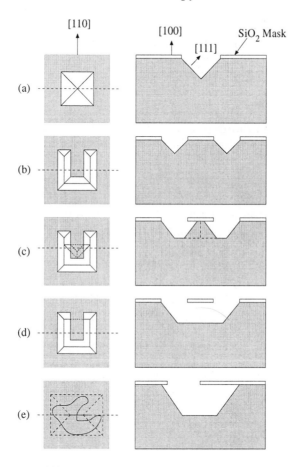

Some examples of anisotropic etching in (100) silicon: (a) typical pyramidal pit, bounded by the (111) planes, etched into (100) silicon with an anisotropic etch through a square hole in an oxide mask; (b) cantilever mask pattern with an anisotropic etch with a slow convex undercut rate; (c) the same mask pattern can result in a substantial degree of undercutting using an etchant with a fast convex undercut rate such as EDP (ethylene diamine pyrochatechol); (d) further etching of (c) produces a cantilever beam suspended over the pit; (e) illustration of the fact that anisotropic etch undercutting converges to predictable shapes after a sufficiently long time. After Petersen (1982).

Unfortunately, there are not as yet any "master equations" that predict etching performance directly from user-controllable factors (an example of a semiempirical etch rate equation can be found in Seidel (1987)). One typically needs to obtain etch rates and characteristics from the literature and by experiment. Also, many of the reported etch rates, etc., are time and usage dependent (i.e., only achieved for "fresh" etchants). Once the etch parameters and etch mixture lifetimes are well defined, most etchants give extremely reproducible results. As mentioned above, the etchant comparison of Williams and Muller (1996) is particularly useful since the entire study was carried out using a consistent set of materials.

ALKALI HYDROXIDE ETCHANTS

The hydroxides of alkali metals (i.e., KOH, NaOH, CeOH, RbOH, etc.) can be used as crystal-orientation-dependent etchants of silicon. The chemistry is still under some debate, but the basic reaction is,

$$\text{Silicon (s)} + \text{Water} + \text{Hydroxide Ions} \rightarrow \text{Silicates} + \text{Hydrogen}$$

The exact reaction is not clear, but none of this debate has much impact on the actual use of the etchants. The reaction sequence appears to be the following (Seidel (1987), Seidel, et al. (1990a)). Silicon atoms at surface react with hydroxyl ions. The silicon is oxidized and four electrons are injected from each silicon atom into the conduction band,

$$\text{Si} + 2\text{OH}^- \rightarrow \text{Si(OH)}_2^{2+} + 4e^-$$

Simultaneously, water is reduced, leading to the evolution of hydrogen,

$$4\text{H}_2\text{O} + 4e^- \rightarrow 4\text{OH}^- + 2\text{H}_2$$

The complexed silicon, Si(OH)_2^{2+}, further reacts with hydroxyl ions to form a soluble silicon complex and water,

$$\text{Si(OH)}_2^{2+} + 4\text{OH}^- \rightarrow \text{SiO}_2(\text{OH})_2^{2-} + 2\text{H}_2\text{O}$$

Thus the overall reaction is,

$$\text{Si} + 2\text{OH}^- + 2\text{H}_2\text{O} \rightarrow \text{SiO}_2(\text{OH})_2^{2-} + 2\text{H}_2$$

For KOH, Seidel (1987) demonstrated that at 72°C the etch rate was maximal at 15 weight percent (wt%) KOH (it should be considered that typical KOH pellets are anhydrous, but absorb H_2O rapidly from air), and for increasing concentrations, the rate decreased with the fourth power of the water concentration (since water is consumed by the reaction). Using 15 wt% KOH, Seidel's reported (100) etch rate was ≈ 55 μm/h at 72°C. In general, concentrations below 20 wt% are not used due to high surface roughness and the formation of potential insoluble precipitates. A more typical concentration of KOH is in the range of 40 wt%. A thorough and useful overview of alkaline etchants, their properties and mechanisms can be found in Seidel, et al. (1990a).

It is important to note that one can also use ammonium hydroxide or the so-called "quaternary ammonium" compounds, which *contain no alkali ions* (these ions, particularly sodium, can be extremely detrimental to metal-oxide-semiconductor (MOS) transistors that may be on fully integrated transducers). These etchants are discussed below.

Formulation	Temp °C	Etch Rate (μm/min)	(100)/(111) Etch Ratio	Masking Films (etch rate)
KOH (44 g) Water, Isopropanol (100 ml)	85	1.4	400:1	SiO_2 (1.4 nm/min) Si_3N_4 (negligible)
KOH (50 g) Water, Isopropanol (100 ml)	50	1.0	400:1	approx. as above
KOH (10 g) Water (100 ml)	65	0.25 to 1.0	-	SiO_2 (0.7 nm/min) Si_3N_4 (negligible)

Table of example alkali hydroxide etchant formulations. Note that isopropyl alcohol can be added as a diluent to increase selectivity, as discussed below. After Petersen (1982).

All of these alkali hydroxide etchants exhibit extremely high selectivity, etching the (111) plane up to 400 times more slowly than the (100) plane. In addition, they can all be dopant modulated (Seidel, et al. (1990b)). The etch rate can be slowed down drastically in regions doped with boron to a concentration of $\geq$ 2×10^{19} cm^{-3}. Apparently, the mechanism of this reduction is that in the heavily doped regions, the width of the space charge region layer at the silicon surface shrinks dramatically, leading to the rapid recombination of electrons generated by oxidation reactions (rather than their confinement to the surface). The reduction in the availability of these electrons limits the reduction of water to form OH⁻ ions that are necessary for the etch reaction to proceed. Seidel, et al. (1990b) present detailed experimental work that led to this hypothesis.

Price (1973) showed that the addition of sufficient *isopropyl alcohol* (IPA), a less polar diluent, to saturate the solution (typically, a layer of IPA stays above the KOH solution) greatly increases the selectivity for (111) versus (100) planes. Although reporting much lower selectivities than other authors, he showed that the (111):(100) etch rate ratios went from 8 without IPA in 40 wt% KOH to 34 with IPA at saturation.

For most micromachining and active circuit processing (e.g., MOS devices), (100) orientation material is used, for which hydroxide etches produce pyramidal pits with 54.74° (111) sidewall angles relative to the surface. It is possible to obtain very low-etched surface roughness for a "mirror-like" finish (this is more difficult with the other anisotropic etchants such as EDP and TMAH, which are described below). In general, KOH etching produces smoother surfaces at low and high molarities, with the maximum roughness occurring at 5 to 6 M (28 to 34 wt%), decreased dramatically with stirring (presumably displacing hydrogen bubbles) and increased temperature (Palik, et al. (1991)). Hillock formation can also be suppressed by the addition of a suitable oxidizing agent (e.g., ferricyanide ions, $Fe(CN)_6^{3-}$), as described by Bressers, et al. (1996). They reported a drastic reduction in hillock formation with the use of 18 mM $K_3Fe(CN)_6$ as an additive to 4 M KOH solutions, used at 70°C.

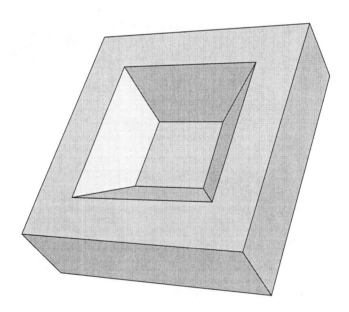

Illustration of pyramidal pit with exposed (111) planes as typically obtained using wet anisotropic etchants without etching to completion of the pyramid.

Using (110) silicon, one can obtain "perfectly" rectangular trenches over considerable distances because the etch rate is so high in the (110) direction relative to the other two planes (Bean (1978)). Tuckerman and Pease (1981) demonstrated the use of such trenches as liquid cooling fins for integrated circuits. One can use "fusion bonding" (explained below) to couple a separate micro heat sink directly to an ordinary silicon wafer (i.e. (100)) on which active circuits could be fabricated. Another useful reference on such devices is Kaminsky (1985).

A commonly asked question is whether or not one can make holes all the way through a wafer using anistotropic etches. The answer is yes, but not with a constant cross-section or high aspect ratio. This is because for (110) wafers, one obtains two perfectly parallel sides and two convergent planes that would effectively "close off" a hole. In the case of the long cooling channels shown below, these planes only appear at the ends of the trenches that were sawed off.

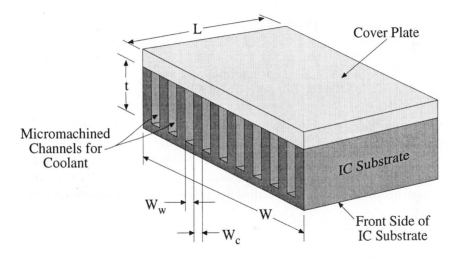

Diagram of a micromachined silicon heat sink incorporated into an integrated circuit (IC). For a 1 cm² silicon IC chip, using water as the coolant, the optimum dimensions are approximately $W_w = W_c = 57\ \mu m$ and $z = 365\ \mu m$. The channels are anisotropically etched into the (110) wafer with a KOH-based etchant. The cover plate is 7740 (Pyrex™) glass anodically bonded to the silicon (see below) forming a fluid-tight seal. Thermal resistances less than 0.1 °C/W were measured, demonstrating extremely high heat transfer capability. Adapted from Tuckerman and Pease (1981).

Further considerations of micromachining with (110) silicon can be found in Bean (1978) and Kendall and deGuel (1985). General discussions of wet etching mechanisms for alkaline etchants can be found in Seidel (1987) and Seidel, et al. (1990a, 1990b). Descriptions of cesium hydroxide as anisotropic silicon etchants can be found in Clark, et al. (1988) and Chambers and Wilkiel (1993), and rubidium hydroxide is discussed in Wang, et al. (1994). Another interesting wet silicon etching paper is Bäcklund and Rosengren (1992).

AMMONIUM HYDROXIDE

As mentioned above, certain hydroxide-based anisotropic etchants for silicon do not incorporate alkali ions that can be detrimental to CMOS integrated circuits (and sometimes prohibited in clean rooms). Ammonium hydroxide (NH_4OH) is one such etchant that has been known about for many years.

Kern (1978) demonstrated the use of NH_4OH (9.7% in H_2O) to achieve 0.11 μm/min (6.6 μm/h) etch rates in (100) Si (temperature range 85 to 92°C). Little further work appears to have been done on this subject until Schnakenberg, et al. (1990) presented their analysis. They explored a variety of concentrations from 1 to 18 wt% NH_4OH at a temperature of 75°C. They noted a maximum (100) silicon etch rate of 30 μm/h but extremely bad hillock formation (surface roughness). They reported that their best results were obtained at 3.7 wt% at 75°C for stirred etch baths, and for this recipe, they demonstrated boron-dependent etch rate modulation at 1.3×10^{20} cm^{-3} with a selectivity of 1:8,000.

Ammonia-based etchants have not been popular for a number of reasons, including their relatively slow etch rate, hillock formation problems, and rapid evaporative losses of ammonia gas (noxious) when heated. As discussed below, tetramethyl ammonium hydroxide in various formulations can be used in its place, with far superior performance and still without alkali metal contamination.

TETRAMETHYL AMMONIUM HYDROXIDE (TMAH)

Tetramethyl ammonium hydroxide (TMAH, $(CH_3)_4NOH$), is one the more useful wet etchant chemistries for silicon. It is safer than EDP (discussed below), can be modified with additives so that it does not etch aluminum, begins to slow down for boron-doping levels above approximately 1×10^{19} cm^{-3}, and is relatively low cost. Due to these features, some micromachining operations are showing a trend away from EDP and toward TMAH as a wet silicon etchant where alkali hydroxides are not suitable. In general, TMAH formulations are reported as a weight percentage of TMAH in water (additives are sometimes used to improve etch properties, as described below).

A potentially significant trade-off with the use of TMAH is that, as for NH_4OH, the surface morphology tends to be rougher than that obtained with the other common etchants, although new formulations (discussed below) appear to control this effect. In addition, the (100):(111) plane selectivity ratios tend to be much lower than for alkali hydroxides, on the order of 10 to 35 for TMAH in the

10 to 40 wt% concentration range (etching at 90°C, with corresponding etch rates of 0.5 to 1.5 µm/min). For typical TMAH solutions, etch rate and surface roughness decrease as the TMAH concentration is increased. Tabata, et al. (1991, 1992) studied these issues and found that at 5 wt%, the surfaces are quite rough due to longer H_2 bubble residence times, becoming quite smooth at approximately 20% (note that the formula given above is 10 wt%, trading off some surface roughness for lack of aluminum etching).

TMAH is in a class of chemicals referred to as *quaternary ammonium hydroxides* (also including tetra*ethyl* ammonium hydroxide, which also etches silicon). TMAH is already present in most clean rooms in "MOS-clean" grade (low sodium) since it is used in most positive photoresist developers (those that do not contain choline). As mentioned above, ordinary ammonium hydroxide (NH_4OH) can be used as a silicon etchant, but the main problem is that it is really a gas dissolved in water. It is volatile and when heated, it must be replenished constantly.

TMAH exhibits useful selectivity for boron etch-stops, the etch rate of TMAH falls off ten times at 10^{20} cm^{-3} boron concentration (conditions: 22 wt% TMAH, 90°C). Etch rate decreases up to 40:1 for 2×10^{20} cm^{-3} boron concentration (note that 2.5×10^{20} cm^{-3} is the solid solubility limit for boron in silicon) were reported by Steinsland, et al. (1995) in 25 wt% TMAH at 80°C. The etch-stop selectivity can also be improved somewhat by the addition of isopropyl alcohol (see Merlos, et al. (1993)).

A very useful property of TMAH is that typical masking layers show excellent resistance to etching. For example, silicon dioxide films exhibit typical etch rates in the range of 0.05 to 0.25 nm/min, and silicon nitride films also offer comparable performance. Some very useful TMAH etch selectivity data for various dielectrics versus (100) silicon were presented by Schnakenberg, et al. (1991a, 1991b) and Ristic, et al. (1994) and is reproduced in the table below. Additional data can be found in Merlos, et al. (1993). Two caveats are appropriate for the tabulated data: 1) TMAH is seldom used at 4 wt% and 2) *it is very difficult to compare different plasma-deposited films* because the actual stoichiometry (and hence etch rate) is extremely process dependent (one should always be extremely careful when interpreting PECVD film results).

Silicon can be dissolved in TMAH solutions and lowers the pH, provides selectivity toward aluminum metallization (but increases surface roughness), and decreases (100) etch rate. A typical TMAH formulation (a modification of the formula described in Reay, et al. (1994)) that provides excellent etch characteristics with minimal aluminum etching is 250 ml TMAH (as obtained from Aldrich Chemical Co., Milwaukee, WI, 25 wt%), 375 ml deionized (DI) water, and 22 g silicon (dissolved in solution). The mechanism underlying minimal attack of aluminum with lowered pH is related to chemical passivation of the aluminum. Ordinarily,

$Al(OH)_3$ forms and "passivates" the aluminum but will dissolve in strong acids or bases (it will form in air but can be removed using HF). With silicon doping of the TMAH solution, the pH is lowered (by consuming OH^-) and a less soluble alumino-silicate is formed at the surface.

Selectivity of TMAH Etchants for Various Dielectrics versus (100) Silicon			
Dielectric	Selectivity 4 wt% TMAH, 80°C	Selectivity (Si-doped, 13.5g/l), 4 wt% TMAH, 80°C	Selectivity 20 wt% TMAH, 95°C
Thermal Silicon Dioxide	5.3×10^3	34.7×10^3	5.2×10^3
Low-Temperature Oxide (LTO)	1.3×10^3	4.2×10^3	2.8×10^3 (360°C LTO) 3.4×10^3 (360°C LTO)
PECVD Oxide	1.4×10^3	4.3×10^3	no value given
LPCVD Silicon Nitride	24.4×10^3	49.3×10^3	38×10^3
PECVD Silicon Nitride	9.2×10^3	18.5×10^3	3.6×10^3
Source	Schnakenberg, et al. (1991)	Schnakenberg, et al. (1991)	Ristic, et al. (1994)

Table comparing published selectivities of TMAH etchants for common dielectrics.

Tabata (1995) reasoned that lowering the pH with acids ((NH_4)$_2CO_2$ or (NH_4)HPO_4) could offer the same protection of aluminum without the difficulty of having to dissolve silicon. He observed that if the pH was lowered from 13 (for 22 wt% TMAH) to 12, the Al etch rate dropped from 1 μm/min to 1 nm/min and similarly for 10 wt% TMAH (lowering the pH from 12.5 to 11.5, with the same three order of magnitude Al etch rate decrease). The corresponding amounts of dissolved silicon required for the same pH changes would be 3 and 1.3 mol/l. The reported hillock size was $\approx$ 5 μm with etch rates > 0.7 μm/min. Thus it seems that the mechanism of aluminum protection in TMAH is indeed driven by lowered pH and is not inherently related to the polysilicic acid formed by silicon dissolution.

An alternative approach to lowering the pH (same effect as dissolving silicon) is to add polysilicic acid directly to the TMAH solution, as demonstrated by Hoffman, et al. (1995). Using a solution of 80 ml of 25 wt% TMAH with 16 g silicic acid and sufficient DI water to bring the volume to 250 ml, they reported an etch rate of 35 to 70 nm/min at 70°C and relatively isotropic etch results, but with little attack of exposed aluminum. While neither the etch rates nor the degree of anisotropy were optimized, the addition of polysilicic acid to prevent aluminum attack is a useful alternative to dissolving silicon.

Hydrogen evolved from the etch reaction can form bubbles that can in turn cause local "micromasking," resulting in hillocking of the etched surface. If a sufficiently high hillock density forms on the surface of the silicon, the (100) etch rate can drop a great deal. It is thus logical to consider the use of oxidizers to consume the hydrogen as it is generated. The addition of oxidizers to such etchants (including the other alkaline etchants) has been investigated by several groups, including Schnakenberg, et al. (1991a), Campbell, et al. (1995), and Schnakenberg, et al. (1991b), with various degrees of success (difficulties included the necessity to carefully control oxidizer concentrations, reduction in etch rates, and etching of aluminum despite loading TMAH with silicon). Klaassen, et al. (1996), described the use of peroxydisulfate oxidizers in TMAH to eliminate hillock formation yet preserve aluminum etch protection via added silicon or silicic acid (adding peroxydisulfates actually increases etch rates on the order of 25%, presumably through elimination of hydrogen masking). The optimum formulation reported was 5 g/l ammonium peroxydisulfate ($(NH_4)_2S_2O_8$) in 5 wt% TMAH solution with 16 g/l dissolved silicon at 80°C for an etch rate of ≈ 0.8 μm/min and a 6 to 8 hour working life when mixed (due to dropping pH over time).

ETHYLENE DIAMINE PYROCHATECHOL (EDP)

Ethylene diamine pyrochatechol (EDP, sometimes referred to as EPW for ethylene diamine pyrochatechol water) is a classic, but hazardous, anisotropic dopant-modulated silicon etch, as described by Finne and Klein (1967), Bassous (1975), and Reisman, et al. (1979). It should be noted that the (100):(111) selectivity of EDP formulations is on the order of 35, much lower than those of alkali hydroxide etchants, but its selectivity for heavy p-type doping is much greater. A typical EDP formulation (Reisman, et al. (1979)) is 1 liter ethylene diamine, 160 g pyrochatechol, 6 g pyrazine, and 133 ml water. Petersen (1982) also gives two formulations, along with etch rates of silicon and masking films, shown in the table below.

Formulation	Temp °C	Etch Rate (μm/min)	(100)/(111) Etch Ratio	Masking Films (etch rate)
Ethylene diamine (750 ml) Pyrocatechol (120 g) Water (100 ml)	115	0.75	35:1	SiO₂ (0.2 nm/min) Si₃N₄ (0.1 nm/min) Au, Cr, Ag, Cu, Ta (negligible)
Ethylene diamine (750 ml) Pyrocatechol (120 g) Water (240 ml)	115	1.25	35:1	as above

Table of EDP formulations, showing etch rates for silicon, (100)/(111) etch rate ratios, and etch rates of masking films. After Petersen (1982).

The basic chemistry of EDP etching (after Finne and Klein (1967)) includes the following steps:

Ionization of Ethylenediamine:

$$NH_2(CH_2)_2NH_2 + H_2O \rightarrow NH_2(CH_2)_2NH_3^+ + OH^-$$

Oxidation - Reduction (oxidation of silicon):

$$Si + 2OH^- + 4H_2O \rightarrow Si(OH)_6^{2-} + 2H_2$$

Chelation of Hydrous Silica:

$$Si(OH)_6^{2-} + 3C_6H_4(OH)_2 \rightarrow [Si(C_6H_4O_2)_3]^{2-} + 6H_2O$$

Heavy ($> 7 \times 10^{19}$ cm^{-3}) boron doping results in a 50 times slowing of etch rate, and this etch is still 35:1 anisotropic for the (111) versus the (100) planes. One can use epitaxially grown silicon that is *in situ* doped for such an etch stop or boron can be diffused into the wafer (the diffusion can be oxide masked). EDP etching is readily masked using SiO$_2$, Si$_3$N$_4$, Au, Cr, Ag, Cu, Ta, and many other materials (this is often a key consideration when choosing etchants). The fact that it does not readily attack SiO$_2$ means one can use very simple (native oxide) masks, and much of the early bulk etching work was done this way.

Some EDP etchants attack aluminum quickly, which can be a major constraint to micromachining with "standard" processes such as foundry CMOS. The formulation given above has one of the lowest aluminum etch rates (400:1 Al versus (100) silicon), as described by Moser (1993) (he provides an excellent review of the use of EDP with standard CMOS as a post-processing step). Moser (1993) published silicon etch rates versus temperature for the EDP formulation given above of 14 µm/h at 70°C, 20 µm/h at 80°C, 30 µm/h at 90°C, and 36 µm/h at 97°C. He also describes a cure for the common problem of etch pits being coated with polymerized Si(OH)$_4$ and the aluminum bond pads with aluminum hydroxide Al(OH)$_3$. Moser's post-EDP etch protocol consists of a 20 s rinse in DI water, a 120 s dip in 5% ascorbic acid solution (vitamin C), a 120 s rinse in DI water, and a 60 s dip in hexane (C$_6$H$_{14}$) helps to prevent any undercut microstructures from sticking during drying.

While such procedures can improve the etching results, EDP mixtures are very corrosive and potentially carcinogenic (as are many materials in typical clean rooms, including photoresist)! EDP etchants require the use of a reflux condenser, are incredibly corrosive, and are usually never allowed in most clean rooms used for "mainstream" integrated circuit fabrication.

HYDRAZINE

As noted in Petersen (1982), hydrazine (H_4N_2)/water mixtures can serve as anisotropic silicon etchants. The recipe of 100 ml hydrazine (H_2N_4) in 100 ml water (and/or isopropanol), used at 100°C, yields an etch rate of 2 μm/min, has no doping dependence, and can be masked with silicon dioxide or aluminum. No (100):(111) etch rate selectivity was reported. A detailed study by Mehregany and Senturia (1988) discussed the 50:50 hydrazine/water mixture in detail. They reported (100) silicon etch rates over the temperature range of 70 to 120°C (the latter being the boiling temperature for the mixture) of ≈ 0.8 to 2 μm/min for heavily antimony doped wafers and 1.5 to 3.3 μm/min for moderately doped samples. The (100):(111) selectivities measured were lower than those for KOH or EDP. At 118°C, silicon dioxide was etched at ≈ 10 nm/h, with the etch rates for silicon nitride, Ag, Au, Ti, and Ta not being measurable. It was also stated that Al, Cu, Zn, and organic polymers are rapidly etched by hydrazine. It should be noted that hydrazine is highly corrosive, potentially explosive, requires the use of a reflux reactor, and is a suspect carcinogen, making it somewhat unattractive as an etchant for micromachining.

AMINE GALLATES

Amine gallate etchants, as described by Linde and Austin (1992), are comprised of a mixture of ethanolamine (high boiling point solvent), gallic acid, water, pyrazine, hydrogen peroxide, and a surfactant. These etchants achieve high etch rates (up to 140 μm/h on (100) Si) and stop at high boron concentrations (> 3×10^{19} cm^{-3}, a lower concentration than it takes to stop EDP). Amine gallates are similar to EDP in terms of etching and masking layers (gentle on SiO_2), but apparently safer. Peroxide and pyrazine can be added to increase the etch rate, but surface roughness also increases. There is not much information on amine gallate etchants in the literature, probably due to the success of alternative etchants such as EDP and TMAH (see below).

An example of amine gallate etchant formulation from Linde and Austin (1992) is 100 g gallic acid, 305 ml ethanolamine, 140 ml water, 1.3 g pyrazine, and 0.26 ml of 10% FC-129™ surfactant (Aldrich Chemical Co., Milwaukee, WI). When used at 118°C, the reported etch rate is 1.7 μm/min, with a (100):(111) selectivity in the range of 50 to 100, and etch rate reduction by ten times for boron doping levels above 10^{20} cm^{-3}. This etch formulation can readily be masked using silicon dioxide, silicon nitride, or a number of metal films (e.g., Au, Cr, Ag, Cu, and Ta).

ULTRASONIC AGITATION IN WET ETCHING

In many typical wet etching applications, two principal non-idealities are often encountered: geometry-dependent etching and surface roughness due to masking by bubbles (as described above in the case of TMAH).

Geometry-dependent etching refers not to crystal-plane-dependent anisotropy, but, rather, to effects of mask feature size and interactions between adjacent etched structures due to their proximity to one another. For example, the etch rates for long, narrow grooves in (110) silicon is strongly influenced by the width of the grooves and their pitch, as described by Ohwada, et al. (1995). Similar effects are observed in plasma/RIE etching and are generally attributed to aperture effects and local reactant depletion ("loading") as features crowd together, respectively.

Ohwada, et al. (1995) used ultrasonic agitation at 28 kHz or 28, 45, and 100 kHz alternated at $\approx$ 1 ms intervals to improve local agitation in KOH etching of (110) silicon such that feature size and spacing had almost no effect (for the triple-frequency case) on etch rate (for test grooves of 1 to 10 μm in width). This can be of great importance when varying features are desired to etch to a uniform depth. These investigators theorized that the improved results seen in the switched frequency case are due to the fact that movement of the standing waves established in the fluid by the ultrasonic energy helps to "stir" the fluid at the reaction sites.

Surface roughness ("hillock" formation) due to micromasking is often seen in wet etch processes wherein the products of the etch reaction include gases (typically H_2 for wet etching of Si). Several approaches to mitigating this effect have been tried, including the use of surfactants, the use of ultrasonic agitation and preemptive chemical elimination of the bubbles, as discussed above with respect to TMAH etching. Ohwada, et al. (1995) also noted that their use of triple-frequency (switched) ultrasonic agitation (rather than chemical means) essentially completely eliminated surface roughness in KOH etching.

ETCH-STOP LAYERS FOR DOPANT-SELECTIVE ETCHANTS

As discussed above for individual etchants, highly p-doped (p++) silicon regions greatly attenuate the etch rate. Selective p++ doping, typically done using a gaseous or solid boron diffusion source with a mask (such as silicon dioxide) to select doped regions, can be used to define specific regions of the silicon that remain, while the bulk is etched away. A classic example of this type of "lost wafer" process is that of Najafi, et al. (1985) used to fabricate needle-like probes for recording the electrical activity of neural cells. It should be noted that a limitation of diffusing in the dopant is the maximum depth that is practically achievable (on the order of 15 μm).

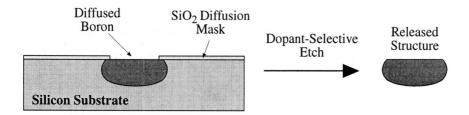

Illustration of the use of heavy boron doping with a dopant-selective etch to form free structures defined by the extent of the boron diffusion at a dopant-stopping concentration.

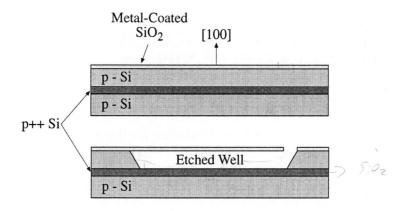

An example of the use of a buried etch-stop layer to fabricate movable cantilevers. After Petersen (1982). The silicon is anisotropically etched from under the patterned, metal-coated silicon dioxide to release the beams. EDP etching automatically terminates on the buried p++ layer.

Surface or "buried" p++ etch-stop layers can be epitaxially grown and used very successfully with EDP, TMAH, and KOH-type etchants. It may be necessary to add some germanium to reduce strain in a boron-doped p++ epitaxial layer, since boron is introduced substitutionally into the Si lattice, and since boron is a smaller atom, the layer is under tensile stress. Germanium is a larger atom than silicon, and helps to compensate. Such buried boron-doped silicon layers can be used to form the bases of etched pits or to fabricate membranes using back-side etching, as illustrated below.

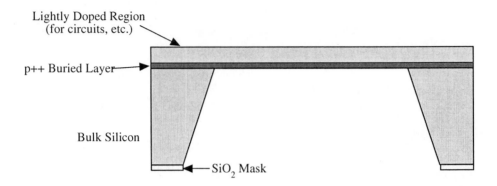

Illustration of the use of a buried etch-stop layer to fabricate a membrane with precisely controlled thickness.

5.2 WET ETCHING OF GALLIUM ARSENIDE AND RELATED III-V COMPOUNDS

GaAs wet etching is similar to that of Si in that (111) planes etch slower than others; however, all (111) planes are not equal since some are entirely Ga and others entirely As atoms. The usual notation for this is (111)A = arsenic, reactive face, and (111)B = gallium, passive face. Many oxidizers will serve as GaAs etchants: peroxides, halogens such as bromine, $K_3Fe(CN)_6$, $Ce(SO_4)_2$, and $KMnO_4$, for example. The oxides produced are amphoteric (dissolve in acidic or basic solution) or, in some cases, organic solvents such as methanol. Gallium (111)B faces etch slower than others and mesa structures (if properly aligned) can have two opposite edges with gently sloping walls and two with undercut walls. This behavior is seen with most *zincblende* structures (i.e., all III-V semiconductors).

It is useful to compare etch rates for different crystal planes in Si and III-V compounds: diamond-like (e.g., Si) {100} > {110} > {111} and zincblendes (e.g., GaAs): {111}A > {100} > {110} > {111}B. Note that some hydroxide etches (e.g., KOH) effectively swap the {100} and {110} relative etch rates for III-V compounds.

A useful overview of GaAs micromachining and mechanical devices is available from the Materials Science Division of Uppsala University (Hjort (1993)). Several micromachining processes (wet etching, dopant selective etching, fusion bonding, etc.) are described for GaAs. Three example GaAs etchants are 1) $Br_2:CH_3OH$ (1:99) with well-defined (111) planes but rough bottoms if deeper than ≈ 10 μm, 2) $H_3PO_4:H_2O_2:CH_3OH$ (1:1:3) with less well-defined (111) planes but smooth bottoms, and 3) $H_2PO_4:H_2O_2:H_2O$ (3:1:50) with similar behavior to the second formula.

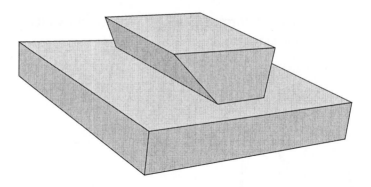

Illustration of a mesa structure possible in III-V compounds (see Runyan and Bean (1990)).

An etchant system with $\approx$ 10:1 to 100:1 selectivity between AlGaAs and GaAs is citric acid:H_2O_2:H_2O (one possible composition is 5 g:2 ml:5 ml) and can be useful for sacrificial etches of epitaxial layers, for example. (It is noteworthy that since Al is generally more reactive than Ga, most selective etchants etch $Al_xGa_{(1-x)}As$ faster than GaAs.) This etchant is discussed in Juang, et al. (1990) for GaAs/$Al_{0.3}Ga_{0.7}As$. They obtained optimum selectivity ($\approx$ 95:1 GaAs to AlGaAs etch ratio) using a 10:1 ratio of citric acid (50% by weight) and H_2O_2 (30%) at room temperature (note that the GaAs must be semi-insulating for this etchant to work as described). This etch is safe and environmentally friendly. One can readily fabricate AlGaAs membranes using such an etch system. It is also worth noting that very high selectivity ($\approx 10^7$) of $Al_xGa_{(1-x)}As$ over GaAs can be obtained using an HF etchant when x, the Al fraction, exceeds $\approx$ 0.4 to 0.5 (see Konagai, et al. (1978)). Other useful GaAs and general III-V semiconductor wet etch references include Kern (1978), Shaw (1981), Juang, et al. (1990), and Yablonovitch, et al. (1987).

5.3 ELECTROCHEMICAL MODULATION OF WET ETCHING

Injection of holes to permit Si to be oxidized to hydroxides can be controlled by external application of a potential, and this approach can be used with various etchants to obtain different effects. An electrochemical circuit is formed in a solution, with the silicon held positive (anodic) relative to a platinum counterelectrode. Typically, HF/H_2O solution is used for very smooth silicon etching (referred to as "electropolishing"). The positive applied voltage provides a source of holes at the silicon surface that attracts OH⁻ ions; this causes the silicon to oxidize, and since the oxide is very quickly dissolved by HF, the reaction continues. Silicon nitride works quite well as a masking layer for such HF etches, in particular, LPCVD films. As discussed below, if the applied potential is reduced below a certain point, *porous* silicon is formed in place of a smoothly etched surface (this can be very useful in some situations).

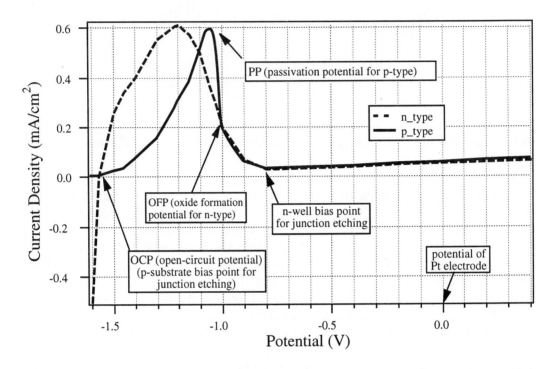

Plots representing the I-V characteristics of n- and p-type (100) silicon in 40% KOH solution at 60°C. After Kloeck, et al. (1989).

In a general electrochemical etching system, one can examine the I-V characteristics of the surface being etched (relative to an inert counterelectrode in the solution) in order to better understand the (electro)chemical mechanisms. For a typical etchant used in an electrochemically modulated mode (involving oxidation and subsequent dissolution of silicon), the I-V characteristics for n- and p-type materials are similar to those shown above. For both doping types, there is an *open circuit potential* (OCP) at which there is nearly zero current flow and the silicon etches just as if it were not in a circuit at all (i.e., just dropped into the etchant solution). As the potential applied between the silicon and the solution is made more positive (anodic current flow), more holes are supplied to the surface silicon atoms, speeding up the oxidation of Si into $Si(OH)_2^{2+}$, which subsequently reacts further with water and is dissolved. As the applied potential is made more positive, eventually the *passivation potential* (PP) is reached where SiO_2 is formed, effectively passivating the surface and stopping etching. For etchants like KOH, TMAH, etc., this means that one can electrically modulate the etching up and then completely *off* if desired. If HF/H_2O is used, the etch rate can be increased without a passivation limit since SiO_2 dissolves readily in HF. It is worth noting that Al and Ti also exhibit such passivation potentials. Another interesting point is that the potentials used to control such etch-stop phenomena can sometimes be generated from mono-lithic photovoltaic cells, as demonstrated by Lapadatu, et al. (1995) (however, this approach will neither be area efficient nor generally applicable).

5.3.1 DIODE JUNCTION ETCH-STOP

Another approach to electrochemical etch modulation is to form a p-n junction on the surface of a p-type wafer and set things up so that etching stops at the n-type layer's surface (i.e., when the diode is destroyed). A potentiostat is used to control the voltages in the system. The potentiostat essentially is a voltage regulator, using feedback from the reference electrode to control the voltage between the wafer and the solution (the reference electrode scheme helps to account for resistive voltage drops through the bulk solution since the reference electrode is physically very close to the wafer).

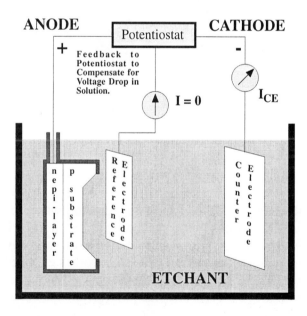

A standard three-electrode system for electrochemical etch-stops. After Kloeck, et al. (1989).

As shown above, the p-type silicon floats at its OCP (or is held there electrically to prevent it from shifting due to potential leakage currents, but this is not illustrated above) and etches quickly. The p-n junction is held in reverse bias (positive potential applied to the n-type silicon) so that when the diode is etched away, the positively biased n-type silicon is directly exposed to the solution and stops etching The bias voltage is applied to the n-type silicon, and the p-type silicon is either biased near the OCP or allowed to float. The figure above shows the simpler case, with the p-type silicon floating.

The *n-type silicon is biased well above its passivation potential* so that as soon as the p-type silicon is etched away, the exposed n-type immediately passivates and the etching stops. In electronic terms, a positive potential is applied to the n-type silicon, effectively reverse-biasing the p-n junction. Thus, the voltage drop is primarily across the p-n junction and the p-type silicon sits at roughly its open-circuit potential until the diode is destroyed, at which point the voltage appears directly between the n-type silicon and the solution and etching stops. By stopping etching at a well-defined junction, one can achieve very accurate thickness control of membranes (see Kloeck, et al. (1989)). This is routinely done in industry (i.e., using n-epitaxial silicon on p-type, as shown in the above diagram).

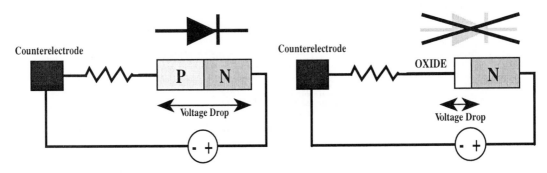

Illustration of diode junction etching.

Ashruf, et al. (1997) described such an etch-stop wherein no electrical connections to any of the structures being etched were required. Their approach was to make use of a Au/Cr/n-Si/TMAH electrochemical cell to provide the necessary voltage to passivate the n-silicon region when it becomes exposed by dissolution of the p-silicon beneath it. A drawback to this process is that the sides of the wafers or dice being etched must be electrically insulated for the etch stop to work properly.

JUNCTION ETCH-STOP USING TMAH AND STANDARD CMOS

Using TMAH, the junction etch-stop technique can be used to postprocess prefabricated CMOS chips (with little or no etching of Al bond wires and no damage to standard ceramic DIP packages) as received from a foundry (Reay, et al. (1994)).

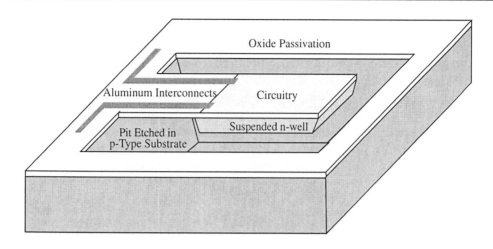

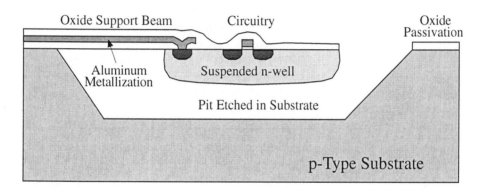

Example of electrochemically controlled TMAH etching of standard CMOS. Courtesy R. Reay and E. Klaassen.

Reay, et al. (1994) biased the p-type wafer near its OCP (rather than letting it float) to make sure the etching proceeded (even a relatively high-resistance leakage path across the p-n junction can stop the etching of the p-type silicon). Thermally and electrically isolated single-crystal silicon islands (including active circuitry) could be formed on fully packaged CMOS chips using this approach. The standard n-wells in the CMOS process were used to form the n-type part of each diode. The n-wells that were to be undercut were biased to -0.8 V (more positive than the PP) relative to a platinum counterelectrode and the p-type substrate was biased at -1.6 V (near its OCP). Areas to be etched away were defined by superimposing device *active area*, *contact cut*, *via cut,* and *pad* openings in the CAD layout. Etching was carried out in silicon-doped TMAH (discussed above) to prevent attack of aluminum bond-pads and other structures (the added Si scavenges OH^- ions, lowering pH and thus reducing Al etching).

A variant of this process was demonstrated by Olgun, et al. (1997) in which very low thermal mass p+ silicon/n polysilicon thermopiles were fabricated. By following the electrochemically controlled n-well etch with an EDP etch to remove the n-wells, only the p+ regions remained suspended beneath the silicon dioxide supports. Schneider, et al. (1997) also demonstrated a similar process, but where both p- and n-type MOSFET's could be integrated onto the same suspended silicon island. This was accomplished through the use of deep n-wells to completely surround any desired p-wells, which were thereby protected during the electrochemically modulated KOH etch used.

5.3.2 PHOTON-PUMPED ELECTROCHEMICAL ETCHING

Another interesting approach to modulation of wet etching is to photon-pump to generate the holes to drive the etching of silicon. Peeters, et al. (1994) described one such technique using photon pumping in a KOH system to generate the carriers across a p-n junction. Lehmann and Föll (1990) and Lehmann (1996) used photo-generated carriers to supply the necessary holes to the bases of prefabricated pits whose sharp bases served "high field points" to "focus" the holes and to obtain very highly anisotropic etch results. HF was used as the etchant, and its anodic I-V characteristic has two important ranges such that below a critical current density (J_{PSL}), porous silicon is formed, and above it, electropolishing occurs. They used anodically biased n-type silicon and demonstrated that a broad range of hole morphologies could be etched. Extremely deep, high-aspect-ratio holes (> 70:1) and trenches were obtained (remarkable for a wet etch that is not crystal plane dependent). This is not a trivial process to undertake, but allows formation of porous silicon as well (under voltage control) (see Lehmann and Föll (1990)).

Lehmann (1996) described the application of this technology to the formation of arrays of 2 μm pores 165 μm deep that were insulated with a dielectric layer and then coated with doped polysilicon to form dense capacitors (specific capacitance ≈ 4 μF•V/mm^3). This approach may only offer significant benefits in certain applications such as this (where high-aspect-ratio dry etching cannot be used to obtain equivalent performance). Another useful reference for this type of etching is Lehmann (1993).

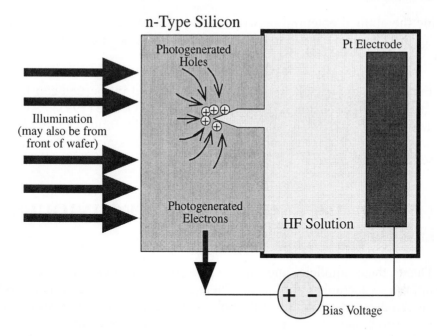

Illustration of photon-pumped electrochemical etching used to form high-aspect-ratio holes in silicon. After Lehmann and Föll (1990).

5.4 POROUS SILICON FORMATION

If silicon etching is performed in very concentrated HF (i.e., 48% HF, 98 wt% ethanol) or solutions otherwise deficient in OH⁻, with an *anodic* electrochemical bias on the silicon, the silicon is not fully oxidized during etching, producing a brownish film of porous silicon. The porous silicon is still a single-crystal, but sponge-like since it is permeated by voids. Its single-crystal nature can be exploited by growing epitaxial Si over porous silicon. Porous silicon can have interesting optical properties, including fluorescence and electroluminescence (with an appropriate electrode). For example, after 6 hours in 40% HF, a yellow film is formed that glows bright red under illumination with appropriate wavelengths (an Ar laser at 488 or 514.5 nm).

The density of the porous layer is controllable. The higher the applied current density during etching, the lower the density of the porous silicon (e.g., one can vary this over time to get buried lower density layers). Again, silicon nitride works well as a mask for this.

Anderson, et al. (1994) describe a method for selectively forming porous silicon between two low-stress silicon nitride layers, starting with a sandwiched polysilicon layer. They demonstrated chambers surrounded by porous plugs by changing the applied potential between the level at which electropolishing occurs (forming the chamber) and porous silicon formation occurs (forming the porous plugs).

Further details of porous silicon formation and properties can be found in Beale, et al. (1985), Canham (1990), and Lehmann and Gösele (1991). An interesting example of its use as a sacrificial layer can be found in Steiner, et al. (1993). The formation and use of porous polycrystalline silicon has also been studied, and is discussed in Anderson, et al. (1994).

5.5 OTHER BULK WET ETCH TECHNIQUES AND MATERIALS

Due to their simplicity and low cost, wet etch techniques are likely to remain a popular micromachining approach. There are a number of such methods beyond those described above, and several apply to materials other than silicon or III-V compound semiconductors. Potentially useful examples of these techniques are presented below.

5.5.1 ANISOTROPIC WET ETCHING OF POROUS ALUMINUM

It is well known that aluminum can be anodically oxidized in certain dilute acid solutions (chromic, oxalic, phosphoric, sulfuric, etc.) and that a porous aluminum oxide forms due to dissolution of the oxide as it forms. Columnar "cells" of aluminum oxide 40 to 300 nm in diameter form during oxidation, with central pores much smaller than this that form due to electrochemical etching of the oxide from the bottom of the pores, which is enhanced by the high electric field there.

Tan, et al. (1995), demonstrated high-aspect-ratio aluminum microstructures that were fabricated from pure aluminum substrates by forming a porous anodic oxide on the entire substrate and masking regions of the oxide using patterned Cr, sealing the pores in those regions. The unmasked anodic oxide regions were rapidly etched in a weak acid, apparently by rapid lateral etching between the pores. Their process began cleaning in trichloroethane, acetone and 2-propanol (in an ultrasonic bath) and electropolishing the substrates in a solution of 130 ml H_2O, 40 g CrO_3, 260 ml 85% H_3PO_4, and 80 ml 98% H_2SO_4 at 80°C at a current density of 0.2 A/cm^2 for 3 min. Sulfuric, oxalic and phosphoric acids in the 2 to 10% concentration range were used for anodization relative to a Pt counterelectrode at a current density of 60 mA/cm^2. For 5% H_2SO_4 anodizing solution at 20°C and 60 mA/cm^2 (not regulated, constant 24 V voltage used), the anodic film formation rate was ≈ 1 µm/min. After deposition and patterning of the Cr mask layer, the exposed porous anodic oxide was removed by etching in 5% sulfuric acid solution at 40°C (no electrical bias for strictly chemical etching) and resulted in vertical etch rates of ≈ 2 μm/min, with less than 1 nm/min lateral etch rate.

5.5.2 ANISOTROPIC WET ETCHING OF QUARTZ

Quartz (single-crystal SiO_2) has a hexagonal lattice, with no center or plane of symmetry, and is an electrical insulator with very useful piezoelectric properties. The photolithographic micromachining of quartz dates back at least to the late 1960's, when the first quartz tuning forks were fabricated for wrist watches (see Staudte (1968) and Staudte (1973)).

Quartz can be anisotropically etched using wet etchants that attack SiO_2, such as HF, ammonium flouride (NH_4F), or saturated ammonium bifluoride (NH_4HF_2). A complete discussion of this is presented by Danel, et al. (1990), Danel and Delapierre (1991), and Hedlund, et al. (1993). Typically NH_4HF_2 is used, at a temperature of 82°C (see Ueada, et al. (1985) and Danel, et al. (1990)). In terms of the crystallographic axes of quartz, the etch rates can be as high as 75 μm/h along the z-axis, many fold higher than for the x- or y-axes, providing considerable anisotropy (it should be noted that the resulting etch profiles are dependent upon the type of etchant and etching conditions). Thus, most work reported to date has been carried out using z-axis quartz wafers (z-axis perpendicular to the wafer surface). A table of etch rates versus etchant compositions is given below (note that these data are for a temperature of 25°C, while most etching reported in the literature was carried out at temperatures of ≈ 80°C), as presented by Vondeling (1983) in a paper containing an overview of fluoride-based etchants for quartz. Vondeling (1983) also discussed the effects of etchant composition and etching conditions on the resulting surface roughness.

Etchant Formula (mol/l)	x-Axis Etch Rate (μm/hr)	y-Axis Etch Rate (μm/hr)	z-Axis Etch Rate (μm/hr)
10.9 HF	0.02	< 0.005	9.6
7.2 HF + 4 NH_4F	0.025	0.005	2.55
5.4 NH_4NH_2	0.015	0.015	1.1
5.4 NH_4HF_2 + 1.8 NH_4F	0.015	0.015	0.75

Table of quartz etch rates for various etchant formulations, along x-, y-, and z-axes at 25°C. After Vondeling(1983).

Generally, the etching is masked using thin metal patterns (e.g., 200 nm Au/50 nm Cr, as used by Toshiyoshi, et al. (1993)), which can be deposited, then patterned using lift-off, wet etching, or ion milling with a photoresist mask. The metal mask can later be patterned again to define the drive/sense electrodes for quartz transducers.

A number of micromachined quartz devices have been demonstrated, such as a torsional differential pressure sensor (Ueda, et al. (1985), a magnetically driven, force-balanced accelerometer (Danel, et al. (1990)), a tuning-fork resonator (Clayton, et al. (1989)), an optical chopper (Toshiyoshi, et al. (1993)), and micro-galvanometers (Ueda, et al. (1985) and Sugiyama, et al. (1991)).

Toshiyoshi, et al. (1993) demonstrated a process (illustrated below) to fabricate z-axis quartz microactuators of various geometries that would provide maximal mechanical movement. They began with 100 μm thick, z-axis quartz substrates, on which 50 nm Cr and 200 nm Au were deposited on both sides by sputtering. Using a photoresist mask, the metals were patterned to form etch masks. Before the quartz was etched, a new photoresist layer was applied to each side and patterned, allowing the metal to be formed into the desired electrode shapes after quartz etching. The quartz etching was carried out in saturated NH_4HF_2 at 82°C, after which the pre-patterned photoresist was used as a mask to ion mill the metal into the final electrode shapes. For some actuator designs, they reported displacements of up to 170 μm p-p with 120 V AC applied.

A useful review of the design of quartz microstructures, with discussion of the mechanical analysis and modeling required for resonator design (in particular, modal analysis), can be found in Clayton, et al. (1989).

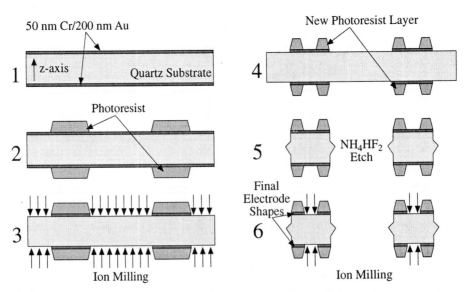

Illustration of a quartz micromachining process. After Toshiyoshi, et al. (1993).

5.5.3 ION-IMPLANT-ASSISTED WET ETCHING

Ion implantation at sufficiently high doses can be used to alter the stoichiometry of a substrate in desired regions, making those regions susceptible to chemical attack. This approach can yield anisotropic etch results in materials for which this is otherwise impractical, such as sapphire.

Sapphire, or single-crystal Al_2O_3 can be grown on (100) silicon by CVD (chemical vapor deposition) or obtained as whole wafers (this is the inverse of the process of growing epitaxial silicon on sapphire that is sometimes used for fabricating active circuits for high radiation environments). Ishida, et al. (1995) demonstrated that regions of single-crystal Al_2O_3 that were implanted with a dose of 3×10^{15} cm^{-2} of silicon atoms (at 80 kV), forming aluminosilicate ($Al_2O_3 \bullet SiO_2$) could readily be etched in aqueous HF solution, while the nonimplanted regions were left untouched. This approach may have utility in the fabrication of flow channels and other structures in sapphire wafers, which are otherwise fairly difficult to micromachine due to its chemical inertness. It is important to note, however, the reported etch depths achievable were only ≈ 0.1 μm deep.

5.5.4 ION-TRACK DAMAGE-ASSISTED WET ETCHING

It is known that the damage to single-crystal materials that occurs due to the transit of high-energy ionizing radiation can be used to selectively etch damaged regions. Since ionizing radiation from beam sources can be highly collimated, it is possible to facilitate anisotropic etching of many materials with bombardment with such radiation.

For example, there is considerable interest in micromachining single-crystal quartz due to its excellent piezoelectric, optical and mechanical properties, yet anisotropic etching that is not determined by crystal planes has not been easy to achieve. Hjort, et al. (1996) reported the use of bombardment with heavy ions (^{129}Xe or ^{197}Au) to form ion tracks. Impinging relativistic (≥ 10 MeV/nucleon) ions form localized plasmas along their paths, resulting in highly anisotropic "tubes" of damaged material, ≈ 10 nm in diameter, with depths of up to centimeters). They precoated Z- or AT-cut quartz wafers with a 200 nm Au/Cr layer to be used as a mask and irradiated the wafers through that layer at doses between 10^8 and 10^{10} cm^{-2}. The mask was then patterned photolithographically with wet etching, and the exposed damaged quartz was etched using a 20 M KOH solution at 143°C (near boiling) for up to 16 hours. The resulting structures showed aspect ratios on the order of 100 at depths up to several hundred microns. It was also noted that, after etching, a 1,200°C anneal "heals" tracks left in the desired structures, although the piezoelectric coefficient would be greatly reduced at temperatures above ≈ 570°C. It is important to consider that, for such processes, the entire surface is damaged by the radiation, and it does not seem feasible to protect some regions selectively.

General discussion of nuclear track damage as a micromachining tool can be found in Fischer and Spohr (1983).

5.5.5 ONE-SIDED WAFER ETCHING

It is often necessary to etch only *one* side of a wafer at a time, and many home-brew contraptions have been built and many people have used things like waxes and epoxies in an effort to protect one side (quite often such materials are very hard to completely remove). Kung, et al. (1991) described an inexpensive apparatus for clamping a wafer and exposing only the side to be etched (note that many O-ring materials leave residues on wafers).

They also presented a simple and useful automatic leveling circuit that keeps the etching chamber full as some of the water boils off (a common problem with heated aqueous etchants). The auto leveling circuit measures the resistance of the etch bath and can provide additional water as necessary to keep the resistance (and hence solution concentration) constant. The comparator, through the transistor output buffer, opens a solenoid valve when more water is needed (i.e., when the fluid resistance increases past a preset threshold. It is worth pointing out that this idea can be generalized to closing a loop around a fluid bath with some type of sensor (conductivity, pH, ion, etc.) and automatically controlling that parameter.

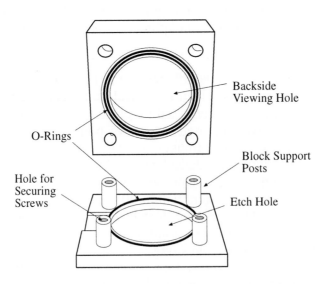

Illustration of a one-sided wafer etcher design. After Kung, et al. (1991).

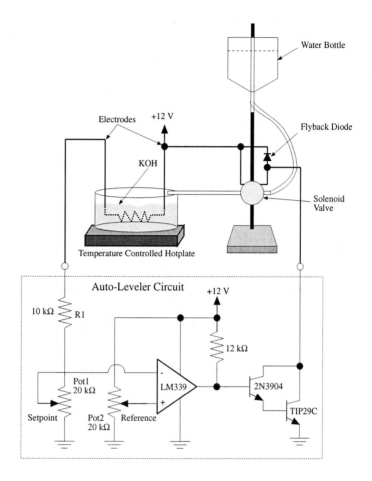

Example of an automatic aqueous etchant leveler system. After Kung, et al. (1991).

A similar O-ring-based one-sided etching chuck design, complete with optical end-point detection (accomplished by shining light through the wafer during etching and measuring the transmitted light) was described by Brugger, et al. (1997).

5.5.6 GENERAL WET-ETCHING CONCEPTS

General theory of crystalline etching (Gibbs (1928)) states that the equilibrium form of a crystal is that which minimizes the surface energy (i.e., the energy that can be liberated by chemical reaction there). A useful discussion of the use of free energy theorems to predict the shapes of etched or growing crystals can be found in Jaccodine (1962). There are numerous etchants for a multitude of materials described in the literature, but very few of them have been used for micromachining (e.g., micromachining of single-crystal quartz has been investigated in detail, as described above). It is likely that some interesting materials and etchants will be "rediscovered" by studying literature of other disciplines.

5.6 VAPOR-PHASE DRY ETCHING

In addition to plasma-based methods (described below), dry etching can be achieved spontaneously with suitably reactive gases/vapors. In some cases, this approach yields the desired release properties of plasma/RIE etching without the need for complex and expensive equipment.

Although the fact that a family of fluorine-containing compounds (noble gas fluorides and interhalogens) will readily etch silicon (with nearly infinite selectivity to masking layers such as SiO_2) has been known for more than a decade, it is only recently that interest in the use of these techniques for micromachining has picked up. One key reason for this, aside from avoiding plasma processing equipment, is that the reactions are at least theoretically quite controllable through temperature and the partial pressure of the reactant species (typically in an inert diluent gas), as described by Ibbotson, et al. (1984). In contrast, the many variables in plasma etching may or may not be readily controllable (via feed gas composition, pressure and power, which in turn influence substrate temperature, reactant species concentration, etc., in a coupled manner). Significant differences between the two types of etching, however, are that the spontaneous non-plasma reactions are all isotropic (nearly perfectly so, which can be very advantageous in some situations).

5.6.1 XENON DIFLUORIDE ETCHING

A non-plasma, *isotropic* dry etch process for silicon is possible using XeF_2 (Winters and Coburn (1979) and provides very high selectivity for aluminum, silicon dioxide, silicon nitride, and photoresist. These properties make it an extremely useful etchant for post-processing CMOS integrated circuits, although the etched surfaces produced are extremely rough. Used originally for exposing the undersides of MOS transistors by etching away the underlying substrate silicon (Hecht, et al. (1985)), the applicability of XeF_2 to micromachined sensors and actuators was demonstrated by Hoffman, et al. (1995). They showed that with a simple bell-jar setup run at 1 torr, XeF_2 could be sublimed from its solid form at room temperature and that this etch has excellent selectivity for CMOS process layers. The etch reaction, as discussed in Chang, et al. (1995), is (approximately),

$$2XeF_2 + Si \rightarrow 2Xe + SiF_4$$

where only the silicon is in a solid phase. This reaction proceeds by non-dissociative adsorption of XeF_2 at the silicon surface, dissociation of fluorine, reaction to form the adsorbed SiF_4 product, and desorption of the product and residual xenon.

Hoffman, et al. (1995) used the "open" layer (superposition of all cut layers) in standard MOSIS 2 μm CMOS dice to define regions to be etched and reported typical etch rates of 1 to 3 μm/min (as high as 40 μm/min in some instances, as reported by Chang, et al. (1995)). Since the process does not attack aluminum, the CMOS dice can be etched once already mounted and bonded in a ceramic DIP package. A key point is that the etched surfaces have a granular structure (10 μm and smaller feature size), making it unsuitable for situations where smooth surfaces are required. In addition, unless the reaction rates are controlled or modulated, the heat generated by this exothermic reaction may adversely affect some microstructures. In addition, an important safety concern is that XeF_2 reacts with water (even moisture in air) to form Xe and HF (the latter can also unintentionally etch silicon dioxide in addition to being toxic).

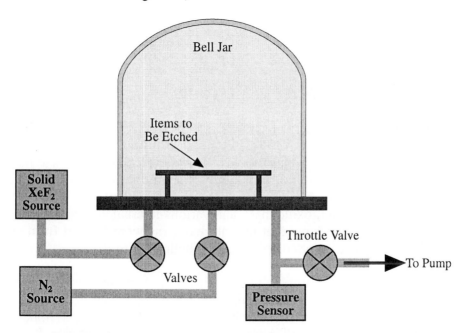

Illustration of an XeF_2 etch system. After Hoffman, et al. (1995). During etching, the XeF_2 valve is opened and the pump is throttled to achieve a pressure of 1 torr to sublimate the XeF_2. Etching is done in cycles to allow generated heat to dissipate. After etching is completed, N_2 is used to purge the system.

Chang, et al. (1995) noted that careful dehydration of the samples to be etched was necessary to prevent the formation of a silicon fluoride polymer (perhaps similar to Teflon™), which can dramatically slow or stop etching. This was carried out using a 120°C bake for > 5 min prior to etching. They also noted that the reaction is quite exothermic (generating $\approx$ 1 W/cm^2 of etch area) and that the reaction rate may be considerably lowered unless the etching is pulsed. They used

1 minute etch intervals at 2.5 torr XeF_2, with pumping down to vacuum ($\approx$ 20 mtorr). With such a protocol, they noted considerable improvements in etch rate over continuous etching at 2.5 torr.

In terms of masking films, Pister (1995) reported that XeF_2 does not appear to etch several useful materials, including photoresist, thermal silicon dioxide, PSG, BPSG, Al, Au, TiNi alloy, silicon nitride, and acrylic. More recently, Chu, et al. (1997) confirmed that there was no measurable etch rate for Al, Cr, TiN, stoichiometric LPCVD silicon nitride, thermal silicon dioxide, PECVD silicon carbide, and photoresist. They measured some etching of Ti and W, with selectivities to Si etching given as Ti:Si = 85:1 and Mo:Si = 6:1. It should be noted that most metals form passivated and non-volatile fluorides at their surfaces, preventing etching.

Hoffman, et al. (1995) demonstrated aluminum hinges, piezoresistive accelerometers, and complex fold-to-assemble structures realized through CMOS post-processing.

5.6.2 INTERHALOGEN ETCH CHEMISTRIES

Köhler, et al. (1996) described a method to avoid the extremely rough silicon surfaces that are formed using XeF_2 etching, yet still retaining its advantage of being a simple-to-implement dry etch process. They developed a process to fabricate smooth and accurately shaped molds in silicon for the fabrication of microlenses through the molding of PMMA in them. In order to carry out the silicon etching, they used a thermal silicon dioxide mask and various interhalogen gases (BrF_3 and ClF_3) with a xenon diluent (note that the interhalogens were formed from single-element feed-gases). The reported results were obtained with BrF_3.

BrF_3 is formed from bromine and fluorine and reacts with exposed silicon, forming SiF_4 and elemental bromine, which can be reused. The mixture for which optimum results were obtained was 7 mbar (5.3 torr) of bromine, 21 mbar (15.8 torr) of fluorine, and 980 mbar (735 torr) of xenon (as also reported in Köhler, et al. (1995)). Compared to silicon etched in pure fluorine, the surface roughness was reduced from $\approx$ 150 nm to <40 nm (reportedly indistinguishable from the unetched, polished wafer surfaces). The etch results were nearly perfectly isotropic. Wang, et al. (1997) measured etch rates for thin-films in BrF_3 and reported selectivities with respect to silicon for several materials: LPCVD silicon dioxide:Si = 3,000:1; silicon nitride (depends on Si concentration) = 400 to 800:1; hard-baked AZ4400 and AZ1518 photoresist = 1,000:1; and Al, Cu, Au, and Ni = > 1,000:1.

Such interhalogen etches could be quite useful in a number of silicon etching applications, and it is likely that they can be used to post-process CMOS integrated circuits, for example. However, the complexities and dangers of working with halogen gases must be given serious consideration.

5.6.3 OTHER VAPOR-ETCHING METHODS

It is likely that several other etching methods for silicon are possible, particularly using strongly reactive vapors of acids. In addition, these methods should be applicable to non-silicon materials used in micromachining, as discussed below. For example, Huang, et al. (1996) reported the use of HF vapor to etch sputter-deposited ZnO piezoelectric layers.

5.7 PLASMA/REACTIVE ION ETCHING

In this class of "dry" etching reactions, external energy in the form of radio frequency (RF) power drives chemical reactions (i.e., takes the place of elevated temperatures or very reactive chemicals). Energetic ions supply the necessary energy such that reactions can be achieved at low temperatures in a plasma that might otherwise require temperatures upwards of 1,000°C. For example, many such reactions can operate in the temperature range of 150 to 250°C, and some can even be run at room temperature. This is a very powerful type of process that can achieve impressive anisotropy (aspect ratio) in cases where bombardment of the surface by ions (generally perpendicular to the wafer) drive the etch reaction. The full spectrum of isotropic through anisotropic etches is available using these methods.

Dry etches were originally developed in the early 1970's partly because silicon nitride turned out to be a nearly ideal passivation layer for MOS devices, but there was no fully satisfactory wet etchant available, particularly when Al was exposed. RF energy applied to a pair of plates accelerates stray electrons, increasing their kinetic energy to levels at which they can break chemical bonds in the reactant gases upon impact, forming ions and additional electrons. With ongoing input of RF energy into the chamber, electron/molecule collisions continue to yield ions and electrons, while exposed surfaces within the chamber absorb or neutralize these species. After a number of RF cycles, a steady-state discharge is reached in which the generation and loss processes are balanced.

At this point the discharge is characterized by a central glow or bulk region and dark or sheath regions between the bulk region and the electrodes (it is the sheath regions where nearly all of the potential drop occurs between the electrodes, analogous to the case of a depletion region). The bulk region is semi-neutral, with nearly equal numbers of negative and positively charged particles. The higher mobility electrons can exit the bulk region more readily than ions in response to the RF cycles, giving rise to the high-field sheath regions that form to retard the further departure of electrons from the bulk region (maintaining charge neutrality).

The DC component of the sheath field (sheath bias) accelerates positive ions originating in the bulk region and thus gives rise to the bombardment of the wafers by the resulting highly energetic ions. The DC bias will occur on its own (as described above), or it can be controlled using a separate DC power supply (typically 0 to 600 V). This produces a "bias-sputter" or "bias-etch" configuration and allows some control of the electron and ion impact energies on the wafers (often done to reduce plasma damage to acceptable levels).

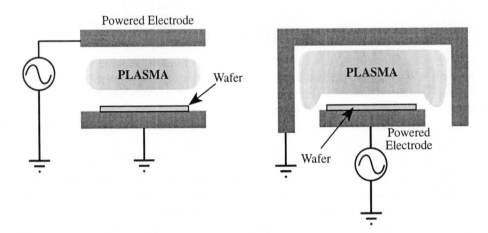

Conceptual illustration of a plasma etcher (left) (grounded wafer, symmetrical electrodes) and a reactive ion etcher (right) (powered wafer, grounded surface area much greater than powered electrode). The reactive ion etcher (RIE), with 1) the wafer on the powered electrode, 2) asymmetric electrodes, and 3) relatively low operating pressures (10^{-3} to 10^{-1} torr versus 10^{-1} to 10^{+1} torr for plasma etching) generates relatively higher energy ions perpendicular to the wafer surface.

Some examples of Si etch chemistries (reactant gases) are $CClF_3$ (Freon™ 13) + Cl_2, $CHCl_3$ + Cl_2, SF_6, NF_3, CCl_4, CF_4 (Freon™ 14) + H_2, and C_2ClF_5 (Freon™ 115). It should be noted that chlorofluorocarbon (CFC) etchants are generally being phased out (by law) since they are harmful to the Earth's ozone layer.

As an example of plasma/RIE etch mechanisms, the SF_6/Freon™ 115 (C_2ClF_5) etch chemistry commonly used to anisotropically etch silicon is examined. For SF_6, dissociation, ionization, and attachment reactions release the fluorine free radicals, which carry out the bulk of the silicon etching (a similar but much larger series of reactions is possible for C_2ClF_5). The SF_6 reactions are:

Dissociation Reactions

$$e^- + SF_6 \rightarrow SF_5 + F + e^-$$
$$e^- + SF_5 \rightarrow SF_4 + F + e^-$$
$$e^- + SF_4 \rightarrow SF_3 + F + e^-$$

Ionization Reactions

$$e^- + SF_6 \rightarrow SF_5^+ + F + 2e^-$$
$$e^- + SF_6 \rightarrow SF_3^+ + F_2 + F + 2e^-$$
$$e^- + SF_4 \rightarrow SF_3^+ + F + 2e^-$$

Attachment Reactions

$$e^- + SF_6 \rightarrow SF_5^- + F$$
$$e^- + SF_4 \rightarrow SF_3^- + F$$
$$e^- + SF_4 \rightarrow SF_3^+ + F + 2e^-$$

Fluorine free radicals from the dissociation of SF_6 are mainly responsible for the desired silicon etching,

$$SiF_x + F \rightarrow SiF_{x+1} \text{ where } x = 0 \text{ to } 3$$

and it is believed that silicon is removed from the wafer surface mainly as SiF_4. Unlike chlorine- and bromine-based processes, the fluorine plasma silicon etching reactions proceed spontaneously, not requiring ion bombardment. Thus the fluorine free radicals result in high etch rates but by themselves produce etch profiles that are nearly isotropic (the lateral etch rate is nearly equal to the vertical rate).

The role of the C_2ClF_5 is to introduce a polymeric deposition process in parallel with the etch process. In regions of low ion bombardment (such as the sidewalls of etched holes), a fluorine-rich fluorocarbon polymer layer (6 to 10 nm) forms and greatly inhibits the lateral silicon etching. At the vertical surfaces, where ion bombardment is the highest, the fluorocarbon layer is carbon-rich and less than 2 nm thick and here the silicon etch rate is substantial (> 250 nm/min). The C_2ClF_5 in addition to the SF_6 is not sufficient to form the etch-retarding sidewall layer and additional carbon from photoresist mask erosion is required. For deep silicon etches, a relatively thick (many microns) photoresist layer is required (e.g., AZ4660 positive-working resist [Hoechst] can be used). By adjusting the composition of the reactant mixture, anisotropy can be controlled, and algorithmic approaches to the control of etch characteristics can be determined, as demonstrated by Jansen, et al. (1995a, 1995b).

For all such dry etch processes, the amount of exposed silicon in a given area, as well as the geometries of etched features, can locally affect etch depths through a number of factors. These include variations in the consumption of reactants (which must diffuse into the regions being etched), changes in the amount of ion bombardment (e.g., less for small, deep pits) at the etched surfaces, and potentially diffusion of reaction products away from the etched features. These phenomena are sometimes taken into account by designing mask patterns appropriately, based on experimental etch results.

Many dry silicon etch chemistries do not attack the dielectrics and metals used in CMOS processing, and if isotropic, they can be used to undercut structures to form bridges and cantilevers. For example, Linder, et al. (1991) discussed various structures that could be fabricated by undercutting aluminum thin-film regions using plasma etching with an SF_6/O_2 chemistry. They reported 1.3 μm/min silicon etch rate with > 300:1 selectivity for aluminum and nearly fully isotropy (0.8:1 undercut:depth ratio).

5.7.1 DOPANT-DEPENDENT PLASMA ETCH ISOTROPY

Contrary to popular belief, it is possible (in some cases) to modulate silicon plasma etch anisotropy via local dopant concentrations. Li, et al. (1995) demonstrated that a Cl-based plasma etch can be used to etch lightly doped p- or n-type silicon anisotropically and heavily n-doped silicon isotropically. By forming buried n+ layers beneath a lightly doped epitaxial layer, they were able to selectively undercut structures above the buried n+ regions.

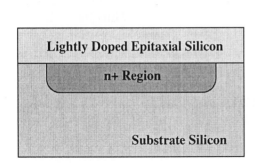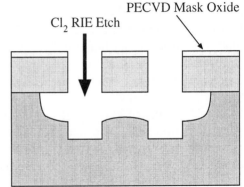

Illustration of dopant-dependent plasma etch isotropy. After Li, et al. (1995).

While a definitive explanation for this effect is not yet available, it has been proposed that since the heavy n-doping raises the Fermi level, the energy barrier for charge transfer to chemisorbed Cl atoms is reduced, enhancing the reaction. In addition, the more ionic nature of the Si-Cl bond formed on the heavily n-doped silicon facilitates Cl chemisorption and penetration into the Si lattice. For undoped, or lightly doped silicon, Cl etching will not take place without bombardment by energetic ions, but the enhanced etching of heavily n-doped silicon will occur without this bombardment, facilitating the desired undercutting effect. For implanted As^+ ions, the minimum dose for this effect to appear was 1×10^{15} cm^{-2} (for a peak carrier concentration of 8×10^{19} cm^{-3}, an extremely high doping level). It is noteworthy that Sb cannot be used as the dopant for this application since the required threshold dose is greater than its solid solubility limit in silicon (2 to 5×10^{19} cm^{-3}).

Li, et al. (1995) demonstrated the combination of this approach with a standard bipolar process to fabricate an integrated accelerometer, but results were not reported.

5.7.2 HIGH-ASPECT-RATIO DRY ETCHING METHODS

The ability to etch deep, anisotropic structures in silicon is of considerable interest in the micromachining community for a variety of applications, including fabricating deep fluidic channels and single-crystal mechanical structures. There are several approaches to obtaining deep etching with high anisotropy, and at least two commercial etchers designed for this purpose are currently available.

DEEP REACTIVE ION ETCHING (DRIE)

A very high-aspect-ratio silicon etching method referred to as deep reactive ion etching (DRIE) relies on a high-density (inductively coupled) plasma source and an alternating process of etching and protective polymer deposition to achieve aspect ratios of up to 30:1 (sidewall angles $90 \pm 2°$), with photoresist selectivities of 50 to 100:1, silicon dioxide selectivities of 120 to 200:1, and etch rates on the order of 2 to 3 μm/min (see Klaassen, et al. (1995) and Bhardwaj, et al. (1997)). The practical maximum etch depth capability of this approach is on the order of 1 mm, and precise etch depths can readily be obtained using buried SiO_2 etch-stop layers (e.g., formed by bonding an oxidized wafer to a second wafer).

The concept of alternating between etching and polymer deposition is described in the German patent of Lärmer and Schilp (1994). They describe the etching step using SF_6/Ar with a substrate bias of -5 to -30 V so that the cations generated in the plasma are accelerated nearly vertically into the substrate being etched. After etching for a short time, the polymerization process is started. A mixture of trifluoromethane (CHF_3) and argon was used (although other fluorocarbon gases such as C_4F_8 combined with SF_6 tend to be utilized in current commercial equipment), and all exposed surfaces (sidewalls and horizontal surfaces) are coated with a Teflon™-like (polymerized CF_2) polymer layer ≈ 50 nm thick. If ion bombardment, due to a small applied bias voltage, is used during the polymerization step, the formation of polymer on the horizontal surfaces can essentially be prevented. The etching step is then repeated, and the polymer deposited on the horizontal surfaces is rapidly moved due to the ion bombardment and the presence of reactive fluorine radicals. Commercial etchers of this type are now available from Surface Technology Systems, Ltd., Redwood City, CA, and Plasma-Therm, Inc., St. Petersburg, FL.

CRYOGENIC DRY ETCHING

Cryogenic cooling of the wafer can greatly enhance anisotropy of etching. Commercial machines have appeared on the market using this approach (e.g.,

Alcatel, San Jose, CA). By cooling the chuck to liquid nitrogen temperatures (77 K) and using a helium gas flow under the wafer to transfer heat, the wafer's temperature can be maintained at cryogenic temperatures during etching. Apparently, the mechanism is condensation of the reactant gas(es) on the sidewalls of the etched structures (condensing gas at the bottoms of the structures is removed by ion bombardment). A potentially important issue with cryogenic dry etching is that if microstructures become thermally isolated due to the etching, cryogenic temperatures (and hence high aspect ratios) may not be maintained locally.

Using pure SF_6, the Alcatel machine can yield aspect ratios on the order of 30:1 and is capable of etching all of the way through a full-thickness silicon wafer. For useful examples of the application of cryogenic dry etching to micromachining, see Murakami, et al. (1993), or Esashi, et al. (1995).

Both the DRIE and cryogenic approaches have great promise (especially for micromechanics) and yield devices of single crystal silicon with extremely high aspect ratios and uniform, well-defined mechanical properties.

MAGNETICALLY CONTROLLED DRY ETCHING

Furuya, et al. (1993) showed very high-aspect-ratio results for oxygen RIE etching polyimide using a titanium etch mask. They could attain nearly 7:1 aspect ratios with extremely smooth sidewalls. They used magnetically controlled reactive ion etching (MC-RIE) to obtain high-density, low-energy plasmas. This involves setting up a magnetic field parallel to the cathode surface (i.e., the plate on which the wafers sit), forcing the plasma density to vary with the applied magnetic field.

They noted that, ordinarily, if one adds fluorine-containing reactant gases to an oxygen RIE or plasma etch of polyimide, the etch rates rise, but unfortunately, the Ti, Si, Cr, or other mask layers are etched quite rapidly. They circumvented this by incorporating the fluorine into the polyimide rather than adding it as a reactant gas, thereby providing a very localized source of fluorine with only oxygen in the feed gas. Unfortunately, the fluorinated polymers they used are not commercially available (Murakami, et al. (1993) synthesized them for their own use at NTT labs). Another very useful feature of these materials is that they are very transparent and useful for optical applications.

They observed that the Ti mask etching was governed by ion bombardment (as the plasma density goes up, the ion flux goes up but its kinetic energy goes down) and the etch rate peaks and then tapers off. The polyimide etch rate increases monotonically with plasma density, suggesting that availability of reactant species is the rate-limiting step. Therefore, they chose to etch at a field strength above the Ti etch rate peak, increasing the polyimide etch rate. It should be noted that the incorporated fluorine method may work even without MC-RIE.

THERMALLY ASSISTED ION-BEAM ETCHING

Berenschot, et al. (1996) reported the use of thermally assisted ion beam etching to fabricate high-aspect-ratio (on the order of 20:1) structures in Teflon™ (polytetrafluoroethylene, or PTFE). Ion-beam etching (using argon ions) was used to obtain a highly directional ion flux and low operating pressure (below 1 mtorr) and, in combination with radiant heating from a halogen incandescent light source, provided the high aspect ratios reported. Masking was accomplished through the use of shadow masks of silicon or the use of a thin intermediate layer of Al or Ti to enhance adhesion of photoresist, which is then patterned as usual. Etch rates on the order of 1.5 μm/min were reported.

5.7.3 VARIABLE ANISOTROPY ETCH PROCESSES

Single-crystal silicon microstructures such as cantilevers, suspended beams, etc., can be realized using a combination of anisotropic and isotropic dry etches. By switching between them during the process, it is possible to form undercut structures. In the process described by Shaw, et al. (1993a, 1993b, 1994) and Zhang and MacDonald (1991, 1992), a silicon wafer was coated with 2.5 μm of PECVD silicon dioxide, and patterned resist was used to pattern this masking oxide using magnetron ion etching. This was followed by an anisotropic Cl_2/BCl_3 RIE etch to form the trenches and the deposition of a thin (0.3 μm) PECVD silicon dioxide layer to protect the sidewalls. The bottoms of the oxide-coated trenches were opened with a CF_4 RIE etch, followed by a second anisotropic Cl_2/BCl_3 RIE etch to deepen the trenches. The beam structures thus formed were then undercut using an isotropic SF_6 etch, after which aluminum is sputtered to form electrostatic electrodes.

There are many possible variations on such processes wherein combinations of anisotropic and isotropic etches are used to obtain the desired geometries. Naturally, there are geometric limitations on structures that can reasonably be undercut, generally restricting the structures obtained to assemblies of uniform cross sectional beams. A variant of the process that can produce two levels of electrically isolated, suspended microstructures was reported by Hofmann and MacDonald (1997).

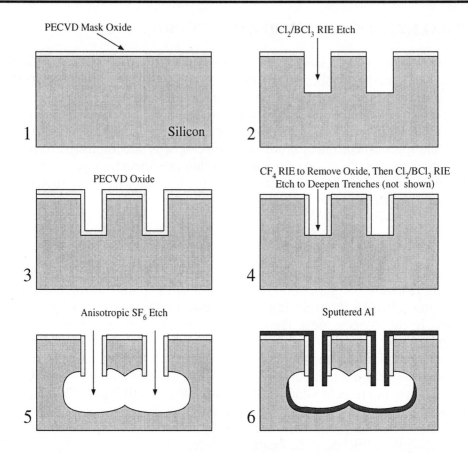

Illustration of a combination anisotropic/isotropic silicon etch to form released, single-crystal silicon mechanical structures (not to scale). Drawn after text description in Shaw, et al. (1993).

Such single-crystal silicon structures could be combined with active circuitry if they were formed on prefabricated, circuit-bearing wafers as a post-processing step. Shaw and MacDonald (1996) demonstrated the fabrication of single crystal electrostatic actuators on wafers containing bipolar/MOS operational amplifiers and showed that such post-processing did significantly alter the properties of the preexisting circuits. Their mechanical structures generally had silicon beam dimensions of ≈ 1 µm in width and 10 µm in height. After the silicon etching, the resulting structures were coated with a CVD silicon dioxide, followed by sputtering and patterning of aluminum to form conductive sidewalls for the electrostatic actuators.

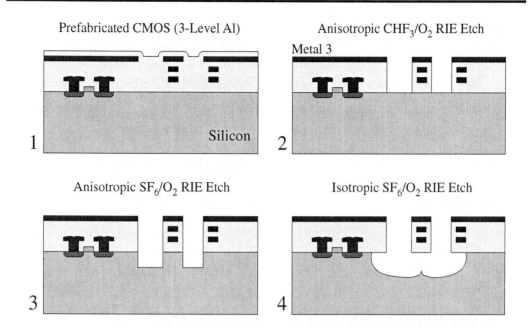

Combination anisotropic/isotropic etch to release laminated aluminum/silicon diox-ide beams using conventional prefabricated CMOS integrated circuits. After Fedder, et al. (1996).

Fedder, et al. (1996) demonstrated the use of a similar approach of combined anisotropic/isotropic dry etching to fabricate laminated aluminum/silicon dioxide mechanical structures using the top level metal as an etch mask. They used an anisotropic silicon dioxide etch (47 sccm CHF_3 + 3 sccm O_2, 25 mtorr, 200 W, 44 nm/min etch rate), to remove the silicon dioxide, followed by an anisotropic silicon etch (50 sccm SF_6 + 12.5 sccm O_2, 150 mtorr, 100 W, 0.5 μm/min etch rate) and finally an isotropic silicon etch (50 sccm SF_6 + 5 sccm O_2, 50 mtorr, 100 W, 1 μm/min etch rate) to release the mechanical structures. While no circuits were combined with the mechanical structures (nor were any test circuit results reported to verify lack of damage from the post-processing), the released mechanical devices demonstrated were fabricated directly from standard CMOS wafers, indicating that full integration should be readily achievable.

In addition, a variable anisotropy method combining deep reactive ion etching with XeF_2 vapor etching was demonstrated by Toda, et al. (1997).

Another method of obtaining variable anisotropy in etching was demonstrated by Juan and Pang (1995a, 1995b). They performed high-aspect-ratio reactive ion etching and followed it with the diffusion of a heavy boron etch-stop into the silicon. This provided high-aspect-ratio structures that could withstand wet etching (using EDP) that would only take place in areas where the heavily doped silicon was first removed by dry etching. A variation of this process, used for fabricating selectively anchored micromechanical structures was reported by Weigold and Pang (1997). Another example of a combination of dry and wet etching was presented by Ensell (1995), who demonstrated the fabrication of undercut single-crystal silicon beams from (111) wafers. This process involved the use of dry etching to define the shapes of the beams, oxidation to protect the sidewalls of the beams (the oxide at the bases of the dry-etched features was removed using another dry etch), and lateral undercutting using a two-step HNA/KOH etch chemistry.

DRY ETCHING OF NON-SILICON SEMICONDUCTORS

A variety of dry etching processes exist for non-silicon semiconductors, such as III-IV compounds. For example, high-aspect-ratio etching methods also have been demonstrated for GaAs, as described by Zhang and MacDonald (1993).

5.8 LASER-DRIVEN BULK PROCESSING

5.8.1 LASER DRILLING

Without special ambient gases, powerful enough laser beams can be used to simply ablate silicon (a localized, rapid, thermal evaporation process). This technology has been used to drill holes through wafers (silicon and silicon-on-sapphire drilling are both well documented in the literature) with the hope of fabricating electrical feed-throughs (Anthony (1982a, 1982b)). Drilling of silicon was reported with light at a wavelength of 1.06 μm from an Nd:YAG laser (neodymium-doped yttrium aluminum garnet). Short pulses (typically 100 ns) were used to ablate rather than melt the silicon. Holes as small as 8 μm in diameter have successfully been drilled through wafers 100 to 200 μm in thickness (aspect ratio = 12.5:1 to 25:1), but unfortunately considerable damage may be caused (cracking, silicon ejecta, dislocation defects, vacancies, etc.), but it is likely that this could be improved if desired.

Laser drilling methods can be slow (depends on equipment) but in any case it is not a parallel process (analogous to a pen plotter). However, even visually transparent materials can be micromachined in this way (i.e., glass or sapphire) as long as the appropriate wavelength, power, and pulse timing is selected. One can also diffuse dopants into the holes to make diodes and drive the junction in beyond the damaged region (this might be useful for interconnects since the diodes could be self-insulating from the substrate and conductive through the substrate).

5.8.2 LASER ANNEALING

Similar to laser drilling, but with lower energy, larger spot sizes can be used to anneal silicon. Variants of this approach may prove useful for reducing stresses in micromechanical structures, although this has not yet been reported in the literature. It should be noted that, in practice, most blanket annealing is done using rapid thermal processing (e.g., using high-power infrared lamps) rather than lasers, which are generally too slow and inefficient.

5.8.3 LASER-DRIVEN ETCHING

Another way to selectively add external energy for micromachining is to use laser beams to drive chemical reactions (Ehrlich and Tsao (1989)), referred to as *laser-assisted chemical etching* (LACE). It takes much less delivered energy to drive chemical reactions for etching or deposition than to vaporize silicon. In certain ambient gases (e.g., Cl_2 at 100 torr), the application of laser energy to silicon surfaces can result in very high local etch rates, such as 100 μm/s.

Direct heating of the silicon speeds the reaction locally, and if the appropriate wavelength is chosen (e.g., 500 nm for Cl_2), free radicals (e.g., Cl^*) can be formed by photolysis as in plasma/RIE etching. The laser beam may be scanned across the silicon using galvanometers or other deflection methods and some form of amplitude modulation to control etching (e.g., sequentially removing unit volume elements or *voxels*). This process can be used to make via holes, channels and very complex structures (undercut, overhanging structures cannot be etched, however).

Etch resolutions approaching 1 μm^3 have been achieved in some materials and voxel rates of 5×10^4 per second have been attained using acousto-optic scanners, a microscope objective scanned across the wafer, beam-blanking optics, and a 15 W CW argon-ion laser at 488 nm (Bloomstein and Ehrlich (1991)). The focal plane can be lowered in steps that correspond to one side of a voxel after one 2-D etching step is completed. Using the published numbers of Bloomstein and Ehrlich (1992), a 900 mW incident laser power, for 1 μm^2 spot (assuming no losses) gives a power density of 90 kW/cm^2 (only 55 kW/cm^2 is required to melt the Si). In this case, the reaction is not diffusion-limited because of the small spot size, but is limited by the rate of collision of Cl_2 molecules with the surface (therefore the speed can be increased by increasing Cl_2 pressure, and scales roughly linearly with it). The spot size of the laser can also be increased to speed up the etch rate, but more laser power is then required. Bloomstein and Ehrlich (1991) predicted that a ten times speedup is possible.

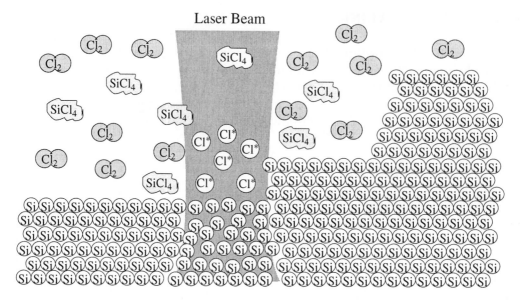

Illustration of laser-assisted chemical etching (LACE) in Cl_2. After Bloomstein and Ehrlich (1991). The Cl^* symbols represent the highly reactive chlorine radicals formed locally by the laser beam.

Times to remove large volumes of silicon can sometimes become prohibitive. For example, to remove a 10×10 μm cube, it takes ≈ 0.02 s, a 100×100 μm cube takes ≈ 20 s, and a 1×1 mm cube takes ≈ 5.6 h. Even with the predicted ten times speedup, this will still be fairly slow, but still potentially faster than electric discharge machining at these scales. The Cl_2 process is extremely selective (1,000:1) for silicon over SiO_2, so buried (continuous) channels can be etched *under* a SiO_2 layer (potentially useful for fluidics, etc.). However, the Cl_2 and $SiCl_4$ must be able to diffuse to the opening of the channels at the edge of the die/wafer, limiting channel length.

It is noteworthy that a similar etch process for GaAs was previously demonstrated by Osgood, et al. (1984), in which CF_3Br was photodissociated using a 193 nm light from a pulsed excimer laser. This produced various volatile organic species such as $Ga(CF_3)_3$, $GaBr(CF_3)_2$, etc., and resulted in etch rates of 1 μm/min, but with a buildup of reaction products that required continuous UV laser irradiation to remove.

In addition to the dry etch processes discussed above, a laser-driven wet etching process was described by Ade, et al. (1987). They used a dilute HF solution (10% by volume) to locally etch cylindrical holes in n-type (10 Ω•cm) (100) silicon using UV excitation. A frequency-doubled Ar-ion laser was used to obtain a 257 nm light, focused down to the desired hole diameter (15 to 18 μm).

Using a power density of 10 W/cm^2, they were able to etch 20 to 25 μm deep holes with a 1 μm/min etch rate. A similar liquid-based, laser-driven etch system for GaAs and AlGaAs was demonstrated by Podlesnik, et al. (1984) and Willner, et al. (1989). Podlesnik, et al. (1984) used various ratios of $H_2SO_4:H_2O_2:H_2O$ (e.g., 1:1:100) to obtain a wide variety of etch rates. In n-type GaAs, using 257 nm light at 10 W/cm^2, they obtained etch rates on the order of 10 μm/min. They demonstrated the etching of 1 μm via holes through a 100 μm thick GaAs substrate using a power of 100 mW/cm^2. By using a mixture of $HNO_3:H_2O$ (1:20) or $HF:H_2O$ (1.5:20), and using a 257 nm laser power density of 3 kW/cm^2, they achieved etch rates of ≈ 20 and 500 μm/min, respectively. In addition, Willner, et al. (1989) used a mixture of $HF:HNO_3:H_2O$ at ratios of 4:1:50 and a 257 nm laser power density of 3 kW/cm^2 to obtain GaAs and AlGaAs etch rates on the order of 750 μm/min. Further information on such photoelectrochemical etching of GaAs and InP can be found in Khare, et al. (1992, 1993a, 1993b).

Again, LACE is not a parallel process and not fast enough for most manufacturing applications. Nonetheless, it may have utility in specialty micromachining or for making molds, etc. (see Müllenborn, et al. (1995) for additional discussion of its applications in silicon micromachining).

6. "SURFACE" (ADDITIVE) PROCESSES

This section contains a discussion of processes that add material (sometimes to act as spacers, later to be removed, in the case of *sacrificial layers*) above the surface of the substrate. Many of them are more common in typical integrated circuit fabrication than bulk wet etches or the more "exotic" micromachining processes. In this section many additive methods are presented, including a wide variety of thin-film deposition and etching processes, electroplating, epitaxial growth, and others.

6.1 THIN-FILM PROCESSES

6.1.1 NON-METALLIC THIN-FILMS FOR MICROMACHINING

Probably the most common dielectric thin-films in semiconductor fabrication facilities are silicon dioxide and silicon nitride, and polysilicon is also typically available. Combination films, such as silicon oxynitrides are also becoming more readily available. Other interesting, but less common, films include SiC, SiB, C-diamond, and others, as discussed below. Some useful properties of the common thin-films (and some metals for comparison) are given in the table below. A key warning with respect to this is that these values can only be treated as *approximations*! They are all extremely process-dependent.

Key parameters for thin-films in micromachining applications are temperature of deposition (trying not to destroy underlying materials/devices), intrinsic stress/strain (and gradients) of the film (can warp microstructures), step coverage, resistance to micromachining etchants, resistance to ion penetration (particularly for "wet" transducers), and uniformity/pinholes (defects will cause breakdown or failure during etching). These parameters vary for a given type of film depending on the type of deposition used and the exact settings for each type of deposition. This variability is a non-trivial issue, especially in production settings. The intrinsic stresses in films are very dependent on the process and exact parameter used, discussed for each of the materials below. For information on resistance to wet etchants, see discussions in the preceding sections dealing with each of the etchants themselves.

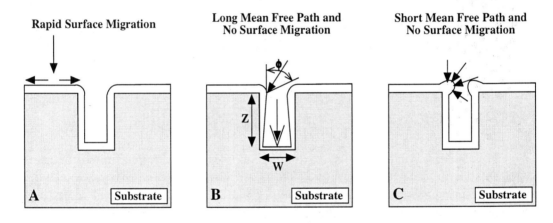

Illustration of possible step coverage profiles. After Adams (1983): A) rapid surface migration process (before reaction), yielding uniform coverage since reactants adsorb and move, then react; B) long mean free path process and no surface migration, with reactant molecule arrival angle determined location on features (local "field of view" effects are important); and C) short mean free path process with no surface migration, yielding nonconformal coating.

Deposition processes all have characteristic deposition properties, and a critical one is step coverage (the ability of the film to conform to steps in the substrate or surface features). Of course, sometimes choosing a thin-film for structural or masking purposes is not so much a decision about optimal properties, but, rather, *how best to use available equipment.*

Material	Coefficient of Thermal Expansion 10^{-6}/K	Young's Modulus GPa	Thermal Conductivity W/m·K	Density g/cm^3
Si	2.33 [2] 2.6 [4]	190 [2] 162 [4]	149 [2] 170 [4]	2.3 [2] 2.42 [4]
SiO$_2$	0.4 [4]	92 (sputtered) [1] 67 (dry) [1] 57 (wet) [1] 70 (bulk) [1]	1.4 [4]	2.66 [4] 2.3 (plasma) [6]
Si$_3$N$_4$	2.8 [4]	146 (CVD) [1] 130 (sputtered) [1]	18.5 [4]	3.44 [4] 2.9 to 3.1 (LPCVD) [6] 2.4 to 2.8 (PECVD) [6]
Poly-Si	2.33 [4]	150 [5]	20 to 30 [3]	2.33 [4]
Polyimide (PIQ L200, Hitachi)	2.0 [7]	8.63 [7]	-	-
Polyimide (PIQ 3200, Hitachi)	54 [7]	2.95 [7]	-	-
Aluminum	23.0 [4]	69 [4]	236 [2] 234 [4]	2.7 [2] 2.692 [4]
Gold	14.3 [4]	80 [4]	318 [4]	19.4 [4]
Platinum	8.9 [4]	147 [4]	73 [4]	21.4 [4]
Nickel	12.8 [4]	210 [4]	90.9 [4]	9.04 [4]

Table of material properties for general comparison (to be used cautiously, since some are for bulk materials). From Petersen (1978) [1], Petersen (1982) [2], Lenggenhager (1994) [3], Riethmüller and Benecke (1988) [4], Tang, et al. (1990) [5], Sze (1988) [6], and Suh, et al. (1996) [7]. A diverse database of such properties that illustrates their wide variations can be found at http://mems.isi.edu/mems/materials.

SILICON DIOXIDE

Silicon dioxide is certainly the most commonly available dielectric in silicon fabrication laboratories. It can be grown *in situ* or deposited with or without dopants (still insulating even if doped). Thermally grown SiO_2 is commonly used as a MOS gate insulator. If deposited SiO_2 is phosphorus-doped, it is referred to as *phosphosilicate glass*, "P-glass," or PSG, often used as a final passivation layer. If boron doping is used, it is referred to as *borosilicate glass* or BSG. Glass that is doped with a combination of phosphorus and boron is often called BPSG or low-temperature oxide (LTO) and has excellent low-temperature reflow properties (allowing high-aspect-ratio surface features to be "smoothed over" or *planarized*).

Silicon dioxide can also be used as an insulator between multi-level metals, an etch mask, an implantation or diffusion mask, or as part of a transducer structure itself. It is optically transparent over a wide range of wavelengths, making it useful for a large variety of micromachined optical devices.

As mentioned above, silicon dioxide can be grown by thermal oxidation or deposited by a large number of techniques, including chemical vapor deposition (CVD), low-pressure chemical vapor deposition (LPCVD), plasma-enhanced chemical vapor deposition (PECVD), sputtering, etc. As for most thin films, its properties are very dependent on the technique by which it is made. The common direct (*in situ*) growth and deposition techniques are listed below:

• Direct (thermal) SiO_2 growth:

Oxygen ambient: $Si + O_2 \rightarrow SiO_2$ ("dry" oxidation)

Steam ambient: $Si + 2H_2O \rightarrow SiO_2 + 2H_2$ ("wet" oxidation)

• SiO_2 deposition (CVD, etc.) example reactions:

Silane + oxygen: $SiH_4 + O_2 \rightarrow SiO_2 + 2H_2$

Tetraethoxysilane (TEOS) decomposition: $Si(OC_2H_5)_4 \rightarrow SiO_2 +$ by-products

Dichlorosilane + nitrous oxide: $SiCl_2H_2 + 2N_2O \rightarrow SiO_2 + 2N_2 + 2HCl$

Only low-temperature oxides can be used over Al metallization (temperature below $\approx 350°C$ to prevent melting of the Al), so, in general, PECVD or silane LTO processes are used in these situations. Note that only for PECVD can the stress really be *controlled* at will (i.e., one can make it compressive, near neutral, or tensile, with some compromises on other properties).

Deposition Type	PECVD	$SiH_4 + O_2$	TEOS	$SiCl_2H_2 + N_2O$	Native Oxide (thermal)
Typical Temp.	200°C	450°C	700°C	900°C	1,100°C
Composition	$SiO_{1.9}(H)$	$SiO_2(H)$	SiO_2	$SiO_2(Cl)$	SiO_2
Step Coverage	Varies (Adams says nonconformal)	Nonconformal	Conformal	Conformal	Conformal
Thermal Stability	Loses H	Densifies	Stable	Loses Cl	Excellent
Density (g/cm^3)	2.3	2.1	2.2	2.2	2.2
Refractive Index	1.47	1.44	1.46	1.46	1.46
Stress MPa	300 comp to 300 tens	300 tens	100 comp	300 comp	300 comp
Dielectric Strength (10^6 V/cm or 10^2V/μm)	3 to 6	8	10	10	10
Etch Rate (nm/min) (100:1 H_2O:HF)	40	6	3	3	≈ 3

Table of silicon dioxide properties for several deposition methods. After Adams (1983).

SILICON NITRIDE

Silicon nitride is an extremely useful dielectric material, but it can only be effectively wet etched using a boiling phosphoric acid (H_3PO_4 mixture), generally necessitating the use of plasma etching to pattern it. It is useful as a passivation layer (good barrier for H_2O and alkali ions), a capacitor dielectric, a structural material and as a mask for etching, selective oxidation of Si, etc. For short optical path lengths (where its large losses are acceptable), it has reasonable optical transparency. Since silicon nitride's refractive index is different from silicon dioxide, it can be used to make antireflective coatings and dielectric mirrors and filters, as discussed in the Optical Transducers chapter.

The common CVD/LPCVD silicon nitride deposition reactions are:

$$3SiH_4 + 4NH_3 \rightarrow Si_3N_4 + 12H_2$$

$$3SiCl_2H_2 + 4NH_3 \rightarrow Si_3N_4 + 6HCl + 6H_2$$

For typical micromachining applications, low-stress LPCVD silicon nitride is deposited so that it is silicon rich (high refractive index). It is possible to obtain nearly zero stress films in this way.

Deposition Type	LPCVD	PECVD
Typical Temp.	700 to 800°C	< 250 to 350°C
Composition	$Si_3N_4(H)$	SiN_xH_y
Si/N Ratio	0.75	0.8 to 1.2
% H	4 to 8	20 to 25
Refractive Index	2.01	1.8 to 2.5
Density (g/cm³)	2.9 to 3.1	2.4 to 2.8
Resistivity (Ω•cm)	10^{16}	10^6 to 10^{15}
Dielectric Strength (10^6 V/cm or 10^2 V/μm)	10	5
Energy gap (eV)	5	4 to 5
Stress (MPa)	1,000 tens (can be ≈ zero for Si rich films)	200 comp to 500 tens

Table of LPCVD and PECVD silicon nitride properties. After Adams (1983).

Common PECVD silicon nitride deposition reactions are:

- Silane and N_2O in Ar plasma: $SiH_4 + 4N_2O \rightarrow SiO_2 + 4N_2 + 2H_2O$

- Silane in NH_3 in Ar plasma: $SiH_4 + NH_3 \rightarrow SiNH + 3H_2$
  ("theoretical" reaction: $3SiH_4 + 4NH_3 \rightarrow Si_3N_4 + 12H_2$ at 700 to 900°C)

- Silane in nitrogen plasma: $2SiH_4 + N_2 \rightarrow 2SiNH + 3H_2$

PECVD nitrides are not stoichiometric (Si_xN_y) and incorporate hydrogen. This may not be a problem, but it may affect etchant resistance, for example. The plasma drive frequency used has a significant effect on stoichiometry and stress in PECVD silicon nitrides: 13.56 MHz yields $\approx$ 400 MPa *tensile* stress and frequencies in the range of 50 kHz yield $\approx$ 200 MPa *compressive* stress. Some newer PECVD systems (such as those from Surface Technology Systems, Ltd., Redwood City, CA.) use rapid switching between two frequencies to obtain nearly stress-free films. If the switching speed is slow relative to the deposition rate, one obtains stacked layers of alternating stress properties. If the switching speed is fast enough, one can obtain sub-monolayer deposition of each stress state, yielding a more uniform film. By controlling the relative duty cycle of the two oscillators, stress can be readily controlled.

Refractive index is a very useful diagnostic tool for examining LPCVD and PECVD films: high indices indicate silicon-rich films, while low indices generally indicate oxygen impurities.

The sodium barrier efficiency of silicon nitride has been tested with CVD silicon nitride by evaporating radioactive $Na^{22}Cl$ on the surface, using a 600°C, 22 h drive-in, and then counting the radioactivity as the nitride is etched away. Typical penetrations were less than 5 nm (Dalton and Drobeck (1968), Franz and Langheinrich (1969)). Similar experiments were conducted by Kwon (1986) on PECVD silicon nitride using a 40 h boil in 0.9% saline, an 80 h room temperature saline soak, and a 10 h 300°C "drive-in." Using Auger electron microscopy and secondary ion mass spectroscopy (SIMS), it was shown that ion penetration was limited to the top 3 nm of the silicon nitride.

SILICON CARBIDE

Silicon carbide is also a very useful dielectric material due to its extreme hardness and resistance to chemical attack (it can, however, readily be chemically patterned). It can be deposited using PECVD techniques or grown in situ. A typical deposition reaction is,

$$SiH_4 + CH_4 \rightarrow SiC + 4H_2$$

Amorphous, hydrogenated silicon carbide (a-SiC:H) can be deposited using PECVD methods and, as for all PECVD films, properties can vary widely, dependent on exact deposition conditions, as described in Tong, et al. (1993) and El Khakani, et al. (1994). Due to its high corrosion resistance and good mechanical properties (if the film is prepared properly), silicon carbide can be used in many micromachining applications (see Klumpp, et al. (1994) and Flannery, et al. (1997a)), but unfortunately, it is not available in all integrated circuit laboratories.

Flannery, et al. (1997a) demonstrated the use of a dual-frequency (13.5 MHz and 187 kHz) plasma deposition system (STS 310 PC, Surface Technology Systems, Ltd., Redwood City, CA) to prepare PECVD a-SiC:H films with a broad range of properties. The best reported film was deposited at a temperature of 350°C and a pressure of 1,600 mtorr from CH_4 (1400 sccm) and 2% SiH_4 in Ar (2840 sccm) with 100 W of high-frequency power in 2 s pulses and 150 W of low-frequency power in 4 s pulses. The resulting film properties included a stress level of -80 MPa (compressive); a volume resistivity of 2×10^{11} Ω•cm; and no measurable etch rates in KOH (45 wt%, 80°C), HF (49%, 25°C), or aqua regia (1:3 70% HNO_3 and 40% HCl at 25°C). The film could readily be etched ($\approx$ 70 nm/min) in an SF_6-based plasma. This particular PECVD a-SiC:H film was used to fabricate electrochemical sensors that could operate continuously in HF, sealed fluidic channels for transporting corrosive chemicals, for protecting a standard silicon pressure sensor (with bond wires), and as an etch mask for glass. Typical ranges of properties for a-SiC:H films from Flannery (1997b) and El Khakani, et al. (1994), respectively, are density 1.5 to 2.4 and 2.4 g/cm^3, and Young's modulus 100 and 88 to 153 GPa.

Polycrystalline SiC can also be grown from a polysilicon base by reaction with hydrocarbon fragments from propane gas at 1,360°C, as demonstrated by Fleischman, et al. (1996). They grew films of poly-SiC on the order of 2 μm thick on 2 μm poly-Si films (deposited on oxidized silicon wafers). The poly-SiC was then patterned using a patterned 0.5 μm Al mask, reactive ion etching in CHF_3/O_2 (97%:3%) (with an SiC etch rate of 40 nm/min), removal of the Al mask in HF, and release by etching the polysilicon layer with KOH. It is important to note that since these films were grown and not deposited, they were unable to form SiC anchors to the substrate, and, instead, relied on regions wherein the poly-Si was not fully undercut as anchors. Also, as mentioned above and described in Tong, et al. (1992), single-crystal silicon carbide can be grown epitaxially on silicon.

POLYCRYSTALLINE DIAMOND

A variety of processes have been developed for the deposition of polycrystalline diamond films on a variety of substrates. The high hardness and thermal conductivity of diamond make this material attractive for a number of applications, although the high temperatures used for deposition (on the order of 900°C) and highly stressed films (several hundred MPa or even GPa) have limited its use in micromachining.

Aslam and Shulz (1995) describe some applications of diamond to microstructures. Their hot-filament CVD process uses CH_4 and H_2, brought in the vicinity of a Ta filament held at 2,200 to 2,400°C. At a pressure of 50 torr and a substrate temperature of 890°C, polycrystalline diamond films can be grown to thicknesses of 3 to 4 μm. They also described a selective diamond growth process in which photoresist loaded with microscopic diamond crystals was used to selectively nucleate further CVD diamond growth. This special photoresist was spun onto Cr-coated silicon wafers, baked and patterned. An electroplated Cr template was then formed around the remaining photoresist regions, after which the wafers were placed into the CVD reactor. The photoresist vaporized, leaving behind the diamond particles, which acted as nucleation sites for diamond growth. At 900°C, 4 to 6 μm thick diamond films were grown. After growth, the Cr template could be removed. The authors noted that the relatively high vapor pressure of Cr at 900°C led to reactor contamination problems, and suggested that other template matals be investigated.

POLYSILICON

Polysilicon is useful not only as a gate material for MOSFETs, but also for high-value resistors, conductors, ohmic contacts to crystalline silicon, mechanical structures (broadly used in the micromachining community), piezoresistors, etc. Typically, polysilicon is deposited using LPCVD by pyrolyzing silane in a low-pressure reactor at 600 to 650°C (usually the silane is diluted to 1 to 2% in nitrogen, below the concentration at which it spontaneously burns in air),

$$SiH_4 \rightarrow Si + 2H_2$$

At lower temperatures, deposition rates are too slow to be practical. At higher temperatures, the reaction occurs in the gas phase as well as on the surface, so one obtains poor, non-adherent films. Poly-Si can also be deposited using PECVD methods.

The structure of the poly-Si is strongly influenced by dopants, deposition temperature, and subsequent thermal processing. This is, of course, critical for transducer applications. For example, at 605°C, an amorphous film is obtained, while at 630°C, a columnar architecture forms.

Poly-Si can be doped in situ by adding phosphine, arsine, or diborane (phosphine and arsine decrease the deposition rate significantly, while diborane increases it). Conductivity is a strong function of temperature for *in situ* doped poly-Si, and it is critical to take this into account if it is being used over a wide temperature range. Poly-Si can also be doped by implantation or diffusion (the former is used most often in MOS circuit manufacture).

Poly-Si can also be sputter-deposited (doped as desired) and, as demonstrated by Abe and Reed (1996), can be annealed to form fine-grained, low-stress films for micromechanical structures. A useful reference on the mechancial properties of surface micromachined polysilicon is Kahn, et al. (1996).

OTHER SEMICONDUCTORS AND THIN-FILM TRANSISTORS

It is readily possible to deposit a variety of other semiconductors using sputtering, evaporation, CVD, PECVD, or other methods. Using suitable materials (e.g., Te, CdS, CdSe, other II-VI compounds, amorphous Si, and polycrystalline Si) and processes, it is possible to fabricate thin-film transistors (TFT's) on a variety of substrates (e.g., glass, plastics, paper, etc.). Such devices could be used to enable whole new classes of micromachined sensors and actuators with on-board circuitry, and require only simple vacuum processing methods, and some (e.g., CdSe) can be deposited by ordinary thermal evaporation onto room temperature substrates. Despite their potential utility, there appears to be no work to date on combining TFT's with micromachined sensors or actuators, except displays and perhaps some early optical sensor arrays.

Efforts to develop TFT technologies, begun in the early 1960's, and despite the successful demonstrations of large-scale active matrix displays, optical sensor arrays, and circuits, they have only recently (1980's) become accepted in volume manufacturing of TFT-driven liquid crystal displays (see the Optical Transducers chapter). A detailed discussion of TFT fabrication is beyond the scope of this book, and the reader is referred to the outstanding historical review of these technologies by Brody (1984). A more recent review, comparing the relative merits of CdSe, amorphous and polycrystalline Si TFT's can be found in Brody (1992). In addition, an example description of a process to fabricate complementary CdSe/In and CdSe/Ge:Cu TFT's on glass (or other) substrates can be found in Doutreloigne, et al. (1991).

ORGANIC COMPOUNDS

Many organic compounds can be deposited via PECVD, and carbon-based polymers are sometimes involved in anisotropic plasma/RIE etches as sidewall protection layers. Such films can be deposited from many organic precursors (such as Freon™ gases used in plasma/RIE etching).

Organic films can be used to improve mechanical properties of surfaces or to form hydrophobic regions. For example, Jansen, et al. (1994) demonstrated the deposition and patterning of fluoropolymers using PECVD with CHF_3 as the source gas. They also discussed spin-on and evaporation deposition methods, and the applications of the polymers in micromachining. Man, et al. (1996) reported the use of a Faraday cage to create a field-free region in a parallel-plate plasma system within which fluorocarbon films $(CF_2)_x$ (polymerized tetraflouoroethylene, or PTFE) could be grown conformally to thicknesses of 10 to 20 nm. They used tetrafluoroethylene (C_2F_4) or decafluorobutane (C_4F_{10}) as the source gases and fabricated a stainless steel mesh Faraday cage above the specimen being coated to protect it from ion bombardment that could remove the polymer. They observed very good release properties for structures after direct immersion in DI water. Elevated temperature (Arrhenius) accelerated testing, predicted a lifetime of > 10 yr at 150°C, and repetitive contact tests indicated continued hydrophobicity after 10^8 contacts.

Similarly, Smith, et al. (1997) reported on the use of trifluoromethane (CHF_3) in a plasma system to deposit similar $(CF_2)_x$ polymers, but used the system's inherent ground grid within the chamber to ensure that there was minimal ion bombardment of the substrate without using a Faraday cage. They observed that the coefficient of friction (apparently the dynamic coefficient) was reduced from 1.0 to 0.07 for polysilicon micromotors, comparing favorably for the coefficient of bulk Teflon™ (PTFE) of 0.04.

SPUTTERED INORGANIC THIN-FILMS

Sputtered films of inorganic compounds can be deposited, but care must be taken to ensure that the resulting films have the desired stoichiometry. This approach is very common for film deposition for optical devices, typically using MgF_2, CaO, LiF, SiO (silicon monoxide), and indium-tin oxide (ITO, a transparent, conductive film often used in LCD displays). When sputtering nonconductive materials RF-driven sputtering is used.

A very useful description of sputter deposition of piezoelectric lead-zirconate-titanate (PZT, or $PbZr_xTi_{1-x}O_3$) films for micromachined transducer applications is provided in Mescher, et al. (1995). They used RF sputtering from a mixed lead, zirconium, and titanium oxide target with a ratio of zirconium to titanium oxide of 52:48 and a 5% molar excess of lead oxide to compensate for proportionally less lead incorporated into the deposited films. They used a RF power level of 100 W for a 2 inch target in an argon ambient with 10% oxygen and a substrate temperature of 450°C (maintained by radiant heating). To transform the film from the as-deposited pyrochlore phase to the desired perovskite structure, they used a rapid thermal annealer in pure oxygen and reported an optimal anneal temperature of 700°C for only two seconds (they also described furnace annealing, but reported poor performance with that approach).

Sakata, et al. (1996) also reported on sputtered PZT with a piezoelectric coefficient of 100 pC/N, comparable to bulk PZT materials. They used a multi-element metal target (Pb/Sr/Ti) and controlled the stoichiometry of the films through the relative areas of the target elements (1:8:9). The process used magnetron sputtering in an argon/oxygen ambient, followed by a 600 s, 700°C anneal in pure oxygen and a 2.5 to 10 V/µm poling step at 200°C. The film stress increased during annealing to ≈ 400 MPa, and the measured Young's modulus was 75 GPa.

SPIN-ON NON-METALLIC THIN-FILMS

Spin-on dielectrics are available and are applicable to micromachining. Most spin-on dielectrics are glasses (spin-on glass or "SOG"), and "prelayers" of SOG can be used to reduce stress in reflowable films such as PSG and BPSG.

A typical spin-on-glass (SOG) formula, after Quenzer, et al. (1996) is: 40 g ethanol, 0.6 g HCl, 40 g tetraethylorthosilicate (TEOS), and 14 g H_2O, mixed together in this sequence and heated at 60°C for 1 h. This recipe yields an SOG that is only suitable for thin (300 nm) layers. The SOG is annealed at 420°C in air to decompose the organic materials and densify the film.

Spin-on organic layers are also commonly available (e.g., photoresist, polyimide, etc.), as well as fluorocarbons such as the cyclic perflouro polymer reported by Matsumoto, et al. (1997).

6.1.2 WET ETCHING OF NON-METALLIC THIN-FILMS

There are hundreds of wet etchants (some well characterized, others not) for a variety of materials in the literature, and a large number are commercially available.

• **Silicon dioxide** is typically wet etched with HF in various aqueous concentrations and is commonly buffered with ammonium fluoride since the etch rate depends on the fluoride concentration and it is consumed by the reaction:

$$SiO_2 + 6HF \rightarrow H_2SiF_6 + 2H_2O$$

The commonly used form of buffered HF (BHF) is 10:1 HF/NH_3F, also known as buffered oxide etch (BOE). HF does not etch Si or GaAs appreciably, but HF will flourinate the surface of oxide-free silicon.

Alkali hydroxides (i.e., NaOH, KOH, etc.) will etch SiO_2 slowly, as well as photoresist and the substrate if it is Si or GaAs (but only if the hydroxides are present in high concentrations; not the case, e.g., for positive photoresist developers).

It is sometimes desirable to etch SiO_2 without etching Al, for example, when the oxide is used as a sacrificial layer for Al-based micromechanical structures. Gennissen and French (1997) describe a suitable approach using highly concentrated HF solutions. They noted that etching of Al in HF is governed by the concentration of H_3O^+ ions, which is low in concentrated HF since it is a weak acid:

$$2Al + 6H_3O^+ \rightarrow 3Al^{3+} + H_2 \uparrow + 3H_2O$$

They demonstrated the use of 73% HF to achieve a thermal SiO_2 etch rate of 1.5 μm/min versus 2.2 nm/min for Al (680:1 selectivity). To aid in release (by maintaining a liquid film on the released microstructures when they are removed from the etchant bath), one can add isopropyl alcohol with considerable reduction in selectivity. The authors noted that water rinses should not be used after release, since the dilution of the remaining HF will immediately increase the H_3O^+ concentration, attacking the Al rapidly (they used multiple isopropyl alcohol rinses, followed by a cyclohexane rinse with sublimation to release the structures).

• **Silicon nitride** is typically wet etched using phosphoric acid (H_3PO_4), which will attack SiO_2 at 1/40th the rate, allowing it to be used as a mask. "Hot" (often boiling, with a reflux condenser) H_3PO_4 must be used to get acceptable etch rates (on the order of 1 to 10 nm/min). HF can be used, but it is very slow. Plasma/RIE etching is far more practical and thus is more common (see below).

• **Organic films** are generally etched using strong oxidizers, chosen depending on the composition of the films. Sulfuric acid/hydrogen peroxide ("Piranha") will etch most organic compounds (including, e.g., photoresist). In general, when the above etchants fail, the compound is either resistant due to halogenation, incorporates considerable quantities of inorganic materials, or has some other chemical resistance (oxygen plasma etching often succeeds in such cases).

• **Polysilicon** is generally etched with the same chemistries as for silicon (KOH, EDP, TMAH) but with faster etch rates (e.g., Ristic, et al. (1994) reported 77 μm/h etch rate for polysilicon versus 64 μm/h for (100) silicon in 20 wt% TMAH at 95°C). Increased etch rates for poly-Si over single-crystal Si are probably due to increased etch rates at grain boundaries. As for single-crystal silicon, dopants can also affect etch rates of polysilicon.

6.1.3 DRY ETCHING OF NON-METALLIC THIN-FILMS

Dry etching allows etching in situations that may be impossible to address with wet etchants (e.g., undercutting aluminum structures by dry etching the underlying silicon or photoresist). It can eliminate sticking problems with wet etching of microstructures that are often drawn down toward the substrate by capillary forces when drying after release. Many of these processes can yield high-aspect-ratio features. Some examples of etchant chemistries are given below:

• **Silicon dioxide:** CF_4 (Freon™ 14) + O_2 (10%), CHF_3 (Freon™ 23), C_2F_6 (Freon™ 116) or C_3F_8 (Freon™ 118).

• **Silicon nitride:** CF_4 (Freon™ 14) + O_2 (4%), CHF_3 (Freon™ 23), C_2F_6 (Freon™ 116) or SF_6 + He.

• **Silicon Carbide:** SF_6, CF_4 or NF_3.

• **Organic films:** O_2 is used to "burn" them to H_2O, CO, and CO_2 (this process is often referred to as *plasma ashing*). This is often the best way to remove charred or overbaked photoresist.

• **Polysilicon:** $CClF_3$ (Freon™ 13) + Cl_2, $CHCl_3$ + Cl_2, SF_6, NF_3, CCl_4, CF_4 (Freon™ 14) + H_2 or C_2ClF_5 (Freon™115). Dry etching of polysilicon is also modulated by dopants. For example, n-type doping of polysilicon increases etch rates in chlorine plasmas (Williams and Muller (1996)).

It is important to note that many of the etchant gases shown above are chlorinated fluorocarbons, *known to damage the ozone layer of the atmosphere.* Thus most etch processes tend to emphasize SF_6 as a fluorine source since it is not a "greenhouse gas."

6.1.4 METALLIC THIN-FILMS FOR MICROMACHINING

There is a large variety of metallic thin-films that are potentially useful in micromachining applications, such as conductors, resistors, mechanical elements, transducer elements, etc. Due to its popularity in the fabrication of conventional integrated circuits, by far the most common metal available in semiconductor fabrication facilities is aluminum, and it has been used extensively in micromachining.

There are a variety of metal deposition techniques including resistive evaporation, electron-beam evaporation, sputter deposition, magnetron sputter deposition, CVD, laser-assisted CVD, and electrodeposition (the latter two are covered in separate sections below). Each approach is applicable to certain metals, and there are many trade-offs when choosing one for a given application. This section covers the common thin-film deposition processes, while laser and electroplating methods are covered in separate sections below.

RESISTIVE EVAPORATION

In resistive evaporation, an electrically heated filament of the desired metal or a refractory metal "boat" containing the desired metal is used to evaporate a film onto the wafer(s). (This is sometimes referred to as "thermal" evaporation, but all evaporation processes are thermal in nature.)

Nearly any material can be evaporated (if it can be raised to a temperature at which a vapor pressure of $\approx 10^{-3}$ to 10^{-2} torr can be generated). The melting point of a metal is not necessarily a criterion for suitability, since some metals sublimate (although many melt before evaporating). The major limitation of resistive evaporation is the limiting temperature of the tungsten "boat" typically used ($\approx 1400°C$), but this is quite suitable for Au, Al, Mg, etc., and Cr is just at the limit of feasibility.

Resistive evaporation is essentially a vapor condensation process. The metal atoms stick as soon as they arrive at a surface, not allowing for much movement that can help with step coverage and limiting stress. The substrate temperature has a great influence on the final stress by controlling the amount "stuck" metal atoms can move around. For unheated silicon substrates, evaporated films will generally be tensile (low melting point metals such as Al and Au tend to have low stresses, and as the melting point increases, stress tends to increase as well). If the substrate is heated to a temperature comparable to the metal's melting point, stress will tend to decrease since this is an epitaxial growth situation (with metal atoms able to move to their lowest-energy positions, for minimal stress). It should be noted that adhesion of evaporated films is often poorer than sputtered films. Typical deposition rates (material dependent) are on the order of 0.5 to 5 nm/s, and there can be substantial substrate heating due to direct infrared illumination.

ELECTRON-BEAM EVAPORATION

An electron beam (8 to 10 keV, generally 100 to 200 mA) is scanned over a target, generating a vapor for deposition. Target temperatures, controlled via electron-beam power (typically 1 kW into a 1 mm^3 volume), focus, and degree of target cooling, can reach $\approx 2,800°C$. This allows for the deposition of the same metals possible with resistive evaporation, as well as Ti, Pt, Pd, and refractory metals such as W (very useful since it has a nearly identical thermal expansion coefficient to Si), Mo, and Ta. High deposition rates are possible with electron (e)-beam evaporation, generally $\approx$ ten times faster than CVD or sputtering (typically 5 to 200 nm/s), and there is generally less substrate heating than for resistive evaporation. Metals and certain dielectrics can be deposited (this is seldom done in the same system). As for resistive evaporation, stresses are generally tensile for unheated substrates, and increase with the melting temperature of the metal.

At higher voltages, x-rays generated can damage silicon and necessitate anneal steps for MOS devices. These x-rays originate from "soft" metals such as Al or Mg, with energies of only 100 to 200 eV. They generally do not pose a health hazard, unless the Cu underlying the target (used for its high thermal conductivity) is bombarded and K-α x-rays are generated (this requires ≈ 30 kV acceleration voltage, and most electron-beam evaporators use ≤ 10 kV). In addition, secondary electrons are produced and, depending upon the degree of shielding provided, can reach the substrate, potentially causing damage.

An *ion gun* can be used with evaporation to drive the film to compressive rather than tensile stresses. Ions striking the surface impart extra energy to the metal atoms on the surface, effectively making their arrangement more compact than that achieved with simple evaporative condensation. The amount of this compression is controlled by the energy and flux of ions emitted by the gun. For metals of high melting points, and thus high tensile stresses, the co-bombardment of the surface with ions of well-controlled energies and fluxes can provide the ability to effectively null the tensile stresses in evaporated films.

SPUTTER DEPOSITION

In sputter deposition, inert ions (e.g., Ar^+) are accelerated using DC or RF drive through a potential gradient so that they bombard a target, generating ejected clusters of the target material from the recoil of their incident momentum. Almost any material can be sputtered if a sufficiently high-energy plasma can be generated (for dielectrics, RF sputtering is required). There are several methods to obtain the necessary plasma: *DC glow discharge* (simple hardware, only for soft metals, low-energy plasma), *planar RF* (two parallel plates, RF applied to target, works for dielectrics as well), *planar magnetron* (where a magnetic field creates semicircular electron trajectories, increasing interaction with gas molecules, creating more ions) and *cylindrical magnetron*, or *S-gun* (where a magnetically induced closed "race-track" trajectory is created for electrons, forming a high-energy plasma region near the target, away from the substrate).

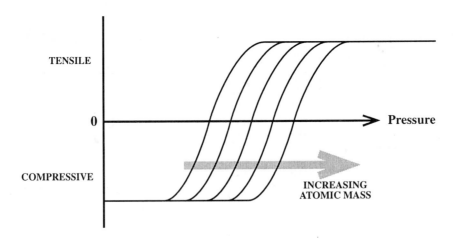

Illustration of relationship among pressure, atomic mass of the metal, and stress in sputtered films.

Sputtering provides a flux of more energetic atoms than in evaporation, allowing them more surface mobility and also knocking off (resputtering) some that are already deposited. Thus, sputtering has better step coverage and stress control, but a trade-off is that lift-off is more difficult (due to improved step coverage).

Sputtering is an excellent method to deposit alloys (same composition as target) and dielectric compounds. For example, Walker, et al. (1990) describe sputter deposition of TiNi alloy ("shape memory alloy," as described in the Mechanical Transducers Section). Mescher, et al. (1995) and Sakata, et al. (1996) describe the use of sputtering to deposit piezoelectric lead zirconate titanate (PZT) thin films. Araki and Okabe (1996) describe sputtered NdTbFeB thin-film magnets.

Negative aspects of sputtering include the fact that there is some incorporation of "background" gas (i.e., Ar) and that there may be considerable heating of substrate due to secondary electrons emitted from the target.

Typically, sputter-deposited metal films are under *tensile* stress, sometimes nearing or exceeding the yield strength (cracking). As seen from the literature for Cr, Ti, Ni, Mo, Ta, etc. (probably true for most, if not all metals), at low working pressures the stress changes to *compressive*. The pressure at which the transition occurs increases with the atomic mass of the metal being deposited. An excellent reference on stresses in sputtered films is Thornton and Hoffman (1977).

COMPARISON OF EVAPORATION AND SPUTTERING

Since evaporation and sputtering are the main methods for depositing metallic films in micromachining, it is useful to compare their properties in terms of step coverage, shadowing, and composition of the resulting films (the stresses typically encountered with each type of deposition were discussed above).

STEP COVERAGE

Evaporation is a "directional from the source" deposition process, and thus has poor step coverage (however, this is very useful for lift-off processes, discussed below). Sputtering is also directional, but from a spatially distributed source, so that the molecules maintain some directionality for only roughly their mean free path (typically ≈ 1 cm at a pressure of ≈ 10 mtorr). For planar RF or planar magnetron sputtering, sidewall coating is nearly equal to surface coating (since the substrate is immersed in the plasma). For S-gun sputtering, the substrate is removed from the plasma and the sidewalls are coated less thickly than the surface.

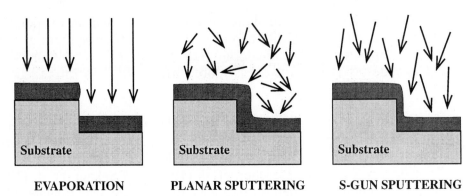

| EVAPORATION | PLANAR SPUTTERING | S-GUN SPUTTERING |

Illustration highlighting the differences in step coverage between evaporation (left), planar sputtering (center), and S-gun sputtering (right).

SHADOWING

The directionality of evaporation leads to shadow effects that can be a problem in some cases, but it can also offer advantages in micromachining. To avoid shadowing effects (for good step coverage), wafers can be mechanically scanned in front of an evaporation source using a rotating *planetary* stage. The shadowing effect can also be deliberately utilized to create features smaller than the minimum lithographic resolution, as illustrated below.

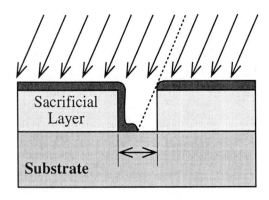
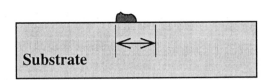

Illustration of the use of shadowing to produce features finer than the lithographic resolution. At left, a sacrificial layer (typically photoresist) is used to deposit metal at a known angle. At right, after removal of the sacrificial layer, a finer structure remains than the original line width. This approach is similar to standard "lift-off" deposition (discussed below), but in that case, an overhanging "lip" of photoresist is used to aid removal of the unwanted metal.

Another related technique is that of shadow masking, in which a physical template is held in close proximity to the substrate to pattern deposited metal (again, generally using a directional process such as evaporation). The template generally consists of a thin membrane with holes made in it such that only metal with a "line-of-site" path through the holes can arrive at the substrate. Burger, et al. (1995) demonstrated micromachined shadow masks made using a combination of wet and dry etching. They etched the mask patterns 25 μm into the surface of a (100) silicon wafer using RIE, coated the wafer with LPCVD silicon nitride, and then etched from the back side using KOH until the bottoms of the RIE-etched holes were visible. Following removal of the silicon nitride, the shadow masks were used with an e-beam evaporator to form patterned metallizations on the sides and bottoms of 380 μm deep pyramidal pits (etched using KOH into (100) silicon). The mask may also be a part of a micromachined structure, and Burger, et al. (1995) also demonstrated this using undercut silicon nitride membranes over deep pyramidal pits as *in situ* shadow masks. Paranjape, et al. (1997) also reported using a patterned membrane, undercut using a TMAH etch, with shadow mask deposition to pattern metal in the resulting etched pit.

FILM COMPOSITION

Since evaporation is a high-temperature process for the source material, the compositions of source compounds may change on the way to the substrate (e.g., $2SiO_2 \rightarrow 2SiO + O_2$, $2Si_3N_4 \rightarrow 2Si_3N_2 + N_2$, $ZnS \rightarrow Zn + S$).

Dual-beam evaporation (two independently controlled electron beams and two separate targets) can be used to control stoichiometry (as for Ti/W alloys) when the two materials have very different vapor pressures.

Sputtering generally *does not* cause changes in composition and is preferred for compounds and alloys where the composition is critical.

CHEMICAL VAPOR DEPOSITION OF METALS

Direct chemical vapor deposition (CVD) from volatile metal compounds provides good step coverage since it is not "directional." Generally, thermal energy drives the reaction that yields the desired metal as a product. CVD processes generally occur at high temperatures (600 to 800°C) unless they are plasma enhanced (PECVD). This may be an important consideration when planning the thermal budget of a process (in terms of when CVD steps must be carried out so as not to damage previously fabricated structures). Primarily W, but also Mo, Ta, and Ti have been deposited using CVD (also Al in research applications).

SELECTIVE METAL CVD

This class of reactions is based on chemistry that replaces silicon atoms with those of the desired metal. A well-known example is selective deposition of tungsten using low-pressure CVD , described by the following reaction,

$$2WF_6 + 3Si \rightarrow 2W + 3SiF_4$$

In practice, one exposes silicon or polysilicon to the WF_6 gas stream only where W is desired, typically using SiO_2 as a masking layer. An interesting use of such a process for fabricating microstructures (loops, wires, etc.), was presented by Busta, et al. (1987). MacDonald, et al. (1989) demonstrated a wide variety of selective CVD W structures deposited on silicon, thin-film polysilicon or metals, and silicon-implanted SiO_2.

Manginell, et al. (1996) demonstrated the selective deposition of platinum on electrically heated, silicon nitride encapsulated, surface micromachined polysilicon beams. Deposition was carried out using platinum acetylacetonate as the reactant. This compound was vaporized by heating to 140°C in a nitrogen purged, atmospheric pressure deposition chamber. Deposition of Pt occurs only on surfaces heated above 450°C, and the beams on which Pt was desired were heated electrically by DC or pulsed power. DC heating tended to produce rougher films (mass transport limited), while pulsed heating yielded smoother layers. (When used as catalytic hydrogen sensors, the rougher, thicker films were more sensitive.)

ADHESION LAYERS FOR METALS

Quite often it is necessary to deposit a thin (10 to 30 nm) adhesion layer of a reactive metal such as Ti, Hf, Cr, etc., beneath a relatively non-reactive metal that would otherwise not adhere. This is absolutely necessary for Au, Pt, and other low-reactivity metals, but selective removal of the adhesion layer can result in selective release of some regions of metal (see Smela, et al. (1995)).

A typical "trilayer" is Ti/Au/Ti at 30 nm/500 nm/30 nm, with the top Ti layer included to promote adhesion of the next layer to be deposited above it (when vias through the layer above the trilayer are desired, the Ti must be removed if electrical contact is desired. Since some of the adhesion metals can alloy with the metal being deposited (e.g., Ti/Au), diffusion barrier layers such as Pt or Pd are sometimes included (if not, Au can also diffuse through Ti and into GaAs or Si substrates).

6.1.5 WET ETCHING OF METALLIC THIN-FILMS

Wet etching, for patterning deposited films, is possible for certain metals, but only if the etchant does not attack desirable structures consisting of other materials (i.e., underlying dielectrics).

• **Aluminum** can be etched by strong acids or bases (would react with water if it was not self-passivating with its oxide, Al_2O_3).

$$2Al + 6NaOH \rightarrow 2Na_3AlO_3 + 3H_2$$

$$2Al + 6HCl \rightarrow 2AlCl_3 + 3H_2$$

A typical Al etchant (parts by volume) is acetic acid:nitric acid:phosphoric acid 20:3:77 (this mixture will etch GaAs, so $HCl:H_2O$ is used 1:2 instead when this is not desired). If the Al_2O_3 on the Al is too thick, a small amount of HF in the etchants will allow etching to proceed.

• **Gold** can be etched with aqueous KI_3 solution, an opaque, black etchant with corrosive vapors (due to excess I_2). The opacity of this etchant makes end-point detection quite difficult. A typical KI_3 etchant recipe is 4 g KI, 1g I_2 in 40 ml H_2O. It is generally used in the temperature range of 20 to 50°C.

• **Platinum** can be etched using *aqua regia*, which is nitric acid:hydrochloric acid in a 1:3 ratio (this mixture will also etch gold).

• **Chromium** can be etched using etchants such as 5 g ceric ammonium nitrate, 4 ml nitric acid (70%) in 5 ml H_2O, generally used at 25°C.

• **Copper** can be etched using diluted nitric acid (5 ml nitric acid (70%) in 5 ml H_2O) at 25°C. Aqueous solutions of ferric chloride ($FeCl_3$) are often used as well, and are commercially available as printed-circuit board etchants.

• **Nickel** etching can be done using 25 ml nitric acid (70%) and 15 g ammonium persulfate in 100 ml H_2O at 25°C. As for copper, ferric chloride solutions are also useful etchants.

• **NiCr** can be etched using 10 g ceric ammonium nitrate and 10 ml nitric acid (70%) in 100 ml H_2O at 25°C.

• **Palladium** can be etched using 3 g potassium dichromate, 260 ml phosphoric acid (85%), and 20 ml hydrochloric acid (37%) in 120 ml H_2O at 25°C.

• **Titanium** can be etched in dilute HF or HF/H_2O_2.

• **TiW** (a thin-film resistive material with sheet resistance of $\approx$ 25 to 200 Ω/square and temperature coefficient of resistance (TCR) of $\approx$ 0 to 50 ppm/°C [sometimes much higher]) can be etched in 30% hydrogen peroxide at 40°C.

• **TaN** (a thin-film resistive material with sheet resistance of $\approx$ 25 to 100 Ω/square and TCR of $\approx$ -50 to -100 ppm/°C) can be etched using 50 ml of nitric acid (70%) and 100 ml of hydrofluoric acid (49%) in 50 ml of H_2O at 25°C.

The above etching formulas (except that for Al) can be found in the "Thin Film Products and Services" literature of American Technical Ceramics (Huntington Station, NY). For considerably more extensive listings of etchants for metals and non-metals, the reader is referred to Vossen and Kern (1978) and Williams and Muller (1996).

6.1.6 DRY ETCHING OF METALLIC THIN-FILMS

Dry etching of many metals is possible, but Al is almost the only metal dry etched in the mainstream semiconductor industry. For Al, the etch process must be able to break through Al_2O_3 layer (self-passivation) and argon carrier gas can be used for this purpose via sputtering that breaks through the oxide. Some example reactants for Al etching include BCl_3, CCl_4, HBr + Ar, HCl + Ar, BCl_3 + Cl_2 + $CHCl_3$ + N_2, etc.

For gold, $C_2Cl_2F_4$ or $C_2Cl_2F_4$ + O_2 can be used (the latter has a higher etch rate), but such etching is generally not permitted in etchers that are used to fabricate active circuits due to concerns about gold contamination. Other metals for which there are available dry etch processes are: molybdenum, tungsten, titanium, Ti:W, and several silicides (i.e., Ti, Ta, and W). There is a large and comprehensive table of etch chemistries in Runyan and Bean (1990). Cross-contamination via the etch chamber, not only for gold, is often a serious problem when chambers are shared with "mainstream" processes. Etch-rate data and selectivities for several dry etches for metals are given in Williams and Muller (1996).

There are many structures that can be built using dry etching of thin-films and/or bulk material (i.e., Si) that cannot readily be fabricated any other way (e.g., by dry etching sacrificial layers to avoid sticking problems with microstructures that are released via wet etching).

6.1.7 LIFT-OFF PATTERNING

A simple (and old-fashioned) method to pattern metallic films is simply to allow metal to adhere to the substrate only in regions where it is ultimately desired (as opposed to "etch back" processes wherein the metal is blanket deposited and later patterned). Lift-off is particularly useful for metals that cannot be etched without attacking/destroying the substrate, such as Ir.

Lift-off processes take advantage of the fact that the step coverage of most metal deposition methods (except CVD and electrodeposition) is quite limited (worse for evaporation), so the metal cannot overcoat steep or undercut steps. A "negative" mask (i.e., patterned photoresist) is fabricated for the metal (exposing only underlying regions where metal is wanted) and the edges of the mask are undercut, allowing discontinuous metal regions on the substrate and on the mask to be formed. The "sacrificial" mask, is then dissolved away, "lifting off" the unwanted metal.

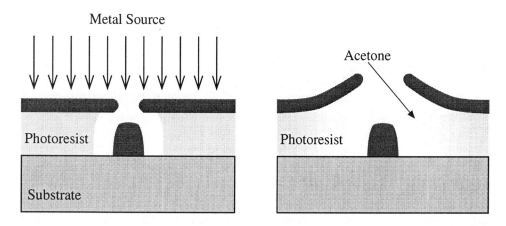

Illustration of a photoresist-based lift-off metal deposition process used to fabricate metallization patterns. Metal is deposited by evaporation and adheres to the substrate where photoresist was exposed and developed away. After evaporation, the photoresist is dissolved in acetone, lifting off the undesired metal.

A typical process involves the use of standard photoresist that is processed by soaking briefly in chlorobenzene to form an overhang "lip" that forms the discontinuous metal layers and at the same time allows solvent (or etchant) access to the sacrificial layer. The mechanism of this type of lift-off structure formation (as described in Halverson, et al. (1982)) is 1) chlorobenzene penetrates into the resist to a certain depth (controlled by the duration of its exposure to it) and removes residual solvent and low-molecular-weight resins and 2) the penetrated photoresist layer develops at a slower rate and thus forms the overhanging lip. An example is the use of AZ1350J photoresist (see Hatzakis, et al. (1980), a classic lift-off paper): spin on and pre-bake at 70°C for 15 min, expose and soak in chlorobenzene (prior to development) for ≈ 5 min (soak time controls lip shape). This process is extremely sensitive to timing, temperature, and the water content of the spun-on photoresist. It is interesting to note that Brumfield and Walker (1958) reported that Jello™ could be used as a lift-off mask for aluminum deposition. They stated that Jello™ was painted on wafers, allowed to dry, patterned by hand, and lifted off in water. They did extensive testing, noting that "various Jellos were evaluated with lemon giving best results."

An alternative is to utilize various combinations materials, such as a bilayer of two that will allow the lower layer to be undercut, forming the necessary overhanging structures. These materials do not need to be photosensitive. An example is the use of sputtered silicon above a layer of aluminum. The silicon and aluminum can be patterned by dry etching and the aluminum undercut using a brief wet etch. After deposition of the desired material, the aluminum can be completely removed using a standard etchant (i.e., acetic acid:nitric acid:phosphoric acid), lifting off the unwanted metal and silicon. Using different etchants, the roles of the two layers can be reversed.

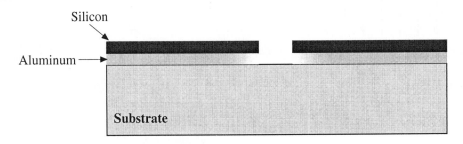

Illustration of polysilicon/aluminum bilayer lift-off structure, with the aluminum undercut to form overhang structures.

6.2 LASER-DRIVEN DEPOSITION

A laser can supply energy to drive a deposition reaction ("LCVD"), allowing metal to be deposited in specific locations. For example, Gilgen, et al. (1987) demonstrated laser-driven patterned deposition of W from $W(CO)_6$, Mo from $Mo(CO)_6$, and Pt from Pt-bis-hexafluoro acetylacetonate using an Ar^+ laser operating in the range of 350 to 360 nm, at power levels of 0.15 to 1 MW/cm^2. As another example, Bloomstein and Ehrlich (1992) discussed the laser-driven deposition of thin layers of Pt from $Pt(PF_3)_4$ and Co from $Co_2(CO)_{10}$ to provide metallization on top of three-dimensional subtractive structures fabricated using LACE.

Spectacular three-dimensional structures can also be fabricated using LCVD (Westberg, et al. (1991), Wallenberger (1995)). They are obtained by scanning the focal point of a laser beam using various mechanical or optical means. Once deposition is started on a surface, the beam can be moved away from it if the direction growth is desired.

Westberg, et al. (1991) demonstrated the fabrication of various non-planar boron structures using LCVD decomposition of BCl_4 at a pressure of $\approx$ 16 torr and a reactant gas flow velocity of 2.1 cm/s, using an Ar^+ laser at a wavelength of 514.5 nm. The reaction used was,

$$2\ BCl_3(g) + 3\ H_2(g) \rightarrow 2\ B(s) + 6\ HCl$$

The process was used to make crystalline (ß-form) or amorphous B structures, depending on the delivered laser power (100 and 50 kW/cm^2 for crystalline and amorphous deposition, respectively). Growth rates were 5 μm/s for crystalline and 2 μm/s for amorphous. Further work on LCVD boron structures can be found in Wallenberger and Nordine (1992).

Single-crystal and polycrystalline Si can also be laser deposited from SiH_4, using a process such as that reported by Boman, et al. (1992) and Westberg, et al. (1993). They fabricated Si rods and more complex 3-D structures using 6% SiH_4 in Ar with a CW Ar^+ laser with a 13 μm spot size and a wavelength of 514.5 nm. The deposition reaction was identical to that used for CVD polysilicon deposition,

$$SiH_4(g) \rightarrow Si(s) + 2\ H_2(g)$$

Boman, et al. (1992) and Westberg, et al. (1993) also demonstrated a micro-solenoid made by growing tungsten on a silicon fiber. With their approach, apparently W could only be laser grown from WF_6 on silicon surfaces, due to the involvement of the Si in the reaction. They used a 6:1 mixture of WF_6/H_2 at 300 torr and the same laser setup as for the Si deposition. For tungsten LCVD, the reactions were,

$$2WF_6(g) + 3Si(s) \rightarrow 2W(s) + 3\ SiF_4 \text{ (initial reaction)}$$

$$WF_6(g) + 3H_2(g) \rightarrow W(s) + 6HF(g) \text{ (once started)}$$

Lehmann and Stuke (1991) demonstrated millimeter-scale, 3-D Al structures achieved from $(CH_3)_3N\bullet AlH_3\bullet N(CH_3)_3$ using LCVD with an Ar^+ laser at a wavelength of 514.5 nm. Ten micron Al lines with an average spacing of 150 μm were deposited on a polycarbonate form. After deposition, the polycarbonate was dissolved in liquid chloroform (CH_3Cl), leaving behind free-standing Al structures. They used this method to fabricate complex structures as tall as 7 mm. Later work by Lehmann and Stuke (1994) demonstrated the ability to grow similar complex structures from insulating aluminum oxide. They used precursors of trimethylamine alane ($AlH_3\bullet N(CH_3)_3$) and oxygen or N_2O. With oxygen at concentrations in the 30 to 35% range, optically transparent aluminum oxide was deposited. They achieved growth rates of 10 to 80 μm/s for aluminum oxide rods between 3 and 20 μm in diameter.

Several other materials have been deposited by LCVD. For example, Krchnavek, et al. (1987) demonstrated the laser-driven deposition of Zn from a diethyl-Zn precursor using a 272 nm, frequency-doubled Ar^+ laser at a power density of 15 kW/cm^2. In addition, Wallenberger and Nordine demonstrated the deposition of C from using a 5 W yttrium-aluminum-garnet (YAG) laser, using methane and ethylene precursor gases. Many other CVD processes can theoretically be adapted to LCVD, with the principal difference being that in LCVD, the thermal energy to drive the reaction is supplied locally rather than globally. While the LCVD approach has the potential for making complex structures, it is a *serial* process by nature, and thus limited in throughput.

6.3 ELECTRODEPOSITION (ELECTROPLATING)

Electroplating is a very useful process for producing additive metal structures above a substrate and can be done in a beaker with simple and inexpensive equipment (often as a processing step done outside of the clean room). The object to be plated is metallized and/or masked to define regions on which metal is to be deposited and is maintained at a negative potential (it is the *cathode*) relative to an inert (generally Pt) positive counterelectrode (*anode*). The electroplating solution (generally aqueous) contains a *reducible* form of ion of the desired metal. By biasing the surfaces to be plated at a negative potential, electrons are supplied to the surface of the exposed metallic conductors and metal ions in the solution are reduced at the surface (one or more electrons are added to each ion so that it becomes a metal atom), depositing metal there. The metal ions must also bond to the surface for plating to be effective (an intermediate layer may be required for some combinations of plated metal and substrate). Two excellent electroplating references are Harrison and Thompson (1973) and Bockris and Reddy (1970).

A large number of metals can be electroplated, such as Au, Ag, Cu, Ni, Pt, Pd, and several others. Two examples of reduction processes, showing different mechanisms, are shown below:

• **Copper** is typically plated from copper sulfate solution. The reaction is a one-step reduction:

$$Cu^{2+} + 2e^- \rightarrow Cu(s)$$

• **Gold** is typically plated from a cyanide solution and the reaction is a two-step process.

a) A fast equilibrium between the gold cyanide ion and a neutral, reducible species:

$$Au(CN)_2^- \leftrightarrow AuCN + CN^-$$

b) A slow charge transfer reaction:

$$AuCN + e^- \rightarrow Au(s) + CN^-$$

In an interesting analogy to anisotropic wet etching (whereby the slowest etching crystal planes are exposed), in electroplating, the fastest growing crystal planes "grow themselves out of existence," exposing the slowest growing planes, as illustrated below. To some extent, the morphology of the electroplated metal can be controlled by varying the current density used for electroplating and/or by the use of small quantities of additives. With names such as *grain refiners, hardeners,* and *brighteners,* these are trace quantities of compounds that are added to a plating solution to modify the deposited metal (e.g., trace quantities of cobalt as a "hardener"). These are generally present in commercial plating solutions (unless they are specified to be "pure" plating solutions) and are typically proprietary.

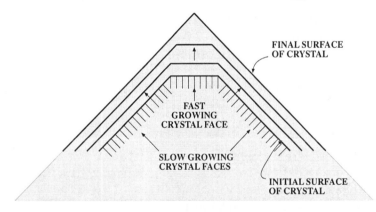

Illustration of how crystal planes growing at different rates during electroplating interact, with the fastest growing faces eventually disappearing. After Bockris and Reddy (1970).

Another common class of additives are *conducting salts* that do not participate in the electrodeposition reactions, but increase the conductivity of the solution, increasing the voltage drop at the interfaces to the cathode (object being plated) by reducing the voltage drop through the bulk solution. In addition, *surfactants* in the plating solution can greatly increase the uniformity of the plating thickness and morphology across a wafer. A good one is PFOS (perfluorooctyl sulfonic acid at a concentration of 30 ppm, as described by Missel, et al. (1980)).

Examples of plating solutions (not tested by the author) that are useful for micromachining include:

• **Gold** (Andricacos, et al. (1977))

Potassium gold cyanide (dicyanoaurate) ($KAu(CN)_2$)	20 g/l
Potassium citrate ($K_3C_6H_5O_7 \bullet H_2O$)	150 g/l
Potassium phosphate (dibasic) (HK_2O_4P)	40 g/l

(used with a current density of 100 mA/cm^2)

• **Nickel** (Andricacos, et al. (1977))

Nickel sulfate ($NiSO_4 \bullet 6H_2O$)	330 g/l
Nickel chloride ($NiCl_2 \bullet 6H_2O$)	45 g/l
Boric acid (H_3BO_3)	38 g/l

(used at 60°C with a current density of 100 mA/cm^2)

• **Nickel-Iron** (Frazier and Allen (1993))

Nickel sulfate ($NiSO_4 \bullet 6H_2O$)	200 g/l
Nickel chloride ($NiCl_2 \bullet 6H_2O$)	5 g/l
Ferrous sulfate ($FeSO_4 \bullet 7H_2O$)	8 g/l
Boric acid (H_3BO_3)	25 g/l
Sodium saccharin ($C_7H_4NNaO_3S \bullet 2H_2O$)	3 g/l
Sulfuric acid (H_2SO_4) to bring the pH to 2.5	

(used with a current density of 5 to 10 mA/cm^2)

(Note that other Ni-Fe, or Permalloy™ electroplating references include Liu, et al. (1995), Castellani, et al. (1978), Anderson and Grove (1981), Komaki (1993), and Grande and Talbot (1993).)

• **Copper** (Frazier and Allen (1993))

Copper sulfate ($CuSO_4 \bullet 5H_2O$)	120 g/l
Sulfuric acid (H_2SO_4)	100 g/l

(used with a current density of 10 mA/cm^2)

Liakopoulos, et al. (1996) demonstrated the electroplating of thick **CoNiMnP** structures, followed by their magnetization using an external rare-earth permanent magnet. Their electroplating formula, used at a current density of 10 to 20 mA/cm^2 at room temperature, consisted of:

Cobaltous chloride ($CoCl_2 \bullet 6H_2O$)	23.8 g/l
Nickel chloride ($NiCl_2 \bullet 6H_2O$)	23.8 g/l
Manganese sulfate ($MnSO_4 \bullet H_2O$)	3.38 g/l
Sodium chloride ($NaCl$)	23.4 g/l
Boric acid ($B(OH)_3$)	24.7 g/l
Sodium phosphate ($NaH_2PO_2 \bullet H_2O$)*	4.4 g/l

Sodium lauryl sulfate [wetting agent] ($C_{12}H_{25}NaO_4S$)   0.20 g/l
Sodium saccharin ($C_7H_4NNaO_3S \cdot 2H_2O$)   1 g/l
* Possible error in the manuscript

Complete recipes for electroplating NiFe, NiCo, and NiFeMo high-permeability magnetic films are also described in Taylor, et al. (1997).

Electroplating should always be done in a fume hood. Note that many plating solutions are *cyanide-containing compounds.* These solutions should be kept away from strong acids that could unintentionally lead to the release of HCN. HCN can be formed and may escape into the air. CN^- binds so tightly to hemoglobin in blood that oxygen cannot be transported, potentially leading to death.

Electroplating of photoresist can also be carried out from aqueous solutions (see Vidusek (1989), Kersten, et al. (1995), and Linder, et al. (1996)). The use of such a resist (PEPR 2400, a product of Shipley, Ltd., Newton, MA) to fabricate through-wafer interconnects was demonstrated by Linder, et al. (1996). The PEPR 2400 resist was deposited from an aqueous solution onto 200 to 300 nm Cu, Al, or Au seed layers at a current density of 10 mA/cm² (100 to 300V applied), with the deposition self-terminating at a thickness of 5 to 18 μm as a function of a thickness controlling additive (PEPR 2400 TC), bath temperature, and applied voltage. The electroplated photoresist was soft-baked at 105°C, exposed (Linder, et al. (1996) describe optimum exposure conditions for patterning within pyramidal pits, taking into account reflections), and developed. The standard developer is 1 wt% sodium bicarbonate in water at 35°C (≈ 1 μm/min development rate), but it attacks aluminum. Although these researchers developed a developer that does not suffer from this problem, it was not reported. Further information on electroplated photoresist can be found in Heschel and Bouwstra (1997).

6.3.1 ELECTRODEPOSITION MECHANISMS

The electrochemical mechanisms underlying electrodeposition exhibit a saturable J-V characteristic (current density, J, is used, rather than the total current, I). Above a certain applied potential, the total current depositing metal plateaus at J_{lim}. Further increases in voltage above this point do increase the current, but the excess current ends up driving unwanted reactions (i.e., electrolysis of water). The excess voltage above the saturation current density is called the *overpotential.* If the overpotential is large, porous or "black" metal can be formed (see below). Thus, for a given electroplating chemistry, there is generally a recommended current density that, for a specified temperature, provides a nearly optimal deposition rate. Often, these recommended current densities do not simply correspond to the maximum rate, but also take into account the morphology of the plated metal. For more information about electrode reactions, the reader is directed to the Chemical and Biological Transducers chapter.

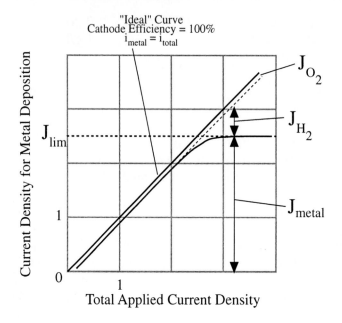

Normalized plot showing that of the total applied current density, only a fraction (this fraction, expressed as a percentage, is referred to as the "cathode efficiency") is available for metal ion reduction, with the balance being consumed by unwanted reactions such as water hydrolysis. Modified after Andricacos, et al. (1977).

6.3.2 DC ELECTROPLATING

Direct current electroplating is the "basic" method discussed above. Typically, a DC power supply is used (potentiostatic) and the voltage is adjusted to obtain the desired current density. An alternative approach is to regulate the current (amperostatic), compensating for changing impedances of the electrodes during plating to maintain a more constant current density.

6.3.3 PULSED ELECTROPLATING ("PEP")

An alternative to DC electroplating is to pulse the current on and off at a specified duty cycle (e.g., typical pulse parameters for gold deposition are 1 kHz pulse frequency with a 10% duty cycle), referred to as pulsed electroplating (or PEP). To obtain a comparable plating rate (usually in μm/min for a constant surface area), one therefore needs higher peak currents than for DC plating. A benefit of this approach is that diffusion of reactant species from the bulk solution *in between current pulses* can "recharge" the region closest to the cathode where the reactants are depleted by each plating pulse. This can raise the reactant concentration at the interface (normally quite depleted during DC plating).

Pulsed electroplating has several benefits: 1) DC current favors the growth of existing metal "grains" (small crystals), while pulse current favors the nucleation of new grains (much "finer-grained" metal deposits are possible with PEP); 2) stress control is possible by adjusting the duty cycle; 3) control of crystal orientation through PEP can improve electrical and wear properties of plated films; and 4) PEP allows the electroplating process to provide better deposition in corners or spaces that would otherwise not receive the same coating thickness (i.e., PEP has higher "throwing power").

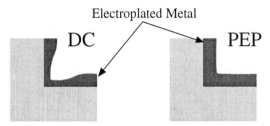

Illustration of increased throwing power of pulsed electroplating.

Its disadvantages, generally not significant, include 1) inclusions such as gases and organic contaminants, which can be increased with PEP over those with DC plating, and 2) current distributions may theoretically be less uniform than for DC plating (leading to poorer thickness uniformity across a wafer). However, with respect to the latter issue, in practice, the opposite is typically the case, especially with good agitation. For additional information, useful references on PEP include Andricacos, et al. (1977) and Grimmett, et al. (1993).

6.3.4 AGITATION FOR ELECTROPLATING

Excellent agitation is key to achieving good uniformity and also increases the maximum current that can be used since it "pumps" reactants from the bulk solution into the regions near the cathode. This is true for any plating technique except, perhaps, where diffusion-limited effects are desired (such as plating "black" metal layers). Typically, propeller-type agitation is used to obtain good solution turnover at the cathode surface.

6.3.5 "BLACK" METAL FILMS

If electroplating is carried out at a sufficiently high current density so that a large reactant concentration gradient is created near the cathode(s), there is a tendency for dendritic growth to occur. High field points formed by the "tallest" crystallites favor continued plating there. The dendrites can "reach" out into regions of higher concentration, further favoring their continued growth.

If the current density is increased even further, the structure of the dendritic metal can become so fine that it is better described as "porous" and is often optically black. "Platinum black" is commonly deposited on microelectrodes used to record bioelectric signals because it can greatly increase the effective surface area of the microelectrode, decreasing its electrical impedance by two (or more) orders of magnitude (this reduces Johnson noise but keeps the signal recording site localized in space). Various "black" films are also used as infrared absorbers for infrared detectors since they are extremely efficient at converting the incident radiation into heat (they perform well in "black body" absorbers). It is interesting to note that such "black" metal films can be as little as 5% metal by volume.

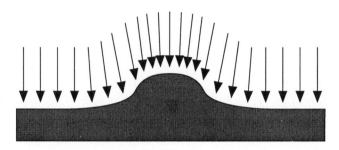

Illustration of preferential electroplating at high field points. After Bockris and Reddy (1970).

A typical platinum black plating solution (used with a current density of 30 mA/cm^2) (Lang, et al. (1992)) is:

Chloroplatinic acid (H_2PtCl_6 (anhydrous))	2 g
Lead acetate ($Pb(CH_2COOH)_2 \cdot 3H_2O$)	16 mg
Water	58 ml

"Black" metal films can also be deposited by evaporation of metal in an atmosphere of low-pressure nitrogen (Lang, et al. 1992) or other gases such as argon. For pressures on the order of 10 torr, metal atoms are scattered by the background gas. Some stick to each other, and can be deposited as a porous "snow-like" layer.

6.3.6 "ELECTROLESS" PLATING

Electroless plating of metals makes use of chemical reducing agent, rather than electrical reduction, to deposit metals from solution. Electroless plating solutions are set up so that they are nearly unstable (i.e., the temperature is adjusted to just below the point at which all of the metal ions would be spontaneously reduced [precipitate]). Exposed metal surfaces (sometimes pre-treated with a catalyst such as Pd) induce the reduction reaction, which continues to proceed locally after initiation. Examples of reducing agents for electroless gold plating are potassium borohydride (KBH_4) or dimethylamine borane (DMAB).

The following is an example of a useful electroless Au formula:

Potassium hydroxide	KOH	11.2 g/l
Potassium cyanide	KCN	13.0 g/l
Potassium borohydride	KBH_4	21.6 g/l
Potassium gold cyanide	$KAu(CN)_2$	5.8 g/l

The above formula must be mixed in the sequence listed and not heated above 85°C. At 75°C, a deposition rate of ≈ 6 µm/h is obtained. Similar approaches exist for copper and other metals, as described in Mallory and Hajdu (1990).

Furukawa, et al. (1993) described a method for direct electroless plating on *bare silicon*, accomplished through an initial roughening step of the silicon, followed by a conventional Pd "seed" layer deposition and plating. Their process consisted of 1) degrease cleaning (weak base solution); 2) silicon surface roughening using HNO_3/HF/DI water solution (18% of 49 wt% HF solution/44% of 70 wt% HNO_3 solution/38%); 3) catalytic deposition of Pd from $PdCl_2$, $SnCl_2$, and HCl solution; 4) removal of residual hydrated Sn complex using HCl; and 5) electroless plating of nickel.

The plating solution used by this group consisted of $NiSO_4$ as the nickel source, NaH_2PO_2 as the reducing agent, and CH_3COONa as a buffer and complexing agent for nickel. The plated film stress could be controlled through the amount of incorporated phosphorus. Their plating solution is commercially available from McGean-Rohco, Inc., Cleveland, OH. Depending on bath temperature and pH, plating rates between 2 and 40 µm/h are possible. They successfully demonstrated structures as narrow as 2.5 µm and as tall as 15 µm (aspect ratio $\approx 6{:}1$) using photoresist templates.

As an alternative to wet seed layer deposition, Bhansali and Sood (1995) demonstrated that, through implantation of Pd ions into patterned polyimide (PIX 3400, used as a thick plating template) on silicon, electroless Cu plating could be seeded either on the silicon alone or on both the silicon and polyimide (they used a CuCN, KOH, and formaldehyde-based plating solution). The dose for silicon-only plating was 1.5×10^{15} cm^{-2} and for both silicon and polyimide was 3.2×10^{16} cm^{-2}. While this process requires a "metal vapor vacuum arc" (MEVVA) implanter that may not be commonly available, it does provide some ideas for future selective electroless plating work.

6.3.7 TEMPLATES FOR PLATING

Template layers (generally sacrificial) formed above a substrate are extremely useful for electroplating (and for electroless plating and selective CVD as mentioned above) and serve to define the regions in which metal will be deposited. Electrical contact to the regions to be electroplated (exposed where the template is removed) must be made. A thin "shorting layer" of metal is deposited over all of the underlying structures, connecting them together electrically. This layer is generally removed upon completion of the electroplating step and removal of the template. In a typical process, the shorting layer is deposited prior to the fabrication of the template, which may be spun-on photoresist, for example. However, if the template layer has a sufficiently low aspect ratio to allow contiguous metal coverage, the shorting layer can be deposited after its fabrication.

For example, in Au electroplating, a Ti/Au or Cr/Au shorting layer is often used. After plating, the Au can be removed using a quick etch with KI_3 solution (short enough that the plated structures are not damaged or undercut), and the Ti can be removed using buffered HF solution (e.g., buffered oxide etch, or "BOE"). Similarly, approaches with various template layers (including polyimides and other organics) can be used with a wide variety of substrates. Very high-aspect-ratio *submicron* structures have been fabricated in this way (e.g., see Romero (1983)).

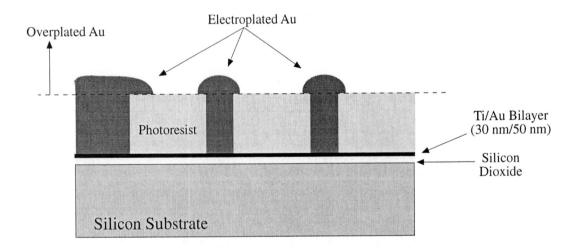

Illustration of the use of a sacrificial photoresist template layer for electroplating gold onto an oxidized silicon substrate. A temporary electrical shorting layer is necessary to interconnect regions where metal deposition is desired. Note that if plating is not halted before the metal exceeds the height of the template, overplating can occur (in some cases, it can be used to advantage, in others it is highly undesirable).

In addition, sacrificial layers (often combined with a template layer) can be used to form air bridges or cavities, the former being used in microwave integrated circuits. An example of a sacrificial/template plating process is illustrated below.

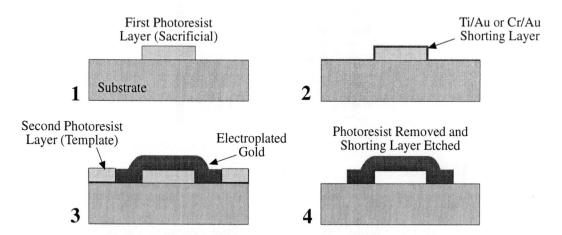

Illustration of the use of a sacrificial photoresist spacer to form an electroplated air bridge. In this case, the shorting layer must have sufficient step coverage to coat the sacrificial structures.

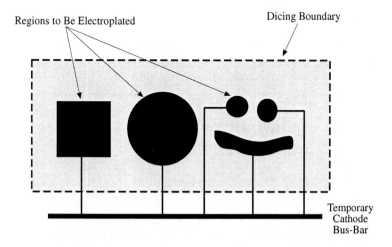

Illustration of the use of temporary shorting traces to interconnect regions to be electroplated.

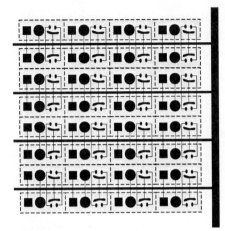

Example of global organization of shorting traces for electroplating, designed for disconnection during dicing.

Another approach to electrically interconnecting regions to be electroplated is to use a photolithographic "shorting trace" that connects the desired regions together and is later removed (i.e., is sawed through during dicing). If "shorting traces" or "temporary bus-bars" are used, the electroplated devices can be arrayed to minimize the number of dicing saw cuts to separate them. This approach is commonly used in the manufacture of printed circuit boards (particularly for board edge connectors).

SELF-SHORTING PLATING PATTERNS

It is possible to make use of the fact that plating processes are fairly isotropic by allowing nearby plated features to merge together (over a template layer or without one) deliberately by "overplating," forming final shapes that would otherwise be very difficult to realize. This approach was demonstrated by Wagner, et al. (1997), who used the natural shorting of electrically isolated metal traces on the substrate with one electrically powered trace to form wedge-shaped structures using electroplating. As plated metal spread out from the driven trace, the nearby traces were eventually shorted to it and began to be plated. This process continued, with the metal deposited over the driven trace being the tallest, and with decreasing height over each of the subsequently merged electrically isolated traces.

If the electrically driven trace in an array of floating traces could be suitably positioned, this approach could be used to fabricate a wide variety of useful 3-D structures, two hypothetical examples of which are illustrated below. By appropriately varying the spacing between the floating traces (Wagner, et al. (1997) used a linear spacing), one could also exert some degree of control over the slopes of the structures.

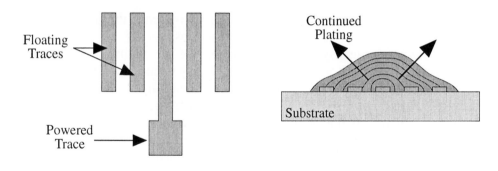

TOP VIEW   CROSS SECTION

Illustration of self-shorting of electrically isolated metal traces with an electrically driven trace during electroplating to form wedge-shaped structures. Adapted from Wagner, et al. (1997).

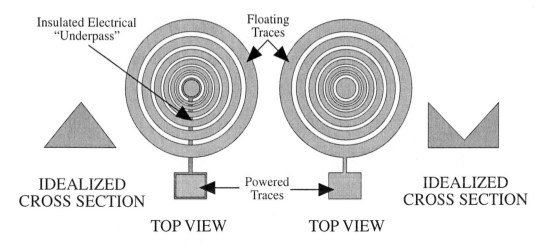

IDEALIZED CROSS SECTION   TOP VIEW   TOP VIEW   IDEALIZED CROSS SECTION

Illustration of extensions of the technique of Wagner, et al. (1997) to fabricate conical and "inverse conical" electroplated structures (for the conical structure, the inner, powered trace must be insulated from the floating traces, requiring two levels of metallization with an intervening insulating layer). The spacing and width of the floating traces could be adjusted to control the shape to some extent. (The idealized cross sections are not drawn to the same scale as the electroplating traces shown in the top views.)

"TEAR-OFF" ELECTROPLATED STRUCTURES

Electroplating templates can be used to form "tear-off" structures that can be pulled free from the substrate and seed layer and used to fabricate out-of-plane features. For example, if a narrow trench (e.g., 5 µm) is etched through a photoresist template down to an electroplating seed layer, the cross section of the electroplated metal rapidly assumes a nearly cylindrical shape once the plating proceeds past the top of the template. Distortion of the photoresist may occur, often helping to make the base of the plated structure more rounded. The resulting structure is attached to the substrate only along a very thin strip originally forming the base of the template trench. If a soft metal, such as gold, is used for the electroplating, these structures can readily be lifted from their thin adhered regions and folded into complex structures. For example, 100 µm diameter Au rods, starting from 5 µm trenches in photoresist, were fabricated in this manner and folded up to form vertical structures several millimeters in height above the substrate (Kovacs, et al. (1992)).

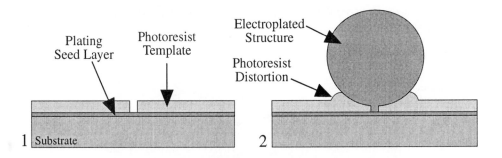

Illustration of the use of deliberate template distortion around a narrow trench to form electroplated metal structures that can be lifted from the substrate and formed into out-of-plane structures.

PHOTOIMAGED PLATING TEMPLATES

Directly photoimaged plating templates can be fabricated using many photosensitive materials (photoresists, polyimide, etc.) as opposed to using templates that require patterning via plasma or RIE etching (such as thick, preformed sheets of organic material), for example. Photoresist is very useful for relatively low-aspect-ratio structures, but can be distorted by stresses of the plating process (plated metal expanding against the patterned resist), resulting in distorted final devices (high-temperature "hard-baking" or UV hardening the photoresist helps to resist distortion). Polyimide is much stiffer than photoresist when fully cured, and is often a better choice for higher aspect ratios.

Allen (1993) and Frazier and Allen (1993) described the use of photosensitive polyimide (Ciba-Geigy Probimide 348) as a template material. Direct patterning was done using UV exposure at 350 mJ/cm^2 for 40 µm thick layers, followed by Ciba-Geigy QZ3301 developer and QZ3312 rinse (ultrasonically assisted development is required if the layers are thicker than 25 μm). The template layer was removed after electroplating using 70°C, 30 wt% KOH solution (this will not work if aluminum is present since KOH attacks it). They demonstrated structures plated from nickel-iron and copper solutions. The plated devices could be released from the substrate by undercutting the titanium or chromium adhesion layer beneath the electroplating seed layer. Optically imaged structures reported by Allen (1993) have achieved aspect ratios of nearly 8:1 and plasma-etch-based template formation is discussed in Allen (1993).

SYNCHROTRON EXPOSED TEMPLATES: THE LIGA PROCESS

Developed by W. Ehrfeld, et al. at Karlsruhe, Germany (see Becker, et al. (1986) and Ehrfeld, et al. (1987, 1988)), a process called "Lithographie, Galvano-formung, Abformung," (LIGA), for lithography, electroplating, and molding, can yield extremely high-aspect-ratio metallic structures. It is a microfabrication process based on template-guided electroplating, which is capable of yielding structures with extremely high aspect ratios (at least 100:1) thanks to the use of extremely well-collimated synchrotron radiation (x-rays) to expose the template layer with nearly perfectly straight sidewalls. Naturally, one needs access to a synchrotron source to carry out this process, but there are many sites with the capabilities (e.g., Brookhaven, Princeton, Stanford, Lawrence Berkeley Laboratory, University of Wisconsin, etc). As discussed in other chapters, a wide variety of devices have been made using the LIGA process, such as a working gear train, magnetic motors, fluidic pumps, etc.

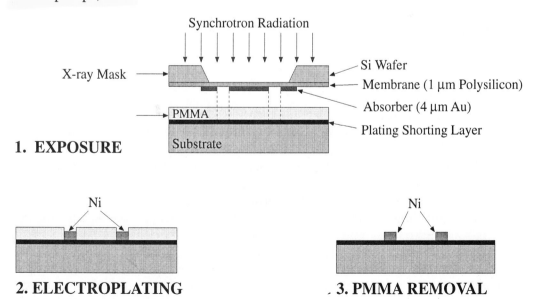

Illustration of the LIGA fabrication process. Adapted from Ehrfeld, et al. (1988) and Guckel, et al. (1990).

As described in Guckel, et al. (1990), the first step is to make a special x-ray mask using a 1 µm thick polysilicon membrane (silicon nitride, particularly low-stress, silicon-rich LPCVD films are also suitable for this purpose), with electroplated Au serving as the x-ray absorber layer. The process flow was as follows: 1) 5 nm Cr and 15 nm Ni was deposited to serve as a plating mask, 2) several 1 to 2 µm layers of polymethyl methacrylate (PMMA) was spun on and baked for a total thickness of ≈ 8 µm, 3) ordinary photoresist (opaque to deep UV) was spun on, pre-baked and exposed, 4) the PMMA was exposed through the patterned photoresist, and 5) the PMMA was developed to obtain nearly vertical sidewalls.

Then a wafer or a precision polished metal substrate must be coated with 10 to 1,000 µm of cross-linked PMMA to serve as the template for subsequent electroplating. The process flow is 1) PMMA was dissolved in MMA and a cross-linking agent EGDA (ethylene glycol dimethacrylate) is added, 2) polymerization was initiated with dimethyl aniline (DMA) and benzoyl peroxide (BPO), 3) PMMA was applied by casting (not spin-on), and 4) the cast PMMA was mechanically polished (this may not be necessary in all applications).

The PMMA was then exposed through the x-ray mask material patterned on a membrane using a highly collimated source such as a synchrotron (x-rays break the cross-links in the PMMA). The PMMA was then developed in a special solvent system (this is the same developing system used to make the x-ray masks), and the resulting structures have extremely high aspect ratios. The developer (a critical process technology) formulated by Ehrfeld, et al. (1988) is:

2-(2-butoxyethoxy)ethanol	60 vol%
tetrahydro-1, 4-oxazin (morpholine)	20 vol%
2-aminoethanol	5 vol%
water	15 vol%

Metal was then electroplated into the PMMA mold. Typically, Ni from a sulfamate plating system is used at 50°C at a current density $J = 50$ mA/cm^2.

Massoud-Ansari, et al. (1996) and Guckel, et al. (1996), reported a multi-level, LIGA-like process for fabricating stacked electroplated nickel structures. The process began with an oxidized silicon substrate with a metallic plating seed layer, to which a preformed PMMA sheet (up to 500 µm thick) was solvent bonded. Synchrotron x-ray exposure was carried out, followed by electroplating and diamond lapping and polishing to the desired final height (350 µm in this case). This process, starting at the bonding of a PMMA sheet can be repeated to form multiple, stacked layers (as long as electroplating can be carried out, requiring electrical continuity between [potentially multiple and electrically isolated from each other] stacked sections or electroless plating).

Molds (for producing large numbers of replicas) can be formed by vacuum casting a suitable resin into the metallic shapes. Metal deposition into the molds can be used to make a succeeding generation of metal replicas. While molding may have been the original intent of the LIGA process developers, it has been rarely used for this purpose thus far. For further information on LIGA, other useful references include Rogner, et al. (1992) and Mohr, et al. (1992). In addition, the combination of LIGA and active circuitry (on an underlying substrate) is discussed in Ehrfeld, et al. (1987, 1993).

RIE OR PLASMA-ETCHED TEMPLATES

RIE-based template patterning has been used successfully in many electroplating applications. For example, spin-on polyimide can be used by hard curing it, depositing a thin Si mask (e.g., a sputtered Si layer of ≈ 300 nm serves as an excellent etch mask [Cr can also be used, but it is typically highly stressed]), patterning the mask, and using oxygen RIE for the high-aspect-ratio template etch. For example, Amoco Ultradel 7505D can be used, but solvent evaporation from thick layers creates a large amount of stress. Presumably, lower solvent-content polyimides would be better suited to this application.

Another approach to etched electroplating template formation is the use of preformed polymer layers that are glued onto the wafer (above the electroplating seed layer). Preformed sheets of polyimide (DuPont Kapton™), available in various thicknesses, including 25, 50, 75, and 125 μm, have been successfully used in this manner, with various adhesive layers including polymethylmethacrylate (PMMA) (in the case of PMMA, vacuum curing with a weight atop the sheet yielded adequate adhesion for electroplating conditions). Aspect ratios on the order of 10:1 have been obtained for structures as tall as 70 μm using this approach. A variety of other materials can potentially be used in such applications as long as they can be patterned to achieve sufficient aspect ratios.

6.3.8 TEMPLATE-FREE LOCALIZED ELECTROPLATING

Localized electroplating to produce 3-D microstructures (columns and helical springs) was demonstrated by Madden and Hunter (1996). They used an x-y-z stepper motor stage to move an electrochemically sharpened, epoxy-coated PtIr wire electrode (de-insulated at its tip) over a Cu or Ni substrate submerged in a plating solution. Electroplating current could be locally controlled, and the electrode moved to direct the formation of various structures. The authors demonstrated Ni columns that were 100 μm tall and 10 μm in diameter, and a five-turn Ni spring with a turn diameter of ≈ 1 mm and a cross-sectional diameter of ≈ 150 μm. Although this process is clearly serial and therefore slower than parallel lithographic methods, it may have applications in rapid prototyping of metallic microcomponents.

6.4 SELECTIVE EPITAXIAL GROWTH

Selective growth of epitaxial silicon (and other materials) can be used to form useful microstructures. Several silicon sources have been used for this purpose, and examples include silicon tetrachloride ($SiCl_4$), dichlorosilane (SiH_2Cl_2), trichlorosilane ($SiHCl_3$), and silane (SiH_4). For the most common chemistry (silicon tetrachloride), the reaction is hydrogen reduction:

$$2H_2 + SiCl_4 \rightarrow Si(s) + 4HCl$$

(Note that if too much HCl builds up over the wafers, the reaction will flow in the opposite direction, etching the Si substrate. Very good process control is required in practice.)

Pits and grooves can be selectively filled with epitaxial Si if HCl is *added* to the gas mixture. It etches the surface epitaxial Si deposits much faster than those below the surface. More selective epitaxial growth is possible if oxide masks are used. Polysilicon grows on the SiO_2 and single-crystal silicon grows on the bare, exposed substrate silicon. If HCl is used, the polysilicon can be removed as it forms, leaving only single-crystal Si mesas. Several other selective epitaxy schemes are possible. For example, Bartek, et al. (1995) demonstrated selective epitaxial silicon growth in confined spaces between two patterned silicon nitride layers, the lower one having vias to allow access to an underlying silicon substrate. The epitaxial silicon would grow upward from those vias and fill the space between the two parallel silicon nitride layers.

Some possible selective epitaxial growth schemes are illustrated below (these have been used extensively for the fabrication of silicon-on-insulator semiconductor devices for high-voltage operation). Silicon can also be grown on some other substrates such as sapphire (Al_2O_3) (several years ago, this was a very popular research area because it offered high performance thanks to isolated circuit regions and reduced parasitics). Such *silicon-on-sapphire* (or SOS) wafers are still used in applications where radiation resistance is critical. (Note that since Si is a cubic lattice and Al_2O_3 is hexagonal, epitaxial growth occurs because a 45° rotation of the Si lattice pattern over R-cut sapphire gives a rough match between every other atom, at the cost of some strain at the interface that somewhat compromises electronic performance.) Silane chemistry is typically used for this, with the following reaction,

$$SiH_4 \rightarrow Si(s) + 2H_2 \text{ (pyrolysis)}$$

Silicon can also be epitaxially grown on GaAs (and vice versa), but the resulting materials are highly strained and generally have many defects (poor electronic properties result). Silicon carbide can also be epitaxially grown on silicon, and may be quite useful in micromachining applications, as described by Tong, et al. (1992).

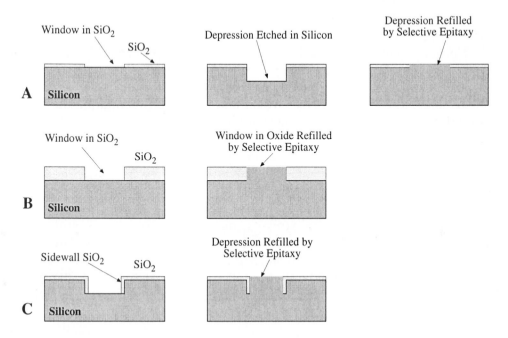

Illustration of three possible schemes of selective epitaxy. After Runyan and Bean (1990).

7. BONDING PROCESSES

Bonding processes are widely used in micromachining applications to allow similar or dissimilar substrates and/or components to be permanently coupled mechanically (and sometimes electrically). There are a variety of processes available, ranging from those that form covalent bonds between the substrates to those using intermediate adhesive layers as simple as commercial glues. In many cases, hermetic sealing is desired, and this requirement is becoming more important as bonded-on cavities are being used for packaging micromachined devices that would be compromised by conventional plastic injection molded packaging (very commonly used with low-cost integrated circuits).

7.1 ANODIC BONDING

There are well-known bonding methods where a metal/glass interface is heated to the point that metal oxides form and "mix" with the glass, forming an excellent bond (this method is used in making hard vacuum seals in vacuum tubes). The problem is that the required temperature can be on the order of 1,000°C. Pomerantz (1968) and Wallis and Pomerantz (1969) showed that high electric fields in *anodic bonding* can substitute for the high temperatures, permitting the bond to be formed at ≈ 400°C. Anodic bonding is therefore a more versatile bonding process that can be used when one wishes to permanently bond a glass sheet to a silicon wafer (i.e., to form a cavity). Positive ions in the glass drift toward the negative electrode, producing a large electric field at the glass/silicon interface. The large field pulls the two surfaces together, facilitating the bonding that would otherwise have to take place at much higher temperatures. The anodic bonding process can be carried out by 1) placing a glass (e.g., Corning 7740) sheet on a bare or oxidized silicon wafer, 2) heating the assembly to ≈ 400°C, and 3) holding the silicon at positive (anodic) potential relative to the glass (a power supply on the order of 1.2 kV is typically used).

To help minimize thermal stresses, one should use glasses with thermal expansion coefficients close to that of silicon, such as Corning 7740 (Pyrex™) and 7070, Schott 8330 and 8329, as well as 7570 (Iwaki) types. A key point for anodic bonding is that while several references in the literature state that anodic bonding must be done in vacuum, this is not the case (atmospheric bonding works well).

If cavities are to be formed by bonding glass over pre-etched pits, the pressure within the pits will reflect the ambient pressure and temperature during bonding (thus a wide range of cavity pressures is possible) so that the final pressure can be predicted via the ideal gas law if the cavity volume, sealing pressure, and temperature are approximately taken into account using the ideal gas law.

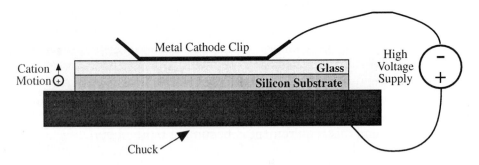

Illustration of a typical anodic bonding apparatus.

Thin (on the order of 500 nm) metal lines passing under the glass do not interfere with the hermeticity of the seal. However, the metal used must withstand the temperatures used. For thicker lines, one can planarize with, for example, LTO and then chemomechanically polish (see Sooriakumar, et al. (1993)).

A potential problem when anodically bonding glass to substrates containing prefabricated circuits is the possibility of electrostatic discharge damage, although this can be mitigated through the use of static protection circuits, metallic shields, and pre-etched cavities in the glass above the location of the underlying circuitry. Also, as discussed below, silicon-to-silicon bonding with thin deposited glass layers can be accomplished at far lower voltages.

7.1.1 ANODIC BONDING USING DEPOSITED GLASS

The anodic bonding process can also be used to bond silicon to silicon using an intermediate glass layer that can be sputtered, evaporated, or deposited using spin-on glass. Esashi, et al. (1990) used a sputtered glass film (i.e., Pyrex™) for this purpose, and reported deposition of 0.5 to 4.0 µm layers of 7740 (Pyrex™) and 7570 (Iwaki) glasses using magnetron RF sputtering in 30% O_2/70% Ar at 6 mtorr. As usual, the negative side of the power supply was connected to the glass-covered wafer and the positive side (anodic) to the bare silicon. It is important to note that for this type of anodic bonding, the minimum bonding voltages were only 30 to 60 V (versus roughly 1 kV for bulk glass-to-silicon bonding). This process worked for bonding glass-coated Si to bare Si, Al-coated glass, and indium-tin-oxide-(ITO) coated glass, and could have broad utility in micromachining applications.

Evaporated glass can also be used for anodic bonding, and is often more appropriate than sputtered glass since the deposition rates have been reported to be as much as three orders of magnitude higher (see de Reus and Lindahl (1997)). Krause, et al. (1995) used Shott 8329 glass deposited using electron-beam evaporation to achieve deposition rates as high as 4 µm/min. They observed large compressive stresses in the films (enough to cause 50 to 60 µm of bowing in standard 4 in. silicon wafers) after deposition. Annealing at 500°C for 10 min greatly reduced the stress, and this temperature was recommended because it is higher than that used for the bonding and allows out-gassing to occur that could otherwise cause problems during bonding. Cracking during the anneal step limited glass thickness to ≈ 10 µm (more than adequate for most applications). Pre-cleaning in organic solvents (not specified) was found to be critical to bond quality. Silicon-to-silicon bonds with 7 µm glass layers were carried out at 150 V at 450°C. The bond strengths were at least as high as bulk silicon-to-glass anodic bonds, reaching values up to 40 MPa. de Reus and Lindahl (1997) further discussed this approach, and reported that anodic bonding could be initiated with voltages as low as 15 V for 5 µm thick glass layers, and at temperatures as low as 300°C. They noted that annealing at 340°C for 15 min was suitable for greatly reducing the stress in the evaporated glass layer.

Quenzer, et al. (1996) demonstrated anodic bonding of two silicon wafers using a specially formulated spin-on-glass mixture containing the necessary mobile alkali ions for the bonding to take place. They reported the use of two formulations, of which the most generally useful is 35.5 g methyltriethylorthosilicate (MTEOS), 11.5 g tetraethylorthosilicate (TEOS), 50.0 g silica sol (presumably a silica suspension), 10 g H_2O, 16 g potassium acetate (10% solution in methanol), and 5 g acetic acid, mixed together in that sequence and refrigerated thereafter (the shelf-life under refrigeration is $\approx$ 2 weeks). Using this formulation, 0.6 to 1.5 μm thick layers can be deposited, and are treated using a 450 to 500°C air anneal to decompose the organic components. Anodic bonding was successful at 400 to 420°C at 60 V (for the 1.5 μm SOG film).

7.2 SILICON FUSION BONDING

The *fusion bonding* process involves direct silicon-to-silicon bonding, and is broadly used in industry to fabricate a variety of micromachined structures (as well as substrates for high-voltage circuits where silicon-on-insulator structures are beneficial). Two wafers, with or without thermal SiO_2 layers, can be directly bonded together. The resulting composite has almost no thermal stress because the thermal expansion coefficients of the two substrates are identical. In addition, the bonds have effectively the same mechanical strength as the silicon itself, making for a very robust method of forming cavities, channels, etc., in between wafers.

The exact mechanisms of bond formation are apparently not yet fully understood (see Barth (1990)). However, the basic idea is to make sure the two surfaces have plenty of hydroxyl groups (this can be accomplished in many ways; Barth suggests boiling in nitric acid). This step makes the wafers very hydrophilic. Hydrogen bonds formed when the wafers are brought into contact hold them initially until the (postulated) bonding reaction occurs,

$$Si\text{-}O\text{-}H + H\text{-}O\text{-}Si \ \rightarrow \ Si\text{-}O\text{-}Si + H_2O$$

The bonding process occurs between 300 and 800°C, and sometimes, 800 to 1,100°C anneals in oxygen or nitrogen are used to strengthen the bonds. An alternative process exists that substitutes higher temperatures ($\approx$ 1,100°C) and electrostatic energy for the hydration step, but SiO_2 must be present on both surfaces. Particulates trapped between the wafers can form voids that may be a problem in some applications. They can be visualized using transmission infrared microscopy (as discussed in, e.g., Harendt, et al. (1990)) as a diagnostic technique.

7.3 OTHER BONDING TECHNIQUES

Several other techniques are available for bonding wafers and structures together, and *the simplest ones are often overlooked.* Photoresist can readily be used as an inter-wafer adhesive, and appears to have reasonable stability. PMMA has been used to glue sheets of polyimide and other materials to various substrates. Various waxes are available as temporary adhesives or protective layers, and are sometimes used to protect one surface of a wafer during wet etching. "Hardware store" adhesives of various sorts have also been employed in micromachined structure fabrication.

In some of these cases, the adhesive layers are difficult to remove (although removal may not be necessary in certain applications) either grossly or selectively, but oxygen plasma etching generally gives good results ("dirty" materials can leave behind considerable residual materials, often inorganic in nature). An example of the fabrication of micromachined devices using adhesive bonding can be found in Maas, et al. (1996). Another useful reference is Field and Muller (1990), where the use of a low-temperature (450°C) melting glass to hermetically bond wafers together is described.

7.4 COMPOUND PROCESSES USING BONDING

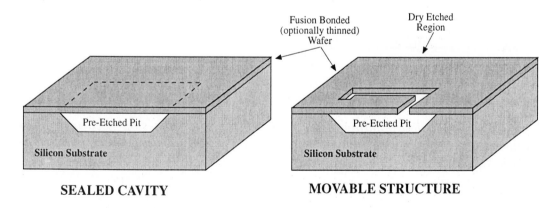

Illustration of the formation of sealed cavities and movable structures using fusion bonding of two wafers.

Several compound processes involving bonding two or more wafers and etching have been demonstrated for fabricating bulk/surface micromachined single-crystal silicon structures. Potential advantages over polysilicon mechanical structures are much greater uniformity of properties and the ability to easily integrate circuits on the structures themselves. However, this is traded off against the need to carry out bulk processing that is typically wet, as well as the bonding step itself. One approach, using fusion bonding combined with "pre-etched" silicon pits, can be used for cavity formation or to make movable structures (examples of such processing can be found in Petersen, et al. (1988, 1991) and Hsu and Schmidt (1994).

Noworolski, et al. (1995) demonstrated a process wherein an underlying "handle" wafer is pre-etched with KOH to form cavities, oxidized and fusion bonded to a p-type substrate with an n-type epi layer (defining the final structure thickness). The p-type region of the top wafer is etched away in an electrochemical etch in KOH, leaving the n-type layer behind on the underlying oxide. If moving structures are desired, plasma etching is carried out through the n-type silicon above the prefabricated cavities. A limitation of this type of process is the maximum practical epi-layer thickness (and hence cost), determining the maximum micro-structure thickness.

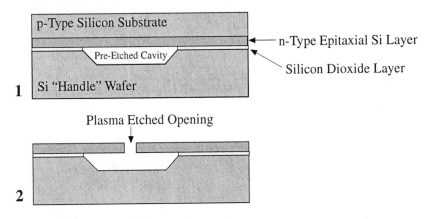

Illustration of fusion bonding and an electrochemical etch-stop used to form a well-defined silicon layer above a cavity-containing "handle" wafer. After No-worolski, et al. (1995).

Klaassen, et al. (1995) demonstrated a similar process wherein cavity-containing "handle" wafers were fusion bonded to top wafers on which both moving and fixed structures could be formed by dry etching. In this process, however, deep reactive ion etching (DRIE) was carried out to achieve extremely high-aspect-ratio mechanical structures without the need for the electrochemical etch-stop process described above. It should be noted that while electrically isolated structures can be formed (anchored to the silicon dioxide between the two bonded wafers), making electrical connections to the individual structures was difficult and required the use of wire bonds. Brosnihan, et al. (1997) demonstrated a technique for forming planar regions in between isolated structures that are silicon nitride lined and back-filled with polysilicon. This allows conductive traces to be formed between isolated structures at the cost of several extra process steps. While not yet demonstrated with on-chip circuitry, the process may alleviate the need for wire bonds between regions. The economics and reliability issues associated with forming a few extra wire bonds versus increasing the process complexity are not yet clear.

Gui, et al. (1997) demonstrated a fairly complex process for fabricating movable, single-crystal silicon mechanical structures using a stack consisting of silicon, oxide, polysilicon, oxide, and silicon. Using anisotropic etching of the top silicon wafer, followed by isotropic undercutting by etching the polysilicon, the structures were released.

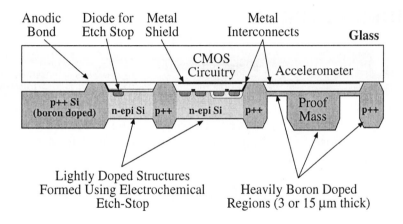

Illustration of compound anodic bond/boron etch stop/electrochemical etch-stop process. After Gianchandani, et al. (1995). CMOS circuitry (center) can be fabricated on n-epitaxial silicon that is protected from the EDP etchant using an appropriate electrochemical bias. Two levels of heavy boron doping provide a second etch-stop mechanism for fabricating mechanical structures such as accelerometers (right).

Another compound-wafer, single-crystal silicon approach was demonstrated by Gianchandani, et al. (1995), which combined heavily doped boron regions with electrically biased n-epi regions (containing CMOS circuitry in some cases) anodically bonded to a glass substrate, and released using an EDP etch (dissolving away most of the original silicon wafer). Shallow (3 μm) and deep (15 μm) boron diffusion steps were used to form thick and thin regions of silicon. They demonstrated capacitive accelerometers and fully functional operational amplifiers on dice processed in this manner.

Flip-chip, or bump bonding is another relevant bonding process in which circuit-bearing dice are transferred onto circuit boards or other substrates using solderable or alloyable metallic "bumps" on the bond pads. The dice are turned upside-down and aligned to matching pads on the substrate, then thermally bonded. This approach, which has been in use for several decades, is useful for simultaneously forming a large number of electrical interconnects, potentially between dissimilar materials (e.g., bump bonding is used to connect HgCdTe infrared focal plane arrays to underlying silicon-based signal processing chips). Singh, et al. (1997) demonstrated the use of such a process to transfer polysilicon microstructures to glass substrates (such transfers to CMOS substrates also appear feasible). Whether it is the circuitry or the micromachined structures that are transferred, this approach allows the two processes to be carried out independently, each with its own optimization paths unconstrained by the other.

8. SACRIFICIAL PROCESSES

Several interesting processes have been demonstrated that rely on "sacrificial" materials that are used as "forms" or "spacers" to make desired shapes and are later removed (some sacrificial layer concepts were presented in the electroplating sections). In a sense, when photoresist is used to define a pattern, it is a sacrificial layer since it is almost always removed. However, in the current context, sacrificial processes refer to those for making three-dimensional (or at least not completely planar) structures.

As described by Howe (1988), CVD phosphosilicate glass (PSG) etches much faster in HF than thermally grown or undoped oxides, making it an attractive sacrificial layer for undercutting polysilicon structures. Fan, et al. (1987) used such a process to realize moving mechanical structures, including pin joints, springs, cranks, etc. Oxidized porous silicon etches even faster, making it another very useful sacrificial layer material (Steiner, et al. (1993)). Aluminum, photoresist, and a number of other materials common in microfabrication laboratories can also be used as sacrificial layers, and can often readily (and selectively) be etched using dry etch methods. A large number of sacrificial processes have been demonstrated, using both wet and dry etching, and the technique can readily be extended to form multiple levels of interconnected, released structures (including trapped moving structures such as rotating and sliding members, as discussed in other chapters, particularly Mechanical Transducers).

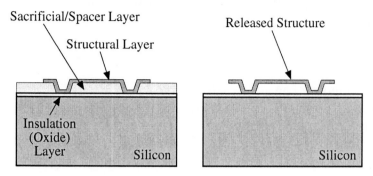

Illustration of the use of a sacrificial layer (e.g., phosphosilicate glass) to form a raised and released microstructure of another material (e.g., polysilicon). After Howe (1988).

8.1 STICKING PROBLEMS DURING WET RELEASE

Attractive capillary forces can be a real problem with wet-released micro-structures because the drying fluid (usually water) tends to pull deformable micro-structures into contact with the substrate and/or each other. (Wetted wafers can stick together with impressive force.)

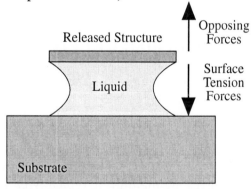

Illustration of surface tension forces in opposition to forces resisting the "pulling down" of a micromechanical structure. After Legtenberg, et al. (1993).

Legtenberg, et al. (1993) studied these phenomena. They studied the capillary forces and showed that once deflected, *van der Waals forces* (attractive and repulsive electrostatic dipole-dipole interactions between molecules) are responsible for the stiction of *hydrophobic surfaces* and *hydrogen bonding* (a particularly strong attraction between a hydrogen atom of one molecule and a pair of unshared electrons of another molecule) is the dominant adhesion mechanism for *hydrophilic surfaces*.

These conclusions were reached by studying various beams (double-clamped) during drying, equating the mechanical forces opposing the surface tension forces:

$$U_{total} = U_{bending} + U_{stretching} + U_{surface\ tension}$$

Hydrogen bonding can occur when there are hydrogen atoms bonded to nitrogen, oxygen, or fluorine. This is a fairly common situation with the compounds used in micromachining. For example, hydrophilic silicon has a large number of surface -OH groups. It is well known that the use of less polar solvents than water, such as methanol, ethanol, etc., greatly reduces this effect.

Illustration of hydrogen bonding between water molecules.

8.1.1 PREVENTING STICKING DURING WET ETCH RELEASE

A wide variety of stiction-minimizing process approaches have been developed for surface micromachining applications. One technique is to remove the liquid in a way that does not allow surface tension to have any detrimental effects, such as freeze-drying or critical point drying, wherein the liquid/solid transitions directly to gas. Another approach is to take advantage of geometry- or process-specific methods to mitigate the forces leading to stiction. Also, one can carefully control the surfaces so that the attractive forces are minimized, typically by chemical means.

PHASE-CHANGE RELEASE METHODS

A variety of phase-change release methods are reviewed in Kim and Kim (1997), all relying on avoiding the surface tension force of drying solvent below microstructures by transitioning it to gas phase either from a solid form or by removing the liquid/gas interface (critical point methods). Takeshima, et al. (1991) discussed the use of tertiary-butyl (t-butyl) alcohol in a freeze-drying scheme wherein the water was replaced by it, the butyl alcohol was frozen and then sublimed under relatively low vacuum. This is a simple and effective method. Similar work with p-dichlorobenzene was done by Kobayashi, et al. (1992) and Lin, et al. (1995), despite that compound's toxicity, higher melting temperature (56°C versus 26°C for butyl alcohol), and longer sublimation times. For both of these compounds, the liquid surrounding the devices to be released (typically polysilicon structures with sacrificial PSG or similar oxide) is changed from the etchant (e.g., HF) to DI water, methanol, and the final sublimating compound. Kim and Kim (1997) describe the use of a Peltier heat pump to control the substrate temperature during the sublimation, good control over rates of change of temperatures.

Mulhern, et al. (1993) demonstrated the use of critical point drying to release micromechanical structures. Critical point drying is very commonly used (and has been for decades) in the preparation of biological specimens for electron microscopy, where one wishes to remove the water without distorting the tissues. The basic idea is to take advantage of the fact that, under the correct conditions (the "supercritical region"), the liquid and vapor phases cease to exist as distinct states. When this occurs, the interface between them is eliminated and, after transitioning directly to the gas phase, the gas can be gently vented without disturbing the structures.

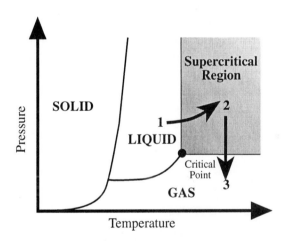

Illustration of the phase relationships in the method of critical point drying. After Mulhern, et al. (1993).

For CO_2, the supercritical region is for temperatures above 31.1°C and pressures above 72.8 atm (1073 psi). Mulhern, et al. (1993) made simple polysilicon structures using an LTO sacrificial layer and the LTO was HF etched and the structures rinsed in DI water without letting them dry. The water was exchanged with methanol by dilution and then transferred to a pressure vessel in which the methanol was replaced by liquid CO_2 at 25°C and 1,200 psi. The contents of the pressure vessel were then heated to 35°C and the CO_2 was vented at a temperature above 35°C (ensuring that it only exists in gaseous form). Their results showed that free-standing poly-Si cantilever beams up to 850 μm in length could be released without sticking, compared to only 80 μm lengths using air drying. In addition, Stowe, et al. (1997) demonstrated critical point drying to release single-crystal silicon cantilevers up to 200 μm long and only 50 nm thick.

Ohtsu, et al. (1996) reported the use of a multi-solvent drying process to release thin, compliant silicon beams (0.3 to 1.2 μm thick with a 0.7 μm gap to the substrate over cantilever dimensions in the hundreds of microns). Their process consisted of replacing the initial DI water (after wet etch release) with methanol, ethyl ether, and finally Fluorinert™ (a perfluorinated hydrocarbon liquid with surface tension approximately five times lower than water).

GEOMETRY/PROCESS-SPECIFIC RELEASE METHODS

A less general approach to reducing sticking following wet etch release is the use of methods specific to given geometries or processes. Such a process was demonstrated by Abe, et al. (1995), who studied the drying of post-release singly and doubly clamped beams. For singly clamped cantilevers, they observed (with DI water rinses) that for the shorter structures, drying occurred from the tip uniformly back to the base; yet for longer cantilevers, the fluid dried in the opposite direction, causing stiction problems. By adding small, narrower width protrusions (with convex corners) from the ends of the long cantilevers, sticking was considerably reduced. They also observed that very rapid, high-temperature drying (500°C for 10 s) greatly improved release yields for surface micromachined structures rinsed in methanol.

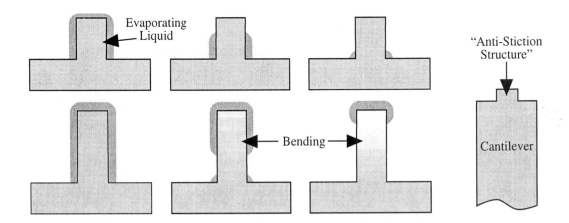

Illustration of wet release drying modes for short cantilevers (top three diagrams, left to right), long cantilevers (bottom three diagrams, left to right) and the "anti-stiction structure" for improving cantilever release yield. After Abe, et al. (1995).

In addition, Liu, et al. (1997) demonstrated the use of magnetic levitation of electroplated NiFe alloy magnetic structures on thin-film cantilevers as a means for improving wet etch release. The sacrificial PSG layer was removed in HF, followed by rinsing in water and a final rinse solution (water, isopropyl or methyl alcohol, acetone, etc.) and drying using heat while applying a magnetic field to pull the cantilevers in the out-of-plane direction. As for the previous example, this may not be a very general approach. Another special purpose approach is that of Suh, et al. (1996, 1997) that took advantage of the tendency of polyimide bilayers to curl out of the substrate plane while the underlying Al sacrificial layer was etched. This naturally provides a mechanical force to oppose that of the drying fluid meniscus beneath each structure, greatly aiding the release process.

SURFACE TREATMENT RELEASE METHODS

Chemical modification of the surfaces to be released, carried out after release but while they are still wet, can also greatly reduce stiction forces. These techniques are discussed below in the Surface Modification section.

8.2 EXAMPLE SACRIFICIAL PROCESSES

A few examples of sacrificial processes are presented below that extend beyond the surface micromachined polysilicon methods discussed above. Some processes use wet etching for release, and there is also a great deal of interest in the use of sacrificial layers that can be removed by dry etching. If carefully done, the structures can be released with no damage to the micromechanical structures unless the etching chamber is vented rapidly enough to cause local turbulence.

8.2.1 SACRIFICIAL LIGA ("SLIGA") PROCESS

Similar to the use of organic sacrificial layers to form electroplated air bridges (as discussed above), Christenson, et al. (1992) discuss the use of sacrificial layers with the LIGA technique to form high-aspect-ratio electroplated metal structures with underlying cavities. They made use of sacrificial layers such as polyimide, deposited silicon dioxide or polysilicon, all of which could be selectively removed without damaging the electroplated Ni microstructures. This process, with a suitable choice of sacrificial layer (deposited at a sufficiently low temperature), was shown to be compatible with prefabricated photodiodes in the silicon substrate (and is likely to be compatible with more complex circuit fabrication processes).

8.2.2 SIMOX AS A SACRIFICIAL LAYER

Wafers with a "buried" silicon dioxide layer (beneath a top layer of silicon) are available and quite useful for micromachining. The basic "SIMOX" wafer

(Separation by Ion Implantation of Oxygen) is made by ion implanting conventional silicon wafers with oxygen (requiring very high implant energies and doses, and currently suffering from relatively low throughput). Under the correct conditions, the implanted oxygen can be made to form a thin (300 to 400 nm) layer of SiO_2 buried beneath 200 to 300 nm of native silicon. Epitaxial silicon of high quality can then be grown from the top silicon layer if desired. By etching through the top silicon layer, the SiO_2 can be undercut, freeing micromechanical structures. Diem, et al. (1993) describe several interesting micromechanical structures fabricated in this way (pressure sensors and accelerometers).

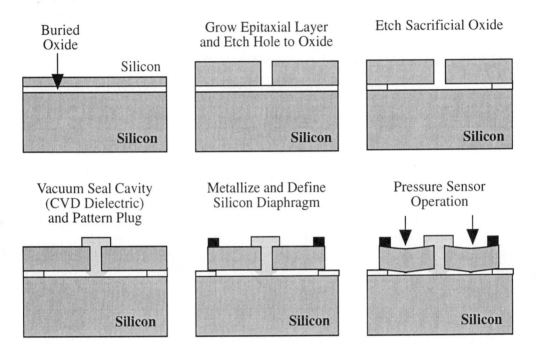

Illustration of sealed-cavity SIMOX pressure sensor formation as demonstrated by Diem, et al. (1993) (not to scale).

8.2.3 VAPOR-PHASE SACRIFICIAL LAYER ETCH

Vapor-phase etching of sacrificial oxides can be accomplished using HF and methanol (CH_3OH), without the need for a complex instrument such as a plasma etcher. Such a process was demonstrated for etching TEOS sacrificial layers beneath polysilicon by Lee, et al. (1997). Using anhydrous HF gas combined with nitrogen passed through a methanol bubbler, they obtained etch rates of 10 to 15 μm/h for partial pressures of 15 and 4.5 torr for HF and methanol, respectively. It should be noted that such processes may not be compatible with all metallization schemes that may be used on the microstructures being released (e.g., Al).

8.2.4 PLASMA ETCH RELEASE OF ORGANIC SACRIFICIAL LAYERS

The Texas Instruments electrostatic micromirror displays (or digital micro-mirror displays [DMD™s], discussed in the Optical Transducers chapter) are released by dry etching organic sacrificial layers such as photoresist (see Hornbeck (1989)), although adequate details to reproduce the process are understandably not given. Storment, et al. (1994)) described a process that is also based on this approach, depositing layers of aluminum above organic (polyimide) sacrificial layers. In both of these examples, the organic layers are removed using an oxygen plasma etch. Basically, what is required is an organic sacrificial layer material that can stand up to the conditions for depositing the micromachined structures above or around it, and still be removable later (without leaving any residue behind). Sputtered silicon can readily be used as such a dry etched sacrificial layer, for example, using fluorine-based etch gases to remove it from under materials such as aluminum.

8.3 TEMPLATE REPLICATION

In template replication processes, molding being a subset thereof, the substrate is not part of the finished devices and is used as a "mold." A great advantage of such processes is that the mold can often be reused many times, allowing for greater investment in its formation than possible on a "one-shot" basis (this is true of macroscopic molding techniques, wherein an injection mold may cost thousands of times more than a finished molded component). Beyond scaling down "conventional" molding, there are several unique "template replication" approaches that are possible with micromachining approaches.

8.3.1 INJECTION MOLDING WITH MICROMACHINED TEMPLATES

In conventional injection molding, molten plastics are injected into metal molds, cooled and removed (generally, the molds are two-part and can be opened after molding). A variant of this process, used to fabricate metallic or ceramic parts, uses plastics ("binders") mixed with 50 or more percent metal or ceramic powder. The mixture is injected into molds, cooled, removed, fired in a reducing atmosphere (e.g., H_2) to remove the binder, and finally sintered in an inert atmosphere near the melting point of the powder, allowing it to consolidate into the final part.

Using LIGA or other precision-machined templates, feature sizes on the order of a few microns are possible using injection molding. For example, Larsson, et al. (1997) demonstrated precision molding using silicon templates formed using anisotropic/isotropic wet etching or deep reactive ion etching. Using an injection molding method developed for compact disc fabrication, they replicated features etched into the silicon by coating it with a Ti/Ni seed layer and electroplating nickel to a thickness between 400 μm and 1 mm in a nickel sulfamate bath. The backsides of the electroplated mold replicas were planarized (method not given), removed from the silicon masters, and used in a commercial injection molding machine. Polycarbonate, with a glass temperature of 144°C was used at an injection temperature of 360°C, a tool temperature of 120°C, and a pressure of 6 MPa. This approach provided a relief depth of ≈ 0.5 μm and a lateral resolution of a few microns, with molding cycle times of less than 10 s. As another example, Lin, et al. (1997) demonstrated the use of anisotropically wet etched pyramidal pits in (100) silicon to form plastic light reflecting pyramids. They carried out the molding by hot embossing thin PMMA sheets with a silicon mold.

Both, et al. (1995) demonstrated the use of LIGA-fabricated structures for precision molding to replicate them, and since molds can often be reused hundreds of times, this approach should prove quite cost-effective in some applications. They fabricated the "positive" (geometry identical to the final, desired objects) mold using conventional LIGA (electroplated Ni) and fabricated a custom molding fixture that was designed for thermal expansion mismatch minimization since the molding is done at up to 250°C (but still requiring alignment to ± 10 μm at forces of up to several tens of kN).

The test structures (accelerometers) required a Ti conducting layer for the anchored regions and a Cu sacrificial layer for the moving regions. Once these layers were patterned, a thick PMMA layer was formed on the substrate. The PMMA was molded using the custom fixture and the remaining thin layer of PMMA in the regions where it was excluded was removed using O_2 RIE etching. Although the authors do not discuss how, apparently the Ti (as well as the Cu) thus exposed was directly used as the plating base, allowing the original mold to be replicated using electroplating. Finally, the Cu sacrificial layer was etched, freeing the movable regions.

8.3.2 PLATING-BASED TEMPLATE REPLICATION

A plating-based template replication process was used by Kiewit (1973) to make metal scribes and chisels for ruling optical gratings. Pyramidal and more complex shapes were made by wet etching (EDP-related compounds) patterns into silicon using patterned SiO_2 masks. Electroless plating was used to fill in the pits with Ni-P or Ni-B alloy. A backing layer of electroplated copper was added and the was silicon etched away using the same etchant, leaving behind the "microtools."

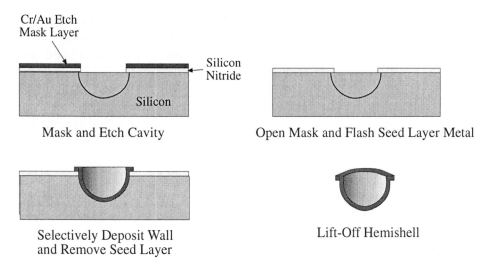

Fabrication sequence for free-standing metal hemishells using an isotropic silicon-etching technique. After Wise, et al. (1979, 1981).

Another example is the hemispherical cavities made by Wise, et al. (1979, 1981) for use as thermonuclear fusion targets. Isotropic etching was used to make hemispherical cavities in silicon, and the cavities were then replicated by either growing oxide or using a conformal metal thin-film layer as a base and then electroplating metal selectively in the cavities (using the substrate as an electrical connection). Typical dimensions of the hemishells were 350 μm diameter with 4 μm thick walls.

As described above, LIGA-derived structures can be used to form extremely high-aspect-ratio templates for electroplating. As an alternative, high-aspect-ratio reactive ion etching to fabricate the molding templates was described by Elders, et al. (1995). They used $SF_6/O_2/CHF_3$ etch chemistries with a 50 nm Cr mask to etch (100) silicon to greater than 20 μm depth with low roughness. They demonstrated electroplating directly onto the silicon substrate using a nickel sulfamate based plating chemistry (in order to make them conductive enough, the silicon wafers were boron doped using B_2H_6 for 1 h at 1,100°C for a sheet resistance of 3.9 Ω/square, after which the silicon was etched in KOH to expose the Ni structures).

The silicon templates fabricated by Elders, et al. (1995) were also used directly as mold templates to emboss 0.75 mm polycarbonate sheets at temperatures of 185 to 200°C (above the polycarbonate's glass transition temperature of 145°C). To facilitate release of the plastic structures, the silicon was plasma-coated with a thin fluorocarbon layer in the same plasma system used to etch the template.

Electroforming is a related plating-based template replication process in which templates are formed in metal and plating is used to replicate them. In some electroforming processes, the template is "peeled" away from the plated replica, and in others it is dissolved away. As described by Siewell, et al. (1985), Hewlett-Packard ink-jet print heads made use of a thin-film, 3-D electroformed nickel nozzle for ink ejection. The electroformed orifice plates are formed using an etched stainless steel template or "mandrel." The reusable mandrel is prepared for electroplating by the lamination of dry film photoresist to its surface, and the subsequent patterning of the resist. Nickel electroplating is then carried using a custom sulfate/chloride nickel plating solution and the electroplated nickel is then peeled away from the mandrel (it is weakly adherent since the steel has a surface oxide). This approach can be used to fabricate a wide variety of metallic objects with micron-scale precision.

8.3.3 CVD-BASED TEMPLATE REPLICATION

Keller and Ferrari (1994) presented a process by which high-aspect-ratio (on the order of 20:1) polysilicon structures could be formed by etching deep trenches in a silicon wafer, growing or depositing a sacrificial silicon dioxide release layer within the trenches, filling the trenches with polysilicon, and etching the sacrificial layer, releasing the template-molded polysilicon. Extensions to this basic process include the addition of lateral polysilicon bridges by lapping and polishing the initial polysilicon deposit and depositing a second layer that is patterned prior to release. A more recent paper by Keller and Howe (1995) describes the process in detail. The process (referred to by the authors as "HEXSIL") is illustrated below.

A simplified description of the process (from both of the above references) proceeds as follows. Plasma (not RIE) etching of the silicon was carried out (400 sccm He_2, 180 sccm Cl_2 at 425 mtorr and 300 W RF power) using a CVD silicon dioxide mask. After each 30 min of etching, the sidewall residue was removed in a "silicon isotropic wet etch." After completing the plasma etch, the silicon isotropic etch was performed again to smooth the sidewalls (in the earlier reference, a thin thermal oxide was grown and removed to aid in smoothing). To optimally shape the sacrificial layer, a trilayer of 5.5 μm PSG, 3.5 μm CVD silicon nitride and 5.5 μm PSG (or one of several variations) was deposited. The structural layer of CVD polysilicon (or CVD silicon nitride or electroless nickel) was then deposited in the trenches. At this point, the CVD polysilicon 1 layer was deposited at 580°C so that it would be very fine-grained, to replicate the template shape well. After the polysilicon 1 layer was annealed in pure nitrogen at 1,000°C for 1 h, it was lapped and polished (slurry of 0.3 μm alumina in glycerin). The polysilicon 2 layer was then deposited and patterned, mechanically coupling the polysilicon 1 structures together. The sacrificial layer of silicon dioxide was then removed in 49% HF and a surfactant, freeing the structures, and the template could be reused at this point.

Bimorph structures were also fabricated by using CVD polysilicon layers with different residual stress levels (achieved by not annealing the second layer).

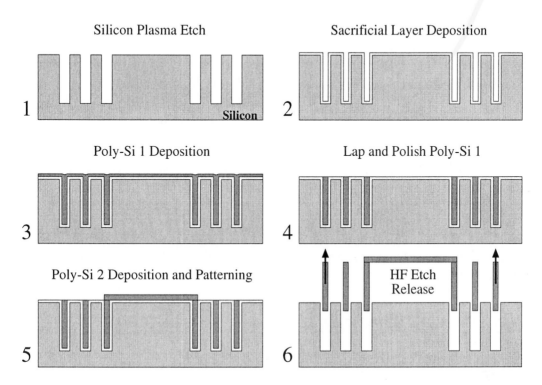

Illustration of template-molded polysilicon process. After Keller and Ferrari (1994).

8.3.4 CERAMIC SLURRY-BASED TEMPLATE REPLICATION

Ceramic materials can be formed by casting slurries and then firing (heating, generally in an oxidizing ambient, to remove organic binders) to obtain desired shapes. Micromachined templates can be used for this purpose. Hirata, et al. (1995) demonstrated the formation of high-aspect-ratio piezoelectric ceramic structures using synchrotron radiation to form high-aspect-ratio cylindrical voids in a copolymer of methyl methacrylate (MMA) and methacryl acid (MAA). Their goal was to create arrays of piezoelectric actuators with low cross-talk for imaging ultrasound applications. Following exposure and development of the polymer (using methyl isobutyl ketone), lead zirconate titanate (PZT) slurry was formed using the template and solidified. The template was then removed using an O_2/CF_4 plasma etch, carefully controlled so as not to raise the substrate temperature, causing damage to the PZT rods being released. The released PZT rods were then fired (or "sintered"), a new polymer layer was cast among them, and they were lapped from both sides to a uniform height ($\approx 100 \ \mu m$, with diameters of $\approx 20 \ \mu m$, for an aspect ratio of ≈ 5), bonded to drive electrode layers on both sides and poled.

Such an approach could be applied to the micromachining of a wide variety of ceramics, and could in principle be done without the need for synchrotron radiation since photoresist and polyimide templates can readily be used for aspect ratios on the order of five for such dimensions.

8.3.5 PREFORMED, ABOVE SUBSTRATE TEMPLATES

Zhang and Kim (1995) fabricated hollow hemispheres above a silicon substrate by attaching sacrificial hemispherical spacers to a silicon wafer and then coating the substrates with LPCVD silicon nitride. They etched from the back side of the wafer to remove the hemispheres, leaving hollow hemispherical shells of silicon nitride. While these authors did not disclose any of their processing methods, a suitable set of materials might be silicon or aluminum (only if low-temperature PECVD silicon nitride were used) hemispheres and KOH or TMAH as an etchant. In addition, the sacrificial release etch can be a different process from that used to etch through the wafer from the back side; front side access holes that are later sealed could be used.

9. SEALED CAVITY FORMATION

There are several possible methods for obtaining sealed cavities in addition to forming them using pre-etched pits and wafer bonding, as discussed above. The basic idea is to form a structure that includes a sacrificial layer that can be removed through narrow access holes, which are then covered over. This approach can be used to form sealed pressure references (see Guckel and Burns (1984)), fluid channels, etc.

The "covering over" step can be done a number of ways, three of which are listed here (obviously, their use depends on the conditions of sealing, i.e., atmospheric pressure, vacuum, etc.):

1) "Simple" approaches include the use of cyanoacrylate, epoxy, photoresist, etc., and are generally limited to atmospheric pressure for application.

2) Thin-film sealing layer can be carried out using CVD, LPCVD, PECVD, sputtered, or evaporated layers.

3) "Reactive" sealing involves reacting the cavity's structural material to form a seal (i.e., oxidizing a polysilicon structure) and sometimes thus clearing the residual reactant gas from the cavity (or allowing it to diffuse out, as occurs for H_2).

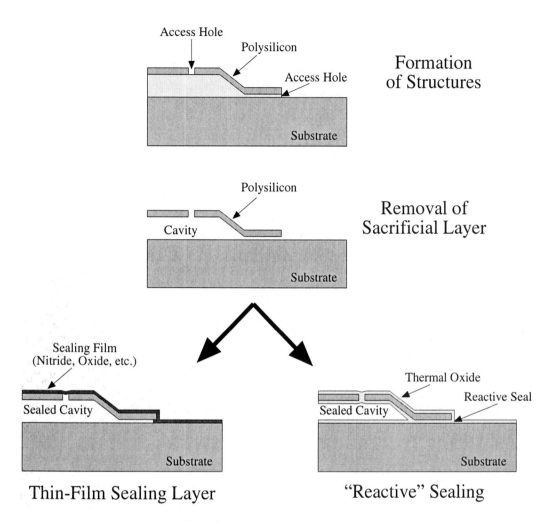

Illustration of the use of a sacrificial layer to form sealed, surface micromachined cavities using either a thin-film sealing layer (left) or a "reactive" seal (right).

9.1 GETTERS FOR SEALED CAVITIES

When sealed cavities are made, as presented above or by various bonding techniques, it is often desirable to remove residual gases that could potentially react with the sealed microstructures over time. In the fabrication of vacuum tubes, a typical approach was to include a *getter* material (typically a highly reactive metal, such as barium) that could be activated after the tube was sealed (by electrically heating a small region where the getter was located). The hot getter would react with all residual gases (except noble gases, such as argon) and eliminate them from the tube cavity, where they could lead to many problems over time.

This approach is being explored for micromachined devices, particularly since sealed cavities are being used experimentally for transducers that may be affected by residual gas (e.g., hot-filament devices that may react with such gases, or inertial sensors, which are affected by squeeze-film damping effects). Minami, et al. (1995) described the use of an integrated non-evaporable getter composed of a thin Ni/Cr heater 2 mm × 6 mm × 50 µm in size, coated locally with a mixture of Ti and a Zr/V/Fe alloy. When electrically heated to ≈ 400°C, the mixture could react with gases trapped during cavity formation. They demonstrated this approach with anodically bonded accelerometer cavities, and showed that the use of the getter to remove residual gases could be used to decrease squeeze-film damping effects on the accelerometer's frequency response.

10. SURFACE MODIFICATION

Surface modification entails the covalent bonding of monolayers (generally) of chemicals to a surface in order to change its surface properties. There are two major classes of surface modification chemistries that are in common use: siloxane chemistries for silicon substrates and thiol chemistries for gold. This approach is finding increasing use in biotechnology applications of micromachining (e.g., making surfaces more compatible with cells, biochemicals, etc.) and is being investigated for wet etch release promotion, mechanical property enhancement, etc. For these approaches to be of use beyond lab bench demonstrations, they must be stable over time and often (e.g., when packaging involves elevated temperatures) thermally. In addition, application-specific properties may be important, such as lack of reaction or degradation in chemical/fluidic applications and resistance to impact or shear in mechanical devices.

The most common surface modification chemistry used in clean rooms is hexamethyldisilazane (HMDS). It is used to enhance adhesion of photoresist to an underlying layer by coating it with methyl (CH_3) groups ("methylating" the surface). The formula for HMDS is,

$$(CH_3)_3 - Si - NH - Si - (CH_3)_3 \quad \text{or} \quad [(CH_3)_3Si]_2NH$$

This is useful to examine not only because it is common, but also because it is a model for a class of surface modification processes that are very useful for micromachined transducers. The chemical modification mechanism is the removal of ligands and/or moisture on the surface of (e.g., an SiO_2 layer) by reacting with them, leaving a surface that is methylated and very compatible with photoresist and other organic materials. Hydroxyl groups are removed from the SiO_2 surface by the following reaction,

$$2SiOH + [(CH_3)_3Si]_2NH \rightarrow 2SiOSi(CH_3)_3 \text{ \{bound to the surface\}} + NH_3$$

Surface moisture is removed by the formation of hexamethyldisiloxane and ammonia (both of which evaporate),

$$H_2O + [(CH_3)_3Si]_2NH \rightarrow [(CH_3)_3Si]_2O + NH_3$$

The SiO_2 surface then is coated with a monolayer of methyl groups.

This type of coupling chemistry can be used to attach a wide variety of molecules (including proteins and DNA) to silicon substrates, etc., that can be coupled to a similar silane compound (or thiol group for attachment to gold). The resulting monolayers are often referred to as self-assembled monolayers, or SAMs. For example, a variety of trichlorosilanes of the form $RSiCl_3$ (where R is an alkyl group) are available to form SAMs on silicon. As described by Srinivasan, et al. (1997), the reaction at the surface involves hydrolysis of the $SiCl_3$ group to form three silanols (SiOH bonds), which condense with other silanols on both oxidized silicon and other reacting precursor molecules. The result is a dense, cross-linked network of siloxanes at the surface, with an oriented array of the alkyl tail groups forming the SAM, which may greatly alter the surface properties of the original surfaces.

Deng, et al. (1995) reported the use of silane-based SAMs to improve the wear characteristics of polysilicon-based micromotors. After removal of the sacrificial low-temperature oxide layers, the polysilicon was hydrated/oxidized in a solution of 70:30 H_2SO_4:H_2O_2 for 10 min at 80°C, followed by rinsing in H_2O, methanol and chloroform (or dichloromethane), followed by reaction with the desired alkyl dichloro- or tricholoro-silane compound. Covalent bonds are formed between the hydroxyl groups on the substrate and the hydrolyzed alkyl silane molecules, and trace adsorbed water on the surfaces allows cross-linking of the alkyl groups through a siloxane network (near to the surface to which they are bound covalently). They investigated several potential SAMs, finding marked improvements in wear and stiction for octadecyltrichlorosilane (OTS) and (3,3,3-trifluororopyl) trichlorosilane (TFP) over a 9 month testing period (9 months and 80 million cycles).

A similar SAM approach (also using OTS) was demonstrated by Houston, et al. (1996) to greatly reduce adhesion forces (nearly four orders of magnitude) and produce SAMs that are thermally stable up to 400°C for short periods (5 min in nitrogen or vacuum). As well as using typical solvent mixtures for the OTS (e.g., 4:1 hexadecane:carbon tetrachloride), they demonstrated the use of chloroform instead of carbon tetrachloride. Approximately one drop of 95% OTS was used per 50 ml of solvent solution, for a concentration of $\approx$ 1 mM. The SAM deposition process used consisted of an H_2O rinse, H_2O_2 soak (15 min), H_2O rinse, isopropyl alcohol rinse (twice), isooctane rinse for water removal (twice), OTS solution soak

(15 min), isooctane rinse (twice), isopropyl alcohol rinse (twice), and a final H_2O rinse. An anneal step at $\approx 100°C$ is then carried out to form cross-links between the OTS alkyl chains. The resulting OTS SAMs were ≈ 2.5 nm in thickness.

More recent work by Srinivasan, et al. (1997) presented an alternate antistiction SAM chemistry using 1H,1H,2H,2H-perflourodecyltrichlorosilane (FDTS, $C_{10}H_4F_{17}SiCl_3$) that avoided the use of chlorinated solvents and provided thermal stability up to 400°C in N_2 (i.e., compatibility with even high-temperature packaging steps for microstructures). A 1 mM FDTS solution in iso-octane was prepared in an N_2-filled glovebox to prevent water vapor in the air from causing bulk polymerization of the FDTS. In a procedure similar to that described above for OTS, polysilicon microstructures with sacrificial silicon dioxide that had been etched away were treated with FDTS. The resulting SAMs provided better antistiction and thermal stability performance than OTS.

While this area needs further study, particularly in terms of long-term performance of such monolayer films, the results to date indicate that useful improvements in stiction and mechanical properties are possible with such surface modifications.

Chemical properties of surfaces may also be modified using the SAM approach. Thomas, et al. (1996) presented an example of improving the selectivity of a chemical sensor (surface acoustic wave device, in this case) through surface modification using a SAM on an Au film. Dendrimers, hyper-branched polymers with regularly repeating branch sequences attached to a central "trunk," have also been used to modify sensor surfaces to confer molecular recognition properties. They confer good chemical sensitivity (comparable to thicker polymer layers) but retain the fast response of SAMs (since binding and other reactions do not have to propagate through thicker layers). Crooks, et al. (1996) demonstrated the use of spheroidal poly(amidoamine) dendrimers to modify gold-coated surface acoustic wave sensors via an intermediate mercaptoundecanoic acid SAM layer. Results for detection of various hydrocarbons were improved over non-dendrimer SAMs.

11. PRINTING AND STEREOLITHOGRAPHY

Various forms of relatively simple printing can be adapted to the fabrication of microstructures with feature sizes comparable to the smallest achievable by conventional microlithography. While such approaches currently cannot be used to fabricate complex microstructures, or those with integrated electronics, they can potentially be used to fabricate microstructures in applications where other methods may not be viable (e.g., fabricating micron- or submicron-scale features on very non-planar surfaces or over very large areas). By repeatedly printing layers that become part of a final structure (or act as sacrificial layers during its fabrication), such printing-based methods can be extended to form 3-D structures (this is referred to as *stereolithography*). While none of these printing methods is likely to displace mainstream microlithography in the near term, one key advantage they have is their low cost in comparison to typical clean room processing.

11.1 SCREEN PRINTING

The basic principle of screen printing is to use a template or "mask" that is removed in certain areas to allow the transfer of "ink" in desired patterns to an underlying substrate (analogous to lift-off deposition of a metal film using a template layer). Typically, the template ink is applied through regions of a screen (historically, silk has been used for this purpose, hence the term "silk screening") that allow it to wick through, while preventing it from penetrating through the remaining regions. The application of this approach to micromachining does not necessarily require any new technology, but alignment accuracy may be considerably less than for conventional optical lithography, particularly if many layers must be accurately superimposed. Conventional screen printing has accuracies that are quite sufficient to pattern relatively large features (on the order of 100 μm or larger) on wafers, such as active polymer layers in biosensors. A key advantage of this approach is that such a layer can be applied patterned in one operation, as opposed to, for example, spinning it on, applying photoresist, patterning the resist, selectively removing the layer, and finally removing the remaining resist. Naturally, however, the material to be deposited must be available in a liquid form that is compatible with screen printing (appropriate viscosity and chemical nature), such as an organic polymer dissolved in a solvent. After screening, nearly all thick film pastes must be fired at 200 to 400°C. Modern screen printing for surface-mount printed circuit board manufacturing has a resolution of $\approx$ 500 μm, with registration capability of $\approx$ 75 to 125 μm

An example of the use of screen printing (and transfer printing, as discussed below) to the fabrication of metal oxide gas sensors (discussed in the Chemical and Biological Transducers chapter) is provided by Golovanov, et al. (1995). They used printing techniques to fabricate a variety of sensors on alumina (Al_2O_3) substrates, including Pt heating traces and SnO_2 sensing layers. The methods used were described by Leppävuori, et al. (1994), and are capable of producing 50 μm features.

11.2 TRANSFER PRINTING

Transfer printing is a very mature technology due to its ubiquitous use in the preparation of books, newspapers, magazines, and the like. The basic principle is to "wet" a plate with raised regions (or regions that attract the ink via surface tension effects, while the remaining regions repel it) and transfer the desired pattern, via an intermediate (unpatterned) carrier, to the object to be inked. Interesting uses of micro- and nanoscale contact/transfer printing are described in Wilbur, et al. (1994). They used microcontact printing to transfer self-assembling monolayers (SAMs) of long chain alkanethiolates to act as nanometer-scale resists. Jackman, et al. (1995), describe the use of similar printing techniques with flexible stamps to form features as small as 25 µm on curved surfaces.

11.3 POWDER-LOADED POLYMERS FOR PRINTING

The technique of "loading" polymer or other initially liquid matrix materials with powders can be used to obtain composite materials with desired mechanical, magnetic, thermal, electrical, chemical, or other properties that can be applied like paints, silk-screened, and sometimes photolithographically patterned. For example, loading a liquid matrix with finely powdered silver has been used for many years to form conductive traces that can be screened or hand painted in the desired shapes. Slowly diffusing (or "depot") forms of chemicals such as pharmaceuticals are often prepared by mixing powders into matrices with varying degrees of permeability to a solvent, generally water, that will be present in the site of use (e.g., implantable devices, catheters, etc.).

As an example of a magnetic application, Lagorce and Allen (1996) demonstrated the loading of Dupont PI-2555 polyimide with strontium ferrite powder to form patternable permanent magnets. Unfortunately, their paper does not provide information as to the optimal amount of powder added nor about additives that were added to improve particle dispersion. They demonstrated application of the ferrite loaded polyimide via 100 µm thick silk-screening (250 µm features) or spin-coating with multiple ≈ 9 µm layers. The spin-coated layers were patterned using a top patterned layer of photoresist and using the photoresist developer to expose regions of the polyimide as the photoresist is developed (200 µm features were achieved with this method).

Piezoelectric powders can also be used to load liquids, as discussed by Clegg, et al. (1997), who loaded epoxy resin with 60% (by volume) bismuth ferrite-doped lead titanate powder (14.1% Ti, 37.3% Pb, 35.1% Bi, and 13.5% Fe, from Morgan Matroc Ltd., Worcestershire, UK), cured the epoxy, and poled the resulting composite piezoelectric material at 3.3 MV/m for 1 h at 65°C. They were able to demonstrate successful opertation of piezoelectric actuators fabricated using this approach.

11.4 THREE-DIMENSIONAL LITHOGRAPHY

Three-dimensional "stereolithography" (or solid free-form fabrication) with polymers is well known for making macroscale prototypes. Typically, optical energy (i.e., laser) is focused and locally hardens a photopolymer, which is built up in layers to form the desired object. For example, Ikuta and Hirowatari (1993) discuss a polymer-based stereolithography technique for microscale devices using a Xenon lamp as the light source, scanning the substrate using an X-Y stage and the focusing lens using a Z-positioner. They achieved a 5 μm spot size for (roughly) $5 \times 5 \times 3$ μm voxels. Unfortunately, their paper gives no information as to the type of polymer, nor any of the details to replicate the work. However, there are some interesting examples of 3-D structures. Another useful reference in this area is Takagi and Nakajima (1993).

11.5 SPATIAL FORMING VIA TRANSFER PRINTING

Taylor, et al. (1995) demonstrated a process that they consider a hybrid between injection molding and stereolithography (referred to as "spatial forming"). They used a customized offset printer to transfer "negative" ink for a template that was later filled with a "positive" ink (binder/metal powder mix) in a manner similar to injection molding.

One negative (template) ink used consisted of polyurethane acrylate resin (Resin 783, Morton Thiokol, Inc., Chicago, IL) with ≈ 20% boron nitride pigment. This ink was transferred using a conventional lithographic printing plate (patterned using an e-beam defined lithographic mask) using a non-wettable silicone coating removed where ink is to stick for transfer. A printing roller was used to pick up the patterned ink from the plate and to transfer it to the (ceramic) substrates. Each ≈ 0.5 μm ink layer was individually UV-cured and the process was repeated until 20 to 30 layers were deposited. At this point, the positive (metal-containing) ink, consisting of an epoxy acrylate (600 Ebecryl™, Radcure Specialties, Louisville, KY) loaded with 50% (vol.) 17-4 PH stainless steel powder (3 μm average grain size), was knifed into the template, UV-cured, and the entire assembly was mechanically planarized. These steps were repeated until ≈ 600 layers were deposited. At this point, the organic binder and negative ink were decomposed in a H_2 atmosphere at ≈ 800°C, followed by a 1,250°C Ar_2 sintering step. Finished parts with heights of up to 250 μm and feature sizes down to 25 μm were realized for "extruded" shapes, and there is considerable discussion of methods for continuously fabricating (i.e., in a conveyor belt fashion) large-scale, interlinked, 3-D structures.

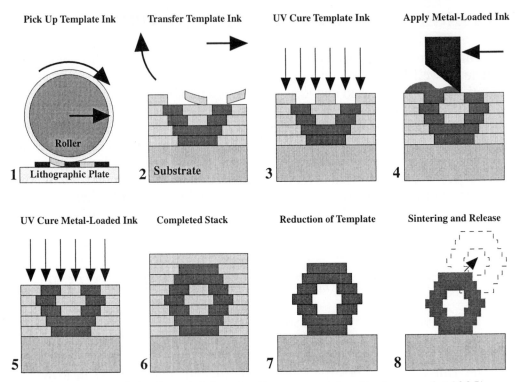

Illustration of the "spatial forming" process. After Taylor, et al. (1995).

12. OTHER MICROMACHINING TECHNIQUES

12.1 SHARP TIP FORMATION

In a variety of applications, such as tunneling transducers, field emission devices, and scanning tunneling or scanning force microscopes, very sharp tips are required. Micromachining methods offer a variety of techniques for achieving this, and some relevant examples are presented below.

12.1.1 SELF-OCCLUDING MASKS

Very sharp field emitter tips can be formed by allowing continued metal deposition (which needs to be nearly perpendicular to the substrate, as obtained with evaporation) onto a template layer to gradually close off holes in which metal is being deposited (Spindt, et al. (1989)). When tip formation is completed, the unwanted metal is simply lifted off by dissolving the template, with extremely (atomically) sharp tips formed within the template. Alternative sharp tip microfabrication processes are described in McGruer, et al. (1991), Branston and Stephani (1991), and Marcus, et al. (1991).

Applications in vacuum microelectronics, including field emission displays, have largely motivated the application of these techniques. A major problem with such tips when used as field emitters has been the lack of uniformity of emission characteristics, and it appears that the solution will be surface modification/control (i.e., surface states cause the variations in device characteristics) or resistive ballasting between multiple parallel tips.

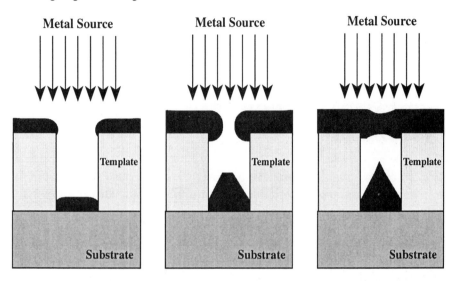

Illustration of Spindt tip formation by allowing holes defined in a sacrificial template layer to be self-occluded by deposited metal, forming atomically sharp tips.

12.1.2 MICROMASKING IN PLASMAS

A well-known effect in plasma and RIE silicon etching is the formation of very sharp needle-like structures (commonly referred to as "grass") through micro-masking of the silicon by particulates from the etcher (often aluminum that is sputtered from the etch chamber walls). Such features are typically undesirable in mainstream semiconductor manufacturing.

Jansen, et al. (1995b) describe a method of deliberately obtaining such structures for use as ultra-sharp tips for scanning probe microscopy and other applications. They suggest using SF_6/O_2 plasma and adjusting the SF_6 and O_2 flows until the surface appears optically black. As well as providing conventional structural etching, this process allows the formation of the micromasked tips through micromasking and the formation of a passivating silicon oxyfluoride surface layer. Increasing O_2 flow creates more vertical sidewalls (by aiding the formation of the passivating layer) and adding CHF_3 results in a more undercut profile (since its breakdown products scavenge the available O_2). They demonstrated a complete range of high-aspect-ratio micromechanical structures with smooth sidewalls and a variety of needle and pillar structures (fabricated using deliberate masking, rather than the random micromasking phenomenon).

12.1.3 WET ETCHING

Another technique for fabricating sharp tips from silicon is to use a light-field mask with small SiO_2 squares oriented along the [110] directions and to undercut them using KOH or a similar anisotropic wet etchant. If etching is stopped just after the SiO_2 squares are released, tips with radii of curvature less than 100 nm can be achieved (Trimmer, et al. (1995)).

12.2 CHEMICAL-MECHANICAL POLISHING AND PLANARIZATION

Chemical-mechanical polishing (CMP) is used quite commonly in "mainstream" integrated circuit fabrication to obtain planar surfaces in between repeated interconnect/dielectric layer depositions. The non-planar regions in the dielectric surfaces are planarized since the etching process depends on both chemistry and mechanical pressure, and can be polished to average roughnesses of less than 2 nm.

CMP is typically carried out using an alkaline, silica-containing slurry in combination with mechanical polishing (as illustrated below) to planarize oxide surfaces. Raised features are etched much faster than flat regions since mechanical forces that enhance the etching are concentrated there.

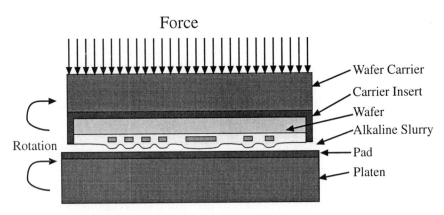

Illustration of the chemical-mechanical polishing (CMP) approach to planarizing integrated circuits. After Nasby, et al. (1996).

Yasseen, et al. (1995) demonstrated CMP of polysilicon (the top layer of prereleased electrostatic motors) in combination with surface micromachining wherein the average surface roughness was reduced from 42 to 1.7 nm, allowing planar diffraction gratings to be etched into the polysilicon.

Nasby, et al. (1996) reported the use of CMP to allow the fabrication of polysilicon micromechanisms in cavities prior to CMOS fabrication, with the CMP used to planarize sacrificial oxide deposited to fill in the trenches. This approach reversed the typical sequence of adding micromechanical structures to CMOS, and removed the need to modify the CMOS process so that the resulting circuits could stand the high temperatures used to deposit the polysilicon ($\approx 600°C$).

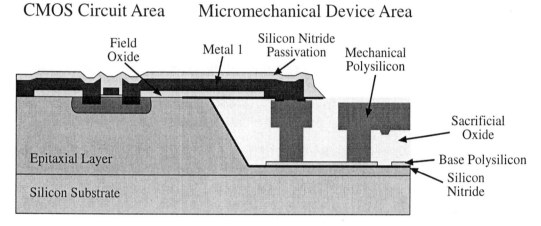

Simplified illustration of Nasby, et al.'s (1996) "buried" micromechanical polysilicon process with CMP planarization and CMOS post-fabrication. After Nasby, et al. (1996).

12.3 ELECTRIC DISCHARGE MACHINING (EDM)

A technique that is often used in the precision fabrication of macroscopic components, electric discharge machining (EDM), is also applicable to micromachining. The basic principle is to draw an electric arc from a negatively-biased, sharp, robust tip from the workpiece, a necessarily conductive material (metal or semiconductor), typically submerged in a dielectric fluid. Regions of the workpiece are selectively eroded away by the discharge, at a rate much faster than the erosion of the tip. Features as small as 5 μm have been realized in metals and silicon, as discussed in Reynaerts, et al. (1997), which also provides an overview of different EDM techniques. Unfortunately, this approach to fabrication is serial and slow, but can be of use for prototyping and, in rare instances, manufacturing.

An interesting application of EDM to microstructure fabrication can be found in Suzuki, et al. (1995). They used EDM with 2 µm resolution to fabricate a 1/1,000th-scale model automobile in Al, which was diamond powder polished to achieve a mean surface roughness of 0.13 µm. A 30 µm thick NiP (95% Ni, 5% P) layer was then plated onto the Al, which was subsequently etched away using KOH. Finally, the NiP shell was plated with 2 µm of Au. Using this technique and various other precision machining methods, an entire model car was fabricated, 4.8 × 1.8 × 1.8 mm in size, complete with operating electromagnetic motor and even a front license plate. The car was able to move at 100 mm/s. The EDM-based shell-fabrication process used to form the model's body was seen to be able to yield impressively accurate small-scale models. However, the time-consuming EDM-fabricated mold is destroyed each time a part is made.

12.4 ABRASIVE POWDER MACHINING

Abrasive powder machining, or "sandblasting," can also be applied to micromachining. A variety of resist materials could conceivably withstand such processing. For example, Little (1984) reported the following recipe for gelatin-based photoresist that is apparently tough enough to withstand abrasive etching with alumina grit. Dissolve 7 g (one packet) of Knox™ unflavored gelatin in 75 ml of water at 40 to 50°C, filter to remove bubbles and add 0.3 g potassium dichromate. Store in an amber bottle in the refrigerator. Shelf life when refrigerated was reported to be ≈ 1 month. This resist was applied to a substrate with an eye dropper at 40°C under yellow safelights and dried at room temperature (3 to 4 h). Exposure was carried out using 365 nm radiation. (Note that Little used this resist on glass and exposed it from the glass side). Development was done in hot water (60°C) for 10 to 20 min, followed by drying and abrasive machining.

12.5 PRECISION MECHANICAL MACHINING

As evidenced by a long history of very high precision "conventionally" machined components, these well-established techniques can be used alone or in combination with the other micromachining techniques described herein. For a number of years, ultraprecision machining techniques have been capable of yielding components with large sizes (several meters) and tolerances that in some cases exceed that of lithography used in micromachining. For example, these errors were on the order of 0.4 µm straightness errors and 1.27 µm displacement errors over 0.4 m for a diamond turning machine in operation at the Lawrence Livermore National Laboratory in 1979, as discussed in Bryan (1979). As described by Donaldson and Patterson (1983), a more accurate diamond turning machine was built at the same laboratory to handle components up to 2.1 m in diameter and weighing up to 4500 kg, with accuracies on the order of 25 nm RMS.

Yamagata, et al. (1996) reported the fabrication of microstructures (such as a 10 μm diameter cylindrical shaft and ≈ 100 μm diameter threaded screw) using precision cutting. Their reported tool resolution was that 0.05 and 0.5 μm per cycle cut depths were used. Friedrich and Vasile (1996) discussed the use of 22 to 100 μm micromilling tools for precision fabrication of molds in PMMA and other materials. They demonstrated wall slopes as high as 89.5° for trench depths as high as 62 μm. Perhaps the most impressive in this area is the use of less than $1,000 worth of bench-top equipment (miniature lathe) to fabricate ultraprecision structures, such as miniature screws 0.01 in. in diameter (254 μm) with a thread depth of 12.7 μm (Sherline Products Inc., San Marcos, CA). An overview of the concepts of precision mechanism design using "conventional" machining can be found in Evans (1989) and Smith and Chetwynd (1992). These examples are useful to provide some perspective and show that lithographic processing is not the only way to achieve high precision, although the parallelism it provides does make it attractive in a wide variety of applications. A variety of other precision mechanical machining approaches are also reviewed in Madou (1997).

12.6 SCANNING PROBE MACHINING

The availability of scanning probe microscopy has opened up the avenue of its use in precision micromachining. While at present a serial process (until parallel probes are developed), this approach does offer the ability to manipulate materials at the atomic level. The use of this approach in micromachining is described by Suda, et al. (1996).

12.7 THERMOMIGRATION

Droplets of metals that alloy with silicon can be "migrated" in straight-line paths from a cold side of a wafer to the hot side (temperature gradient of ≈ 50°C/cm, or 2.75°C across a typical wafer). The basic process is that the molten "alloy zone" migrates toward the hot side, dissolving silicon atoms at the hot side, transporting them through the zone, and depositing them on the cold side. Conductive paths are left in the "wake" of the moving alloy zone, which moves at ≈ 3 μm/min at 1,100°C. Conductive "wires" have been made in this way that reach all the way through a wafer. The idea was to use them to make "double-sided" chips and other structures that depend on through-substrate connections.

One problem is that thermomigration must be carried out at very high temperatures ($\approx$ 1,000°C) and another is that high stresses are induced in the wafers. A potential solution to this problem is to use a CW laser to guide the thermomigration and then follow that with a thermal anneal step to reduce stress. This type of process has potential importance in "stacked-chip" processes for applications where high density is essential. Good references for thermomigration techniques include Cline and Anthony (1978), Mizrah (1980), and Kimerling, et al. (1980).

12.8 PHOTOSENSITIVE GLASS MICROMACHINING

Photosensitive glasses, patterned using standard UV photolithographic techniques, are available for micromachining. As discussed in Trotter (1991) and Madou (1997), a variety of photosensitive mechanisms can be used. For example, in glasses made from mixtures of SiO_2, LiO_2, K_2O, Al_2O_3, CeO_2 and Ag, incident UV light causes photoelectrons to be released from the Ce atoms. These electrons are trapped to form a latent image analogous to that in photographic film. Heating to 500°C releases the electrons, which reduce Ag ions to form islands that in turn nucleate the crystallization of a variety of compounds such as lithium metasilicate, sodium bromide, or sodium fluoride. These crystallized regions etch much faster (15 to 30 times) than the unexposed regions, allowing selective etching.

An example photosensitive glass process is available from Schott Glass, Inc. (Schott Glaswerke, Mainz, Germany). The Foturan® photoetchable glass is exposed to UV light via a standard chromium-on-quartz photolithographic mask. The glass is then heat-treated, and UV-exposed regions crystallize, as explained above. When the exposed glass is etched in HF, the crystallized regions are removed, leaving only the unexposed glass structures behind. The use of Foturan® for micromachining is discussed in Dietrich, et al. (1993). Another photosensitive glass manufactured by Corning, Inc. (Corning, NY), makes use of the fact that during thermal crystallization, the UV-exposed regions physically contract, forcing unexposed regions to bulge outward under pressure. This technique has been used to fabricate arrays of photolithographically defined optical lenses, as discussed in the Optical Transducers chapter. Other photosensitve materials, such as plastics, are discussed in Madou (1997).

12.9 FOCUSED ION-BEAM MICROMACHINING

Focused beams of reactive ions such as fluorine can also be used to micromachine a wide variety of materials. While inherently a serial process, impressive resolution is achievable with focused ion beam, or FIB, etching (see Shaver and

Ward (1985)). For example, an argon beam was used by Davies, et al. (1996) to fabricate components in the size range of 1 to 100 μm, with tolerances on the order of 0.1 μm. They demonstrated the fabrication of beams, shafts, and cantilevers in Al, Ni, stainless steel, etc., and also fabricated disks, gears, and cogs with a resolution down to 50 nm. An example of silicon micromachining with FIB etching can be found in Moore, et al. (1995), who demonstrated the fabrication of submicron gaps at oblique angles through 7 μm thick silicon cantilevers. A variety of other materials have been micromachined using FIB etching, including CVD diamond (Mori, et al. (1995)) and single-crystal diamond (Hunn and Christensen (1994)). Specific applications of FIB etching to diode laser mirrors and other optical components can be found in Elliott, et al. (1988) and Gargin, et al. (1973), respectively.

MICROMACHINING TECHNIQUES REFERENCES

GENERAL REFERENCES ON MICROFABRICATION

Madou, M., "Fundamentals of Microfabrication," CRC Press, Inc., Boca Raton, FL, 1997.

Petersen, K. E., "Silicon as a Mechanical Material," Proceedings of the IEEE, vol. 70, no. 5, May 1982, pp. 420 - 457.

Ristic, Lj. [ed.], "Sensor Technology and Devices," Artech House, London, 1994.

Runyan, W. R., and Bean K. E., "Semiconductor Integrated Circuit Processing Technology," Addison-Wesley Publishing Company, Inc., 1990.

Sze, S. M. [ed.], "VLSI Technology," McGraw-Hill, Inc., New York, NY, 1983.

SPECIFIC REFERENCES

Abe, T., and Reed, M. L., "Low Strain Sputtered Polysilicon for Micromechanical Structures," Proceedings of IEEE International Workshop on Micro Electro Mechanical Systems, San Diego, CA, Feb. 11 - 15, 1996, pp. 258 - 261.

Abe, T., Messner, W. C., and Reed, M. L., "Effective Methods to Prevent Stiction During Post-Release-Etch Processing," Proceedings of the IEEE Micro Electro Mechanical Systems Conference, Amsterdam, Netherlands, Jan. 29 - Feb. 2, 1995, pp. 94 - 99.

Ade, R. W., Harstead, E. E., Amirfazli, A. H., Cacouris, T., Fossum, E. R., Prucnal, P., and Osgood, R. M., Jr., "Silicon Photodetector Structure for Direct Coupling of Optical Fibers to Integrated Circuits," IEEE Transactions on Electron Devices, vol. ED-34, no. 6, June 1987, pp. 1283 - 1289.

Adams, A. C., "Dielectric and Polysilicon Film Deposition," Chapter 3 in "VLSI Technology," Sze, S. M., [ed.], McGraw-Hill, Inc., New York, NY, 1983, pp. 93 - 129.

Allen, M. G., "Polyimide-Based Process for the Fabrication of Thick Electroplated Microstructures," Proceedings of Transducers '93, the 7th International Conference on Solid-State Sensors and Actuators, Yokohama, Japan, June 7 - 10, 1993, Institute of Electrical Engineers, Japan, pp. 60 - 65.

Anderson, N. C., and Grove, N.C., Jr., U.S. Patent No. 4,279,707, issued 1981.

Anderson, R. C., Muller, R. S., and Tobias, C. W., "Porous Polycrystalline Silicon: A New Material for MEMS," Journal of Microelectromechanical Systems, vol. 3, no. 1, Mar. 1994, pp. 10 - 18.

Andricacos, P. C., Cheh, H. Y., and Linford, H. B., "Application of Pulsed Plating Techniques to Metal Deposition: Part IV - Microthrowing Power of Gold Deposition," Interim Report AES Project 35, Plating and Surface Finishing, Sept. 1977, pp. 44 - 46.

Anthony, T. R., "Diodes Formed by Laser Drilling and Diffusion," Journal of Applied Physics, vol. 53, no. 12, Dec. 1982, pp. 9154 - 9164.

Anthony, T. R., "Forming Feedthroughs in Laser-Drilled Holes in Semiconductor Wafers by Double-Sided Sputtering," IEEE Transactions on Components, Hybrids, and Manufacturing Technology, vol. CHMT-5, no. 1, Mar. 1982, pp. 171 - 180.

Araki, T., and Okabe, M., "(Nd, Tb)-Fe-B Thin Film Magnets Prepared by Magnetron Sputtering," Proceedings of IEEE International Workshop on Micro Electro Mechanical Systems, San Diego, CA, Feb. 11 - 15, 1996, pp. 244 - 249.

Ashruf, C. M. A., French, P. J., Sarro, P. M., Nagao, M., and Esashi, M., "Fabrication of Micromechanical Structures with a New Electrodeless Electrochemical Etch Stop," Proceedings of Transducers '97, the 1997 International Conference on Solid-State Sensors and Actuators, Chicago, IL, June 16 - 19, 1997, vol. 1, pp. 703 - 706.

Aslam, M., and Schulz, D., "Technology of Diamond Microelectromechanical Systems," Proceedings of Transducers '95, the 8th International Conference on Solid-State Sensors and Actuators, Stockholm, Sweden, June 25 - 29, 1995, vol. 2, pp. 222 - 224.

Bäcklund, Y., and Rosengren, L., "New Shapes in (100) Si Using KOH and EDP Etches," Journal of Micromechanics and Microengineering, vol. 2, no. 2, June 1992, pp. 75 - 79.

Bartek, M., Gennissen, P. T. J., French, P. J., and Wolffenbuttel, R. F., "Confined Selective Epitaxial Growth: Potential for Smart Silicon Sensor Fabrication," Proceedings of Transducers '95, the 8th International Conference on Solid-State Sensors and Actuators, Stockholm, Sweden, June 25 - 29, 1995, vol. 1, pp. 91 - 94.

Barth, P., "Silicon Fusion Bonding for Fabrication of Sensors, Actuators and Microstructures," Sensors and Actuators, vols. A21 - A23, Feb. - Apr. 1990, pp. 919 - 926.

Bassous, E., "Fabrication of Novel Three-Dimensional Microstructures by the Anisotropic Etching of (100) and (110) Silicon," IEEE Transactions on Electron Devices, vol. ED-25, no. 10, Oct. 1978, pp. 1178 - 1185.

Bassous, E., U.S. Patent No. 3,921,916, issued 1975.

Beale, M. I. J., Chew, N. G., Uren, M. J., Cullis, A. G., and Benjamin, J. D., "Microstructure and Formation Mechanism of Porous Silicon," Applied Physics Letters, vol. 46, no. 1, Jan. 1, 1985, pp. 86 - 88.

Bean, K. E., "Anisotropic Etching of Silicon," IEEE Transactions on Electron Devices, vol. ED-25, no. 10, Oct. 1978, pp. 1185 - 1193.

Becker, E. W., Ehrfeld, W., Hagmann, P., Maner, A., and Munchmeyer, D., "Fabrication of Microstructures with High Aspect Ratios and Great Structural Heights by Synchrotron Radiation Lithography, Galvanoforming, and Plastic Moulding (LIGA Process)," Microelectronic Engineering, vol. 4, no. 1, May 1986, pp. 35 - 56.

Berenschot, E., Jansen, H., Burger, G.-J., Gardeniers, H., and Elwenspoek, M., "Thermally Assisted Ion Beam Etching of Polytetrafluoroethylene: A New Technique for High Aspect Ratio Etching of MEMS," Proceedings of IEEE International Workshop on Micro Electro Mechanical Systems, San Diego, CA, Feb. 11 - 15, 1996, pp. 277 - 284.

Bhansali, S., and Sood, D. K., "Selective Seeding of Copper Films on Polyimide Patterned Silicon Substrate, Using Ion Implantation," Proceedings of Transducers '95/Eurosensors IX, Stockholm, Sweden, June 25 - 29, 1995, vol. 1, pp. 95 - 98.

Bhardwaj, J., Ashraf, H., and McQuarrie, A., "Dry Silicon Etching for MEMS," Proceedings of the 191st Meeting of the Electrochemical Society, Microstructures and Microfabricated Systems III Symposium, Montréal, PQ, May 4 - 9, 1997, vol. 97-5, pp. 118 - 130.

Bloomstein, T. M., and Ehrlich, D. J., "Laser Deposition and Etching of Three-Dimensional Microstructures," Proceedings of Transducers '91, the 1991 International Conference on Solid-State Sensors and Actuators Digest of Technical Papers, IEEE Press, San Francisco, CA, June 24 - 27, 1991, pp. 507 - 511.

Bloomstein, T. M., and Ehrlich, D. J., "Stereo Laser Micromachining of Silicon," Applied Physics Letters, vol. 61, no. 6, Aug. 1992, pp. 708 - 710.

Bockris, J. O'M., and Reddy, K. N., "The Electrogrowth of Metals on Electrodes," Section 10.2 in "Modern Electrochemistry," Plenum Press, New York, NY, 1970, pp. 1173 - 1231.

Bogenschütz, A. F., Krusemark, W., Löcherer, K.-H., and Mussinger, W., "Activation Energies in the Chemical Etching of Semiconductors in HNO_3-HF-CH_3COOH," Journal of the Electrochemical Society: Solid State, vol. 114, no. 9, Sept. 1967, pp. 970 - 973.

Boman, M., Westberg, H., Johansson, S., and Schweitz, J.-A., "Helical Microstructures Grown by Laser Assisted Chemical Vapour Deposition," Proceedings of the IEEE Workshop on Micro Electro Mechanical Systems '92, Travemünde, Germany, Feb. 4 - 7, 1992, pp. 162 - 167.

Both, A., Bacher, W., Heckele, M., Müller, K. D., Ruprecht, R., and Strohrmann, M., "Molding Process with High Alignment Precision for the LIGA Technology," Proceedings of the IEEE Micro Electro Mechanical Systems Conference, Amsterdam, Netherlands, Jan. 29 - Feb. 2, 1995, pp. 186 - 190.

Branston, D. W., and Stephani, D., "Field Emission from Metal-Coated Tips," IEEE Transactions on Electron Devices, vol. 38, no. 10, Oct. 1991, pp. 2329 - 2333.

Bressers, P. M. M. C., Kelly, J. J., Gardeniers, J. G. E., and Elwenspoek, M., "Surface Morphology of p-Type (100) Silicon Etched in Aqueous Alkaline Solution," Journal of the Electrochemical Society, vol. 143, no. 5, May 1996, pp. 1744 - 1750.

Brody, T. P., "CdSe - The Ideal Semiconductor for Active Matrix Displays," Proceedings of the SPIE Conference on High-Resolution Displays and Projection Systems, San Jose, CA, Feb. 11 - 12, 1992, vol. 1664, pp. 2 - 13.

Brody, T. P., "The Thin-Film Transistor - A Late Flowering Bloom," IEEE Transactions on Electron Devices, vol. ED-31, no. 11, Nov. 1984, pp. 1614 - 1628.

Brosnihan, T. J., Bustillo, J. M., Pisano, A. P., and Howe, R. T., "Embedded Interconnect and Electrical Isolation for High-Aspect-Ratio, SOI Inertial Instruments," Proceedings of Transducers '97, the 1997 International Conference on Solid-State Sensors and Actuators, Chicago, IL, June 16 - 19, 1997, vol. 1, pp. 637 - 640.

Brugger, J., Beljakovic, G., Despont, M., Biebuyck, H., de Rooij, N. F., and Vettiger, P., "High-Yield Wafer Chuck for Single-Sided Wet Etching of MEMS Structures," Proceedings of Transducers '97, the 1997 International Conference on Solid-State Sensors and Actuators, Chicago, IL, June 16 - 19, 1997, vol. 1, pp. 711 - 713.

Brumfield, D., and Walker, P., 1958, reference unknown.

Bryan, J. B., "Design and Construction of an Ultraprecision 84 Inch Diamond Turning Machine," Precision Engineering, vol. 1, no. 1, July 1979, pp. 13 - 17.

Burger, G. J., Smulders, E. J. T., Berenschot, J. W., Lammerink, T. S. J., Fluitman, J. H. J., and Imai, S., "High Resolution Shadow Mask Patterning in Deep Holes and Its Application to an Electrical Wafer Feed-Through," Proceedings of Transducers '95, the 8th International Conference on Solid-State Sensors and Actuators, Stockholm, Sweden, June 25 - 29, 1995, vol. 1, pp. 573 - 576.

Busta, H. H., Feinerman, A. D., Ketterson, J. B., and Cueller, R. D., "Strings, Loops and Pyramids - Building Blocks for Microstructures," 1987 IEEE Micro Robots and Teleoperators Workshop, Hyannis, MA, Nov. 1987, pp. 9/1 - 9/5.

Campbell, S. A., Cooper, K., Dixon, L., Earwaker, R., Port., S. N., and Schiffrin, D. J., "Inhibition of Pyramid Formation in the Etching of Si p(100) in Aqueous Potassium Hydroxide-Isopropanol," Journal of Micromechanics and Microengineering, vol. 5, no. 3, Sept. 1995, pp. 209 - 218.

Canham, L. T., "Silicon Quantum Wire Array Fabrication by Electrochemical and Chemical Dissolution of Wafers," Applied Physics Letters, vol. 57, no. 10, Sept. 3, 1990, pp. 1046 - 1048.

Carns, T. K., Tanner, M. O., and Wang, K. L., "Chemical Etching of $Si_{1-x}G_{ex}$ in $HF:H_2O_2:CH_3COOH$," Journal of the Electrochemical Society, vol. 142, no. 4, Apr. 1995, pp. 1260 - 1266.

Castellani, E. E., Powers, J. V., and Romankiw, L. T., U.S. Patent No. 4,102,756, issued 1978.

Chambers, F. A., and Wilkiel, L. S., "Cesium Hydroxide Etching of (100) Silicon," Journal of Micromechanics and Microengineering, vol. 3, no. 1, Mar. 1993, pp. 1 - 3.

Chang, F. I., Yeh, R., Lin, G., Chu, P. B., Hoffman, E., Kruglick, E. J. J., Pister, K. S. J., and Hecht, M. H., "Gas-Phase Silicon Micromachining with Xenon Difluoride," Microelectronic Structures and Microelectromechanical Devices for Optical Processing and Multimedia Applications, Oct. 24, 1995, Austin, TX, in Proceedings of the SPIE, July 1995, vol. 2641, pp. 117 - 128.

Christenson, T. R., Guckel, H., Skrobis, K. J., and Jung, T. S., "Preliminary Results for a Planar Microdynamometer," Technical Digest of the Solid-State Sensor and Actuator Workshop, Hilton Head Island, SC, June 22 - 25, 1992, pp. 6 - 9.

Chu, P. B., Chen, J. T., Yeh, R., Lin, G., Huang, J. C. P., Warneke, B. A., and Pister, K. S. J., "Controlled Pulse-Etching with Xenon Difluoride," Proceedings of Transducers '97, the 1997 International Conference on Solid-State Sensors and Actuators, Chicago, IL, June 16 - 19, 1997, vol. 1, pp. 665 - 668.

Clark, L. D., Jr., Lund, J. L., and Edell, D. J., "Cesium Hydroxide (CsOH): A Useful Etchant for Micromachining Silicon," Technical Digest of the IEEE Solid State Sensor and Actuator Workshop, Hilton Head Island, SC, 1988, p. 5.

Clayton, L. D., Eernisse, E. P., Ward, R. W., and Wiggins, R. B., "Miniature Crystalline Quartz Electromechanical Structures," Sensors and Actuators, vol. 20, nos. 1 - 2, Nov. 1989, pp. 171 - 177.

Clegg, W. W., Jenkins, D. F. L., and Cunningham, M. J., "The Preparation of Piezoceramic-Polymer Thick-Films and Their Application as Micromechanical Actuators," Sensors and Actuators, vol. A58, no. 3, Mar. 1997, pp. 173 - 177.

Cline, H. E., and Anthony, T. R., "Migration of Fine Molten Wires in Thin Silicon Wafers," Journal of Applied Physics, vol. 49, no. 4, Apr. 1978, pp. 2412 - 2419.

Crooks, R. M., Bergbreiter, D. E., Bruening, M. L., Wells, M., Zhou, Z., Ricco, A. J., and Osbourn, G. C., "Versatile Materials for Use as Chemically Sensitive Interfaces in SAW-Based Sensor Arrays," Proceedings of the 1996 Solid-State Sensor and Actuator Workshop," Hilton Head Island, SC, June 3 - 6, 1996, pp. 19 - 22.

Dalton, J. V., and Drobeck, J., "Structure and Sodium Migration in Silicon Nitride Films," Journal of the Electrochemical Society, vol. 115, 1968, pp. 865 - 868.

Danel, J. S., and Delapierre, G., "Quartz: A Material for Microdevices," Journal of Micromechanics and Microengineering, vol. 1, no. 4, Dec. 1991, pp. 187 - 198.

Danel, J. S., Michel, F., and Delapierre, G., "Micromachining of Quartz and its Application to an Acceleration Sensor," Sensors and Actuators, vol. A23, nos. 1 - 3, Apr. 1990, pp. 971 - 977.

Davies, S. T., Hayton, D. A., and Tsuchiya, K., "Fabrication of Microstructures by Ion Beam Micromachining," Microlithography and Metrology in Micromachining II, Austin, TX, Oct. 14 - 15, 1996, in Proceedings of the SPIE - The International Society for Optical Engineering, vol. 2880, Oct. 1996, pp. 248 - 255.

de Reus, R., and Lindahl, M., "Si-to-Si Wafer Bonding Using Evaporated Glass," Proceedings of Transducers '97, the 1997 International Conference on Solid-State Sensors and Actuators, Chicago, IL, June 16 - 19, 1997, vol. 1, pp. 661 - 664.

Delapierre, G., "Micro-Machining: A Survey of the Most Commonly Used Processes," Sensors and Actuators, vol. 17, nos. 1 - 2, May 1989, pp. 123 - 138.

Deng, K., Collins, R. J., Mehregany, M., and Sukenik, C., "Performance Impact of Monolayer Coating of Polysilicon Micromotors," Proceedings of the IEEE Micro Electro Mechanical Systems Conference, Amsterdam, Netherlands, Jan. 29 - Feb. 2, 1995, pp. 368 - 373.

Diem, B., Delaye, M. T., Michel, F., Renard, S., and Delapierre, G., "SOI (SIMOX) as a Substrate for Surface Micromachining of Single Crystalline Sensors and Actuators," Proceedings of Transducers '93, the 7th International Conference on Solid-State Sensors and Actuators, Yokohama, Japan, June 7 - 10, 1993, Institute of Electrical Engineers, Japan, pp. 233 - 236.

Dietrich, T. R., Abraham, M., Diebel, J., Lacher, A., and Ruf, J., "Photoetchable Glass for Microsystems: Tips for Atomic Force Microscopy," Journal of Micromechanics and Microengineering, vol. 3, no. 4, Dec. 1993, pp. 187 - 189.

Donaldson, R. R., and Patterson, S. R., "Design and Construction of a Large, Vertical Axis Diamond Turning Machine," Proceedings of the SPIE Conference on Contemporary Methods of Optical Manufacturing and Testing, San Diego, CA, Aug. 24 - 26, 1983, SPIE vol. 433, pp. 62 - 67.

Doutreloigne, J., de Baets, J., de Rycke, I., de Smet, H., van Calster, A., and Vanfleteren, J., "The Electrical Performance of a Complementary CdSe:In/Ge:Cu Thinfilm Transistor Technology for Flat Panel Displays," Solid-State Electronics, vol. 34, no. 2, Feb. 1991, pp. 143 - 147.

Ehrfeld, W., Bley, B., Götz, F., Hagmann, P., Maner, A., Mohr, J., Moser, H. O., Münchmeyer, D., Schelb, W., Schmidt, D., and Becker, E. W., "Fabrication of Microstructures Using the LIGA Process," 1987 IEEE Micro Robots and Teleoperators Workshop, Hyannis, MA, Nov. 9 - 11, 1987, pp. 11/1 - 11/11.

Ehrfeld, W., Götz, F., Schelb, W., and Schmidt, D., "Method of Producing Microsensors with Integrated Signal Processing," U.S. Patent No. 5,194,402, issued Mar. 16, 1993.

Ehrfeld, W., Götz, F., Münchmeyer, D., Schelb, W., and Schmidt, D., "LIGA Process: Sensor Construction Techniques via X-Ray Lithography," Technical Digest, IEEE Solid-State Sensor and Actuator Workshop, Hilton Head Island, SC, June 6 - 9, 1988, pp. 1 - 4.

Ehrlich, D. J., and Tsao, J. Y. [Eds.], "Laser Microfabrication: Thin Film Processes and Lithography," Academic Press, Inc., Boston, MA, 1989.

El Khakani, M. A., Chaker, M., Jean, A., Boily, S., Kieffer, J. C., O'Hern, M. E., Ravet, M. F., and Rousseaux, F., "Hardness and Young's Modulus of Amorphous a-SiC Thin Films Determined by Nanoindentation and Bulge Tests," Journal of Materials Research, vol. 9, no. 1, Jan. 1994, pp. 96 - 103.

Elders, J., Jansen, H. V., Elwenspoek, M., and Ehrfeld, W., "DEEMO: A New Technology for the Fabrication of Microstructures," Proceedings of the IEEE Micro Electro Mechanical Systems Conference, Amsterdam, Netherlands, Jan. 29 - Feb. 2, 1995, pp. 238 - 243.

Elliott, R. A., DeFreez, R. K., Puretz, J., Orloff, J., and Crow, G. A., "Focused-Ion-Beam Micromachining of Diode Laser Mirrors," Communications Networking in Dense Electromagnetic Environments, Los Angeles, CA, Jan. 14 - 15, 1989, in Proceedings of the SPIE - The International Society for Optical Engineering, vol. 876, Jan. 1988, pp. 114 - 120.

Ensell, G., "Free Standing Single-Crystal Silicon Microstructures," Journal of Micromechanics and Microengineering, vol. 5, no. 1, Mar. 1995, pp. 1 - 4.

Esashi, M., Nakano, A., Shoji, S., and Hebiguchi, H., "Low-Temperature Silicon-to-Silicon Anodic Bonding with Intermediate Low Melting Point Glass," Sensors and Actuators, vol. A23, nos. 1 - 3, Apr. 1990, pp. 931 - 934.

Esashi, M., Takinami, M., Wakabayashi, Y., and Minami, K., "High-Rate Directional Deep Dry Etching for Bulk Silicon Micromachining," Journal of Micromechanics and Microengineering, vol. 5, no. 1, Mar. 1995, pp. 5 - 10.

Evans, C., "Precision Engineering: An Evolutionary View," Cranfield Press, Bedford, U.K., 1989.

Fan, L. S., Tai, Y. C., and Muller, R. S., "Pin Joints, Gears, Springs, Cranks, and Other Novel Micromechanical Structures," Proceedings of Transducers '87, Fourth International Conference on Solid-State Sensors and Actuators Digest of Technical Papers, IEEE Press, Tokyo, Japan, June 2 - 5, 1987, pp. 849 - 852.

Fedder, G. K., Santhanam, S., Reed, M. L., Eagle, S. C., Guillou, D. F., Lu, M. S.-C., and Carley, L. R., "Laminated High-Aspect-Ratio Microstructures in a Conventional CMOS Process," Proceedings of IEEE International Workshop on Micro Electro Mechanical Systems, San Diego, CA, Feb. 11 - 15, 1996, pp. 13 - 18.

Field, L. A., and Muller, R. S., "Fusing Silicon Wafers with Low Melting Temperature Glass," Sensors and Actuators, vol. A23, nos. 1 - 3, Apr. 1990, pp. 935 - 938.

Finne, R. M., and Klein, D. L., "A Water-Amine Complexing Agent System for Etching in Silicon," Journal of the Electrochemical Society, vol. 114, no. 9, Sept. 1967, pp. 965 - 970.

Fischer, B. E., and Spohr, R., "Production and Use of Nuclear Tracks: Imprinting Structure on Solids," Review of Modern Physics, vol. 55, no. 4, Oct. 1983, pp. 907 - 948.

Flannery, A. F., Mourlas, N. J., Storment, C. W., Tsai, S., Tan, S. H., and Kovacs, G. T. A., "PECVD Silicon Carbide for Micromachined Transducers," Proceedings of Transducers '97, the 1997 International Conference on Solid-State Sensors and Actuators, Chicago, IL, June 16 - 19, 1997, vol. 1, pp. 217 - 220.

Flannery, A. F., personal communication, 1997.

Fleischman, A. J., Roy, S., Zorman, C. A., Mehregany, M., and Matus, L. G., "Polycrystalline Silicon Carbide for Surface Micromachining," Proceedings of IEEE International Workshop on Micro Electro Mechanical Systems, San Diego, CA, Feb. 11 - 15, 1996, pp. 234 - 238.

Franz, I., and Langheinrich, W., "Distribution of Sodium in Silicon Nitride," Solid State Electronics, vol. 12, no. 3, Mar. 1969, pp. 145 - 150.

Frazier, A. B., and Allen, M. G., "Metallic Microstructures Fabricated Using Photosensitive Polyimide Electroplating Molds," Journal of Microelectromechanical Systems, vol. 2, no. 2, June 1993, pp. 87 - 94.

Friedrich, C. R., and Vasile, M. J., "Development of the Micromilling Process for High-Aspect-Ratio Microstructures," Journal of Microelectromechanical Systems, vol. 5, no. 1, Mar. 1996, pp. 33 - 38.

Furukawa, S., Miyajima, H., Mehregany, M., and Liu, C. C., "Electroless Plating of Metals for Microelectromechanical Structures," Proceedings of Transducers '93, the 7th International Conference on Solid-State Sensors and Actuators, Yokohama, Japan, June 7 - 10, 1993, Institute of Electrical Engineers, Japan, pp. 66 - 69.

Furuya, A., Shimokawa, F., Matsuura, T., and Sawada, R., "Micro-Grid Fabrication of Fluorinated Polyimide by Using Magnetically Controlled Reactive Ion Etching (MC-RIE)," Proceedings of the IEEE Micro Electro Mechanical Systems Conference, Fort Lauderdale, FL, Feb. 1993, pp. 59 - 64.

Gargin, H. L., Garmire, E., Somekh, S., Stoll, H., and Yariv, A., "Ion Beam Micromachining of Integrated Optics Components," Applied Optics, vol. 12, no. 3, Mar. 1973, pp. 455 - 459.

Gennissen, P. T. J., and French, P. J., "Sacrificial Oxide Etching Compatible with Aluminum Metallization," Proceedings of Transducers '97, the 1997 International Conference on Solid-State Sensors and Actuators, Chicago, IL, June 16 - 19, 1997, vol. 1, pp. 225 - 228.

Gianchandani, Y. B., Ma, K. J., and Najafi, K., "A CMOS Dissolved Wafer Process for Integrated P++ Microelectromechanical Systems," Proceedings of Transducers '95/Eurosensors IX, Stockholm, Sweden, June 25 - 29, 1995, vol. 1, pp. 79 - 82.

Gibbs, J. W., "Collected Works," Longmans, Green & Co., New York, NY, 1928.

Gilgen, H. H., Cacouris, T., Shaw, P. S., Krchnavek, R. R., and Osgood, R. M., Jr., "Direct Writing of Metal Conductors with Near-UV Light," Applied Physics B, vol. 43, 1987, pp. 55 - 66.

Golovanov, V., Solis, J. L., Lantto, V., and Leppävuori, S., "Different Thick-Film Methods in Printing of One-Electrode Semiconductor Gas Sensors," Proceedings of Transducers '95, the 8th International Conference on Solid-State Sensors and Actuators, June 25 - 29, 1995, Stockholm, Sweden, vol. 2, pp. 874 - 877.

Grande, W., and Talbot, J., "Electrodeposition of Thin Films of Nickel-Iron: I. Experimental," Journal of the Electrochemical Society, vol. 140, no. 3, Mar. 1993, pp. 669 - 674.

Grimmett, D., Schwartz, M., and Nobe, K., "A Comparison of DC and Pulsed Fe-Ni Alloy Deposits," Journal of the Electrochemical Society, vol. 140, no. 4, Apr. 1993, pp. 672 - 677.

Guckel, H., and Burns, D. W., "Planar Processed Polysilicon Sealed Cavities for Pressure Transducer Arrays," Proceedings of the IEEE International Electron Devices Meeting, San Francisco, CA, Dec. 9 - 12, 1984, pp. 223 - 225.

Guckel, H., Christenson, T. R., Skrobis, K. J., Denton, D. D., Choi, B., Lovell, E. G., Lee, J. W., Bajikar, S. S., and Chapman, T. W., "Deep X-Ray Lithographies for Micromechanics," Technical Digest of the 1990 Solid-State Sensor and Actuator Workshop, Hilton Head Island, SC, June 4 - 7, 1990, pp. 118 - 122.

Guckel, H., Mangat, P. S., Emmerich, H., Massoud-Ansari, S., Klein, J., Earles, T., Zook, J. D., Ohnstein, T., Johnson, E. D., Siddons, D. P., and Christenson, R. T., "Advances in Photoresist-Based Processing Tools for 3-Dimensional Prescision and Micro Mechanics," Proceedings of the Solid-State Sensor and Actuator Workshop, Hilton Head Island, SC, June 3 - 6, 1996, pp. 60 - 63.

Gui, C., Jansen, H., de Boer, M., Berenschot, J. W., Gardeniers, J. G. E., and Elwenspoek, M., "High Aspect Ratio Single Crystalline Silicon Microstructures Fabricated with Multi Layer Substrates," Proceedings of Transducers '97, the 1997 International Conference on Solid-State Sensors and Actuators, Chicago, IL, June 16 - 19, 1997, vol. 1, pp. 633 - 636.

Halverson, R. M., MacIntyre, M. W., and Motsiff, W. T., "The Mechanism of Single-Step Liftoff with Chlorobenzene in a Diazo-Type Resist," IBM Journal of Research and Development, vol. 26, no. 5, Sept. 1982, pp. 590 - 595.

Harendt, C., Graf, H.-G., Penteker, E., and Höfflinger, B., "Wafer Bonding: Investigation and in situ Observation of the Bond Process," Sensors and Actuators, vol. A23, nos. 1 - 3, Apr. 1990, pp. 927 - 930.

Harrison, J. A., and Thompson, J., "The Electrodeposition of Precious Metals; A Review of the Fundamental Electrochemistry," Electrochemica Acta, vol. 18, no. 11, 1973, pp. 829 - 834

Hatzakis, M., Canavello, B. J., and Shaw, J. M., "Single-Step Optical Lift-Off Process," IBM Journal of Research and Development, vol. 24, no. 4, July 1980, pp. 452 - 460.

Hecht, M. H., Vasquez, R. P., and Grunthaner, F. J., "A Novel X-ray Photoelectron Spectroscopy Study of the Al/SiO_2 Interface," Journal of Applied Physics, vol. 57, no. 12, June 1985, pp. 5256 - 5261.

Hedlund, C., Lindberg, U., Bucht, U., and Söderkvist, J., "Anisotropic Etching of Z-Cut Quartz," Journal of Micromechanics and Microengineering, vol. 3, no. 2, June 1993, pp. 65 -73.

Heschel, M., and Bouwstra, S., "Conformal Coating by Photoresist of Sharp Corners of Anisotropically Etched Through Holes in Silicon," Proceedings of Transducers '97, the 1997 International Conference on Solid-State Sensors and Actuators, Chicago, IL, June 16 - 19, 1997, vol. 1, pp. 209 - 212.

Hirata, Y., Okuyama, H., Ogino, S., Numazawa, T., and Takada, H., "Piezoelectric Composites for Micro-Ultrasonic Transducers Realized with Deep-Etch X-Ray Lithography," Proceedings of the IEEE Micro Electro Mechanical Systems Conference, Amsterdam, Netherlands, Jan. 29 - Feb. 2, 1995, pp. 191 - 196.

Hjort, K., "Gallium Arsenide Micromechanics," PhD Thesis, Acta Universitatis Upsaliensis, Upsalla, Sweden, 1993.

Hjort, K., Thornell, G., Spohr, R., and Schweitz, J.-Å., "Heavy Ion Induced Etch Anisotropy in Single Crystalline Quartz," Proceedings of IEEE International Workshop on Micro Electro Mechanical Systems, San Diego, CA, Feb. 11 - 15, 1996, pp. 267 - 271.

Hoffman, E., Warneke, B., Kruglick, E., Weigold, J., and Pister, K. S. J., "3D Structures with Piezoresistive Sensors in Standard CMOS," Proceedings of the IEEE Micro Electro Mechanical Systems Conference, Amsterdam, Netherlands, Jan. 29 - Feb. 2, 1995, pp. 288 - 293.

Hofmann, W., and MacDonald, N. C., "Fabrication of Multiple-Level Electrically Isolated High-Aspect-Ratio Single Crystal Silicon Microstructures," Proceedings of the Tenth Annual Workshop of Micro Electro Mechanical Systems, Nagoya, Japan, Jan. 26 - 30, 1997, pp. 460 - 464.

Hornbeck, L. J., "Deformable-Mirror Spatial Light Modulators," in Spatial Light Modulators and Applications III, in Proceedings of the SPIE - the International Society for Optical Engineering, 1990, San Diego, CA, Aug. 7 - 8, 1989, vol. 1150, pp. 86 - 102.

Houston, M. R., Maboudian, R., and Howe, R. T., "Self-Assembled Monolayer Films as Durable Anti-Stiction Coatings for Polysilicon Microstructures," Proceedings of the 1996 Solid-State Sensor and Actuator Workshop," Hilton Head Island, SC, June 3 - 6, 1996, pp. 42 - 47.

Howe, R. T., "Surface Micromachining for Microsensors and Microactuators," Journal of Vacuum Science and Technology B, vol. 6, no. 6, Nov./Dec. 1988, pp. 1809 - 1813.

Hsue, C. H. and Schmidt, M. A., "Micromachined Structures Fabricated Using A Wafer-Bonded Sealed Cavity Process," Technical Digest of the Solid-State Sensor and Actuator Workshop, Hilton Head Island, SC, June 13 - 16, 1994, pp. 151 - 155.

Huang, Y., Zhang, H., Kim, E. S., Kim, S. G., and Jeon, Y. B., "Piezoelectrically Actuated Microcantilever for Actuated Mirror Array Application," Proceedings of the 1996 Solid-State Sensor and Actuator Workshop, Hilton Head Island, SC, June 3 - 6, 1996, pp. 191 - 195.

Hunn, J. D., and Christensen, C. P., "Ion Beam and Laser-Assisted Micromachining of Single-Crystal Diamond," Solid State Technology, vol. 37, no. 12, Dec. 1994, pp. 57 - 60.

Ibbotson, D. E., Mucha, J. A., Flamm, D. A., and Cook, J. M., "Plasmaless Dry Etching of Silicon with Fluorine-Containing Compounds," Journal of Applied Physics, vol. 56, no. 10, Nov. 1984, pp. 2939 - 2942.

Ikuta, K., and Hirowatari, K., "Real Three Dimensional Micro Fabrication Using Stereo Lithography and Metal Molding," Proceedings of the IEEE Micro Electro Mechanical Systems Conference, Fort Lauderdale, FL, Feb. 7 - 10, 1993, pp. 42 - 47.

Ishida, M., Kim, H., Kimura, T., and Nakamura, T., "A New Etching Method for Single Crystal Al_2O_3 Film on Si Using Si Ion Implantation," Proceedings of Transducers '95/Eurosensors IX, Stockholm, Sweden, June 25 - 29, 1995, vol. 1, pp. 87 - 90.

Jaccodine, R. J., "Use of Modified Free Energy Theorems to Predict Equilibrium Growing and Etching Shapes," Journal of Applied Physics, vol. 33, no. 8, Aug. 1962, pp. 2643 - 2647.Jackman, R. J., Wilber, J. L., and Whitesides, G. M., "Fabrication of Submicrometer Features on Curved Substrates by Microcontact Printing," Science, vol. 269, no. 5224, Aug. 4, 1995, pp. 664 - 666.

Jansen, H. V., Gardeniers, J. G. E., Elders, J., Tilmans, H. A. C., and Elwenspoek, M., "Applications of Fluorocarbon Polymers in Micromechanics and Micromachining," Sensors and Actuators, vols. A41 - A42, nos. 1 - 3, Apr. 1994, pp. 136 - 140.

Jansen, H., de Boer, M., and Elwenspoek, M., "The Black Silicon Method VI: High Aspect Ratio Trench Etching for MEMS Applications," Proceedings of IEEE International Workshop on Micro Electro Mechanical Systems, San Diego, CA, Feb. 11 - 15, 1996, pp. 250 - 257.

Jansen, H., de Boer, M., Legtenberg, R., and Elwenspoek, M., "The Black Silicon Method: A Universal Method for Determining the Parameter Setting of a Fluorine-Based Reactive Ion Etcher in Deep Silicon Trench Etching with Profile Control," Journal of Micromechanics and Microengineering, vol. 5, no. 2, June 1995, pp. 115 - 120.

Jansen, H., de Boer, M., Otter, B., and Elwenspoek, M., "The Black Silicon Method IV: The Fabrication of Three-Dimensional Structures in Silicon with High Aspect Ratios for Scanning Probe Microscopy and Other Applications," Proceedings of the IEEE Micro Electro Mechanical Systems Conference, Amsterdam, Netherlands, Jan. 29 - Feb. 2, 1995, pp. 88 - 93.

Juan, W. H., and Pang, S. W., "A Novel Etch-Diffusion Process for Fabricating High Aspect Ratio Si Microstructures," Proceedings of Transducers '95, Stockholm, Sweden, June 25 - 29, 1995, vol. 1, pp. 560 - 563.

Juan, W. H., and Pang, S. W., "High-Aspect-Ratio Si Etching for Microsensor Fabrication," Journal of Vacuum Science and Technology A, vol. 13, no. 3, part 1, May - June 1995, pp. 834 - 838.

Juang, C., Kuhn, K. J., and Darling, R. B., "Selective Etching of GaAs and $Al_{0.3}Ga_{0.7}As$ with Citric Acid/Hydrogen Peroxide Solutions," Journal of Vacuum Science and Technology B, vol. 8, no. 5, Sept./Oct. 1990, pp. 1122 - 1124.

Kahn, H., Stemmer, S., Nandakumar, K., Heuer, A. H., Mullen, R. L., Ballarini, R., and Huff, M. A., "Mechanical Properties of Thick, Surface Micromachined Polysilicon Films," Proceedings of IEEE International Workshop on Micro Electro Mechanical Systems, San Diego, CA, Feb. 11 - 15, 1996, pp. 343 - 348.

Kaminsky, G., "Micromachining of Silicon Mechanical Structures," Journal of Vacuum Science and Technology, vol. B3, no. 4, July/Aug. 1985, pp. 1015 - 1024.

Keller, C., and Ferrari, M., "Milli-Scale Polysilicon Structures," Technical Digest of the 1994 Solid-State Sensor and Actuator Workshop, Hilton Head Island, SC, June 13 - 16, 1994, pp. 132 - 137.

Keller, C. G., and Howe, R. T., "Hexsil Bimorphs for Vertical Actuation," Proceedings of Transducers '95, the 8th International Conference on Solid-State Sensors and Actuators, Stockholm, Sweden, June 25 - 29, 1995, vol. 1, pp. 99 - 102.

Kendall, D. L., and deGuel, G. R., "Orientations of the Third Kind: The Coming of Age of (110) Silicon," Micromachining and Micropackaging of Transducers, Elsevier, Amsterdam, Netherlands, 1985.

Kern, W., "Chemical Etching of Silicon, Germanium, Gallium Arsenide and Gallium Phosphide," RCA Review, vol. 39, June 1978, pp. 278 - 308.

Kersten, P., Bouwstra, S., and Petersen, J. W., "Photolithography on Micromachined 3D Surfaces Using Electrodeposited Photoresists," Sensors and Actuators, vol. A51, no. 1, Oct. 1995, pp. 51 - 54.

Khare, R., Hu, E. L., Brown, J. J., and Melendes, M. A., "Micromachining of III-V Semiconductors Using Wet Photoelectrochemical Etching," Journal of Vacuum Science and Technology, vol. B11, no. 6, Nov./Dec. 1993, pp. 2497 - 2501.

Khare, R., Hu, E. L., Reynolds, D., and Allen, S. J., "Photoelectrochemical Etching of High Aspect Ratio Submillimeter Waveguide Filters from n^+ GaAs Wafers," Applied Physics Letters, vol. 61, no. 24, Dec. 14, 1992, pp. 2890 - 2892.

Khare, R., Young, D. B., Snider, G. L., and Hu, E. L., "Effect of Band Structure on Etch-Stop Layers in the Photoelectrochemical Etching of GaAs/AlGaAs Semiconductor Structures," Applied Physics Letters, vol. 62, no. 15, Apr. 12, 1993, pp. 1809 - 1811.

Kiewit, D. A., "Microtool Fabrication by Etch Pit Replication," Review of Scientific Instruments, vol. 44, no. 12, Dec. 1973, pp. 1741 - 1742.

Kim, J. Y., and Kim, C.-J., "Comparative Study of Various Release Methods for Polysilicon Surface Micromachining," Proceedings of the Tenth Annual Workshop of Micro Electro Mechanical Systems, Nagoya, Japan, Jan. 26 - 30, 1997, pp. 442 - 447.

Kimerling, L. C., Leamy, H. J., and Jackson, K. A., "Photoinduced Zone Migration (PIZM) in Semiconductors," in Proceedings of the Symposium on Laser and Electron Beam Processing of Electronic Materials, The Electrochemical Society, vol. 80-1, 1980, p. 242.

Klaassen, E. H., Petersen, K., Noworolski, J. M., Logan, J., Maluf, N. I., Brown, J., Storment, C., McCulley, W., and Kovacs, G. T. A., "Silicon Fusion Bonding and Deep Reactive Ion Etching; A New Technology for Microstructures," Digest of Technical Papers from Transducers '95/Eurosensors IX, Stockholm, Sweden, June 25 - 29, 1995, vol. 1, pp. 556 - 559.

Klaassen, E. H., Reay, R. J., Storment, C., Audy, J., Henry, P., Brokaw, A. P., and Kovacs, G. T. A., "Micromachined Thermally Isolated Circuits," Proceedings of the 1996 Solid-State Sensor and Actuator Workshop, Hilton Head Island, SC, June 3 - 6, 1996, pp. 127 -131.

Kloeck, B., Collins, S., de Rooij, N., and Smith, R. L, "Study of Electrochemical Etch-Stop for High-Precision Thickness Control of Silicon Membranes," IEEE Transactions Electron Devices, vol. 36, no. 4, Apr. 1989, pp. 663 - 669.

Klumpp, A., Schaber, U., Offereins, H. L, Kühl, K., and Sandmair, H., "Amorphous Silicon Carbide and Its Application in Silicon Micromachining," Sensors and Actuators, vol. A41, nos. 1 - 3, Apr. 1994, pp. 310 - 316.

Kobayashi, D., Hirano, T., Furuhata, T., and Fujita, H., "An Integrated Lateral Tunneling Unit," IEEE Micro Electro Mechanical Systems Workshop, Travemünde, Germany, 1992, pp. 214 - 219.

Köhler, U., Guber, A. E., Bier, W., Heckele, M., and Schaller, Th., "Fabrication of Microlenses by Combining Silicon Technology, Mechanical Micromachining and Plastic Molding," Miniaturized Systems with Micro-Optics and Micromechanics, San Jose, CA, Jan. 30 - 31, 1996, in Proceedings of the SPIE - The International Society for Optical Engineering, Jan. 1996, vol. 2687, pp. 18 - 22.

Köhler, U., Guber, A., and Bier, W., "Plasmaless Silicon Etching with Mixtures of Bromine and Fluorine," 11th European Symposium on Fluorine Chemistry, Bled, Slovenia, Sept. 1995, p. 22.

Komaki, K., "Analysis of Domain Structure and Permeability of Electrodeposited NiFe Strip Films," Journal of the Electrochemical Society, vol. 140, no. 2, Feb. 1993, pp. 529 - 533.

Konagai, M., Sugimoto, M., and Takahashi, K., "High Efficiency GaAs Thin Film Solar Cells by Peeled Film Technology," Journal of Crystal Growth, vol. 45, 1978, pp. 277 - 280.

Kovacs, G. T. A., Cutkosky, M., Maluf, N. I., and Storment, C. W., "Out-of-Plane Microactuator Demonstration," Semiannual DARPA Technical Report for the Period Aug. 1 - Dec. 15 1992, Stanford University.

Kovacs, G. T. A., Petersen, K., and Albin, M., "Silicon Micromachining: Sensors to Systems," Analytical Chemistry, vol. 68, 1996, pp. 407A - 412A.

Krause, P., Sporys, M., Obermeier, E., Lange, K., and Grigull, S., "Silicon to Silicon Anodic Bonding Using Evaporated Glass," Proceedings of Transducers '95, the 9th International Conference on Solid-State Sensors and Actuators, Stockholm, Sweden, June 25 - 29, 1995, vol. 1, pp. 228 - 231.

Krchnavek, R. R., Gilgen, H. H., Chen, J. C., Shaw, P. S., Licata, T. J., and Osgood, R. M., Jr., "Photodeposition Rates of Metal from Metal Alkyls," Journal of Vacuum Science and Technology, vol. B5, no. 1, Jan./Feb. 1987, pp. 20 - 26.

Kung, J. T., Karanicolas, A. N., and Lee, H.-S., "A Compact, Inexpensive Apparatus for One-Sided Etching in KOH and HF," Sensors and Actuators, vol. A29, no. 3, Dec. 1991, pp. 209 - 215.

Kwon, O.-H., "A Microelectrode with CMOS Multiplexer for an Artificial Ear," Doctoral Dissertation in Electrical Engineering, Stanford University, Technical Report No. G909-3, Oct. 1986.

Lagorce, L. K., and Allen., M. G., "Micromachined Polymer Magnets," Proceedings of IEEE International Workshop on Micro Electro Mechanical Systems, San Diego, CA, Feb. 11 - 15, 1996, pp. 85 - 90.

Lang, W., Kühl, K., and Sandmaier, H., "Absorbing Layers for Thermal Infrared Detectors," Sensors and Actuators, vol. A34, no. 2, Sept. 1992, pp. 243 - 248.

Lapadatu, D., De Cooman, M., and Puers, R., "A Double Sided Capacitive Miniaturised Accelerometer Based on Photovoltaic Etch Stop Technique," Proceedings of Transducers '95, the 8th International Conference on Solid-State Sensors and Actuators, Stockholm, Sweden, June 25 - 29, 1995, vol. 2, pp. 546 - 549.

Lärmer, F., and Schilp, P., "Method of Anisotropically Etching Silicon," German Patent No. DE 4,241,045, issued 1994.

Larsson, O., Öhman, O., Billman, Å., Lundbladh, L., Lindell, C., and Palmskog, G., "Silicon Based Replication Technology of 3D-Microstructures by Conventional CD-Injection Molding Techniques," Proceedings of Transducers '97, the 1997 International Conference on Solid-State Sensors and Actuators, Chicago, IL, June 16 - 19, 1997, vol. 2, pp. 1415 - 1418.

Lee, J. H., Lee, Y. I., Jang, W. I., Lee, C. S., and Yoo, H. J., "Gas-Phase Etching of Sacrificial Oxides Using Anhydrous HF and CH_3OH," Proceedings of the Tenth Annual Workshop of Micro Electro Mechanical Systems, Nagoya, Japan, Jan. 26 - 30, 1997, pp. 448 - 453.

Legtenberg, R., Elders, J., and Elwenspoek, M., "Stiction of Surface Microstructures after Rinsing and Drying: Model and Investigation of Adhesion Mechanisms," Proceedings of Transducers '93, the 7th International Conference on Solid-State Sensors and Actuators, Yokohama, Japan, June 7 - 10, 1993, Institute of Electrical Engineers, Japan, pp. 198 - 201.

Lehmann, O., and Stuke, M., "Generation of Three-Dimensional Free-Standing Metal Micro-Objects by Laser Chemical Processing," Applied Physics A (Solids and Surfaces), vol. A53, 1991, pp. 343 - 345.

Lehmann, O., and Stuke, M., "Three-Dimensional Laser Direct Writing of Electrically Conducting and Isolating Microstructures," Materials Letters, vol. 21, Oct. 1994, pp. 131 - 136.

Lehmann, V., "Porous Silicon - A New Material for MEMS," Proceedings of IEEE International Workshop on Micro Electro Mechanical Systems, San Diego, CA, Feb. 11 - 15, 1996, pp. 1 - 6.

Lehmann, V., "The Physics of Macropore Formation in Low Doped n-Type Porous Silicon," Journal of the Electrochemical Society, vol. 140, 1993, pp. 2836 - 2843.

Lehmann, V., and Föll, H., "Formation Mechanism and Properties of Electrochemically Etched Trenches in n-Type Silicon," Journal of the Electrochemical Society, vol. 137, no. 2, Feb. 1990, pp. 653 - 659.

Lehmann, V., and Gösele, U., "Porous Silicon Formation: A Quantum Wire Effect," Applied Physics Letters, vol. 58, no. 8, Feb. 25, 1991, pp. 856 - 858.

Lenggenhager, R., "CMOS Thermoelectric Infrared Sensors," Dissertation for the Degree of Doctor of Natural Sciences, Swiss Federal Institute of Technology, Zurich, Switzerland, DISS. ETH no. 10744, 1994.

Leppävuori, S., Väänänen, J., Lahti, M., Remes, J., and Uusimäki, A., "A Novel Thick-Film Technique, Gravure Offset Printing, for the Realization of Fine-Line Sensor Structures," Sensors and Actuators, vol. A42, nos. 1 - 3, Apr. 15, 1994, pp. 593 - 596.

Li, Y. X., French, P. J., Sarro, P. M., and Wolffenbuttel, R. F., "Fabrication of a Single Crystalline Silicon Capacitive Lateral Accelerometer Using Micromachining Based on Single Step Plasma Etching," Proceedings of the IEEE Micro Electro Mechanical Systems Conference, Amsterdam, Netherlands, Jan. 29 - Feb. 2, 1995, pp. 398 - 403.

Liakopoulos, T. M., Zhang, W., and Ahn, C. H., "Electroplated Thick CoNiMnP Permanent Magnet Arrays for Micromachined Magnetic Device Applications, "Proceedings of IEEE International Workshop on Micro Electro Mechanical Systems, San Diego, CA, Feb. 11 - 15, 1996, pp. 79 - 84.

Lin, G., Kim, C.-J., Konishi, S., and Fujita, H., "Design, Fabrication and Testing of a C-Shape Actuator," Proceedings of Transducers '95/Eurosensors IX, Stockholm, Sweden, June 25 - 29, 1995, vol. 2, pp. 416 - 419.

Lin, L., Shia, T. K., and Chiu, C.-J., "Fabrication and Characterization of IC-Processed Brightness Enhancement Films," Proceedings of Transducers '97, the 1997 International Conference on Solid-State Sensors and Actuators, Chicago, IL, June 16 - 19, 1997, vol. 2, pp. 1427 - 1430.

Linde, H., and Austin, L., "Wet Silicon Etching with Aqueous Amine Gallates," Journal of the Electrochemical Society, vol. 139, no. 4, Apr. 1992, pp. 1170 - 1174.

Linder, C., Tschan, T., and de Rooij, N. F., "Deep Dry Etching Techniques as a New IC Compatible Tool for Silicon Micromachining," Proceedings of Transducers '91, the 1991 International Conference on Solid-State Sensors and Actuators Digest of Technical Papers, IEEE Press, San Francisco, CA, June 24 - 27, 1991, pp. 524 - 527.

Linder, S., Baltes, H., Gnaedinger, F., and Doering, E., "Photolithography In Anisotropically Etched Grooves," Proceedings of IEEE International Workshop on Micro Electro Mechanical Systems, San Diego, CA, Feb. 11 - 15, 1996, pp. 38 - 43.

Little, W. A., "Microminiature Refrigeration," Review of Scientific Instruments, vol. 55, no. 5, May 1984, pp. 661 - 680.

Liu, C., Tsao, T., and Tai, Y.-C., "A High-Yield Drying Process for Surface Microstructures Using Active Levitation," Proceedings of Transducers '97, the 1997 International Conference on Solid-State Sensors and Actuators, Chicago, IL, June 16 - 19, 1997, vol. 1, pp. 241 - 244.

Liu, C., Tsao, T., Tai, Y.-C., Leu, T.-S., Ho, C.-H., Tang. W.-L., and Miu, D., "Out-of-Plane Permalloy Magnetic Actuators for Delta-Wing Control," Proceedings of MEMS '95, Amsterdam, Netherlands, IEEE, Jan. 29 - Feb. 2, 1995, pp. 7 - 12.

Maas, D., Büstgens, B., Fahrenberg, J., Keller, W., Ruther, P., Schomburg, W. K., and Seidel, D., "Fabrication of Microcomponents Using Adhesive Bonding Techniques," Proceedings of IEEE International Workshop on Micro Electro Mechanical Systems, San Diego, CA, Feb. 11 - 15, 1996, pp. 331 - 336.

MacDonald, N. C., Chen, L. Y., Yao, J. J., Zhang, Z. L., McMillan, J. A., Thomas, D. C., and Haselton, K. R., "Selective Chemical Vapor Deposition of Tungsten for Microelectromechanical Structures," Sensors and Actuators, vol. 20, nos. 1 - 2, Nov. 1989, pp. 123 - 133.

Madden, J. D., and Hunter, I. W., "Three-Dimensional Microfabrication by Localized Electrochemical Deposition," Journal of Microelectromechanical Systems, vol. 5, no. 1, Mar. 1996, pp. 24 - 32.

Mallory, G. O., and Hajdu, J. B. [eds.], "Electroless Plating: Fundamentals and Applications," American Electroplaters and Surface Finishers Society, Orlando, FL, 1990.

Man, P. F., Gogoi, B. P., and Mastrangelo, C. H., "Elimination of Post-Release Adhesion in Microstructures Using Thin Conformal Fluorocarbon Films," Proceedings of IEEE International Workshop on Micro Electro Mechanical Systems, San Diego, CA, Feb. 11 - 15, 1996, pp. 55 - 60.

Manginell, R. P., Smith, J. H., Ricco, A. J., Moreno, D. J., Hughes, R. C., Huber, R. J., and Senturia, S. D., "Selective, Pulsed CVD of Platinum on Microfilament Gas Sensors," Proceedings of the 1996 Solid-State Sensor and Actuator Workshop," Hilton Head Island, SC, June 3 - 6, 1996, pp. 19 - 22.

Marcus, R. B., Ravi, R. S., Gmitter, T., Busta, H. H., Niccum, J. T., Chin, K. K., and Liu, D., "Atomically Sharp Silicon and Metal Field Emitters," IEEE Transactions on Electron Devices, vol. 38, no. 10, Oct. 1991, pp. 2289 - 2293.

Massoud-Ansari, S., Mangat, P. S., Klein, J., and Guckel, H., "A Multi-Level LIGA-Like Process for Three-Dimensional Actuators," Proceedings of IEEE International Workshop on Micro Electro Mechanical Systems, San Diego, CA, Feb. 11 - 15, 1996, pp. 285 - 289.

Mastrangelo, C. H., and Tang, W. C., "Semiconductor Sensor Technologies," Chapter 2 in "Semiconductor Sensors," Sze, S. M. [ed.], John Wiley and Sons, New York, NY, 1994, pp. 17 - 95.

Matsumoto, Y., Yoshida, K., and Ishida, M., "Fluorocarbon Film for Protection from Alkaline Etchant and Elimination of In-Use Stiction," Proceedings of Transducers '97, the 1997 International Conference on Solid-State Sensors and Actuators, Chicago, IL, June 16 - 19, 1997, vol. 1, pp. 695 - 698.

McGruer, N. E., Warner, K., Singhal, P., Gu, J. J., and Chan, C., "Oxidation-Sharpened Gated Field Emitter Array Process," IEEE Transactions on Electron Devices, vol. 38, no. 10, Oct. 1991, pp. 2389 - 2391.

Mehregany, M., and Senturia, S. D., "Anisotropic Etching of Silicon in Hydrazine," Sensors and Actuators, vol. 13, no. 4, Apr. 1988, pp. 375 - 390.

Merlos, A., Acero, M., Bao, M. H., Bausells, J., and Esteve, J., "TMAH/IPA Anisotropic Etching Characteristics," Sensors and Actuators, vols. A37 - A38, June - Aug. 1993, pp. 737 - 743.

Mescher, M., Abe, T., Brunett, B., Metla, H., Schlesinger, T. E., and Reed, M., "Piezoelectric Lead-Zirconate-Titanate Actuator Films for Microelectromechanical Systems Applications," Proceedings of the IEEE Micro Electro Mechanical Systems Conference, Amsterdam, Netherlands, Jan. 29 - Feb. 2, 1995, pp. 261 - 266.

Minami, K., Moriuchi, T., and Esashi, M., "Cavity Pressure Control for Critical Damping of Packaged Micro Mechanical Devices," Proceedings of Transducers '95, the 8th International Conference on Solid-State Sensors and Actuators, Stockholm, Sweden, June 25 - 29, 1995, vol. 1, pp. 240 - 243.

Missel, L., Duke, P., and Montelbano, T., "Square Profile Gold by Pulse Plating," Semiconductor International, Feb. 1980, pp. 67 - 78.

Mizrah, T., "Joining and Recrystallization of Si Using the Thermomigration Process," Journal of Applied Physics, vol. 51, no. 2, Feb. 1980, pp. 1207 - 1210.

Mohr, J., Bley, P., Stohrmann, M., and Wallrabe, U., "Microactuators Fabricated by the LIGA Process," Journal of Micromechanics and Microengineering, vol. 2, no. 4, Dec. 1992, pp. 234 - 241.

Moore, D. F., Burgess, S. C., Chiang, H.-S., Klaubert, H., Shibaike, N., and Kiriyama, T., "Micromachining and Focused Ion Beam Etching of Si for Accelerometers," Micromachining and Microfabrication Process Technology, Austin, TX, Oct. 23 - 24, 1995, in Proceedings of the SPIE - The International Society for Optical Engineering, Oct. 1995, vol. 2639, pp. 253 - 258.

Mori, Y., Ino, T., Tokura, H., and Yoshikawa, M., "Micromachining of CVD Diamond Films Using a Focused Ion Beam," Proceedings of the 3rd International Conference on the Applications of Diamond Films and Related Materials, Gaithersburg, MD, Aug. 21 - 24, 1995, vol. 1, pp. 233 - 240.

Moser, D., "CMOS Flow Sensors," Doctoral Thesis, Swiss Federal Institute of Technology, Zurich, Switzerland, 1993.

Moser, D., Parameswaran, M., and Baltes, H., "Field Oxide Microbridges, Cantilever Beams, Coils and Suspended Membranes in SACMOS Technology," Sensors and Actuators, vols. A21 - A23, Feb. - April 1990, pp. 1019 - 1022.

Mulhern, G. T., Soane, D. S., and Howe, R. T., "Supercritical Carbon Dioxide Drying of Micro-structures," Proceedings of Transducers '93, the 7th International Conference on Solid-State Sensors and Actuators, Yokohama, Japan, June 7 - 10, 1993, Institute of Electrical Engineers, Japan, pp. 296 - 299.

Müllenborn, M., Dirac, H., Petersen, J. W., and Bouwstra, S., "Fast 3D Laser Micromachining of Silicon for Micromechanical and Microfluidic Applications," Proceedings of Transducers '95, the 8th International Conference on Solid-State Sensors and Actuators, Stockholm, Sweden, June 25 - 29, 1995, vol. 1, pp. 166 - 169.

Murakami, K., Wakabayashi, Y., Minami, K., and Esashi, M., "Cryogenic Dry Etching for High Aspect Ratio Microstructures," Proceedings of the IEEE Microelectromechanical Systems Conference, Fort Lauderdale, FL, Feb. 1993, pp. 65 - 70.

Najafi, K., Wise, K. D., and Mochizuki, T., "A High-Yield IC-Compatible Multichannel Recording Array," IEEE Transactions on Electron Devices, vol. ED-32, no. 7, July 1985, pp. 1206 - 1211.

Nasby, R. D., Sniegowski, J. J., Smith, J. H., Montague, S., Barron, C. C., Eaton, W. P., and McWhorter, P. J., "Application of Chemical-Mechanical Polishing to Planarization of Surface-Micromachined Devices," Proceedings of the 1996 Solid-State Sensor and Actuator Workshop," Hilton Head Island, SC, June 3 - 6, 1996, pp. 48 - 53.

Noworolski, J. M., Klaassen, E., Logan, J., Petersen, K., and Maluf, N., "Fabrication of SOI Wafers with Buried Cavities Using Silicon Fusion Bonding and Electrochemical Etchback," Proceedings of Transducers '95/Eurosensors IX, Stockholm, Sweden, June 25 - 29, 1995, vol. 1, pp. 71 - 74.

Ohtsu, M., Minami, K., and Esashi, M., "Fabrication of Packaged Thin Beam Structures by an Improved Drying Method," Proceedings of IEEE International Workshop on Micro Electro Mechanical Systems, San Diego, CA, Feb. 11 - 15, 1996, pp. 228 - 233.

Ohwada, K., Negoro, Y., Konaka, Y., and Oguchi, T., "Groove Depth Uniformization in (110) Si Anisotropic Etching by Ultrasonic Wave and Application to Accelerometer Fabrication," Proceedings of the IEEE Micro Electro Mechanical Systems Conference, Amsterdam, Netherlands, Jan. 29 - Feb. 2, 1995, pp. 100 - 105.

Olgun, Z., Akar, O., Kulah, H., and Akin, T., "An Integrated Thermopile Structure with High Responsivity Using Any Standard CMOS Process," Proceedings of Transducers '97, the 1997 International Conference on Solid-State Sensors and Actuators, Chicago, IL, June 16 - 19, 1997, vol. 2, pp. 1263 - 1266.

Osgood, R. M., Jr., Gilgen, H. H., and Brewer, P., "Summary Abstract: Low-Temperature Deposition and Removal of Material Using Laser-Induced Chemistry," Journal of Vacuum Science and Technology, vol. 2, no. 2, Apr. - June 1984, pp. 504 - 505.

Palik, E. D., Glembocki, O. J., Heard, I., Jr., Burno, P. S., and Tenerz, L., "Etching Roughness for (100) Silicon Surfaces in Aqueous KOH," Journal of Applied Physics, vol. 70, no. 6, Sept. 15, 1991, pp. 3291 - 3300.

Parameswaran, M., Baltes, H. P., and Robinson, A. M., "Polysilicon Microbridge Fabrication Using Standard CMOS Technology," Digest of Technical Papers, IEEE Solid-State Sensors and Actuators Workshop, Hilton Head Island, SC, June 6 - 9, 1988, pp. 148 - 150.

Parameswaran, M., Baltes, H. P., Ristic, Lj., Dhaded, A. C., and Robinson, A. M., "A New Approach for the Fabrication of Micromechanical Structures," Sensors and Actuators, vol. 19, no. 3, Sept. 1989, pp. 289 - 307.

Paranjape, M., Giacomozzi, F., Landsberger, L., Kahrizi, M., Margesin, B., Nikpour, B., and Zen, B., "A Micromachined Angled Hall Magnetic Field Sensor Using Novel In-Cavity Patterning," Proceedings of Transducers '97, the 1997 International Conference on Solid-State Sensors and Actuators, Chicago, IL, June 16 - 19, 1997, vol. 1, pp. 397 - 400.

Peeters, E., Lapadatu, D., Puers, R. and Sansen, W., "PHET, An Electrodeless Photovoltaic Electrochemical Etchstop Technique," Journal of Microelectromechanical Systems, vol. 3., no. 3, Sept. 1994, pp. 113 - 123.

Petersen, K. E., "Dynamic Micromechanics in Silicon: Techniques and Devices," IEEE Transactions on Electron Devices, vol. ED-25, no. 10, Oct. 1978, pp. 1241 - 1250.

Petersen, K. E., Barth, P., Poydock, J., Brown, J., Mallon, J., Jr., and Bryzek, J., "Silicon Fusion Bonding for Pressure Sensors," Digest of Technical Papers, IEEE Solid-State Sensors and Actuators Workshop, Hilton Head Island, SC, June 6 - 9, 1988, pp. 144 - 147.

Petersen, K. E., Gee, D., Pourahmadi, F., Craddock, R., Brown, J., and Christel, L., "Surface Micromachined Structures with Silicon Fusion Bonding," Proceedings of Transducers '91, the 1991 International Conference on Solid-State Sensors and Actuators Digest of Technical Papers, IEEE Press, San Francisco, CA, June 24 - 27, 1991, pp. 397 - 399.

Pister, K. S. J., personal communication, 1995.

Podlesnik, D. V., Gilgen, H. H., and Osgood, R. M., Jr., "Deep-Ultraviolet Induced Wet Etching of GaAs," Applied Physics Letters, vol. 45, no. 5, Sept. 1, 1984, pp. 563 - 565.

Pomerantz, D. I., "Anodic Bonding," U.S. Patent No. 3,397,279, issued Aug. 13, 1968.

Price, J. B., "Anisotropic Etching of Silicon with KOH-H_2O-Isopropyl Alcohol," in Semiconductor Silicon, Huff, H. R. and Burgess, R. R. [eds.], Electrochemical Society Proceedings, Princeton, NJ, 1973, p. 339.

Quenzer, H. J., Dell, C., and Wagner, B., "Silicon-Silicon Anodic-Bonding with Intermediate Glass Layers Using Spin-On Glasses," Proceedings of IEEE International Workshop on Micro Electro Mechanical Systems, San Diego, CA, Feb. 11 - 15, 1996, pp. 272 - 276.

Reay, R. J., Klaassen, E. H. and Kovacs, G. T. A., "Thermally and Electrically Isolated Single-Crystal Silicon Structures in CMOS Technology," IEEE Electron Device Letters, vol. 15, no. 10, Oct. 1994, pp. 399 - 401.

Reisman, A., Berkenblit, M., Chan, S. A., Kaufmann, F. B., and Green, D. C., "The Controlled Etching of Silicon in Catalyzed Ethylene-Diamine-Pyrochatechol-Water Solutions," Journal of the Electrochemical Society: Solid-State Science and Technology, vol. 126, no. 8, Aug. 1979, pp. 1406 - 1415.

Reynaerts, D., Heeren, P.-H., and van Brussel, H., "Microstructuring of Silicon by Electro-Discharge Machining (EDM) - Part I: Theory," Sensors and Actuators, vol. A60, nos. 1 - 3, May 1997, pp. 212 - 218.

Riethmüller, W., and Benecke, W., "Thermally Excited Silicon Microactuators," IEEE Transactions on Electron Devices, vol. 35, no. 6, June 1988, pp. 758 - 763.

Rogner, A., Eicher, J., Munchmeyer, D., Peters, R.-P., and Mohr, J., "The LIGA Technique - What Are the Opportunities?," Journal of Micromechanics and Microengineering, vol. 2, no. 3, Sept. 1992, pp. 133 - 140.

Romero, G., "Fabrication of Submicrometer Gold Lines Using Optical Lithography and High-Growth-Rate Electroplating," IEEE Electron Device Letters, vol. EDL-4, no. 7, July 1983, pp. 210 - 212.

Sakata, M., Wakabayashi, S., Goto, H., Totani, H., Takeuchi, M., and Yada, T., "Sputtered High $|d_{31}|$ Coefficient PZT Thin Film for Micro Actuators," Proceedings of IEEE International Workshop on Micro Electro Mechanical Systems, San Diego, CA, Feb. 11 - 15, 1996, pp. 263 - 266.

Schnakenberg, U., Benecke, W., and Lange, P., "TMAHW Etchants for Silicon Micromachining," Proceedings of Transducers '91, the 1991 International Conference on Solid-State Sensors and Actuators Digest of Technical Papers, IEEE Press, San Francisco, CA, June 24 - 27, 1991, pp. 815 - 818.

Schnakenberg, U., Benecke, W., and Löchel, B., "NH_4OH-Based Etchants for Silicon Micromachining," Sensors and Actuators, vol. A23, nos. 1 - 3, Apr. 1990, pp. 1031 - 1035.

Schnakenberg, U., Benecke, W., Löchel, B., Ullerich, S., and Lange, P., "NH_4OH Based Etchants for Silicon Micromachining: Influence of Additives and Stability of Passivation Layers," Sensors and Actuators, vol. A25, nos. 1 - 3, Oct. 1990 - Jan. 1991, pp. 1 - 7.

Schneider, M., Müller, T., Häberli, A., Hornung, M., and Baltes, H., "Integrated Micromachined Decoupled CMOS Chip On Chip," Proceedings of the Tenth Annual Workshop of Micro Electro Mechanical Systems, Nagoya, Japan, Jan. 26 - 30, 1997, pp. 512 - 517.

Seidel, H., "The Mechanism of Anistotropic Silicon Etching and Its Relevance for Micromachining," Proceedings of Transducers '87, Record of the 4th International Conference on Solid-State Sensors and Actuators, Tokyo, Japan, June 2 - 5, 1987, pp. 120 - 125.

Seidel, H., Csepregi, L., Heuberger, A., and Baumgärtel, H., "Anisotropic Etching of Crystalline Silicon in Alkaline Solutions I: Orientation Dependence and Behavior of Passivation Layers," Journal of the Electrochemical Society, vol. 137, no. 11, Nov. 1990, pp. 3612 - 3626.

Seidel, H., Csepregi, L., Heuberger, A., and Baumgärtel, H., "Anisotropic Etching of Crystalline Silicon in Alkaline Solutions II: Influence of Dopants," Journal of the Electrochemical Society, vol. 137, no. 11, Nov. 1990, pp. 3626 - 3632.

Shaver, D. C., and Ward, B. W., "Semiconductor Applications of Focused Ion Beam Micromachining," Solid State Technology, vol. 28, no. 12, Dec. 1985, pp. 73 - 78.

Shaw, D. W., "Localized GaAs Etching with Acidic Hydrogen Peroxide Solutions," Journal of the Electrochemical Society: Solid-State Science and Technology, vol. 128, no. 4, Apr. 1981, pp. 874 - 880.

Shaw, K. A., Adams, S. G., and MacDonald, N. C., "A Single-Mask Lateral Accelerometer," Digest of Technical Papers, Transducers '93, Yokohama, Japan, June 7 - 10, 1993, pp. 210 - 213.

Shaw, K. A., and MacDonald, N. C., "Integrating SCREAM Micromachined Devices with Integrated Circuits," Proceedings of IEEE International Workshop on Micro Electro Mechanical Systems, San Diego, CA, Feb. 11 - 15, 1996, pp. 44 - 48.

Shaw, K. A., Zhang, Z. L., and MacDonald, N. C., "SCREAM I: A Single Mask, Single-Crystal Silicon Process for MicroElectroMechanical Structures," Proceedings of the 1993 Micro Electro Mechanical Systems Workshop - MEMS '93, Fort Lauderdale, FL, Feb. 7 - 10, 1993 pp. 155 - 160.

Shaw, K. A., Zhang, Z. L., and MacDonald, N. C., "SCREAM I: A Single Mask, Single-Crystal Silicon, Reactive Ion Etching Process for MicroElectroMechanical Structures," Sensors and Actuators, vol. A40, 1994, pp. 210 - 213.

Siewell, G. L., Boucher, W. R., and McClelland, P. H., "The ThinkJet Orifice Plate: A Part with Many Functions," Hewlett-Packard Journal, May 1985, pp. 33 - 37.

Singh, A., Horsley, D. A., Cohn, M. B., Pisano, A. P., and Howe, R. T., "Batch Transfer of Microstructures Using Flip-Chip Solder Bump Bonding," Proceedings of Transducers '97, the 1997 International Conference on Solid-State Sensors and Actuators, Chicago, IL, June 16 - 19, 1997, vol. 1, pp. 265 - 268.

Smela, E., Inganäs, O., and Lundrström, I., "Differential Adhesion Method for Microstructure Release: An Alternative to the Sacrificial Layer, Proceedings of Transducers '95, the 8th International Conference on Solid-State Sensors and Actuators, Stockholm, Sweden, June 25 - 29, 1995, vol. 2, pp. 350 - 351.

Smith, B. K., Sniegowski, J. J., and LaVigne, G., "Thin Teflon-Like Films for Eliminating Adhesion in Polysilicon Microstructures," Proceedings of Transducers '97, the 1997 International Conference on Solid-State Sensors and Actuators, Chicago, IL, June 16 - 19, 1997, vol. 1, pp. 245 - 248.

Smith, S. T., and Chetwynd, D. G., "Foundations of Ultraprecision Mechanism Design," Gordon and Breach Science Publishers, Philadelphia, PA, 1992.

Sooriakumar, K., Haeberle, R. J., and Meitzler, A. H., "A Technique for Anodically Bonding Glass Wafers with Metal Surface Patterns to Silicon Wafers," Proceedings of Transducers '93, the 7th International Conference on Solid-State Sensors and Actuators, Yokohama, Japan, June 7 - 10, 1993, Institute of Electrical Engineers, Japan, pp. 191 - 193.

Spindt, C.A., Holland, C. E., Brodie, I., Mooney, J. B., and Westerberg, E. R., "Field-Emitter Arrays to Vacuum Fluorescent Display." IEEE Transactions on Electron Devices, vol. 36, no. 1, part 2, Jan. 1989, pp. 225 - 228.

Srinivasan, U., Houston, M. R., Howe, R. T., and Maboudian, R., "Self-Assembled Fluorocarbon Films for Enhanced Stiction Reduction," Proceedings of Transducers '97, the 1997 International Conference on Solid-State Sensors and Actuators, Chicago, IL, June 16 - 19, 1997, vol. 2, pp. 1399 - 1402.

Staudte, J. H., "Micro-Resonators in Electronics," Proceedings of the 22nd Annual Frequency Control Symposium, Atlantic City, NJ, Apr. 22 - 24, 1968, pp. 226 - 231.

Staudte, J. H., "Subminiature Quartz Tuning Fork Resonator," Proceedings of the 27th Annual Frequency Control Symposium, Cherry Hill, NJ, June 12 - 14, 1973, pp. 50 - 54.

Steiner, P., Richter, A., and Lang, W., "Using Porous Silicon as a Sacrificial Layer," Journal of Micromechanics and Microengineering, vol. 3, no. 1, Mar. 1993, pp. 32 - 36.

Steinsland, E., Nese, M., Hanneborg, A., Bernstein, R. W., Sandmo, H., and Kittilsland, G., "Boron Etch-Stop in TMAH Solutions," Proceedings of Transducers '95, the 8th International Conference on Solid-State Sensors and Actuators, Stockholm, Sweden, June 25 - 29, 1995, vol. 1, pp. 190 - 193.

Storment, C. W., Borkholder, D. A., Westerlind, V., Suh, J. W., Maluf, N. I., and Kovacs, G. T. A., "Flexible, Dry-Released Process for Aluminum Electrostatic Actuators," Journal of Micro-electromechanical Systems, vol. 3, no. 3, Sept. 1994, pp. 90 - 96.

Stowe, T. D., Yasumura, K., Kenny, T. W., Botkin, D., Wago, K., and Rugar, D., "Attonewton Force Detection Using Ultra-Thin Silicon Cantilevers," Applied Physics Letters, vol. 71, no. 2, July 14, 1997, pp. 288 - 290.

Suda, M., Nakajima, K., Furuta, K., Mitsuoka, Y., Sakuhara, T., and Ataka, T., "Electrochemical and Optical Processing of Micro Structures by Scanning Probe Microscopy (SPM)," Proceedings of IEEE International Workshop on Micro Electro Mechanical Systems, San Diego, CA, Feb. 11 - 15, 1996, pp. 296 - 300.

Sugiyama, N., Yamazaki, D., and Ueda, T., "A Quartz Galvanometer for Optical Scanning in a Laser Printer Application," Proceedings of Transducers '91, the 1991 International Conference on Solid-State Sensors and Actuators, San Francisco, CA, June 24 - 27, 1991, pp. 734 - 737.

Suh, J. W., Glander, S. F., Darling, R. B., Storment, C. W., and Kovacs, G. T. A., "Combined Organic Thermal and Electrostatic Omnidirectional Ciliary Microactuator Array for Object Positioning and Inspection," Proceedings of the 1996 Solid-State Sensor and Actuator Workshop, Hilton Head Island, SC, June 3 - 6, 1996, pp. 168 - 173.

Suh, J. W., Glander, S. F., Darling, R. B., Storment, C. W., and Kovacs, G. T. A., "Organic Thermal and Electrostatic Ciliary Microactuator Array for Object Manipulation," Sensors and Actuators, vol. A58, no. 1, Jan. 1997, pp. 51 - 60.

Suzuki, H., Ohya, N., Kawahara, N., Yokoi, M., Ohyanagi, S., Kurahashi, T., and Hattori, T., "Shell-Body Fabrication for Micromachines," Journal of Micromechanics and Microengineering, vol. 5, no. 1, Mar. 1995, pp. 35 - 40.

Tabata, O., "pH-Controlled TMAH Etchants for Silicon Micromachining," Proceedings of Transducers '95/Eurosensors IX, Stockholm, Sweden, June 25 - 29, 1995, vol. 1, pp. 83 - 86.

Tabata, O., Asahi, R., Funabashi, H., and Sugiyama, S., "Anisotropic Etching of Silicon in $(CH_3)_4NOH$ Solutions," Proceedings of Transducers '91, the 1991 International Conference on Solid-State Sensors and Actuators Digest of Technical Papers, IEEE Press, San Francisco, CA, June 24 - 27, 1991, pp. 811 - 814.

Tabata, O., Asahi, R., Funabashi, H., Shimaoka, K., and Sugiyama, S., "Anisotropic Etching of Silicon in TMAH Solutions," Sensors and Actuators, vol. A34, no. 1, July 1992, pp. 51 - 57.

Takagi, T., and Nakajima, N., "Photoforming Applied to Fine Machining," Proceedings of IEEE Micro Electro Mechanical Systems Conference, Fort Lauderdale, FL, Feb. 7 - 10, 1993, pp. 173 - 178.

Takahata, K., Aoki, S., and Sato, T., "Fine Surface Finishing Method for 3-Dimensional Micro Structures," Proceedings of IEEE International Workshop on Micro Electro Mechanical Systems, San Diego, CA, Feb. 11 - 15, 1996, pp. 73 - 78.

Takeshima, N., Gabriel, K. J., Ozaki, M., Takahashi, J., Horiguchi, H., and Fujita, H., "Electrostatic Parallelogram Actuators," Proceedings of Transducers '91, the 1991 International Conference on Solid-State Sensors and Actuators, IEEE Press, San Francisco, CA, June 24 - 27, 1991, pp. 63 - 66.

Tan, S., Reed, M., Han, H., and Boudreau, R., "High Aspect Ratio Microstructures on Porous Anodic Aluminum Oxide," Proceedings of the IEEE Micro Electro Mechanical Systems Conference, Amsterdam, Netherlands, Jan. 29 - Feb. 2, 1995, pp. 267 - 272.

Tang, W. C., Nguyen, T.-C., H., Judy, M. W., and Howe, R. T., "Electrostatic-Comb Drive of Lateral Polysilicon Resonators," Sensors and Actuators, vols. A21 - A23, Feb. - Apr. 1990, pp. 328 - 331.

Taylor, C. S., Cherkas, P., Hampton, H., Frantzen, J. J., Shah, B. O., Tiffany, W. B., Nanis, L., Booker, P., Salahieh, A., and Hansen, R., "'Spatial Forming' A Three Dimensional Printing Process," Proceedings of the IEEE Micro Electro Mechanical Systems Conference, Amsterdam, Netherlands, Jan. 29 - Feb. 2, 1995, pp. 203 - 208.

Taylor, W. P., Schneider, M., Baltes, H., and Allen, M. G., "Electroplated Soft Magnetic Materials for Microsensors and Microactuators," Proceedings of Transducers '97, the 1997 International Conference on Solid-State Sensors and Actuators, Chicago, IL, June 16 - 19, 1997, vol. 2, pp. 1445 - 1448.

Thomas, R. C., Ricco, A. J., Yang, H. C., Dermody, D., and Crooks, R. M., "Chemical Class Specificity Using Self-Assembled Monolayers on SAW Devices," Proceedings of the 1996 Solid-State Sensor and Actuator Workshop," Hilton Head Island, SC, June 3 - 6, 1996, pp. 28 - 31.

Thornton, J. A., and Hoffman, D. W., "Internal Stresses in Titanium, Nickel, Molybdenum, and Tantalum Films Deposited by Cylindrical Magnetron Sputtering," Journal of Vacuum Science and Technology, vol. 14, no. 1, Jan./Feb. 1977, pp. 164 - 168

Toda, R., Minami, K., and Esashi, M., "Thin Beam Bulk Micromachining Based on RIE and Xenon Difluoride Silicon Etching," Proceedings of Transducers '97, the 1997 International Conference on Solid-State Sensors and Actuators, Chicago, IL, June 16 - 19, 1997, vol. 1, pp. 671 - 674.

Tong, L., Mehregany, M., and Matus, L. G., "Silicon Carbide as a New Micromechanics Material," Technical Digest of the Solid-State Sensor and Actuator Workshop, Hilton Head Island, SC, June 22 - 25, 1992, pp. 198 - 201.

Tong, L., Mehregany, M., and Tang, W. C., "Amorphous Silicon Carbide Films by Plasma-Enhanced Chemical Vapor Deposition," Proceedings of the IEEE Micro Electro Mechanical Systems Conference, Fort Lauderdale, FL, Feb. 1993, pp. 242 - 247.

Toshiyoshi, H., Fujita, H., Kawai, T., and Ueda, T., "Piezoelectrically Operated Actuators by Quartz Micromachining for Optical Applications," Proceedings of the IEEE Micro Electro Mechanical Systems Workshop, Fort Lauderdale, FL, Feb. 7 - 10, 1993, pp. 133 - 138.

Trimmer, W., Ling, P., Chin, C.-K., Orton, P., Gaugler, R., Hashimi, S., Hashimi, G., Brunett, and Reed, M., "Injection of DNA into Plant and Animal Tissues with Micromechanical Piercing Structures," Proceedings of the IEEE Micro Electro Mechanical Systems Conference, Amsterdam, Netherlands, Jan. 29 - Feb. 2, 1995, pp. 111 - 115.

Trotter, D. M., Jr., "Photochromic and Photosensitive Glass," Scientific American, vol. 264, no. 4, Apr. 1991, pp. 56 - 61.

Tuckerman, D. B., and Pease, R. F. W., "High-Performance Heat Sinking for VLSI," IEEE Electron Device Letters, vol. EDL-2, no. 5, May 1981, pp. 126 - 129.

Ueda, T., Koshaka, F., and Yamazaki, D., "Quartz Crystal Micromechanical Devices," Proceedings of the Third International Conference on Solid-State Sensors and Actuators, Transducers '85, Philadelphia, PA, June 11 - 14, 1985, pp. 113 - 116.

Vidusek, D. A., "Electrophoretic Photoresist Technology: An Image of the Future-Today," Circuit World, vol. 15, no. 2, Jan. 1989, pp. 6 - 10.

Vondeling, J. K., "Fluoride-Based Etchants for Quartz," Journal of Materials Science, vol. 18, no. 1, Jan. 1983, pp. 304 - 314.

Vossen, J. L., and Kern, W., Thin Film Processes, Academic Press, Inc., New York, NY, 1978.

Wagner, B., Reimer, K., Maciossek, A., and Hofmann, U., "Infrared Micromirror Array with Large Pixel Size and Large Deflection Angle," Proceedings of Transducers '97, the 1997 International Conference on Solid-State Sensors and Actuators, Chicago, IL, June 16 - 19, 1997, vol. 1, pp. 75 - 78.

Walker, J. A., Gabriel, K. J., and Mehregany, M., "Thin-Film Processing of TiNi Shape Memory Alloy," Sensors and Actuators, vols. A21 - A23, Feb. - Apr. 1990, pp. 243 - 246.

Wallenberger, F. T., "Rapid Prototyping Directly from the Vapor Phase," Science, vol. 267, no. 5202, Mar. 3, 1995, pp. 1274 - 1275.

Wallenberger, F. T., and Nordine, P. C., "Strong, Small Diameter, Boron Fibers by LCVD," Materials Letters, vol. 14, no. 4, Aug. 1992, pp. 198 - 202.

Wallis, G., and Pomerantz, D. I., "Field Assisted Glass-Metal Sealing," Journal of Applied Physics, vol. 40, no. 10, Sept. 1969, pp. 3946 - 3949.

Wang, T., Surve, S., and Hesketh, P. J., "Anisotropic etching of silicon in rubidium hydroxide," Journal of the Electrochemical Society, vol. 141, no. 9, Sept. 1994, pp. 2493 - 2497.

Wang, X.-Q., Yang, X., Walsh, K., and Tai, Y.-C., "Gas-Phase Silicon Etching with Bromine Trifluoride," Proceedings of Transducers '97, the 1997 International Conference on Solid-State Sensors and Actuators, Chicago, IL, June 16 - 19, 1997, vol. 2, pp. 1505 - 1508.

Weast, R. C. [ed.], "CRC Handbook of Chemistry and Physics," CRC Press, Inc., Boca Raton, FL, 1988.

Weigold, J. W., and Pang, S. W., "A New Frontside-Release Etch-Diffusion Process for the Fabrication of Thick Si Microstructures," Proceedings of Transducers '97, the 1997 International Conference on Solid-State Sensors and Actuators, Chicago, IL, June 16 - 19, 1997, vol. 2, pp. 1435 - 1438.

Westberg, H., Boman, M., Johansson, S., and Schweitz, J.-Å., "Free-Standing Silicon Micro-structures Fabricated by Laser Chemical Processing," Journal of Applied Physics, vol. 73, no. 11, June 1993, pp. 7864 - 7871.

Westberg, H., Boman, M., Johansson, S., and Schweitz, J.-Å., "Truly Three Dimensional Structures Microfabricated by Laser Chemical Processing," Proceedings of Transducers '91, the 1991 International Conference on Solid-State Sensors and Actuators Digest of Technical Papers, IEEE Press, San Francisco, CA, June 24 - 27, 1991, pp. 516-519.

Wilbur, J. L., Kumar, A., Kim, E., and Whitesides, G. M., "Microfabrication by Microcontact Printing of Self-Assembled Monolayers," Advanced Materials, vol. 6, nos. 7 - 8, July - Aug. 1994, pp. 600 - 604.

Williams, K. R., and Muller, R. S., "Etch Rates for Micromachining Processing," Journal of Microelectromechanical Systems, vol. 5, no. 4, Dec. 1996, pp. 256 - 269.

Willner, A. E., Podlesnik, D. V., Gilgen, H., and Osgood, R. M., Jr., "Ultrafast Aqueous Etching of Gallium Arsenide," Proceedings of the Materials Research Society Symposium on Photon, Beam, and Plasma Stimulated Chemical Processes at Surfaces, Boston, MA, Dec. 1 - 4, 1986, pp. 403 - 410.

Willner, A. E., Ruberto, M. N., Blumenthal, D. J., Podlesnik, D. V., and Osgood, R. M., Jr., "Laser Fabricated GaAs Waveguiding Structures," Applied Physics Letters, vol. 54, no. 19, May 8, 1989, pp. 1839 - 1841.

Winters, H. F., and Coburn, J. W., "The Etching of Silicon with XeF_2 Vapor," Applied Physics Letters, vol. 34, no. 1, 1979, pp. 70 - 73.

Wise, K. D., Jackson, T. N., Masnari, N. A., Robinson, M. G., Solomon, D. E., Wuttke, G. H., and Rensel, W. B., "Fabrication of Hemispherical Structures Using Semiconductor Technology for Use in Thermonuclear Fusion Research," Journal of Vacuum Science and Technology, vol. 16, 1979, pp. 936 - 939.

Wise, K. D., Robinson, M. G., and Hillegas, W. J., "Solid-State Process to Produce Hemisperical Components for Inertial Fusion Targets," Journal of Vacuum Science and Technology, vol. 18, no. 3, Oct. 1981, pp. 1179 - 1182.

Wu, S.-Y., "A Hybrid Mass-Interconnection Method by Electroplating," IEEE Transactions on Electron Devices, vol. ED-25, no. 10, Oct. 1978, pp. 1201 - 1203.

Yablonovitch, E., Gmitter, T., Harbinson, J. P., and Bhat, R., "Extreme Selectivity in the Lift-Off of Epitaxial GaAs Films," Applied Physics Letters, vol. 51, no. 26, Dec. 28, 1987, pp. 2222 - 2224.

Yamagata, Y., Mihara, S., Nishioki, N., and Higuchi, T., "A New Fabrication Method for Micro Actuators with Piezoelectric Thin Film Using Precision Cutting Technique," Proceedings of IEEE International Workshop on Micro Electro Mechanical Systems, San Diego, CA, Feb. 11 - 15, 1996, pp. 307 - 311.

Yasseen, A. A., Smith, S. W., Mehregany, M., and Merat, F. L., "Diffraction Grating Scanners Using Polysilicon Micromotors," Proceedings of the IEEE Micro Electro Mechanical Systems Conference, Amsterdam, Netherlands, Jan. 29 - Feb. 2, 1995, pp. 175 - 180.

Zhang, H., and Kim, E. S., "Dome-Shaped Diaphragm Microtransducers," Proceedings of the IEEE Micro Electro Mechanical Systems Conference, Amsterdam, Netherlands, Jan. 29 - Feb. 2, 1995, pp. 256 - 260.

Zhang, Z. L., and MacDonald, N. C., "An RIE Process for Submicron, Silicon Electro-Mechanical Structures," Proceedings of Transducers '91, the 1991 International Conference on Solid-State Sensors and Actuators Digest of Technical Papers, IEEE Press, San Francisco, CA, June 24 - 27, 1991, pp. 520 - 523.

Zhang, Z. L., and MacDonald, N. C., "A RIE Process for Submicron Silicon Electromechanical Structures," Journal of Micromechanics and Microengineering, vol. 2, no. 1, Mar. 1992, pp. 31 - 38.

Zhang, Z. L., and MacDonald, N. C., "Fabrication of Submicron High-Aspect-Ratio GaAs Actuators," Journal of Microelectromechanical Systems, vol. 2, no. 2, June 1993, pp. 66 - 73.

CITED INDUSTRY REFERENCES AND SUPPLIERS

Source of Unusual Silicon Substrates

Virginia Semiconductor, 1501 Powhatan St., Fredericksburg, VA 22401, Phone: (703) 373-2900.

Sources of Glass and Quartz Wafers

Hayward Quartz, 4190 Technology Drive, Fremont CA 94538, Phone: (510) 657-9605, Fax: (510) 657-6404.

Hoya Electronics, 960 Rincon Circle, San Jose, CA 95131, Phone: (408) 435-1450.

Source of Xenon Difluoride

Harris Specialty Chemicals, PCR Division, Inc., P.O. Box 1466, Gainsville, FL 32602, Phone: (352) 376-8246.

Source of Photosensitive Polyimides

Olin Microelectronic Materials, Inc., 1025 S. 52nd St., Tempe, AZ 85281, Phone: (602) 829-1010.

Source of Piezoelectric Powders and Other Ceramics

Morgan Matroc Ltd., Bewdley Road Stourport-on-Severn, Worcestershire DY13 8QR England, Phone: 01299 827000, Fax: 01299 827872, http://www.morganmatroc.com/

Source of Thin-Film Products and Services

American Technical Ceramics, Inc., One Norden Lane, Huntington Station, NY 11746-2102, Phone: (516) 547-5700, Fax: (516) 547-5748.

Precision Machining

Sherline Products, Inc., 170 Navajo Street, San Marcos, CA 92069-2593, Phone: (619) 744-3674, Fax: (619) 744-1574, http://www.sherline.com/sherline

Source of Electroplatable Photoresist

Shipley, Inc., 455 Forest St., Marlborough, MA 01752-3001, Phone: (508) 481-7950, Fax: (508) 485-9113.

1. INTRODUCTION

Numerous micromachined mechanical transducers have been developed or demonstrated, and many of the truly "micromachined" sensor devices being shipped today are mechanical (as opposed to sensors based on conventional integrated circuits that, through inherent mechanisms, sense light, temperature, etc.). This chapter focuses on the phenomena that can be sensed or acted upon with micromachined mechanical devices, but the reader should be aware that important discussion of mechanical transduction is presented in other chapters, since it appears as an intermediate mechanism in many devices. Also, some categories of mechanical devices, such as valves, flow restrictors, pumps, etc., are covered in the separate Microfluidic Devices chapter.

An important addition to "conventional" uses of common semiconductors (mainly silicon) began in the 1970's as some engineers realized that silicon possessed remarkable mechanical properties. The first mechanical uses of silicon were in the area of strain gauges, where their extremely high sensitivities compared to competing methods quickly won them market share. This was followed by the gradual development of a large variety of applications, and today the dominant micromachined mechanical transducers are pressure sensors and accelerometers. Petersen (1982) provides an excellent overview of the wide variety of silicon applications in mechanical devices that had already been studied by the early 1980's. (This classic review article is probably the most cited paper in the field of micromachining.)

This chapter begins with a concise review of basic mechanics, followed by a discussion of basic mechanisms. The remainder of the chapter progresses through sensing mechanisms, sensor designs, actuation mechanisms, and specific actuators, including resonators and electrical relays.

2. BASIC MECHANICS

This section is provided to familiarize the reader with some of the basic concepts and terminology of mechanical design and analysis. It is not intended to replace a thorough review of mechanics, and one is encouraged to refer to Gere and Timoshenko (1990) or Popov (1968). However, as a convenient reference, and as an aid to understanding the mechanical scaling laws applicable to microstructures, some of the most relevant concepts have been included.

2.1 AXIAL STRESS AND STRAIN

When a force is applied to a surface, that surface is said to be *stressed*. The average value of this stress equals the loading force, F, divided by the area, A, over which it is applied,

$$\sigma = \frac{F}{A} \quad \text{(typically in N/m}^2\text{, or Pa)}$$

Forces that act perpendicular to the surface are called *axial* or *normal* forces and produce axial or normal stresses. By convention, tensile forces, which pull on a surface, are given a positive sign. Compressive stresses, which push on a surface, are negative. It should be noted that σ is used to represent normal strain and τ to represent shear stress, as discussed below.

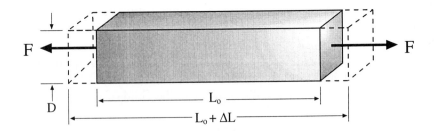

Illustration of the relationship of an applied force to a beam and the resulting dimensional change. Note that the assumption here is that there is no resultant dimensional change in D, which is generally not the case, as explained below. The strained beam is longer by ΔL from its original length, L_o. Courtesy K. Honer.

When subjected to a stress, materials literally get pushed (or pulled) out of shape. *Strain*, ε, is a measure of this deformation (within the elastic limit for a material), and equals the change in length, ΔL, divided by the original length, L_o, of an object,

$$\varepsilon = \frac{\Delta L}{L_o}$$

(It should be noted that the term *microstrain* is often used, and simply refers to $\varepsilon \times 10^{-6}$. This is convenient since in many practical situations, strains are in the 1 to 100 microstrain range.)

Most materials of interest obey Hooke's law; that is, they deform linearly with load. Since load is proportional to stress and deformation is proportional to strain, stress and strain are linearly related. The proportionality constant that relates them is known as the *elastic modulus* or *Young's modulus* of a material, and is usually given the symbol, E,

$$E = \frac{\text{stress}}{\text{strain}} = \frac{\sigma}{\varepsilon} \quad \text{(typically in N/m}^2\text{)}$$

This parameter is relatively constant below the plastic flow point of a given material. The higher the elastic modulus of a material, the less it deforms for a given stress, and thus the stiffer it is. For example, an incompressible material (i.e., one that would not deform under any stress) would have an infinite Young's modulus,

$$E = \lim_{\frac{\Delta L}{L} \to 0} \frac{\frac{F}{A}}{\frac{\Delta L}{L}} = \infty$$

while a "soft" material would deform considerably for a given amount of stress, so its modulus of elasticity would be quite small.

For example, E for Si is 190 GPa (1 Pa = 1 n/m²), 73 GPa for SiO_2 (quartz), and 1,035 GPa for diamond. The Young's moduli of several other materials used in micromachining are given in the Micromachining Techniques chapter. It should be pointed out that the elastic moduli for crystalline materials is dependent on their orientation. However, the calculation of orientation-dependent moduli is beyond the scope of this book and one can usually use average values for initial calculations to good effect. In addition, for most materials, the moduli for tension and compression are equal, but their maximum values of stress in these states may not be equal.

2.2 SHEAR STRESS AND STRAIN

Shear stress is stress due to force applied in parallel to surfaces of an object, as opposed to normal force applied to surfaces for axial stress. Shear stress is denoted by the symbol, τ, to distinguish it from axial stress,

$$\tau = \frac{F}{A} \quad \text{(typically in N/m}^2\text{)}$$

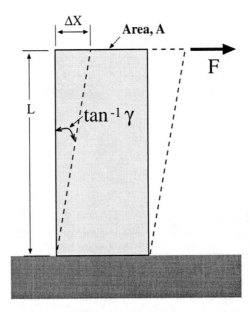

Illustration of shear strain, γ, of a rectangular element resulting from a load parallel to the top surface. In order to balance the load on the element, the support anchor is also applying a load, F, in the opposite direction, as well as vertical forces directed up on the right side and down on the left (not shown). Courtesy K. Honer.

Shear strain, γ, is slightly different from axial strain. Whereas axial strain is a measure of linear deformation, shear strain can be thought of as related to an angle that a deformed element's sides make with respect to its original shape (as illustrated above). As in the axial case, shear strain is linearly proportional to shear stress. However, the proportionality constant is different. This new constant, G, is called the *shear modulus* of elasticity,

$$G = \frac{\text{shear stress}}{\text{shear displacement angle (rad)}} = \frac{\tau}{\gamma} = \frac{\frac{F}{A}}{\frac{\Delta X}{L}} \quad \text{(typically in N/m}^2\text{)}$$

For isotropic materials (those having identical properties in every direction, which is not the case for most single-crystal materials), the shear modulus, G, is related to the elastic modulus, E, by,

$$E = 2G(1+\mu) = 3K(1-2\mu)$$

where μ is *Poisson's ratio* and K, the *bulk modulus*, is defined as the ratio of hydrostatic stress to volume compression,

$$K = \frac{\text{hydrostatic stress}}{\text{volume compression}} = \frac{\frac{F}{A}}{\frac{\Delta V}{V}} \quad \text{in N/m}^2$$

The bulk modulus of a material represents its volume change under uniform pressure. In general, solids are less compressible (larger K) than liquids due to their rigid atomic lattices (e.g., for water, $K = 2.0 \times 10^9$ N/m^2, for aluminum, $K = 7 \times 10^{10}$ N/m^2, and for steel, $K = 14 \times 10^{10}$ N/m^2 (from Giancoli (1989)).

2.3 POISSON'S RATIO

When a material is subjected to an axial load, it deforms in the direction of the load. However, it may also deform in directions perpendicular to the load as shown in the figure below. When subjected to a tensile load, the length of an object will typically increase and its girth will decrease. When compressed, its length will decrease and its girth will increase.

In this situation, there are two strains, one axial (ε_l) and one transverse (ε_t),

$$\varepsilon_a = \frac{\Delta L}{L_o} \quad \text{and} \quad \varepsilon_t = \frac{\Delta D}{D_o}$$

and as illustrated below, the longitudinal strain is tensile and the transverse strain is compressive (ε_l and ε_t will usually be of the opposite sign). Note that the sign convention is $\varepsilon_l > 0$ for tension and $\varepsilon_t < 0$ for compression. *Poisson's ratio* is the ratio of the transverse strain to the axial strain,

$$\nu = \frac{\text{transverse strain}}{\text{longitudinal strain}} = -\frac{\varepsilon_t}{\varepsilon_a} = -\frac{\dfrac{\Delta D}{D_o}}{\dfrac{\Delta L}{L_o}}$$

and ν (or μ) is always defined as a positive value (as noted above, ε_l and ε_t will usually be of the opposite sign). Typical values of Poisson's ratio are 0.2 to 0.5 for most materials. For most metals, Poisson's ratio is ≈ 0.3. Rubbers have a Poisson's ratio closer to 0.5, which corresponds to a conservation of volume. Cork has a Poisson's ratio near zero, which is why it was chosen as a plug for wine bottles (i.e., it will not expand when strained during insertion).

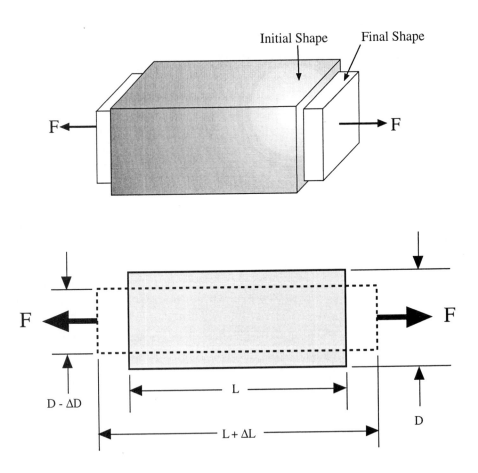

Illustration of Poisson's ratio. In addition to straining in the direction of the load, a material also deforms in directions perpendicular to the load. Courtesy K. Honer.

2.4 COMMONLY USED DEFLECTION EQUATIONS FOR MICROSTRUCTURES

Most micromechanical structures are based on deflection of rectangular cross-sectional beams, torsion of rectangular cross-sectional beams, or deflection of membranes. This is to be expected due to the fact that the thin films from which the structures are fabricated tend to take on such cross sections when patterned. Naturally, the degree to which they are actually "rectangular" depends on the anisotropy of the etch used to form them, but to a first approximation, the rectangular assumption is usually satisfactory. The following subsections present the most common equations for linear deflection of these three basic structures.

2.4.1 STATIC BEAM EQUATIONS

Many common microstructures are based on deflection of beams. They are used in the suspension of rigid plates or by themselves as cantilever devices and are a natural choice for bearing-less motion. For the standard beam equations to be valid, the beams must have lengths at least an order of magnitude greater than their other dimensions and the deflections must be small relative to their lengths as well.

Most micro-beams are constrained at one or two points and subjected to either a point load or an evenly distributed load. The equations that describe the load versus deflection of a beam for these boundary conditions are shown below, for a point load, F, in N, or a distributed load, ρ, in N/m.

Point Load	Distributed Load

$$y(x) = \frac{F}{6EI}(3x^2L - x^3)$$

$$y(x) = \frac{\rho x^2}{24EI}(6L^2 - 4Lx + x^2)$$

$$\sigma_{max} = \frac{FLt}{2I}$$

$$\sigma_{MAX} = \frac{\rho L^2 t}{4I}$$

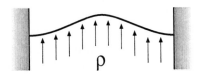 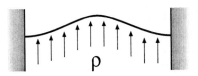

$$y(x) = \frac{Fx}{48EI}(3Lx - 4x^2)$$

$$y(x) = \frac{\rho x^2}{24EI}(L - x)^2$$

{for $x \le L/2$ away from a support}

$$\sigma_{MAX} = \frac{FLt}{8I}$$

$$\sigma_{MAX} = \frac{\rho L^2 t}{12I}$$

where,
L = length of the beam, in m
t = thickness of the beam, in m
I = bending moment of inertia, which for a beam of a rectangular cross section is given by,

$$I = \frac{1}{12}wt^3 \quad (\text{in } m^4)$$

where w is the width of the beam, in m. It is important to note that no matter what the boundary conditions or loading, the deflection is proportional to the load. This is a consequence of a linear material and small deflections. The same is true for torsional structures and membranes.

2.4.2 STATIC TORSION EQUATIONS

In addition to bending, beams may be twisted about their axes. This is known as *torsion* and is illustrated in the figure below.

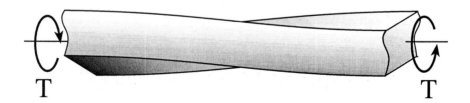

Illustration of torsion of a rectangular cross-sectional beam about its primary axis. Courtesy K. Honer.

The constitutive equation for torsional structures deflecting through an angle θ (in radians) is,

$$\theta = \frac{TL}{KG}$$

where T is the applied torque, and K is a constant depending on geometry.

For a circular beam, K is given by,

$$K = \frac{1}{2}\pi r^4$$

where r is the radius of the beam.

For a rectangular beam of dimensions x_0 and y_0, K is given by,

$$K = \frac{x_o y_o^3}{16}\left[\frac{1}{3} - 0.21\frac{y_o}{x_o}\left(1 - \frac{y_o^4}{12x_o^4}\right)\right] \quad \text{for } x_0 > y_0$$

There are equations in the literature for other shapes. Useful references are Hopkins (1987), Roark and Young (1989), Avallone and Avallone (1996), and Gere and Timoshenko (1990).

2.4.3 STATIC PLATE EQUATIONS

Plates are similar to beams except that plates have widths comparable to their lengths and are typically much thinner than their lengths or widths. In addition, in plates there may be stress gradients in the z-direction (thickness). For these structures, Poisson's ratio becomes important as the lateral strain serves to stiffen the plates. Thus a plate will have less curvature than a beam under equivalent load — approximately $(1 - v^2)$ as much for a plate bending in only one direction. One way of looking at this is that for a simple beam being deflected downward, material closer to the bottom surface of the beam will expand, while the upper part will contract. A plate can be considered to be an array of parallel beams fused together, and their interaction opposes these dimensional changes, resulting in greater resistance to bending.

Membranes are generally thinner than plates and typically have fixed (constrained) boundaries (e.g., a drumhead). The stress in the z-direction is typically uniform (not always the case in microstructures). Membranes usually deflect in two directions. Under uniform pressure, P, in N/m², a circular membrane will deflect as follows:

$$\delta_{max} = \delta_{center} = \frac{3Pr^4(1 - v^2)}{16Et^3}$$

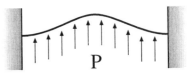

The above equation assumes that the membrane is under zero initial stress. This is usually not the case. Residual tensile stresses will tend to reduce the maximum deflection, while residual compressive stresses may result in buckling even in the absence of a load. Micromachined membranes are usually intentionally under moderate tensile stresses.

2.5 DYNAMICS

In the above sections, the loads on the structures were assumed to be constant over time, and no such assumption is made here. The basics of dynamic response for microstructures is described to illustrate that a static analysis may be insufficient to ensure a safe design.

Any lumped mechanical structure can be modeled as a simple mass on a spring. For this case, the governing equation is,

$$m\frac{d^2x}{dt^2} + b\frac{dx}{dt} + kx = F_{external}$$

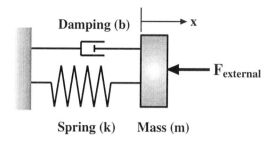

where,

m = mass, in kg
b = damping coefficient, in (N•s)/m
k = spring constant, in N/m
$F_{external}$ = applied force, in N = (kg•m)/s^2

Damping arises from any losses that the system experiences when it moves. It can be due to aerodynamic drag of the mass or from small plastic deformations taking place in the spring, among other things. The external load can be any force, such as gravity, electrostatic attraction, or thermal expansion. If, however, $F_{external}$ has the form of a sine wave, the movement of the mass will also be sinusoidal at the same frequency (since this is a linear system). The magnitude and relative phase of this sine wave is a strong function of frequency for such a second-order system (a typical log-log plot of this dependence, called a *Bode plot*, is shown below).

Analyzing this second-order system, one obtains a second-order (low-pass filter) force-to-displacement response of the form (where s is the Laplace variable),

$$H(s) = \frac{\dfrac{1}{m}}{s^2 + \left(\dfrac{b}{m}\right)s + \dfrac{k}{m}}$$

The electrical analog of this system is an RLC circuit, as illustrated below. This circuit has an electrical transfer function (volts-to-volts) given by,

$$H(S) = \frac{V_o}{V_i} = \frac{\dfrac{1}{LC}}{s^2 + \left(\dfrac{1}{RC}\right)s + \dfrac{1}{LC}} = \frac{\omega_o^2}{s^2 + \dfrac{\omega_o}{Q}s + \omega_o^2}$$

with a natural frequency and quality factor, Q, (a measure of how "lossless" the system is in terms of energy — high Q systems tend to resonate for long periods of time once energized) given by,

$$Q = \frac{\text{energy stored per cycle}}{\text{energy lost per cycle}} = \omega_o RC \quad \text{where} \quad \omega_o = \frac{1}{\sqrt{LC}}$$

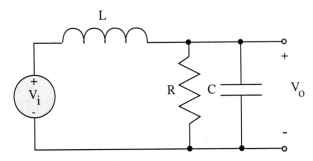

Schematic of the electrical RLC analog of a second-order mechanical system.

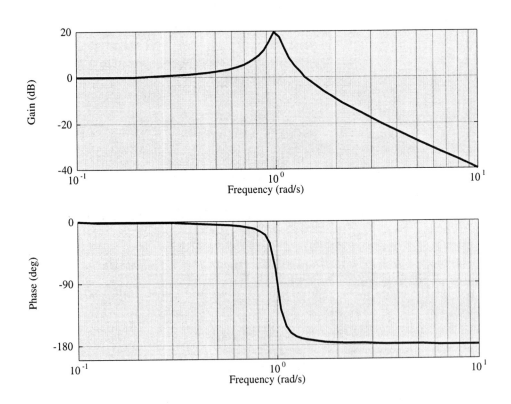

Typical gain and phase response plots (a Bode plot) for a second-order system.

Three important parameters for a second-order system are DC gain, natural or resonant frequency, and quality factor. These can be calculated for mechanical and electrical systems as shown in the table below.

Parameter	Mechanical System	Electrical System
DC Gain	$\dfrac{1}{k}$	1
Natural Frequency, ω_o	$\sqrt{\dfrac{k}{m}}$	$\dfrac{1}{\sqrt{LC}}$
Quality Factor, Q	$\omega_o \dfrac{m}{b} = \sqrt{\dfrac{km}{b^2}}$	$\sqrt{\dfrac{R^2C}{L}}$

Table showing key parameters for mechanical and electrical second-order systems.

While not discussed in detail here, it should be noted that for torsional systems, there is a similar governing equation,

$$I\frac{d^2\theta}{dt^2} + b\frac{d\theta}{dt} + k\theta = T_{external}$$

where I is the second moment of inertia.

For frequencies below the natural frequency, the system will respond with close to the DC gain. At the natural frequency, the response is Q times the DC response and is 90° out of phase with the input force (or voltage). At frequencies above the natural frequency, the response falls off by 40 dB per decade and phase approaches a phase of -180°.

Tuning forks have a very high Q, which is why they resonate at predominately one frequency. This is good for deliberate resonators, as discussed in Section 7, but it brings up a potential material failure problem for other mechanical structures. A comprehensive static analysis, assuming a maximum force, may indicate a safe amount of deflection and hence stress. However, at resonance the deflection, and hence stress, are Q times as large as at steady state. Increasing the damping of the system helps to ensure that out of control oscillations do not cause the structure to exceed its maximum stress levels.

2.6 THERMAL NOISE

As discussed above, damping is desirable unless a structure should deliberately resonate. Unfortunately, anywhere damping is present there is also noise. This is directly analogous to the electronic case, wherein pure LC circuits (infinite resistance) can have infinite Q in theory. In practice, resistive damping limits Q quite severely. These electrical resistances all generate Johnson noise (thermal noise), which is flat in spectral density ("white"), but is generally shaped by the transfer function of the circuit it is in.

In the case of mechanical noise, the molecules in any material at a temperature above absolute zero are constantly vibrating. This vibration produces small random motions in microstructures. When the position of a microstructure in a sensing mechanism, this random vibration shows up as noise analogous to Johnson noise in electronic circuits. The equipartition theorem states that the value of this noise is,

$$\frac{1}{2}k\langle x^2 \rangle = \frac{1}{2}m\langle v^2 \rangle = \frac{1}{2}k_b T$$

where,

$\langle x^2 \rangle$ = mean squared average displacement

$\langle v^2 \rangle$ = mean squared average velocity

k_b = Boltzmann's constant = 1.38066×10^{-23} J/K (note that here k_b is used, rather than k, to avoid confusion with the spring constant, k)

T = temperature in K

(Note that this is an idealized model, derived based on the assumption of a thermally driven, statistical ensemble of non-interacting particles, e.g., a gas. In this case, it should be treated as an approximation.)

In terms of the spectral density of the noise force, one can use Nyquist's relation to obtain,

$$F_{noise} = \sqrt{4k_b bT} \quad (\text{in } \frac{N}{\sqrt{Hz}})$$

where simply replacing the damping, b, with the electrical resistance, R, allows the formula to be used for electrical noise. In fact, the same techniques used to model circuits can be used to model noise in mechanical structures by taking advantage of the parallelism between mechanical structures and circuits. The output noise of a mechanical system is white noise filtered by the force-to-displacement response of the system and the electrical transfer function in the electronic case. Noise issues are discussed in more detail in Section 7 below, and the reader is also referred to Gabrielson (1993, 1995) and Motchenbacher and Connelly (1993).

3. MECHANICAL PROPERTIES OF MATERIALS

3.1 MATERIAL FAILURE

As indicated above, when a material is stressed, it undergoes a strain proportional to the stress. This is true up to a point. If the stress is above the *yield stress*, the material will deform significantly more and will be permanently deformed when the stress is removed. If, however, the stress is above the *ultimate stress*, the material will fail completely and break into separate sections if not constrained.

For some materials, such as many metals, the ultimate stress is significantly greater than the yield stress. These materials will bend before they break and are referred to as *ductile*. Materials such as silicon, however, have ultimate stresses that are virtually the same as their yield stresses. These materials will break suddenly and without warning. These materials are *brittle*. The table below shows the yield strengths for some common materials, with more examples provided in the Micro-machining Techniques chapter.

Material	Yield Strength (MPa)
Al	170
Steel	2,100
W	4,000
Si	7,000
Quartz (SiO_2)	8,400
Diamond	53,000

Brief table of example yield strengths of materials. From Petersen (1982).

Determining when a structure will fail is not as simple as summing up the contributions from all loads and checking to see if they exceed the yield stress. For example, some materials fail at lower stresses in shear, others in tension, making the sign of the applied load particularly important. However, checking to make sure stresses are below the yield stress is a good way to eliminate grossly erroneous designs quickly.

A material can also fail even if it is never exposed to a load greater than its yield strength. Over time, cyclic loads can cause *fatigue* at stresses significantly below yield values. When a material fatigues, tiny cracks appear as a result of local stress and grow with each cycle until the material breaks. For some materials used in micromachined structures, fatigue failures have been documented and for others, such as the < 100 nm metal films used in various mechanical light modulators, large strains can be applied without failure over trillions of cycles (this is discussed in the Optical Transducers chapter). General rules for the fatigue properties of microscale materials are not yet clear, but several groups are actively investigating these issues. Connally and Brown (1991, 1992) and Brown, et al. (1993, 1997) used deliberate stress concentration regions on resonators to show that the presence of moisture during mechanical cycling had a major influence on crack growth and failure. The effect of stress level on fatigue life was also demonstrated, consistent with intuition from macroscopic devices.

3.2 GENERAL MATERIALS CONSIDERATIONS

It is critical to note that the commonly available tabulated values of mechanical properties of materials are generally derived from bulk specimens, and thus may not be very relevant to the materials and scales used in micromachined devices. For example, one can find wide variations in published values of parameters such as Young's moduli for polysilicon, and these are likely to be different due more to the process-specific nature of polysilicon properties than to experimental errors. Also, as mentioned above, crystalline materials have anisotropic mechanical properties. Even if the properties were consistent, since the Young's moduli for many of the materials of interest are very high, there are practical issues that interfere with measurements of these materials, such as compliance of measuring tools.

Micromachined structures are often laminates. In macroscopic composites, it is well known that the overall properties are not simply a linear interpolation between the properties of the individual constituent materials (e.g., fiberglass behaves neither like glass fibers nor like epoxy, nor like anything in between). This is a key concept to keep in mind when analyzing the mechanical (and other) properties of microstructures.

There will be considerable ongoing benefit to more basic science work in the area of mechanical and other properties of thin-film and small-scale materials as the micromachining industry continues to expand. Such efforts have, to some extent, been hampered by the fact that process parameters greatly influence these properties and may be difficult to reproduce on one set of equipment, let alone on the equipment of others. In a great many cases, improved knowledge of material properties would greatly reduce the amount of "cut-and-try" engineering that is common in micromachining endeavors despite the emergence of modeling tools.

3.3 MECHANICAL CHARACTERIZATION OF THIN-FILMS

For many micromachined transducers and structures, the mechanical properties of applied thin-films can be critical to their success, as discussed in Ristic, et al. (1994). For example, stresses or stress gradients can cause thin-film structures (such as surface micromachined polysilicon or metal devices) to warp to the point that they are useless. It is often also necessary to determine other mechanical properties such as Young's moduli, Poisson's ratios, etc. This section is concerned with direct and indirect means of measuring or estimating these parameters.

3.3.1 STRESS MEASUREMENT

It should be noted that stresses can be due to the deposition process(es) themselves or to the post-deposition process history of a thin-film. With appropriately designed experiments, it is usually possible to determine the stage at which such stresses arise.

The typical macroscopic method for measuring stresses in thin-films is to quantify optically the differential "bow" or curvature of a wafer before and after the deposition of the film. Commercial instruments that operate on this principle can provide stress resolutions in the MPa range without difficulty. A useful discussion of this technique is provided in Flinn (1988). Microscale methods for measuring such stresses can make use of a variety of methods, including deflection of thin-film structures freed by removal of a sacrificial layer, external mechanical probing of free regions of a film, pressure-based membrane-deflection methods, or resonant frequency measurement.

UNIFORM STRESSES

For films wherein the stress is uniform in the direction perpendicular to the substrate surface, a free region of the film will relax to a non-stressed state, with an accompanying dimensional change.

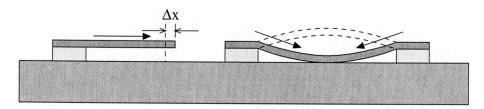

Illustration of (left) dimensional change (expansion) of a simple cantilever with a uniform, compressive residual stress when released, and (right) the buckling of a doubly supported beam in the same circumstances.

For a simple cantilever, the length, L, will change an amount ΔL, and this can potentially be measured directly relative to a fixed landmark or vernier on the substrate, as demonstrated by Fan, et al. (1988). If the Young's modulus of the film, E, is known, the stress, ε, can be estimated (since ΔL/L is the strain, σ) using,

$$\varepsilon = \frac{\sigma}{E}$$

As discussed in Ristic, et al. (1994), this type of measurement is difficult with films having large Young's moduli since the strains are very small. However, one advantage of this approach is that both compressive and tensile stresses can be measured. Test structures that geometrically amplify the strains somewhat can be used in some cases, but the total deflections still generally remain small for reasonable device sizes (see Allen, et al. (1987) and Mehregany, et al. (1987)). If both high-resolution lithography and electron microscopy are available, such measurement approaches may be useful for stress estimates.

For estimating compressive stress, a simple technique is to fabricate doubly supported beams of various lengths and determining which of them have buckled at a given stress level. As demonstrated by Guckel, et al. (1985), the stress can be estimated in this case using the equation for the critical buckling strain,

$$\varepsilon_{CR} = \frac{\pi^2 t^2}{KL^2}$$

where t is the beam thickness, in m, and K is a constant with value $3 < K < 12$, where the low and high bounds represent a doubly supported beam and a cantilever beam, respectively. In this case, the constant K needs to be determined, as well as the Young's modulus, in order to obtain a stress value from the estimated strain.

For estimating tensile stress, the "ring-and-beam" test structure (illustrated below) introduced by Guckel, et al. (1988a, 1988b) and Guckel and Burns (1990) can be used. When released from the substrate so that the structure is supported at the ends of two supporting members, the film will relax to relieve residual tensile stress by contracting. As the ring contracts, the central beam will be compressed and, if the compressive forces are sufficient, will buckle. The critical value of strain required for this buckling is given by,

$$\varepsilon_{CR} = \frac{1}{3G} \left(\frac{\pi t}{2 R_{CR}} \right)^2$$

where R_{CR} is the ring radius, in m, and G is the geometry-dependent ratio of strain in film to strain in crossbar. With a scaled array of cantilever and/or ring-in-beam test structures, the approximate value of the residual film stress can be determined by observing (typically by optical microscopic methods, such as interference microscopy) which length beams have buckled.

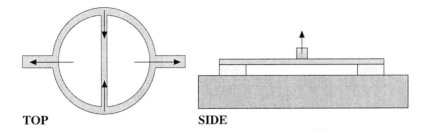

TOP   **SIDE**

Illustration of top and side views of "ring-and-beam" test structures for measuring tensile residual stresses. The geometry used converts tension on the ring into compressive forces on the central beam, causing it to buckle. After Guckel, et al. (1988a, 1988b) and Guckel and Burns (1990).

NONUNIFORM STRESSES (STRESS GRADIENTS)

While it is generally possible to design microstructures that are tolerant of uniform residual stresses, the more serious problem is stress *gradients*. During processing (deposition, thermal cycling, etc.), such gradients can be formed in thin films. Such variations in stress lead to internal bending moments that cause the films to deform ("curl") when released (converting the stress gradient into a strain gradient). As described in Fan, et al. (1990), the net bending moment is given by,

$$M = \int_0^t \sigma(y) \, w \left(y - \frac{t}{2} \right) dy$$

where,
  y = height above the bottom of the film, in m
$\sigma(y)$ = stress at height y, in N/m^2
  w = width of structural member in which the moment is present, in m
  t = thickness of film, in m

In practice, when such gradients are present, they are generally seen as upward or downward curvature of released structures, and if severe enough, they may preclude their use.

For a simple cantilever, the vertical deflection at a distance x from the support, $\delta(x)/x$, is given by (Roark and Young (1989)),

$$\frac{\delta(x)}{x} = K + \frac{\left(1 - v^2\right)}{2EI} Mx$$

where,

K = constant determined by boundary conditions at the support end

E/(1 - v^2) = biaxial modulus of the film (compensating Young's modulus for stiffening of the beam during stretching)

I = moment of inertia of the cantilever about the z-axis, in m^4

M = internal bending moment defined above, in N•m

Such deflections can be measured using optical and other microscopic means, and can theoretically be used to estimate the value of M.

Fan, et al. (1990) proposed the use of Archimedian spirals as test structures for measuring the average values of such gradients.

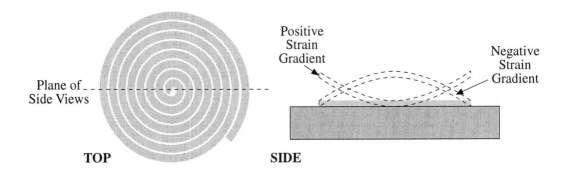

Illustration of a spiral test structure for stress gradients, showing (at right) views of the relaxed states of the spiral for positive and negative strain gradients (viewed as slices through the spiral along the plane indicated at left).

If the stress gradient is positive (increasingly tensile with increasing height above the substrate), the spiral will assume an open bowl shape. If the stress gradient is negative (increasingly compressive with increasing height above the substrate), the spiral will assume a dome shape. Note that to cover stress gradients of both polarities, test spirals with anchors to the substrate at the center and outermost turn must be included so that their relaxed states are above the substrate. In theory, one can also extract information as to the magnitude of the strain gradient by measuring the height of the free end of the spiral, its rotation, and contraction of the diameter of the spiral (unfortunately, these parameters may be quite tedious to measure). In practice, however, spiral structures are most convenient as indicators of the presence and sign of significant residual stress gradients. For further information on stress gradient effects on a variety of thin-film microstructures, the reader is referred to Howe (1994).

3.3.2 MEASUREMENT OF OTHER MECHANICAL PROPERTIES

DIRECT MEASUREMENT

As is commonly done for macroscopic specimens, key mechanical properties of materials (Young's modulus, Poisson's ratio, and tensile strength) can be measured directly by applying carefully controlled, uniform stress to a specimen and measuring the resulting dimensional changes. By measuring the resulting dimensional changes in the directions of, and perpendicular to, the applied stress, the Young's modulus and Poisson's ratio can be computed.

Sharpe, et al. (1996, 1997) reported on a relatively simple technique for measuring these parameters for thin-film specimens. Their approach is based on the fabrication of a released, doubly supported cantilever beam of the thin-film to be tested. Markers of different reflectivity from the thin-film under test are fabricated on the beam, and laser interferometry is used to measure their displacements relative to each other. As described in Sharpe, et al. (1996, 1997), test specimens were fabricated by depositing the test layer on a silicon wafer and back-side etching to form the released, doubly supported membrane. In this example, the authors used two layers of polysilicon, one 2.0 μm thick and the other 1.5 μm thick, both deposited by LPCVD and annealed at 1,050°C for 1 hour to reduce the grain size and stress. Prior to deposition of the polysilicon, the substrates were coated with silicon nitride and PSG (phosphosilicate glass) on both sides. To release the beams, the wafers were cut into individual dice, the back-side polysilicon and PSG were removed, the back-side silicon nitride was patterned, the top side of the wafer was protected by waxing it to a carrier, and the silicon bulk was etched using an unspecified wet etchant. With the bulk etching complete, the PSG beneath the polysilicon test film was removed using HF and the die was removed by dissolving the wax in an organic solvent. Key to practical handling of the final dice was the inclusion of support strips of bulk silicon that prevent forces from being applied to the test beam before it is mounted in the stress test assembly. Once a test die was mounted (with adhesive), the support strips were cut using a diamond saw.

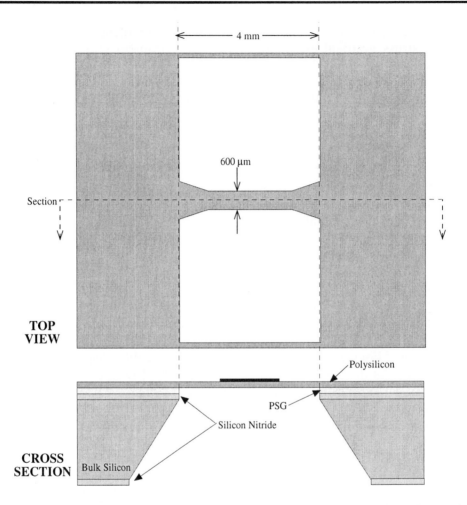

Illustration (not to scale) of a bulk micromachined, doubly supported cantilever with removable support strips (at top and bottom of top view) for handling prior to stress testing. Adapted from Sharpe, et al. (1997).

The measurements were made using an air-bearing supported mechanism with a piezoelectric translator and a load cell. Movement of optical markers on the test specimen's surface was detected by measuring the changes in fringe patterns from laser light diffracted from each edge of a marker on the polysilicon's surface. It should be noted that although Sharpe, et al. (1997) reported only on the use of this direct method on polysilicon, it could, in principle, be applied to any thin-film material from which a doubly supported cantilever could be fabricated.

Sharpe, et al. (1997) also present an excellent summary of the mechanical properties of polysilicon, with average values that are in agreement with many previously reported results. The reader should be warned that for any such measurements, process-, doping-, and thermal-history-dependent variations must be considered when making comparisons. The results Sharpe, et al. (1997) presented were derived from polysilicon fabricated using the Microelectronics Center of North Carolina's "MUMPS" polysilicon foundry process. Those values were Young's modulus = 169 ± 6.15 GPa, Poisson's ratio = 0.22 ± 0.011, and tensile strength = 1.20 ± 0.150 GPa. It should be noted that their paper also summarizes similar results from 13 other references and brief comments on the methods used. An excellent discussion of the structural and mechanical properties of polysilicon deposited using various methods can be found in Krulevitch (1994).

INDIRECT MEASUREMENT

As alternatives to direct measurement of mechanical properties of thin films, several indirect methods are available. Several of these are reviewed in Sharpe, et al. (1997) and Ristic, et al. (1994). For example, if a cantilever can be electrostatically (easiest for conductive films) or photothermally driven (e.g., with a laser beam) with a sinusoidal waveform and its vibration amplitude is measured (typically using a deflected laser beam), a plot of amplitude versus frequency will show a characteristic resonance at a frequency f_R. From this information, the Young's modulus of the film can be estimated using,

$$E = \frac{2\pi f_R^2 A L^4 \rho}{3.52 I}$$

where,
A = cross-sectional area, in m^2
L = cantilever length, in m
ρ = film density, in kg/m^3
I = moment of inertia of the cantilever, in m^4

A doubly supported beam can also be used in a similar fashion, with the resonant frequency determined by the tension in the beam.

4. BASIC MECHANISMS AND STRUCTURES

As a building block for fabricating sensors or machines on a microscale, mechanisms are often necessary to couple energy between actuators and sensors and/or the outputs of a mechanical system. A useful definition of "mechanism" (from Mehregany, et al. (1988) is the "means for transmitting, controlling, or constraining relative movement." It is possible to make mechanisms such as joints, linkages, gears, sliders, hinges, etc., using micromachining techniques.

The three major process examples discussed below all rely on some combination of oxide or PSG sacrificial layers and polysilicon. This reflects the fact that most of the micromachined mechanisms to date have been fabricated using surface micromachining. It would be more difficult to use such bulk methods to fabricate, for example, rotating structures.

4.1 IN-PLANE ROTARY MECHANISMS

An interesting example of early surface micromachined mechanisms is a paper by Mehregany, et al. (1988). This paper contains a good general discussion of "kinematic pairs." *Higher kinematic pairs* are defined as those in contact over a point, line, or a curve (e.g., meshed gear teeth) and *lower kinematic pairs* are those in contact over a larger area (e.g., ball joint, screw, etc.). The fabrication of gear trains, turbines with flow channels, and grippers or "tongs" was accomplished using a fairly general purpose process.

The basic process (illustrated below) began with the growth of thick (4.0 µm) thermal SiO_2 on a silicon substrate (it is interesting to note that this would introduce considerable wafer bow due to the compressive stress in thermal SiO_2). An RIE etch (using CHF_3 plasma) was done all the way through the thick oxide in some regions (for flow channels) and only part way through in other regions to later define the shapes of annular bearings and such structures. This was followed by the deposition of 4.5 µm of low-stress LPCVD polysilicon at 630°C over the patterned oxide. Then 2.0 µm of CVD oxide (TEOS) was deposited over the polysilicon and patterned with a CHF_3 plasma to form a mask for polysilicon etching. The polysilicon thus exposed was then patterned using a $Cl_2/CFCl_3/Ar$ RIE etch, followed by the removal of the remaining etch-mask oxide. Next 1.2 µm of CVD oxide (TEOS) was blanket deposited and patterned for a subsequent etch to define constraining members. The next step was to RIE etch (using CHF_3) through the exposed oxide (made in the first step) down to the silicon substrate to allow for attachment of the constraining members (also defining fluid flow channels). A second 3.0 µm thick polysilicon layer was then deposited and patterned using a $Cl_2/CFCl_3/Ar$ RIE etch to form constraining members. Finally, the mechanisms were released in an HF wet etch.

It should be noted that the release etch often took 6 to 8 hours at room temperature in 10:1 buffered HF but only 40 min for 1:1 HF:water. Miniature "tongs," requiring a large area to be undercut by the sacrificial etch) took 48 hours to release. In later designs, many researchers included several perforations through large polysilicon structures to allow increased access of the etchant to the sacrificial layer(s). Also noteworthy is that no annealing step was required to reduce stress in the polysilicon, although this is commonly used in more recent processes of this type. In addition, current processes use much thinner layers.

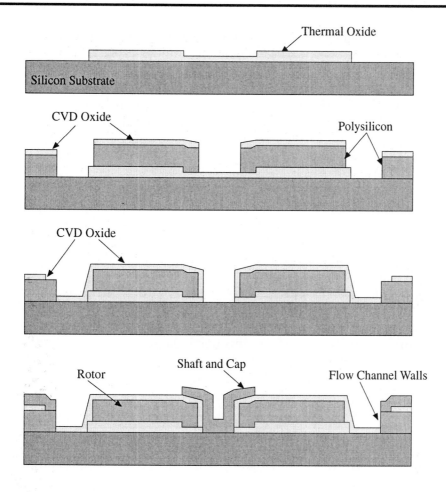

Illustration of a process flow used to surface micromachine rotating mechanisms. After Mehregany, et al. (1988).

Microturbines 125 μm in diameter were successfully fabricated and run with air pressure, but their speed could not be accurately measured using 2,000 frame-per-second cameras. Estimated speeds were greater than 15,000 RPM, and while this sounds very fast, it should be pointed out that a typical Cox Tee-Dee™ 0.028 size (0.28 in^3 displacement) internal combustion model airplane engine will operate at 28,000 RPM under castor-oil power, many motorcycle engines "red-line" at ≈ 14,000 RPM, and automotive turbo chargers can reach 50,000 RPM. Gear trains with annular bearings were successfully demonstrated, with reduction ratios between 1.4:1 and 1:1. Under high-speed operation, the constraining shaft "caps" often failed by breaking off. In addition, in-plane grippers (or "tongs") that were fabricated in this manner were manually actuated by pushing on a "handle."

Publishing simultaneously, Fan, et al. (1988b) demonstrated a wide variety of movable mechanisms fabricated using a very similar process. They used PSG as the sacrificial layer and the two-layer polysilicon approach was basically identical, as was the wet etch release using HF. Fan, et al. (1988b) did use a single process to fabricate constrained and unconstrained structures (i.e., anchored to the substrate or not), only calling for extra masking and patterning steps between layers.

The process used by Fan, et al. (1988b) began with the deposition of a 1.5 μm thick phosphorus-doped (8% by weight) LPCVD SiO_2 (i.e., PSG) (at 450°C) on a silicon wafer, followed by the patterning and etching of the PSG to form contacts to the substrate for anchors. A 1.0 μm layer of undoped LPCVD polysilicon was then deposited at 630°C, and this first polysilicon layer was patterned using a CCl_4 plasma. At this point, an extra photolithographic step was used to define areas of the first polysilicon layer that were selectively undercut 2.0 μm using buffered HF etching. Then a 0.5 μm thick layer of PSG was deposited, followed by photolithography and patterning to form contacts to substrate (etching all the way through the PSG layers) for a second polysilicon layer. A separate photolithography and plasma etch patterning step was then carried out to form contacts between the two polysilicon layers. At this point, the second 1.0 μm thick undoped LPCVD polysilicon layer was deposited and patterned using CCl_4 plasma, with extra etch time allowed to remove all polysilicon from step regions. The wafer was then annealed for 1 hour at 1,000°C in N_2 to reduce stress in polysilicon. As a final step, the structures were released in 5:1 buffered HF over 6 hours.

Structures demonstrated by Fan, et al. (1988b) included: 1) fixed-axle pin joints (anchored to substrate), which allow a member to rotate about a fixed point on the silicon; 2) self-constraining pin joints (free to move relative to substrate), which allows a member to rotate about a point on another member (which may in turn be movable); 3) multiple-joint cranks, formed from a combination of several self-constraining joints and one fixed axle pin joint; 4) sliders, which are sliding plates constrained to slide along one direction by overlapping guides, implemented with or without end stops; 5) gear/slider combinations, which are sliders that can be actuated by gears meshing with pins along their sides; 6) cranks constrained by slots with central pins; 7) spring axles, which were constrained spiral springs between polysilicon 1 and 2, used for mechanical energy storage; and 8) beam springs, which were constrained linear springs between polysilicon 1 and 2.

Their experiments yielded an estimated Young's modulus for their polysilicon of 169 GPa and a fracture strength of 2 to 3 GPa.

The above examples have all been surface micromachined rotary devices. Perhaps one of the most ambitious, yet potentially most practical, applications of *bulk micromachined* turbines is the project underway at MIT to realize a microscale gas turbine for power generation (Epstein, et al. (1997)). The proposed device would ultimately be 1 cm in diameter and 3 mm thick, and would be fabricated from SiC and theoretically produce 10 to 20 W of electrical power (using a coupled generator) while only using 10 g/h of H_2 fuel. Despite technical hurdles such as speeds as high as 2 million RPM, a key incentive driving this development is that the power densities of hydrogen or hydrocarbon fuels is far greater than that possible with any known batteries.

All of these mechanisms are severely limited by friction due to sliding contact at interfaces to the substrate and other members. Considerable research effort is being expended to mitigate this problem (as discussed below), but to some extent, this problem can be expected to remain. It is interesting to note that in nature, sliding contact mechanisms are virtually nonexistent at the microscale.

4.2 OUT-OF-PLANE MECHANISMS

The mechanisms described above are capable of moving only in the two-dimensional (2-D) plane of the wafer's surface. In order to provide access to the third dimension, Pister, et al. (1992) demonstrated the fabrication of movable hinges and "fold-up" structures, again using a very similar fabrication process.

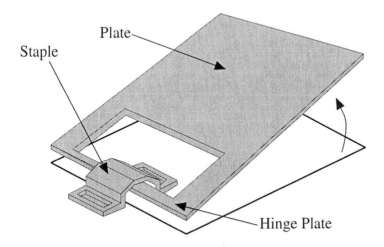

Illustration of a surface micromachined out-of-plane hinge that can be fabricated using a two-layer polysilicon micromachining process. After Pister, et al. (1992).

The hinges were fabricated using a two-polysilicon layer, sacrificial PSG process, which was very similar to those described above (however, it is described here since it represents a more recent process).

Their process began with the deposition of the lower sacrificial PSG (0.5 to 2.5 μm) on bare silicon by LPCVD. This was followed by the deposition of the lower, undoped polysilicon (1 to 2 μm) by LPCVD and another temporary doped PSG layer. The combination was annealed at 950°C in an N_2 ambient not only to remove stress in the polysilicon, but also to dope it using the phosphorus in the PSG. The top (doped) PSG layer was then removed in 5:1 buffered HF and the polysilicon was patterned using a CCl_4, O_2 and He_2 plasma to form most of the movable structures. The second sacrificial PSG layer was then deposited, and both sacrificial PSG layers were patterned using a CCl_4, CHF_3 and He_2 plasma to define contacts between the top polysilicon layer (deposited next) and either the substrate or the lower polysilicon layer. A second temporary doped PSG layer was then deposited and used in a subsequent annealing/doping step for the top polysilicon, which was then patterned using a plasma process as above. The final step was the release of the mechanisms in 49% HF (1 min), followed by a rinse in deionized (DI) H_2O and air drying.

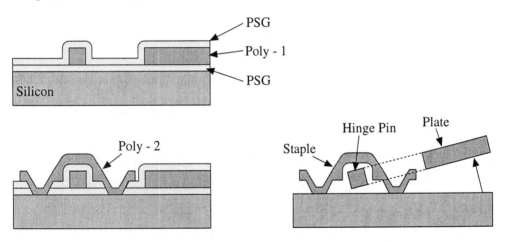

Cross-sectional illustration of the formation of a hinge structure using a two-level polysilicon process. After Pister, et al. (1992).

Only three masks were used to form a rich variety of structures, which were manually rotated (a time-consuming serial process) into position after release. Some structures could be "locked" in place either by locking structures or by depositing an overcoating of PECVD SiO_2. As illustrated below, three basic hinge types (based on whether or not they were anchored to the substrate or to other polysilicon members, and on their direction of rotation) could be fabricated, being defined simply through the layout of lithographic masks.

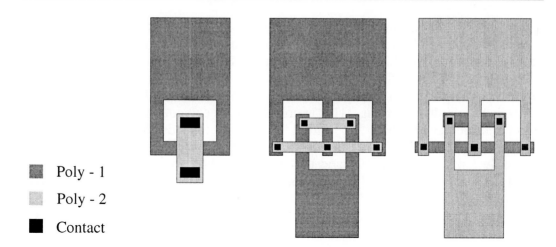

Poly - 1

Poly - 2

Contact

Illustration of three basic hinge types available in a two-level polysilicon surface micromachining process. After Pister, et al. (1992).

They demonstrated a variety of demonstration devices, including hot-wire anemometers (200 µm long, 2 µm wide polysilicon "hot wires" with electrical contacts through serpentine springs), a conformal micro-probe, a "tissue growth dynamometer" (for measuring forces exerted by growing embryonic tissues), and a parallel-plate gripper (pulling the "handle" in turn pulled on 1 mm long "tendons" that pulled the normally closed gripper plates apart). More recently, Yeh, et al. (1995) demonstrated a variety of more complex "fold-up" structures for exploratory work in fabricating articulated manipulators for miniaturized robotic systems.

4.3 STRUCTURAL MEMBERS

While not strictly considered mechanisms by themselves, structural members are critical in connecting between mechanisms, supporting other structures, etc. In the processes discussed above, such members can readily be fabricated from lengths of polysilicon. They can be made using one or both layers of polysilicon (the latter being possible if a contact cut is made in the inter-polysilicon sacrificial layer prior to deposition of the upper layer).

Rather than using solid polysilicon, however, Judy and Howe (1993a, 1993b), developed a process for fabricating hollow polysilicon beams. Their process is more complicated than the "typical" single-layer polysilicon processes often used for electrostatic actuators, but offers a significant advantage: increased "stiffness-to-mass ratio," leading to higher resonant frequencies for a given geometry (allowing for higher-frequency operation). The basic process (details covered in Judy and Howe (1993a, 1993b), and illustrated below) was essentially to pattern a polysilicon/PSG/polysilicon sandwich to define the top and bottom of each beam, to add sidewalls using a blanket polysilicon deposition and anisotropic etch, to create access holes for HF etching of the now trapped PSG, and to release the structures. It is interesting to note that the PSG layer beneath the lowest released polysilicon layer was dimple etched using a timed HF etch to form replica dimples in the polysilicon to reduce sticking. This approach can be extended to make other beam structures such as "I," "C," "U," and "L" beams, or the sidewalls themselves can be used (see Judy and Howe (1993b)). The final HF etch to release the devices took only 1 min with 49% HF, and it is surprising that tubes 2×2 µm $\times$ 200 µm could be cleared that fast. (It was proposed that unintentional access for the HF was provided at grain boundaries in the polysilicon, since beams without access holes were also cleared of oxide during the etch.) For further details on this type of processing, the reader is referred to Judy (1994).

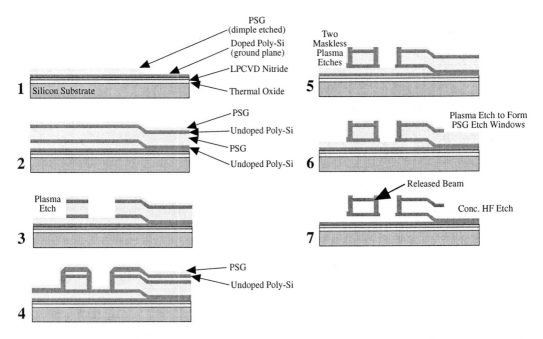

Illustration of a hollow beam, surface micromachining process for electrostatic actuators. After Judy and Howe (1993b).

4.4 BISTABLE MECHANISMS

Bistable mechanisms have also been demonstrated using micromachining. Devices such as compressively stressed membranes, can "snap" between two stable states to provide mechanical latching. This concept has considerable promise in terms of reducing the power required to hold a mechanical structure in a particular state (e.g., this would be very useful in implantable drug delivery systems). A good example of such structures can be found in Huff, et al. (1991), but very few bistable micromechanical devices have been realized to date.

4.5 SELF-ASSEMBLY

While still in early stages of development, various groups have discussed the concept of self-assembly of three-dimensional (3-D) microstructures. On the nanoscopic scale, nature makes use of this approach extensively in biological systems (e.g., the self-assembly of complex proteins from individual subunits). In these cases, non-covalent bonding forces reversibly hold individual components into larger aggregates. In these cases, random (Brownian) motion provides the input energy to drive the subunits into their correct orientations. Considerable discussion of this subject, as well as the large variety of non-self-assembled biological systems can be found in Alberts, et al. (1994), Darnell, et al. (1991), and many other molecular biology texts.

A variety of self-assembly processes have been demonstrated on a microscopic scale in an effort to eliminate the need for manual or robotic assembly of micromachined structures for multiple components. To date, the two methods used are the use of random motion to drive mechanical interaction/latching or chemical bond assembly mechanisms. Cohn, et al. (1991) demonstrated purely mechanical self-assembly of hexagonal silicon dice into ordered arrays, using an underlying vibrating diaphragm to provide energy. As an alternative to assembling structures from non-connected components, mechanical self-assembly of coupled, hinged polysilicon members has been demonstrated. The use of water turbulence during rinsing, or directed air jets, to drive hinged polysilicon structures into their final positions was described by Pister, et al. (1992). Latching mechanisms for microstructures have been described in several publications, including Yeh, et al. (1995) and Han, et al. (1992). Fan, et al. (1997) used on-chip scratch-drive actuators (later used as positioners) to power the folding of hinged polysilicon structures into a 3-D shape above the substrate.

The use of surface tension to promote self-assembly was demonstrated by Hosokawa, et al. (1996). They used thin-film structures of polysilicon, polyimide, and nickel with a large rounded semicircle and a sharp pointed end connected to the cut line of the semicircle. The authors took advantage of the fact that large surface tension forces acted at the sharp points when the objects were floated in water, drawing them together to form clusters when they were agitated using a magnet. Green, et al. (1995) made use of the surface tension of molten solder to drive self-assembly of microstructures. Bowden, et al. (1997) demonstrated the use of controllable regions of hydrophobic and hydrophilic polydimethylsiloxane (PDMS) structures to drive their self-assembly while floating in water. PDMS is naturally hydrophobic, but can be made hydrophilic by oxidation in an O_2 plasma. After oxidation, hydrophobic surfaces can be regenerated by cutting away regions of PDMS (or by covering them with tape prior to oxidation). Using externally powered swirling of the water bath, a variety of different shapes were self-assembled into ordered arrays. Yeh and Smith (1994a, 1994b) demonstrated the fluid-powered self-assembly of GaAs light-emitting diodes (LEDs) into etched holes in a silicon substrate.

As an alternative to such self-assembly techniques, the batch transfer of microcomponents from one substrate or assembly to another could be used to construct more complex micromachines. Such a technique was demonstrated by Cohn, et al. (1996) using the transfer of separately fabricated polysilicon structures to a single-crystal silicon substrate. In principle, this approach could be extended to realize multi-level transferred structures, but it does not appear to have the ability to replace the folding of hinged structures, for example, in terms of achievable object complexity. However, such batch-transfer approaches do not require any further interaction with the resulting microstructures.

5. MECHANICAL SENSORS

As mentioned above, there is a tremendous variety of direct mechanical sensors that have and could be micromachined. In order to study this vast area, it makes sense to begin with a discussion of the basic mechanical sensing mechanisms (e.g., for force, displacement, strain, etc.) and then see how these mechanisms can be applied to realize a wide range of micromachined sensors.

5.1 SENSING MECHANISMS

The following table summarizes the commonly used mechanical sensing mechanisms in micromachined devices. Important considerations when choosing such a mechanism include the need for local (or even monolithically integrated) circuitry, whether or not the transduction mechanism is DC-responding (piezoelectric sensors are the only common variety that are generally not), temperature coefficients, long-term drift, overall system complexity, and others.

Mechanism	Parameter Sensed	Needs Local Circuits?	DC Response?	Complex System?	Linearity	Issues
Metal Strain Sensor	strain	NO	YES	+	+++	• low sensitivity • very simple
Piezoresistive Strain Sensor	strain	NO	YES	+	+++	• temperature effects can be significant • easy to integrate
Piezoelectric	force	NO	NO	++	++	• high sensitivity • fabrication can be complex
Capacitive	displacement	YES	YES	++	poor	• very simple • extremely low temperature coefficients
Tunneling	displacement	YES	YES	+++	poor	• sensitive to surface states • drift performance not yet proven
Optical	displacement	NO	YES	+++	+++	• rarely employed in mechanical microsensors

Table comparing some of the major properties of the mechanical sensing mechanisms commonly used in micromachined devices.

5.1.1 RESISTIVE AND PIEZORESISTIVE STRAIN SENSORS

Strain sensors are an integral part of many micromachined devices, serving to measure strain or, indirectly, displacement of structures. A *strain gauge* is a conductor or semiconductor that is fabricated on or bonded directly to the surface to be measured. Changes in gauge dimensions result in proportional changes in resistance in the sensor. This is partly due to stretching (changes in dimension) and partly due to the *piezoresistive effect*, discovered by Lord Kelvin in 1856. As might be expected, the sensitivity of gauges can be quite different, depending on their design. Strain gauges of all types can be very linear over considerable ranges of strain, making them attractive in a variety of applications.

In general, the sensitivity is expressed by the *gauge factor* (dimensionless),

$$ GF = \frac{\text{relative resistance change}}{\text{strain}} = \frac{\dfrac{\Delta R}{R}}{\dfrac{\Delta L}{L}} = \frac{\Delta R}{\varepsilon R} $$

(here longitudinal strain, ε_l, is used). One can use partial derivatives to derive a general expression for the gauge factor in terms of the physical parameters of the strain gauge. This begins with the derivation of a relation between resistance and changes in its underlying parameters, as seen in,

$$R = \frac{\rho L}{A} \quad \text{in } \Omega$$

where,
ρ = resistivity, in $\Omega \cdot cm$
L = length, in cm
A = cross-sectional area, in cm^2

Differentiating the resistance equation, one obtains,

$$dR = \frac{\rho}{A} dL + \frac{L}{A} d\rho - \frac{\rho L}{A^2} dA$$

which can be divided by the above equation for resistance to obtain,

$$\frac{dR}{R} = \frac{dL}{L} + \frac{d\rho}{\rho} - \frac{dA}{A}$$

It must be noted that from this point on in the derivation it becomes geometry-specific, and a cylindrical wire is assumed for simplicity. It is useful to use Poisson's ratio to express the relative dimensional change in diameter, D, versus length, L,

$$\nu = -\frac{\varepsilon_t}{\varepsilon_1} = -\frac{\frac{\Delta D}{D}}{\frac{\Delta L}{L}} \approx -\frac{\frac{dD}{D}}{\frac{dL}{L}}$$

where, for a cylinder, the area and diameter are related through,

$$A = \frac{\pi D^2}{4} \quad \text{and} \quad \frac{dA}{A} = \frac{2dD}{D}$$

which in turn allows Poisson's ratio to be written and rearranged to obtain,

$$\frac{dA}{A} = -2\nu \frac{dL}{L}$$

finally allowing the differential form of the resistance to be written,

$$\frac{dR}{R} = (1+2v)\frac{dL}{L} + \frac{d\rho}{\rho}$$

where the first term represents the dimensional effect and the second term represents the piezoresistive effect (change in resistivity of the material of the strain gauge). From this, the gauge factor can be expressed in terms of these parameters as,

$$GF = \frac{\frac{dR}{R}}{\frac{dL}{L}} = \frac{\frac{dR}{R}}{\varepsilon_1} = (1+2v) + \frac{\frac{d\rho}{\rho}}{\varepsilon_1}$$

Naturally, such a derivation can be carried out for strain gauges with non-cylindrical shapes.

As shown in the table below, the gauge factors of different types of strain gauges can be vastly different, due mainly to whether or not they have a significant piezoresistive effect (as do the semiconductor types).

Type of Strain Gauge	Gauge Factor
Metal Foil	1 to 5
Thin-Film Metal	≈ 2
Bar Semiconductor	80 to 150
Diffused Semiconductor	80 to 200

Table comparing the gauge factors of different types of strain gauges.

METALLIC STRAIN GAUGES

For metals, ρ does not vary significantly with strain (as long as the cross-sectional dimensions are much larger than the grain size), and v is typically in the range of 0.3 to 0.5, for gauge factors on the order of two. However, in practice, macroscopic metal strain gauges often have gauge factors higher than this, so it appears that some piezoresistive effect and/or change in total wire volume comes into play. Whether or not this would be helpful in micromachined strain gauges is most likely a moot point since the much larger gauge factors of piezoresistive strain gauges make them nearly ubiquitous in this domain.

Metal strain gauges may be made from thin wires or metal films (thin-film strain gauges) that may be directly fabricated on top of microstructures. Thin-film metal strain gauges are easier to fabricate (photolithographically) and allow for more complex shapes. They are generally built on flexible plastic substrates (sometimes self-adhesive) and can be glued onto a surface.

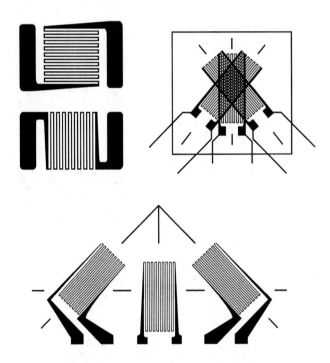

Illustration of typical metal strain gauge designs. Three strain gauges arranged as shown in the lower part of the illustration allow σ_x, σ_y and τ_{xy} to be resolved at a given location (this is a standard configuration). After Norton (1989).

SEMICONDUCTOR STRAIN GAUGES

In semiconductor strain gauges, the piezoresistive effect is very large, leading to much higher gauge factors. P-type silicon has a gauge factor up to 200 and n-type has a negative gauge factor, down to $\approx$ -140. Strain gauges can be locally fabricated in bulk silicon through ion implantation or diffusion, or the entire substrate can be used as the sensor. Unfortunately, these semiconductor strain gauges also have much higher-temperature coefficients of resistivity, making temperature compensation more important. (For example, one can use a Wheatstone bridge with a reference strain gauge that is not deformed to compensate, as long as all bridge elements are isothermal.)

The detailed theory of piezoresistivity is not covered herein, but good explanations can be found in Kanda (1982), Yamada, et al. (1982), and Middelhoek and Audet (1989). In short, the effective mobilities of majority carriers in a piezoresistive material are affected by stress (this effect is highly orientation dependent). For p-type materials, the effective mobility of holes decreases so the resistivity increases. For n-type materials, the effective mobility of electrons increases so the resistivity decreases. The observed change in mobility results from the strain-induced distortions of the energy band structure, and this can be calculated quite accurately if necessary.

While the strong temperature dependence of the gauge factor for single-crystal semiconductor strain gauges makes their use sometimes more difficult, polycrystalline and amorphous silicon are useful alternatives (which are not anisotropic). The total resistance of polycrystalline silicon is determined by the resistance of the silicon grains and that of the grain boundaries, the latter being the most important aspect. Within the grains, the resistivity behaves essentially like that of the single-crystal material, and so as the temperature increases, the mobility decreases and the resistivity *increases*. At grain boundaries, depletion regions develop due to charge trapping, and here as the temperature increases, more carriers can overcome these boundaries, *decreasing* the resistivity. By balancing these effects (i.e., changing the dose of ion implant doping), the net temperature coefficient can be adjusted to nearly zero. It is worth pointing out that silicon (and polysilicon and amorphous silicon) are centrosymmetric and not piezoelectric (unless stressed). Piezoresistive behavior is completely different from piezoelectric properties (discussed below).

5.1.2 PIEZOJUNCTION EFFECT

A related effect to piezoresistance is the *piezojunction effect,* which is a marked shift in the I-V characteristic of a p-n junction when mechanical stresses are applied to it. One can also fabricate pressure sensitive tunnel diodes, MOSFETs, MESFETs, etc. While this effect has been discussed in the micromachining literature, little use has been made of it in micromachined transducers. Friedrich, et al. (1997) presented a piezojunction-based strain sensor where reverse I-V characteristics, dominated by band-to-band tunneling, were modulated by applied strain.

5.1.3 PIEZOELECTRIC EFFECT

Piezoelectricity is a phenomenon in which a mechanical stress on a material produces an electrical polarization and, *reciprocally*, an applied electric field produces a mechanical strain. The Curies discovered the effect in 1880, and the first useful applications were made by Cady in 1921 with his work on quartz resonators (see Cady (1964)). The effect can certainly be used to sense mechanical stress (or, indirectly, displacement, etc.) and as an actuation mechanism. However, a key potential limitation of this transduction mechanism is that the piezoelectric effect produces a DC charge (polarization), but not a DC current. Thus such transducers are inherently incapable of providing a DC response. The limited low-frequency response of piezoelectric devices is primarily due to parasitic charge leakage paths, and can be significantly improved through micromachining and directly coupling piezoelectric outputs to MOSFET gates (see Chen, et al. (1982, 1984), discussed below in the accelerometer section).

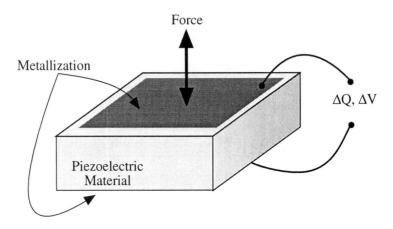

Illustration showing the generation of incremental charge (and hence voltage) on metallized electrodes on opposite sides of a slab of piezoelectric material, in response to the application of force.

Centrosymmetric crystals such as silicon and germanium are not piezoelectric. If such materials are strained, the effective centers of the positive and negative charges do not move with respect to each other, preventing the formation of dipoles as required for piezoelectricity. Thus, materials whose crystal structures lack centers of symmetry are required for the piezoelectric effect to be possible. Thus, to fabricate silicon-based piezoelectric transducers, a suitable material must be deposited on the devices. (It is possible to make centrosymmetric crystals such as silicon exhibit a weak piezoelectric effect by applying a uniaxial stress, but the effect is too small for any practical applications.) The III-V and II-VI compounds (e.g., GaAs, CdS, ZnO, etc.) are not centrosymmetric and have bonds that are partly covalent and partly ionic in nature, and thus are piezoelectric. Materials such as CdS and ZnO can be deposited by co-evaporation or sputtering, with the sputtered ZnO being more common approach.

Piezoelectricity, pyroelectricity, and ferroelectricity all derive from one single physical cause: the existence of an electric polarization vector, **P**. As a rule of thumb, if a crystal is piezoelectric, almost always it will be pyroelectric and ferroelectric too (there are very few exceptions in some exotic materials). This points out another factor limiting the use of piezoelectric sensing for low frequencies and DC, since most suitable materials exhibit considerable temperature errors due to their pyroelectric behavior. This effect can, however, be mitigated through the use of compensation capacitors made from the same piezoelectric material but left unstrained (Chen, et al. (1982, 1984)).

In terms of sensitivity to stress, piezoelectric materials are commonly characterized by the charge sensitivity coefficients, d_{ij}, (in units of C/N), which relates the amount of charge generated at the surfaces of the material (of area A) on the i axis to the applied force, F, on the j axis,

$$\Delta Q_i = d_{ij} \Delta F_j = d_{ij} \Delta \sigma A$$

From this, the voltage change across the conductive plates (at a spacing x) can be written as,

$$V = \frac{Q}{C} = \frac{Qx}{\varepsilon_o \varepsilon_r A} \quad \rightarrow \quad \Delta V_i = \frac{d_{ij} \Delta F_j x}{\varepsilon_o \varepsilon_r A}$$

The piezoelectric effect is reversible, such that the application of a voltage ΔV gives rise to a corresponding force, ΔF, and resulting dimensional change ΔL. This is commonly used in piezoelectric actuators. Typical values for ΔL vary between 10^{-10} and 10^{-7} cm/V. Thus, to obtain displacements on the order of micrometers (μm), voltages exceeding 1,000 V are often necessary, unless stacked actuators or mechanical motion amplification methods are used.

Material	Type	Piezoelectric Constant pC/N	Relative Permittivity (ε_r)
Quartz	single crystal	d_{33} = 2.33 [2, 3]	4.5 [2], 4.0 [3]
Polyvinylidene fluoride (PVDF)	polymer	d_{31} = 20, d_{32} = 2, d_{33} = -30 [2] d_{31} = 23 [1] d_{33} = 1.59 [3]	12 [1, 2]
Barium titanate (BaTiO$_3$)	ceramic (Perovskite crystal)	d_{31} = 78 [1, 2] d_{33} = 190 [3]	1,700 [1, 2] 4,100 [3]
Lead zirconate titanate (PZT)	ceramic	d_{31} = 110 [1, 2] d_{33} = 370 [3]	1,200 [1] 300 to 3,000 [3]
Zinc oxide (ZnO)	metal oxide	d_{33} = 246 [4]	1,400 [4]

Table of relevant properties of piezoelectric materials. From Hauptmann (1991) [1], Fraden (1997) [2], Gardner (1994) [3], and Bernstein, et al. (1997) [4].

Common piezoelectric materials with applications in micromachining include quartz (basis of quartz crystal oscillators and resonators); polyvinylidene fluoride (PVDF); lead zirconate titanate (PZT); perovskite crystals such as barium titanate , lithium niobate, etc.; III-V compounds such as GaAs, GaP; and II-VI compounds such as ZnO, ZnS, ZnSe, etc. Relevant properties of several piezoelectric materials are given in the table above.

PZT is a ceramic with a high value for the piezoelectric strain constant. However, it is somewhat difficult to deposit as a thin film. Polyvinylidene fluoride (PVDF) and ZnO are most often used in the microfabrication of piezoelectric transducers.

PVDF, which is a carbon-based polymer, is usually deposited as a spin cast film from a dilute solution in which PVDF powder has been dissolved. As for most piezoelectric materials, processing after deposition greatly affects the behavior of the PVDF film. For example, heating and stretching can increase or decrease the piezoelectric effect. PVDF and most other piezoelectric films require a polarization after deposition (called *poling*). This is done by the application of a large electric field for a few hours using electrodes deposited on both sides of the film. The net polarization (which is closely related to the piezoelectric strain constant) at the end of this process depends on the time integral of the applied field up to a saturation level. It is important to note the electro-spray PVDF deposition method of Asahi, et al. (1993) discussed in the Optical Transducers chapter. Alternatively, Swartz and Plummer (1979) described the attachment of a prefabricated PVDF sheet to an array of MOSFETs using epoxy (in their application, ultrasound receivers, acoustic energy impinging upon the PVDF generated charges that were sensed by the MOS-FETs). These approaches eliminate the need for a poling step, and may make PVDF a more attractive thin-film piezoelectric material than the others.

ZnO is the most common piezoelectric material used in microfabrication. Unlike PVDF, it can be sputter-deposited as a polycrystalline thin film with its c-axis (along which piezoelectricity is strongest) perpendicular to the surface of the substrate. Pure Zn is usually sputtered in an O_2/Ar plasma to form ZnO. ZnO has also found broad applications as a pyroelectric material.

The required high voltages make piezoelectric materials poor actuators for displacements in the micron regime, but they are precise actuators on the sub-nanometer scale (the reverse argument shows that small displacements will cause large detectable voltages making piezoelectric materials very sensitive sensors). Piezoelectric actuation is ideal, however, for scanning tunneling and scanning force microscopes, where small, precise displacements are required. Piezoelectric materials are also very useful in micromachined transducers such as surface acoustic wave (SAW) devices, accelerometers, microphones, etc.

5.1.4 CAPACITIVE SENSING

Perhaps one of the most important, and oldest, precision sensing mechanisms is capacitive (or electrostatic). The physical structures of capacitive displacement sensors are extremely simple (one or more fixed plates, with one or more moving plates). The inherent nonlinearity of most capacitive sensors is often overshadowed by their simplicity and very small temperature coefficients. With the monolithic integration of signal conditioning circuitry, the additional problem of measuring often miniscule capacitance changes in the face of large parasitics is mitigated. Several potential capacitive sensing modes are illustrated below.

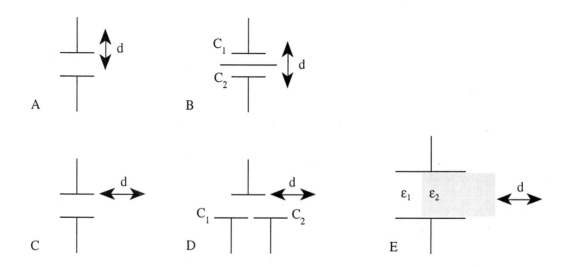

Illustration of four different possible capactive sensing modes. After Cobbold (1974). In case A, the distance between the plates is varied, which leads to an extremely nonlinear transfer function. In case B, an intermediate plate is moved relative to two fixed plates, providing a fixed total capacitance and lending itself to differential measurements. Cases C and D represent two modes (single-ended and differential, respectively) where the overlap area of the plates is varied with position, providing far greater linearity (impeded to some extent, nonetheless by fringing field effects). Another mode of operation, varying the dielectric constant between the plates (e.g., by moving a dielectric slab between them) is illustrated as case E.

The basic parallel-plate capacitor equation (assuming no fringing fields) is,

$$C = \frac{\varepsilon_o \varepsilon_r A}{d} \quad \text{in F}$$

where,
ε_o = dielectric constant of free space = 8.854188×10^{-14} F/cm
ε_r = relative dielectric constant of the material between the plates
A = overlapping plate area, in cm
d = plate separation, in cm

Similarly, as is often the case in micromachined structures, for n dielectric layers of a relative dielectric constant, ε_{ri}, the overall capacitance is,

$$ C = \frac{\varepsilon_o A}{\left(\dfrac{d_1}{\varepsilon_{r1}} + \dfrac{d_2}{\varepsilon_{r2}} \cdots \dfrac{d_n}{\varepsilon_{rn}} \right)} \quad \text{in F} $$

Capacitive sensor structures are relatively simple to fabricate. As illustrated above, one can vary d, ε, or A, providing very nonlinear (in the former two cases) or quite linear position-to-capacitance transfer functions (in the latter case). While macroscopic capacitive transducers of nearly any imaginable shape can be (and largely have been) implemented, this is not the case for micromachined devices. Membrane-type capacitive devices (e.g., microphones or pressure sensors) are straightforward to fabricate, but they are extremely nonlinear, since d varies. Comb-type capacitors are commonly used in surface micromachined devices, and are theoretically based on varying the overlapping area for greater linearity. However, at such scales (particularly for surface micromachined devices), fringing fields can become very significant or even dominant. Thus, the parallel-plate capacitor equation is only useful for first-order estimates at best. Varying the dielectric constant between the plates (e.g., by moving a slab of a different dielectric constant from the ambient between the plates) appears not to have been employed in micromachined devices often, and there seems to be little incentive to do so given the relative ease of using the other two modes. One noteworthy exception is the class of humidity and chemical sensors in which the dielectric constant of a sensitive layer is varied in relation to the concentration of analyte.

Despite these potentially difficult issues, a key redeeming feature of capacitive transducers is their near lack of a temperature coefficient (as long as the material in the gap has a low-temperature coefficient of its dielectric coefficient, e.g., as air and vacuum do). According to Baxter (1997), the temperature coefficient of the dielectric constant of air at 1 atm and 20°C is ≈ 2 ppm/°C for dry air and 7 ppm/°C for moist air. However, the change in dielectric coefficient of air versus pressure is more sizable at 100 ppm/atm. If the dielectric between the plates is a gas at a stable pressure (or vacuum), the dominant (and often minor) thermal effect on capacitance is due to differential thermal expansion of the structures themselves. An additional advantage of capacitive sensing is the fact that it is non-contact. A classic paper on capacitive transduction is Foldvari and Lion (1964), showing many basic configurations of mechanical to capacitive devices (although the circuits discussed, e.g., the diode twin-T network, are dated in specific implementation, they are still very relevant). Furthermore, a book entirely dedicated to capacitive sensing is Baxter (1997).

While capacitive transduction is inherently less noisy than resistive (with its attendant thermal, or Johnson noise), the noise performance of capacitive versus piezoresistive approaches is not always better, particularly since surface micromachining approaches often result in extremely small (femto or attofarad) capacitances. In such cases, the necessary interface electronics (and Brownian motion of the extremely low-mass capacitive structures) often supplies enough noise to eliminate any potential signal-to-noise ratio (SNR) advantage of capacitive sensing (one needs only examine current commercial piezoresistive versus capacitive accelerometers to verify this).

Changing capacitance can be measured using a number of well-known circuit techniques, such as 1) charge-sensitive amplifiers, 2) charge-redistribution techniques, 3) impedance measurements (measuring the impedance in a bridge or other configuration), 4) RC oscillators (making the unknown capacitance the time-constant determining capacitance in an oscillator and measuring the frequency), and 5) direct charge coupling (e.g., use the moving plate as the gate of a field-effect transistor (FET)). In general, these circuits can (and often must) be integrated with the capacitive sensors themselves, or at least positioned nearby to minimize the effects of parasitic capacitances. Examples of capacitive sensor interface circuits for micromachined sensors can be found in Park and Wise (1983), Smith, et al. (1986), Kung, et al. (1988), and Kung and Lee (1992).

Such electrostatic devices are also capable of being used as actuators, but are very nonlinear in this mode, as discussed below.

5.1.5 TUNNELING SENSING

Tunneling transduction, as discussed in the Optical Transducers chapter, is extremely sensitive due to the exponential relationship of tunneling current, I, to the tip/surface separation ,

$$I = I_o e^{\left(-\beta \sqrt{\phi} z\right)}$$

where,

I_o = scaling factor, dependent on materials, tip shape, etc.

β = conversion factor, typical value = 10.25 eV$^{-1/2}$/nm

ϕ = tunnel barrier height in electronvolts (eV), typical value = 0.5 eV

z = tip/surface separation in nanometers (nm), typical value = 1 nm

While extremely nonlinear, the sensitivity of this displacement sensing approach can be harnessed by using closed-loop feedback to linearize the system. Also, while imaging microscopy applications call for extremely (atomically) sharp tunneling tips, this is not necessary for displacement sensing applications since one atom or another will be closest to the tunneling current sink and will thus dominate. To date, there have been limited applications of tunneling transduction in micromachined systems (note that Grade, et al. (1996) and Scheeper, et al. (1997) discuss the issues involved in manufacturing such devices).

5.2 MICROMACHINED MECHANICAL SENSORS

In this section, a sampling of the wide variety of micromachined mechanical sensors is presented, beginning with strain gauges and progressing to more complex devices. Again, the reader is encouraged to explore the large and growing literature discussing a multitude of other examples.

5.2.1 MICROMACHINED STRAIN GAUGES

The examples of micromachined strain gauges discussed here were chosen because they demonstrate strain measurement where it would be difficult or impossible using fabrication methods other than micromachining.

IMPLANTABLE STRAIN GAUGES

Angell, et al. (1980) fabricated microminiature silicon strain gauges for implantation in animals to study the forces in tissues that lead to *decubitus ulcers* ("bed sores") in patients who are bedridden for prolonged periods. The overall length of the devices was 1.7 mm, and they include silicon loops at each end for passing through a fine surgical suture, used to anchor them to the tissues. Piezoresistive elements were formed through a masked diffusion, and wet etching was used to define the overall shape of the gauges, with an overall thickness of 60 μm. The active region (location of the piezoresistor) was thinned to 30 μm using a back-side etch.

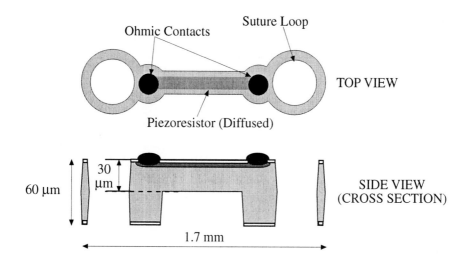

Illustration of a micromachined implantable strain gauge. After Angell (1980).

PENETRATING MICRO-STRAIN-GAUGE PROBE

As part of an ongoing effort at the University of Michigan to realize penetrating probes for recording from and stimulating brain tissue, Najafi and Hetke (1990) demonstrated a microprobe with an on-board strain gauge. A key question was whether or not the probes were strong enough to penetrate the overlying tissues of the brain (*arachnoid* and *pia*, and the very tough *dura mater*). The probes should be made as thin as possible (minimizing damage to cells) while remaining strong enough to penetrate the tissues.

The probe shapes were defined using a boron diffusion etch-stop and an EDP etch, being etched away from the starting wafer. Some versions of the probes carry on-board electronics to amplify electrical activity of brain cells (this is discussed in the Biological Transducers chapter). To measure the forces required for tissue penetration, Najafi and Hetke (1990) outfitted their probes with phosphorus-doped polysilicon piezoresistive sensors, pushed them through various tissues, and recorded the deflections in real time. The change in resistance is proportional to the deflection of the probe (bending or buckling), which is proportional to the force on the probe. As expected, the change in resistance was nearly perfectly linear with deflection and did not exhibit hysteresis.

The authors measured the fracture stress for various probe widths and, knowing Young's modulus, they were able to determine the maximum strain and hence maximum deflection allowable for the probes. An important result is that the actual fracture stresses for small silicon "needles" is five to six times higher than that for bulk silicon. By measuring the actual deflection (and hence stress) using the strain gauges, they were able to see how close they came to the fracture stress of each type of probe tested. Even for the very tough dura, using probes only 30 μm thick, the stresses were still an order of magnitude lower than the fracture stress. For "softer" tissues, only 15 μm thick probes were required.

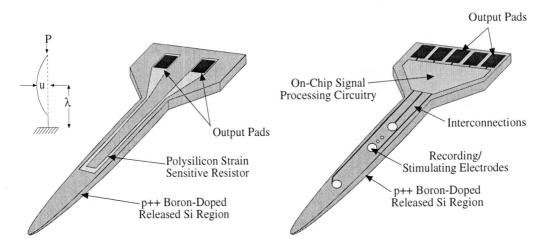

Illustration of (left) strain-gauge equipped test probe and (right) typical neural recording/stimulation probe fabricated using similar methods, but including electrodes and signal processing circuitry. Adapted from Najafi and Hetke (1990).

Similar silicon-substrate probes, using dry etching for shape definition, were developed by Kane, et al. (1995). The probes were used to stimulate mechanical sensors in the cornea of the eye and in human skin. Probes of lengths between 0.5 and 2 mm were fabricated. Using an on-board piezoresistive strain gauge to provide feedback to a macroscopic voice-coil actuator, they generated closed-loop-controlled forces up to 3 mN with a measurement resolution of 10 μN.

SINGLE-CELL STRAIN GAUGES

Using a combination of XeF_2 bulk etching of standard CMOS devices and manual assembly using tungsten microprobes, Lin, et al. (1997) demonstrated a micromachined strain gauge that could measure the contractile force of a single heart cell. Signals from polysilicon piezoresistive strain gauges were amplified using an on-chip preamplifier to allow sensing of contraction forces. During the experiments, a cell was manually attached to "microclamps" on strain-sensitive deflectable members. Including an off-chip amplifier, the system deflection gain was 2.4 V/μm, and forces of up to 32 μN were measured from single cells (the noise floor was 1 μN).

RESONANT STRAIN GAUGES

In direct analogy to the use of increased string tension to tune instruments such as guitars, the resonant frequency of a doubly supported beam can be altered by applied strain. This approach has been used to fabricate pressure sensors (discussed in Section 5.2.4) and accelerometers (see Burns, et al. (1995) and Roszhart, et al. (1995)). A stand-alone resonant strain gauge with optical excitation and readout was demonstrated by Zook, et al. (1995). Their designs were a merged combination of a photodiode, an electrostatically deflected resonator beam and an upper cap. The cap and beam were surface-micromachined using polysilicon with sacrificial silicon dioxide layers. In operation, light entered through the cap layer, passing through the beam on its way to the photodiode below. The photogenerated potential would electrostatically attract the resonator beam, which in turn would alter the optical interference in the cavity formed by the three structures. With such deflection of the beam, the photovoltage would decrease, allowing the beam to return to its original state. If the correct phase relationship between the photovoltage and the beam's position was maintained, self-generating oscillations would ensue when the device was illuminated, and the frequency at which reflected light was modulated could be used as the optical readout of the applied strain. A variety of electrostatically excited resonant strain gauges have also been implemented, and an interesting example can be found in Yoshida, et al. (1995), with a useful discussion of relevant design issues in Gui, et al. (1995).

5.2.2 ACCELEROMETERS

Accelerometers find application in diverse areas, including automotive uses such as deploying air bags, navigation, activity detection for pacemakers, machine monitoring, etc. Micromachined accelerometers are rapidly opening up new application areas, thanks to their reduced cost/volume and improved performance in many (but not all) designs. For example, MacDonald (1990) is a useful overview of automotive applications (although the review is somewhat dated due to the rapid advance of micromachining technology, it is still useful as a reference point). Interesting general overviews of sensors for automotive applications (including accelerometers) are found in Sulouff (1991) and Giachino and Miree (1995).

It is worth noting that the automotive environment, as described by Giachino and Miree (1995), includes temperatures in the range of -40 to 150°C, mechanical shocks up to 50 g, vibration up to 15 g, electromagnetic fields up to 200 V/m, and exposure to a variety of chemicals (e.g., fuels, oil, brake fluid, transmission fluid, antifreeze fluid, salt spray, etc.). Needless to say, even for transducers that do not need physical ports in their packaging, such as accelerometers, these harsh conditions make the transition from laboratory demonstration to practical application quite challenging.

(Note that in the context of accelerometers, the unit "g" is used to represent the acceleration due to the Earth's gravity, which is 9.80665 m/s^2 at 45° latitude at sea level, and should not be confused with grams.)

Range	± 1 g	antilock braking (ABS)/traction control system (TCS)
	± 2 g	vertical body motion
	± 40 g	wheel motion
	± 50 g	air bag deployment
	± 100°/s	steering feedback
Accuracy	± 2 %	5% at temperature extremes
Cross-Axis Sensitivity	< 1 to 3 %	all applications
Shock Survivability	> 500 g	1 m drop onto concrete
Frequency Response	0 to 5 Hz	vertical motion
	0.5 to 50 Hz	horizontal motion (up to 1 kHz for air bags)
Temperature Range	-40 to 85°C	most applications
	-40 to 125°C	under hood

Table of typical automotive accelerometer specifications. After MacDonald (1990).

BASIC ACCELEROMETER CONCEPTS

One can develop accelerometers to measure linear acceleration, defined by,

$$a = \frac{d^2x}{dt^2} = \frac{dv}{dt} \quad \text{where} \quad v = \frac{dx}{dt}$$

and where x is the linear displacement in meters (m), and v is the linear velocity in meters per second (m/s).

In addition, one can measure angular acceleration, defined by,

$$\alpha = \frac{d^2\theta}{dt^2} = \frac{d\omega}{dt} \quad \text{where} \quad \omega = \frac{d\theta}{dt}$$

and where θ is the angular displacement in rad, and ω is the angular velocity in rad/s.

For each of the axes in three dimensions, this corresponds to three linear and three angular accelerations, for a total of six acceleration components. Typically, angular acceleration is measured using linear accelerometers at known positions relative to the rotational axis of interest. Thus the discussion herein will be limited to linear accelerometer designs.

In general, accelerometers sense acceleration by using a *proof mass*, or *seismic mass* on which external acceleration can act, and measure the displacement of the proof mass, the force exerted by the proof mass against the frame, or the force required to keep it in place (the latter case, an *active* accelerometer, requires a closed-loop control system). These designs rely upon measuring the force generated by the acceleration acting on the proof mass directly, m, or indirectly. This force is simply $F = ma$. There are many ways to measure the force or displacement, including strain gauges, capacitors, surface acoustic wave devices, strain-sensitive resonant beams, magnetometers, optical detectors (interferometers, etc.), tunneling sensors, etc.

In practice, a typical passive accelerometer has the basic components illustrated below: a proof mass, a spring, a damping element, and a hard limiter. In this case, the displacement of the proof mass is measured by equating the force generated by acceleration of the mass to the displacement of a spring,

$$F = ma = kx_{rel}$$

where k is the stiffness of the spring, or the spring constant, and x_{rel} is the displacement relative to the accelerometer frame (as opposed to x, displacement of the frame itself). The spring also serves to return the proof mass back to its null position when acceleration ceases.

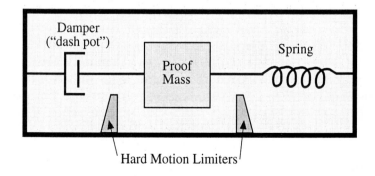

Illustration of the basic components of an accelerometer.

In order to control the frequency response characteristics of the accelerometer, a damper (or "dash pot") is placed in series with the assembly. This element provides a force that is proportional to the velocity of the proof mass relative to the frame, v_{rel}, as,

$$F_{damper} = bv_{rel} = b\frac{dx_{rel}}{dt}$$

where b is the viscous resistance coefficient of the damper, or damping coefficient, as discussed above in Section 2.5. It should be noted that in practice this damping is provided by trapped gas (e.g., air) around the proof mass (squeeze-film compressive damping). This gives an overall second-order force equation of,

$$F = m\frac{d^2x}{dt^2} = kx_{rel} + b\frac{dx_{rel}}{dt}$$

As mentioned in Section 2.5, the mechanical resonant frequency and quality factor are given by,

$$\omega_o = \sqrt{\frac{k}{m}} \quad \text{and} \quad Q = \omega_o\frac{m}{b} = \frac{\sqrt{mk}}{b}$$

and the displacement magnitude versus acceleration frequency (ω) for a sinusoidal excitation acceleration is given by,

$$|x(\omega)| = \frac{1}{\sqrt{\left(1 - \frac{\omega^2}{\omega_o^2}\right)^2 + \frac{\omega^2}{Q^2\omega_o^2}}}\left(\frac{m}{k}\right)\left|\frac{d^2x}{dt^2}\right|$$

In terms of design, the above equations provide considerable insight. For a second-order system, ideal ("flat") transient response occurs when $Q = 1/2$. This case is referred to as critical damping (note that the damping factor ζ is sometimes used, where $2Q\zeta = 1$). This is controlled via the proof mass, the spring constant, and the damping coefficient. In the majority of passive micromachined accelerometer designs being manufactured at present, the spring constant corresponds to the stiffness of a silicon beam (or beams) and the damping coefficient corresponds to squeeze-film effects in gaps between the proof mass and the supporting structure.

While the beam geometry and material composition generally allows a great deal of control over the spring constant, squeeze-film effects are more difficult to design (or model) and are sometimes handled on a purely empirical basis. At low frequencies and large gaps, squeeze-film effects produce a damping, or loss, while at high frequencies with narrow gaps, an elastic spring effect is produced. (It should be noted that squeeze-film effects can be totally removed by operating a device in vacuum, and this is often a useful technique to allow direct measurement of the otherwise obscured properties of many micromachined devices.) In studying the micromachined accelerometer designs discussed below, it is extremely useful to identify the structural elements that control these important design parameters.

FORCE-BALANCED ACCELEROMETER CONCEPTS

Passive accelerometers rely on spring effects (generally compliant members) whereas active rebalance use some form of closed-loop force generation mechanism to hold the proof mass in place (the required force generation signal is thus proportional to the acceleration of the proof mass). Active rebalance schemes used in the past have included magnetic and electrostatic sensing and torquing (a large number of the devices have been developed prior to the advent of micromachining). In both cases, the spring constant of the rebalancing force is a prime determinant of an accelerometer's sensitivity. Feedback that is proportional to the integral of the acceleration signal can be used to *electronically* control the damping of a feedback-based accelerometer (similarly, feedback can be used to control the bandwidth).

STRAIN-GAUGE ACCELEROMETERS

One of the earliest examples of a micromachined piezoresistive strain-gauge accelerometer (or, in fact, *any* micromachined accelerometer) is the device made by Roylance and Angell (1979) for use in biomedical implants to measure heart wall accelerations. The accelerometers were fabricated from one silicon wafer and two 7740 Pyrex™ glass wafers anodically bonded to the silicon to form an enclosure for the proof mass. To allow for the excursions of the silicon proof mass, cavities were isotropically etched into the glass. The displacement of the proof mass was sensed using a diffused piezoresistor in the bending beam connecting the proof mass to the supporting rim of silicon. The overall dimensions of the devices were $2 \times 3 \times 0.6$ mm.

The basic process used to fabricate the accelerometers began with the etching of some alignment holes in the silicon (necessary for later double-sided alignment, illustrated below), followed by the growth and patterning of 1.5 μm of thermal oxide on the silicon. Two diffusions were then carried out, one for 10 Ω/square p+ contacts and one for 100 Ω/square piezoresistors (there was no metallization on the silicon). A thick, densified oxide (most likely LTO, but this information was not given in the paper) was deposited on the front side of the wafer, and the back-side oxide was patterned for KOH etch, which was then carried out to define the shape of the bending beam. The oxide on the front side of the wafer was then patterned and used with a KOH etch to form the air gap for the proof mass. The glass was then metallized (with an Au/Cr etch mask that was etched in 30% HNO_3, 70% HF at 48°C) to form the necessary cavities, after which the Au/Cr mask was removed and Al metallization was deposited and patterned on the glass to form bond pads. The silicon wafer, sandwiched between the two glass wafers, was then anodically bonded and diced.

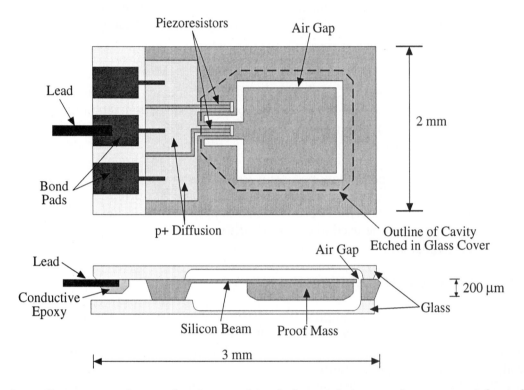

Overall structure of an early micromachined piezoresistive accelerometer. Adapted from Roylance and Angell (1979). Note that the upper piezoresistor is not subject to strain. (Not to scale.)

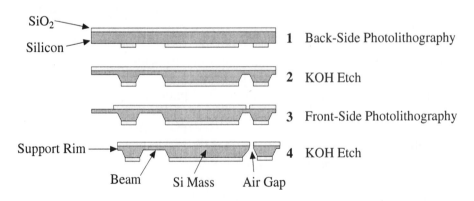

Procedure for double-sided etching to release the silicon accelerometer beam. Adapted from Roylance and Angell (1979).

These devices were used to demonstrate the detection of accelerations down to 0.001 g, allowing cardiac accelerations to be measured directly. They reported a full-scale range of $\pm$ 200 g, a sensitivity of 50 $\mu V/(g \bullet V_{supply})$, an off-axis sensitivity of 10%, a piezoresistive effect temperature coefficient of -0.2 to -0.3%/°C (i.e., the variation in sensitivity versus temperature), and a resonant frequency of 2,330 Hz. Their paper also presents a thorough analysis of the mechanics (static and dynamic) and SNR limitations of their accelerometers.

This approach provides a good example of the use of both front-side and back-side bulk micromachining to obtain more complex shapes than those available with single-sided etching. As will be seen from the examples below, modern designs rely on better controlled processes to decrease the physical dimensions of the devices, provide tighter device-to-device matching, and carefully tune their performance. However, at least for bulk micromachined devices, there are still many similarities, such as the use of double-sided processing and (often) anodic bonding to a glass substrate.

The illustration below shows a modern, bulk micromachined, strain gauge accelerometer and exhibits the refinements such as the use of silicon fusion bonding, design for overrange and damping, and optimized strain-gauge positioning. The silicon fusion bonding is used to form the precisely controlled upper layer of silicon over a predefined cavity in the underlying wafer, where the cavity forms a damping gap (shown as "Gap 2" in the illustration). The strain gauge is positioned farther back onto the supporting frame than one could guess using intuition (i.e., one might place it on the beam itself), since it was determined through finite element modeling that the maximum stress concentration is located this far back from the beam's anchor point. In addition, self-test capability is available on some units, provided through a thermally expanded (under electrical control) beam, which can buckle and push the proof mass downward, simulating a known acceleration. A similar accelerometer, with electrostatic self-test capability has also been demonstrated (Allen, et al. (1990)).

A wide variety of more complex piezoresistive accelerometers have been fabricated. For example, a three-axis device, fabricated using anisotropic bulk etching of silcon was presented by Takao, et al. (1995). The integration of piezoresistive accelerometer elements with commercial CMOS are discussed in Riethmüller, et al. (1991) and Seidel, et al. (1995). In general, however, current high-volume commercial piezoresistive accelerometers are passive designs with off-chip electronics (this is typically dictated by yield issues, although in some cases a major limiting factor is the fact that the manufacturers do not have CMOS fabrication capabilities in-house).

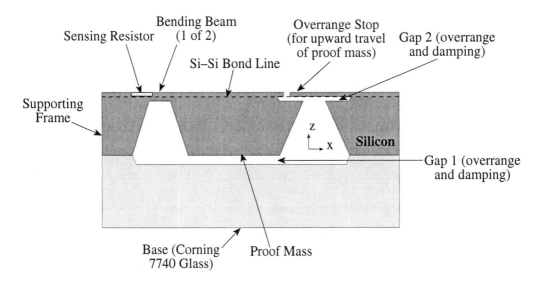

Cross-sectional view of a modern strain-gauge-type accelerometer manufactured by Lucas NovaSensor. Courtesy of K. Petersen.

CAPACITIVE ACCELEROMETERS

Several micromachined accelerometers have been fabricated that use capacitive sensing as the displacement sensing transduction mechanism. In theory, one can modify a strain-gauge-based accelerometer design to obtain a capacitive design by using the proof mass as the moving plate of a capacitor with a fixed opposite plate. In practice, the designs are quite different. Not only do the designs shown above vary the capacitor gap (versus area) for a very nonlinear response, but they would also require exquisite control over device-to-device gap dimensions. Some capacitive designs have been implemented in such a passive (open-loop) fashion. A good example of an open-loop capacitive accelerometer design based on a movable electroplated, torsional/differential capacitor is that of Cole (1991) (this device is commercially available from Silicon Designs, Inc., Issaquah, WA).

This design made use of an electroplated, asymmetrical torsional capacitor plate that would rotate under acceleration, varying the ratio of capacitance measured at two underlying sensing plates on the substrate. The sense elements used were on the order of 1×0.6 mm $\times 5$ μm thick, and the net capacitances sensed were on the order of 150 fF. Sensitivity of these devices can be controlled by varying the torsion bar length and width (for a 25 g device, they were 8 μm wide, 100 μm long, and 5 μm thick). An important design note is that the single-point support used at the pedestal minimized effects of thermal expansion coefficient mismatches between the plate material and the substrate.

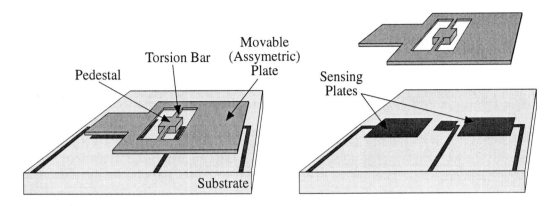

Illustration of an asymmetrical torsional plate design for a capacitive accelerometer. Adapted from Cole (1991). In this case, the torional plate is fabricated using Ni electroplating.

These accelerometers are packaged for commercial sale with a separate interface integrated circuit (IC) that produces a pulse rate output proportional to acceleration. The IC includes a modulator, demodulator, and sense amplifier to measure the changes in plate capacitance, as well as a sigma-delta A/D converter and clock generator. The sensor die and the interface IC are mounted on a substrate (presumably ceramic). The entire assembly is packaged in a 20 pin leadless chip carrier (LCC) package for an overall size of $8.9 \times 8.9 \times 2.8$ mm. Another implementation of the Cole design is the two-chip accelerometer developed by Ford Microelectronics, Inc. and described by Spangler and Kemp (1995).

An interesting discussion of the design issues involved in a parallel moving-plate design for a passive capacitive accelerometer is presented in Peeters, et al. (1991), and a useful reference of the suspension of such moving plates is Aine (1991). In the case of Aine (1991), epitaxial silicon on both sides of a wafer is used to form the suspension when undercut, with a proof mass that is the full thickness of the original wafer. Rather than epitaxial silicon, shallow boron diffusions were used to form similar suspensions by Yazdi and Najafi (1997), resulting in a potentially navigation-grade (μg sensitive) capacitive accelerometer.

Roessig, et al. (1997) presented early work on the fabrication of resonant accelerometers using surface micromachined polysilicon mechanisms. With these designs, the resonant frequency of a pair of comb-drive resonators is shifted when accelerations move a proof mass, coupling force into the resonators. Since one resonator is placed under tensile load and the other under compression, this approach provides first-order temperature compensation. To date, sensitivities as high as 45 Hz/g have been demonstrated with 68 kHz resonant frequencies. While the overall utility of this approach is not yet clear, it should be noted that to measure frequency to 1 hertz resolution requires a 1 second counting period (or a phase-locked frequency multiplier), which illustrates the limited signal frequency response of such accelerometers (which may, however, be adequate for navigation and other low-frequency applications).

FORCE-BALANCED CAPACITIVE ACCELEROMETERS

For several possible geometries, it is possible to apply electrostatic rebalancing forces to the moving mass of an accelerometer and to obtain the acceleration signal directly from the rebalance signal. The force-balancing concept has been in use for many years in macroscopic accelerometers and geophones, since it provides inherent linearization and other benefits such as the capability for built-in self-test (with the test signal applied as an electrical offset to the force-balance loop).

This approach, implemented with off-chip circuits, was used with a bulk micromachined capacitive accelerometer as presented by Rudolf, et al. (1987) for use in aerospace applications. They implemented a force-balanced system with electrically switchable ranges of ±1 g (0 to 3 kHz bandwidth), ±0.1 g (0 to kHz), ±0.01 g (0 to 1 kHz) and ± 0.001 g (0 to 100 Hz), and demonstrated μg resolution.

As discussed by Goodenough (1991) and Sherman, et al. (1992), Analog Devices is now producing the ADXL-50, a ± 50 g accelerometer for air bag deployment, which has been one of the most publicized micromachined devices to date. The device, packaged in a standard TO-100 IC package, contains a micromachined polysilicon suspended comb structure and complete on-chip BiMOS signal conditioning circuitry. The basic polysilicon surface micromachining technology is the same as that discussed in the Micromachining Techniques chapter (release via sacrificial oxide). As illustrated below, a 2 μm thick polysilicon movable proof mass/central capacitor plate with a large number of ≈ 100 μm polysilicon fingers is positioned such that the fingers are located between alternating sets of fixed fingers at a spacing of 1 μm (with the structure held 1 μm above the substrate). Each of the two sets of fixed fingers collectively forms a capacitor plate. The capacitance measurement used to sense displacement of the proof mass is differential measurement between the proof mass fingers and both sets of fixed fingers. If the proof mass is centered, the two capacitances are equal. Any displacement away from center causes one capacitance to rise while the other decreases.

The difference in capacitances is measured by applying one of two complementary (180° out of phase) 1 MHz square waves to each of the fixed plates and by measuring the voltage on the proof mass plate. If the two capacitances are equal, the measured voltage should theoretically be zero. An on-chip amplifier buffers this voltage and it is then demodulated to remove any residual components of the 1 MHz carrier. The resulting DC voltage is amplified further and used as a feedback signal, through a 3 MΩ resistor, applied to the proof mass plate in order to keep it centered electrostatically. The two fixed plates are held at 0.2 and 3.4 V DC levels, respectively, and these are the potentials against which the feedback voltage acts. The proof mass is held at 1.8 V, midway between these two voltages, if no acceleration is acting on it Thus, when the power is on, the proof mass is held in the center between the two fixed plates, and the voltage required to hold it there by feedback is used as the acceleration output signal. Due to the high gain of the feedback loop, the proof mass generally does not move by more than $\approx$ 10 nm.

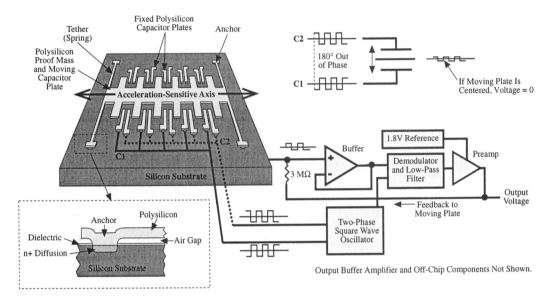

Illustration of the structure (left) and circuitry of the Analog Devices ADXL-50 integrated, surface micromachined accelerometer. The position of the movable, acceleration-sensitive capacitor plate/proof mass is sensed using a two-phase clock scheme, and a feedback signal is applied to the plate to keep it stationary. The feedback voltage is proportional to the applied acceleration and is the output signal. DC bias voltages of 0.2 V for C1 and 3.4 V for C2 are not shown. Adapted from Goodenough (1991) and the ADXL-50 datasheet (Analog Devices, Inc., Norwood, MA).

The entire chip is 3 × 3 mm in size, with the surface micromachined region taking up $\approx$ 0.6 mm^2 ($\approx$ 7%) and the 3 MΩ resistor using $\approx$ 0.4 mm^2 ($\approx$ 4%). The overall system response is from DC to 10 kHz, with a typical 6.6 mg/$\sqrt{Hz}$ noise density ($\approx$ 0.7 g RMS for full bandwidth operation, although the bandwidth is typically reduced using external filtering components in practice). The mechanical system itself (open-loop) has a resonance of $\approx$ 24 kHz, with a Q of 3 to 4. A more sensitive version, the ADXL-05 has a softer mechanical spring system, for a resonant frequency of $\approx$ 12 kHz, the same Q, and a typical noise density of 0.5 mg/$\sqrt{Hz}$ (see Chau, et al. (1995)).

Development of the ADXL-50 required merging the polysilicon microma-chining process with the existing 4 μm analog BiCMOS process, and this was certainly not trivial. Once done, however, volume production was undertaken, and this design (as well as its descendants) is available as a commodity item. In newer versions, the noise of these devices has been considerably decreased.

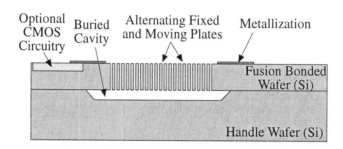

Cross-sectional illustration of a deep reactive ion etched, single-crystal silicon, force-balanced capacitive accelerometer. Adapted from van Drieënhuizen, et al. (1997).

Deep reactive ion etching with buried-cavity, fusion-bonded wafers has also been used to form high-aspect-ratio, force-balanced capacitive accelerometers (Mohan, et al. (1996) and van Drieënhuizen, et al. (1997)). Comb-type finger arrays were etched into the upper silicon wafer, with the etch stopping either on the silicon dioxide at the bond interface or in the pre-etched buried cavities at the interface. A suspended proof mass could move capacitive fingers relative to fixed fingers in separated actuation and sensing regions (the top view of these structures is similar to that of electrostatic comb-drive devices such as the ADXL-50 accelerometer discussed above). The high aspect ratio achievable with this approach (≈ 25) allows for very high off-axis stiffness, large capacitances (≈ 3 pF) and a relatively heavy proof mass (43 μg) for improved thermal noise over surface micromachined designs. A separate CMOS control circuit was used, including an oscillator, phase detector, integrator, phase shifter, and output amplifiers. Performance measurements of the two-chip system showed a bandwidth of 1 kHz, a sensitivity of 700 mV/g, and a dynamic range of 44 dB (corresponding to a resolution of 35 mg for a 5 g full-scale device or 7 mg for a 1 g full-scale device). Since electronic noise was dominant, the estimated noise floor with improved circuitry was in the sub-mg range.

The deep reactive ion etching process is CMOS compatible, allowing circuitry to be included on the accelerometer devices (Mohan, et al. (1996)), although the large capacitances available do not necessitate this, as demonstrated by the use of an off-chip processing circuit. However, an important potential problem with such designs is that in order to electrically isolate electrode and circuit regions, they must be separated from the bulk silicon by etching down to the underlying silicon dioxide layer. While resulting in excellent isolation, this necessitates forming electrical interconnects after the etching has been carried out. This has been done using wire bonding (van Drieënhuizen, et al. (1997)), but this may not be desirable in all manufacturing settings. As an alternative interconnect method, Brosnihan, et al. (1997) demonstrated the use of a technique for forming planar regions in between isolated structures that are silicon nitride lined and back-filled with polysilicon. This allows conductive traces to be formed between isolated structures at the cost of several extra process steps, with the relative trade-offs with respect to wire bonding not clear at present.

PIEZOELECTRIC ACCELEROMETERS

Macroscopic piezoelectric accelerometers are quite common, although they generally offer no useful DC response due to charge leakage and pyroelectric (temperature-dependent) effects. A classic use of micromachining technology to compensate for these factors and to realize a piezoelectric accelerometer with near DC response was that of Chen, et al. (1982, 1984) (this appears to be the first accelerometer demonstrated that includes monolithic amplification circuitry). ZnO thin-film piezoelectric elements were coupled to the gates of MOSFETs to nearly eliminate leakage paths (due to direct coupling). Unstrained ZnO compensation capacitors in series with the strained sense capacitors served to nearly cancel the pyroelectric effect. They measured decay times for stress-induced piezoelectric signals of ≈ 7 days using this approach.

Their fabrication sequence consisted of a modified PMOS process (this also could easily be done today using a CMOS process). Following field oxide growth, thin gate/sensor thermal oxide growth, threshold adjust implant, and polysilicon gate deposition, a heavy boron implant was carried out to define the source/drain regions, to dope the polysilicon gates, and to form an etch-stop for a later EDP etch to free the cantilevers. ZnO was magnetron-sputtered and etched using dilute acetic and phosphoric acids, followed by sputtered SiO_2, NiCr/Al metallization, and a final sputtered SiO_2 layer. The cantilevers were etched using either a direct EDP etch (with the ≈ 1 μm p++ region defining the cantilever's thickness) or etched from both sides of the wafers using patterned SiO_2 masks rather than the p++ etch stop to define the cantilever (thus being as thick as the substrate). Using a region of unetched silicon at the free end of a beam for a proof mass, they reported a sensitivity of 1.5 mV/g and a flat frequency response from 3 Hz to 3 kHz. Noise was reported as having a flat spectral distribution and magnitude of 0.05 g for a 3.63 Hz bandwidth (or roughly 26 mg/√Hz, for a full bandwidth noise level of 1.4 g). Further work on this type of accelerometer (and discussion of resonant and force-rebalance designs) appears in Motamedi (1987).

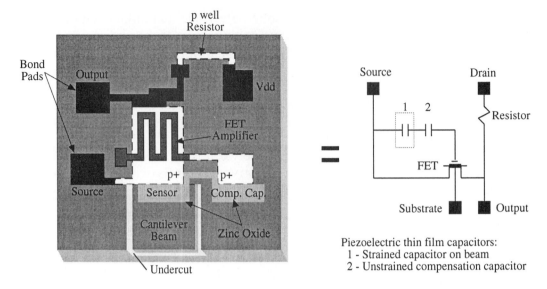

Physical structure and schematic of a micromachined piezoelectric (or so-called "PI-FET") accelerometer. Adapted from Chen, et al. (1982).

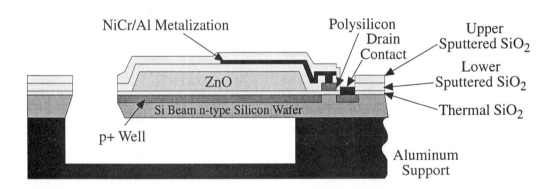

Cross-sectional view of a PI-FET accelerometer. After Chen, et al. (1982). The beam as shown would be defined by double-sided wafer etching, although the p++ region defined by ion implantation could also be used to define thinner beams. (Not to scale.)

DeVoe and Pisano (1997) demonstrated a surface micromachined implementation of piezoelectric accelerometers (also using ZnO as the sensing material). They fabricated polysilicon beams (released using an HF etch of a sacrificial layer) with a silicon nitride top layer, a sputtered ZnO piezoelectric layer, and a top platinum electrode layer. The multilayer stack was patterned using ion-milling, after which the polysilicon was dry etched using a chlorine plasma. Considerable attention was paid to balancing the residual stresses between the layers to prevent unwanted curvature of the beams (primarily due to the -80 MPa stress of the ZnO layer). They demonstrated 2% linearity over a full-scale range of $\pm$ 25 g, with a sensitivity of 0.21 fC/g. Naturally, the use of such thin-film multimorph structures has some key limitations, including the need for careful stress control and the potential for large deflections with temperature changes (with degradation of sensitivity and linearity). Nonetheless, the work presented demonstrated that piezoelectric accelerometers can be constructed in miniscule volumes using surface micromachining techniques.

TUNNELING ACCELEROMETERS

As discussed above, tunneling is an extremely sensitive means of sensing displacement at the sub-nanometer level. In practice, this approach requires closed-loop control to be practical, so all tunneling accelerometers implemented to date have required some form of this. Rockstad, et al. (1993) developed a series of mechanical transducers that utilize the tunneling effect. Electrostatic actuation is used to maintain a constant tunneling current in the face of applied acceleration so the tip does not actually move appreciably, but the voltage applied to the electrostatic plates varies in direct proportion to the acceleration. The devices were fabricated by stacking multiple wafers (often glued together using photoresist) that were patterned using laser-printer-generated masks. As mentioned above, since the tunneling tips for non-imaging applications need not be particularly sharp, they were able to realize functional tunneling-based sensors with relatively simple fabrication methods.

Descriptions of more recent efforts to fabricate tunneling-based accelerometers using a variety of approaches can be found, for example, in Rockstad, et al. (1995), Wang, et al. (1996), Grade, et al. (1996), Scheeper, et al. (1997), and Yeh and Najafi (1997).

One of the key concerns with tunneling accelerometers is that despite their high sensitivity, there is the potential for long-term drift. Whether or not this would actually be an issue in a production device, there is considerable effort to address it. Grade, et al. (1997) demonstrated that the dominant source of this low-frequency noise in tunneling accelerometers is due to thermal expansion mismatch. This result is encouraging in the sense that the effect can potentially be mitigated through improved designs.

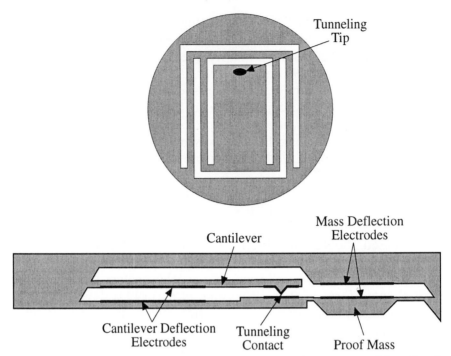

Illustration of a bulk micromachined tunneling accelerometer. After Rockstad, et al (1993).

LATCHING ACCELEROMETERS

Ciarlo (1992) demonstrated micromachined accelerometers that latch when a threshold acceleration is reached (providing a mechanical memory). There may be many applications of such accelerometers to monitor the shipping of delicate equipment, for example (i.e., for handling damage claims) if they could be conveniently interrogated. At present, no commerically available micromachined latching accelerometers appear to exist.

ACCELEROMETER SWITCH ARRAYS

As for a large number of micromachined transducers, there are many potential benefits to fabricating *arrays* of accelerometers on a single chip, including increased dynamic range and redundancy. Dynamic range enhancement with arrays is possible if the accelerometer being monitored at a given time is in its optimal operating range (i.e., most linear). By using an array of scaled accelerometers with overlapping ranges, it is possible to always locate one that is in its optimal range, providing a greater overall dynamic range than possible with a single accelerometer. This is loosely analogous to the use of a scaled sensor or actuators in biological tissues, referred to as *recruitment* (in muscle, e.g., more actuator units and those of increased strength are brought to bear as more force is needed).

Threshold accelerometers, or *g-switches,* are basically switches that close when there is sufficient acceleration acting on the proof mass to bend a beam and to make an electrical contact.   In 1972, the first micromachined (plated gold) accelerometer arrays were demonstrated (Frobenius, et al. (1972)). The state of the art in this area in 1987, as well as practical problems with existing designs, was reviewed by Robinson, et al (1987).

Loke, et al. (1991) fabricated arrays of such accelerometers with different sensitivities, acting electrically like an array of acceleration comparators. They used silicon dioxide cantilevers of different lengths in an array to obtain different acceleration thresholds. Their process consisted of depositing a low-stress CVD oxide 2.2 μm thick on a silicon substrate, followed by a 1,000°C anneal in nitrogen for 30 min for stress relief. Thermal evaporation was then used to deposit 20 nm Cr and 100 nm Au, which was patterned to serve as a mask for etching the underlying oxide at 50°C in 10:1 buffered oxide etch (this is a higher temperature than normal for densified CVD oxide). The remaining Au/Cr was then patterned to form metallization for the arrays (including switch contacts on the cantilevers, interconnects, and bond pads), and the silicon was EDP etched to form cavities for accelerometer movement. Using a Cr/Au mask, pits were etched into 7740 Pyrex™ glass, the metal mask was stripped, and 20 nm Cr and 1,000 nm Au were deposited and patterned on the glass to form the switch contacts. The silicon cantilevers were then coated with protective photoresist and diced, followed by resist removal and O_2 plasma cleaning of all surfaces. The glass and silicon wafers were then glued together with epoxy and diced. Electrostatic drive was used to pre-test the switches, and it is interesting to note that despite the use of 100 MΩ current-limiting resistors, welding tended to occur. Accelerations from 400 to 20,000 g could be measured with the arrays.

Another example of such an array of accelerometer switches was presented by Noetzel, et al. (1995), who formed laterally deflectable beams above a substrate using electroplating of Au above a patterned sacrificial layer (covered by the plating seed layer of Cr/Au) such as Al. Using a photoresist template, they formed long beams with rectangular proof masses positioned at various points along them. The ends of the beams were positioned so that if they deflected sufficiently, they would touch matching electroplated Au contacts, completing the electrical circuit. A potential advantage of this approach is that the fabrication technology is very low cost. However, a high degree of control over the electroplated metal's mechanical properties would be required if consistency between arrays and between fabrication runs was required.

MULTI-AXIS ACCELEROMETERS

It is often necessary to measure accelerations along multiple axes. While it is certainly possible simply to integrate two or more discrete, single-axis accelerometers into a single package to achieve this, there is considerable interest in the development of monolithic multi-axis designs. Aside from potential cost and size advantages, a key feature of fabricating multi-axis accelerometers on a single die is that their alignment is extremely precise, since it is defined lithographically.

An example of a bulk micromachined, piezoresistive, fully integrated three-axis accelerometer is the design presented by Takao, et al. (1997). They integrated all of the necessary signal processing electronics for the three axes into a CMOS die, with a suspended central proof mass released through a post-CMOS combination of wet and dry etching. Accelerations were sensed by measuring induced strains in the four silicon suspension beams. They reported z-, x-, and y-axis sensitivities of 192, 23, and 23 mV/g; nonlinearities of < 1%, < 1.5%, and < 1.5%; and resonant frequencies of 1.5, 1.25, and 1.25 kHz, respectively. Unfortunately, 25% cross-axis errors were measured for the x- and y-axes, but this could apparently be reduced computationally.

Surface micromachining has also been used to fabricate multi-axis accelerometers, as exemplified by a design described by Lemkin, et al. (1997). They used standard polysilicon structural layers, combined with on-chip CMOS circuitry, to achieve simultaneous multi-axis force balancing of a single proof mass. They reported a noise floor for all three axes of 0.7 mg/$\sqrt{Hz}$, z-, x-, and y-axis full-scale ranges of ±5.5, ±11, and ±11 g; resonant frequencies of 5.2, 6.6, and 6.6 kHz, respectively. Cross-coupling between axes was no worse than -36 dB.

Given market demand for multi-axis accelerometers, it is likely that fully integrated designs will be considerably refined in the near future.

5.2.3 MICROMACHINED GYROSCOPES

A *gyroscope* is a device that measures rotation rate or whole angle rotation. As such, it has many applications in transportation such as navigation, adaptive control of braking and acceleration, etc. Macroscopic gyroscopes fall into two broad categories: nonmechanical (optical) or mechanical. Nonmechanical gyroscopes make use of optical rings in which light beams are directed in counterrotating directions. Doppler shifts of the light beams are detected when the gyroscope frame rotates (this is the *Sagnac effect*). While macroscopic mechanical gyroscopes often use a spinning disk to create an inertial reference, micromachined gyroscopes typically use vibrating structures because of the difficulty of micromachining rotating parts with sufficient mass to be useful. Vibratory gyroscopes make use of the fact that if a mechanical member is vibrating along one reference axis, rotation of the gyroscope frame will couple some of the vibrational energy into another axis (or two, depending on the rotation direction).

This coupling is due to the *Coriolis effect,* which is the force exerted upon a mass, m, moving (i.e., vibrating) at velocity $\vec{v}$, when it is subjected to a rotation $\vec{\Omega}$, and it is given by,

$$\vec{F}_c = 2m\vec{v} \times \vec{\Omega}$$

For vibratory gyroscopes, two degrees of mechanical freedom are thus required, one for excitation of the vibration and an orthogonal one into which the Coriolis effect will couple force. In many types of vibrating gyroscope design, it is also critical to design mechanisms with isoelastic properties for matched frequency responses in both axes and with high quality factors (Q values) since energy coupled by the Coriolis effect is amplified by this factor. In designs where this is a goal, symmetrical structures with low loss mechanisms (such as acoustic radiation or mechanical losses) are highly desirable.

There are three general operating modes for mechanical gyroscopes (with the latter two being the most suitable for micromachined implementations):   1) *whole angle mode* — allow free movement and measure angle of deflection; 2) *open-loop vibration mode* — set up fixed vibration modes and measure the deviation (Coriolis acceleration) of these modes to determine rotation rate; and 3) *force-to-rebalance mode* — set up fixed vibration modes, but operate in a closed-loop mode (as a deviation is measured, it is continuously driven to zero by feedback circuitry).

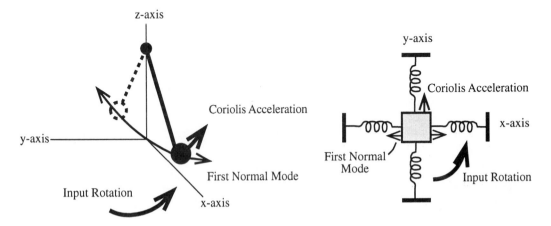

Foucault pendulum (whole angle mode) (left) and normal mode (open-loop vibration) (right) models of gyroscopes. In the former case, a pendulum is allowed to swing initially in the first normal mode along the x-axis, and rotation causes energy to be coupled into the y-axis, causing the pendulum to precess. After Putty and Najafi (1994).

As discussed by Putty and Najafi (1994), there are five classes of vibratory gyroscopes. Vibrating prismatic beams are essentially beams that are fixed at one end and vibrated along one direction, with coupled vibration sensed in the orthogonal direction. Tuning forks are, as the name implies, structures with two identical tines and a common base, with the two tines driven differentially and the resulting Coriolis force detected as a twisting force about the base of the assembly (an example of a micromachined quartz tuning fork gyroscope can be found in Söderkvist (1990)). Dual accelerometer designs use a pair of accelerometers driven 180° out of phase (as for tuning fork designs) in a direction orthogonal to their acceleration sensing axis, and rotation-induced vibrations are sensed along their sensitive axes. Vibrating shells are symmetrical structures in the shape of a hollow cylinder, ring or similar structure, where vibrational energy is stored in an elliptically shaped mode that can precess in a manner analogous to the Foucault pendulum illustrated above.

Greiff, et al. (1991) demonstrated a micromachined gyroscope (along the lines of a vibrating prismatic beam design) using a gimbal structure as illustrated below. Their gyro was driven in force-to-rebalance mode (measure force required to re-balance the structure) for improved linearity and extended bandwidth. The outer gimbal structure was driven to oscillation (3 kHz) at a constant amplitude. Coriolis force acting upon the inner gimbal is sensed and driven to zero by electrodes bridging above the device and other electrodes buried in the substrate. The moving parts of the device were undercut using an EDP anisotropic etchant, with the entire gimbal structure defined by a high concentration boron diffusion etch-stop. They reported a resolution of 4.0°/s in a 1 Hz bandwidth and demonstrated measurement of change in output signal capacitance on the order of tens of attofarads.

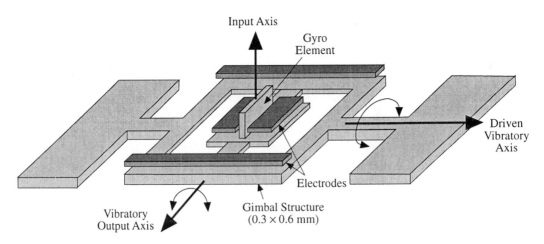

Illustration of a micromachined force-to-rebalance gyroscope. After Greiff, et al. (1991).

A piezoelectrically driven and sensed tuning fork gyro prototype was demonstrated by Voss, et al. (1997). Their design, using AlN piezoelectric material, was fabricated using bulk micromachining of bonded SOI-wafers. While actual angular rate responses were not published, the design is worth mention because the bulk silicon forming the tuning fork had a significantly larger inertial mass than any of the surface micromachined designs demonstrated, and thus could potentially provide a greater sensitivity. It is also worth noting that the authors used laser ablation to tune the two resonant modes of the tuning fork.

Putty and Najafi (1994) described an electroplated ring (hollow shell design) gyroscope built above integrated active circuits, as illustrated below. The symmetry of the device provided two identical resonance modes with the same natural frequency.

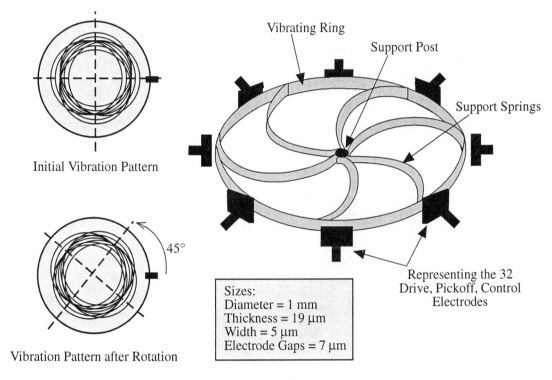

Initial Vibration Pattern

45°

Vibration Pattern after Rotation

Vibrating Ring

Support Post

Support Springs

Representing the 32 Drive, Pickoff, Control Electrodes

Sizes:
Diameter = 1 mm
Thickness = 19 μm
Width = 5 μm
Electrode Gaps = 7 μm

Illustration of micromachined resonant ring gyroscope. After Putty and Najafi (1994).

The balanced and isolated support of the post allowed for high Q and long time constant devices for high performance applications. Thirty-two electroplated capacitive electrodes were oriented around the vibrating ring to drive, sense, and thus control its vibration. On-chip CMOS circuitry converted the ring vibration into a buffered output voltage (in the future, more circuits could be integrated).

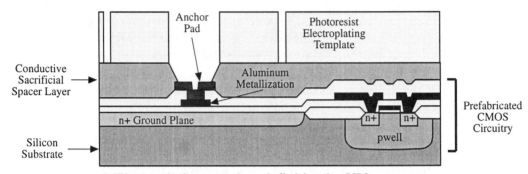

Anchor Pad

Photoresist Electroplating Template

Conductive Sacrificial Spacer Layer →

Aluminum Metallization

Prefabricated CMOS Circuitry

n+ Ground Plane

n+   n+

pwell

Silicon Substrate →

1. Electroplating template definition by UV exposure.

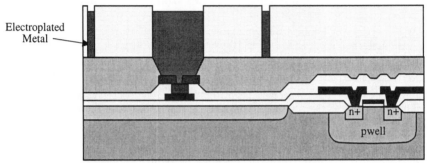

Electroplated Metal →

n+   n+

pwell

2. Electroplating of sensor element.

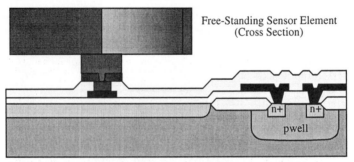

Free-Standing Sensor Element (Cross Section)

n+   n+

pwell

3. Template and sacrificial spacer layer removal.

Fabrication process for electroplated ring gyroscope with on-chip active circuits. Adapted from Putty and Najafi (1994).

The fabrication process (illustrated above) was begun with the active circuit base prefabricated. The exposed Al metallization of the underlying CMOS circuits was passivated and an Al interconnect for the ring anchor pad was deposited and patterned. An unspecified conductive sacrificial spacer was deposited and patterned to define movable portions of the sensor and to serve as a base layer for electroplating the ring structure. A template layer was deposited and patterned using UV lithography, followed by the electroplating of the sensor (ring) structure and fixed electrodes. This was followed by removal of the template and sacrificial layer.

Their measured results included a Q of 2,000 and a resolution of 0.5°/s in a 10 Hz bandwidth. A follow-up paper by Sparks, et al. (1997) discusses improvements to the design, which was then implemented as a two-chip set (gyro resonator with simple CMOS buffers/amplifier and a more complex mixed-signal CMOS control IC, vacuum packaged in a ceramic dual in-line carrier). The reported performance for the packaged gyro system includes a rate range of $\pm$ 100°/s, bias drift < $\pm$ 10°/s, sensitivity drift < $\pm$ 3%, nonlinearity < 0.2% , noise level < 0.5°/s RMS, bandwidth of 25 Hz, and a supply current of 35 mA at 5 V.

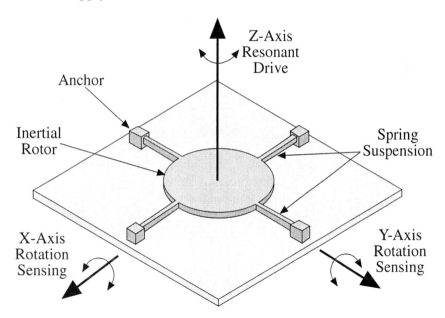

Illustration of the operating principle of a two-axis micromachined gyroscope with z-axis resonant drive. Rotation about either the x- or y-axis results in a Coriolis angular acceleration, and thus a tilting oscillation in the other of the two in-plane axes. Adapted from Juneau, et al. (1997).

Multi-axis micromachined gyroscopes have also been demonstrated and offer the same advantage of lithographically precise inter-axis alignment available in multi-axis accelerometers. Juneau, et al. (1997) described a polysilicon-based, surface micromachined, two-axis gyroscope with on-chip circuitry. Their design made use of a z-axis resonantly driven angular rotor, suspended by four orthogonal springs. Any rotation of the substrate around the x-axis would induce a Coriolis angular acceleration about the y-axis, in turn causing a tilting oscillation of the rotor about the y-axis. Similarly, rotations of the substrate about the y-axis result in oscillation about the x-axis. Different sense modulation frequencies were used for each of the two axes, with two pickup electrodes for each. Variations in processing produced devices with well-matched modes and low noise but with worse cross-axis sensitivity, and devices with poorly matched modes, high noise, and better cross-axis sensitivity. A proposed means to avoid the trade-off nature of these parameters was to employ closed-loop operation in future designs.

Generally speaking, gyroscope sensitivity is related to the moment of inertia of the moving element. Miniaturization reduces this moment, and therefore micro-machined gyroscopes suffer from sensitivity problems (as do micromachined accelerometers) compared to macroscopic competitors. Due to the broad interest in gyroscopes in consumer applications including vehicle stabilization, video camera stabilization, and personal navigation, improving them is likely to remain a topic on which a considerable amount of energy will be expended. It is likely that micromachining approaches will succeed, although very low cost gyroscopes made using injection molding, metal stamping, and other conventional technologies currently dominate the marketplace.

5.2.4 MECHANICAL PRESSURE SENSORS

Pressure sensors represent another fairly mature and commercially viable area for micromachined mechanical sensors. Examples of a wide variety of designs will be discussed below, following a brief description of the units of measure and basic operating principles. Only mechanical pressure sensors are discussed here (thermal conductivity, or Pirani-type pressure sensors are discussed in the Thermal Transducers chapter).

There is an enormous range of application areas for sensing pressure in liquids and gases. Examples include automotive manifold air pressure and other systems (tires, hydraulics, oil, etc.), environmental (heating, ventilation, and air conditioning) control, aerospace systems, medical (arterial blood pressure, etc.), and many others. Micromachining technology has been able to make large inroads into this area due to improved performance, reduced size, and (sometimes) reduced cost. However, as for most micromachined devices, particularly in commodity markets, the cost of the sensor die itself is often just a fraction (sometimes as little as 20%) of that of the total tested and packaged sensor.

PRESSURE MEASUREMENT UNITS

There are many unusual units for measuring pressure, but the preferred (SI) unit is the Pascal (Pa), which is 1 N/m^2. The reader will undoubtedly encounter most of these units when researching this topic, and thus a table of conversion factors is given below.

Unit	Conversion to Pa
Atmosphere	1.01325×10^5 Pa
Bar	10^5 Pa
Inch of mercury (0°C)	3,386.4 Pa
Millimeter of mercury (0°C) or torr	133.32 Pa
Inch of Water (0°C)	284.8 Pa

Table of pressure conversion factors relative to the SI unit, the Pascal.

Most pressure sensors used today do not use the old-fashioned "fluid barometer" principle wherein the height of a column of liquid is measured as an indicator of pressure, but, instead, they use sealed gas or vacuum-filled cavities (these are referred to as *aneroid* pressure sensors). The basic operating principle of such an aneroid pressure sensor is to couple the pressure to be measured to one surface of a membrane and to measure its deflection (knowing the pressure-to-deflection transfer function). Determining or designing the transfer function between pressure and the output signal is not always trivial, but for basic membranes, there are analytically derived solutions that can serve as at least first-order design guides (again bearing in mind that many of the membranes used in practice will be composites of thin-film layers). Microphones and related acoustic sensors are also pressure sensors, typically with high sensitivities and no appreciable DC responses. To measure the displacement, any suitable sensing technique can be used, including strain gauges, capacitance, piezoelectric (if no DC response is required), optical, tunneling, etc. Usually, strain gauge or capacitive sensing is used in commercial pressure sensors.

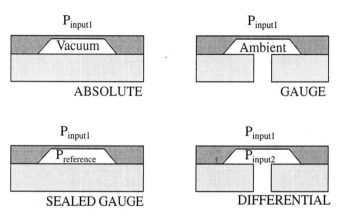

Types of pressure sensor designs commonly implemented in micromachined form. After Bryzek, et al. (1991).

As illustrated above, pressure sensors can be built to measure pressure relative to a sealed reference cavity (e.g., containing a vacuum) or differentially using two input ports. It should be noted that for sealed cavity designs, a vacuum is the preferred since there will be no temperature-dependent pressure changes in the reference pressure.

As discussed in van Mullem, et al. (1991), for a circular clamped membrane, the deflection of the center can be related to the pressure, P, and other physical (design) parameters by,

$$\frac{Pr^4}{Eh^4} = \frac{16}{3(1-v^2)}\left(\frac{y}{h}\right) + \frac{7-v}{3(1-v)}\left(\frac{y}{h}\right)^3$$

where,
P = applied pressure, in Pa
r = diaphragm radius, in m
E = Young's modulus, in Pa
h = diaphragm thickness, in m
v = Poisson's ratio
y = deflection at the diaphragm's center, in m

As for many electronic systems, such a system has linear (*small signal*) and nonlinear (*large signal*) operating regions. For small signal operation (y < h/2), the first (linear) term of the equation approximates the behavior of the diaphragm and the cubic term dominates for larger deflections. *Corrugated* membranes can provide both increased net deflection for equivalent loads and increased linear operating range (useful references are di Giovanni (1982) and Jerman (1990)). Such membranes have been used for many years in "macroscopic" devices and have more recently been adopted in micromachining applications. The deflection/pressure relationship then becomes,

$$\frac{Pr^4}{Eh^4} = A_p\left(\frac{y}{h}\right) + B_p\left(\frac{y}{h}\right)^3$$

where,

$$A_p = \frac{2(q+2)(q+1)}{3\left(1-\left(\frac{v}{q}\right)^2\right)} \quad \text{and} \quad B_p = \frac{32}{q^2-9}\left(\frac{1}{6} - \frac{3-v}{(q-v)(q+3)}\right)$$

and q is the "corrugation quality factor," which is defined for *sinusoidal corrugations* (generally only approximated in micromachining applications) by,

$$q = \sqrt{\frac{s}{L}\left(1 + 1.5\left(\frac{H}{h}\right)^2\right)}$$

where,
 s = corrugation arc length
H = corrugation depth
L = corrugation spatial period

(Note that q = 1 for a flat diaphragm and that the exact corrugation shape has little influence on q so that sinusoidal approximations for rectangular cross sections are generally reasonable.)

   Van Mullem, et al. (1991) give an excellent description of the relative performance improvements available through the use of corrugations. They fabricated test membranes using a simple sacrificial layer process wherein the sacrificial layer (aluminum) was etched only part way through to the substrate and then used as a template for forming polyimide membranes. They pointed out that for the types of micromachining processes employed, they could achieve a maximum diaphragm quality factor of ≈ 10 (limited by minimum diaphragm thickness and etching aspect ratios), but that this provides an extension of the linear operating range over *two orders of magnitude*. Additional discussion of the use of corrugated silicon diaphragms can be found in Jerman (1990) (in this case, the diaphragms were fabricated by etching the corrugations into silicon, diffusing in an etch stop, and then bulk etching to form the membrane).

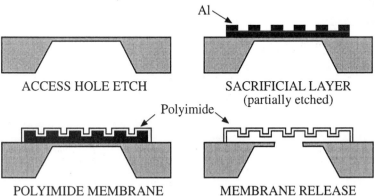

Sacrificial layer process for fabricating corrugated membranes. After van Mullem, et al. (1991).

   Most commercial diaphragm-type pressure sensors do not as yet use corrugated membranes, but they achieve impressive performance nonetheless. It may, however, be worth the extra effort to take advantage of this approach in new designs.

PIEZORESISTIVE PRESSURE SENSORS

A typical structure of a piezoresistive pressure sensor is a planar silicon diaphragm formed by electrochemical or dopant-selective anisotropic etching (bulk micromachining). The majority of currently shipping micromachined pressure sensors are of this type. Piezoresistors are placed near the edge(s) of the membrane, and, within the linear operating range of the sensor, provide an electrical output that is proportional to the deflection of the membrane and hence the pressure (the piezoresistors in these devices actually measure strain at the edges of the membrane).

In practice, four piezoresistors are used, arranged in a Wheatstone bridge so as to maximize the output signal. In addition, temperature compensation and trim resistors are often included (note that this type of design applies to accelerometers and other piezoresistance-based sensors as well in some cases). A typical circuit configuration is illustrated below. Even an uncompensated piezoresistive pressure sensor can provide accuracies on the order of 1% over a 55°C operating temperature span. However, temperature coefficients can be a major problem for automotive and other application areas where wide temperature variations are routine.

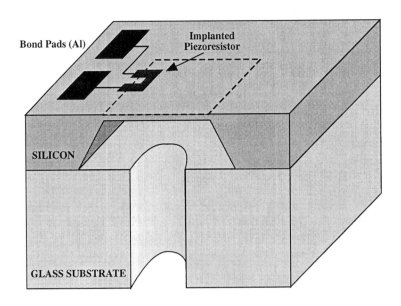

Simplified diagram of a typical piezoresistive bulk micromachined diaphragm silicon pressure sensor (differential configuration shown).

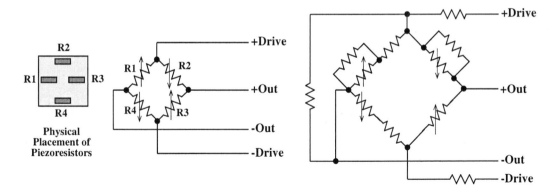

Illustration of simple (left) and temperature-compensated (right) bridges as typically used in piezoresistive pressure sensors. After Bryzek, et al. (1991).

For low-pressure operation, membrane thickness must be decreased to maintain usable deflections; yet membrane stress increases at a far greater rate, ultimately limiting this type of scaling. For such applications, the membrane is typically stiffened in sections (by adding *bosses*) to limit overall diaphragm deflection and thereby limiting the stress in the diaphragm. The use of bossed membranes to obtain 500 times overpressure protection in an micromachined pressure sensor is discussed in Christel, et al. (1990), and further discussion of bossed designs can be found in Mallon, et al. (1990).

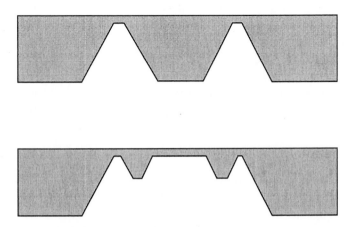

Examples of bossed membranes that are typical in low-pressure bulk micromachined pressure sensors. After Bryzek, et al. (1991).

INTEGRATED PIEZORESISTIVE PRESSURE SENSORS

An interesting example of a commercially available integrated piezoresistive pressure sensor is the Motorola MPX5100 series (Motorola, Inc., Phoenix, AZ). A previous bipolar circuit process (unfortunately not designed around a (100) wafer orientation) was adapted to incorporate a bulk micromachined membrane fabricated using back-side etching. On-chip temperature compensation and (active) signal conditioning circuitry (bipolar circuitry) are included on some devices in the form of three operational amplifiers. These devices and their descendants are now in volume production and provide on the order of ± 1% accuracy over a temperature range of -40 to +100°C. The processes are not published but can be inferred to be quite similar to the bulk micromachined pressure sensors described above, starting with the bipolar process steps.

SURFACE MICROMACHINED PIEZORESISTIVE PRESSURE SENSORS

Guckel (1991) described one of the few attempts to realize fully surface micromachined pressure sensors. To date, there still do not appear to be any shipping in volume despite the potential cost savings achievable with the smaller surface area micromachined structures. The idea behind this work was to make such a purely surface micromachined pressure sensor largely of a single material to avoid thermal mismatch problems. A surface machined cavity ("pill box") approach was chosen (as illustrated below), providing an inherent overpressure stop.

Almost perfectly planar oxide-filled cavities were formed in the starting wafer, to be later etched out to form the sealed reference cavity, taking advantage of the fact that for each micron of SiO_2 grown thermally from silicon, theoretically 0.43 μm of silicon must be consumed. Approximately 750 nm of thermal SiO_2 was grown (masked with silicon nitride) in regions that would form the cavities, using up approximately one-half as much silicon. The oxide was then stripped in HF, and another 750 nm was grown. Since the oxide is roughly half as dense as the silicon, this second oxidation ensured that the oxide layers in the cavities were nearly flush with the wafer's surface. It should be noted that the formation of the initial cavities on the order of 0.3 μm deep could have been done using a plasma etch or other process as well, as long as they did not produce rough surfaces.

Piezoresistive elements (doped polysilicon) were formed, sandwiched between silicon nitride films on top of the polysilicon plates that were fabricated above the cavities. An implant dose of 1.75×10^{15}/cm^2 at 80 keV results in a near zero temperature coefficient of resistance (TCR) for the (p-type) polysilicon resistors. The sacrificial oxide beneath the plates was then removed and the resulting empty cavities were vacuum sealed using a reactive sealing technique (as discussed in the Micromachining Techniques chapter).

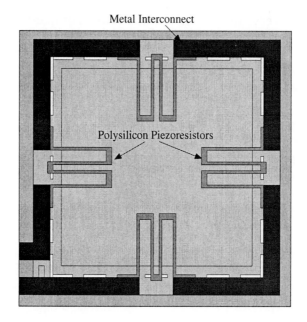

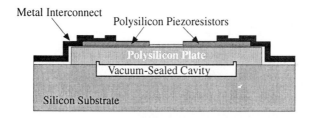

Illustration of a surface micromachined pressure transducer. After Guckel (1991).

The pressure transducers fabricated in this way apparently worked quite well, but their quantitative performance was not reported in Guckel (1991). However, referring to Burns (1988), it can be seen that devices with full-scale ranges of 5 to 500 psi were fabricated. Example sensitivities were in the range of 0.54 to 1.3% of full scale for devices ranging from 5 to 25 psi full scale, with linearity on the order of 0.2% at fixed temperatures and a temperature coefficient of 143 ppm/°C.

CAPACITIVE PRESSURE SENSORS

Capacitive sensing mechanisms can readily be employed to realize pressure sensors, despite the fact that this mechanism is inherently nonlinear (since the capacitance is inversely proportional to gap width). As for other applications of capacitive sensing in mechanical transducers, the near zero temperature coefficient of the approach is very attractive. Ford Motor Company introduced the first silicon capacitive manifold air pressure sensor in 1979 (termed the "SCAP"), followed by micromachined capacitive pressure sensors for implantable/clinical use (with and without monolithic circuitry) reported by Sander, et al. (1980), Ko, et al. (1982), Lee and Wise (1982), and a number of others since that time.

Since the membranes employed are typically clamped at all edges, the capacitance of such membrane structures is *not* given simply by the parallel-plate capacitor equation, but it can be used as a crude starting point,

$$C = \frac{\varepsilon_o A}{d}$$

which gives a ΔC in terms of change in gap width, Δd, of,

$$\Delta C = -\frac{\varepsilon_o A}{d^2} \Delta d$$

For non-parallel plates, it is necessary to integrate the local gap height, $d(x,y)$ over the entire plate area,

$$\Delta C = C_o - \int_y \int_x \frac{\varepsilon}{d_o - d(x,y)}\, dxdy$$

where d_o is the initial (parallel) gap width. Quite often, finite element or other numerical approaches are used for this purpose (or linearized small-signal solutions may be used for small membrane deflections).

An interesting example of a micromachined capacitive pressure sensor is an acutely implantable blood pressure sensor developed at the University of Michigan in the 1980's (see Chau and Wise (1988) and Ji, et al. (1991)). They fabricated the pressure transducer membranes using boron-selective EDP etching of structures that were anodically bonded to a glass substrate pre-etched to form cavities for lead attachment and metallized to form the lower capacitor plate and interconnections to a separate CMOS processor chip. As illustrated below, the membrane formation process was carried out by pre-etching depressions into a silicon wafer, oxidizing and patterning the oxide, and performing a deep boron diffusion to define the membrane frame. This was followed by a shallower boron diffusion to form the membrane and the deposition of a silicon dioxide/silicon nitride electrical insulating layer on the surface of the diaphragm (all other dielectric materials were removed). The silicon wafer was anodically bonded to a metallized glass wafer containing the opposite capacitor plates, bond pads, and etched grooves to accept interconnect wires. The silicon wafer was then etched away in EDP, leaving behind the membranes.

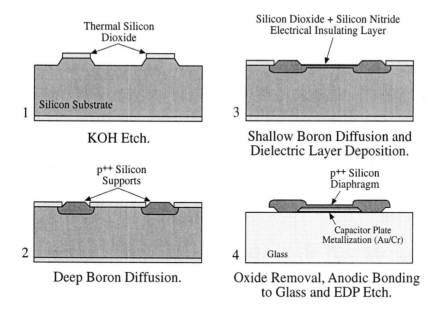

Process sequence for boron-selective etch capacitive membrane pressure sensor. After Chau and Wise (1988).

For $290 \times 550 \times 1.5$ µm sensor diaphragms with a capacitor gap of 2 µm, the zero pressure capacitance was 490 fF at 37°C. The pressure sensitivity at that point was 0.41 fF/mm Hg (3.1 aF/Pa), increasing to 1.39 fF/mm Hg (10.4 aF/Pa) at a pressure of 500 mm Hg (66.7 kPa, less than 1 atm). The measured capacitance was used as the frequency-determining capacitance in a simple Schmitt-trigger oscillator whose output signal was transmitted out as a modulation of the supply current. Thus only two wires were required to energize and read out pressure from the devices. The nonlinearly scaled output frequency was digitized by counting pulses and linearized using a compensation look-up table stored in a nonvolatile memory. The original processor chip was implemented in a 10 µm E/D (enhancement/depletion) NMOS process, but the later work was done using a 3 µm CMOS process, resulting in an idle current of < 10 µA.

As illustrated below, the sensor and circuit dice were mounted on the same glass substrate. The completed assemblies were 350 µm $\times$ 100 µm $\times$ 1 mm, including electronics.

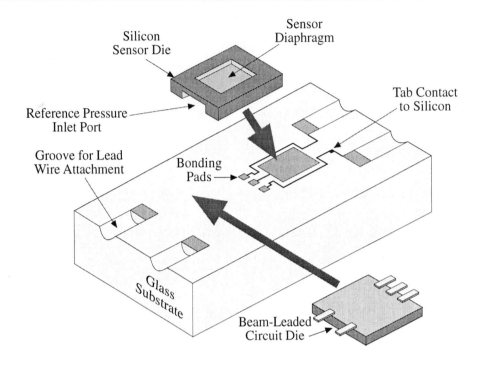

Illustration of the overall construction of a catheter-based in vivo pressure sensor. After Chau and Wise (1988).

Zhang and Wise (1995) and Chavan and Wise (1997) described a capactive pressure sensor fabricated using a similar boron diffusion/sacrificial silicon/glass substrate process, but they fabricated the sensors in scaled arrays. The scaled array concept allows the dynamic range of the array to greatly exceed that of any individual sensor (this approach can be applied to a wide variety of scalable sensors and actuators).

Another interesting biomedical application of capacitive pressure sensors is in intraocular (within the eye) pressure monitoring (useful in diseases such as glaucoma, wherein the pressure in the anterior chamber of the eye becomes abnormally high). As discussed by Collins (1967, 1970), passive implantable pressure sensors with wireless readout capabilities were realized using miniature capacitive pressure sensors and inductors to form pressure-sensitive tuned LC resonators. The resonant frequency of the sensor could be determined using a so-called "grid-dip" oscillator (a circuit in which the oscillation frequency is determined by a tunable LC resonant tank and in which the amplitude can be diminished by inductive coupling of a nearby resonator of appropriate frequency). By scanning the grid-dip oscillator's frequency, the resonant frequency of the pressure sensor (and hence the pressure) was measured via the location of the maximum attenuation of the oscillator.

The original designs of Collins (1967, 1970), typically a 5 mm diameter by 1 to 2 mm thick and cylindrically shaped, were fabricated by hand. On both sides of a hollow cylindrical glass support, thin glass or mylar membranes were attached, each connected to flat, 50 μm thick copper spiral coils that were positioned in close proximity within the sensor. When external pressure deflected the glass membranes inward, the two coils were brought closer together, increasing their mutual coupling and decreasing the resonant frequency.

More recently, Rosengren, et al. (1994) demonstrated such resonant pressure sensors (proposed by Bäcklund, et al. (1990)) made using bulk micromachined silicon pressure sensors and external wire coils. The pressure-sensing membranes were fabricated using KOH etching of high-resistivity silicon wafers (> 14 kΩ•cm) to prevent coupling losses, and the membrane region was doped to 0.1 Ω•cm to form one of the capacitor plates. The other capacitor plate was formed using a second high-resistivity oxidized silicon wafer to which the membrane-bearing wafer was fusion bonded (forming a sealed vacuum reference cavity). Once the pressure sensors were diced, a 5 mm diameter, 6 to 12 turn coil of 50 μm diameter gold wire was formed around each sensor die, and electrically connected in parallel with the sensing capacitor. Each assembly was then sealed in silastic (Dow Corning 3140 RTV). Measurements on earlier designs indicated a pressure sensitivity of 1 kHz/mmHg over the range of 0 to 80 mmHg, a quality factor (Q) of $\approx$ 30, and a resonant frequency of $\approx$ 1 MHz. The authors predicted that, by using the low-resistivity silicon, the quality factor would be > 100 and the resonant frequency could be increased to 40 MHz.

An additional example of an integrated pressure sensor with on-board telemetry was presented by Carr, et al. (1995). They used a single-transistor Hartley oscillator, an on-chip inductor, and the capacitive pressure sensor to form an externally powered, pressure-tuned LC oscillator.

Further examples of capacitive pressure sensors, but with fully monolithic circuitry, can be seen in papers such as Suzuki, et al. (1985, 1987), Kress, et al. (1990), Schorner, et al. (1990), Kung and Lee (1991), Shöneberg, et al. (1991), Kandler, et al. (1992), Dudaicevs, et al. (1993, 1995).

PRESSURE SWITCHES

In several applications, the measured pressure needs only to be compared against a fixed reference level and an on or off ("one-bit" D/A conversion) output produced. Such needs can be addressed by pressure switches, as long as their overall cost is lower than, for example, a linear pressure sensor and a comparator circuit. Huff, et al. (1991) demonstrated pressure switches based on bistable membranes that can be in only two states, buckled upward or downward. This mechanically derived hysteresis proves to be very useful for threshold pressure switches.

Their process involved pre-etching cavities in a silicon substrate, fusion bonding and thinning a top wafer, patterning the switch membranes, and providing electrical contacts. A key point involves the formation of the initially upward buckled membranes. During fusion bonding at 1,000°C in O_2, oxygen is trapped in the preformed cavities (approximately 0.8 atm at room temperature). Huff, et al. (1991) heated the bonded wafers in a ramp from 600 to 1,050°C to expand the trapped gas sufficiently to load the silicon past its yield point and plastically deform the membrane, resulting in an upward buckled state.

Further examples of micromachined pressure switches can be found in Allen (1987), de Bruin, et al. (1990), and Hiltmann, et al. (1997).

RESONANT PRESSURE SENSORS

Typical piezoresistive or capacitive pressure sensors generally require fairly high precision analog circuitry to condition and amplify their small output signals. Those types of pressure sensors may therefore be subject to drift and noise due to imperfections in those circuits. Resonating structures make use of the fact that measurement of frequency is one of the most robust and high precision methods available, and also that their resonant frequency is generally not a function of the imperfections of the electronics. Petersen, et al. (1991) described a very high precision (100 ppm accuracy over automotive temperature range) resonant beam pressure sensor fabricated using silicon fusion bonding.

The resonating pressure sensor was fabricated by the following process. Shallow pits were etched into n-type substrates, and p-type deflection electrodes were diffused in and above the pits, followed by fusion bonding of a second wafer above the first. The top wafer was then ground and polished down to a thickness of 6 μm. A passivation layer was then formed on the top wafer and the sensing piezoresistors were formed using ion implantation, after which contact holes for metallization to connect to the diffused deflection electrodes were etched. Bond pads and interconnect metallization were then deposited and patterned, followed by etching of the diaphragm from the back of the wafer. Finally, two slots were etched next to the beam to release it over the buried cavity.

The beam was resonated by applying signals as small as a few millivolts between the beam and the deflection electrode and providing feedback from the piezoresistors in the beam. Pressure applied to the underlying wafer's diaphragm increases the tension on the resonating beam (like a guitar string), increasing its resonant frequency. The frequency shift versus pressure for the structure proved to be nearly perfectly linear.

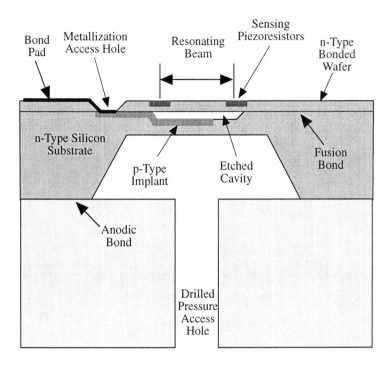

Cross-sectional illustration of a micromachined resonant pressure sensor. After Petersen, et al. (1991).

Further discussion of similar micromachined resonant pressure sensors can be found in Ikeda, et al. (1990a, 1990b), Parsons, et al. (1992), Stemme and Stemme (1990, 1992), Welham, et al. (1995), and Burns, et al. (1995). Another type of resonant pressure sensor relies on the dependence of resonant frequency on the pressure-dependent viscosity of a gas film trapped between two membranes. Such devices were described by Andrews, et al. (1993) and Gutierrez, et al. (1997). The latter group described a micromachined resonant pressure sensor with a range of 10^{-6} to 10^{-1} torr.

5.2.5 MICROPHONES

Microphones are transducers that convert acoustic energy into electrical energy and are essentially "leaky" pressure transducers, since it is undesirable for them to sense DC pressure changes (to prevent overloading of the diaphragms with barometric pressure changes). At the same time, they are required to be as sensitive as possible. In general, microphones are ubiquitous in communications devices, consumer electronics, etc. Micromachined microphones could be attractive for volume and cost-sensitive applications such as hearing aids and surveillance devices, where competition on a cost basis is less difficult. It is important to note that in the conventional size domain for microphones, with volumes as small as ≈ 0.25 cm^3, there are outstanding devices available for a fraction of a dollar, making it difficult for micromachined versions to compete.

As illustrated below, microphones generally consist of a diaphragm that is caused to vibrate by impinging waves of acoustic pressure. The enclosure behind the diaphragm serves to define its amplitude response, and must include a pressure equalization port of some kind to prevent the microphone from operating as a static (DC) pressure sensor. A back plate, acoustic holes, and a back chamber allow the response of the microphone to be tuned (the back chamber provides an acoustic compliance in a manner analogous to those in loudspeakers). Although the example shown below is of a condenser microphone (discussed below), most microphones have these components in some form.

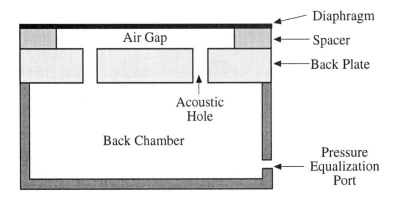

Schematic cross-sectional view of an idealized capacitive or "condenser" microphone to illustrate general aspects of the microphone structure. After Scheeper, et al. (1994).

Micromachining has enabled fabrication of piezoelectric, piezoresistive, capacitive, and FET microphones with diaphragm areas less than 1 mm^2. A thorough overview of micromachined microphones has been written by Scheeper, et al. (1994).

PIEZORESISTIVE MICROPHONES

Piezoresistive microphones make use of a diaphragm incorporating four piezoresistors in a Wheatstone bridge configuration. Piezoresistive microphones typically have sensitivities in the range of 25 μV/Pa and frequency responses in the range of 100 Hz to 5 kHz. Despite this relatively low sensitivity, an advantage of piezoresistive transduction is relatively low output impedance. An example of such a device is shown below (Schellin and Hess (1992).

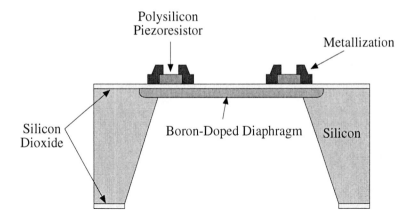

Illustration of the cross section of a piezoresistive microphone. After Schellin and Hess (1992). Note that the microphone requires a vented back chamber to operate (not shown).

CAPACITIVE MICROPHONES

Capacitive or "condenser" microphones are by far the most common type of silicon microphone to date, and consist of a variable gap capacitor. To operate, such microphones need to be biased with a DC voltage (to form a surface charge). Capacitive micromachined microphones typically have sensitivities in the range of 0.2 to 25 mV/Pa, capacitances between 1 and 20 pF, and frequency responses in the range of 10 Hz to 15 kHz. A potential disadvantage of capacitive sensing is decreased sensitivity for high frequencies (for a given area) due to air-streaming resistance of the narrow air gap. However, published data seem to indicate comparable responses to micromachined microphones realized using other transduction methods.

The illustration below shows some examples of the many types of micromachined capacitive microphones that have been implemented. These examples are cited in Scheeper, et al. (1994) and include Hohm (1986), Bergqvist and Rudolf (1990), Bergqvist, et al. (1991), Bourouina, et al. (1992), Scheeper, et al. (1992), and van der Donk (1992). Another recent example can be found in Horwath, et al. (1995). In addition, micromachined capacitive microphones for underwater use, or *hydrophones*, were presented by Bernstein, et al. (1994).

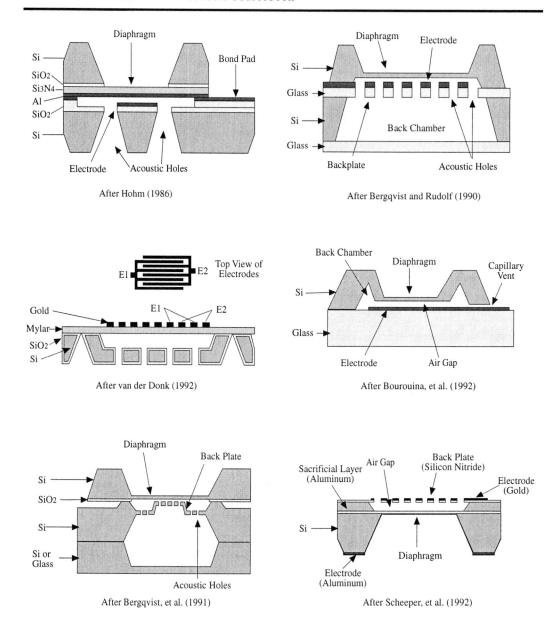

Illustrations of example micromachined capacitive microphones. After authors shown beneath each figure, and adapted from Scheeper, et al. (1994).

A polyimide membrane, capacitive microphone with an integrated CMOS preamplifier was described by Pedersen, et al. (1997). The process used relied only on low-temperature (< 300°C) processing (e.g., polyimide deposition/curing and metallization), and was thus feasible as a CMOS post-processing method.

Despite the inclusion of an on-chip amplifier, however, the reported sensitivity of 2.5 mV/Pa was not outside the range achievable with passive capacitive microphones.

Capacitive microphones optimized for ultrasound applications have also been micromachined, and an example with many background references is presented in Jin, et al. (1997).

Electret-type microphones are also capacitive transducers, but the electret material can store a permanent charge, eliminating the need for external DC biasing. Most commercial microphones of this type use Teflon™ electret materials. Hsieh, et al. (1997) demonstrated a micromachined Teflon™ electret microphone fabricated using spin-on Teflon™ (DuPont AF 1601S). To form the permanently stored charge in the electret layer, electrons were implanted into the Teflon™ using a back-lighted thyratron, which can produce pulsed electron beams of millimeter diameters and keV energies. Charge densities of 1×10^{-5} to 6×10^{-4} C/m^2 were obtained using this method, and at room temperature, the charge was stable for greater than 1 year. The reported microphone sensitivity was 0.2 mV/Pa, with a bandwidth greater than 8 kHz.

PIEZOELECTRIC MICROPHONES

Piezoelectric microphones make use of a piezoelectric material that is mechanically coupled to the diaphragm. Piezoelectric micromachined microphones typically have sensitivities in the range of 50 to 250 µV/Pa and frequency responses in the range of 10 Hz to 10 kHz. A disadvantage of piezoelectric microphones is their relatively high noise level. An example of such a design is shown below (Royer, et al. (1983)) in which a thin silicon membrane supports a ZnO piezoelectric layer and electrodes.

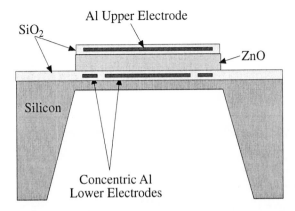

Illustration of the cross section of a micromachined piezoelectric microphone. After Royer, et al. (1983). Note that the microphone requires a vented back chamber to operate (not shown).

A more recent ZnO-based design was presented by Lee, et al. (1996) in which a cantilever, rather than a diaphragm supported on all sides, was used to increase the compliance of the structure. Careful stress management during fabrication resulted in flat cantilevers (less than 20 µm deflection) even with relatively large sizes (2 × 2 mm, for a 4.5 µm thickness). Not only did this microphone have one of the highest measured sensitivities for a micromachined device (30 mV/Pa), but it was also demonstrated as a micro-speaker, with a sound output level of 100 dB SPL (sound pressure level) at 4.8 kHz with a 6 V drive signal.

Bernstein, et al. (1997) described the use of an array of sol-gel deposited lead zirconate titanate (PZT) piezoelectric transducers as an underwater acoustic imager. The fabrication process is noteworthy because it illustrates the use of sol-gel deposition of piezoelectric materials. After oxidizing silicon wafers and patterning the oxide, a heavy boron dose was diffused in all but the oxide-protected areas where through-wafer via holes would later be etched. The wafers were then re-oxidized, and a Ti/Pt lower electrode layer was deposited. This was followed by deposition of the PZT by spinning on a sol-gel mixture of lead acetate trihydrate, zirconium n-propoxide and titanium isopropoxide in a glacial acetic acid solvent. The sol-gel was spun on, dried at 150°C to remove the solvent, rapidly heated at 400°C to remove residual organics, and pre-annealed at 600°C (6 min) to densify the layer and prevent further shrinkage. The final anneal was done at 700°C for 1 hour. The PZT deposition yielded an 0.25 µm layer, and the process was repeated many times to obtain a useful total thickness. The PZT was then wet-etched using a photoresist mask (etchant not described). The top insulator was formed using spin-on polyimide, which was patterned to allow contacts to the PZT to be formed by a Ti/Pt interconnect layer that was then deposited and patterned. Finally, an EDP back-side etch was used to form membranes defined on the actuator side by the p++ diffused boron layer. The PZT films were then poled at room temperature at 50 V DC bias (10 V/µm for the 5 µm thick PZT layer) for 2 minutes.

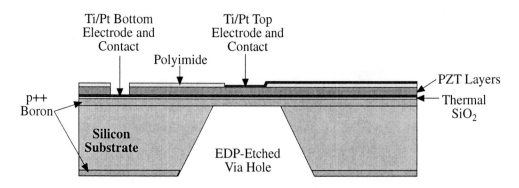

Illustration of a micromachined underwater microphone, with sol-gel deposited PZT piezoelectric transducer layer. Adapted from Bernstein, et al. (1997).

The resulting 4 µm PZT films had a relative dielectric constant, ε_r, of 1,400 and a piezoelectric coefficient, d_{33}, of 246 pC/N. The 8×8 arrays of transducers fabricated were used both to emit and to receive acoustic energy in water (the frequency range was 0.3 to 2 MHz) and successfully yielded acoustic imagery using an external acoustic source and lens.

These arrays demonstrated the use of spin-on, sol-gel PZT deposition for micromachined transducers. The principal drawbacks of this approach is the lengthy fabrication process due to the repetitive thin-layer depositions required, as well as the necessary high-temperature anneals. The latter fact unfortunately makes combining such PZT films with active circuits very difficult because the high-temperature anneals would have to be done before all CMOS steps (particularly metallization) were done, but the presence of lead in the film would likely prevent introduction of the wafers into "clean" metallization and etching systems.

MOVING-GATE FET MICROPHONES

Micromachined microphones have been constructed in which a field-effect transistor (FET) is the transduction element (this approach may also be applicable to other forms of mechanical transduction, but has seldom been used). The membrane of the microphone is biased and forms a moving gate relative to a fixed source and drain. Published results for FET microphones give sensitivities in the range of 0.2 to 6 mV/Pa and frequency responses in the range of 100 Hz to 30 kHz. An advantage of FET microphones is the low output impedance, since the preamplifier is integrated within the microphone. A disadvantage is the absence of a bias element to define a stable gate potential (allowing temperature drift) and relatively high noise levels. An illustration of a bulk micromachined, two-wafer design for such a microphone is shown below (Kuhnel (1991)).

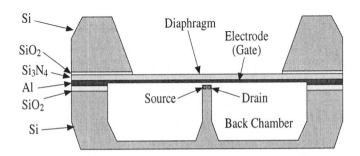

Cross-sectional illustration of a moving-gate field-effect microphone. After Kuhnel (1991).

5.2.6 TACTILE SENSORS

The idea of tactile sensors is to provide "sensory" feedback regarding contact with objects for robotic end effectors (grippers, etc.). This feedback can be very useful in such tasks as precision assembly. It is interesting to note that, while studies of normal tactile sensing in humans and primates indicate that sensing shear forces are critical for precision manipulation, the vast majority of researchers fabricating tactile sensors have concentrated on normal forces. Ideally, such sensors would perform the sensory functions of skin, providing information on contact, force, shear, velocity, 3-D shape, slip, and thermal properties. To have accuracies comparable to human skin, the sensors must be able to resolve features smaller than 1 mm. However, this is a greatly oversimplified statement. For example, in day-to-day situations, we can certainly resolve steps on the order of tens of microns (e.g., by moving a fingertip across a page of printed output from a laser printer, one can feel the raised print). Thus the successful and "human-like" application of micromachined tactile sensors for fine manipulation requires careful attention to system aspects beyond the sensor itself.

Tactile sensors are among the oldest applications of micromachining technology, dating back to the mid-1970's. The major categories of tactile sensors are piezoelectric, piezoresistive, capacitive, and optical. An excellent comprehensive overview of the various technologies is given in Nicholls and Lee (1989). Many of the sensor methods described below suffer from hysteresis (i.e., material "memory"), and this is an issue of significant interest in this area of research. Another important issue is temperature compensation, since the object contacted may be at a temperature other than that of the sensor and may thus locally heat or cool it.

PIEZOELECTRIC TACTILE SENSORS

Polyvinylidene fluoride (PVDF) is commonly used for these sensors. As mentioned in Section 5.1.3, it must be pre-treated or "poled" (with a high electric field) to make it piezoelectric, but once prepared, it has a wide dynamic range, good linearity, and excellent mechanical properties (durability, flexibility, etc.). Like all piezoelectric devices, these tactile sensors have no DC response; consequently, they are typically used to detect differential excitations. The lack of a DC response may rule out the use of such tactile sensors in many robotics applications. Esashi, et al. (1990) fabricated a thin, flexible square mesh of 64 piezoelectric sensors that would theoretically be quite rugged (airtight, waterproof) and could be wrapped around curved contours such as robotic fingers. Other examples of such devices can be found in de Rossi, et al. (1991) and Kolesar, et al. (1992).

RESISTIVE TACTILE SENSORS

Resistive tactile sensors have a DC response and thus are useful for shear and bending stress sensors, pressure sensors, etc. A single piezoresistive element in a Wheatstone bridge is a common implementation, but this will not necessarily provide adequate temperature compensation (it depends on where the bridge elements are relative to the transducer element). Some examples of micromachined piezoresistive tactile sensors can be found in French, et al. (1988) and Pati, et al. (1988). Sugiyama, et al. (1990) demonstrated a 32×32 element piezoresistive normal force sensing tactile sensor with on-chip CMOS scanning circuitry. Their 16,000 transistor integrated tactile sensor could generate force images (using external cancellation of offsets) and had a sensitivity of 1 to 2 mV•cm^2/Kg ($\approx$ 10 to 20 µV/kPa).

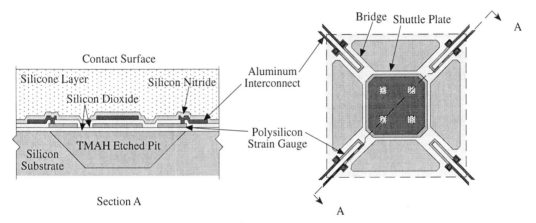

Illustration of a micromachined, CMOS-compatible shear and normal force tactile sensor. After Kane, et al. (1996). Courtesy B. J. Kane.

Kane, et al. (1996) demonstrated a micromachined CMOS-compatible normal and two-axis shear-sensing tactile sensor based on TMAH-undercut square shuttle plates (80×80 µm) of LTO supported at their four corners by LTO-encased polysilicon strain gauges. After the shuttle plates were fully undercut, a preformed elastomer layer was glued to the sensing surfaces to form as a mechanical interface and a protection layer. By algebraically combining the output signals from the four strain gauges, the two shear components and normal force could be independently resolved. The sensors exhibited sensitivities of 12 mV/kPa to shear stresses and 51 mV/kPa to normal stresses, with a noise floor (including external electronics) of 3.6 mV and minimum resolvable stresses (SNR = 1) of 0.29 and 0.07 kPa, respectively. The mechanical natural frequency of the system, including the elastomer layer, was $\approx$ 40 Hz.

As a simple alternative to piezoresistive sensors, conductive elastomer (containing conductive particles) can be used. In such materials, the overall conductivity goes up with applied pressure, and can be reasonably linear and certainly very inexpensive. The basic concept of a conductive rubber normal force tactile sensor is illustrated below (this approach is used in some keyboards). Shear sensors can also be realized with this approach if one electrode is split, and the resistance is measured differentially (this method could potentially be extended to resolve both in-plane shear components). This approach may be readily applicable to micromachined devices via silk screening or similar printing methods (see the Micromachining Techniques chapter for further information on this approach), allowing such conductive materials to be added above prefabricated active circuitry.

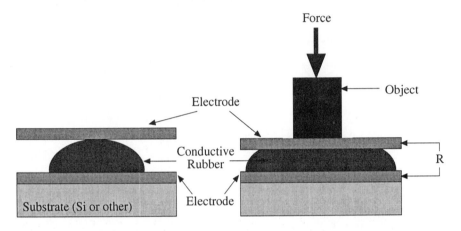

Illustration showing the basic concept of a conductive rubber tactile sensor. After Nicholls and Lee (1989).

CAPACITIVE TACTILE SENSORS

As for other mechanical transduction applications, capacitive methods can readily be applied to tactile sensors. The capacitance in such devices can be varied either by deflecting the electrodes with respect to each other, by moving a piece of dielectric into or out of the space between the plates, or by measuring the capacitance between fixed plates on the sensor and the object(s) being sensed.

Novak (1989) reported a capacitive sensor (moving electrode type) that detects both 2-D shear and normal forces simultaneously. Wolfenbuttel and Regtien (1990) described a surface micromachined polysilicon bridge capacitive tactile sensor with local signal conditioning circuitry.

Suzuki, et al. (1990a, 1990b) described a boron-doped, EDP-etched capacitive tactile sensor array wherein the capacitive elements were anodic bonded to a glass substrate, as illustrated below. Their 32 × 32 element array (1.6 × 1.6 cm overall size) was fabricated with two boron diffusions (one deep and one shallow), followed by the anodic bonding and release steps. The thick center plate for the sense capacitor was supported by thinner beams. The authors reported a sensitivity of 0.27 pF/g/element (maximum force 1 g/element) and a temperature coefficient of < 30 ppm/°C. More recently, De Souza and Wise (1997) reported the use of a similar process to fabricate such tactile sensors at spatial resolutions up to 500 pixels per inch.

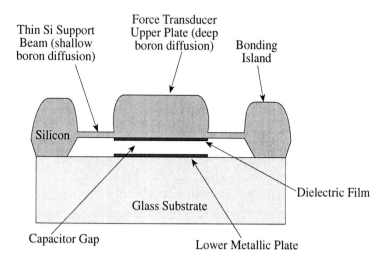

One element of a micromachined capacitive tactile sensor.   After Suzuki, et al. (1990a, 1990b).

Chu, et al. (1995) demonstrated a capacitive tactile sensor with the ability to resolve shear and normal forces (three axes). They made use of a bulk micromachining process to fabricate circular membranes with four underlying electrodes. Using a silicon protrusion above the membrane, forces applied were translated into capacitance changes that could be decomposed into the three axes of force. The authors reported sensitivities of 0.13 pF/g for normal forces and 0.32 pF/g for shear forces.

OPTICAL TACTILE SENSORS

There are a variety of ways to measure force and/or shear using optical means. One approach, discussed by Chao and Neudeck (1990), is to modulate the transmitted light through an optical fiber by bending it under pressure. Kawashima and Aoki (1987) have proposed a 2-D array of optical sensors that recovers a pressure distribution using CT (computerized tomography) decoding algorithms. This sensor is rugged (able to withstand over a Newton of force), flexible, and simple. It requires considerable external signal processing to decode its results, however. King and White (1985) and White (1987) described a tactile sensor based on the deformation of a elastomer membrane with pyramidal protrusions against an edge-lighted transparent sheet. As the protrusions press against the lighted sheet, light is reflected outward in relation to the pressure applied, and can be detected using a solid-state camera, for example. While none of these approaches was micromachined, these basic principles could be applied in such implementations.

OTHER TACTILE SENSORS

Several other types of tactile sensors have been fabricated, including those based on electric field measurement (Jiang, et al. (1991)); tunneling current changes as a function of applied pressure; and magnetoelastic sensing (materials modulate an incident magnetic field with applied stress). In addition, Wang et al. (1991) designed a PMOS ring oscillator whose frequency is modulated by forces acting on pressure-sensitive transistors.

5.2.7 BIOLOGICAL MECHANOSENSORS

It has long been critical for living organisms to perceive their external environment through mechanical means (among others, such as optical, chemical, etc.). Even a cursory study of physiology brings to light the fact that the simplest of organisms are really sensor-driven systems.

BIOLOGICAL ACCELEROMETERS

In a wide variety of vertebrate animals, the sense of balance is obtained through a multi-axis accelerometer system designed to determine the orientation of the organism's head relative to the gravitation vector. More precisely, these *vestibular* organs function as inclinometers, measuring one's angle relative to the direction considered to be "down." In many organisms, fluid moving in tubes provides this function, and in humans, three-axis tilt sensing is provided by the *semicircular canals,* the *utricle* and the *saccule,* located near the inner ear organ, the *cochlea* (discussed below). Fluid movement is sensed by shear-responsive hair cells that transduce this signal into neural impulses.

Gravity sensing is even found in single-celled organisms. For example, Fenchel and Finlay (1984) discuss gravity sensing in the ciliated protozoan *Loxodes striatus*, which must locate its optimum oxygen concentration in water by swimming up or down. Thus, it requires gravity sensors, which are provided in the form of specialized organelles known as Müller bodies, containing granules of Ba or Sr salts (apparently the sulphates). Apparently, these 2 to 4 μm diameter gravity sensors (one to four per organism) are directly coupled to the cilia that propel the organism, providing direct orientation-to-position feedback. Interestingly, as observed by Gabrielson (1995), if these sensors were only four times smaller, thermal noise would make it impossible for the organisms to differentiate up from down.

BIOLOGICAL TACTILE SENSORS

The skin, constituting the largest organ of a human, is a vast array of mechanical and thermal sensors. The major mechanical sensory modalities are described as *discriminative touch* (recognition of shape, size, and texture), *proprioception* (sense of position and movement of limbs and body), *nociception* (pain), and *thermal sense*. This is a complex and only partly understood subject, and more than a simple introduction is beyond the scope of this book.

Considering only mechanical sensors in the skin, there are four major receptor types, differentiated by the types of stimuli they respond to and thus their physical construction. As for most of the nervous system, the information output of these sensors appears to be encoded in frequency-modulated streams of electrical impulses referred to as action potentials. The sensors are located in the skin itself, and the signals they generate are transmitted quite far before being processed at a simple level in the spinal cord and finally in the brain. The four basic types of mechanoreceptors in the skin are:

1) Meissner's corpuscles — touch, rapidly adapting
2) Pacinian corpuscles — flutter, rapidly adapting
3) Ruffini corpuscles — vibration, rapidly adapting
4) Merkel's receptors — steady skin indentation, slowly adapting

These receptors differ in the rates at which they slow down their action potential frequencies in response to a mechanical input signal (rapidly adapting receptors slow down their firing rates quickly and thus can respond to rapidly changing inputs). In addition, bare nerve endings (without visible, specialized sensory structures) in the skin respond to pricking pain (as well as to thermal insults to the skin). These latter sensors can be as small as 300 to 500 nm in diameter.

The reader interested in further information on this subject is referred to Kandel, et al. (1991), which contains excellent overviews of the sensory systems themselves, as well as the coding and processing of sensory information.

BIOLOGICAL ACOUSTIC SENSORS

Another major mechanical input system is the organs of hearing, a complex series of structures. The central structure in the human auditory system is the *cochlea*, which is effectively a mechanical spectrum analyzer. The other components function to couple acoustic energy to the cochlea and to route information from it to other parts of the nervous system. The external ear gathers acoustic energy, and mechanical impedance matching between the eardrum and the oval window (entry point of acoustic energy into the cochlea) of the cochlea is carried out by three bones known as the *malleus, incus,* and *stapes* (moving inward from the eardrum).

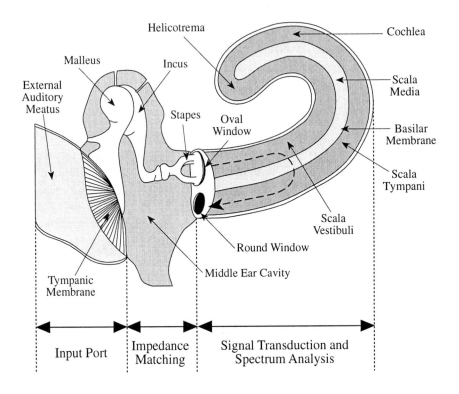

Illustration of the anatomy of (left to right) the outer, middle, and inner ear of the human, with functions indicated in engineering terms below. Adapted from Kelly (1991).

The *basilar membrane* in the fluid-filled cochlea oscillates in response to mechanically coupled acoustic energy applied to the oval window. *Hair cells* in the basilar membrane convert its mechanical movements into electrical/chemical neural signals that are routed to the brain.

The basilar membrane is narrow and stiff near the entry point for sounds (the oval window) and becomes progressively wider and more compliant near the end of the spiral. The basilar membrane also has taut cross striations akin to the strings in a piano. The stiff end of the cochlea responds to high-frequency sounds, decreasing in frequency to the end of the spiral (the *apex*), which senses low-frequency sounds. Thus the hair cells at different locations along the basilar membrane are tuned by virtue of their position. (It should be noted that a micromachined analog of the cochlea, meant as a mechanical spectrum analyzer, was demonstrated by Haronian and MacDonald (1995).)

The structure of the hair cells also varies along the length of the cochlea. They are short and stiff near the oval window and long and floppy at the apex. In addition, the hair cells have resonant properties, due to a coupled electrical/mechanical feedback system within them. The combination of these tuning effects results in extremely high-frequency resolution at the level of the hair cells. Interestingly, hair cells are also capable of *generating* acoustic energy, and acoustic emissions from the ear have been studied for some time.

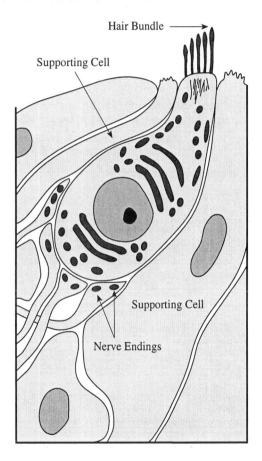

Illustration of a hair cell and surrounding supporting cells, positioned in the basilar membrane. Adapted from Kelly (1985).

It would be quite difficult to micromachine as versatile an acoustic sensor as the ear (120 dB dynamic range, exquisite frequency selectivity, long operating lifetime, very low power), but it is certainly worth studying for inspiration.

6. MECHANICAL ACTUATORS

6.1 ACTUATION MECHANISMS

By definition, mechanical actuators convert electrical (or other) energy into mechanical energy. Unfortunately, there is no "perfect" actuator technology, but, rather a series of trade-offs in terms of fabrication complexity, environmental robustness, range of motion, available force, etc. The "ideal" actuator would use little power, have a high mechanical efficiency, be robust to mechanical/environmental conditions, be capable of fast motion if necessary, have a high power-to-mass ratio, and have a linear proportionality between force/torque/speed, etc., and a control signal. In practice, not all of these issues would matter (e.g., one might require a slow, but strong actuator that is always shielded from temperature variations, such as some types of muscle fibers).

There have recently been several interesting surveys of actuator technologies, and the reader is referred to Fujita and Gabriel (1991), Hollerbach, et al. (1991), and Hunter and Lafontaine (1992) for useful comparisons and a multitude of literature references on the subject. Two general comparison tables based on the first and third of these references are shown below.

Type of Motor	Torque/Mass (N•m/kg)	Power/Mass (W/kg)
Sarcos Dextrous Arm (electrohydraulic)	120	600
McGill/MIT Electromagnetic Motor	15	200
Polyacrylic Acid/Polyvinyl Alcohol Polymeric Actuator	17	6
NiTi Shape Memory Alloy	20	6
Human Biceps Muscle	20	50
Burleigh Piezoelectric Inchworm	3	0.1

Table of general comparisions of (macroscopic) robotics motors and human skeletal muscle, in terms of torque/mass and power/mass ratios. After Hollerbach, et al. (1991).

Type of Actuator	Stress (MPa)	Strain (%)	Strain Rate (Hz)	Power Density (W/kg)	Efficiency %
Electrostatic (macroscopic composite)	0.04	> 10	> 1	> 10	> 20
Cardiac Muscle (human)	0.1	> 40	4	> 100	> 35
Polymer (polyacrylic acid/polyvinyl alcohol)	0.3	> 40	0.1	> 5	30
Skeletal Muscle (human)	0.35	> 40	5	> 100	> 35
Polymer (polyaniline)	180	> 2	> 1	> 1,000	> 30
Piezoelectric Polymer (PVDF)	3	0.1	> 1	> 100	< 1
Piezoelectric Ceramic	35	0.09	> 10	> 1,000	> 30
Magnetostrictive (Terfenol-D)	70	0.2	1	> 1,000	< 30
Shape Memory Alloy (NiTi bulk fiber)	> 200	> 5	3	> 1,000	> 3

Table of linear actuator materials. After Hunter and Lafontaine (1992). The authors noted that the values provided did not always represent optimal materials, as development is active in many of these categories and that the power needed for accessory systems, such as cooling, were not included in the calculations.

This section is organized by actuation mechanism, not by specific actuator function, since relatively few useful stand-alone actuators have been demonstrated. In general, the actuators form part of an overall functional unit, and these are described throughout the chapters in the appropriate domains of operation.

6.1.1 ELECTROSTATIC ACTUATION

The fundamental actuation principle behind electrostatic actuators is the attraction of two oppositely charged plates, as was quite familiar to Benjamin Franklin. In principle, despite the nonlinear force-to-voltage relationship in such actuators, they are very low power and simple to fabricate. They have been used extensively in micromachined devices, since it is relatively simple to fabricate closely spaced gaps with conductive "plates" on opposite sides (see Muller (1990)). In many cases, the delivered output power and efficiency are far less than those predicted by theory (often due to fringing fields, surface leakage, etc.), but electrostatic actuators are still extremely important in many applications (particularly when the actuator merely needs to move itself, as opposed to other objects). An example of this was seen above in the case of the ADXL-50 micromachined accelerometer, where electrostatic actuation was used to keep a movable polysilicon proof mass centered between sensing capacitor plates.

In estimating the force generated by an electrostatic actuator, one can begin with *Coulomb's law*, which gives the force between two point charges,

$$F_{elec} = \frac{1}{4\pi\varepsilon_r\varepsilon_o} \frac{q_1 q_2}{x^2}$$

where q_1 and q_2 are the two charges in coulombs and x is the distance separating them. If there are more than two charges, it is necessary to determine the force between each charge pair and to superimpose these vectors to find the resultant force. For most realistic electrostatic actuators, this unfortunately becomes quite complex, although suitable finite element and other computational methods exist.

For first-order approximations in many simple geometries, one can sometimes start with a parallel-plate capacitor approximation, but for most actuator shapes (i.e., cantilevers), this only holds for *very* small angles. For a parallel-plate capacitor with plate area, A, (neglecting fringing fields), the energy stored at a given voltage, V, is given by,

$$W = -\frac{1}{2}CV^2 = -\frac{1}{2}\frac{\varepsilon_r\varepsilon_o A V^2}{x}$$

and the force between the plates is,

$$F = \frac{dW}{dx} = +\frac{1}{2}\frac{\varepsilon_r\varepsilon_o A V^2}{x^2}$$

(It is interesting to note that in all capacitors, the voltage applied tends to crush the inter-plate dielectric.) From this, it is clear that the force versus distance and force versus voltage relationships are nonlinear, although in some cases they can be linearized through closed-loop control.

ELECTROSTATIC CANTILEVER ACTUATORS

A useful analysis of the relationship of applied drive voltage and deflection in a micromachined cantilever beam (intended as an optical modulator in a display system, as discussed in the Optical Transducers chapter) can be found in Petersen (1978a). The basic structure of the actuator is illustrated below.

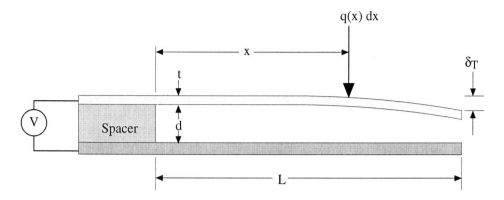

Illustration of an electrostatically deflected cantilever structure showing variable definitions for analysis. After Petersen (1978a).

Following Petersen's derivation, the goal is to obtain a relationship between applied drive voltage and tip deflection. From the mechanical engineering theory, it is known that a concentrated load at a position, x, from the fixed end of a cantilever beam of width, w, results in a tip deflection, δ_T, given by,

$$(d\delta)_T = \frac{x^2}{6EI}(3L - x)wq(x)\,dx$$

where the electrostatic force, q(x), at a distance x, is,

$$q(x) = \frac{\varepsilon_o}{2}\left(\frac{V}{d - d(x)}\right)^2$$

and where,
E = Young's modulus of the cantilever
I = moment of inertia of the cantilever
L = beam length
x = distance of force (load) from the fixed end of the beam
d = gap between cantilever and deflection electrode

The total tip deflection can be found by integrating the above equation from the fixed end to the tip of the beam (x = L),

$$\delta_T = w\int_0^L \frac{(3L - x)}{6EI}x^2 q(x)\,dx$$

To make the solution of the integral possible, one can assume a square-law curvature of the beam at any point along its length,

$$\delta(x) \approx \left(\frac{x}{L}\right)^2 \delta_T$$

This in turn yields a normalized load, F, required to produce a specified tip deflection,

$$F \equiv \frac{\varepsilon_o w L^4 V^2}{2EId^3} = 4\Delta^2 \left(\frac{2}{3(1-\Delta)} - \frac{\tanh^{-1}\sqrt{\Delta}}{\sqrt{\Delta}} - \frac{\ln(1-\Delta)}{3\Delta} \right)^{-1}$$

where Δ is the deflection at the tip (δ_T/d).

This equation for normalized load is plotted below versus normalized deflection and shows that the relationship between tip deflection and applied voltage is indeed extremely nonlinear, and that once deflection exceeds a threshold voltage, the position of the tip is unstable, and the beam spontaneously deflects all the way down.

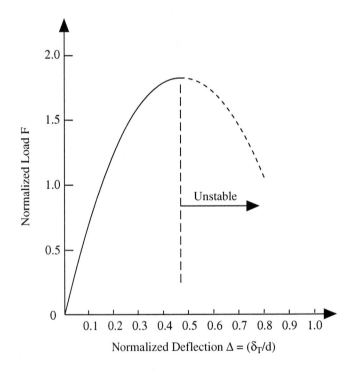

Plot showing the normalized load, F, versus the normalized deflection, Δ, for an electrostatically deflected cantilever. Adapted from Petersen (1978a). Beyond a certain threshold deflection, spontaneous collapse of the cantilever occurs.

The threshold voltage is approximately given by,

$$V_{th} \approx \sqrt{\frac{18EId^3}{5\varepsilon_o L^4 w}}$$

and appeared to be fairly accurate in correlating with Petersen's experimental results. Petersen also noted that the effective Young's modulus/moment of inertial product for such a composite structure (his were a layer of metal above an insulating layer of silicon dioxide) could be approximated to first order (and used for the EI product in the above equations) by,

$$(EI)_{eff} \approx \left(\frac{wt_1^3}{12}\right) \left(\frac{4 + 6\frac{t_2}{t_1} + \frac{E_1 t_1}{E_2 t_2}}{1 + \frac{E_1 t_1}{E_2 t_2}} \right)$$

where, t_1, t_2, E_1, and E_2 are the thicknesses and Young's moduli, respectively, of the lower (1) and upper (2) layers of the beam. This illustrates a point made above: most micromachined structures need to be treated as composite materials. Further, for such a cantilever beam structure, the first resonant frequency (bending) can be estimated using,

$$f_{R1} = \frac{3.52}{2\pi} \sqrt{\frac{EI}{vL^4}}$$

As can be seen from this reasonably tractable example (keeping in mind that fringing fields, which can be quite important at these aspect ratios, were ignored), the analyses of electrostatic structures can be rather involved. Analysis of more complex structures, such as the torsional or comb-drive electrostatic actuators discussed below, is often done using numerical approaches.

TORSIONAL ELECTROSTATIC ACTUATORS

Torsional electrostatic actuators have, in some cases, advantages over cantilever designs, particularly since they are (if supplied with dual deflection electrodes) able to deflect in two directions, rather than only one. As for the cantilever designs, they can be fabricated using a number of processes, including polysilicon with sacrificial oxide, electroplated metal with sacrificial organic layer, sputtered aluminum with sacrificial organic layer, etc.

As well as the Texas Instruments torsional electrostatic actuators discussed in Hornbeck (1995), a similar all-aluminum electrostatic actuator technology was developed in an academic setting, and all process details are available in Storment, et al. (1994). In both processes, thin aluminum torsional members (150 nm or thinner) determine most of the mechanical properties of actuators consisting of much thicker aluminum plates suspended above deflection electrodes from aluminum "posts." The process avoids wet release problems through a dry (oxygen plasma) release of the aluminum structures by removing the underlying sacrificial polyimide on which they are built. These approaches are designed to allow direct integration of aluminum actuators with prefabricated CMOS control circuits. None of the process steps entails temperatures above the 300 to 350°C limit for prefabricated (conventional) CMOS. (CMOS that is to be integrated with polysilicon structures generally must have a modified process flow to allow for the high-temperature deposition of the polysilicon prior to bond-pad and interconnect metallization). Since all of the applications of torsional electrostatic actuators to date appear to be in the optical modulator area, the reader is referred to the Optical Transducers chapter, where this material is covered in detail.

ELECTROSTATIC COMB DRIVES

Comb-drive-type electrostatic actuators make use of large numbers of fine interdigitated "fingers" that are actuated by applying a voltage between them. If the fingers are relatively thin compared to their lengths and widths, the attractive forces are mainly due to fringing fields, not parallel-plate fields. Unlike the cantilever and, to a large extent, the torsional actuators discussed above, comb-drive actuators can generate relatively large movements in the plane of the substrate. In addition, the asymmetry of the fringing fields (due to air or vacuum above, and conductive substrate below) leads to considerable out-of-plane, or "levitation" forces, as described by Tang, et al. (1990). Additional ground-plane electrodes can reduce this effect where it is not desired. As for all electrostatic actuators, there is an inherent ability to carry out capacitive position sensing using the actuator (or other structures coupled to it).

A key distinction between these types of electrostatic actuators is that in the case of the comb drive, the displacement-to-voltage relationship is much more linear. In comb drives, the capacitance is varied through changing the area, not the gap width, and since the capacitance is linearly related to area, the *displacement will vary as the square of the applied voltage*. A good reference on the theory of comb-drive actuators is Tang, et al. (1990).

This type of actuator design lends itself to fabrication using the polysilicon/sacrificial oxide surface micromachining processes discussed above, and to any other processes that are capable of producing structures that have freely moving regions that may be small vertically (typically one or two microns) but can be spread out over a considerable area of the substrate.

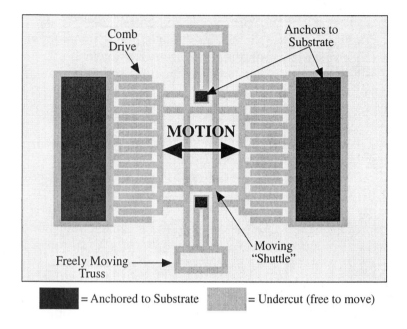

Illustration of the comb-drive electrostatic actuator concept. After Judy and Howe (1993a, 1993b). Note that the anchors to the substrate may or may not be insulated from it electrically, allowing potentials to be applied as needed.

Comb-drive actuators have been used to drive many electrostatic actuators, such as the electrostatically controlled microgripper of Kim, et al. (1990, 1992), discussed below. There have been several devices constructed wherein the moving structure's resonant frequency is used to advantage, such as high-Q mechanical filters (or oscillator elements).

FEEDBACK STABILIZATION OF ELECTROSTATIC ACTUATORS

In some situations, the highly nonlinear voltage-to-position relationships in electrostatic actuators can be a significant drawback. However, several techniques are available to apply feedback to stabilize and linearize them. As explained above, electrostatic comb drives can be stabilized using an extra set of combs to measure the position of the shuttle and to provide an appropriate bias voltage to return the shuttle to a desired position. For example, this approach is employed in the Analog Devices ADXL series of accelerometers (see Goodenough (1991) and Sherman, et al. (1992)). Torsional and cantilever type actuators can be stabilized using out-of-plane feedback via optical means, as demonstrated by Honer, et al. (1995, 1996) for characterizing aluminum torsional devices, although this technique is not likely to be useful in fully integrated devices.

As a general concept for "passive" stabilization, Seeger and Crary (1997) showed, through simulations, that placing a capacitor (conventional or MOS) in series with an electrostatic actuator can serve to stabilized it. The effect of the series capacitance is such that as the actuator begins to collapse, the voltage division between the two capacitors forces the voltage across the actuator to decrease as it closes (since its capacitance then increases). This provides local, negative feedback and considerable stabilization. The major trade-off with using this approach is that since a voltage divider is used, there is an inherent increase in the overall voltage that must be applied to obtain actuation.

ELECTROSTATIC ROTARY MICROMOTORS

Electrostatic actuation has also been used to implement rotary motor structures. The basic idea is to create a central freely moving rotor with surrounding capacitive plates that can be driven in correct phase to cause the rotor to turn. Relatively high speeds are possible, and recent work has advanced the modeling and design of these structures. Unfortunately, these motors were the subject of a good deal of hype in the popular press, and few, if any, practical applications have resulted. However, it is not unreasonable to expect that some useful applications for these devices will materialize, particularly if their lifetimes can be extended to many tens of thousands of hours (one promising application area is optical scanning, as discussed in Yasseen, et al. (1995)). Approaches to lubricating such motors using chemical surface modification are discussed in the Micromachining Techniques chapter.

These devices have generally been fabricated using sacrificial oxide/polysilicon processes as discussed above. Fan, et al. (1988a), presented the first paper on such micromotors, describing a variety of motors with rotor diameters in the range of 60 to 120 μm. These designs used six stator poles and eight rotor poles (or "teeth"). Using voltages of 200 to 300 V, and a three-phase drive scheme, they achieved rotation rates of 50 to 500 RPM. Further information on this subject, showing the evolution of the designs and their analyses can be found in Lober and Howe (1988), Fan, et al. (1989), Mehregany, et al. (1989, 1990a, 1990b), Tai and Muller (1989), and Tai, et al. (1989). In addition, Hackett, et al. (1991) described the fabrication of electrostatically actuated motors for RF switches on GaAs (discussed below in Section 7.2).

As described in Trimmer and Jebens (1989a, 1989b), *harmonic*, or "wobble" motors are designed around the principle of a rotor turning in a slightly larger stator ring, such that it "wobbles" around the central axis as it turns. An interesting advantage of such motors is that in an ideal case, the motor operates purely by rolling, without sliding friction. Also, since the rotor can closely approach the stator, large electrostatic forces can be developed. As the rotor rolls around the stator ring, each turn causes only a fractional rotation of the rotor about its own axis, leading to an effective "gear reduction" or ratio between drive frequency and rotation frequency. As described by Mehregany, et al. (1990a, 1990b, 1991), for a wobble motor with a central bearing (as shown above), this ratio, n, is given by,

$$n = \frac{r_b}{\delta}$$

where r_b is the radius of the bearing, and δ is the bearing clearance (gap between bearing and rotor).

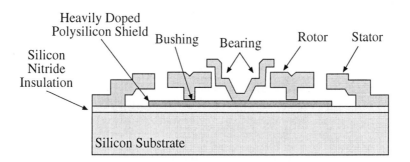

Illustration of a typical surface micromachined electrostatic motor, in this case of the harmonic, or "wobble motor" type. Adapted from Mehregany, et al. (1991). The top view shows the complete motor, and the bottom view shows a cross section.

Trimmer and Jebens (1989a, 1989b) demonstrated a version made by conventional machining. Jacobsen, et al. (1989a, 1989b) demonstrated a variety of microscale wobble motors made using direct assembly of microscale components, electric discharge machining, cylindrical photolithographic etching, and coextrusion of metal and plastic. Mehregany, et al. (1990a, 1990b, 1991, 1992) demonstrated surface micromachined, polysilicon wobble motors and provided detailed analyses of their theory and performance. More recently, electroplating has been used to fabricate electrostatically driven wobble motors. Fan and Woodman (1995a) and Fan, et al. (1995) presented a two-level electroplating process using a low-stress copper plating process with photoresist templates. They fabricated millimeter-scale motors for actuation of rotating magnetic media and read/write heads for future disk drives.

As a demonstration of conversion of the outputs of linear actuators to rotary motion, Garcia and Sniegowski (1995) presented a "microengine" in which the outputs of two orthogonal electrostatic comb drives (driven with square waves 90° out of phase) were coupled to an output gear via two drive linkages. They fabricated the devices using a sacrificial oxide, three-level polysilicon process, and demonstrated operation at angular speeds between 30 and 300,000 RPM. They also completed life tests exceeding 12 million rotations at 1,200 RPM (corresponding to 7 days).

ELECTROSTATIC LINEAR MICROMOTORS

In addition to rotary micromotors, electrostatic linear motors have also been micromachined, as proposed by Trimmer and Gabriel (1987). Hoen, et al. (1997) fabricated such a motor assembled from two separately fabricated silicon structures: a stator consisting of two levels of metallization (insulated using PSG) and electro-plated nickel posts (to align to the translator) and a translator fabricated from silicon using the "SCREAM" variable anisotropy etching process. They demonstrated displacements of up to 8 μm (total displacement of 16 μm) with 1 μm step sizes, using only 6 V of applied bias (4 μm displacements were achieved with only 4 V, below the 5 V levels often used for logic circuits). Hoen, et al. (1997) also provide a description of the underlying theory of operation for such motors.

Lateral comb drives can also be used to fabricate a class of linear micromotor, the *linear vibromotor,* based on repeated oblique impacts to a slider in order to periodically overcome static friction and cause stepwise movement (Lee, et al. (1993)). Only short duration (a few microseconds) forces are required for this purpose. A single slider can be outfitted with opposing sets of impactors that can provide *push/pull motion.*

As discussed by Daneman, et al. (1996), the quadratic response of comb drives leads to a primary driving frequency (fundamental) proportional to the product of the DC and AC voltage components. Electrostatic comb drives can be driven at resonance, amplifying the original electrostatic force by the quality factor (they reported 30 to 100 in air at 1 atm). Several comb-drive cycles are required for the amplitude to increase, so Daneman, et al. (1996) ran them in bursts of four to five cycles and demonstrated such motors for use in positioning optical elements (see the Optical Transducers chapter and Daneman, et al. (1995)) and achieved linear motion velocities of > 1 mm/s. (In general, it appears that optical applications of such actuators may be one of the first areas in which they become practical.)

An interesting micromachined actuator capable of very precise, stepwise linear motion is the so-called "scratch drive" actuator (Akiyama and Shono (1993)). As illustrated below, the basic concept is to utilize a flexible, conductive plate with a small bushing at one end. When a voltage is applied between the plate and a buried conductor on the substrate, the plate buckles down, pushing the bushing forward a small distance. When the voltage is removed, asymmetries in the friction between the bushing and the insulator surface result in some degree of "rectification" of motion, producing a net movement of the plate. The cycle can be repeated, for continous, stepwise linear motion.

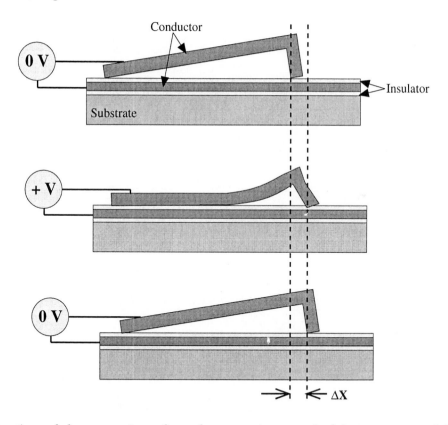

Illustration of the operation of an electrostatic scratch drive actuator. Adapted from Fukuta, et al. (1997).

In a recent example, Fukuta, et al. (1997) reported the use of 70 μm long, 50 μm wide, 1 μm thick, and 1.5 μm bushing height scratch drive actuators. The actuators were operated with ± 180 V pulses at 250 Hz. The relatively high forces generated were sufficient to deform considerably larger polysilicon structures, which were heated using electric current to permanently reshape them thermally in their new orientations.

ELECTROSTATIC MICROGRIPPERS

Kim, et al. (1990, 1992) demonstrated a polysilicon electrostatic microgripper that could achieve a 10 µm movement with only 20 V applied voltage. The microgripper consisted of a 7 × 5 mm silicon die, a 1.5 mm long support cantilever of boron-doped silicon substrate silicon (protruding from the die), and a 400 µm long polysilicon overhanging gripper extending from the end of the support cantilever. The fabrication process used is illustrated below.

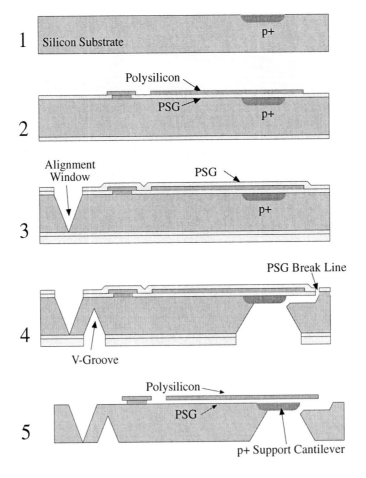

Electrostatic microgripper fabrication process. After Kim, et al. (1992). Note that in the final stage (5), some regions of the polysilicon are undercut and free to move, and others are anchored to the subsrate.

ELECTROSTATIC RELAYS AND SWITCHES

Applications of micromachining to the fabrication of relays and RF switches are discussed in detail below in Section 7.2. Electrostatic actuation has been one of the more successful means for driving these devices due to its extremely low power requirements, although most current devices still require relatively large voltages.

6.1.2 THERMAL ACTUATION

As discussed in the Thermal Transducers chapter, there are a number of possible thermal actuation means based on the expansion of solids or fluids. A large number of them have been exploited in micromachined devices.

THERMAL EXPANSION OF SOLIDS

Thermal expansion of materials can be readily applied to the actuation of microstructures. An interesting example is the thermal-expansion-driven gripper design demonstrated by Keller and Howe (1997). The gripper was fabricated using the HEXSIL silicon mold replication process (described in the Micromachining Techniques chapter). The 8×1.5 mm $\times 40$ μm, normally closed gripper could be opened 35 μm with a 75 mW power input, and silicon dioxide pegs as small as $1 \times 4 \times 40$ μm could be manipulated and placed in 4×4 μm holes. In addition, an example of the use of single-crystal silicon beams as thermal expansion actuators can be found in Klaassen, et al. (1995). As for all thermal expansion driven actuators, careful design to maximize the generated strain and the thermal isolation of the expanding region are key to achieving realistic operating power levels.

BIMORPH THERMAL ACTUATORS

One of the basic thermal actuation schemes is to use the difference in thermal coefficients of expansion of two bonded materials, referred to as *thermal bimorph* actuation. A heater is typically sandwiched between the two "active" materials and, when electrically driven, causes them to expand differentially. Advantages of this approach include nearly linear deflection-versus-power relationships and environmental ruggedness (e.g., these actuators can be run in liquids of sufficiently low thermal conductivities). Disadvantages include high power, low bandwidth (determined by thermal time constants), and more complex construction than simple electrostatic actuators. Despite their disadvantages, these actuators have been used extensively. The basic principles of thermal bimorph actuators (differential thermal expansion) are discussed in the Thermal Transducers chapter. In addition, the underlying theory is discussed in a classic paper, Timoshenko (1925). It should be noted that micromachined bimorphs can be constructed from a wide variety of materials, both organic and inorganic.

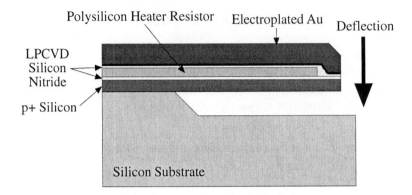

Cross-sectional illustration of a thermal bimorph actuator where electroplated gold and a layer of epitaxial silicon form the main bimorph structure (with thin polysilicon and silicon nitride layers interspersed). After Riethmüller and Bencke (1988). (Not to scale.)

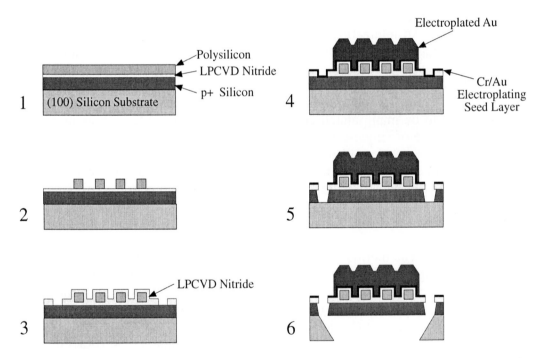

Gold/silicon thermal bimorph actuator process. After Riethmüller and Benecke (1988).

Reithmüller and Benecke (1988) demonstrated a thermal bimorph actuator based on the differential thermal expansion of gold and silicon. Their actuators used up to 200 mW per actuator to achieve deflections up to 100 µm. The fabrication process used to realize these devices is illustrated below. They began with the epitaxial growth of a 4 μm thick p+ silicon layer (1.3×10^{20} cm^{-3}) on p-type (100) wafers. A thin LPCVD silicon nitride electrical insulating layer was then deposited, followed by the deposition of an 0.5 μm polysilicon layer and its doping via ion implantation. The polysilicon was then patterned using dry etching to form the heaters, followed by the deposition of another LPCVD silicon nitride layer, which was patterned to form contact vias and to define the geometry of the actuators. A Cr seed layer was deposited, followed by the electroplating of a 1.8 μm gold layer to serve as the top layer of the bimorph and the electrical interconnects. Finally, the Cr/Au layer was ion-milled, and the silicon etched in EDP to undercut the bimorph cantilevers.

Yang and Kim (1995) demonstrated a multilayer thermal cantiliver actuator (actually a trilayer structure rather than a bimorph) that made use of two independent polysilicon heater layers with a central silicon dioxide layer. Such an actuator could be deflected up or down, based on which of the polysilicon layers was heated by passing a current through it. Two of these actuators were combined with a tension band of silicon nitride to form a bistable cantilever structure that could be changed from an "up" to a "down" state with thermal drive, but with no holding power. Electrical heating pulse times as low as 2 μs were used, for a maximum mechanical switching frequency greater than 200 Hz.

THERMAL ARRAY ACTUATORS

Ataka, et al. (1993a, 1993b) presented an interesting paper demonstrating the use of an array of thermal bimorph actuators to move objects by ciliary motion. An array of 256 (total) actuators (128 facing each direction) were actuated with a total average power of 1 W. A $2.6 \times 1.5 \times 0.26$ mm piece of silicon weighing 2.4 mg was moved using this array. This early ciliary actuator array was limited such that objects could only be moved along a line. Fujita, et al. (1996) described such thermal acutators, as well as pneumatic devices (discussed below) and the strategies required to operate them to obtain useful macroscopic motions of carried objects.

Suh, et al. (1996) demonstrated omnidirectional ciliary actuators using a similar thermal bimorph actuation scheme, coupled with electrostatic actuators. The actuators, as illustrated below, were fabricated using two layers of different polyimides, Hitachi PIQ-L200 and PIQ-3200 (with coefficients of linear thermal expansion of 2.0 and 54 ppm/°C, respectively). PECVD silicon nitride stiffening layers, TiW heating resistors, and aluminum electrostatic plates were embedded between the two polyimide layers. The devices were fabricated atop a sacrificial aluminum layer that, in other regions where it was protected with a PECVD silicon nitride layer removed after release, also served as the bond-pad and interconnect layer. When the actuators were released using a standard aluminum wet etchant, they deflected upward from the substrate (since they were cured at a higher temperature, and the upper polyimide layer had the larger thermal expansion coefficient), aiding the release process.

The actuators, in groups of four (1 × 1 mm in size for each set of four) to allow four different basic motion directions, were fabricated in an 8 × 8 element array, for a total of 256 actuators. Each actuator element dissipated 16.7 mW for a 5 V drive signal, with a theoretical lifting capacity of 76 µN. Through proper sequencing, small objects could be moved across the array in arbitrary directions. Objects tested included a 3 × 3 mm silicon chip (8.6 mg), a #4-40 stainless steel nut (0.16 g), and an 8-pin plastic dual in-line packaged IC (0.5 g). Step sizes were as small as ≈ 3 µm. Electrostatic hold-down of the actuators after thermal actuation was also demonstrated with voltages as low as 100 V (but 500 V was required for purely electrostatic pull-in). Demonstrations of omnidirectional, vectored motions of components, as well as strategies for operation of such arrays, were presented by Konishi and Fujita (1995) and Bohringer, et al. (1997).

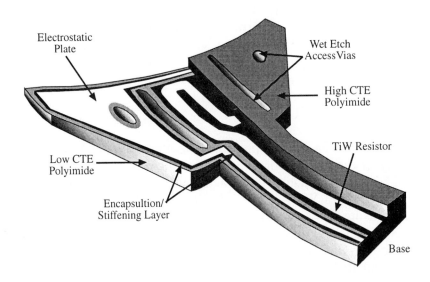

Illustration of one-quarter of an omnidirectional, thermal/electrostatic ciliary actuator. An embedded TiW heating resistor drives the device in thermal mode, and an aluminum electrostatic plate in its tip can be used to hold it down via electrostatic force relative to the substrate. Courtesy J. W. Suh, from Suh, et al. (1996).

Suh, et al. (1995) also demonstrated multi-segmented thermal bimorph actuators fabricated in the same manner. By thermally isolating regions of a bimorph via removal of selected regions of material, and using one TiW heater per segment, it was shown that individual segments could be actuated with only moderate cross-talk. The use of the resistance of the TiW heaters to provide temperature (and, indirectly, position) feedback during heating was demonstrated (the TCR of the TiW heaters was measured to be 361 ppm/°C). Using analog computation circuits, the resistance of a given heater could be determined dynamically during heating. This approach could not only serve to provide the aforementioned position feedback, but also to potentially indicate the additional heat loss from an actuator contacting an object.

DIELECTRIC LOSS HEATING OF THERMAL BIMORPHS

Most dielectrics, when exposed to RF energy, exhibit losses that convert some of that energy into heat. This can serve as an indirect way of heating thermal bimorph structures. Dielectric losses are proportional to the square of the applied RF electric field (directly proportional to applied RF power). The thermal energy generated is given by,

$$P_{\text{thermal}} = \pi f \varepsilon_r'' \varepsilon_o |E_o|^2 \quad \text{in W/m}^3$$

where,
  f = excitation frequency, in Hz
  ε_r'' = imaginary part of relative permittivity (may be a function of frequency)
  E_o = electric field, in V/m

Rashidian and Allen (1993) demonstrated such thermal bimorph actuators based on a layer of the copolymer of vinyldiene fluoride and trifluoroethylene (PVDF-TrFE, which is also a piezoelectric material) on top of a layer of polyimide. This combination of materials exhibits one of the largest differences in thermal expansion coefficient reported for a micromachined structure (PVDF-TrFE has an α_L of 140×10^{-6}°C^{-1} and Dupont polyimide PI2611D has an α_L of 3×10^{-6}°C^{-1}). As illustrated below, they fabricated thermal bimorphs with a lower layer of polyimide and a layer of PVDF-TrFE sandwiched between two aluminum electrodes across which RF power could be applied.

RF power (1 to 100 MHz, 0 to 16 V peak) was applied and large deflections (250 to 900 μm) were achieved over millimeter-scale (length) cantilevers. The authors reported relatively large deflections using RF power inputs on the order of 1 mW (as compared to hundreds of milliwatts for directly heated devices). However, since most previously published thermal bimorph actuators used materials with much lower expansion coefficient mismatches, a fair comparison would have to take that into account. These devices could be remotely actuated by RF power and have been shown to work under water (using a wire-wound inductor connected to the beam drive electrodes and a second one nearby to couple power in, similar to a transformer with physically separated primary and secondary windings).

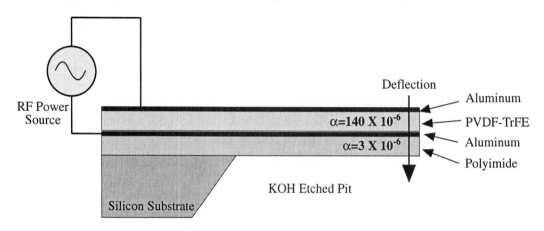

Simplified cross-sectional view of a dielectric loss heated thermal bimorph actuator. After Rashidian and Allen (1993).

Dielectric loss heating of thermal bimorphs could turn out to be a very useful way of generating mechanical force in micromachined structures, particularly if they are relatively isolated from each other.

VOLUME EXPANSION AND PHASE-CHANGE ACTUATORS

Rather than making use of linear expansion of solids, it is possible to construct micromachined actuators that take advantage of volume expansion. A typical approach is to form a cavity with a sealed fluid (e.g., air, water vapor, liquid water, etc.) that can be heated and thus expanded. If part of the cavity (i.e., one wall) is compliant, it will deform under pressure and generate mechanical force. This approach can generate large deflections and forces, but, like most thermal actuation schemes, uses a great deal of power and has low bandwitdh due to thermal time constants. In addition, the formation of sealed, fluid-containing chambers may complicate a fabrication process. Most of these actuators are more properly described as phase-change devices, discussed below.

Phase-change thermal actuators involve thermally changing the phase of a substance to expand its volume and to create pressure (and hence mechanical force). For example, one can change liquids such as water from the liquid to gas phase by heating, generating bubbles that can be used to power mechanisms. Lin, et al. (1991) presented some such "bubble-powered" devices.

As discussed in the Microfluidic Devices chapter, an example of a thermopneumatic valve is the so-called "fluistor," which is commercially available from Redwood Microsystems (Menlo Park, CA) and is described in Zdeblick, et al. (1994). This device makes use of a resistive heater to cause a trapped fluid to vaporize, pushing a silicon diaphragm against a valve seat. At present, both normally open and normally closed valves are available.

Yang, et al. (1997) demonstrated surface micromachined "bellows" actuators wherein a circular, folded thin-film membrane structure is used to obtain extended deflections over simple membranes. The membranes were fabricated using LPCVD silicon nitride structural layers with polysilicon sacrificial layers. By alternating layers of polysilicon and silicon nitride, with appropriate vias made to allow nitride-to-nitride connections, bellows can be fabricated. They released the actuators by using a KOH etch to form vias from the back side to access the bellows and then a final TMAH etch to remove the polysilicon (TMAH provides much higher etching selectivity for polysilicon over nitride, therefore it is used after the initial KOH etch, which stops on a thermal silicon dioxide layer beneath the bellows). For 800 μm diameter bellows, they demonstrated deflections greater than 50 μm for a pressure of 20 psi. The measured deflections achieved with the bellows were more than three times those available with single-layer membranes, although with some-what lower burst pressures (due to failures at the interlayer joints). The authors also demonstrated the use of a separate glass chip with a thin-film heater to seal water into the cavity beneath the bellows and to actuate it thermopneumatically.

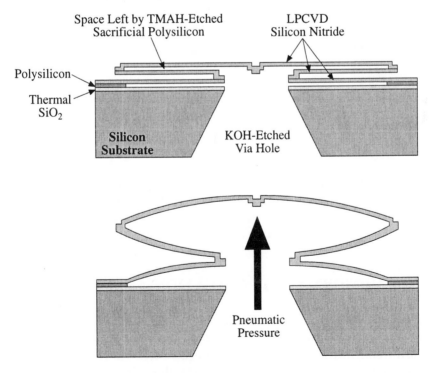

Illustration (not to scale) of a micromachined "bellows" actuator, showing the structural components (top) and deflection of the bellows when pressure is applied inside it. Adapted from Yang, et al. (1997).

Sniegowski (1993) demonstrated an expanding vapor-powered piston actuator that moves parallel to the plane of the substrate. According to Sniegowski (1993), in a comparable substrate surface area, surface-tension-based actuators can provide forces that can be two orders of magnitude greater than those with other actuation schemes. Unfortunately, however, they must be operated in liquid environments, which may limit their maximum operating speeds (through viscosity) and efficiencies (due to the thermal conductivities of the liquids).

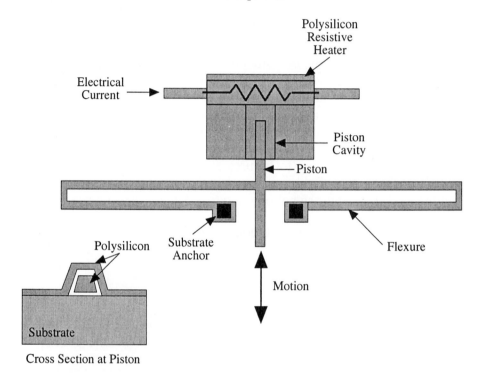

Simplified illustration of a bubble- (surface tension) driven actuator. After Sniegowski (1993). Bubbles of water vapor are thermally formed by the polysilicon heater and expand within the piston cavity, pushing the piston outward. When heating is stopped, the bubble in the piston cavity collapses, allowing the piston to return.

6.1.3 SHAPE MEMORY ALLOY ACTUATION

There are a few materials, notably titanium/nickel alloys, that can exhibit considerable changes in their length (contraction) when heated, and they are collectively referred to as *shape memory alloys* (SMAs). As reviewed in Hunter and Lafontaine (1992) and Krulevitch, et al. (1996), the original observations, made in the early 1950's were that some alloys, once mechanically deformed, would return to their original, undeformed state when heated. Since they are conductive, they can be heated simply by passing a current through them. As discussed below, deformation causes the materials to transition from one crystal phase to another, and this process can be reversed by heating. Their advantages include relatively "linear" control and very high stress (> 200 MPa), and if strains are kept below 2%, they can operate over millions of cycles. Disadvantages include the need for special alloys (often with annealing temperatures that are incompatible with prefabricated control circuitry) and high power consumption (the efficiency is ≈ 3%).

Alloys that have SMA properties include Au/Cu, In/Ti, and Ni/Ti, with the latter being the most commonly used and the alloy discussed herein. The SMA effect is due to a temperature-dependent phase transition between *martensite* (predominantly rhombohedral) and *austensite* (higher symmetry) phases of the alloy. In the martensite phase, the NiTi can be easily deformed plastically and will contract when heated, causing it to return to the austensite phase. As expected for a thermally driven actuator, the time constants of movement are determined by thermal time constants, which can be considerably reduced for thin-film micromachined structures.

Currrently, there are some research groups investigating thin-film NiTi as an actuator for microvalves and other devices, some examples of which are provided in Johnson (1991). Further information on the fabrication and etching of thin-film NiTi layers can be found in Walker and Gabriel (1990) and Jardine, et al. (1994), and Krulevitch (1994) and Krulevitch, et al. (1996) (the last reference provides a thorough review of background work as well as comparisons to other actuator mechanisms for microstructures).

6.1.4 PNEUMATIC/HYDRAULIC ACTUATION

The use of fluid pressure as an actuation power source is another potentially viable method for micromachined devices. The requirement of the micromechanical structures would then be to operate as a valve, fluidic amplifier, or simply as a flow structure. The reader is also referred to the Micromachined Fluidic Devices chapter for more information.

Several devices have been constructed using this general idea, either valving a gas flow or using it to support objects like an "air hockey" table (see, e.g., Pister, et al. (1990)). An important consideration with such actuators is the source of the operating pressure, which may be a pressure vessel or some sort of pump (miniaturization of which may not be possible). Another interesting example is that of Field, et al. (1991), who demonstrated liquid-powered gears integrated with microscale flow channels.

Konishi and Fujita (1993) described an actuator array based on the idea of using electrostatic actuators (valves) to control the flow of gas. Their goal was to use the pneumatic force to move small objects above an array of such actuators. The basic idea was to make valves that could reside in either of four states: venting gas to both sides, to one side, to the opposite side, or completely closed. Their states could be controlled with two electrodes at each valve that would control the position of polyimide valve flaps containing a third electrode. By pulsing the gas pressure and controlling the valves in between pulses, they were able to obtain directionally controlled pulses of gas to move objects.

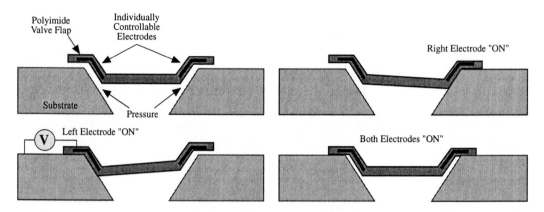

Illustration showing the concept of electrostatically gated pneumatic valves capable of directing gas flow in both directions (top left), to the left (top right), to the right (lower left) and turning gas flow off (lower right).

The authors used a process based on a combination of etch pit replication (to form the basic shape of the valve actuators) on an SOI wafer, an electroplated metallic sacrificial layer (copper), and polyimide valve actuators with sandwiched gold electrodes. The basic process began with 525 μm Si substrate with buried 2 μm oxide and 20 μm thick (100) single-crystal Si top layer (the buried oxide would later serve as an etch-stop). Thermal oxide was grown and HF used to etch the locations of the top surface etch pits for use as electroplating templates, with the pits themselves being etched using KOH. A 100 nm Cr/Cu seed layer was then deposited by evaporation, followed by the application and patterning of a 4 μm photoresist mask and electroplating of a 2 to 3 μm Cu sacrificial layer. At this point, a 1 μm layer of polyimide was deposited to form the lower layer of the valve flaps, followed by lift-off deposition of the 10 nm Cr, 150 nm Au, 100 nm Ni electrostatic electrodes, and the subsequent deposition of another 1 μm thick polyimide layer. Interestingly, the polyimide sandwich was patterned using a relatively thick 7 μm photoresist layer and an O_2 plasma etch to remove the polyimide down to the sacrificial Cu layer (photoresist would not typically be used for this since it erodes very quickly in an O_2 plasma). A 2 μm layer of silicon nitride on the back side of the wafer (when it was deposited was not clearly stated) was then patterned and used, with KOH etching stopping at the buried oxide layer to form a large diaphragm on which the actuators would be located (another omission is the description of how the front side polyimide was protected from the KOH). Finally, the sacrificial Cu and buried oxide layer at the valve sites was wet etched, freeing the actuators.

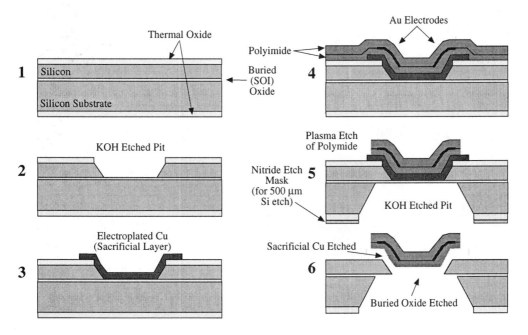

Simplified illustration of the fabrication process for electrostatically gated pneumatic valves used in a micromachined object transport system. After Konishi and Fujita (1993).

Arrays of 63 actuators were fabricated in an overall area of 2×3 mm. Actuation voltages reported were 90 V for 100 μm wide valves with 2 μm gaps. Pulsating (≈ 1 Hz) air pressure (≈ 2 kPa, or 0.3 psi) was applied from the back side and not only provided pulsatile driving power for objects being moved, but undoubtedly also helped to prevent sticking of the valves. A $1 \times 1 \times 0.3$ mm piece of silicon (700 μg) could be moved by the array.

6.1.5 PIEZOELECTRIC ACTUATION

As discussed in Section 5.1.3 above, the piezoelectric effect can be used in both sensing and actuation. As an actuation mechanism, the electrically induced strain is approximately proportional to the applied electric field. Potential advantages include high stress (tens of MPa), high bandwidth, high energy density, and potentially lower operating voltages than electrostatics (this is geometry-dependent). Disadvantages include more complex fabrication (sometimes including a poling step) and much smaller dimensional change under drive (strains on the order of 0.1% or less). In order to obtain greater displacements despite this, piezoelectric bimorphs can be constructed. In macroscopic devices, multilayer piezoceramics are commonly constructed to optimize displacement characteristics. For an interesting discussion of the potential uses of piezoelectric actuators that takes into account the small dimensional variations, see Robbins, et al. (1991).

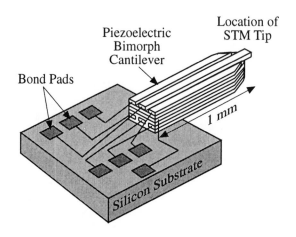

Illustration of a micromachined, multilayer piezoelectric scanning tunneling microscope tip. After Akamine, et al. (1990).

An interesting application of piezoelectric actuation where small strains are very useful is a micromachined, piezoelectrically deflected scanning tunneling microscope (STM) tip (Akamine, et al. (1990)). The current carrying tunneling tip was fabricated on a piezoelectric cantilever capable of deflecting the tip up and down by distances less than 10^{-10} m. The piezoelectric actuation was achieved using a ZnO film sandwiched between aluminum electrodes. The cantilever itself consisted of two sets of these sandwiched layers: the first one for deflecting the tip in and out of the plane (for sensing the tunneling current) and the other for scanning the tip in a planar fashion to form an image. Using multiple electrodes to actuate the ZnO in selective regions, complex tip motions could be achieved, as illustrated below.

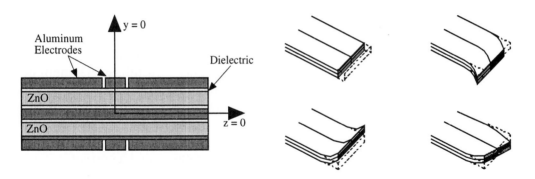

Illustration of the cross section (left) and achievable tip motions of a micrmomachined piezoelectric STM tip. After Akamine, et al. (1990). Note that the individual electrodes allow control of specific regions of the ZnO piezoelectric layers.

6.1.6 MAGNETIC ACTUATORS

A wide variety of micromachined magnetic actuators have been realized, including those that depend on external magnetic fields and those that generate their own, via on-chip coils or even permanent magnets. These actuators are discussed in the Magnetic Transducers chapter.

6.1.7 CHEMICAL ACTUATORS

There are a number of potential chemical actuation techniques that might be applicable to micromachined devices. These materials are attractive from the point of view that they can generate large strains (> 40% for some polymer actuators) and be relatively efficient ($\approx$ 30%), but their drawbacks generally include the need for a wet environment and questionable long-term stability. Further information on chemical actuators is presented in the Chemical Transducers chapter.

6.1.8 HYBRID ACTUATION SCHEMES

A potentially important approach to overcoming the limitations of individual actuation schemes is to use more than one approach at once in hybrid. For example, electrostatic actuation and thermal bimorph actuation can be combined to yield a high-force thermal actuator that can switch to a low-power electrostatic hold-down mode (demonstrations of this are described in Suh, et al. (1996), and Sun, et al. (1996), and a device capable of this mode of operation was described by Lin, et al. (1995)). As another example, magnetic and electrostatic forces can be combined to achieve a combination of the large movements of the magnetic approach, with a low-power electrostatic hold-down state (this is demonstrated in Judy and Muller (1996), discussed in detail in the Optical Transducers chapter). Undoubtedly, several other actuation methods will be combined to obtain similar benefits.

6.1.9 BIOLOGICAL ACTUATORS

Some of the most versatile and efficient microactuators in existence are those in living organisms (two excellent general references for this are Darnell, et al. (1991) and Alberts, et al. (1994)). Advantages to molecular actuators include incredible diversity of function, low cost, nanoscale size, self-assembly, and available existence proof. Their disadvantages include the need for chemical "food" (typically), finite operating lifetime, requirement for specific environmental conditions (i.e., 37°C or another organism's normal body or cell temperature), and a current lack of tools to make use of them.

There are three major classes of biological actuators: *cytoskeleton/microtubules* (within the individual cell, controlling its shape and movements of its organelles), *cilia/flagella* (propulsion for the whole cell), and *muscle* (contraction of the entire cell under external control via other cells) of *skeletal* (voluntary control) plus *cardiac* and *smooth* (involuntary control) types. These actuators evolved over billions of years, and are encoded in multiple gene families. The specific DNA is very similar ("highly conserved") between primitive organisms and humans. This generally indicates that the approach that evolved is a fundamental and unique one. All of these actuators are constructed from polypeptides (long strings of coupled amino acids, synthesized under the control of DNA via transfer-RNA (t-RNA), and messenger-RNA (m-RNA)). At some point, the energy for the movements of these actuators is derived from food, via ATP (adenosine triphosphate), an essential energy storage molecule. Specific enzymes catalyze the local oxidation of ATP, allowing specific interactions to occur between the different actuator proteins. Such interactions are generally the sliding of filaments past each other by means of making and breaking particular molecular associations between the proteins.

CYTOSKELETON AND MICROTUBULES

The cellular "skeleton," or cytoskeleton, and subcomponents called microtubules, are only part of the actuators within cells. They are really structural members that are either moved against other structures or have smaller structures moving along them. They are comprised of 7 nm diameter *actin microfilaments*, 24 nm diameter *microtubules,* and 10 nm diameter *intermediate filaments.* Their existence (assembly or disassembly) within a cell is under tight control by the cell's nucleus. For example, during cell division (*mitosis*), complex webs of microtubules assemble to pull the duplicated chromosomes apart. In other cases, the same materials are around in permanent forms (cilia, etc.).

Microtubules are self-assembled from two repeating subunits: α - and β-*tubulin. Microtubule associated proteins* (MAPs) are also involved in control of self-assembly and do not actually form part of the "structure." Typically (with an important exception being some outgrowths of neural cells), *microtubule organizing centers* (MTOCs) are the physical "starting points" for microtubules, but their function is poorly understood at present.

Many intracellular components (organelles, proteins, etc.) are transported within cells by movement along microtubules, not by diffusion. For example, since protein synthesis only occurs at *ribosomes* near the cell body, proteins must be transported to where they are needed, and this can be *several meters* in cells with long extensions such as neurons. Molecular "motors" using ATP "fuel" do this work of transporting molecules along the microtubules. One of them is *kinesin* (molecular weight 380,000, length $\approx$ 70 nm). Kinesin can either "stick" to the microtubules and move things along them or "stick" to another surface (i.e., a glass or silicon substrate) and move things relative to the substrate (a mode that may be useful for micromachined actuators), as illustrated above.

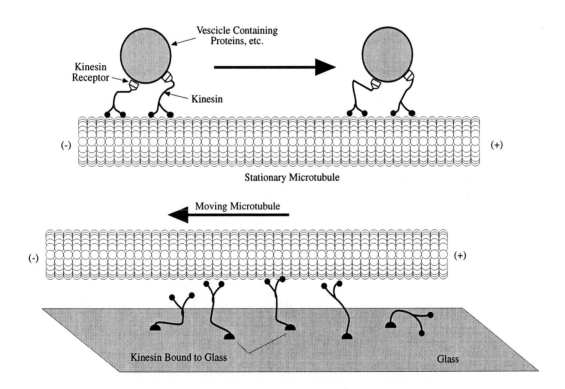

Illustration of two modes of kinesin-based transport, showing (top) protein-filled vescicles being transported along a stationary microtubule by kinesin and (bottom) stationary kinesin molecules moving a microtubule. After Darnell, et al. (1991).

CILIA AND FLAGELLA

Cilia and *flagella* are slender (250 nm in diameter) cylindrical actuators that are used by cells to propel themselves or to move substances relative to them. The main differences between cilia and flagella are that flagella are longer and are used in different ways to generate movement. Arrays of cilia form the major "conveyor belt" systems in our bodies and they move things along in an organized way (e.g., moving mucus along the respiratory passages, as opposed to the peristaltic motion in the intestines, where the smooth muscles in the intestinal walls cause large-scale rhythmic contractions).

Cilia move by a "whiplike" power stroke, followed by a recovery stroke. Flagella move by propagating waves of constant amplitude from the base (at the cell) to the tip (analogous to making waves in a rope by moving it up and down). However, despite their functional differences, cilia are nearly identical in structure. They consist of a bundle of microtubules (the *axoneme*) surrounded by an extension of the cell's membrane.

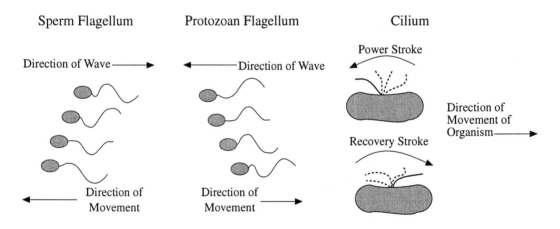

Illustration of the operating modes of flagella (left and center) and cilia (right). After Darnell, et al. (1991).

A typical axoneme consists of nine pairs of A- and B-tubules in a circle, surrounding a pair of single microtubules. Each A-tubule has *dynein* "arms" protruding from it that can "push" the B-tubule of the adjacent A-B pair, sliding it toward the tip of the axoneme. *Nexin* "connectors" (apparently "elastic" links) between each A-B pair and its neighbors, as well as radial "spokes" to the central pair constrain the relative movements of the pairs, converting the pushing of the dynein into bending of the axoneme.

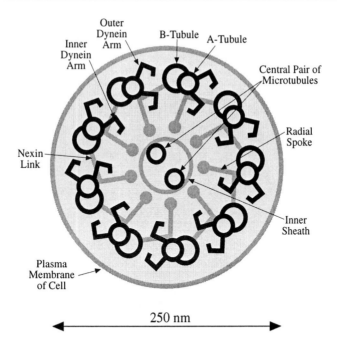

Inner Dynein Arm

Outer Dynein Arm

B-Tubule

A-Tubule

Central Pair of Microtubules

Radial Spoke

Nexin Link

Inner Sheath

Plasma Membrane of Cell

250 nm

Illustration of the structure of a typical cilium or flagellum, showing the central axoneme and outer extension of the cell's plasma membrane. After Alberts, et al. (1994).

MUSCLE

Muscles are organs of higher animals (i.e., cooperating groups of cells) where specialized cells (as opposed to sections of cells, as seen above) team up to generate force. Muscles are involved in circulation, locomotion, eating, digestion, reproduction, etc. Muscle tissue under voluntary control is referred to as *skeletal muscle*, that in the heart is termed *cardiac muscle*, while that under autonomic control is referred to as *smooth muscle*. Despite some stuctural and functional differences, the basic mechanisms discussed below apply to all types of muscle.

Muscles make use of a similar mechanism of large protein molecules moving against each other in an organized way, and use ATP for fuel. The two key molecules in muscle are *actin* and *myosin*. Myosin has a "ratcheting" head that allows it to generate traction against fixed actin filaments, causing them to move relative to each other. Myosin has a long, fibrous "tail" that allows many myosin molecules to aggregate into "bands" that fit in between fixed actin filaments.

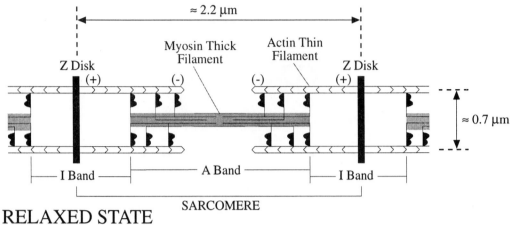

RELAXED STATE

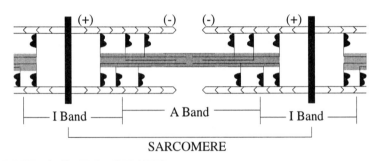

CONTRACTED STATE

Illustration of the structure of the sarcomere (the basic muscle unit), showing (top) the relaxed and (bottom) the contracted states. After Darnell, et al. (1991).

One can couple actin to a substrate such as silicon or glass and get myosin-coated objects to move along the substrate in the presence of ATP. Like microtubules, the actin fibers are polarized, so there is a defined direction along which myosin can move.

The contraction of muscle is controlled by the release of calcium ions (Ca^{2+}) from a specialized reservoir called the *sarcoplasmic reticulum* (a membrane structure within the muscle cell). Neural command signals from nerves cause (indirectly) the release of Ca^{2+} into the muscle fibers, allowing contraction to occur.

It is interesting to study the overall hierarchy of physical structure in skeletal muscle (illustrated below) that shows how multitudes of micron-scale actuators can be arranged to form macroscopic actuators with central control mechanisms.

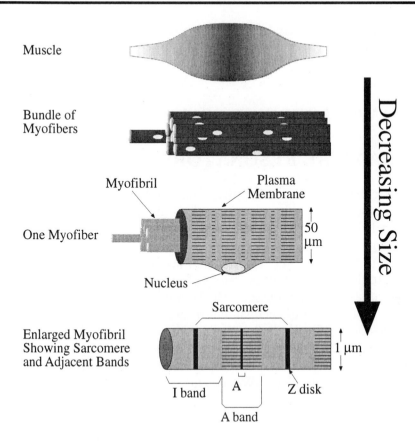

Muscle

Bundle of
Myofibers

Myofibril

Plasma
Membrane

One Myofiber

50
μm

Nucleus

Decreasing Size

Sarcomere

Enlarged Myofibril
Showing Sarcomere
and Adjacent Bands

1 μm

I band   A   Z disk

A band

Illustration of the overall hierarchical structure of skeletal muscle. After Darnell, et al. (1991).

The different types of muscle have differences in function and structure, as mentioned above. In skeletal muscle, as previously discussed, myofibers are so-called syncitia, in which many cells share a single nucleus. In cardiac muscle (which must contract billions of times over the lifetime of a human), there are separate muscle cells, each with its own nucleus. These cells are coupled together by structures called *desmosomes*, which provide mechanical linkage, as well as electrical signaling channels between cells, allowing them to synchronize their beating. Finally, smooth muscle, in the gut and other regions of the body, is not under conscious control. These are also separately nucleated cells, but with a much more disorganized internal structure of the actin and myosin.

It seems that the application of biological molecules in micromachining is still in its infancy, and can bridge the gap between micron-scale and nanometer scale structures. It is certainly worth studying molecular biology as a source of inspiration and perhaps tools for engineering. Molecular structures have evolved into some amazingly optimized forms and possess unique properties, including self-assembly, which one cannot presently replicate using other technologies.

7. MECHANICAL CIRCUIT COMPONENTS

7.1 MECHANICAL RESONATORS

Mechanical resonators, for applications in precision frequency generation and filters, have been used at least as far back as the late 1940's (see Mason (1948) and Roberts and Burns (1949)). (It should be noted that mechanical resonators for other applications are discussed in several other sections of this book, most notably as sensors in the Chemical and Biological Transducers chapter, and a thorough review of piezoelectric resonators can be found in Benes, et al. (1995).) As reviewed in Hathaway and Babcock (1957) and Börner (1965), machined metal components with a variety of mechanical-to-electrical transducers were typically used in early versions of such devices, and their primary application was as intermediate frequency (IF) filters for heterodyne radios. A large variety of both torsional and flexural devices was developed and successfully employed for many years. Surface acoustic wave (SAW) resonators also have been widely used (see Smith, et al. (1969, 1972) for a good overview of their analysis and design).

Currently, macroscopic ceramic resonators (a good early description of these devices can be found in Sauerland and Blum (1968)) are used in high volumes for microprocessor clock generation (where absolute accuracy and stability is not critical) and for IF filters. Based on the use of PZT ceramics (see, for example, Kulcsar (1961) and Srivastava, et al. (1995)), these devices have been broadly applied due to their low cost and relatively high performance (e.g., Q values can range from 200 to 1,500). Typical resonant frequencies are the AM radio IF of 455 kHz, 4.5 MHz for television audio, and 10.7 MHz IF used in commercial FM radio and other communications applications. A variety of micromachined designs have been fabricated, as reviewed in Howe (1994a).

7.1.1 CANTILEVER RESONATORS

In the early 1960's, a group at Westinghouse laboratories began to develop micromachined mechanical resonators (Nathanson and Wickstrom (1965), Nathanson, et al. (1965, 1967), and Newell (1968)), which were probably among the earliest micromachined transducers. The motivation for this work was that while tuning fork oscillators (e.g., quartz) had been greatly reduced in volume, no monolithic fabrication method was available for obtaining high-quality on-chip resonators, as discussed in Newell (1964). They focused on flexural cantilever resonators and, as discussed below, implemented both singly and doubly clamped versions.

The resonant frequency for a singly clamped cantilever (this equation also holds for the fundamental mode of resonance for a doubly clamped cantilever) of length, L, and thickness, t, is given by,

$$f_R = \frac{1}{2\pi}\sqrt{\frac{k}{m}} = \frac{1.03}{2\pi}\frac{tv}{L^2}$$

where the longitudinal acoustic velocity is given by,

$$v = \sqrt{\frac{E}{\rho}}$$

(For silicon, using a density of 2.33 g/cm^3 and a Young's modulus of 130 GPa, one obtains an acoustic velocity of 7.47×10^3 m/s.)

and where,
  k = beam spring constant
  m = beam mass
  t = beam thickness
  L = beam length
  E = Young's modulus of beam material
  ρ = density of beam material

From the resonant frequency equation, it can be seen that the width of the cantilever is not a frequency-determining factor (increasing the width increases both the mass, m_r, and the spring constant k_r, linearly). Thus, for a given material, the key to increasing the resonant frequency of a cantilever is by maximizing the thickness relative to the length (short, thick beams intuitively provide higher resonant frequencies).

As discussed by Newell (1968), the deflection of such a cantilever due to gravity, δ_g, is given by,

$$\delta_g = \frac{3}{2}\left(\frac{WL^3}{Ewt^3}\right)g = \frac{3}{2}\left(\frac{\rho L^4}{Et^2}\right)g \approx \frac{0.38}{f_R^2}$$

where,
 W = cantilever weight
 w = cantilever width
  g = force of gravity = 9.83 m/s^2 (at sea level)

Thus, the magnitude of this gravitational deflection, relative to that of the desired resonance (which increases by a factor of approximately Q per g (force of gravity) of acceleration) decreases as the square of the resonant frequency, making it a difficulty primarily in low-frequency cantilever resonators. The gravitationally induced resonant frequency shift (depending on the orientation of the cantilever relative to the DC gravitation force) is in the range of,

$$\frac{\Delta f}{f_R} \approx \pm \frac{1}{6 L f_R^2}$$

As mentioned above, mechanical filters are widely used as radio IF sections (as band-pass elements) and as low-cost alternatives to quartz crystals in clock circuits. Low-frequency oscillator/filter applications can generally be dealt with using mixed analog/digital techniques that were not available at reasonable cost in the past (e.g., low-frequency clocks derived by digitally dividing a higher frequency, and band-pass filters using switched capacitor filters, which are clocked using a higher-frequency resonator). Thus gravitational effects are generally not of concern, as the primary applications of micromachined mechanical resonators appear to be at higher frequencies where high-volume commercial applications exist.

In terms of the RMS thermal noise produced by such micromechanical resonators, the noise spectrum is flat ("white" noise) but shaped by the force-to-displacement transfer function of the resonator such that it peaks at f_R. (This is a general point; i.e., for a mechanism, the displacement noise is white noise shaped by the structure's force-to-displacement transfer function.) As mentioned above, in such a cantilever structure, with noise displacement δ_N, one can equate the potential energy of thermal agitation (of the molecules comprising the cantilever) to the stored potential energy in the cantilever of effective spring constant, k, for each degree of freedom as,

$$\frac{1}{2} k \langle \delta_N^2 \rangle = \frac{1}{2} k_b T$$

where $\langle \delta_N^2 \rangle$ is the average displacement spectral density over all frequencies, and again noting that k_b is Boltzmann's constant. It can be shown that, for a simple cantilever, δ_N is given by,

$$|\delta_N| = \left(\frac{k_b T}{k} \right)^{\frac{1}{2}} = \left(\frac{k_b T}{M} \right)^{\frac{1}{2}} \left(\frac{1}{2 \pi f_R} \right) \approx \left(\frac{2 k_b T L^3}{E w t^3} \right)^{\frac{1}{2}} \quad \text{in} \quad \frac{m}{\sqrt{Hz}}$$

where M is the effective mass of the cantilever.

A more accurate expression for a simple mass-spring-damper system such as a cantilever is,

$$|\delta_N| = \sqrt{\frac{2k_bT}{\pi f_R kQ\left(1 - \left(2 - \frac{1}{Q^2}\right)\left(\frac{f}{f_R}\right)^2 + \left(\frac{f}{f_R}\right)^4\right)}} \quad \text{in} \quad \frac{m}{\sqrt{Hz}}$$

In general, to obtain the displacement noise in a given frequency range, it is necessary to integrate the above equation over the frequencies of interest. If $f \ll f_o$, this simplifies to a form where integration is reduced to multiplying by the square root of the desired frequency range,

$$|\delta_N| \approx \sqrt{\frac{2k_bT}{\pi f_R kQ}} = \sqrt{\frac{k_bT}{2\pi^3 f_o^3 MQ}} \quad \text{in} \quad \frac{m}{\sqrt{Hz}}$$

From these equations it can be seen that as such resonators are reduced in size through micromachining, their masses will decrease and thermal noise will increase. This is a very important result of scaling and is a serious consideration for a broad variety of mechanical transducers. Similar equations can be derived for other micromachined structures where the effective spring constant can be derived or estimated. It turns out that thermal noise of this nature can be a major problem, and this is apparent in commercial products such as some of the Analog Devices ADXL series integrated accelerometers (wherein the active sensing regions are polysilicon thin-film structures). It should also be noted that noise/drift can also be generated through the adsorption and desorption of contaminating molecules on the surface of the structures, but that this particular noise source can probably be mitigated through careful process control.

Another factor affecting such resonators is the viscous damping of the surrounding gas (or liquid) medium, which is inversely related to Q. For vacuum operation, the viscous damping is zero, and Q is not infinity due to intrinsic damping due to energy dissipation in the resonator material itself (much less in single-crystal materials, e.g., than in amorphous or polycrystalline).

For pressures much below 1 atm ($P < 0.04/w$, in Pa), the mean free path of gas molecules is sufficiently long that they do not interact with each other significantly. Under these conditions, they exchange momentum with the resonator structure at a rate proportional to the difference in velocity between the gas molecules and the resonator (Newell (1968)). In this regime, damping is proportional to gas pressure and Q can be estimated, using,

$$Q = \left(\frac{\pi}{2}\right)^{\frac{3}{2}} \rho t f_R P \left(\frac{RT}{M_M}\right)^{\frac{1}{2}}$$

where P is the pressure and M_M is the molar mass $\approx$ 28.9 g/mol for air, which can be simplified, by subsituting the resonant frequency equation and assuming that air is the surrounding gas, to,

$$Q \approx 93 \left(\frac{t}{L}\right)^2 \left[\frac{(E\rho)^{\frac{1}{2}}}{P}\right]$$

which illustrates the strong dependence of Q on the ambient gas pressure. For most currently practical micromachined resonators (with widths greater than 0.4 μm), gases can be treated as a viscous fluid (molecules interact with each other) and Stoke's law can be employed to compute Q by taking into account the damping force. For a simple cantilever that is far away from any stationary structures (i.e., no squeeze-film damping effects), this gives (not taking into account dissipative losses in the cantilever),

$$Q = \frac{(E\rho)^{\frac{1}{2}} w t^2}{24 v L^2} \quad \text{for pressures } P > \frac{0.04}{w} \quad \text{in Pa}$$

(noting that 1 atm $\approx 10^5$ Pa).

If squeeze-film damping effects are present, Q is further reduced. Using the simplified model in Newell (1968) (assuming that the mass of the air molecules is small relative to the mass of the cantilever and that flow velocities are subsonic), one can derive a new Q value (for cases when the gap between the resonant cantilever and the substrate, d, is less than approximately one-third the width, w),

$$Q \approx \frac{(E\rho)^{\frac{1}{2}} w t^2}{v L^2} \left(\frac{d}{w}\right)^3$$

Similar concepts, although with potentially more complex analyses, apply to other resonator geometries, such as doubly supported cantilevers, drum-head membranes, etc.

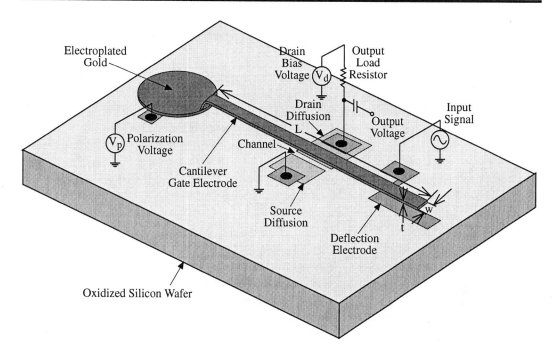

Illustration of a resonant gate field-effect transistor. Adapted from Nathanson, et al. (1967).

The Westinghouse team developed a field-effect transistor structure wherein the "gate" was an electroplated gold cantilever. The so-called resonant gate transistor (RGT), illustrated above, was operated by placing a bias, or "polarization" voltage on the cantilever and electrostatically deflecting it using an electrode on the substrate. While not fully described in the available references, the fabrication process for the RGTs made use of a metal sacrificial layer (above pre-diffused n+ source and drain regions), deposited via a process likely to have been lift-off, onto which the gold cantilever was electroplated.

For the singly clamped cantilever RGT devices, the deflection at resonance, δ_R, is approximated by,

$$\delta_R \approx \eta Q \left(\frac{V_p}{V_{p\,max}} \right) \left(\frac{L^4 \varepsilon_o V_{in}}{t^3 Ed} \right)^{\frac{1}{2}}$$

where,
η = dimensionless deflection electrode geometry dependent constant, giving ratio of deflection at gate to deflection at tip ≈ 0.2
V_p = polarization voltage applied to the cantilever

V_{pmax} = "pull-in voltage," above which the cantilever collapses
V_{in} = input signal used to deflect the cantilever

With V_p/V_{pmax} typically set to ≈ 0.5, assuming E for electroplated gold at 8×10^{10} N/m², and with the relatively high V_p values used of 20 to 70 V, the behavior of the RGT devices could be predicted fairly well. For example, dimensions of L = 500 µm, t = d = 5 µm (note thick gold plating required), $Q \approx 100$ (in air), $f_R \approx 6.5$ kHz, and $\delta_R \approx 1$ µm. Devices could readily be fabricated with resonant frequencies in the range of 1 to 50 kHz. Dual cantilever designs, with the two cantilevers coupled by a short "web" at their bases, were also fabricated to form loosely coupled, multipole band-pass filters.

While the project was eventually abandoned, successful resonators of various types were demonstrated. Difficulties encountered included inadequate control over fabrication steps (primarily the electroplating step) to adequately limit variations in Q and f_R (20 to 30% variations in f_R were observed), difficulty in finding suitable low temperature-coefficient resonator material (likely not a problem with today's technology), difficulty in increasing resonant frequency beyond 50 kHz (potentially solvable using doubly clamped cantilevers, but if so, not using the high temperature-coefficient electroplated gold available to these researchers), and the need for relatively high polarization voltages for the geometries available at the time.

More recently, Clark, et al. (1997) used doubly supported, electrostatically driven and capacitively sensed polysilicon beams to realize a variety of band-pass filters. They demonstrated Butterworth, Chebyshev, and Bessel filters with center frequencies of ≈ 14.5 MHz, Q values in the range of 830 to 1,600, and DC-bias tuning ranges on the order of 500 kHz (it should be noted that the demonstration devices had 13 to 14 dB insertion losses and were tested at 10 µtorr using a turbomolecular pump). Tuning for these devices was accomplished via the super-position of a DC bias voltage, V_p, with the AC signal being filtered (reducing the mechanical spring constant k via the nonlinear electrostatic spring constant that effectively subtracted from it), resulting in a new resonant frequency given by,

$$f_R^* = f_R \left(1 - \frac{V_p C_o}{kd^2} \right)^{\frac{1}{2}}$$

where C_o is the beam/electrode overlap capacitance without bias, and d is the beam/electrode gap. This tuning approach should be applicable to fairly complex filters, which can be constructed by arranging several such resonators in series for the signal path, and individually controlling the DC bias voltage for each one.

Newell (1968) reported silicon-based, doubly clamped cantilever resonators (referred to as "tunistors") with piezoelectric transduction. These devices eliminated the need for an underlying FET structure (and eliminated its noise contribution) and due to the matched the temperature coefficient of the silicon cantilever, virtually eliminated thermal stability problems. For such designs, the deflection of the center of the cantilever at resonance, δ_R, is approximated by,

$$\delta_R \approx \left(\frac{3Q\eta d_{31}}{Es_{11}}\right)\left(\frac{L}{t}\right)^2 V_{in}$$

where d_{31} is the piezoelectric coefficient of the transducing film, and s_{11} is the compliance of transducing film.

The doubly clamped silicon resonators were fabricated using cadmium sulfide as the piezoelectric film ($d_{31} \approx 5.2 \times 10^{12}$ C/N and $s_{11} = 2.1 \times 10^{-11}$ m^2/N) and resonant frequencies and Q values achieved were on the order of 250 kHz and 500, respectively (in air). The measured temperature coefficients of frequency were on the order of 40 ppm/°C. Although micromachining technology available at the time somewhat impeded the formation of the resonators, the major difficulties encountered in further developing these resonators were in depositing suitable piezoelectric films.

A cantilever-based, surface micromachined piezoelectric resonator was described by Lau, et al. (1996). Their fabrication process began with the deposition and patterning of a 300 nm spin-on-glass sacrificial layer, followed by a 150 nm LPCVD silicon nitride mechanical support layer, and a 2 μm RF-sputtered ZnO piezoelectric layer. After the ZnO was patterned, a 400 nm Al electrode layer was evaporated and patterned, followed by a second 600 nm Al layer (for step coverage over the 2 μm ZnO layer). Finally, the silicon nitride was patterned and the cantilevers were released using vapor-phase HF etching. Such a resonator with dimensions of $\approx 40 \times 60$ μm was shown to have a resonant frequency of 531 MHz and a Q of 738.

Prak, et al. (1992) described methods for deliberately exciting only specific resonance modes of doubly clamped cantilevers. The underlying principle is the deliberate shaping of the excitation (and perhaps sensing) devices on the cantilevers (e.g., piezoelectric deflection films patterned along the cantilevers) to deliberately drive desired modes. While these principles are general, the authors demonstrated them with electrostatically actuated resonators and showed relatively selective excitation of modes 1, 2, and 3 in a silicon beam.

7.1.2 LATERAL RESONATORS

More recently, lateral flexural resonators fabricated from polysilicon have been demonstrated that have the potential to be scaled up to the standard radio intermediate frequencies. As described in Nguyen and Howe (1993a, 1994) and Nguyen (1995), standard sacrificial oxide released surface micromachined polysilicon structures are employed to realize a wide variety of mechanical filter configurations. In vacuum, Q values approaching 100,000 appear to be achievable with this approach.

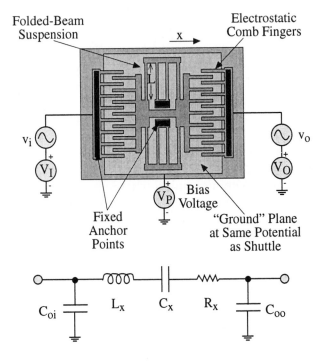

Illustration of a two-port, folded-beam lateral comb-drive resonator. Adapted from Nguyen (1995), with RLC equivalent circuit.

As illustrated above, folded-beam lateral comb-drive resonators can readily be fabricated in polysilicon using a standard sacrificial layer process (example processes are discussed in the Micromachining Techniques chapter). The resonant frequency of such a two-port device is given by (Nguyen (1995)),

$$f_R = \frac{1}{2\pi} \left[\frac{2Et\left(\dfrac{w}{L}\right)^3}{\left(M_p + \dfrac{M_t}{4} + \dfrac{12M_b}{35} \right)} \right]^{\frac{1}{2}}$$

where,
w = width of suspension beams

t = thickness of suspension beams
 L = length of suspension beams (see illustration above)
M_p = mass of shuttle
M_t = mass of folding trusses
M_b = total mass of suspension beams

Also, the DC-bias tuning approach used with doubly clamped beam resonators by Clark, et al. (1997) (discussed in Section 7.1.1 above) can be used in this case as well (the same equation for the new resonant frequency, f_R', described in that section is valid).

For a 100 kHz resonator with typical values of 2 µm thick polysilicon and 2 µm comb finger gaps, Nguyen (1995) gives equivalent circuit component values of C_x = 0.5 fF, L_x = 200 kH, R_x = 500 kΩ and C_{oi} = C_{oo} = 15 fF, for a Q value of 40,000 (note that Cx, Lx and Rx depend upon the DC bias across the transducer, which sets its electrical-to-mechanical transfer function). In comparison, a typical 100 kHz quartz crystal (NT cut) has equivalent circuit component values of C_x = 8 pF, L_x = 365 H, and R_x = 2 kΩ, for a Q value of 3,380 (from Kemper and Rosine (1969), with a more recent published Q of 5,100 for a 90 kHz quartz crystal from Malik (1995)). In both cases, the value of Q was computed, using,

$$Q = \sqrt{\frac{L_x}{C_x R_x^2}}$$

Referring to the resonant frequency equation above, it can be noted that the temperature coefficient of f_R, TC_{fr}, derives from the temperature dependence of the Young's modulus, TC_E, and the thermal expansion coefficient of the polysilicon in the direction of flexing (i.e., change in the variable t), TC_t. Nguyen (1995) reported a -10 ppm/°C temperature coefficient of frequency, with a derived temperature coefficient of polysilicon's Young's modulus of -22.5 ppm/°C (where $TC_E = 2TC_{fr} - TC_t$).

Similar to the cantilever case discussed above, the noise displacement of a lateral resonator can be derived by equating the potential energy of thermal agitation to the stored potential energy in the mechanism. For an effective spring constant, k, this yields a root-mean-squared noise displacement similar in form, but with a new effective mass and spring constant. As for the cantilever case, the noise is Q times larger at resonance due to the shaping of the white spectrum of the thermal noise by the force-to-displacement relationship of the mechanism). The thermal noise itself is fully modeled by the equivalent resistance in the above circuit, and the shaping of its spectrum is similarly modeled by the effects of the other components.

The equivalent resistor value, R_x, (representing the only noise source in the electrrical model) is given by,

$$R_x = \frac{\sqrt{kM}}{Q\left[V_p\left(\frac{\partial C}{\partial x}\right)\right]^2}$$

where,
  k = effective spring constant
  M = effective mass of mechanism
  V_p = applied DC bias voltage
  C = comb capacitance
  x = displacement

Further analysis shows that R_x, and hence the total system noise, can be minimized by minimizing the gaps in the comb-drive fingers (maximizing the partial derivative to capacitance with respect to displacement), which is achievable through process improvements (lithography and dry etching).

Yao and MacDonald (1995) described a single-crystal silicon lateral resonator fabricated using a variable anisotropy dry etching process (described in the Micro-machining Techniques chapter). With structural members on the order of 0.25 µm wide by 2 to 3 µm thick and up to 200 µm long, a resonator was demonstrated with f_R = 0.96 MHz, Q = 4,370 (in vacuum), and an electrical tuning range of -60 kHz for a 35 V tuning voltage.

Nguyen (1995) and Nguyen and Howe (1993a) described the integration of CMOS (a standard process with the exception that W metallization and $TiSi_2$ contact barriers were used so that it could withstand the high temperatures associated with polysilicon processing) with the polysilicon micromechanisms. A fully inte-grated oscillator was demonstrated, combining the resonator mechanism, a transre-sistance amplifier, and an output buffer in a 500 × 500 µm space. Thermally isolated silicon nitride membranes were also used to support some resonator designs and allowed them to be locally heated to reduce shifts in resonant frequency with temperature In addition, multi-resonator filters were implemented with more complex transfer functions, such as the third-order 300 kHz band-pass filter (Q = 590 and stop-band rejection was greater than 38 dB) presented by Wang and Nguyen (1997). As for the macroscopic multi-stage mechanical filters developed in the 1940's and thereafter, it is likely that far more complex filters can be synthesized using this approach.

The use of local thermal annealing for tuning such devices was described by Wang, et al. (1997). By pulsing relatively high currents through regions of polysilicon lateral polysilicon resonators (applied through their anchors to the substrate), they reported frequency trims of up to 2.7% and Q increases of several fold. Lee and Cho (1997) demonstrated electrical tuning via the addition of two extra sets of comb-drive arrays with varied finger lengths. They reported the ability to reduce the resonant frequency up to 3.3%.

7.1.3 MEMBRANE RESONATORS

To fabricate much higher frequency (microwave) micromachined resonators, one can apply bulk acoustic resonator techniques, as discussed by Ruby and Merchant (1994) and Ruby (1996). Their devices consisted of back-side released silicon nitride membranes supported by bulk silicon, with a reactively sputtered aluminum nitride piezoelectric layer deposited thereon. In the frequency range of 1.5 to 7.5 GHz, the authors demonstrated Q values as high as 1,300. They also demonstrated the use of on-chip thin-film heaters for frequency control and/or stabilization. A similar device, also using an aluminum nitride piezoelectric layer on a silicon nitride membrane, was fabricated by Lutsky, et al. (1996) using a somewhat more complex (yet potentially CMOS-compatible) fusion bonding/electrochemical etch-stop/dry etch method. With this approach, a structure similar to that of Ruby and Merchant (1994) was realized, but with buried, atmosphere-vented cavities beneath the resonator membranes. This design provided the acoustic reflecting surface required in addition to the membrane to form a resonator, as opposed to the need for a separate underlying substrate in earlier devices. With 200 μm thick circular membranes, resonant frequencies of 1.36 and 3.28 GHz were seen, with a Q value of 210 for the first resonance.

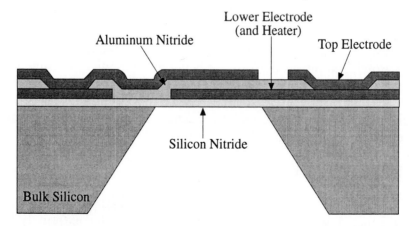

Illustration of a micromachined, microwave frequency bulk acoustic resonator based on a piezoelectric layer fabricated above a back-side released dielectric membrane. Adapted from Ruby and Merchant (1994).

Again, since SNR is a critical parameter in electronic systems, a similar noise analysis can be done for membrane resonators once an effective spring constant is determined. In general, the concepts discussed above with respect to micromachined resonators can be applied to a wide variety of other microstructures. The concept of looking at noise spectrum to examine the mechanical frequency transfer function of a microstructure can be broadly applied. Two excellent references on thermal-mechanical noise analyses are Gabrielson (1993, 1995).

7.2 MECHANICAL RELAYS AND RF SWITCHES

This section covers relays and RF switches, both of which are based on the principle that contacts are physically in contact or separated by an insulating gap (gas or vacuum), depending on the desired state (on or off) of the device. Such electrically controlled switches have a wide variety of applications in electronic circuits (and have been used in the past to construct logic circuits without using any active components). It is important to note that RF switches are somewhat distinct from general purpose relays in that the former are optimized for impedance and parasitic parameters required for higher frequencies, while the latter are often optimized for relatively high currents. While this distinction is fairly arbitrary, in the current literature, it seems to be used consistently.

7.2.1 GENERAL PURPOSE RELAYS

Macroscopic relays are broadly applied in electronic switching where high resistance, low capacitance isolation is necessary, and low on-resistance is advantageous, such as in precision electronic instruments. No solid-state switch (e.g., CMOS transmission gate, JFET, etc.) has yet demonstrated the outstanding off-state isolation, low cross-talk, and on-resistance of conventional, electromagnetic relays. However, a growing number of efforts are being made to micromachine mechanical relays for inclusion on IC substrates. The bulk of the work done to date has been on the use of electrostatic drive for the relays (due to the low power required), although electromagnetic and thermal drive schemes have been demonstrated (thermally driven RF switches are discussed below).

ELECTROSTATICALLY DRIVEN RELAYS

Petersen (1978) described micromachined relays and switches that were electrostatically actuated, with the aim of taking advantage of the extremely low power dissipations achievable with this approach. As illustrated below, a bulk micromachining process was used to fabricate the devices, beginning with the growth of a thermal SiO_2 layer on silicon wafers with a buried p+ layer and more lightly doped upper epitaxial layer (used as a sacrificial in regions where the relays were formed). This was followed by the deposition of a Cr/Au layer and the patterning of it and the underlying SiO_2. Two layers of patterned photoresist were then added to form sacrificial layers of two heights (required to provide a raised area for the electrical contacts, as shown below). A thin Cr/Au layer was then deposited to serve as a seed layer for the electroplating of thick Au, which was done using a third, thicker photoresist layer as a template. The photoresist layers and seed layer were then removed, and the wafers were etched in EDP to form the pits beneath the moving cantilever electrodes. The finished devices were actuated by applying a DC control voltage between the lower Cr/Au layer and the buried p+ silicon layer, which served to draw the cantilever electrodes downward, bringing the Au contacts together.

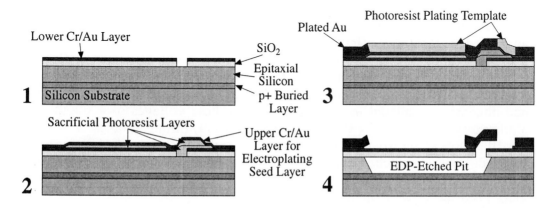

Cross-sectional diagrams of a single-contact micromechanical switch at various stages during the fabrication procedure. (1) After first metal etch and oxide etch. (2) After evaporation of Au-Cr plating base. (3) After selective Au plating through photoresist plating template. (4) Finished structure after photoresist stripping, removal of excess plating base, and EDP etch. After Petersen (1978a).

As discussed in Section 6.1.1 above, the approximate threshold voltage, v_{th}, for such a design is given by Petersen (1978a),

$$V_{th} \approx \sqrt{\frac{18EId^3}{5\varepsilon_o L^4 b}}$$

where,
E = effective Young's modulus for the composite SiO_2/Cr/Au beam
I = effective moment of inertia for the beam
d = gap between cantilever and deflection electrode
L = beam length
b = beam width

The resonant frequency of this type of design, taking into consideration the mass of the electroplated Au contacting bar at its tip is,

$$f_R \approx \frac{1}{2\pi} \sqrt{\frac{3EI}{L^3 (M + 0.23m)}}$$

where M is the mass of contacting bar and m is the mass of cantilever beam.

These two equations can be studied to illustrate the design trade-offs between threshold voltage (the lower the better, in general) and resonant frequency (higher values of f_R correspond to faster relay response. Unfortunately, many designs with low threshold voltages are also relatively slow (note that the gap width, however, can be reduced without affecting f_R, but this may introduce stiction problems during fabrication or use). Typical relay dimensions used by Petersen (1978) and relevant to mechanical properties were 0.35 μm for the SiO_2 layer, 40 nm for the first Cr/Au layer, 5 to 7 μm pit depth and 75 μm long cantilevers. Relays with such dimensions actuated in ≈ 40 μs, with threshold voltages of ≈ 60 V. The cantilevers were deflected (with no switched current) through angles of ≈ 4° for over 10 billion cycles (roughly 12 days at 10 kHz) without any gross failures being observed. With a switched current density of 5×10^4 A/cm^2 (below the level of gross electromigration effects), 2 million cycles were achieved prior to failure. Contact resistances were on the order of 5 Ω.

Drake, et al. (1995) demonstrated a bulk micromachined, electrostatically actuated relay with a polysilicon actuator. As illustrated below, their design made use of electroplated gold "deflection bumps" to push the polysilicon actuator away from mating contacts on a glass cap wafer. When actuated, the actuator pulled upward and made contact with gold metallization on the glass cap, completing the circuit. Their fabrication process began with the growth of a silicon dioxide layer on the silicon substrate; next its patterning to allow the polysilicon to contact the substrate; followed by the deposition, annealing, and patterning of the polysilicon. Gold metallization was then deposited (presumably via electroplating, although this was not discussed in the paper) to form the lower contact and deflection bumps. Following the patterning of the underlying silicon dioxide, anisotropic bulk etching was carried out to form the deflection cavity beneath the actuator. The back side of the silicon wafer was then metallized to serve as a contact to the actuator. A matching glass cap wafer was processed via the deposition and patterning of metallization (upper contacts) and etching to form cavities to allow clearance surface features of the silicon substrate. The two substrates were then bonded together using a proprietary metal sealing technique that was not described in the paper. Measured performance for these relays were actuation voltages in the range of 50 to 100 V, on-resistances (including wiring and contacts) of ≈ 2.3 Ω, actuation lifetimes on the order of 100 million cycles, and closure times of ≈ 20 μs.

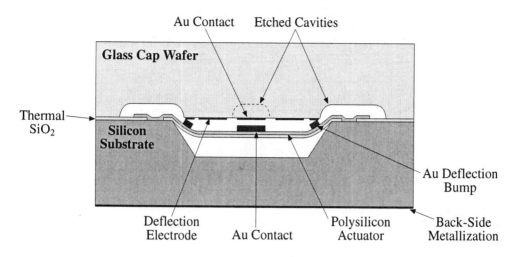

Illustration of a bulk micromachined, electrostatically actuated relay. Adapted from Drake, et al. (1995).

Grétillat, et al. (1995) demonstrated a surface micromachined polysilicon-based electrostatic relay. They fabricated three-level, doubly supported polysilicon beams, released by undercutting them by removing a sacrificial layer. The relays were actuated by applying a potential between the upper polysilicon layer (the actuator) and the substrate. In this design, the electrical contacts were both polysilicon. They reported switching voltages in the range of 41 to 54 V (corresponding to 350 and 250 μm long beams, respectively). The authors note switching voltages in the range of 1 to 10 V, but these voltages are those required to transition from closed to open, while still under bias (demonstrating a relatively small hysteresis). Closure times were on the order of 10 μs, lifetime was demonstrated to be > 10^9 cycles (whether this was tested with or without switched current was not clear), and contact resistances were not given.

Another polysilicon-based, surface micromachined relay, but with mercury wetted contacts (an approach often used in macroscopic relays) was demonstrated by Saffer, et al. (1995). They used the MCNC (Microelectronics Center of North Carolina, Research Triangle Park, NC) foundry polysilicon process (MUMPS) to fabricate structures wherein a long (300 to 500 μm) polysilicon cantilever was deflected laterally (in the plane of the wafer) to contact a mercury droplet ≈ 10 μm in diameter. The mercury was deposited by condensing it from vapor onto a patterned Au pad on the polysilicon (this metal layer is available through MCNC). The resulting Au/Hg alloy (amalgam) led to preferential deposition of the mercury on the pad. Due to a curved actuator beam, relatively large deflections (≈ 30 μm) were achieved with a drive voltage of 60 V, with only 5 μm required to close the contacts. The total resistance of the closed relays (including wiring and contacts) was between 1.9 and 3.2 kΩ, with a current capabiltiy of > 10 mA. It should be noted that these resistance levels were larger than those obtained with commercial CMOS analog switches, but this design could potentially handle larger currents.

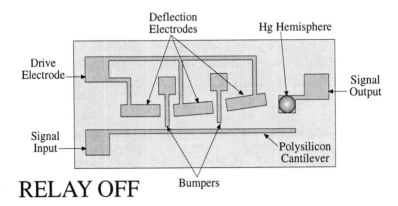

RELAY OFF

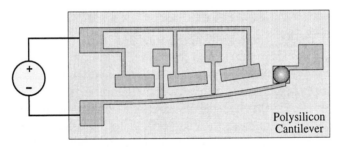

RELAY ON

Illustration of a surface micromachined, mercury-contact electrostatic relay in the off state (top) and on state (bottom). Adapted from Saffer, et al. (1995).

Zavracky, et al. (1997) and Majumder, et al. (1997) described a surface micromachined electrostatic relay, consisting of an electroplated Ni beam, released by etching a sacrificial Cu layer. Gold electrode contacts were used, with electroplated Au bumps contacting sputter-deposited Au electrodes. Measured contact resistances have been in the range of $\approx$ 0.1 to 5 Ω, with threshold voltages of 30 to 300 V, off impedances > 10^{12} Ω, and lifetimes exceeding 10^9 cycles (cold switched, with no current flowing) in a nitrogen environment. At 5 mA current levels, lifetimes on the order of 10^6 cycles were possible, extending to as many as 10^7 cycles at 10 μA. It is likely that for practical use, the actuation voltages will need to be reduced to values closer to typical IC power supply rails and the operating lifetimes will need to be extended. However, even preliminary results show that the extremely high isolation impedances desired for many instrumentation applications can be achieved.

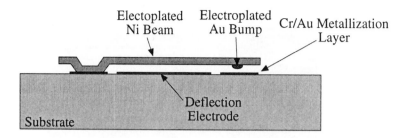

Cross-sectional illustration of a surface micromachined, electrostatic relay, fabricated by electroplating Au bump contacts and Ni beams above a sacrificial copper layer (not shown). Adapted from Zavracky, et al. (1997).

MAGNETICALLY DRIVEN RELAYS

Since the bulk of miniature relays sold at present are magnetically driven, it stands to reason that researchers would attempt to micromachine relays using this operating principle. An important (and obvious) consideration for relays is that, in general, they should use less power to actuate than the power they switch. Most miniature relays are reed-relays, wherein a magnetic field causes two Fe:Ni beams ("reeds") to contact each other, so as to reduce the magnetic flux in the gap between them. These devices are very simple, and since the mass of the beams is generally quite small, require only small magnetic fields (or coil currents). These relays are more efficient than larger solenoid-type devices. Unfortunately, high efficiency is not always possible in micromachined magnetic relays, particularly those with lithographically fabricated coils, due to their relative lack of efficiency.

Rogge, et al. (1995) demonstrated a sacrificial, two-stage LIGA-based magnetic relay with a Permalloy core and electroplated copper coil turns (with underlying sputtered copper pass-throughs beneath the core). Their 4 × 2.3 mm relay was fabricated on an Al_2O_3 substrate beginning with the sputter deposition and patterning of an 0.1/1.0 µm Cr/Au layer to form the bond pads and the bottoms of the coils (the pass-throughs that carry current underneath the core). This was followed by the sputter deposition and patterning of a 5.0 µm Ti sacrificial layer, application of a 120 µm PMMA layer, and LIGA processing to form a high aspect ratio template for formation of the magnetic core. The core was then electroplated (Ni or Ni/Fe, with reported coercivities of 7,500 and 200 A/m, saturation magnetizations of 0.8 and 1.2 T, and relative permeabilities of 12 and 400, respectively), followed by the removal of the first PMMA and application of a second, thicker PMMA layer (170 µm). At this point, a 0.1/0.3 µm Cr/Au plating seed layer was sputter deposited and patterned, defining the regions where copper plating would occur during the final plating step. Once completed, the PMMA was removed and the Ti sacrificial layer was etched away.

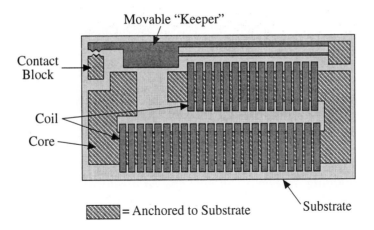

Illustration of monolithic, micromachined magnetic relay. Adapted from Rogge, et al. (1995).

For a relay of such a design, the electromagnetic force generated by the "keeper" moving toward the core is given by,

$$F_{MAG} = \left(\frac{NI\mu_o\mu_r}{\dfrac{A_{gap}}{A_{core}}L_{core} + 2d\mu_r} \right)^2 \frac{A_{gap}}{\mu_o}$$

with the optimum value of A_{gap} for maximum force, given by,

$$A_{gapopt} = 2\frac{A_{core}}{L_{core}}\mu_r d$$

where,

N = number of turns

I = coil current

μ_o = permeability of free space = $4\pi \times 10^{-7}$ T•m/A

μ_r = relative magnetic permeability of core

A_{gap} = cross-sectional area of gap

A_{core} = cross-sectional area of core

L_{core} = path length of magnetic flux (being careful not to confuse this use of the symbol "L" with that for inductance)

d = gap width

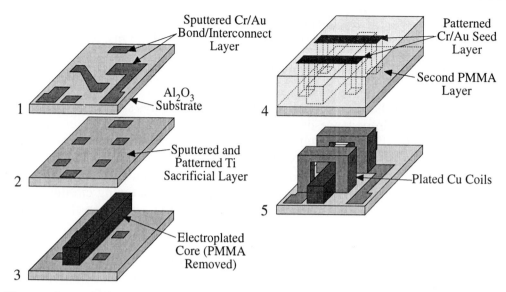

Illustration of the process steps used to fabricate the micromachined magnetic relay using a two-stage LIGA process. Adapted from Rogge, et al. (1995).

Due to the relatively small number of turns possible with monolithically fabricated coils, the switching currents for this design were quite high, at 1 A and 45 mA, with zero-gap forces of 0.22 and 250 mN for the Ni and NiFe, respectively. Since the maximum demonstrated switched current was 1 mA, it is not clear whether or not these relays can switch more power than that used to control them (unfortunately, the coil resistance was not given).

As opposed to the above "in plane" motion relay, Taylor, et al. (1996a, 1996b) demonstrated a monolithic "vertical" actuation micromachined magnetic relay. Magnetic flux generated in a planar meander coil with interdigitated NiFe core draws downward a magnetic plate that is supported by four compliant members at its corners. This plate then shorts out two contacts, which can be opened by turning off the drive current in the coil. These microrelays were fabricated using a polyimide template electroplating process (the material used for the magnetic plate was not given). Actuation occurred with a coil current of 600 mA (320 mW, for a coil resistance of $\approx 0.9\ \Omega$), the contact resistance was $< 30\ \Omega$, and the maximum current that could be electromagnetically switched was 60 mA. The relays were tested over $\approx 17,000$ actuation cycles. (It should be noted that Taylor, et al. (1996a) contains useful references to several other efforts to fabricate micromachined relays.) A more recent paper by Taylor and Allen (1997) describes improved devices (normally open and normally closed), with minimum actuation currents of 180 mA (for a power of 33 mW), carrying 1.2 A of contact current. Lifetime testing has shown that for currents of ≈ 2.5 mA, these relays can survive $> 300,000$ cycles without degradation (although these current levels are far below those needed

to actuate the relays). The off-state resistance was measured to be $> 10^{10} \Omega$, and the pull-in and release times were on the order of 1 to 5 and 2.5 ms, respectively.

7.2.2 RF SWITCHES AND SWITCHED CIRCUITS

Several groups have developed and demonstrated relays specifically designed for RF applications (impedance matched to transmission lines and potentially capacitively coupled). These RF switches have a broad variety of applications where active and passive components can be switched into or out of RF circuits. Example applications are digitally controlled antenna matching circuits, transmit/receive switches, phase shifters for phased array radars, input filters, tuning circuits, and a number of others. In this section, the switches are not separated on the basis of actuation mechanism.

Hackett, et al. (1991) and Larson, et al. (1991a, 1991b) presented an assortment of micromachined, electrostatically actuated RF switching devices fabricated on GaAs substrates. The fabrication process they employed began with GaAs substrates with a pre-implanted active layer, with metallization and ohmic contacts patterned and alloyed into the surface. A 200 nm layer of CVD silicon nitride was then deposited and patterned to remain only in the areas where electrostatic actuators would be fabricated, increasing the breakdown voltage to the substrate. Using a sequence of thin-films and electroplated gold layers, a wide variety of cantilever switches and rotary motor/rotary switch devices were fabricated, in parallel with the active devices on the substrate. For the cantilever switches, characterized over 2 to 45 GHz, reported off-isolation was 35 dB, insertion loss was 0.4 dB, and actuation voltages were in the range of 80 to 200 V.

A family of electrostatic RF switches was demonstrated by Yao and Chang (1995), using PECVD silicon dioxide as the structural material, polyimide as the sacrificial layer, and evaporated gold for contacts and metallization. They presented results that showed a minimum actuation voltage of 28 V, closure time of $\approx$ 30 µs, cycle test survival beyond 6.5×10^{10} cycles, and a DC resistance of 0.22 Ω. Over the frequency range of 100 MHz to 4 GHz, the insertion loss for the switch was $\approx$ 0.1 dB, and isolation was $\geq$ 50 dB. It is particularly noteworthy that the lack of failure over such a large number of mechanical cycles was achieved using a 2 µm PECVD silicon dioxide structural layer for the cantilevers.

Goldsmith, et al. (1995, 1996) and Randall, et al. (1996) fabricated and tested a family of membrane-deflection-based capacitive RF switches for microwave applications. A key advantage of capacitive switches over their resistive counterparts are the potential to make use of non-conductive anti-stiction coatings on the electrodes. Also, as for all electrostatically actuated devices, the drive power can be extremely small. As described in Randall, et al. (1996), the switches consisted of 300 nm thick, circular Al or Au membranes (fabricated with diameters ranging from 50 to

400 µm) suspended 2 to 4 µm above a dielectric-insulated electrode. A 100 nm silicon nitride film on the bottom electrode allows DC deflection voltages to be applied without interference with the AC-coupled RF signal passing through the switch. For such a switch, a critical parameter is the off/on impedance ratio, which defines the off-isolation, and is given by,

$$\frac{Z_{off}}{Z_{on}} = \frac{C_{on}}{C_{off}} = \frac{\varepsilon_R d_{air} + d_{ins}}{d_{ins}} \approx \frac{\varepsilon_R d_{air}}{d_{ins}}$$

where,
ε_R = relative dielectric constant of insulating layer
d_{ins} = thickness of insulating layer
d_{air} = height of air gap between top membrane and insulating layer

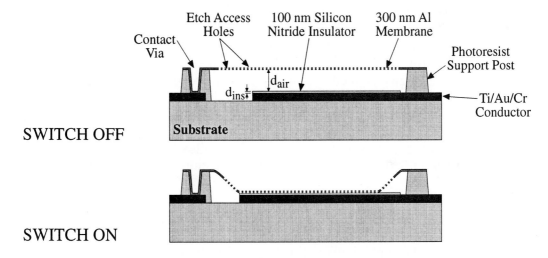

Illustration of a capacitively coupled, electrostatically actuated membrane-type RF switch. Adapted from Goldsmith, et al. (1996).

For typical film properties and thicknesses used, this gives an impedance ratio of ≈ 100:1, which is more than adequate for most microwave switching applications. As described in Randall, et al. (1996), the fabrication process used a series of steps that are compatible with a variety of possible substrates (e.g., silicon, silicon-on-sapphire, or GaAs). They began by lift-off depositing a trilayer of 20 nm of Ti, 550 nm of Au, and 50 nm of Cr, follwed by the deposition and patterning of 100 nm of PECVD silicon nitride for the electrode insulator. Gold was then electroplated to form the transmission lines (presumably using photoresist templates) and a 4 µm (in this case) photoresist sacrificial layer was then deposited and patterned. The membrane Al layer (300 nm thick, in tension at ≈ 20 MPa) was then sputtered onto the sacrificial layer and patterned with a wet etch to form the membrane, as well as a large number of small access holes used so that the dry etch used to release the membrane can access the underlying photoresist. The Al layer also forms electrical connections between the membrane and underlying metallization patterns so that signals can be applied to the membrane. Using a custom-built, microwave-driven downstream etch reactor, an $O_2/NO_2/Ar$ etch process was used to release the membranes. In regions where access holes through the membrane were deliberately omitted, photoresist was left behind to form the support posts for the membrane. Pull-down voltages varied from 50 V (for 50 μm diameter membranes) to < 10 V (for 400 μm membranes). Insertion loss for a 400 μm diameter device was ≈ 0.3 to 0.5 dB at 10 GHz, off/on isolation was 11 dB, and switching times were ≈ 6 μs for closing and ≈ 8 μs for opening the switch.

Several different surface micromachined electrostatic relay designs were presented by Schiele, et al. (1997), using cantilevers, doubly clamped beams, and torsional structures as the actuators. The structural material of the moving structures was a $SiO_2/Au/SiO_2$ thin-film stack. The goal of this group was to fabricate relays suitable for high-frequency signal control, leading them to pursue designs with large gaps (≈ 10 μm) between the contacts when open (to reduce parasitic capacitances). Their designs yielded minimum pull-in voltages of 20 and 60 V for cantilever and doubly clamped designs, respectively. Contact resistances for the cantilever designs were 12 to 80 Ω, and reported lifetimes were > 7,000 cycles. For all of the designs, the maximum contact current was 1 mA, and the lowest reported switching speeds were 2.6 to 20 μs for a cantilever design. Unfortunately, despite the apparent focus on high-frequency applications, no RF test results were reported.

A digitally controlled inductor array, switched using thermal bimorph actuators, was demonstrated by Zhou, et al. (1997). They formed an array of four inductors (15, 29, 47, and 77 nH) in series, with a thermal micro-relay arranged such that it could bypass each inductor with a short-circuit if energized. Rather than selectively undercut the RF components from the top surface of the wafer, they opted to bulk micromachine membranes from the back of the wafer beneath the inductors.

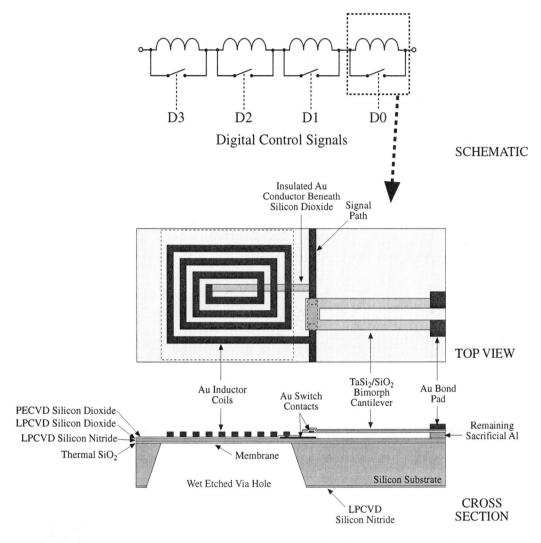

Illustration of a micromachined, digitally controlled inductor array, using thermal bimorph actuators for switching. Adapted from Zhou, et al. (1997). As seen in the schematic (top), one of the series inductor/switch combinations are shown in the top view (center) and cross section (bottom). When a particular switch is actuated, it bypasses the associated inductor.

Their process began with the growth of an 800 nm thermal silicon dioxide layer on (100) wafers, followed by the deposition of a 200 nm LPCVD silicon nitride and 500 nm LPCVD silicon dioxide layers. Two sets of alternating layers of Au (1 μm thick) and Al (1.5 μm thick) were then deposited and patterned, with the Au forming the switch contacts (one layer forming the substrate-level contacts and the other forming the gold contacts on the switch cantilevers) and the Al forming sacrificial layers. The thermal bimorphs for the switches were then formed by depositing and patterning 1.5 μm PECVD silicon dioxide and 400 nm TaSi$_2$ (patterned using SF$_6$ plasma etching) layers. A third 1.5 μm Au layer was then deposited to form the inductor coils and the bonding contacts for the switches. The back-side vias to form the membranes were then etched by patterning the silicon nitride on the backs of the wafers using RIE, followed by a KOH etch. Finally, the switches were released using a standard wet Al etchant. The resulting devices exhibited considerable mutual inductance between the inductors, making the available inductances for the array much larger than the sum of the individual values. However, digital control was demonstrated, with self-resonant frequencies between 1.9 and 4.6 GHz and Q factors between 1.7 and 3.3. The switches (DC-coupled) had measured on-resistances of 0.6 to 0.8 Ω, with thermal closing power requirements of 8 mW (they could be electrostatically held with 20 V) and closure times of 12 μs. The entire switched inductor array was 3.2×0.9 mm in size.

MECHANICAL TRANSDUCERS REFERENCES

GENERAL REFERENCES

Bryzek, J., Petersen, K., Mallon, J. R., Christel, L., and Pourahmadi, F., "Silicon Sensors and Microstructures," Lucas NovaSensor, 1055 Mission Court, Fremont, CA, 1991.

Norton, A. N., "Handbook of Transducers," Prentice-Hall, Inc., Englewood Cliffs, NJ, 1989.

Petersen, K. E., "Silicon as a Mechanical Material," Proceedings of the IEEE, vol. 70, no. 5, May 1982, pp. 420 - 457.

SPECIFIC REFERENCES

Aine, H. E., U.S. Patent No. 5,000,817, issued Mar. 19, 1991.

Akamine, S., Albrecht, T. R., Zdeblick, M. J., and Quate, C. F., "A Planar Process for Microfabrication of a Scanning Tunneling Microscope," Sensors and Actuators, vol. A23, nos. 1 - 3, Apr. 1990, pp. 964 - 970.

Akiyama, T., and Shono, K., "Controlled Stepwise Motion in Polysilicon Microstructures," Journal of Microelectromechanical Systems, vol. 2, no. 3, Sept. 1993, pp. 106 - 110.

Alberts, B., Bray, D., Lewis, J., Raff, M., Roberts, K., and Watson, J. D., "Molecular Biology of the Cell," Third Edition, Garland Publishing, Inc., New York, NY, 1994.

Allen, H. V., "Silicon-Based Micromechanical Switches for Industrial Applications," Proceedings of the IEEE Micro Robots and Teleoperators Workshop, Hyannis, MA, Nov. 9 - 11, 1987, pp. 8/1 - 8/3.

Allen, H. V., Terry, S. C., and de Bruin, D. W., "Accelerometer Systems with Built-in Testing," Sensors and Actuators, vols. A21 - A23, Feb. - Apr. 1990, pp. 381 - 386.

Allen, M. G., Mehregany, M., Howe, R. T., and Senturia, S. D., "Microfabricated Structures for In-Situ Measurement of Residual Stress, Young's Modulus, and Ultimate Strain of Thin-Films," Applied Physics Letters, vol. 51, no. 4, July 27, 1987, pp. 241 - 243.

Andrews, M. K., Turner, G. C., Harris, P. D., and Harris, I. M., "A Resonant Pressure Sensor Based on a Squeezed Film of Gas," Sensors and Actuators, vol. A36, no. 3, May 1993, pp. 219 - 226.

Angell, J. B., "Transducers for in vivo measurement of force, strain and motion," in "Physical Sensors for Biomedical Applications," Neuman, M. R., Fleming, D. G., Cheung, P. W., and Ko, W. H. [eds.], CRC Press, Boca Raton, FL, 1980, pp. 46 - 53.

Asahi, R., Sakata, J., Tabata, O., Mochizuki, M., Sugiyama, S., and Taga, Y., "Integrated Pyroelectric Infrared Sensor Using PVDF Thin-Film Deposited by Electro-Spray Method," Proceedings of Transducers '93, the 7th International Conference on Solid-State Sensors and Actuators, Yokohama, Japan, June 7 - 10, 1993, pp. 656 - 659.

Ataka, M., Omodaka, A., and Fujita, H., "A Biomimetic Micro Motion System - A Ciliary Motion System," Proceedings of Transducers '93, the 7th International Conference on Solid-State Sensors and Actuators, Yokohama, Japan, June 7 - 10, 1993, Institute of Electrical Engineers, Japan, pp. 38 - 41.

Ataka, M., Omodaka, A., Takeshima, N., and Fujita, H., "Fabrication and Operation of Polyimide Bimorph Actuators for a Ciliary Motion System," Journal of Microelectromechanical Systems, vol. 2, no. 4, Dec. 1993, pp. 146 - 150.

Avallone, T. B., III, and Avallone, E. A., [eds.], "Marks' Standard Handbook for Mechanical Engineers," Tenth Edition, McGraw-Hill, New York, NY, 1996.

Bäcklund, Y., Rosengren, L., Hök, B., and Svedbergh, B., "Passive Silicon Transensor Intended for Biomedical, Remote Pressure Monitoring," Sensors and Actuators, vol. A21, nos. 1 - 3, Feb. 1990, pp. 58 - 61.

Baxter, L. K., "Capacitive Sensors: Design and Applications," IEEE Press, New York, NY, 1997.

Benes, E., Gröschl, M., Burger, W., and Schmid, M., "Sensors Based on Piezoelectric Resonators," Sensors and Actuators, vol. A48, no. 1, May 1995, pp. 1 - 21.

Bergqvist, J., and Rudolf, F., "A New Condenser Microphone in Silicon," Sensors and Actuators, vol. A21, nos. 1 - 3, Feb. 1990, pp. 123 - 125.

Bergqvist, J., Rudolf, F., Maisano, J., Parodi, F., and Rossi, M., "A Silicon Condenser Microphone with a Highly Perforated Backplate," Proceedings of Transducers '91, the 1991 International Conference on Solid-State Sensors and Actuators, San Francisco, CA, June 24 - 27, 1991, pp. 266 - 269.

Bernstein, J., Houston, K., Niles, L., Finberg, S., Chen, H., Cross, L. E., Li, K., and Udayakumar, K., "Micromachined Ferroelectric Transducers for Acoustic Imaging," Proceedings of Transducers '97, the 1997 International Conference on Solid-State Sensors and Actuators, Chicago, IL, June 16 - 19, 1997, vol. 1, pp. 421 - 424.

Bernstein, J., Weinberg, M., McLaughlin, E., Powers, J., and Tito, E., "Advanced Micromachined Condenser Hydrophone," Proceedings of the 1994 Solid-State Sensor and Actuator Workshop, Hilton Head Island, SC, June 13 - 16, 1994, pp. 73 - 77.

Bohringer, K. F., Donald, B. R., MacDonald, N. C., Kovacs, G. T. A., and Suh, J. W., "Computational Methods for Design and Control of MEMS Micromanipulator Arrays," IEEE Computational Science and Engineering, vol. 4, no. 1, Jan. - Mar. 1997, pp. 17 - 29.

Börner, M., "Progress in Electromechanical Filters," The Radio and Electronic Engineer, vol. 29, no. 3, Mar. 1965, pp. 173 - 184.

Bourouina, T., Spirkovitch, S., Baillieu, F., and Vauge, C., "A New Silicon Condenser Microphone with a p+ Silicon Membrane," Sensors and Actuators, vol. A31, nos. 1 - 3, Mar. 1992, pp. 149 - 152.

Bowden, N., Terfort, A., Carbeck, J., and Whitesides, G. M., "Self-Assembly of Mesoscale Objects into Ordered Two-Dimensional Arrays," Science, vol. 276, no. 5310, Apr. 11, 1997, pp. 233 - 235.

Brosnihan, T. J., Bustillo, J. M., Pisano, A. P., and Howe, R. T., "Embedded Interconnect and Electrical Isolation for High-Aspect-Ratio, SOI Inertial Instruments," Proceedings of Transducers '97, the 1997 International Conference on Solid-State Sensors and Actuators, Chicago, IL, June 16 - 19, 1997, vol. 1, pp. 637 - 640.

Brown, S. B., Povrik, G., and Connally, J., "Measurement of Slow Crack Growth in Silicon and Nickel Micromechanical Devices," Proceedings of the 1993 Micro Electro Mechanical Systems Workshop, Fort Lauderdale, FL, Feb. 7 - 10, 1993, pp. 99 - 104.

Brown, S. B., Van Arsdell, W., and Muhlstein, L., "Materials Reliability in MEMS Devices," Proceedings of Transducers '97, the 1997 International Conference on Solid-State Sensors and Actuators, Chicago, IL, June 16 - 19, 1997, vol. 1, pp. 591 - 593.

Burns, D. W., "Micromachanics of Integrated Sensors and the Planar Processed Pressure Transducer," Doctoral Dissertation, University of Wisconsin, Madison, 1988.

Burns, D. W., Horning, R. D., Herb, W. R., Zook, J. D., and Guckel, H., "Resonant Microbeam Accelerometers," Proceedings of Transducers '95, the 8th International Conference on Solid-State Sensors and Actuators, Stockholm, Sweden, June 25 - 29, 1995, vol. 2, pp. 659 - 662.

Burns, D. W., Zook, J. D., Horning, R. D., Herb, W. R., and Guckel, H., "Sealed-Cavity Resonant Microbeam Pressure Sensor," Sensors and Actuators, vol. A48, no. 3, May 1995, pp. 179 - 186.

Cady, W. G., "Piezoelectricity," (original edition 1946, McGraw-Hill, New York, NY), revised edition, Dover Publications, New York, NY, 1964.

Carr, W. N., Chamarti, S., and Gu, X., "Integrated Pressure Sensor with Remote Power Source and Remote Readout," Proceedings of Transducers '95, the 8th International Conference on Solid-State Sensors and Actuators, Stockholm, Sweden, June 25 - 29, 1995, vol. 1, pp. 624 - 627.

Chao, H.-L., and Wise, K. D., "An Ultraminiature Solid-State Pressure Sensor for a Cardiovascular Catheter," IEEE Transactions on Electron Devices, vol. 35, no. 2, Dec. 1988, pp. 2355 - 2362.

Chao, J. H. C., and Neudeck, G. W., "Optical Fibre Microbending Sensors by Micromachining Techniques," Electronics Letters, vol. 26, no. 8, Apr. 14, 1990, pp. 513 - 515.

Chau, K. H.-L., Lewis, S. R., Zhao, Y., Howe, R. T., Bart, S. F., and Marcheselli, R. G., "An Integrated Force-Balanced Capacitive Accelerometer for Low-G Applications," Proceedings of Transducers '95, the 8th International Conference on Solid-State Sensors and Actuators, Stockholm, Sweden, June 25 - 29, 1995, vol. 1, pp. 593 - 596.

Chavan, A. V., and Wise, K. D., "A Batch-Processed Vacuum-Sealed Capacitive Pressure Sensor," Proceedings of Transducers '97, the 1997 International Conference on Solid-State Sensors and Actuators, Chicago, IL, June 16 - 19, 1997, vol. 2, pp. 1449 - 1452.

Chen, P.-L., Muller, R. S., and Andrews, A. P., "Integrated Silicon Pi-FET Accelerometers with Proof Mass," Sensors and Actuators, vol. 5, no. 2, Feb. 1984, pp. 119 - 126.

Chen, P.-L., Muller, R. S., Jolly, R. D., Halac, G. L., White, R. M., Andrews, A. P., Lim, T. C., and Motamedi, M. E., "Integrated Silicon Microbeam PI-FET Accelerometer," IEEE Transactions on Electron Devices, vol. ED-29, no. 1, Jan. 1982, pp. 27 - 33.

Christel, L., Petersen, K., Barth, P., Pourahmadi, F., Mallon, J., Jr., and Bryzek, J., "Single-Crystal Silicon Pressure Sensors with 500 x Overpressure Protection," Sensors and Actuators, vols. A21 - A23, Feb. - Apr. 1990, pp. 84 - 88.

Chu, Z., Sarro, P. M., and Middelhoek, S., "Silicon Three-Axial Tactile Sensor," Proceedings of Transducers '95, the 8th International Conference on Solid-State Sensors and Actuators, Stockholm, Sweden, June 25 - 29, 1995, vol. 1, pp. 656 - 659.

Ciarlo, D. R. "A Latching Accelerometer Fabricated by the Anisotropic Etching of (110) Oriented Silicon Wafers," Journal of Micromechanics and Microengineering, vol.2, no.1, Mar. 1992, pp. 10 - 13.

Clark, J. R., Bannon III, F. D., Wong, A.-C., and Nguyen, C. T.-C., "Parallel Resonator HF Micromechanical Bandpass Filters," Proceedings of Transducers '97, the 1997 International Conference on Solid-State Sensors and Actuators, Chicago, IL, June 16 - 19, 1997, vol. 2, pp. 1161 - 1164.

Cobbold, R. S. C., "Transducers for Biomedical Measurements," John Wiley and Sons, New York, NY, 1974.

Cohn, M. B., Kim, C.-J., and Pisano, A. P., "Self-Assembling Electrical Networks: An Application of Micromachining Technology," Proceedings of Transducers '91, the 1991 International Conference on Solid-State Sensors and Actuators, San Francisco, CA, June 24 - 27, 1991, pp. 490 - 493.

Cohn, M. B., Liang, Y., Howe, R. T., and Pisano, A. P., "Wafer-to-Wafer Transfer of Microstructures for Vacuum Packaging," Proceedings of the 1996 Solid-State Sensor and Actuator Workshop, Hilton Head Island, SC, June 3 - 6, 1996, pp. 32 - 35.

Cole, J. C., "A New Sense Element Technology for Accelerometer Subsystems," Proceedings of Transducers '91, the 1991 International Conference on Solid-State Sensors and Actuators, San Francisco, CA, June 24 - 27, 1991, pp. 93 - 96.

Collins, C. C., "Biomedical Transensors: A Review," Journal of Biomedical Systems, vol. 1, no. 1, 1970, pp. 23 - 39.

Collins, C. C., "Miniature Passive Pressure Transensor for Implanting in the Eye," IEEE Transactions on Biomedical Engineering, vol. BME-14, 1967, pp. 74 - 83.

Connally, J. A., and Brown, S. B., "Micromechanical Fatigue Testing," Proceedings of Transducers '91, the 1991 International Conference on Solid-State Sensors and Actuators, San Francisco, CA, June 24 - 27, 1991, pp. 953 - 956.

Connally, J. A., and Brown, S. B., "Slow Crack Growth in Single-Crystal Silicon," Science, vol. 256, no. 5063, June 12, 1992, pp. 1537 - 1539.

Daneman, M. J., Tien, N. C., Solgaard, O., Lau, K. Y., and Muller, R. S., "Linear Vibromotor-Actuated Micromachined Microreflector for Integrated Optical Systems," Proceedings of the 1996 Solid-State Sensor and Actuator Workshop, Hilton Head Island, SC, June 3 - 6, 1996, pp. 109 - 112.

Daneman, M. J., Tien, N. C., Solgaard, O., Pisano, A. P., Lau, K. Y., and Muller, R. S., "Linear Microvibromotor for Positioning Optical Components," Proceedings of the IEEE Micro Electro Mechanical Systems Conference, Amsterdam, Netherlands, Jan. 29 - Feb. 2, 1995, pp. 55 - 60.

Darnell, J., Lodish, H., and Baltimore, D., "Molecular Cell Biology," Second Edition, Scientific American Books, W. H. Freeman and Co., New York, NY, 1991.

de Bruin, D. W., Allen, H. V., Terry, S. C., Jerman, J. H., "Electrically Trimmable Silicon Micromachined Pressure Switch," Sensors and Actuators, vol. A21, nos. 1 - 3, Feb. 1990, pp. 54 - 57.

de Rossi, D., Caiti, A., Bianchi, R., and Canepa, G., "Fine-Form Tactile Discrimination Through Inversion of Data from a Skin-like Sensor," Proceedings of the 1991 IEEE International Conference on Robotics and Automation, Sacramento, CA, Apr. 9 - 11, 1991, pp. 398-403.

De Souza, R. J., and Wise, K. D., "A Very High Density Bulk Micromachined Capacitive Tactile Sensor," Proceedings of Transducers '97, the 1997 International Conference on Solid-State Sensors and Actuators, Chicago, IL, June 16 - 19, 1997, vol. 2, pp. 1473 - 1476.

DeVoe, D. L., and Pisano, A. P., "A Fully Surface-Micromachined Piezoelectric Accelerometer," Proceedings of Transducers '97, the 1997 International Conference on Solid-State Sensors and Actuators, Chicago, IL, June 16 - 19, 1997, vol. 2, pp. 1205 - 1208.

di Giovanni, M., "Flat and Corrugated Diaphragm Design Handbook," Marcel Dekker, Inc., New York, NY, 1982.

Drake, J., Jerman, H., Lutze, B., and Stuber, M., "An Electrostatically Actuated Micro-Relay," Proceedings of Transducers '95, the 8th International Conference on Solid-State Sensors and Actuators, Stockholm, Sweden, June 25 - 29, 1995, vol. 2, pp. 380 - 383.

Dudaicevs, H., Kandler, M., Manoli, Y., Mokwa, W., and Spiegel, E., "Surface Micromachined Pressure Sensors with Integrated CMOS Read-Out Electronics." Proceedings of Transducers '93, the 7th International Conference on Solid State Sensors and Actuators, Yokohama, Japan, June 7-10, 1993, pp. 992 - 994 (also published in Sensors and Actuators, vol. A43, nos. 1 - 3, May 1994, pp. 157 - 163).

Dudaicevs, H., Manoli, Y., Mokwa, W., Schmidt, M., and Speigel, E., "A Fully Integrated Surface Micromachined Pressure Sensor with Low Temperature Dependence," Proceedings of Transducers '95/Eurosensors IX, Stockholm, Sweden, June 25 - 29, 1995, vol. 1, pp. 616 - 619.

Epstein, A. H., Senturia, S. D., Anathasuresh, G., Ayon, A., Breuer, K., Chen, K.-S., Ehrich, F. E., Gauba, G., Ghodssi, R., Groshenry, C., Jacobson, S., Lang, J. H., Lin, C.-C., Mehra, A., Miranda, J. M., Nagle, S., Orr, D. J., Piekos, E., Schmidt, M. A., Shirley, G., Spearing, M. S., Tan, C. S., Tzeng, Y.-S., and Waitz, I. A., "Power MEMS and Microengines," Proceedings of Transducers '97, the 1997 International Conference on Solid-State Sensors and Actuators, Chicago, IL, June 16 - 19, 1997, vol. 2, pp. 753 - 756.

Esashi, M., Shoji, S., Yamamoto, A., and Nakamura, K., "Fabrication of Semiconductor Tactile Imager," Electronics and Communications in Japan, Part 2 (Electronics), vol. J73C-II, no. 1, Jan. 1990, pp. 97 - 104.

Fan, L. S., Muller, R. S., Yun, W., Howe, R. T., and Huang, J., "Spiral Microstructures for the Measurement of Average Strain Gradients in Thin Films," Proceedings of the 1990 IEEE Conference on Micro Electro Mechanical Systems, Napa Valley, CA, Feb. 11 - 14, 1990, pp. 177 - 181.

Fan, L., Wu, M. C., Choquette, K. D., and Crawford, M. H., "Self-Assembled Microactuated XYZ Stages for Optical Scanning and Alignment," Proceedings of Transducers '97, the 1997 International Conference on Solid-State Sensors and Actuators, Chicago, IL, June 16 - 19, 1997, vol. 1, pp. 319 - 322.

Fan, L.-S., and Woodman, S., "Batch Fabrication of Mechanical Platforms for High Density Data Storage," Proceedings of Transducers '95, the 8th International Conference on Solid-State Sensors and Actuators, Stockholm, Sweden, June 25 - 29, 1995, vol. 1, pp. 434 - 437.

Fan, L.-S., Tai, Y.-C., and Muller, R. S., "IC-Processed Electrostatic Micromotors," Proceedings of the IEEE International Electron Devices Meeting, San Francisco, CA, Dec. 12 - 14, 1988, pp. 666 - 669.

Fan, L.-S., Tai, Y.-C., and Muller, R. S., "IC-Processed Electrostatic Micromotors." Sensors and Actuators, vol. 20, no. 1 - 2, Nov. 15, 1989, pp. 41 - 47.

Fan, L.-S., Tai, Y.-C., and Muller, R. S., "Integrated Movable Micromechanical Structures for Sensors and Actuators," IEEE Transactions on Electron Devices, vol. 35, no. 6, June 1988, pp. 724 - 730.

Fan, L.-S., Woodman, S. J., and Crawforth, L., "Integrated Multilayer High Aspect Ratio Milliactuators," Sensors and Actuators, vol. A48, no. 3, May 1995, pp. 221 - 227.

Fenchel, T., and Finlay, B., "Geotaxis in the Ciliated Protozoan Loxodes," Journal of Experimental Biology, vol. 110, 1984, pp. 17 - 33.

Field, L. A., White, R. M., and Pisano, A. P., "Fluid-Powered Rotary Gears and Micro-Flow Channels," Proceedings of Transducers '91, the 1991 International Conference on Solid-State Sensors and Actuators, San Francisco, CA, June 24 - 27, 1991, pp. 1033 - 1036.

Flinn, P. A., "Principles and Applications of Wafer Curvature Techniques for Stress Measurements in Thin Films," Proceedings of the Materials Research Society Symposium: Thin Films: Stresses and Mechanical Properties, Boston, MA, Nov. 28 - 30, 1988 (published in 1989 by the Materials Research Society), vol. 130, pp. 41 - 51.

Foldvari, T. L., and Lion, K. S., "Capacitive Transducers," Instruments and Control Systems, Nov. 1964, pp. 77 - 85.

Fraden, J., "Handbook of Modern Sensors: Physics, Designs, and Applications," AIP Press, Woodbury, NY, 1997.

French, P. J., and Evans, A. G. R., "Polysilicon Strain Sensors Using Shear Piezoresistance," Sensors and Actuators, vol. 15, 1988, pp. 257 - 272.

Friedrich, A. P., Besse, P. A., Ashruf, C. M. A., and Popovic, R. S., "New Piezo-Tunneling Strain Sensor with Very Low Temperature Sensitivity," Proceedings of Transducers '97, the 1997 International Conference on Solid-State Sensors and Actuators, Chicago, IL, June 16 - 19, 1997, vol. 1, pp. 133 - 136.

Frobenius, W. D., Zeitman, S. A., White, H. M., O'Sullivan, D. D., and Hamel, R. G., "Microminiature Ganged Threshold Accelerometers Compatible with Integrated Circuit Technology," IEEE Transactions on Electron Devices, vol. ED-19, no. 1, 1972, pp. 37 - 40.

Fujita, H., and Gabriel, K. J., "New Opportunities for Micro Actuators," Proceedings of Transducers '91, the 1991 International Conference on Solid-State Sensors and Actuators, San Francisco, CA, June 24 - 27, 1991, pp. 14 - 20.

Fujita, H., Ataka, M., and Konishi, S., "Group Work of Distributed Microactuators," Robotica, vol. 14, no. 5, Sept. - Oct. 1996, pp. 487 - 492.

Fukuta, Y., Collard, D., Akiyama, T., Yang, E. H., and Fujita, H., "Microactuated Self-Assembling of 3D Polysilicon Structures with Reshaping Technology," Proceedings of the Tenth Annual Workshop of Micro Electro Mechanical Systems, Nagoya, Japan, Jan. 26 - 30, 1997, pp. 477 - 481.

Gabrielson, T. B., "Fundamental Noise Limits for Miniature Acoustic and Vibration Sensors," Journal of Vibration and Acoustics, Transactions of the ASME, vol. 117, Oct. 1995, pp. 405 - 410.

Gabrielson, T. B., "Mechanical-Thermal Noise in Micromachined Acoustic and Vibration Sensors," IEEE Transactions on Electron Devices, vol. 40, no. 5, May 1993, pp. 903 - 909.

Garcia, E. J., and Sniegowski, J. J., "Surface Micromachined Microengine," Sensors and Actuators, vol. A48, no. 3, May 1995, pp. 203 - 214.

Gardner, J. W., "Microsensors: Principles and Applications," John Wiley and Sons, New York, NY, 1994.

Gere, J. M., and Timoshenko, S., "Mechanics of Materials," Third Edition, PWS Publishers, Boston, MA, 1990.

Giachino, J. M., and Miree, T. J., "The Challenge of Automotive Sensors," Proceedings of the SPIE Conference on Microlithography and Metrology in Micromachining, Austin, TX, Oct. 23 - 24, 1995, SPIE vol. 2640, pp. 89 - 98.

Giancoli, D. C., "Physics for Scientists and Engineers," Second Edition, Prentice-Hall, Inc., Englewood Cliffs, NJ, 1989.

Goldsmith, C., Lin, T.-H., Powers, B., Wu, W.-R., and Norvell, B., "Micromechanical Membrane Switches for Microwave Applications," Proceedings of the 1995 IEEE MTT-S International Microwave Symposium, Orlando, FL, May 16 - 20, 1995, vol. 1, pp. 91 - 94.

Goldsmith, C., Randall, J., Eshelman, S., Lin, T.-H., Denniston, D., Chen, S., and Norvell, B., "Characteristics of Micromachined Switches at Microwave Frequencies," Proceedings of the 1996 IEEE MTT-S International Microwave Symposium, San Francisco, CA, June 17 - 21, 1996, vol. 2, pp. 1141 - 1144.

Goodenough, F., "Airbags Boom When IC Accelerometer Sees 50 g," Electronic Design, vol. 39, no. 15, Aug. 8, 1991.

Grade, J., Barzilai, A., Reynolds, J. K., Liu, C. H., Partridge, A., Kenny, T. W., VanZandt, T. R., Miller, L. M., and Podosek, J. A., "Progress in Tunnel Sensors," Proceedings of the 1996 Solid-State Sensor and Actuator Workshop, Hilton Head Island, SC, June 3 - 6, 1996, pp. 72 - 75.

Grade, J., Barzilai, A., Reynolds, J. K., Liu, C.-H., Partridge, A., Miller, L. M., Podosek, J. A., and Kenny, T., "Low Frequency Drift in Tunnel Sensors," Proceedings of Transducers '97, the 1997 International Conference on Solid-State Sensors and Actuators, Chicago, IL, June 16 - 19, 1997, vol. 2, pp. 871 - 874.

Green, P. W., Syms, R. R. A., and Yeatman, E. M., "Demonstration of Three-Dimensional Microstructure Self-Assembly," Journal of Microelectromechanical Systems, vol. 4, no. 4, Dec. 1995, pp. 170 - 176.

Greiff, P., Boxenhorn, B., King, T., and Niles, L., "Silicon Monolithic Micromechanical Gyroscope," Proceedings of Transducers '91, the 1991 International Conference on Solid-State Sensors and Actuators, San Francisco, CA, June 24 - 27, 1991, p. 966 - 968.

Grétillat, M.-A., Thiébaud, P., Linder, C., and de Rooij, N. F., "Integrated Circuit Compatible Electrostatic Polysilicon Microrelays," Journal of Microelectronics and Microengineering, vol. 5, no. 2, June 1995, pp. 156 - 160.

Guckel, H., "Surface Micromachined Pressure Transducers," Sensors and Actuators, vol. A28, no. 2, July 1991, pp. 133 - 146.

Guckel, H., and Burns, D. W., "Polysilicon Thin Film Process," U.S. Patent No. 4,897,360, issued Jan. 30, 1990.

Guckel, H., Burns, D. W., Tilmans, H. A. C., DeRoo, D. W., and Rutigliano, C. R., "Mechanical Properties of Fine Grained Polysilicon: The Repeatability Issue," Technical Digest of the IEEE Solid-State Sensor and Actuator Workshop, Hilton Head Island, SC, June 6 - 9, 1988, pp. 96 - 99.

Guckel, H., Burns, D. W., Visser, C. C. G., Tilmans, H. A. C., and DeRoo, D., "Fine Grained Polysilicon Films with Built-In Tensile Strain," IEEE Transactions on Electron Devices, vol. ED-35, no. 6, June 1988, pp. 800 - 801.

Guckel, H., Randazzo, T., and Burns, D. W., "A Simple Technique for the Determination of Mechanical Strain in Thin Films with Applications to Polysilicon," Journal of Applied Physics, vol. 57, no. 5, Mar. 1, 1985, pp. 1671 - 1675.

Gui, C., Legtenberg, R., Tilmans, H. A. C., Fluitman, J. H. J., and Elwenspoek, M., "Nonlinearity and Hysteresis of Resonant Strain Gauges," Proceedings of the IEEE Micro Electro Mechanical Systems Conference, Amsterdam, Netherlands, Jan. 29 - Feb. 2, 1995, pp. 157 - 162.

Gutierrez, R. C., Tang, T. K., Stell, C. B., Vorperian, V., and Shcheglov, K., "Bulk Micromachined Vacuum Sensor," Proceedings of Transducers '97, the 1997 International Conference on Solid-State Sensors and Actuators, Chicago, IL, June 16 - 19, 1997, vol. 2, pp. 1497 - 1500.

Hackett, R. H., Larson, L. E., and Melendes, M. A., "The Integration of Micro-Machine Fabrication with Electronic Device Fabrication on III-V Semiconductor Materials," Proceedings of Transducers '91, the 1991 International Conference on Solid-State Sensors and Actuators, San Francisco, CA, June 24 - 27, 1991, pp. 51 - 54.

Han, H., Weiss, L. E., and Reed, M. L., "Micromechanical Velcro," Journal of Microelectromechanical Systems, vol. 1, no. 1, Mar. 1992, pp. 37 - 43.

Haronian, D., and MacDonald, N. C., "A Microelectromechanics Based Artificial Cochlea (MEM-BAC)," Proceedings of Transducers '95, the 8th International Conference on Solid-State Sensors and Actuators, Stockholm, Sweden, June 25 - 29, 1995, vol. 2, pp. 708 - 711.

Hathaway, J. C., and Babcock, D. F., "Survey of Mechanical Filters and Their Applications," Proceedings of the IRE, Jan. 1957, pp. 5 - 16.

Hauptmann, P., "Sensors: Principles and Applications," Prentice-Hall International (UK) Ltd., Hertfordshire, UK, 1991.

Hiltmann, K. M., Schmidt, B., Sandmaier, H., and Lang, W., "Development of Micromachined Switches with Increased Reliability," Proceedings of Transducers '97, the 1997 International Conference on Solid-State Sensors and Actuators, Chicago, IL, June 16 - 19, 1997, vol. 2, pp. 1157 - 1160.

Hoen, S., Merhcant, P., Koke, G., and Williams, J., "Electrostatic Surface Drives: Theoretical Considerations and Fabrication," Proceedings of Transducers '97, the 1997 International Conference on Solid-State Sensors and Actuators, Chicago, IL, June 16 - 19, 1997, vol. 1, pp. 41 - 44.

Hohm, D., and Gerhard-Multhaupt, R., "Silicon-Dioxide Electret Transducer," Journal of the Acoustical Society of America, vol. 75, no. 4, Apr. 1984, pp. 1297 - 1298.

Hohm, D., "Kapazitive Silizium-Sensoren fur Horschallanwendungen," Fortschritt-Berichte VDI, VDI, Dusseldorf, 1986.

Hollerbach, J. M., Hunter, I. W., and Ballantyne, J. A., "A Comparative Analysis of Actuator Technologies for Robotics," in The Robotics Review 2, Khatib, O., Craig, J. J., and Lozano-Perez, T. [eds.], MIT Press, Cambridge, MA, 1991, pp. 301 - 345.

Honer, K. A., Maluf, N. I., Martinez, E., and Kovacs, G. T. A., "A High-Resolution Laser-Based Deflection Measurement System for Characterizing Aluminum Electrostatic Actuators," Proceedings of Transducers '95/Eurosensors IX, Stockholm, Sweden, June 25 - 29, 1995, vol. 1, pp. 308 - 311.

Honer, K. A., Maluf, N. I., Martinez, E., and Kovacs, G. T. A., "Characterizing Deflectable Microstructures via a High-Resolution Laser-Based Measurement System," Sensors and Actuators, vol. A52, nos. 1 - 3, Mar. - Apr. 1996, pp. 12 - 17.

Hopkins, R. B., "Design Analysis of Shafts and Beams," Second Edition, Robert E. Krieger Publishing Company, Malabar, FL, 1987.

Hornbeck, L. J., "Digital Light Processing and MEMS: Timely Convergence for a Bright Future," Proceedings of the SPIE Workshop on Micromachining and Microfabrication '95, Austin, TX, Oct. 23 - 24, 1995, pp. 3 - 21.

Horwath, P., Erlebach, A., Köhler, R., and Kück, H., "Miniature Condenser Microphone with a Thin Silicon Membrane Fabricated on SIMOX Substrate," Proceedings of Transducers '95, the 8th International Conference on Solid-State Sensors and Actuators, Stockholm, Sweden, June 25 - 29, 1995, vol. 2, pp. 696 - 699.

Hosokawa, K., Shimoyama, I., and Miura, H., "Two-Dimensional Micro-Self-Assembly Using the Surface Tension of Water," Proceedings of the Ninth Annual IEEE Workshop on Micro Electro Mechanical Systems, San Diego, CA, Feb. 11 - 15, 1996, pp. 67 - 72.

Howe, R. T., "Applications of Silicon Micromachining to Resonator Fabrication," Proceedings of the 1994 IEEE International Frequency Control Symposium, Boston, MA, May 31 - June 3, 1994, pp. 2 - 7.

Howe, R. T., "Microstructural Design in BIMOS-2C: Polysilicon Properties and Implications for Suspension Design," Workshop on BIMOS-2C Design, Berkeley Sensors and Actuator Center, University of California, Berkeley, CA, Mar. 7, 1994.

Hsieh, W. H., Hsu, T.-Y., and Tai, Y.-C., "A Micromachined Thin-Film Teflon Electret Microphone," Proceedings of Transducers '97, the 1997 International Conference on Solid-State Sensors and Actuators, Chicago, IL, June 16 - 19, 1997, vol. 1, pp. 425 - 428.

Huff, M.A., Nikolich, A. D., and Schmidt, M. A., "A Threshold Pressure Switch Utilizing Plastic Deformation of Silicon," Proceedings of Transducers '91, the 1991 International Conference on Solid-State Sensors and Actuators, San Francisco, CA, June 24 - 27, 1991, pp. 177 - 180.

Hunter, I. W., and Lafontaine, S., "A Comparison of Muscle with Artificial Actuators," Proceedings of the 1992 Solid-State Sensor and Actuator Workshop, Hilton Head Island, SC, June 22 - 25, 1992, pp. 178 - 185.

Ikeda, K., Kuwayama, H., Kobayashi, T., Watanabe, T., Nishikawa, T., Yoshida, T., and Harada, K., "Silicon Pressure Sensor Integrates Resonant Strain Gauge on Diaphragm," Sensors and Actuators, vols. A21 - A23, Feb. - Apr. 1990, pp. 146 -150.

Ikeda, K., Kuwayama, H., Kobayashi, T., Watanabe, T., Nishikawa, T., Yoshida, T., and Harada, K., "Three-Dimensional Micromachining of Silicon Pressure Sensor Integrating Resonant Strain Gauge on Diaphragm," Sensors and Actuators, vols. A21 - A23, Feb. - Apr. 1990, pp. 1007 - 1010.

Jacobsen, S. C., Price, R. H., Wood, J. E., Rytting, T. H., and Rafaelof, M., "A Design Overview of an Eccentric-Motion Electrostatic Microactuator (the Wobble Motor)," Sensors and Actuators, vol. 20, nos. 1 - 2, Nov. 1989, pp. 1 - 15.

Jacobsen, S. C., Price, R. H., Wood, J. E., Rytting, T. H., and Rafaelof, M., "The Wobble Motor: An Electrostatic, Planetary-Armature, Microactuator," Proceedings of the IEEE Workshop on Micro Electro Mechanical Systems, Salt Lake City, UT, Feb. 20 - 22, 1989, pp. 17 - 24.

Jardine, A. P., Madsen, J. S., and Mercado, P. G., "Characterization of the Deposition and Materials Parameters of Thin-Film TiNi for Microactuators and Smart Materials," Materials Characterization, vol. 32, no. 3, Apr. 1994, pp. 169 - 178.

Jerman, J. H., "The Fabrication and Use of Micromachined Corrugated Silicon Diaphragms," Sensors and Actuators, vol. A23, nos. 1 - 3, Apr. 1990, pp. 988 - 992.

Ji, J., Cho, S. T., Zhang, Y., Najafi, K., and Wise, K. D., "An Ultraminiature CMOS Pressure Sensor for a Multiplexed Cardiovascular Catheter," Proceedings of Transducers '91, the 1991 International Conference on Solid-State Sensors and Actuators, San Francisco, CA, June 24 - 27, 1991, pp. 1018 - 1020.

Jiang, J. C., White, R. C., and Allen, P. K., "Microcavity Vacuum Tube Pressure Sensor for Robot Tactile Sensing," Proceedings of Transducers '91, the 1991 International Conference on Solid-State Sensors and Actuators, San Francisco, CA, June 24 - 27, 1991, pp. 238 - 240.

Jin, X. C., Ladabaum, I., and Khuri-Yakub, B. T., "The Microfabrication of Capacitive Ultrasonic Transducers," Proceedings of Transducers '97, the 1997 International Conference on Solid-State Sensors and Actuators, Chicago, IL, June 16 - 19, 1997, vol. 1, pp. 437 - 440.

Johnson, A. D., "Vacuum-Deposited TiNi Shape Memory Film: Characterization and Applications in Microdevices," Journal of Micromechanics and Microengineering, vol. 1, no. 1, Mar. 1991, pp. 34 - 41.

Judy, J. W., and Muller, R. S., "Batch-Fabricated, Addressable, Magnetically Actuated Microstructures," Proceedings of the 1996 Solid-State Sensor and Actuator Workshop, Hilton Head Island, SC, June 3 - 6, 1996, pp. 187 - 190.

Judy, M. W., "Micromechanisms Using Sidewall Beams," Doctoral Dissertation in Electrical Engineering and Computer Sciences, University of California, Berkeley, CA, 1994.

Judy, M. W., and Howe, R. T., "Highly Compliant Lateral Suspensions Using Sidewall Beams," Proceedings of Transducers '93, the 7th International Conference on Solid-State Sensors and Actuators, Yokohama, Japan, June 7 - 10, 1993, Institute of Electrical Engineers, Japan, pp. 54 - 57.

Judy, M. W., and Howe, R. T., "Polysilicon Hollow Beam Lateral Resonators," Proceedings of the IEEE Microelectromechanical Systems Conference, Fort Lauderdale, FL, Feb. 7 - 10, 1993, pp. 265 - 271.

Juneau, T., Pisano, A. P., and Smith, J. H., "Dual Axis Operation of a Micromachined Rate Gyroscope," Proceedings of Transducers '97, the 1997 International Conference on Solid-State Sensors and Actuators, Chicago, IL, June 16 - 19, 1997, vol. 2, pp. 883 - 886.

Kanda, Y., "A Graphical Representation of the Piezoresistance Coefficients of Silicon," IEEE Transactions on Electron Devices, vol. ED-29, no. 1, Jan. 1982, pp. 64 - 70.

Kandel, E. R., Schwartz, J. H., and Jessell, T. M. [eds.], "Principles of Neural Science," Third Edition, Elsevier, New York, NY, 1991.

Kandler, M., Manoli, Y., Mokwa, W., Spiegel, E., and Vogt, H., "A Miniature Single-Chip Pressure and Temperature Sensor," Journal of Micromechanics and Microengineering, vol. 2, no. 3, Sept. 1992, pp. 199 - 201.

Kane, B. J., Cutkosky, M. R., and Kovacs, G. T. A., "CMOS-Compatible Traction Stress Sensor for Use in High-Resolution Tactile Imaging," Sensors and Actuators, vol. A54, nos. 1 - 3, June 1996, pp. 511 - 516.

Kane, B. J., Storment, C. W., Crowder, S. W., Tanelian, D. L., and Kovacs, G. T. A., "Force-Sensing Microprobe for Precise Stimulation Of Mechanosensitive Tissues," IEEE Transactions on Biomedical Engineering, vol. 42, no. 8, Aug. 1995, pp. 745 - 750.

Kawashima, T., and Aoki, Y., "An Optical Tactile Sensor Using the CT Reconstruction Method," Electronics and Communications in Japan, Part 2 (Electronics), vol. 70, no. 10, Oct. 1987, pp. 35 - 43.

Keller, C. G., and Howe, R. T., "HEXSIL Tweezers for Teleoperated Micro-Assembly," Proceedings of the Tenth Annual Workshop of Micro Electro Mechanical Systems, Nagoya, Japan, Jan. 26 - 30, 1997, pp. 72 - 77.

Kelly, J. P., "Auditory System," Chapter 31 in "Principles of Neural Science," Second Edition, Kandel, E. R., and Schwartz, J. H. [eds.], Elsevier, New York, NY, 1985, pp. 396 - 408.

Kelly, J. P., "Hearing," Chapter 32 in "Principles of Neural Science," Third Edition, Kandel, E. R., Schwartz, J. H., and Jessell, T. M. [eds.], Elsevier, New York, NY, 1991, pp. 481 - 499.

Kemper, D., and Rosine, L., "Quartz Crystals for Frequency Control," Electro-Technology, June 1969, pp. 43 - 50.

Kim, C.-J., Pisano, A. P., and Muller, R. S., "Silicon-Processed Overhanging Microgripper." Journal Of Microelectromechanical Systems, vol. 1, no. 1, Mar. 1992, pp. 31 - 36.

Kim, C.-J., Pisano, A. P., Muller, R. S., and Lim, M. G., "Polysilicon Microgripper." Technical Digest of the IEEE Solid-state Sensor and Actuator Workshop, Hilton Head Island, SC, June 4 - 7, 1990, pp. 48 - 51.

King, A. A., and White, R. M., "Tactile Sensing Array Based on Forming and Detecting an Optical Image," Sensors and Actuators, vol. 8, no. 1, Sept. 1985, pp. 49 - 63.

Klaassen, E. H., Petersen, K., Noworolski, J. M., Logan, J., Maluf, N. I., Brown, J., Storment, C., McCulley, W., and Kovacs, G. T. A., "Silicon Fusion Bonding and Deep Reactive Ion Etching: A New Technology for Microstructures," Proceedings of Transducers '95, the 8th International Conference on Solid-State Sensors and Actuators, Stockholm, Sweden, June 25 - 29, 1995, vol. 1, pp. 556 - 559.

Ko, W. H., Bao, M.-H., and Hong, Y.-D., "A High-Sensitivity Integrated-Circuit Capacitive Pressure Transducer," IEEE Transactions on Electron Devices, vol. ED-29, no. 1, Jan. 1982, pp. 48 - 56.

Kolesar, E. S., Jr., Reston, R. R., Ford, D. G., and Fitch, R. C., Jr., "Multiplexed Piezoelectric Polymer Tactile Sensor," Journal of Robotic Systems, vol. 1, no. 9, Feb. 1992, pp. 37 - 63.

Konishi, S., and Fujita, H., "A Conveyance System Using Air Flow Based on the Concept of Distributed Micro Motion Systems," Proceedings of Transducers '93, the 7th International Conference on Solid-State Sensors and Actuators, Yokohama, Japan, June 7 - 10, 1993, Institute of Electrical Engineers, Japan, pp. 28 - 31.

Konishi, S., and Fujita, H., "System Design for Cooperative Control of Arrayed Microactuators," Proceedings of the IEEE Micro Electro Mechanical Systems Conference, Amsterdam, Netherlands, Jan. 29 - Feb. 2, 1995, pp. 322 - 327.

Kress, H.-J., Bantien, F., Marek, J., and Willmann, M., "Silicon Pressure Sensor with Integrated CMOS Signal-Conditioning Circuit and Compensation of Temperature Coefficient," Sensors and Actuators, vol. A25, nos. 1 - 3, Oct. 1990 - Jan. 1991, pp. 21 - 26.

Krulevitch, P. A., "Micromechanical Investigations of Silicon and Ni-Ti-Cu Thin Films," Doctoral Dissertation in Mechanical Engineering, University of California, Berkeley, CA, 1994.

Krulevitch, P. A., Lee, A. P., Ramsey, P. B., Trevino, J. C., Hamilton, J., and Northrup, M. A., "Thin Film Shape Memory Alloy Microactuators," Journal of Microelectromechanical Systems, vol. 5, no. 4, Dec. 1996, pp. 270 - 282.

Kuhnel, W., "Silicon Condenser Microphone with Integrated Field Effect Transistor," Sensors and Actuators, vol. A26, nos. 1 - 3, Mar. 1991, pp. 521 - 525.

Kulcsar, F., U.S. Patent No. 3,006,857, issued Oct. 1, 1961.

Kung, J. T., and Lee, H.-S., "An Integrated Air-Gap-Capacitor Pressure Sensor and Digital Readout with Sub-100 Attofarad Resolution," Journal of Microelectromechanical Systems," vol. 1, no. 3, Sept. 1992, pp. 121 - 129.

Kung, J. T., and Lee, H.-S., "An Integrated Air-Gap-Capacitor Process for Sensor Applications," Proceedings of Transducers '91, the 1991 International Conference on Solid-State Sensors and Actuators, San Francisco, CA, June 24 - 27, 1991, pp. 1010 - 1013.

Kung, J. T., Lee, H.-S., and Howe, R. T., "A Digital Readout Technique for Capacitive Sensor Applications," IEEE Journal of Solid-State Circuits, vol. 23, no. 4, Aug. 1988, pp. 972 - 977.

Larson, L. E., Hackett, R. H., and Lohr, R. F., "Microactuators for GaAs-Based Microwave Integrated Circuits," Proceedings of Transducers '91, the 1991 International Conference on Solid-State Sensors and Actuators, San Francisco, CA, June 24 - 27, 1991, pp. 743 - 746.

Larson, L. E., Hackett, R. H., Melendes, M. A., and Lohr, R. F., "Micromachined Microwave Actuator (MIMAC) Technology - A New Tuning Approach for Microwave Integrated Circuits," Proceedings of the 1991 IEEE MTT-S International Microwave Symposium, Boston, MA, June 10 - 11, 1991, pp. 27 - 30.

Lau, W. W., Song, Y., and Kim, E.-S., "Lateral-Field Excitation Acoustic Resonator for Monolithic Oscillators and Filters," 1996 IEEE International Frequency Control Symposium, Honolulu, HI, June 5 - 7, 1996, pp. 558 - 562.

Lee, A. P., Nikkel, D. J., and Pisano, A. P., "Polysilicon Linear Microvibromotors," Proceedings of Transducers '93, the 7th International Conference on Solid-State Sensors and Actuators, Yokohama, Japan, June 7 - 10, 1993, Institute of Electrical Engineers, Japan, pp. 46 - 49.

Lee, K. B., and Cho, Y.-H., "Frequency Tuning of a Laterally Driven Microresonator Using an Electrostatic Comb Array of Linearly Varied Length," Proceedings of Transducers '97, the 1997 International Conference on Solid-State Sensors and Actuators, Chicago, IL, June 16 - 19, 1997, vol. 1, pp. 113 - 116.

Lee, S. S., Ried, R. P., and White, R. M., "Piezoelectric Cantilever Microphone and Microspeaker," Journal of Microelectromechanical Systems, vol. 5, no. 4, Dec. 1996, pp. 238 - 242.

Lee, Y. S., and Wise, K. D., "A Batch-Fabricated Silicon Capacitive Pressure Transducer with Low Temperature Sensitivity," IEEE Transactions on Electron Devices, vol. ED-29, no. 1, Jan. 1982, pp. 42 - 48.

Lemkin, M. A., Boser, B. E., Auslander, D., and Smith, J. H., "A 3-Axis Force Balanced Accelerometer Using a Single Proof-Mass," Proceedings of Transducers '97, the 1997 International Conference on Solid-State Sensors and Actuators, Chicago, IL, June 16 - 19, 1997, vol. 2, pp. 1185 - 1188.

Lin, G., Kim, C.-J., Konishi, S., and Fujita, H., "Design, Fabrication, and Testing of a C-Shape Actuator," Proceedings of Transducers '95, the 8th International Conference on Solid-State Sensors and Actuators, Stockholm, Sweden, June 25 - 29, 1995, vol. 2, pp. 416 - 419.

Lin, G., Palmer, R. E., Pister, K. S. J., and Roos, K. P., "Single Heart Cell Force Measured in Standard CMOS," Proceedings of Transducers '97, the 1997 International Conference on Solid-State Sensors and Actuators, Chicago, IL, June 16 - 19, 1997, vol. 1, pp. 199 - 200.

Lin, L., Pisano, A. P., and Lee, A. P., "Microbubble Powered Actuator," Proceedings of Transducers '91, the 1991 International Conference on Solid-State Sensors and Actuators, San Francisco, CA, June 24 - 27, 1991, pp. 1041 - 1044.

Lober, T. A., and Howe, R. T., "Surface-Micromachining Processes for Electrostatic Microactuator Fabrication," Technical Digest of the IEEE Solid-State Sensor and Actuator Workshop, Hilton Head Island, SC, June 8 - 9, 1988, pp. 59 - 62.

Loke, Y., McKinnon, G. H., and Brett, M. J., "Fabrication and Characterization of Silicon Micromachined Threshold Accelerometers," Sensors and Actuators, vol. A29, no. 3, Dec. 1991, pp. 235 - 240.

Lutsky, J. J., Naik, R. S., Reif, R., and Sodini, C. G., "A Sealed Cavity TFR Process for RF Bandpass Filters," Technical Digest of the International Electron Devices Meeting, San Francisco, CA, Dec. 8 - 11, 1996, pp. 95 - 98.

MacDonald, G. A., "A Review of Low Cost Accelerometers for Vehicle Dynamics," Sensors and Actuators, vol. A21, nos. 1 - 3, Feb. 1990, pp. 303 - 307.

Majumder, S., McGruer, N. E., Zavracky, P. M., Adams, G. C., Morrison, R. H., and Krim, J., "Measurement and Modeling of Surface Micromachined, Electrostatically Actuated Microswitches," Proceedings of Transducers '97, the 1997 International Conference on Solid-State Sensors and Actuators, Chicago, IL, June 16 - 19, 1997, vol. 2, pp. 1145 - 1148.

Malik, N. R., "Electronic Circuits," Prentice-Hall, Inc., Englewood Cliffs, NJ, 1995, pp. 737 - 739.

Mallon, J. R., Jr., Pourahmadi, F., Petersen, K., Barth, P., Vermeulen, T., and Bryzek, J., "Low-Pressure Sensors Employing Bossed Diaphragms and Precision Etch-Stopping," Sensors and Actuators, vols. A21 - A23, Feb. - Apr. 1990, pp. 89 - 95.

Mason, W. P., "Electromechanical Transducers and Wave Filters," Van Nostrand, New York, NY, 1948.

Mehregany, M., Bart, S. F., Tavrow, L. S., Lang, J. H., Senturia, S. D., and Schlecht, M. F., "A Study of Three Microfabricated Variable-Capacitance Micromotors," Proceedings of Transducers '89, the 5th International Conference on Solid-State Sensors and Actuators, and Eurosensors III, vol. 2, Montreaux, Switzerland, June 25 - 30, 1989, pp. 173 - 179.

Mehregany, M., Bart, S. F., Tavrow, L. S., Lang, J. H., Senturia, S. D., and Schlecht, M. F., "A Study of Three Microfabricated Variable-Capacitance Micromotors," Sensors and Actuators, vol. A21, nos. 1 - 3, Feb. 1990, pp. 173 - 179.

Mehregany, M., Gabriel, K. J., and Trimmer, W. S. N., "Integrated Fabrication of Polysilicon Mechanisms," IEEE Transactions on Electron Devices, vol. 35, no. 6, June 1988, pp. 719 - 723.

Mehregany, M., Howe, R. T., and Senturia, S. D., "Novel Microstructures for the In-Situ Measurement of Mechanical Properties of Thin-Films," Journal of Applied Physics, vol. 62, no. 9, Nov. 1, 1987, pp. 3579 - 3584.

Mehregany, M., Nagarkar, P., Senturia, S. D., and Lang, J. H., "Operation of a Microfabricated Harmonic and Ordinary Side-Drive Motors," Proceedings of the IEEE Workshop on Micro Electro Mechanical Systems, Napa Valley, CA, Feb. 11 - 14, 1990, pp. 1 - 8.

Mehregany, M., Phillips, S. M., Hsu, E. T., and Lang, J. H., "Operation of Harmonic Side-Drive Micromotors Studied Through Gear Ratio Measurements," Proceedings of Transducers '91, the 1991 International Conference on Solid-State Sensors and Actuators, San Francisco, CA, June 24 - 27, 1991, pp. 59 - 62.

Mehregany, M., Senturia, S. D., Lang, J. H., and Nagarkar, P., "Micromotor Fabrication," IEEE Transactions on Electron Devices, vol. 39, no. 9, Sept. 1992, pp. 2060 - 2069.

Middelhoek, S., and Audet, S. A., "Silicon Sensors," Academic Press, Inc., London, UK, 1989, pp. 109 - 125.

Mohan, J., Maluf, N. I., Petersen, K. E., and Kovacs, G. T. A., "An Integrated Accelerometer as a Demonstration of a New Technology Using Silicon Fusion Bonding and Deep Reactive Ion Etching," Late News Poster Session Supplemental Digest of the Solid-State Sensor and Actuator Workshop, Hilton Head Island, SC, June 3 - 6, 1996, pp. 21 - 22.

Motamedi, M. E., "Acoustic Accelerometers," IEEE Transactions on Ultrasonics, Ferroelectrics and Frequency Control, vol. UFFC-34, no. 2, Mar. 1987, pp. 237 - 242.

Motchenbacher, C. D., and Connelly, J. A., "Low Noise Electronic System Design," John Wiley and Sons, New York, NY, 1993.

Muller, R. S., "Microdynamics," Sensors and Actuators, vols. A21 - A23, Feb. - Apr. 1990, pp. 1 - 8.

Najafi, K., and Hetke, J., "Strength Characterization of Silicon Microprobes in Neurophysiological Tissues," IEEE Transactions on Biomedical Engineering, vol. 37, no. 5, May 1990, pp. 474 - 481.

Nathanson, H. C., and Wickstrom, R. A., "A Resonant Gate Transistor with High-Q Bandpass Properties," Applied Physics Letters, vol. 7, no. 4, Aug. 1965, pp. 84 - 86.

Nathanson, H. C., Newell, W. E., and Wickstrom, R. A., "'Tuning Forks' Sound a Hopeful Note," Electronics, vol. 38, Sept. 20, 1965, pp. 84 - 87.

Nathanson, H. C., Newell, W. E., Wickstrom, R. A., and Davis, J. R., Jr., "The Resonant Gate Transistor," IEEE Transactions on Electron Devices, vol. ED-14, no. 3, Mar. 1967, pp. 117 - 133.

Newell, W. E., "Miniaturization of Tuning Forks," Science, vol. 161, Sept. 27, 1968, pp. 1320 - 1326.

Newell, W. E., "Tuned Integrated Circuits – A State-of-the-Art Survey," Proceedings of the IEEE, Dec. 1964, vol. 52, pp. 1603 - 1608.

Newell, W. E., Wickstrom, R. A., and Page, D. J., "Tunistors - Mechanical Resonators for Microcircuits," Proceedings of the IEEE International Meeting on Electron Devices, Washington, DC, Oct. 1967, p. 24.

Nguyen, C. T.-C., "Micromechanical Resonators for Oscillators and Filters," Proceedings of the 1995 IEEE International Ultrasonics Symposium, Seattle, WA, Nov. 7 - 10, 1995, pp. 489 - 499.

Nguyen, C. T.-C., and Howe R. T., "Design and Performance of Monolithic CMOS Micromechanical Resonator Oscillators," Proceedings of the 1994 IEEE International Frequency Control Symposium, Boston, MA, May 31 - June 3, 1994, pp. 127 - 134.

Nguyen, C. T.-C., and Howe, R. T., "CMOS Micromechanical Resonator Oscillator," Proceedings of the IEEE International Electron Devices Meeting, Washington, DC, Dec. 5 - 8, 1993, pp. 199 - 202.

Nguyen, C. T.-C., and Howe, R. T., "Microresonator Frequency Control and Stabilization Using an Integrated Micro Oven," Digest of Technical Papers of Transducers '93, the 7th International Conference on Solid-State Sensors and Actuators, Yokohama, Japan, June 7 - 10, 1993, pp. 1040 - 1043.

Nicholls, H. R. and Lee, M. H., "A Survey of Robot Tactile Sensing Technology," International Journal of Robotics Research, vol. 8, no. 3, June 1989, pp. 3 - 30.

Noetzel, J., Tønnesen, T., Benecke, W., Binder, J., and Mader, G., "Quasianalog Accelerometer Using Microswitch Array," Proceedings of Transducers '95, the 8th International Conference on Solid-State Sensors and Actuators, Stockholm, Sweden, June 25 - 29, 1995, vol. 2, pp. 671 - 674.

Norton, A. N., "Handbook of Transducers," Prentice-Hall, Inc., Englewood Cliffs, NJ, 1989.

Novak, J. L., "Initial Design and Analysis of a Capacitive Sensor for Shear and Normal Force Measurement," IEEE International Conference on Robotics and Automation, Scottsdale, AZ, May 14 - 19, 1989, pp. 137 - 144.

Park, Y. E., and Wise, K. D., "An MOS Switched-Capacitor Readout Amplifier for Capacitive Pressure Sensors," Record of the IEEE 1983 Custom IC Conference, Rochester, NY, May 23 - 25, 1983, pp. 380 - 384.

Parsons, P., Glendinning, A, and Angelidis, D., "Resonant Sensor for High Accuracy Pressure Measurement Using Silicon Technology," IEEE Aerospace and Electronics Systems Magazine, vol. 7, no. 7, July 1992, pp. 45 - 48.

Pati, Y. C., Friedman, D., Krishnaprasad, P. S., Yao, C. T., Peckerar, M. C., Yang, R., and Marrian, C. R. K., "Neural Networks for Tactile Perception," 1988 IEEE International Conference on Robotics and Automation, Philadelphia, PA, Apr. 24 - 29, 1988, pp. 134-139.

Pedersen, M., Olthius, W., and Bergveld, P., "A Polymer Condenser Microphone on Silicon with On-Chip CMOS Amplifier," Proceedings of Transducers '97, the 1997 International Conference on Solid-State Sensors and Actuators, Chicago, IL, June 16 - 19, 1997, vol. 1, pp. 445 - 446.

Peeters, E., Vergote, S., Puers, B., and Sansen, W., "A Highly Symmetrical Capacitive Micro-Accelerometer with Single Degree-of-Freedom Response," Proceedings of Transducers '91, the 1991 International Conference on Solid-State Sensors and Actuators, San Francisco, CA, June 24 - 27, 1991, pp. 97 - 100.

Petersen, K. E, "Dynamic Micromechanics on Silicon: Techniques and Devices," IEEE Transactions on Electron Devices, vol. ED-25, no. 10, Oct. 1978, pp. 1241 - 1250.

Petersen, K. E., "Micromechanical Membrane Switches on Silicon," IBM Journal of Research and Development, vol. 23, no. 4, July 1978, pp. 376 - 385.

Petersen, K. E., Pourahmadi, F., Brown, J., Parsons, P., Skinner, M., and Tudor, J., "Resonant Beam Pressure Sensor Fabricated with Silicon Fusion Bonding," Proceedings of Transducers '91, the 1991 International Conference on Solid-State Sensors and Actuators, San Francisco, CA, June 24 - 27, 1991, pp. 177 - 180.

Pister, K. S. J., Judy, M. W., Burgett, S. R., and Fearing, R. S., "Microfabricated Hinges." Sensors and Actuators, vol. A33, no. 3, June 1992, pp. 249 - 256.

Pister, K. S., Fearing R., and Howe, R., "A Planar Air Levitated Electrostatic Actuator System," IEEE Micro Electro Mechanical Systems, Napa Valley, CA, Feb. 1990, pp. 67 - 71.

Popov, E. P., "Introduction to the Mechanics of Solids," Prentice-Hall, Inc., Englewood Cliffs, NJ, 1968.

Prak, A., Elwenspoek, M., and Fluitman, J. H. J., "Selective Mode Excitation and Detection of Micromachined Resonators," Journal of Microelectromechanical Systems, vol. 1, no. 4, Dec. 1992, pp. 179 - 186.

Putty, M., and Najafi, K., "A Micromachined Vibrating Ring Gyroscope," Solid State Sensor and Actuator Workshop, Hilton Head Island, SC, June 13 - 16, 1994, pp. 213 - 220.

Randall, J. N., Goldsmith, C., Denniston, D., and Lin, T.-H., "Fabrication of Micromechanical Switches for Routing Radio Frequency Signals," Journal of Vacuum Science and Technology, vol. B14, no. 6, Nov./Dec. 1996, pp. 3692 - 3696.

Rashidian, B., and Allen, M. G., "Electrothermal Microactuators Based on Dielectric Loss Heating," Proceedings of the IEEE Microelectromechanical Systems Conference, Fort Lauderdale, FL, Feb. 1993, pp. 24 - 29.

Riethmüller, W. and Benecke, W., "Thermally Excited Silicon Microactuators," IEEE Transactions on Electron Devices, vol. 35, no. 6, June 1988, pp. 758 - 763.

Riethmüller, W., Benecke, W., Schnakenberg, U., and Wagner, B., "Development of Commercial CMOS Process-Based Technologies for the Fabrication of Smart Accelerometers," Proceedings of Transducers '91, the 1991 International Conference on Solid-State Sensors and Actuators, San Francisco, CA, June 24 - 27, 1991, pp. 416 - 419.

Ristic, Lj., Shemansky, F. A., Kniffin, M. L., and Hughes, H., "Surface Micromachining Technology," Chapter 4 in "Sensor Technology and Devices," Ristic, Lj. [ed.], Artech House, Boston, MA, 1994, pp. 95 - 155.

Roark, R., and Young, W., "Formulas for Stress and Strain," McGraw-Hill, New York, NY, 1989.

Robbins, W. P., Polla, D. L., Tamagawa, T., Glumac, D. E., and Judy, J. W., "Linear Motion Microactuators Using Piezoelectric Thin Films," Proceedings of Transducers '91, the 1991 International Conference on Solid-State Sensors and Actuators, San Francisco, CA, June 24 - 27, 1991, pp. 55 - 58.

Roberts, W. van B., and Burns, L. L., "Mechanical Filters for Radio Frequencies," RCA Review, vol. 10, Sept. 1949, pp. 348 - 365.

Robinson, C., Overman, D., Warner, R., and Blomquist, T., "Problems Encountered in the Development of a Microscale g-Switch Using Three Design Approaches," Digest of Technical Papers, Transducers '87, the 4th International Conference on Solid-State Sensors and Actuators, Tokyo, Japan, June 3 - 5, 1987, pp. 410 - 413.

Rockstad, H. K., Kenny, T. W., Reynolds, J. K., Kaiser, W. J., and Gabrielson, T. B., "A Miniature High-Sensitivity Broad-Band Accelerometer Based on Electron Tunneling Transducers," Proceedings of Transducers '93, the 7th International Conference on Solid State Sensors and Actuators, Yokohama, Japan, June 7-10, 1993, pp. 836 - 839 (also published in Sensors and Actuators, vol. A43, nos. 1 - 3, May 1994, pp. 107 - 114).

Rockstad, H. K., Reynolds, J. K., Tang, T. K., Kenny, T. W., Kaiser, W. J., and Gabrielson, T. B., "A Miniature, High-Sensitivity, Electron Tunneling Accelerometer," Proceedings of Transducers '95, the 8th International Conference on Solid-State Sensors and Actuators, Stockholm, Sweden, June 25 - 29, 1995, vol. 2, pp. 675 - 678.

Roessig, T. A., Howe, R. T., Pisano, A. P., and Smith, J. H., "Surface-Micromachined Resonant Accelerometer," Proceedings of Transducers '97, the 1997 International Conference on Solid-State Sensors and Actuators, Chicago, IL, June 16 - 19, 1997, vol. 2, pp. 859 - 862.

Rogge, B., Schulz, J., Mohr, J., Thommes, A., and Menz, W., "Fully Batch Fabricated Magnetic Microactuators Using a Two Layer LIGA Process," Digest of Technical Papers, Transducers '95, Stockholm, Sweden, June 25 - 29, 1995, vol. 1, pp. 320 - 323.

Rosengren, L., Bäcklund, Y., Sjöström, T, Hök, B., and Svedbergh, B., "A System for Wireless Intraocular Pressure Measurements Using a Silicon Micromachined Sensor," Journal of Micromechanics and Microengineering, vol. 2, no. 3, Sept. 1992, pp. 202 - 204.

Rosengren, L., Rangsten, P., Bäcklund, Y., Hök, B., Svedbergh, B., and Selén, G., "A System for Passive Implantable Pressure Sensors," Sensors and Actuators, vol. A43, nos. 1 - 3, May 1994, pp. 55 - 58.

Roszhart, T. V., Jerman, H., Drake, J., and de Cotiis, C., "An Inertial-Grade, Micromachined Vibrating Beam Accelerometer," Proceedings of Transducers '95, the 8th International Conference on Solid-State Sensors and Actuators, Stockholm, Sweden, June 25 - 29, 1995, vol. 2, pp. 656 - 658.

Royer, M., Holmen, J., Wurm, M., Aadland, O., and Glenn, M., "ZnO on Si Integrated Acoustic Sensor," Sensors and Actuators, vol. 4, no. 3, Nov. 1983, pp. 357 - 362.

Roylance, L. M., and Angell, J. B., "A Batch-Fabricated Silicon Accelerometer," IEEE Transactions on Electron Devices, vol. ED-26, no. 12, Dec. 1979, pp. 1911 - 1917.

Ruby, R., "Micromachined Cellular Filters," Proceedings of the 1996 IEEE MTT-S International Microwave Symposium, San Francisco, CA, June 17 - 21, 1996, pp. 1149 - 1152.

Ruby, R., and Merchant, P., "Micromachined Thin Film Bulk Acoustic Resonators," Proceedings of the 1994 IEEE International Frequency Control Symposium, Boston, MA, May 31 - June 3, 1994, pp. 135 - 138.

Rudolf, F., Jornod, A., and Bencze, P., "Silicon Microaccelerometer," Digest of Technical Papers, Transducers '87, the 4th International Conference on Solid-State Sensors and Actuators, Tokyo, Japan, June 3 - 5, 1987, pp. 395 - 398.

Saffer, S., Simon, J., Kim, C.-J., Park, K. H., and Lee, J. H., "Mercury-Contact Switching with Gap-Closing Microcantilever," Proceedings of the SPIE Micromachined Devices and Components Conference II, Austin, TX, Oct. 14 - 15, 1995, SPIE vol. 2882, pp. 204 - 209.

Sander, C. S., Knutti, J. W., and Meindl, J. D., "A Monolithic Capacitive Pressure Sensor with Pulse-Period Output," IEEE Transactions on Electron Devices, vol. ED-27, no. 5, May 1980, pp. 927 - 930.

Sauerland, F., and Blum, W., "Ceramic IF Filters for Consumer Products," IEEE Spectrum, vol. 5, no. 11, Nov. 1968, pp. 112 - 126.

Scheeper, P. R., Reynolds, J. K., and Kenny, T. W., "Development of a Modal Analysis Accelerometer Based on a Tunneling Displacement Transducer," Proceedings of Transducers '97, the 1997 International Conference on Solid-State Sensors and Actuators, Chicago, IL, June 16 - 19, 1997, vol. 2, pp. 867 - 870.

Scheeper, P. R., van der Donk, A. G. H., Olthuis, W., and Bergveld, P., "A Review of Silicon Microphones," Sensors and Actuators, vol. A44, no. 1, July 1994, pp. 1 - 11.

Scheeper, P. R., van der Donk, A. G. H., Olthuis, W., and Bergveld, P., "Fabrication of Silicon Condenser Microphones Using Single Wafer Technology," IEEE Journal of Microelectromechanical Systems, vol. 1, no. 3, Sept. 1992, pp. 147 - 154.

Schellin, R., and Hess, G., "A Silicon Subminiature Microphone Based on Piezoresistive Polysilicon Strain Gauges," Sensors and Actuators, vol. A32, nos. 1 - 3, Apr. 1992, pp. 555 - 559.

Schiele, I., Huber, J., Evers, C., Hillerich, B., and Kozlowski, F., "Micromechanical Relay with Electrostatic Actuation," Proceedings of Transducers '97, the 1997 International Conference on Solid-State Sensors and Actuators, Chicago, IL, June 16 - 19, 1997, vol. 2, pp. 1165 - 1168.

Schorner, R., Poppinger, M., and Eibl, J., "Silicon Pressure Sensor with Frequency Output," Proceedings of Transducers '89, the 5th International Conference on Solid-State Sensors and Actuators and Eurosensors, Montreux, Switzerland, June 25 - 30, 1989 (also published in Sensors and Actuators, vol. A21, nos. 1 - 3, Feb. 1990, pp. 73 - 78.

Seeger, J. I., and Crary, S. B., "Stabilization of Electrostatically Actuated Mechanical Devices," Proceedings of Transducers '97, the 1997 International Conference on Solid-State Sensors and Actuators, Chicago, IL, June 16 - 19, 1997, vol. 2, pp. 1133 - 1136.

Seidel, H., Fritsch, U., Gottinger, R., Schalk, J., Walter, J., and Ambaum, K, "A Piezoresistive Silicon Accelerometer with Monolithically Integrated CMOS-Circuitry," Proceedings of Transducers '95, the 8th International Conference on Solid-State Sensors and Actuators, Stockholm, Sweden, June 25 - 29, 1995, vol. 1, pp. 597 - 600.

Sharpe, W. N., Jr., Yuan, B., Vaidyananthan, R., and Edwards, R. L., "Measurements of Young's Modulus, Poisson's Ratio, and Tensile Strength of Polysilicon," Proceedings of the Tenth Annual Workshop of Micro Electro Mechanical Systems, Nagoya, Japan, Jan. 26 - 30, 1997, pp. 424 - 429.

Sharpe, W. N., Jr., Yuan, B., Vaidyanathan, K. R., and Edwards, R. L., "New Test Structures and Techniques for Measurement of Mechanical Properties of MEMS Materials," Proceedings of the SPIE Symposium on Micromachining and Microfabrication, Oct. 14 - 15, 1996, vol. 2880, pp. 78 - 91.

Sherman, S. J., Tsang, W. K., Core, T. A., Payne, R. S., Quinn, D. E., Chau, K. H.-L., Farash, J. A., and Baum, S. K., "A Low Cost Monolithic Accelerometer; Product/Technology Update," Proceedings of IEEE International Electron Devices Meeting, San Francisco, CA, Dec. 13 - 16, 1992, pp. 501 - 504.

Shöneberg, U., Schnatz, F. V., Brockherde, W., Kopystynski, P., Melhorn, T., Obermeier, E., and Benzel, B., "CMOS Integrated Capacitive Pressure Transducer with On-Chip Electronics and Digital Calibration Capability," Proceedings of Transducers '91, the 1991 International Conference on Solid-State Sensors and Actuators, San Francisco, CA, June 24 - 27, 1991, pp. 304 - 307.

Smith, M. J. S., Bowman, L., and Meindl, J. D., "Analysis, Design, and Performance of Micropower Circuits for a Capacitive Pressure Sensor IC," IEEE Journal of Solid-State Circuits, vol. SC-21, no. 6, Dec. 1986, pp. 1045 - 1056.

Smith, W. R., Gerard, H. M., and Jones, W. R., "Analysis and Design of Dispersive Interdigital Surface-Wave Transducers," IEEE Transactions on Microwave Theory and Techniques, vol. MTT-20, no. 7, July 1972, pp. 458 - 471.

Smith, W. R., Gerard, H. M., Collins, J. H., Reeder, T. M., and Shaw, H. J., "Design of Surface-Wave Delay Lines with Interdigital Transducers," IEEE Transactions on Microwave Theory and Techniques, vol. MTT-17, no. 11, Nov. 1969, pp. 865 - 873.

Sniegowski, J. J., "A Microactuation Mechanism Based on Liquid-Vapor Surface Tension," Abstracts of Late News Papers from Transducers '93, the 7th International Conference on Solid-State Sensors and Actuators, Yokohama, Japan, June 7 - 10, 1993, Institute of Electrical Engineers, Japan, pp. 12 - 13.

Söderkvist, J., "Design of a Solid-State Gyroscopic Sensor Made of Quartz," Sensors and Actuators, vols. A21 - A23, Feb. - Apr. 1990, pp. 293 - 296.

Spangler, L., and Kemp, C. J., "ISAAC -- Integrated Silicon Automotive Accelerometer," Proceedings of Transducers '95, the 8th International Conference on Solid-State Sensors and Actuators, Stockholm, Sweden, June 25 - 29, 1995, vol. 1, pp. 585 - 588.

Sparks, D. R., Zarabadi, S. R., Johnson, J. D., Jiang, Q., Chia, M., Larsen, O., Higdon, W., and Castillo-Borelley, P., "A CMOS Integrated Surface Micromachined Angular Rate Sensor: Its Automotive Applications," Proceedings of Transducers '97, the 1997 International Conference on Solid-State Sensors and Actuators, Chicago, IL, June 16 - 19, 1997, vol. 2, pp. 851 - 854.

Srivastava, A., Bhalla, A., and Cross, L. E., "PZT Ceramic Compositions Having Reduced Sintering Temperatures and Process for Producing Same," U.S. Patent No. 5,433,917, issued July 18, 1995.

Stemme, E., and Stemme, G., "A Balanced Resonant Pressure Sensor," Sensors and Actuators, vols. A21 - A23, Feb. - Apr. 1990, pp. 336 - 341.

Stemme, E., and Stemme, G., "A Capacitively Excited and Detected Resonant Pressure Sensor with Temperature Compensation," Sensors and Actuators, vol. A32, nos. 1 - 3, Apr. 1992, pp. 639 - 647.

Storment, C. W., Borkholder, D. A., Westerlind, V., Suh, J. W., Maluf, N. I., and Kovacs, G. T. A., "Flexible, Dry-Released Process for Aluminum Electrostatic Actuators," Journal of Microelectromechanical Systems," vol. 3, no. 3, Sept. 1994, pp. 90 - 96.

Sugiyama, S., Kawahata, K., Yoneda, M., and Igarashi, I., "Tactile Image Detection Using a 1k-Element Silicon Pressure Sensor Array," Sensors and Actuators, vols. A21 - A23, Feb. - Apr. 1990, pp. 397 - 400.

Suh, J. W., Glander, S. F., Darling, R. B., Storment, C. W., and Kovacs, G. T. A., "Combined Organic Thermal and Electrostatic Omnidirectional Ciliary Microactuator Array for Object Positioning and Inspection," Proceedings of the 1996 Solid-State Sensor and Actuator Workshop, Hilton Head Island, SC, June 3 - 6, 1996, pp. 168 - 173.

Suh, J. W., Storment, C. W., and Kovacs, G. T. A., "Characterization of Multi-Segment Organic Thermal Actuators," Proceedings of Transducers '95, the 8th International Conference on Solid-State Sensors and Actuators, Stockholm, Sweden, June 25 - 29, 1995, vol. 2, pp. 333 - 336.

Sulouff, R. E., "Silicon Sensors for Automotive Applications," Proceedings of Transducers '91, the 1991 International Conference on Solid-State Sensors and Actuators, San Francisco, CA, June 24 - 27, 1991, pp. 170 - 176.

Sun, X.-Q., Gu, X., and Carr, W. N., "Lateral In-Plane Displacement Microactuators with Combined Thermal and Electrostatic Drive," Proceedings of the 1996 Solid-State Sensor and Actuator Workshop, Hilton Head Island, SC, June 3 - 6, 1996, pp. 152 - 155.

Suzuki, K., "High-Density Tactile Sensor Arrays," Advanced Robotics, vol. 7, no. 3, 1993, pp. 283 - 287.

Suzuki, K., Ishihara, T., Hirata, M., and Tanigwa, H., "Nonlinear Analyses on CMOS Integrated Silicon Pressure Sensor," IEEE International Electron Devices Meeting, Washington, DC, Dec. 1 - 4, 1985, pp. 137 - 140.

Suzuki, K., Ishihara, T., Hirata, M., and Tanigwa, H., "Nonlinear Analyses on CMOS Integrated Silicon Pressure Sensor," IEEE Transactions on Electron Devices, vol. ED-34, no. 6, June 1987, pp. 1360 - 1367.

Suzuki, K., Najafi, K., and Wise, K. D., "A 1024-Element High-Performance Silicon Tactile Imager," IEEE Transactions on Electron Devices, vol. 37, no. 8, Aug. 1990, pp. 1852 - 1860.

Suzuki, K., Najafi, K., and Wise, K. D., "Process Alternatives and Scaling Limits for High-density Silicon Tactile Imagers," Sensors and Actuators, vol. A23, nos. 1 - 3, Apr. 1990, pp. 915 - 918.

Swartz, R. G., and Plummer, J. D., "Integrated Silicon-PVF$_2$ Acoustic Transducer Arrays," IEEE Transactions on Electron Devices, vol. ED-26, no. 12, Dec. 1979, pp. 1921 - 1931.

Tai, Y.-C., and Muller, R. S., "IC-Processed Electrostatic Synchronous Micromotors," Sensors and Actuators, vol. 20, nos. 1 - 2, Nov. 15, 1989, pp. 49 - 55.

Tai, Y.-C., Fan, L.-S., and Muller, R. S., "IC-Processed Micro-Motors: Design, Technology and Testing," Proceedings of the IEEE Workshop on Micro Electro Mechanical Systems, Salt Lake City, UT, Feb. 20 - 22, 1989, pp. 1 - 6.

Takao, H., Matsumoto, Y., and Ishida, M., "A Monolithically Integrated Three Axial Accelerometer Using Stress Sensitive CMOS Differential Amplifiers," Proceedings of Transducers '97, the 1997 International Conference on Solid-State Sensors and Actuators, Chicago, IL, June 16 - 19, 1997, vol. 2, pp. 1173 - 1176.

Takao, H., Matsumoto, Y., Seo, H.-D., Tanaka, H., Ishida, M., and Nakamura, T., "Three Dimensional Vector Accelerometer Using SOI Structure for High Temperature," Proceedings of Transducers '95, the 8th International Conference on Solid-State Sensors and Actuators, Stockholm, Sweden, June 25 - 29, 1995, vol. 2, pp. 683 - 686.

Tang, W. C., Nguyen, H., Judy, M. W., and Howe, R. T., "Electrostatic-Comb Drive of Lateral Polysilicon Resonators," Sensors and Actuators, vol. A21, nos. 1 - 3, Feb. 1990, pp. 328 - 331.

Taylor, W. P., Allen, M. G., and Dauwalter, C. R., "A Fully Integrated Magnetically Actuated Micromachined Relay," Technical Digest of the 1994 Solid-State Sensor and Actuator Workshop, Hilton Head Island, SC, June 3 - 6, 1996, pp. 231 - 234.

Taylor, W. P., Allen, M. G., and Dauwalter, C. R., "A Packaging Compatible Fully Integrated Micromachined Relay," Proceedings of the 1996 International Symposium on Microelectronics, Minneapolis, MN, Oct. 8 - 10, 1996, SPIE vol. 2920, pp. 202 - 207.

Taylor, W. P., and Allen, M. G., "Integrated Magnetic Microrelays: Normally Open, Normally Closed, and Multi-Pole Devices," Proceedings of Transducers '97, the 1997 International Conference on Solid-State Sensors and Actuators, Chicago, IL, June 16 - 19, 1997, vol. 2, pp. 1149 - 1152.

Timoshenko, S. P., "Analysis of Bi-Metal Thermostats," Journal of the Optical Society of America, vol. 11, 1925, pp. 233 - 255.

Trimmer, W. S. N., and Gabriel, K. J., "Design Considerations for a Practical Electrostatic Micro-Motor," Sensors and Actuators, vol. 11, no. 2, Mar. 1987, pp. 189 - 206.

Trimmer, W. S. N., and Jebens, R., "Harmonic Electrostatic Micromotors," Sensors and Actuators, vol. 20, nos. 1 - 2, Nov. 1989, pp. 17 - 24.

Trimmer, W., and Jebens, R., "An Operational Harmonic Electrostatic Motor," Proceedings of the IEEE Workshop on Micro Electro Mechanical Systems, Salt Lake City, UT, Feb. 1989, pp. 13 - 16.

van der Donk, A., "A Silicon Condenser Microphone: Modeling and Electronic Circuitry," Ph.D. Thesis, University of Twente, 1992.

van Drieënhuizen, B. P., Maluf, N. I., Opris, I. E., and Kovacs, G. T. A., "Force-Balanced Accelerometer with mG Resolution, Fabricated Using Silicon Fusion Bonding and Deep Reactive Ion Etching," Proceedings of Transducers '97, the 1997 International Conference on Solid-State Sensors and Actuators, Chicago, IL, June 16 - 19, 1997, vol. 2, pp. 1229 - 1230.

van Mullem, C. J., Gabriel, K. J., and Fujita, H., "Large Deflection Performance of Surface Micromachined Corrugated Diaphragms," Proceedings of Transducers '91, the 1991 International Conference on Solid-State Sensors and Actuators, San Francisco, CA, June 24 - 27, 1991, pp. 1014 - 1017.

Voss, R., Bauer, K., Ficker, W., Gleissner, T., Kupke, W., Rose, M., Sassen, S., Schalk, J., Seidel, H., and Stenzel, E., "Silicon Angular Rate Sensor for Automotive Applications with Peizoelectric Drive and Piezoresistive Read-Out," Proceedings of Transducers '97, the 1997 International Conference on Solid-State Sensors and Actuators, Chicago, IL, June 16 - 19, 1997, vol. 2, pp. 879 - 882.

Walker, J. A., Gabriel, K. J., and Mehregany, M., "Thin-Film Processing of TiNi Shape Memory Alloy," Sensors and Actuators, vol. A21, nos. 1 - 3, Feb. 1990, pp. 243 - 246.

Wang, J., McClelland, B., Zavracky, P. M., Hartley, F., and Dolgin, B., "Design, Fabrication and Measurement of a Tunneling Tip Accelerometer," Proceedings of the 1996 Solid-State Sensor and Actuator Workshop, Hilton Head Island, SC, June 3 - 6, 1996, pp. 68 - 71.

Wang, K., and Nguyen, C. T.-C., "High-Order Micromechanical Electronic Filters," Proceedings of the IEEE Micro Electro Mechanical Systems Workshop, Nagoya, Japan, Jan. 26 - 30, 1997, pp. 25 - 30.

Wang, K., Wong, A.-C., Hsu, W.-T., and Nguyen, C. T.-C., "Frequency Trimming and Q-Factor Enhancement of Micromechanical Resonators via Localized Filament Annealing," Proceedings of Transducers '97, the 1997 International Conference on Solid-State Sensors and Actuators, Chicago, IL, June 16 - 19, 1997, vol. 1, pp. 109 - 112.

Wang, Y., Zheng, X., Liu, L., and Li, Z., "A Novel Structure of Pressure Sensors," IEEE Transactions on Electron Devices, vol. 38, no. 8, Aug. 1991, pp. 1797 - 1802.

Welham, C. J., Gardner, J. W., and Greenwood, J., "A Laterally Driven Micromachined Resonant Pressure Sensor," Proceedings of Transducers '95, the 8th International Conference on Solid-State Sensors and Actuators, Stockholm, Sweden, June 25 - 29, 1995, vol. 2, pp. 586 - 589.

White, R. M., "Tactile Sensor Employing a Light Conducting Element and a Resiliently Deformable Sheet," U.S. Patent No. 4,668,861, issued May 26, 1987.

Wolffenbuttel, M. R., and Regtien, P. P. L., "Design Considerations for a Silicon Capacitive Tactile Cell," Sensors and Actuators, vol. A24, no. 3, Sept. 1990, pp. 187 -190.

Yamada, K., Nishihara, M., Shimada, S., Tanabe, M., Shimazoe, M., and Matsuoka, Y., "Nonlinearity of the Piezoresistance Effect of p-Type Silicon Diffused Layers," IEEE Transactions on Electron Devices, vol. ED-29, no. 1, Jan. 1982, pp. 71 - 77.

Yang, X., Tai, Y.-C., and Ho, C.-M., "Micro Bellow Actuator," Proceedings of Transducers '97, the 1997 International Conference on Solid-State Sensors and Actuators, Chicago, IL, June 16 - 19, 1997, vol. 1, pp. 45 - 48.

Yang, Y.-J., and Kim, C.-J., "Testing and Characterization of a Bistable Snapping Microactuator Based on Thermo-Mechanical Analysis," Proceedings of Transducers '95, the 8th International Conference on Solid-State Sensors and Actuators, Stockholm, Sweden, June 25 - 29, 1995, vol. 2, pp. 337 - 340.

Yao, J. J., and Chang, M. F., "A Surface Micromachined Miniature Switch for Telecommunications Applications with Signal Frequencies from DC up to 4 GHz," Digest of Technical Papers, Transducers 95, Stockholm, Sweden, June 25 - 29, 1995, vol. 2, pp. 384 - 387.

Yao, J. J., and MacDonald, N. C., "A Micromachined, Single-Crystal Silicon, Tunable Resonator," Journal of Micromechanics and Microengineering, vol. 5, no. 3, Sept. 1995, pp. 257 - 264.

Yasseen, A. A., Smith, S. W., Mehregany, M., and Merat, F. L., "Diffraction Grating Scanners Using Polysilicon Micromotors," Proceedings of the IEEE Micro Electro Mechanical Systems Conference, Amsterdam, Netherlands, Jan. 29 - Feb. 2, 1995, pp. 175 - 180.

Yazdi, N., and Najafi, K., "An All-Silicon Single-Wafer Fabrication Technology for Precision Microaccelerometers," Proceedings of Transducers '97, the 1997 International Conference on Solid-State Sensors and Actuators, Chicago, IL, June 16 - 19, 1997, vol. 2, pp. 1181 - 1184.

Yeh, C., and Najafi, K., "Micromachined Tunneling Accelerometer with a Low-Voltage CMOS Interface Circuit," Proceedings of Transducers '97, the 1997 International Conference on Solid-State Sensors and Actuators, Chicago, IL, June 16 - 19, 1997, vol. 2, pp. 1213 - 1216.

Yeh, H.-J., J., and Smith, J. S., "Fluidic Self-Assembly of GaAs Microstructures on Si Substrates," Sensors and Materials, vol. 6, no. 6, 1994, pp. 319 - 332.

Yeh, H.-J., J., and Smith, J. S., "Fluidic Self-Assembly of Microstructures and Its Application to Integration of GaAs on Si," Proceedings of the Seventh Annual IEEE Workshop on Micro Electro Mechanical Systems, Oiso, Japan, Jan. 25 - 28, 1994, pp. 279 - 284.

Yeh, R., Kruglick, E. J. J., and Pister, K. S. J., "Microelectromechanical Components for Articulated Microrobots," Proceedings of Transducers '95, the 8th International Conference on Solid-State Sensors and Actuators, Stockholm, Sweden, June 25 - 29, 1995, vol. 2, pp. 346 - 349.

Yoshida, T., Kudo, T., Kato, S., Miyazaki, S.-I., Kiyono, S., and Ikeda, K., "Strain Sensitive Resonant Gate Transistor," Proceedings of the IEEE Micro Electro Mechanical Systems Conference, Amsterdam, Netherlands, Jan. 29 - Feb. 2, 1995, pp. 316 - 321.

Zavracky, P. M., McGruer, N. E., and Majumder, S., "Micromechanical Switches," Journal of Microelectromechanical Systems, vol. 6, no. 1, Mar. 1997, pp. 3 - 9.

Zdeblick, M. J., Anderson, R., Jankowski, J., Kline-Schoder, B., Christel, L., Miles, R., and Weber, W., "Thermpneumatically Actuated Microvalves and Integrated Electro-Fluidic Circuits," Proceedings of the Solid-State Sensor and Actuator Workshop, Hilton Head Island, SC, June 13 - 16, 1994, pp. 251 - 255.

Zhang, Y., and Wise, K. D., "A High-Accuracy Multi-Element Silicon Barometric Pressure Sensor," Proceedings of Transducers '95, the 8th International Conference on Solid-State Sensors and Actuators, Stockholm, Sweden, June 25 - 29, 1995, vol. 1, pp. 608 - 611.

Zhou, S., Sun, X.-Q., and Carr, W. N., "A Micro Variable Inductor Chip Using MEMS Relays," Proceedings of Transducers '97, the 1997 International Conference on Solid-State Sensors and Actuators, Chicago, IL, June 16 - 19, 1997, vol. 2, pp. 1137 - 1140.

Zook, J. D., Burns, D. W., Herb, W. R., Guckel, H., Kang, J.-W., and Ahn, Y., "Optically Excited Self-Resonant Strain Transducers," Proceedings of Transducers '95, the 8th International Conference on Solid-State Sensors and Actuators, Stockholm, Sweden, June 25 - 29, 1995, vol. 2, pp. 600 - 603.

CITED INDUSTRY REFERENCES AND SUPPLIERS

Open-Loop Capacitive Accelerometers

Silicon Designs, Inc., 1445 NW Mall Street, Issaquah, WA 98027, Phone: (206) 391-8329.

Piezoresistive Pressure Sensors with and without On-Chip Signal Conditioning Circuitry

Motorola, Inc., P.O. Box 20912, Phoenix, AZ 85036, Phone: (800) 441-2447, http://www.mot.com/

Piezoresistive Pressure Sensors

Sensym, Inc., 1804 McCarthy Blvd., Milpitas, CA 95035, USA, Phone: (800)573-6796, http://www.sensym.com/

Lucas NovaSensor, 1055 Mission Court, Fremont, CA 94539, Phone: (800) 962-7364 .

Closed-Loop Capacitive Accelerometers

Analog Devices, Inc., One Technology Way, P.O. Box 9106, Norwood, MA 02062-9106, Phone: (617) 329-4700, http://www.analog.com/

Chapter 4:

OPTICAL TRANSDUCERS

1. INTRODUCTION

This chapter is concerned with micromachined optical transducers: sensors that detect light and "actuators" that emit or modulate light. Compared to micromechanical devices, this is a well-established market, yet new devices continue to emerge rapidly. A powerful commercial driver for optical transducers has been consumer electronics. This market sector has provided a high-volume demand for devices such as light emitting and laser diodes, photodiodes and phototransistors, charge-coupled (and recently CMOS) imagers, and computer displays. With aggressive cost competition, these devices have all moved from expensive laboratory items to relatively low-cost commodities.

Recently, some of the micromachining approaches discussed in the previous chapters have been applied to optical transducers. For example, micromechanical mirror displays are currently being used for projection video display applications, and micromachined electromechanical fiber-optic transceivers have been demonstrated. There are many more potential applications for micromachining in optical transducers in areas of beam scanning, beam positioning, fiber-optic coupling, new light emitter structures, advanced photodetectors, and complete miniaturized optical systems.

There is considerable overlap between the transduction mechanisms used in this application domain and others. For example, thermal transduction is often used as an intermediate means of detecting optical energy (particularly for infrared applications) and ultimately generating an electrical output signal. Similarly, micromechanical actuators are often used in optical beam steering. Thus, the reader is referred to the Thermal Transducers, Mechanical Transducers, and other chapters for further information on such indirect transduction methods.

1.1 THE OPTICAL SPECTRUM

Light, as conventionally defined, refers to the electromagnetic spectrum in the frequency range of $\approx 10^{11}$ (far infrared) to 10^{17} Hz (far ultraviolet). The energy of a single photon, E, is given by,

$$E = h\nu \quad \text{in J or eV}$$

where h is Planck's constant = 6.6261×10^{-34} J•s = 4.1361×10^{-15} eV•s, and ν is the frequency of light in Hz (ν, rather than f, is conventionally used), and the wavelength of the light, λ, can be related to its frequency by,

$$\lambda = \frac{c}{\nu}$$

where c is the speed of light in a vacuum = 2.99792458×10^8 m/s.

The optical spectrum includes relatively long wavelengths (infrared [IR]) up to several microns (longer wavelengths are referred to as radio waves) and relatively short wavelengths (ultraviolet [UV]) down to a few tens of nanometers (shorter wavelengths are referred to as x-rays). It is important to note that "visible" light is defined by a human perspective (other species have visual receptors that have different spectral responses). For example, many insects, and even vertebrates such as some birds can see UV light. In humans, the retina can respond to UV light at wavelengths greater than 300 nm, but most of it is absorbed by the lens of the eye (Hecht (1987)). The response of the normal human eye is shown below.

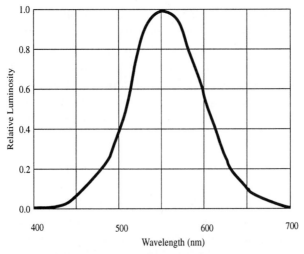

Plot of relative luminosity of various wavelengths, normalized to the wavelength of peak sensitivity in humans (555 nm). After Benson (1992).

The light from the sun, the basis for the evolution of our visual systems, is not "flat" or "white," but has a characteristic spectrum that is in turn modulated by the Earth's atmosphere due to absorptions characteristic of its chemical constituents. Similarly, all light sources normally encountered have emission spectra that are not uniform across large wavelength bands. Those emission spectra are also affected by the media through which the light must pass before reaching the detector (with its own spectral response).

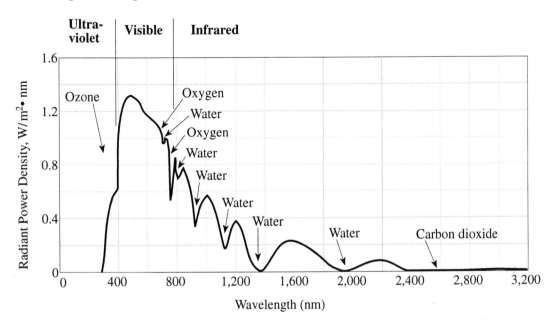

Spectral distribution of solar radiant power density at sea level, showing the ozone, oxygen, water, and CO_2 absorption bands. After Benson (1992).

1.2 UNITS RELEVANT TO OPTICAL TRANSDUCERS

Some of the commonly used units of optical measurement are given below for reference. It should be noted that there are two sets of units for measuring optical output:

Radiometric – measuring pure, raw energy flow, regardless of wavelength, λ

Photometric – measuring energy flow within the wavelength range of human vision

Radiometric units are based on the watt (W) and photometric units are based on the *lumen*. The two are related at $\lambda = 555$ nm, the peak of the human visual response, for which 1 W at 555 nm = 683 lumens (lm). For wavelengths other than

555 nm, the number of lumens is weighted by the "standardized" human visual response (sometimes referred to as *efficacy*). The term "efficacy" is also used to describe the output properties of illumination sources (e.g., a light bulb producing some number of lumens per watt).

The international definition of the lumen is derived from the *candela* (base unit). But the above definition is generally more appealing to engineers since it emphasizes the fact that lumens represent power in a conventional sense and provides a conversion between radiometric and photometric units that is based on the human visual response.

As one might expect the "basic" units of light intensity can be defined using the simple SI units, such as defining radiant flux in watts or flux density as W/m^2.

Luminous intensity: A *candela* (cd) is the luminous intensity, in a given direction, of a source that emits monochromatic radiation of frequency 540×10^{12} Hz (555 nm, or 2.23 eV, corresponding to the peak sensitivity of the human visual system) and that has a radiant intensity in that direction of (1/683) watt per steradian (sr). One candela (cd) per square meter is known as a "nit."

Luminous flux: A *lumen* (lm) is the luminous flux emitted in a solid angle of one steradian by a point source having a uniform intensity of one *candela* (cd•sr). Note that the luminous power output in lumens (photometric) of light emitters outside the human range of perception is, by definition, zero.

$$1 \text{ lm} = (1 \text{ cd})(1 \text{ sr})$$

Illuminance (lx): A *lux* is the illuminance produced by a luminous flux of 1 lumen uniformly distributed over a surface of one square meter $(cd•sr/m^2)$,

$$1 \text{ lx} = \frac{1 \text{ lm}}{m^2} = \frac{1 \text{ cd} • \text{ sr}}{m^2}$$

1.3 BLACKBODY RADIATION

The radiant flux, W_λ, (referred to as the *spectral flux density* or *spectral excitance*) emitted by a body of emissivity, ε, at an absolute temperature, T, is given by the Planck equation, which is notably comprised entirely of fundamental constants. The flux per unit area of emitter, per unit wavelength, centered at a wavelength, λ, is given by,

$$W_\lambda = \frac{\varepsilon(\lambda)\, 2\pi hc^2}{\lambda^5} \frac{1}{\left(e^{\frac{hc}{\lambda kT}} - 1\right)} \quad \text{in W/m}^2$$

where $\varepsilon(\lambda)$ is the emissivity (dimensionless) as a function of wavelength, and k is Boltzmann's constant = 1.38066×10^{-23} J/K.

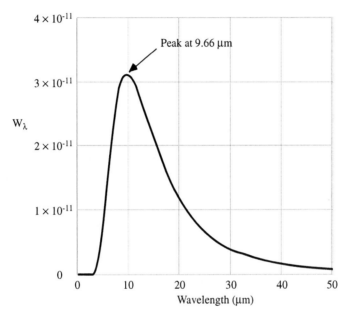

Plot of W_λ versus λ (μm) for a 300 K (room temperature) object assuming $\varepsilon = 1$ (human body temperature corresponds to a peak at $\approx 9.4 \,\mu$m).

The *emissivity* of a body, ε, expresses the degree to which a body emits less efficiently than a blackbody, for which $\varepsilon = 1$. It should be noted that ε usually varies with λ. The wavelength at which the radiant flux is at its maximum is obtained by differentiating the Planck equation with respect to λ and setting it equal to zero, to obtain *Wien's displacement law* (assuming $\varepsilon(\lambda)$ is constant),

$$\lambda_{peak} = \frac{2898}{T} \quad \text{in } \mu\text{m}$$

This relationship of peak wavelength to temperature forms the basis of optical temperature measurements (pyrometry) and for equating colors to temperatures as used in astronomy and photography (e.g., 3,000 K tungsten or 2,600 K "natural" lighting).

The total radiated power per unit area can be determined by integrating the Planck equation from $\lambda = 0$ to ∞, which yields the *Stefan-Boltzmann equation*,

$$W_T = \int_{\lambda=0}^{\infty} W_\lambda d\lambda = \varepsilon\sigma T^4 \quad \text{in W/m}^2$$

where σ is the Stefan-Boltzmann constant $= 5.6697 \times 10^{-12}$ W/cm^2•K^4

The fourth-power dependence on temperature in this equation can be misleading, since the fact that σ is so small means that in typical engineering practice, objects must become very hot before radiating appreciable amounts of energy.

2. OPTICAL SENSORS

2.1 TYPES OF OPTICAL SENSORS

2.1.1 FUNDAMENTAL PRINCIPLE OF OPTICAL DETECTION

One basic principle captures the process of optical detection: *photon absorption causes an electron to undergo an upward energy transition.* This energy transition may be of several possible types:

1) Transition from the valence band to the conduction band of a solid (*photovoltaic* effects).

2) Transition from the conduction band to a vacuum (*photoelectric* effect).

3) Transition to a virtual energy state. After a certain lifetime, the electron returns to its initial energy level (this is the basis of the index of refraction and other polarization effects, e.g., birefringence, the Kerr and Pockels electro-optic effects).

4) Transition to midgap states and back to an initial relaxed level. Photon energy is dissipated by phonons (i.e., heat generation, the basis of many "thermal" optical detectors).

5) Other mechanisms such as excitons (the electron and a hole form a hydrogen-like molecule with a distinct set of energy levels).

2.1.2 DIRECT ELECTRONIC OPTICAL SENSORS

This category of optical sensors consists of those devices wherein the detection of photons *directly* results in an electronic signal (without any intermediate transduction stages such as a conversion of light to heat and a subsequent temperature measurement). There are three basic classes of direct electronic photosensors:

1) *Photoemissive* (Einstein, Nobel prize) — one electron is released from a metal photocathode (in a vacuum) per photon of sufficient energy.

2) *Bulk photoconductive* (no junction) — photons generate carriers that lower the bulk resistance of the material.

3) *Junction-based* — photons generate electron-hole pairs in the depletion region of a semiconductor junction.

All of these photosensors have three basic underlying processes that occur in the transduction process and determine the sensor's final gain: 1) carrier *generation* by incident light, 2) carrier *separation*, transport and/or carrier multiplication, and 3) *collection* of current by an external circuit.

2.1.3 INDIRECT OPTICAL SENSORS

The terms "direct" and "indirect" with respect to optical sensors should not be confused with direct and indirect as used to describe the bandgaps of materials. As mentioned above, for optical sensors direct means direct transduction to an electronic signal, whereas indirect means conversion of optical signals into an intermediate energy form (thermal, chemical, etc.), which is then measured electrically. For example, indirect optical sensors are often used to detect IR light by absorbing it, converting it to heat, which is then measured with what is essentially a thermometer (referred to as a *bolometer* in this case). Other indirect optical sensors might include intermediate states, such as chemical bonds, mechanical forces, etc.

2.1.4 MAJOR SPECIFICATIONS FOR PHOTODETECTORS

Quantum efficiency (η) is defined as the number of carriers generated (but not necessarily collected) per incident photon. The *internal* quantum efficiency refers to the number of carriers generated, while the *external* quantum efficiency accounts for only the carriers collected. As long as the photon energy $h\nu$ exceeds the bandgap of a semiconductor, the internal η is essentially unity.

Gain (symbol varies) is defined as the ratio of the total current that flows in response to photoexcitation to the current that flows in direct response to impinging photons (the primary photocurrent). This is a measure of the carrier multiplication or other mechanisms that go on in the sensor.

Response time is a time constant (in seconds) that corresponds to the response of a photodetector to large signals.

Bandwidth or *frequency response* refers to the frequency range over which the photodetector responds (this is a small-signal parameter), with a cutoff frequency or frequencies (if a low-frequency cutoff exists) defined as the frequency at which the output signal amplitude is reduced by 3 dB.

Responsivity is also an often used figure of merit relating the output signal amplitude to input power, defined as,

$$R_I = \frac{\text{output current}}{\text{optical input power}} = \frac{I_p}{P_{opt}} = \frac{\eta q}{h \nu} = \frac{\eta \lambda(\mu m)}{1.2398} \quad A/W$$

where, q is the charge on the electron = 1.60218×10^{-19} C. An alternative definition is the ratio of the photovoltage to the optical input power is R_V (in V/W), although it is seldom used in the context of semiconductor detectors.

Noise equivalent power (NEP) is the amount of light required to get a signal *equivalent in power to that of the noise* (signal-to-noise ratio = 1) and is defined for a current-output transducer as,

$$NEP \equiv \frac{\text{RMS noise current} \left(\frac{A}{\sqrt{Hz}} \right)}{R_I \left(\frac{A}{W} \right)} \quad \text{in} \quad \left(\frac{W}{\sqrt{Hz}} \right)$$

or, for a voltage-output device as,

$$NEP \equiv \frac{\text{RMS noise voltage} \left(\frac{V}{\sqrt{Hz}} \right)}{R_V \left(\frac{V}{W} \right)} \quad \text{in} \quad \left(\frac{W}{\sqrt{Hz}} \right)$$

(Note that the NEP and the noise current/voltage can be either spectral densities or integrated over a specified bandwidth.)

Photonic detectors (photoconductor, photovoltaic [junction], photoemissive, and photoelectromagnetic detectors), as opposed to indirect optical detectors (bolometers, thermopiles, and pyroelectric detectors), which operate using the generation of carriers, produce either a voltage or current in response (proportional) to the incident optical power. The electrical signal power is proportional to i^2 or v^2, and thus if the incident optical power is proportional to the detector's area, A, the noise current will be proportional to $\sqrt{A}$. (In other words, the noise power is proportional to A, and thus the noise current is proportional to $\sqrt{A}$.) Similarly, for white noise, the noise power is proportional to bandwidth, B, and thus the noise current is proportional to $\sqrt{B}$.

The *detectivity*, D*, is based on the reciprocal NEP, correcting for the proportionality of $NEP \propto \sqrt{A}$ (noting that $NEP \propto \sqrt{B}$), thus,

$$D^* \equiv \frac{\sqrt{A}}{NEP} \quad \text{in} \quad \frac{cm\sqrt{Hz}}{W}$$

For indirect detectors, the current or voltage is not necessarily proportional to the incident optical power, so detectivity cannot be simply defined for them, and comparisons using D* or NEP are risky and potentially misleading. It is safest to compare detectors of the same size, eliminating assumptions regarding noise-versus-area relationships. The plots below allow basic comparison of typical D* values for various photodetectors compared to theoretically ideal D* values at room temperature and 77 K. Theoretical D* values are limited by noise due to emissivity of photons *from* the detector at the temperatures of interest. This is a fundamental thermodynamic process and sets a basic performance limit. For further information, two classic references on detectivity are Jones (1947, 1953).

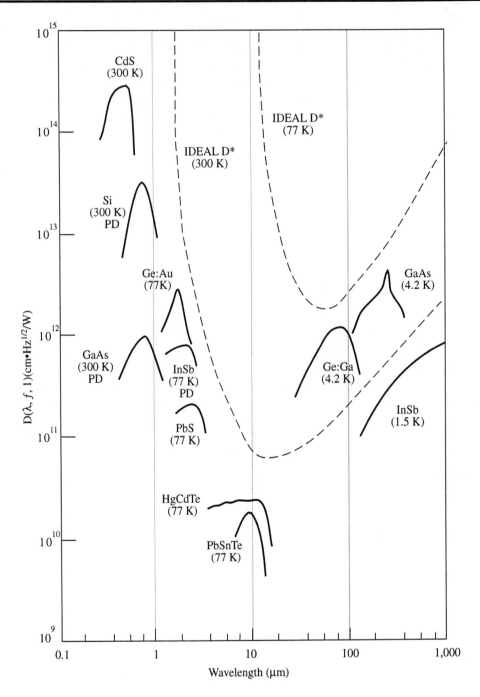

Plots of detectivity, D*, for various photoconductors (and photodiodes, identified as "PD") as a function of wavelength. The dashed curves are the theoretical ideal D* at 77 and 300 K viewing an angle of 2π steradians (not realizable in practice). After Sze (1981).

2.2 DIRECT ELECTRONIC OPTICAL SENSORS

As mentioned above, direct electronic optical sensors operate by direct transduction of incident photons into electronic carriers in the sensor. This is a very broad category of devices, including the familiar photodiodes, phototransistors, etc., and many less common devices. Several types are illustrated in the figure below, and a general comparison table also follows.

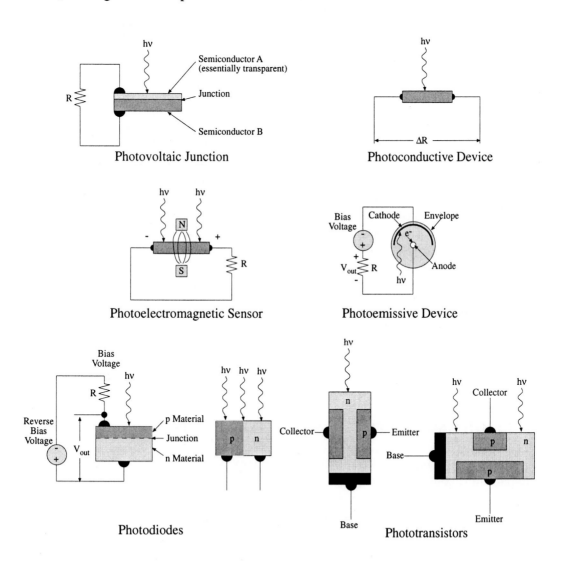

Conceptual illustration of several different types of direct photosensors. After Sze (1981).

Device Type	Gain	Response Time (s)	Typical Temperature
Photomultiplier	$> 10^6$	10^{-7} to 10^{-9}	300 (sometimes cooled)
Photoconductor	1 to 10^6	10^{-3} to 10^{-8}	4.2 to 300
Metal-Semiconductor-Metal Photodetector	1 or less	10^{-10} to 10^{-12}	300
p-n Photodiode	1 or less	10^{-6} to 10^{-11}	300 (sometimes cooled to 77 K)
p-i-n Photodiode	1 or less	10^{-6} to 10^{-9}	300
Metal-Semiconductor Diode	1 or less	10^{-9} to 10^{-12}	300
Avalanche Diode	10^2 to 10^4	10^{-10}	300
Bipolar Phototransistor	10^2	10^{-6} to 10^{-8}	300
Bipolar Photo-Darlington	10^4	10^{-5} to 10^{-6}	300
Field-Effect Phototransistor	10	10^{-7}	300
CCD Cell (Metal-Insulator-Semiconductor Capacitor)	1 or less	10^{-5} to 10^{-8}	300 (sometimes cooled)

Gains and response times of some typical photodetectors (some are optimistic!). After Sze (1981). Note that the CCD cell, and some extrinsic photoconductors, are integrating detectors, and thus the response time figures can be somewhat misleading.

2.2.1 LIGHT ABSORPTION IN SEMICONDUCTORS

Assuming no reflection, if the photon flux density at the surface is Φ_o, then the photon flux at depth x is given by Beer's law,

$$\Phi(x) = \Phi_o e^{-\alpha x} \text{ in photons/(cm}^2\bullet\text{s)}$$

where α is the absorption coefficient, typically in cm^{-1}. (Note that Φ_o can be related to be optical power, I_o, by, $I_o = h\nu\Phi_o$, in W.) It is important to note that α is a strong function of λ and this function varies greatly for each material. The electron-hole pair generation rate at a depth x from the illuminated side of the semicondutor layer (the depletion region in a junction-based device), is given by,

$$G(x) = \alpha \eta \Phi_o e^{-\alpha x} \quad \text{in } cm^{-3} \bullet s^{-1}$$

where n is the internal quantum efficiency (dimensionless, described below).

One certainly needs to consider absorption when designing a photosensor that relies on electron-hole pair generation. It is critical to take into account the depth to which impinging photons are likely to penetrate when designing the physical structure of such a detector. If there is an overlying layer *above* that in which carrier generation is desired, its absorption must also be taken into account. Of the light flux falling onto any air-semiconductor interface, usually 30 to 40% is reflected, resulting in a significant loss. For example, a well-designed antireflective (AR) coating will recover most of this loss (e.g., MgF_2/SiO_2).

If, for example, CMOS-derived optical detectors are realized, a potential absorbing layer above the detectors might be the "scratch mask" or passivation layer, which is typically LTO. Conveniently, such layers tend to be quite transparent, although silicon nitrides tend to be more absorbing than oxides. Typical aluminum metallization and polysilicon layers are opaque to visible light, yet the polysilicon is transparent to longer-wavelength IR light. Thus these layers can be very useful as shields or as sections of Fresnel optics and gratings.

Another important point to note is that most semiconductors have a high refractive index, typically, n = 3 to 3.5, compared to "optical materials" such as glass. The difference in refractive indices across interfaces causes significant refraction and reflection.

2.2.2 BAND STRUCTURE OF PHOTOSENSORS

The bandgap for a semiconductor is the energy difference between the top of the valence band and the bottom of the conduction band, $E_g = E_c - E_v$. These bands are a continuum of allowed energy states for electrons, defining the range of energies in which they can exist. In most semiconductors, the bottom of the conduction band is higher in energy than the top of the valence band, giving a distinct energy bandgap, $E_g > 0$. In some other materials, the conduction and valence bands overlap, so that the top of the valence band is actually higher in energy than the bottom of the conduction band, $E_c < E_v$, so that $E_g < 0$ (a negative bandgap according to the above definition). The consequence of this is that the Fermi energy will lie in the region of overlap between the two bands, making both the conduction and valence bands partially occupied, even at low temperatures. A material with a negative bandgap is usually termed a "semimetal." Semimetals are highly conductive, even at low (cryogenic) temperatures, whereas true semiconductors become nearly perfect insulators as absolute zero is approached. As discussed below, the alloy mercury cadmium telluride ($Hg_{1-x}Cd_xTe$) can be tuned, by varying the alloy fraction, between a conventional semiconductor with a positive bandgap (x = 1, CdTe) to a semimetal with a negative bandgap (x = 0, HgTe), and crosses over between positive to negative bandgaps at ≈ x = 0.06 (6% Cd).

It is important to distinguish between *direct* and *indirect* bandgap materials with respect to optical detection (however, this distinction is not relevant to photoemissive sensors, wherein the emissive material's *work function* is the key parameter). If the minimum energy gap of the solid happens at the k = 0 point in momentum space (i.e., at the Γ point), the crystal is said to have a direct bandgap (e.g., GaAs), otherwise, it is said to be an indirect gap material (e.g., Si, Ge). Because the valence band maxima always occur at Γ, the structure of the conduction band determines what type of bandgap a material has.

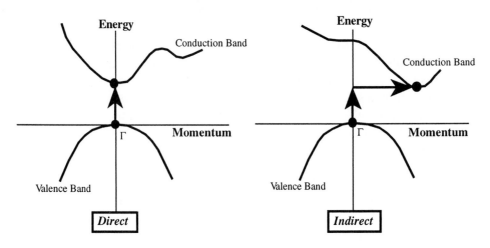

Schematic band diagram for direct and indirect bandgap materials.

A direct bandgap implies that *no* phonon (i.e., change in the carrier momentum) is necessary for an electron to complete an energy transition. This makes the transition a one-particle process, making it very likely to happen. If a phonon is required (change in carrier momentum), the transition becomes a two-particle process and the transition probability drops dramatically.

While both direct and indirect materials have a sharp cutoff in the absorption coefficient at hν = E_g, they have very different slopes in their absorption spectra, as shown below (optical absorption in any material follows the same functional energy dependence as the joint conduction/valence band density of states, which is the basic principle of optical determinations of energy band structure).

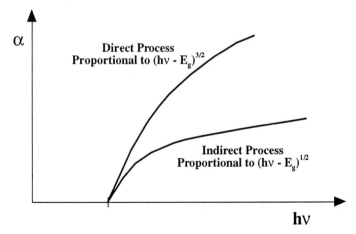

Illustration of the absorption spectra of direct and indirect processes.

2.2.3 QUANTUM EFFICIENCY

The (internal) quantum efficiency of a photodetector is the number of electron-hole pairs generated per absorbed photon (neglecting reflection losses at the surface),

$$\eta = \frac{\left(\dfrac{I_p}{q}\right)}{\left(\dfrac{P_{opt}}{h\nu}\right)}$$

where I_p is the photogenerated current, in A, and P_{opt} is the incident optical power, in W.

The plots below are useful to provide a sense for what theoretical maximum responsivities can be obtained for a given wavelength and internal quantum efficiency, assuming other detector properties are ideal. For near bandgap photons, this is essentially limited to one electron-hole pair. At greater energies (e.g., 3 to 4 eV, deep in the UV region), more than one electron-hole pair can be generated per absorbed photon.

Noting that the responsivity, R_I, is given by,

$$R_I = \frac{I_p}{P_{opt}} = \frac{\eta q}{h\nu} = \frac{\eta q \lambda}{hc}$$

the quantum efficiency as a function of wavelength for a given responsivity is,

$$\eta = \frac{hcR_I}{q\lambda}$$

which is the function plotted as dashed lines (for various values of R_I) below.

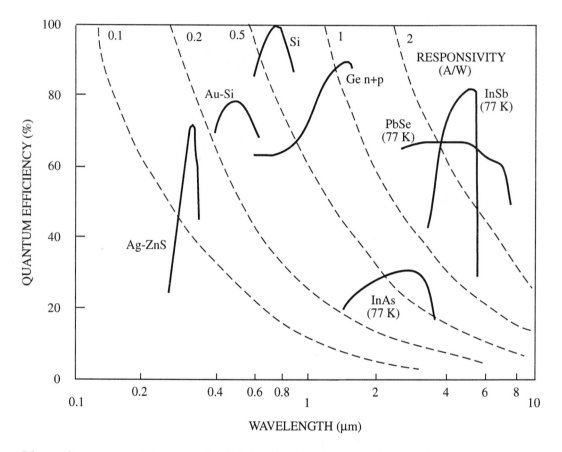

Plots of quantum efficiencies (solid lines) of various semiconductors versus wavelength. Responsivities of hypothetical detectors are shown as dashed lines (assuming that the detector has perfect carrier collection, perfect optical absorption, but imperfect internal quantum efficiency). After Sze (1981).

It should be noted that the often used curves shown above can be quite misleading. They imply that certain materials are inherently poor because of their composition. The inability of certain materials to reach $\eta = 1$ is a result of inefficient electron-hole pair separation and collection (a technological issue) and not generation (a materials issue). As processing techniques and device structures improve, these curves will shift upward.

2.2.4 PHOTOEMISSIVE SENSORS

The basic mechanism of photoemissive sensors is electrons released from a photocathode by elastic collisions with photons of sufficient energy $E = h\nu = hf$ (ν and f simply represent frequency). Gas pressure only effects the backscatter after emission, not the emission process itself; but, in practice, such devices are run at low pressures. The *work function*, ϕ_o (in eV), of the photocathode (negatively biased electron emitter) determines the maximum wavelength that can be detected. This wavelength is obtained simply by equating the work function with the energy of an incoming photon (*Einstein photoelectric effect*),

$$\lambda_{max} = \frac{hc}{\phi_o}$$

The key to fabricating a usable sensor is the appropriate choice of photoemissive material, optimizing quantum efficiency. Pure metals have relatively high work functions and low quantum efficiencies (≈ 0.1 %), so they are rarely used as photocathodes. Typical photocathodes are comprised of evaporated layers of alkali and group V metals (e.g., NaKCsSb) on a metal electrode. *Negative electron affinity* (NEA) photocathodes are also often used in such applications. NEA photocathodes are wide bandgap semiconductors (E_g typically 1 eV or greater), in which the bands at the surface are bent down sufficiently to position the vacuum energy level below the level of the conduction band in the bulk semiconductor. P-type semiconductors are typically used such that the absorption of a photon causes the promotion of an electron from the valence to the conduction band. Once the electron reaches the conduction band, it can escape from the semiconductor surface with no additional energy required. In practice, the wide bandgap semiconductor (e.g., GaAs) is heavily p-doped and then the surface is treated to create defect states that cause the desired band bending at the surface. For GaAs NEA devices, the surfaces are often treated to deposit very thin (monolayer) films of cesium and oxygen for this purpose (the other common alkali metals, Li, Na, K, and Rb, are also effective, but less than Cs). NEA photocathodes should be readily applicable to micromachined photoemissive detectors, should this be desired. An excellent reference on this NEA theory and applications is Escher (1981).

Typical photoemissive sensors obtain additional gain through introducing some noble gas molecules into the vacuum cavity. Emitted electrons can ionize gas atoms, generating further electrons for gain increases of roughly one order of magnitude. The unfortunate side effect of this is that the shot noise is thus increased by more than an order of magnitude, actually reducing the SNR (signal-to-noise ratio).

Multiple extra cathodes, or *dynodes*, can be included in a cascade so that the emitted electrons are multiplied by secondary emissions from each cathode. This is the basis for the photomultiplier tube (PMT). The operation of the PMT is based on this electron cascade mechanism, and not on gas ionization. Each dynode surface is coated with a low work function material such as an alkali metal oxide, as used for some basic photoemissive sensors. Each successive group of emitted electrons is accelerated into another of the extra cathodes (or "dynodes"), resulting in cascade amplification. Such photomultiplier tubes can resolve single photon events and maintain nanosecond resolution, achieved by focusing the electrons to avoid a large spread in transit path lengths.

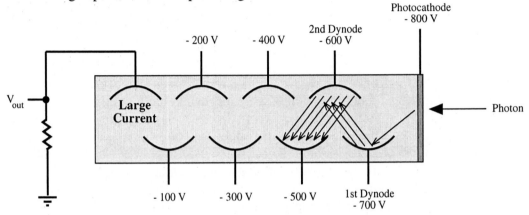

Illustration of the basic principle of photomultiplier tubes.

For n dynodes with an average electron gain N, the overall gain is,

$$G = N^n$$

The major noise source for photoemissive detectors is dark current, i_{dark}, which is due to thermionic emission (unrelated to incident radiation), and is given by the *Richardson-Dushman equation*,

$$i_{dark} = aAT^2 e^{\left(-\frac{\phi_o}{kT}\right)}$$

where,
a = Richardson's constant = 1.2×10^6 A/(m²•K²) in vacuum
A = cathode area, in m²
T = temperature, in K

Another significant noise source is shot noise, due to the discrete nature of arriving electrons and fluctuations in their arrival rate, usually a standard Poisson process. It is important to note that for photomultipliers, electrons considered to be noise (i.e., thermionically emitted) are also amplified by the gain, G. Naturally, in a real system, noise is also added externally by detection electronics such as the load resistor shown above (although running the current into a "virtual ground" solid-state transresistance stage may provide lower noise).

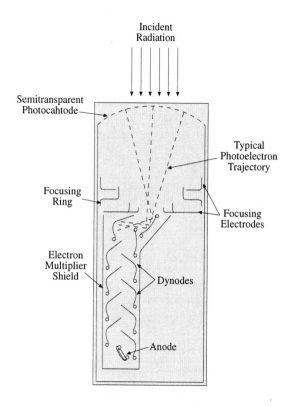

Typical physical construction of a photomultiplier tube.

PMT operation is inherently a much less noisy process than avalanche multiplication in a semiconductor (see below). At any temperature, a PMT has a better SNR than an avalanche photodiode (APD) (discussed below). In an APD, avalance multiplication is randomly initiated in space, but in a PMT, the multiplication process occurs semi-coherently at the dynodes. This coordinated multiplication process is less noisy than a randomly initiated one. In addition, single-carrier multiplication is inherently less noisy than a bipolar (electron-hole pair) multiplication process. As yet, a decent solid-state analog to the PMT has not been invented.

As illustrated below, microchannel devices include a continuous cascade of "cathodes" along slightly angled channels, so that electron multiplication can be obtained as an incident electron passes down the channel. The cathodes are actually fabricated by depositing a thin film of low work function material down the entire length of each channel. Voltage is dropped along the channels, effectively making one long continuous amplification cathode equivalent to a multitude of infinitesimally small discrete cathodes. Arrays of such microchannels (microchannel plates) are used in so-called "starlight-scopes" and other night-vision equipment. Microchannel plates were used by Tektronix for extra gain from an impinging electron beam to manufacture the world's fastest production analog oscilloscope, the Model 7104, reaching a 3 dB bandwidth of 1 GHz (with digital oscilloscopes eclipsing the analog models, this pinnacle of oscilloscope engineering is no longer manufactured, but is still used).

At present, microchannel plates are made by "pulling" large numbers of parallel glass capillaries and cutting the fused arrays at an appropriate angle, followed by cathode film deposition. Micromachined photomultipliers have not as yet been realized, despite some efforts such as that of Horton, et al. (1990). One of the problems is the need for very high-aspect-ratio microchannels at angles other than 90° to the substrate's surface (possibly done using the deep reactive ion etching techniques discussed in the Micromaching Techniques chapter). Another potential approach to fabricating such structures is to use LIGA to form electroplated nickel molds, replicate them with spin-on glass and reverse electroplate the mold away, leaving glass microchannels, as described by Liu, et al. (1997). Another key problem to be solved prior to fabricating useful devices is the need for uniform deposition of cathode materials within the microchannels after they are fabricated.

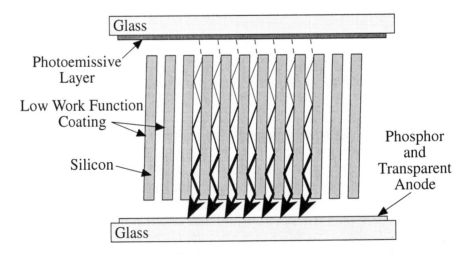

Conceptual illustration of a micromachined silicon microchannel plate (similar to "first-generation" image intensifier tubes).

If higher microchannel aspect ratios than pulled glass tubes can be obtained, the potential for higher resolution and thinner microchannel plates could yield such things as night-vision glasses that are more like normal eyeglasses in size and weight than the bulky devices available today. For further information on photomultipliers and microchannel plates, the reader is referred to Carruthers (1994).

2.2.5 PHOTOCONDUCTIVE SENSORS

This type of optical sensor does not have any form of semiconductor junction. The basic idea is to use a region of semiconductor as a "light-operated variable resistor." Two ohmic contacts are made and the resistance (modulated by incident photons) is measured across them.

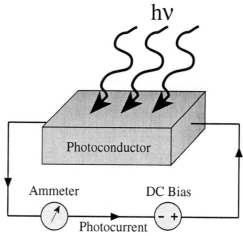

Illustration of the basic principle of photoconductive optical sensor operation.

Resistance *falls* (nonlinearly) with increasing illumination since conductivity is increased by the photogenerated carriers. A classic example of this is the common CdS cell often used in consumer goods to automatically adjust the brightness of LED (light emitting diode) and vacuum fluorescent displays when ambient lighting changes. It is worth noting that the use of the "archaic" CdS cell is actually increasing rapidly. The major application is energy-efficient lighting, and the reason these sensors are preferred is that they are one of the few types of photodetectors that can handle full 120/240 V AC across them (eliminating the need for low-voltage control electronics).

GAIN OF PHOTOCONDUCTIVE SENSORS

A key measure of photoconductive optical sensor performance is the *photocurrent gain*, the ratio of the number of carriers that flow through the device to the number generated per photon (the longer a photogenerated carrier exists prior to recombining the larger this gain). Long-lived carriers contribute more to an increase in conductivity since they permit increased current to flow until they recombine. If the carriers are swept out of the photoconductor quickly, they can make less of a contribution. While this concept is discussed in several textbooks (e.g., Sze (1981)), commonly presented derivations and equations can be misleading or erroneous in practical situations. For example, the photocurrent gain presented by Sze (1981) is,

$$\text{Photocurrent gain} = \frac{\mu_n E \tau}{L} = \frac{\tau v_d}{L} = \frac{\tau}{t_r}$$

where,
μ_n = electron mobility, in cm^2/(V•s)
E = V/L = electric field within the photoconductor, in V/cm
L = length of the photoconductor slab, in cm
$v_d = \mu_n E$ = drift velocity, in cm/s
τ = carrier lifetime, in s
$t_r = L/v_d$ = transit time, in s

is actually only valid for electric fields of zero (not practically useful) and predicts that the photocurrent gain can be increased linearly with the applied field without limit (not true). While it is appealing because it illustrates that carrier lifetime is important and that a short carrier transit time can increase gain, this equation does not properly model real photoconductive devices.

The above equation assumes only a single carrier type, yet electrons and holes are always generated in pairs. A space charge arises due to unequal carrier mobilities, with holes drifting out more slowly and additional carrier (electron) injection occurring at the contacts to maintain neutrality, providing the actual photocurrent gain. If $\mu_n = \mu_p$, the maximum photocurrent gain is unity.

In addition, the above equation assumes that carriers only recombine within the photoconductor, yet, in practice, most carriers are swept out to the contacts by the electric field and thus recombine by means of the external circuit. At a sufficent field where all photogenerated carriers are swept out of the photoconductor, the photocurrent gain saturates. *Typically, photoconductors are operated in this saturated region for high-speed applications.*

It can be shown that the maximum (saturated) photocurrent gain is a function of the ratio of electron and hole mobilities (Beneking (1982)),

$$\text{Maximum photocurrent gain} = \frac{1}{2}\left(1 + \frac{\mu_n}{\mu_p}\right)$$

Thus, in order to obtain significant photocurrent gain, one needs to choose a material for which $\mu_n \gg \mu_p$. For example, GaAs gives $\mu_n \approx 20\,\mu_p$, for a photocurrent gain of ≈ 10.5, while silicon only has $\mu_n = 2.5\,\mu_p$ for a maximum gain of 1.75.

While carrier lifetime certainly can influence photoconductor signal bandwith, it should be noted that for saturated photoconductors, the transit time (which is usually shorter than the lifetime) can be the principal bandwidth determining factor. For example, in a saturated GaAs photoconductor, with lifetimes on the order of 10 ns, fall times of less than 1 ns can be exhibited.

Another key photocurrent gain mechanism is carrier trapping. The trapping (see "Extrinsic Photoconductors" below) and subsequent delayed emission of a carrier back into its original band (distinct from recombination) greatly reduces that carrier's effective mobility, increasing photoconductive gain. Enormous photoconductive gains can thus be produced in GaAs and CdS, for example, with the fall time (and hence bandwidth) limited to the trap emission time constant (which can exceed several milliseconds). While this greatly compromises bandwidth, it provides an integration effect through these traps, making such devices very sensitive and tunable over most of the visible and IR range through a suitable choice of dopants.

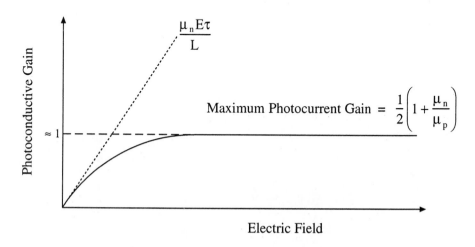

Illustration showing the photocurrent saturation effect in realistic photoconductors (solid line) under the assumption of $\mu_n = \mu_p$, and the approximation from Sze (1981).

In terms of photoconductive optical sensor design, it is generally useful to choose materials with large carrier mobility ratios; short slab lengths, L, to make the transit time short and the bandwidth high (also reducing the resistance and hence thermal noise); and thick photoconductor layers to produce more photocurrent and to increase the fraction of incident light intercepted (increasing the thickness only to the point at which the bulk of the incoming photons have been absorbed).

The maximum practical applied voltage depends on the type of photoconductor, but generally only needs to be adequate to saturate the photoconductor. If the material has a high intrinsic conductivity (giving unwanted *dark current*), one may have excessive power dissipation,

$$P_d = \frac{V^2}{R} = \frac{V^2}{\left(\dfrac{\rho L}{A}\right)} = \frac{V^2 \sigma A}{L} \quad \text{in watts}$$

where,
R = resistance of photoconductor at operating point, in Ω
ρ = resistivity of material, in $\Omega \cdot cm$
A = cross-sectional area of photoconductor, in cm^2
σ = conductivity of material, in $(\Omega \cdot cm)^{-1}$

In order to make the transit time short (to improve bandwidth), the inter-electrode gap, L, can be reduced, and to increase the active area of the photoconductor, the inter-electrode gap can be made to cover a long distance. This leads to the interdigitated electrode designs (and serpentine gaps) seen in most CdS cells.

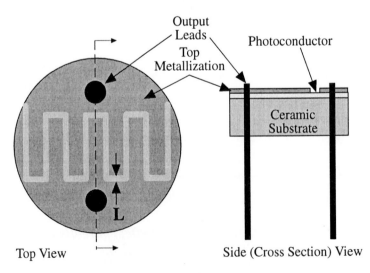

Illustration of a typical CdS photocell.

PHOTOCONDUCTIVE SENSOR SMALL-SIGNAL MODEL

For photoconductive sensors, there is no junction capacitance that is internal to the device. A basic small-signal circuit model (Sze (1981)) is shown below (note that significant parasitic capacitances may exist in practice, but these are dependent on implementation and not shown below).

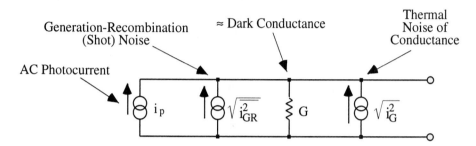

Small-signal circuit model of a photoconductive sensor, not considering parasitic capacitances between the contacts and elsewhere. In addition, loading effects of external circuits are not considered above. After Sze (1981).

It can be shown (Sze (1981)) that for an RMS optical input power, P_{opt}, the small-signal AC signal current for an input frequency ω can be written as,

$$i_p \approx \left(\frac{q\eta P_{opt}}{h\upsilon}\right)\left(\frac{\tau}{t_r}\right)\left(\sqrt{\frac{1}{1+\omega^2\tau^2}}\right)$$

which corresponds to a first-order low-pass filter with a cutoff frequency of,

$$f_c = \frac{2\pi}{\tau}$$

The two noise current sources in the model are the generation-recombination (shot) noise and the thermal (Johnson) noise generated by the conductance, G. For further information, a classic paper on the gain-bandwidth product for photoconductors is Redington (1959), and an example of a full analysis of sensitivity and dynamic range for a high bit-rate optical receiver can be found in Forrest (1985).

INTRINSIC PHOTOCONDUCTORS

These materials are "pure" (no intentional dopants) and electron-hole pairs are generated when the photon energy is greater than the bandgap, E_g, as illustrated below. If this type of carrier generation is the dominant mechanism, the devices are referred to as *intrinsic* detectors. In such detectors, the semiconductor is chosen based (typically) on the basis of the *maximum* wavelength to be detected, given by equating the photon energy ($E_p = h\nu = hc/\lambda$) to the bandgap (to determine the maximum wavelength at which a photon can move an electron into the conduction band),

$$\lambda_{MAX} = \frac{hc}{E_g} \approx \frac{1.24}{E_g} \quad (\lambda \text{ in } \mu m)$$

where E_g is the bandgap, in eV.

For semiconductors, the conductivity is given by,

$$\sigma = q\left(\mu_n n + \mu_p p\right)$$

where,
q = charge on the electron = 1.60218×10^{-19} C
μ_n, μ_p = electron and hole mobilities, in cm^2/(V•s)
n, p = density of electrons and holes, in cm^{-3} (for intrinsic materials, n = p = n_i)

and the light-induced change in conductivity is,

$$\Delta\sigma = qG\left(\mu_n \tau_n + \mu_p \tau_p\right)$$

where τ_n, τ_p is the electron and hole lifetimes, in s, and G is the generation rate (note different usage of "G" from conductance as above), in cm^{-3}•s^{-1}.

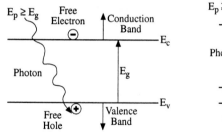

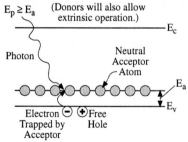

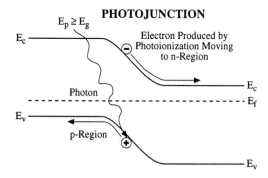

Comparative band diagrams for intrinsic-, extrinsic- and photojunction-type optical sensors. After Cobbold (1974).

Semiconductor	Bandgap (eV) 300 K	Bandgap (eV) 0 K	λ_{max} (µm) 300 K
BN	7.500	-	0.165
C	5.470	5.480	0.227
ZnS	3.680	3.840	0.337
GaN	3.360	3.500	0.369
ZnO	3.350	3.420	0.370
Alpha-SiC	2.996	3.030	0.414
CdS	2.420	2.560	0.512
GaP	2.260	2.340	0.549
BP	2.000	-	0.620
CdSe	1.700	1.850	0.729
AlSb	1.580	1.680	0.785
CdTe	1.560	-	0.795
GaAs	1.420	1.520	0.873
InP	1.350	1.420	0.919
Si	1.120	1.170	1.107
GaSb	0.720	0.810	1.722
Ge	0.660	0.740	1.879
PbS	0.410	0.286	3.024
InAs	0.360	0.420	3.444
PbTe	0.310	0.190	4.000
InSb	0.170	0.230	7.294
Sn	-	0.082	15.122 @ 0 K

Table of various semiconductors in order of increasing λ_{max}. From Sze (1981).

Various materials can be used and give quite different responses, as indicated in the table above. Photoconductive detectors are sometimes cooled to increase SNR by reducing noise, with the primary mechanism being reduction of intrinsic carriers. Note that E_g increases as temperature decreases, thus low-temperature operation causes the usable λ_{MAX} to fall (slightly).

Photoconductive detectors can be three orders of magnitude more sensitive than photovoltaic junctions and, under DC illumination, can be *one million times more sensitive* than simple photoemissive detectors (i.e., equal to photomultipliers). It is also important to note that the effective bandgaps shown in the table above are for high-purity "intrinsic" materials (in practice, there is no way to produce truly intrinsic semiconductors), and can be reduced using dopants (to form extrinsic photoconductors).

EXTRINSIC PHOTOCONDUCTORS

For operation at wavelengths beyond about 10 μm, it is possible to make use of another carrier generation mechanism than that available in most intrinsic photo-conductors (except narrow bandgap materials or semimetals). It should be noted, however, that an extrinsic photoconductor level in a wide bandgap host material is no better than a narrow bandgap material without a dopant. Impurities are added to the photoconductive materials to provide energy level(s) close to the conduction or valence bands. Incident photons cause electrons to be transferred between the valence band and the new energy level or the new energy level and the conduction band. An example of the effects of such doping is provided in the table below.

Material and Dopant	E_g or E_A (eV)	λ_{max} (μm)
Pure Ge	0.66 (E_g)	1.9
Ge with Hg	0.083 (E_A)	15
Ge with Zn	0.029 (E_A)	43

Examples of the effects of doping on energy levels and maximum wavelengths on Ge photoconductors. From Cobbold (1974).

For the intrinsic materials with small E_g the intrinsic carrier density at room temperature can be much greater than that from photogeneration. For extrinsics with small E_a, cooling is usually required to keep the impurity atoms from becoming thermally ionized. In some cases, cryogenic temperatures are used (liquid nitrogen or even helium). Cooling always reduces the noise and thus raises D^*. Cooling is an absolute requirement for many of these detectors. In many cases, their responses are severely deteriorated if operated above their nominal temperatures. In contrast, indirect thermal detectors (e.g., Golay cells or thermopiles, as discussed below) have less drastic decreases in performance if uncooled. If cooling is unavailable, thermal detectors are generally the best for $\lambda > 3$ μm. A wide variety of detector types and operating temperatures are compared for wavelengths of 1.5 to 40 μm in the figure below.

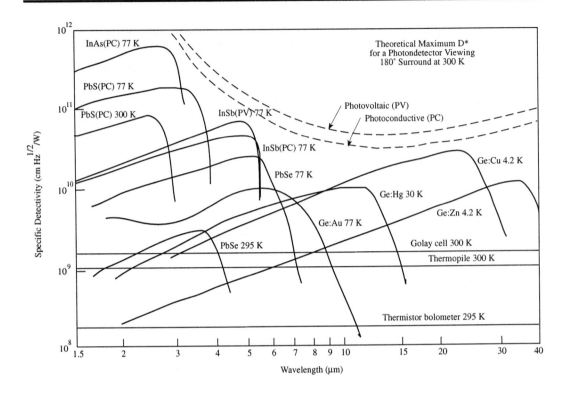

Plots of detectivities of various detectors, including Golay cells, thermopiles, and bolometers for comparison. After Cobbold (1974). Photovoltaic and photoconductive devices are identified as "PC" and "PV," respectively.

CADMIUM SULFIDE AND CADMIUM SELENIDE

CdS and CdSe are very useful in the visible light region. The most common photoconductive cell (lots of commercial applications) is CdS (sometimes CdSe). These are the sensors that are used to dim clock radio displays, dashboard lights, etc., as the ambient light falls. The spectral responses of CdS and CdSe can readily be adjusted via their composition (typically by mixing the two compounds). The responses of various CdS and CdSe compositions are illustrated below.

Typically, one can obtain 100:1 to 10,000:1 resistance variations with illumination going from dark to some maximum level. At low light levels, the trap states in CdS integrate charge (much like a charge-coupled device [CCD], discussed below) and provide excellent sensitivity and dynamic range up to the point where the states saturate. The linearity of CdS cells is quite good, but falls with increasing illumination (this is gain saturation). The temperature coefficient of CdS (and some other photoconductors) is a function of the specific material *and* the illumination level. Response times are on the order of seconds for low light and milliseconds for strong light. In addition, CdS cells exhibit an interesting hysteresis or "light history" effect wherein the conductance depends on the sensor's previous exposure to light and the duration of that exposure. This is caused by slow carrier emission from the long lifetime trap states that are responsible for the devices' high sensitivities at low light levels, as mentioned above.

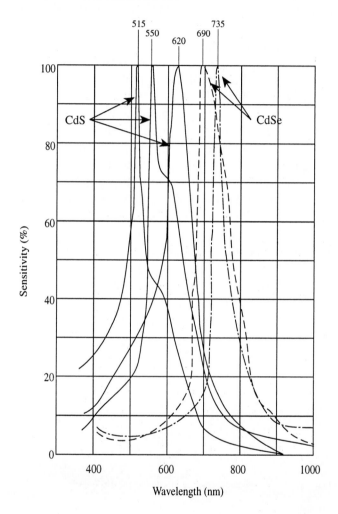

Spectral responses of various CdS and CdSe compositions. After Norton (1993).

Typical commercial CdS or CdSe cells use evaporated photoconductor patterns on ceramic substrates, often with hermetic packaging complete with an optical window. A water-based slurry of CdS can also be used to fabricate such devices (an electrolytic jet spray technique can be used, followed by a sintering step). Following sintering to form the CdS film, the metal electrodes can be deposited using evaporation or sputtering. This could be done on a silicon (or other) substrate without damaging or shorting out prefabricated on-chip circuitry if appropriate underlying insulating materials were deposited (and process temperatures kept low enough). In general, however, junction-based photosensors are used for monolithic detectors because they usually do not require *any* additional processing.

INTEGRATED CdSe PHOTOCONDUCTIVE SENSOR ARRAYS

Capon, et al. (1993) demonstrated a photolithographically fabricated 200 dot-per-inch (dpi) document image sensor (for facsimile machines) based on the fabrication of a one-dimensional array of CdSe photoconductors on glass. Their process began with a Corning 7059 glass substrate on which 150 nm TiW was sputtered (patterned using H_2O_2) to act as a metallization layer and also as a light shield. A 400 nm PECVD SiO_2 layer was then deposited, followed by a 50 nm Al_2O_3 and a 300 nm CdSe film (apparently evaporated) that was etched in 27% HCl. The upper level TiW metallization was then sputter-deposited and patterned by lift-off. The CdSe was then "activated" (thermally doped and recrystallized to increase sensitivity to a useful level) at 450°C in N_2 while in contact with doped CdS powder (CdS + 10 wt.% $CdCl_2$ + 10 wt.% Cu). The bond-pad interconnect level was fabricated using a sputtered TiW/Au patterned by lift-off, followed by the application of a 25 μm polyimide spacer layer. The 256-element arrays were connected to off-chip amplifiers and the performance was shown to be quite impressive (output signals were more than three orders of magnitude larger than comparable photodiode-based sensors).

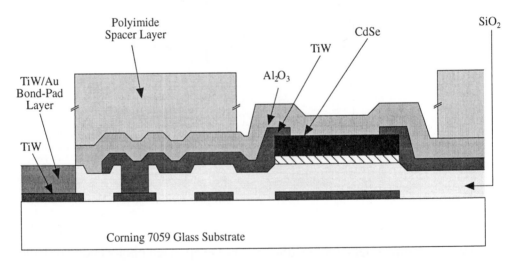

Cross-sectional structure of a microlithographically fabricated CdSe imager array on a glass substrate. After Capon, et al. (1993).

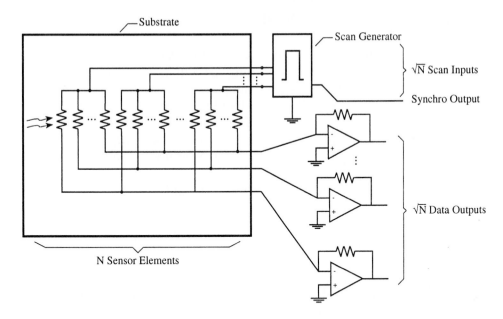

Block diagram of an integrated CdSe sensor array, showing a multiplexing scheme. After Capon, et al. (1993).

LEAD SULFIDE AND LEAD SELENIDE

PbS and PbSe are very useful as IR photodetectors. The spectral response of PbS at 300 K (room temperature) is 1 to 3 μm, and 1 to 4.5 μm at 77 K (liquid nitrogen). For PbSe, the 300 K response is 1 to 4.5 μm and 1 to 6.8 μm at 77 K. PbS or PbSe detectors are commercially available with thermoelectric coolers built into the same package. An interesting subject for research consideration is the feasibility of including thermoelectric cooling devices on a monolithic photoconductive detector.

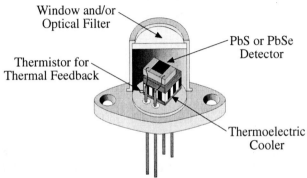

Illustration of PbS or PbSe detector with built-in thermoelectric cooler. After Cobbold (1974).

MERCURY CADMIUM TELLURIDE

Mercury cadmium telluride ("mer-cad-tel") $Hg_xCd_{(1-x)}Te$ is a stoichiometric semiconducting alloy. HgTe is a II-VI semimetal with a negative bandgap. CdTe is a II-VI semiconductor with a positive bandgap. By adjusting the alloy fraction of Hg and Cd, one can reach all of the bandgap range in between, including bandgaps as low as 0.1 to 0.2 eV, useful for long-wavelength IR detection. The spectral response of these alloys can be tuned over the range of 2 to 20 μm. $Hg_xCd_{(1-x)}Se$ has similar properties, but with a narrower useful wavelength range.

Mercury cadmium telluride is used for a large number of focal plane detector arrays due to its high performance as an IR detector (e.g., for heat-seeking missiles), despite the need to work with its toxic constituents. Typically, the $Hg_xCd_{(1-x)}Te$ elements are formed on a separate substrate and then bonded to a silicon substrate (usually "flip-chip" style, with solder bumps formed on one of the two substrates, aligned, and reflowed) that contains the necessary signal-processing circuitry. Large numbers of pixels can be addressed in this manner without a separate external connection to each one. Another two-chip approach, demonstrated by Wu (1978), used electroplated interconnects that were deliberately allowed to bridge the gap between a $Hg_xCd_{(1-x)}Te$ sensor array and a silicon signal-processing chip.

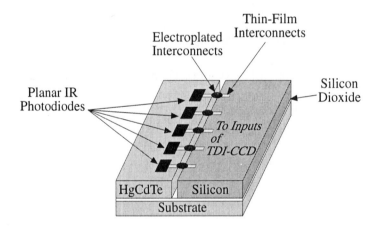

Gap-bridging electroplated connections between a $Hg_xCd_{(1-x)}Te$ sensor array and a silicon companion chip (CCD-based). After Wu (1978).

GENERAL PURPOSE SEMICONDUCTORS

Silicon germanium alloys are promising as photoconductors due to their good performance and the fact that they can be grown epitaxially on ordinary silicon substrates. As for all semiconducting alloys, the spectral response of SiGe can be tuned by varying the relative proportions of Si and Ge. These structures are typically fabricated using molecular beam epitaxy (MBE) methods. A good overview of this topic can be found in Kasper and Schäffler (1991). Silicon and GaAs can also be useful as photoconductors. The two are compared in Constant, et al. (1988b), with the focus on high-gain GaAs devices, and an example of their use in spectroscopic systems is presented in Constant, et al. (1988a).

METAL-SEMICONDUCTOR-METAL PHOTOCONDUCTIVE SENSORS

Metal-semiconductor-metal (MSM) photodetectors (Auston (1983)) consist of two metal lines configured to form a microstrip (typically 50 Ω characteristic impedance) over a semiconductor substrate, separated by a narrow gap (typically 1 to 10 μm). Impinging photon pulses can cause the exposed semiconductor to conduct briefly across the gap, transmitting a pulse from one microstrip to the other. Due to their coplanar structure, the gap capacitance is quite low, leading to relatively fast devices (switching times 100 ps to 1 ns). The structures can be improved by interdigitating the two electrodes (Slayman and Figueroa (1981)), and $\approx$ 50 ps response times have been obtained.

These devices can be built on GaAs to get better results than for silicon, and diamond film provides even better performance (lower dark current). With response times down to $\approx$ 2 ps in some cases, GaAs MSMs are the fastest semiconductor photodetectors available at present. Typically, high-power lasers are required to activate such sensors, since a large number of carriers must be generated to result in the large reduction in gap resistance generally hoped for (down to a few ohms).

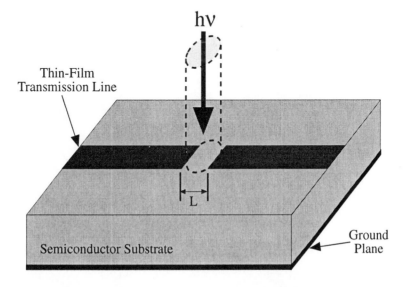

Illustration of the cross section of a metal-semiconductor-metal photodetector. An alternative geometry is to use a thin photoconductive film on an insulating substrate.

A key feature of MSM devices is that they only require a *single* lithographic step and can readily be integrated into a variety of processes. Other useful references, illustrating a variety of such devices, include Auston (1975), Lee (1977), Leonberger and Moulton (1979), Auston, et al. (1980), DeFonzo (1981), Degani, et al. (1981), Figueroa and Slayman (1981), and Darling (1990). These devices have also been integrated with monolithic amplifiers, such as a four-channel integrated detector described by Wada, et al. (1986), with 1.5 Gbit/s data rates and 110 V/W responsivity or a single-channel version with 2 Gbit/s data rate and 400 V/W responsivity presented by Hamaguchi, et al. (1987). Another interesting paper on integrated GaAs MSM detectors is Pohjonen and Andersson (1990).

2.2.6 JUNCTION-BASED PHOTODETECTORS

Junction-based photodetectors have a p-n junction in which the photogeneration of carriers occurs as it does for photoconductors. There is a high electric field across the depletion region (since this is where most of the voltage across the diode is dropped) that serves to separate generated electron-hole pairs (quasi-neutral regions in a detector are essentially loss regions).

The commonly referred to "photovoltaic effect" is the photocurrent generated within the depletion region charging up the open-circuited depletion capacitance. If the junction is shorted (i.e., current flows into a ground or virtual ground), the voltage is lost. Thus the photovoltaic effect is simply an external circuit-based effect.

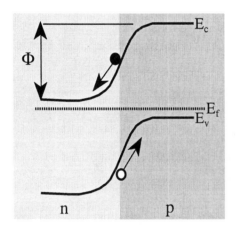 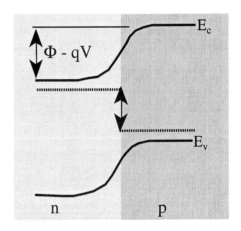

Generation of a voltage at a p-n junction by the creation of an electron-hole pair.

Incoming photons with greater than bandgap energy generate electron-hole pairs that can be swept away by the built-in electric field at the junction. Carriers generated in the neutral regions diffuse into the depletion region and are collected by the built-in field. An electron and hole can diffuse on average of one diffusion length (L_e and L_h, respectively) before they recombine. Thus most of the useful photogeneration occurs in the depletion region (or very nearby so that carriers can diffuse into it). The charge separation results in a potential difference V across the junction that tends to decrease the built-in potential barrier (i.e., forward-bias the junction).

The I-V characteristic of a photodiode is described by,

$$I = I_S \left(e^{\frac{qV}{kT}} - 1 \right) - I_{SC}$$

where I_s is the saturation current, consisting of carriers that can overcome the barrier, Φ, (illustrated in the figure above) with their own thermal energy, and I_{sc} is the short-circuit current, in A.

The bandgap (essentially the barrier height at degenerate doping levels) is the highest open circuit voltage that may be achieved. Illumination shifts the I-V characteristic for a photojunction downward by an amount I_{SC}, as illustrated below.

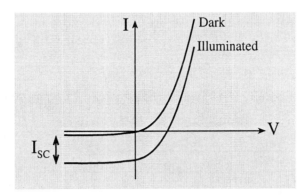

Illustration showing shift in photodiode I-V characteristic when illuminated.

The quantum efficiency of a photodiode is generally ≤ 1. If the photon energy is high enough to generate hot carriers that in turn excite secondary pairs by impact ionization, a quantum efficiency > 1 may be obtained. The dependence of quantum efficiency on photon energy is illustrated below for Ge.

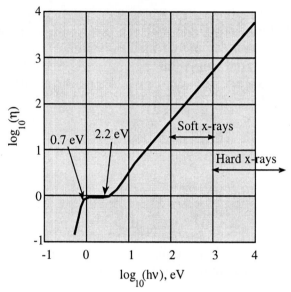

Dependence of quantum efficiency on incident photon energy in germanium. In Ge, the gap voltage is 0.7 eV. At 2.2 eV and above (the point at which the second conduction band is reached), the quantum yield exceeds 1.

PHOTODIODE DESIGN ISSUES

Depletion region width is a critical factor in the design and use of photojunction devices. The cases of wide and narrow depletion regions are compared below:

Wide depletion region:

• Higher quantum efficiency (η) since incoming photons are more likely to generate electron-hole pairs

• Lower junction capacitance, C_j

• Longer transit time, t_r (this slows response, but high reverse bias helps)

Narrow depletion region:

• Lower η

• Larger C_j (note that these devices are typically run with a significant reverse bias, minimizing this effect)

• Shorter t_r (helps to speed response, but increased capacitance may dominate)

• Larger dark currents due to higher dopant levels

Photodiode speed is primarily determined by three factors: 1) diffusion time of carriers generated outside the depletion region into it (reduced by making the depletion region close to the surface), 2) drift time in the diffusion region (reduced by making it only wide enough to absorb maximally but not too thin so that capacitance goes up), and 3) junction capacitance (reduced by strong reverse bias or intrinsic region). Photodiode designs have been refined over many years, and the details are beyond the scope of this book. However, a variety of designs are illustrated below to exemplify their wide variations.

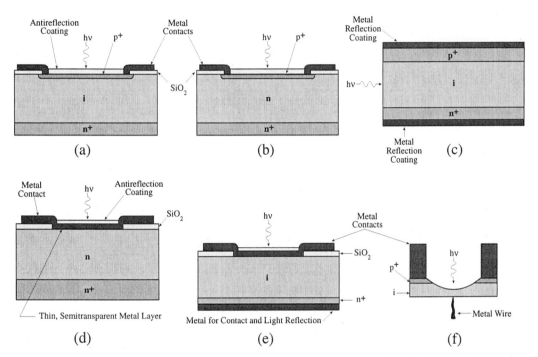

Some examples of junction photodiode designs: a) p-i-n diode, b) p-n diode, c) p-i-n diode with illumination parallel to junction, d) metal-semiconductor (Shottky barrier) diode, e) metal-i-n diode, and f) semiconductor point-contact diode. Adapted from Sze (1981).

The optimum trade-off between the three speed-limiting factors for maximum speed occurs when the depletion layer transit time is on the order of one-half of the modulation period (Sze (1981)). A sensitivity versus speed trade-off occurs if the depletion region is so large that transit times become a big issue, but this is seldom a problem in practice since each 1 μm of depletion region only adds ≈ 10 ps of transit time.

Junction-based photosensors can be operated in photoconductive or photovoltaic (unbiased) mode. As mentioned above, the diode is usually strongly reverse-biased when optimum response speed is required.

PHOTOVOLTAIC OPERATION

As mentioned above, in this mode of operation the photocurrent is allowed to charge up the junction capacitance, and thus a large-valued load resistance is used to allow the photovoltage to develop. Photovoltaic operation is generally not as fast as the photoconductive mode.

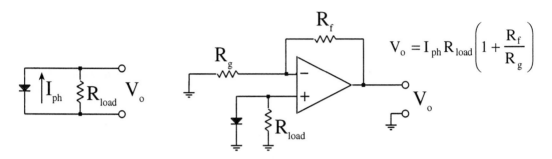

Example of photovoltaic-mode circuits. A photodiode is shown at the left connected to a large-valued resistive load, R_{load}. The photodetector output can be amplified and buffered using an operational amplifier configured as a noninverting voltage amplifier, as shown at the right.

PHOTOCONDUCTIVE OPERATION

In the photoconductive mode of operation, strong reverse bias (> 5 V) is typically used to widen the depletion region (thus lowering C_j) and to decrease the transit time, both of which increase speed. As noted earlier, this is not a different mechanism at work, just the photojunction operating into a low load resistance. With p-i-n (p-type/intrinsic/n-type) photodiodes, nanosecond response times are readily achievable. A typical circuit for photoconductive operation is shown below.

Minority carriers produced near the depletion region have a marked effect on this reverse current by adding to it. Carriers generated within one diffusion length of the depletion region will be pulled into it by the electric field and transported across it, adding to the collected current as if they were generated in the depletion region itself. This is not a major factor in p-i-n photodiodes, wherein the intrinsic region occupies most of the structure. In p-n junction photodiodes, without any explicitly created intrinsic regions, this extra carrier generation may constitute a sizable portion of the overall current detected.

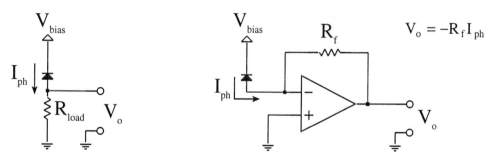

$$V_o = -R_f I_{ph}$$

Example of photoconductive-mode circuits. A photodiode is shown at the left connected to a small-valued resistive load, R_{load}. The photodetector output can be amplified and buffered using an op-amp in transresistance mode, as shown at the right, where the photodiode load is the virtual ground at the op-amp input.

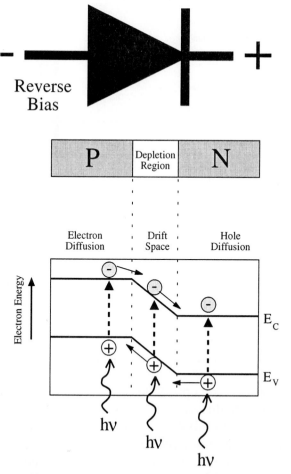

Diagram illustrating the operation of a photodiode under reverse bias (photoconductor mode). After Sze (1981).

PHOTODIODE SMALL-SIGNAL MODEL

For photojunction devices, the small-signal equivalent model is more complex than that for photoconductive devices. Photojunctions have significant capacitance effects due to the junction itself, unlike junctionless photoconductors. A basic small-signal circuit model for a photodiode (Sze (1981)) is shown below. The value of R_L essentially determines whether the photodiode is operated in photoconductive (low R_L) or photovoltaic (high R_L) mode.

As for photoconductive sensors, the small-signal frequency response of this circuit is a first-order low-pass. This low-pass transfer function means that the SNR decreases with increasing signal frequency. An important factor is that for the photodiode, the cutoff frequency can be modulated through C_j (which can be controlled via externally applied voltage bias).

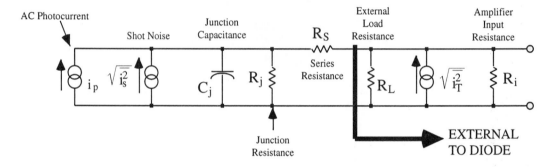

Small-signal circuit model of a photodiode, including external resistance effects from the load resistance, R_L, and the amplifier input resistance, R_i. External thermal noise is shown as i_T. After Sze (1981).

INTEGRATION OF PHOTODIODES WITH STANDARD ACTIVE CIRCUIT PROCESSES

Integrating photodiodes in an active circuit process (e.g., CMOS, bipolar, BiCMOS) is readily achievable, but with a key design compromise over discrete devices. High-speed integrated circuit processes use thin, highly doped regions and low supply voltages. Good optical detectors require thick, lightly doped regions and high voltages. Despite this, several groups have implemented integrated photodiode detectors (mainly in CMOS), typically referred to as *active pixel arrays*. There is great interest in these devices for video camera applications, since they are cheaper and can include more processing circuitry than the charge-coupled device (CCD) alternatives. The CMOS arrays also provide higher possible frame rates and random access capabilility (generally, CCDs are read out serially).

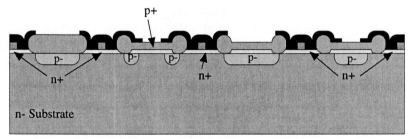

Illustration of integrated photojunctions in CMOS. After Kramer, et al. (1992).

Extremely wide dynamic range is achievable using weak inversion (sub-threshold) MOSFET loads for the photodiodes, providing a typical dynamic range of six to seven decades, compared to four to six decades for a CCD array. It is worth noting that the dynamic range of the subthreshold load/photodiode pixel is limited by that of the subthreshold circuitry, not the photodiode itself, which may operate over more than eight decades of range.

Ricquier and Dierickx (1994) demonstrated a 256×256 pixel CMOS imager using a 2.4 µm CMOS process with square pixel sizes of ≈ 26 µm per side. Their array included an 8-bit A/D converter to provide direct digital outputs. These devices had large variations in pixel gains (so-called *fixed-pattern noise* or FPN) due to variations in threshold voltages. There are ways to improve FPN due to threshold voltage variations through improved circuit design of off-chip correction.

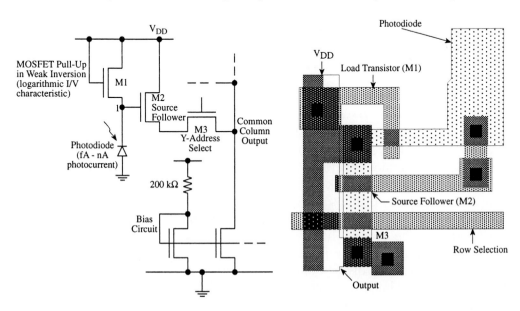

CMOS imager pixel architecture schematic (left) and layout (right). After Ricquier and Dierickx (1994).

More recent examples of active pixel arrays include those of Kramer, et al. (1992), Denyer, et al. (1993), Mendis, et al. (1994), Denyer (1995), Fossum (1995), Nakamura, et al. (1995), Nixon, et al. (1996), and Panicacci, et al. (1996). Many of these authors addressed the FPN problem and were seeking to approximate CCD performance with their CMOS imagers.

ULTRAVIOLET-OPTIMIZED PHOTODIODES

Photodiodes are generally operated in the visible and IR ranges, and they can be optimized for UV applications. An example of this can be found in Bolliger, et al. (1995) and Bolliger (1995), who discussed the use of a modified bipolar process to realize specialized UV photodiodes. Through the use of multiple shallow implants and an integral Fabry-Perot interference filter using multiple evaporated Si and SiO_2 layers, UV-selective photodiode structures were obtained. The reported sensitivity was 90 mA/W at 310 nm for a 1 mm^2 diode with a concentrating lens in the package.

P-I-N PHOTODIODES

The discussion of photodiode operation above applies equally well to the p-i-n diode, which is fabricated with an intrinsic (undoped) region between the p- and n-doped regions. The p-i-n structure allows much higher reverse bias to be applied, providing higher electric fields and shorter transit times (on the order of 10 ps/μm). The width of the intrinsic region can be optimized for the desired wavelength range and the required speed. These diodes are often selected for detector applications when a discrete device can be used and a high bandwith at a reasonable cost is desired. An additional benefit of the intrinsic region is that the leakage current is greatly reduced compared to standard junctions.

METAL-SEMICONDUCTOR (SCHOTTKY) PHOTODIODES

Metal-semiconductor photodiodes are simple structures fabricated by depositing a very thin metal layer ($\approx$ 10 nm) (and generally an antireflection coating, e.g., 50 nm ZnS) on a semiconductor, forming a Schottky diode. (For reference, it should be noted that in Schottky diodes, the metal contact takes the role of the missing doping polarity of the semiconductor. In other words, if a Schottky diode was formed with an n-type substrate, the metal contact would act as the "p-type" material and form the anode.) Illumination passes (in most cases) through the thin metal layer to interact with the junction and substrate (if able to penetrate). This general class of devices is divided into two functional groups: UV/visible detectors acting as conventional photocarrier generation diodes and IR detectors making use of a principle known as *internal photoemission*.

For the UV/visible applications, the metal-semiconductor diodes are treated as general purpose photodiodes in terms of circuit connections and operating conditions. As long as the incoming photons are able to penetrate the metal electrode and to be absorbed near the semiconductor surface (for UV the effective absorption lengths, $1/\alpha$, are very small, on the order of 0.1 µm), the photodiode can be used effectively in this mode. For these applications, the choice of metal is not particularly critical, and Au or Pt are common choices. These devices are capable of very high-speed operation. For example, Wang, et al. (1983) presented a GaAs Schottky-barrier photodiode with a 16 ps impulse response (20 GHz 3dB bandwidth) and Wang and Bloom (1983) demonstrated one with a 5.4 ps impulse response (100 GHz bandwidth).

If a contact metal is chosen that places the Fermi level at $\approx$ 0.3 to 0.4 eV, metal-semiconductor photodiodes can be operated in internal photoemission mode. These devices are typically made by depositing a thin Pt layer on Si and annealing it to form platinum silicide (Pd_2Si is also useful for the 1 to 3 µm band). The internal photoemission process, described in Stotlar (1994) leads to the generation of electrons or holes only (analogous to vacuum photoemission), which can cross the barrier into the semiconductor. These devices are often fabricated in imaging arrays and, when operated at 77 K, they can produce IR images in the range of 1 to 5 µm (3 to 5 µm for PtSi).

An example PtSi imaging array, with an underlying charge-coupled imager structure (discussed below), was presented by Kosonocky, et al. (1985) (this paper also contains a review and comparison of Schottky-barrier IR focal plane arrays up to that date). They fabricated an array of 160×244 pixels with dimensions of 80×40 µm in a die with overall dimensions of 1.5×1.2 cm. They found that the imager's response was maximized for PtSi thicknesses between 1.5 and 2.0 nm and a SiO_2 antireflection coating (on the back side of the substrate) thicknesses on the order of 550 nm. An Al mirror was deposited on the front side of the photodiodes to reflect incident IR, applied through the back of the substrate, back through the junctions. When operated at 77 K (cooled using liquid N_2), they achieved 30 frames per second operation (compatible with television rates) at a noise level equivalent to a temperature difference of < 0.1 K.

AVALANCHE PHOTODIODES

Avalanche photodiodes (APDs) are used where the inherently high gain of avalanche processes is required. They are operated with large reverse bias voltages (typically 80 to 100 V, but can rarely be thousands of volts) and *avalanche multiplication* of photocurrents takes place, providing internal current gain. The current gain is obtained when photogenerated (or otherwise sourced) carriers gain enough energy from the electric field to generate secondary carriers through impact ionization of valence electrons into the conduction band (creating free holes in the valence band). These secondary carriers are also accelerated by the electric field and participate in further impact ionizations, providing gain.

As described in Stillman and Wolfe (1977), the output current is thus the primary photocurrent times the avalanche multiplication factor, M, which is dependent on the bias voltage. The impact ionization coefficients of electrons and holes, $\alpha_n(E)$ and $\beta_p(E)$ (in cm^{-1}), respectively, represent the reciprocal of the average distance that a carrier will travel (for a given electric field) before generating an additional electron-hole pair through impact ionization (the impact ionization process can occur for both carrier types). It turns out that for high performance APDs, it is desirable to have one of the two coefficients equal to zero, and for a depletion region width, W, the avalanche multiplication factor for electrons is then given (with $\beta_p = 0$) by,

$$M_n = e^{\left(\int_{x=0}^{x=W} \alpha_n dx\right)} = e^{(\alpha_n W)}$$

The time course of a current pulse resulting from a single electron is equivalent to the electron's inherent transit time through the depletion region, τ_n, plus the hole transit time, τ_p, making it roughly twice the duration with avalanche gain as without, and is *independent of the degree of multiplication* (100 GHz bandwidths are feasible for certain APDs). Thus (for the case when one of the impact ionization coefficients is equal to zero) there is no gain-bandwidth product limitation for avalanche multiplication. The reason that one of the coefficients should ideally be zero is illustrated by the case where they are equal. While the gain can be very high in this case, the pulse width resulting from a single carrier becomes very long (infinitely long when avalanche *breakdown* occurs), representing a gain-bandwidth relationship (in addition, noise is greater for cases where one coefficient is non-zero). In practice, neither ionization coefficient is zero, and the avalanche multiplication factor is given by,

$$M = \frac{\left[1 - \left(\frac{\beta_p}{\alpha_n}\right)\right] e^{\left\{\alpha_n W \left[1 - \left(\frac{\beta_p}{\alpha_n}\right)\right]\right\}}}{1 - \left(\frac{\beta_p}{\alpha_n}\right) e^{\left\{\alpha_n W \left[1 - \left(\frac{\beta_p}{\alpha_n}\right)\right]\right\}}}$$

The gain-bandwidth product for values of $M > \alpha_n/\beta_p$ and high-frequency operation is maximized for small depletion region widths, W (in cm), and high electron velocities, and is given by (not considering parasitic RC effects),

$$M(\omega)\omega = \cfrac{1}{N\tau\left(\cfrac{\beta_p}{\alpha_n}\right)} = \cfrac{1}{N\left(\cfrac{W}{v_n}\right)\left(\cfrac{\beta_p}{\alpha_n}\right)}$$

where N is the empirical coefficient varying between 1/3 for $\alpha_n/\beta_p = 1$ and 2 for $\alpha_n/\beta_p = 10^{-3}$, and v_n is the electron velocity, in cm/s.

For silicon devices, α_n is typically ten or more times greater than β_p, contributing to good noise performance and high achievable bandwidth. In germanium, on the other hand, the coefficients are very close, making it a poor material for APDs, but still it is often used in the 1.0 to 1.6 μm range for optical communications.

The high reverse bias used with APDs requires very uniform doping and sometimes elaborate doping profiles for "perfect" junction edges (to prevent breakdown, and also to ensure that the field-dependent ionization coefficients do not exhibit local variations).

Quantum efficiencies in APDs can be > 90% at the peak wavelength of response, so the primary photocurrent generation mechanism is highly efficient. Unfortunately, avalanche gain comes along with noise that is much worse than for photomultipliers, since non-photogenerated carriers are also subject to avalanche multiplication.

There is a wide variety of APD designs, largely motivated by lightwave communications applications. An APD subtype of note is the SAGM, or separate absorption gain multiplication type, wherein photocarrier generation and gain regions of the devices are physically separate. Thorough overviews of avalanche photodiode technologies, including underlying principles and device structures, can be found in Stillman and Wolfe (1977) and Kaneda (1985).

PHOTOTRANSISTORS

The bipolar phototransistor is essentially an ordinary bipolar junction transistor (BJT) with a large base-collector junction as a light-collecting element. A base terminal is sometimes provided, but seldom used. Base current, $I_B = I_{ph}$, is optically generated and forward biases the base-emitter junction. The base current is multiplied by the transistor's β.

A key trade-off for bipolar phototransistors is that designing for high gain and high speed reduces the efficiency of optical collection. Thus such devices are generally used for switching applications and have poor optical sensitivity compared to photodiodes.

PhotoFETs have also been demonstrated in a variety of materials, yet have not been used in high volume commercially. Examples include a photo-JFET (Bandy and Linvill (1973) and GaAs photo-MESFETs (Edwards (1980) and Bethea, et al. (1983) (the latter achieving 12 ps rise times).

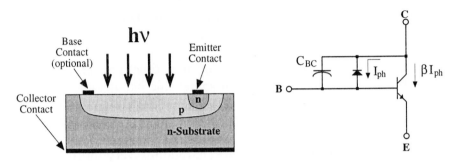

Illustration of the cross section (left) and equivalent circuit (right) of a bipolar phototransistor.

PHOTO-DARLINGTON TRANSISTORS

A photo-Darlington transistor is simply a bipolar phototransistor driving second, "ordinary" BJT in a Darlington configuration. In this case, the photocurrent is multiplied by both transistors' β values, and if they are the same, the net photocurrent gain is $\approx \beta^2$. Photo-Darlingtons are even slower than simple bipolar phototransistors.

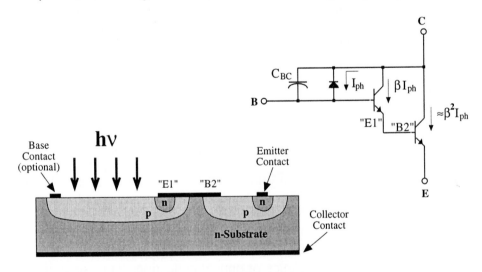

Illustration of the cross section (left) and equivalent circuit (right) of a bipolar photo-Darlington transistor.

SOLAR CELLS

Solar cells are essentially "giant" photodiodes operated in the photovoltaic mode, generating sufficient photocurrent to produce useful output power. The important parameters in a solar cell are the "conversion efficiency" (fraction of incident optical power available at its output terminals) and the "power output" (magnitude of the output power). The limiting factor is the optical spectrum of the sun (note that this is quite different at the Earth's surface versus in space). A lower bandgap intercepts a wider portion of the solar spectrum but results in a lower open circuit voltage and a larger saturation current.

Optimization is key for solar cells. For space applications where there is no atmospheric absorption, the optimum bandgap is 1.6 eV. For higher atmospheric absorption, the bandgap shifts down to about 1.4 eV (near that of GaAs and CdTe). Another important parameter is minority carrier lifetime, τ. A larger lifetime implies that more carriers can be collected before they recombine.

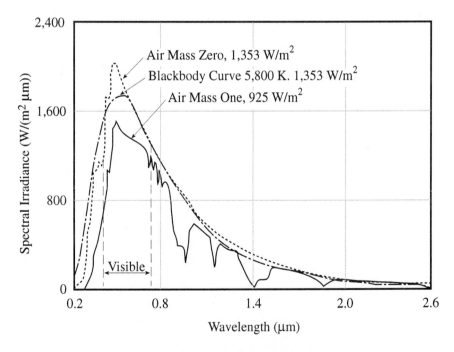

Plot of the solar power spectrum, showing that the solar blackbody radiation curve (the effective surface temperature of the sun is 5800 K) is altered by the absorption of the atmosphere. After Sze (1981).

Important applications for solar cells are in space hardware, alternative energy sources and consumer goods ("solar-powered calculators," etc.), or distributed sensor systems where direct power supplies are not feasible (and where batteries need to be supplemented or recharged). Solar radiation at the Earth's distance from the sun is 1,353 W/m^2 (roughly the same power into 1 m^2 as the maximum available from a normal 15 ampere (A) household outlet, around 1,700 W). Assuming optimum loading conditions, the theoretical maximum efficiency of a solar cell is 31%, assuming E_g = 1.35 eV (III-V semiconductors), making $\approx$ 419 W available per square meter of cell area.

There are a multitude of solar cell types and much design effort has been expended to optimize them for power output, cost, etc. Solar cell types include single-crystal, polycrystalline, thin-film (polycrystalline), etc. Since most of these devices do not rely on micromachining, the focus here will be on an example that does, and shows the benefits of the approach.

Cuevas, et al. (1990) developed a family of extremely efficient solar cells that make use of micromachining technology to enhance their performance. Micromachining techniques greatly increased their efficiency by recovering light reflected by the necessary metal lines on the surface. Pyramidal pits were etched in arrays on the collecting surface. Pure ammonium hydroxide was used as the anisotropic etchant. This was followed by a mild isotropic etch (HNO$_3$:HF at 100:1 for 1 min) used to "sharpen" the crests of the ridges. (It should be noted that *random* pyramidal surface texturing can also be used to increase efficiency.) These micromachined solar cells are commercially available from Sunpower, Inc., (Sunnyvale, CA).

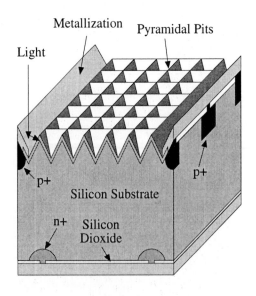

Illustration of a micromachined solar cell. After Cuevas, et al. (1990).

ANOMALOUS PHOTOVOLTAIC EFFECT

A high-voltage photovoltaic effect is observed in some thin-layer semiconductors. The voltage can be many times larger than the bandgap potential. Little is
understood about this phenomenon, and its magnitude depends on the way the film
growth was performed. It has been observed in GaAs, PbS, CdTe, Si, and Ge. A
voltage of up to 500 V can develop in CdTe at an illuminance of 0.1 lux. Voltages
of up to 100 V have been observed in thin-film Si structures. The simplest model
for the anomalous photovoltaic effect uses the concept of summation of photovoltages
generated in an array of "microcells" defined by microscopic grain boundaries. It
would certainly be of interest to explore micromachining structures that deliberately
make use of this effect.

2.2.7 CAPACITIVE PHOTOSENSORS

Probably the simplest and commonest photodetector (used in camcorders
and other consumer applications) is the *metal-insulator-semiconductor* (MIS) device,
otherwise referred to as a charge-coupled-device (CCD) cell. The basic structure,
as shown below, is a metal gate above a dielectric and a semiconductor substrate
below. This forms a MOS capacitor, the charge on which arises from photogenerated
carriers. Generally, the gate oxide for MIS photodetectors is made as thin as
possible.

A bias is applied to place the underlying semiconductor in an inversion state
(i.e., positive gate voltage for p-type substrate) so that electron-hole pairs generated
by incident photons will be separated. The accumulated charge can be shifted
between potential wells created below the MIS structures, forming a basis for the
CCD arrays used for imaging. A key point is that this charge transfer has an
extremely high efficiency (> 99.99% in typical devices). It should be noted that
this direct charge transfer is not comparable to switched capacitor charge transfer
(closing a switch between two capacitors and allowing charge to move until equilibrium is reached), which has a maximum efficiency of 50%.

CHARGE-COUPLED IMAGE SENSORS

CCD arrays are, as mentioned above, arrays of MIS sensors arranged so that
photogenerated charge can be stored and transferred between elements by an appropriate variation of control voltages applied to surface electrodes. The CCD was
developed as a way to increase the density of photosensor arrays that could be
integrated for imaging purposes and as a memory/signal processing approach. The
former is now by far its dominant use.

MIS structures are arranged in groups (usually of three if older metal-gate
technology is used) that are electrically interconnected and addressed together in a

way that creates localized potential wells that can be shifted in one direction, moving the photogenerated charge with them. While a three-phase system guarantees unidirectional charge transfer, this can also be guaranteed using a two-phase system with overlapping gates (i.e., in a two-level polysilicon process, as illustrated below), reducing the CCD cell size further than for a three-phase system.

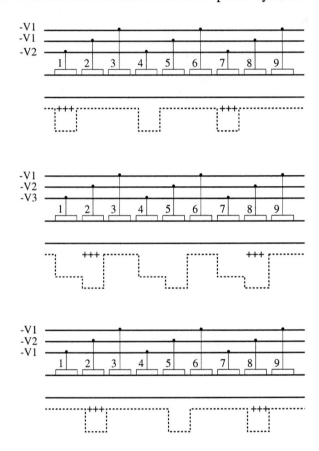

Illustration of operating mode of (older style) a three-phase CCD array showing the mechanism of the charge transfer between depleted MOS capacitors. Adapted from Millman (1979).

To manufacture color CCDs (such as those in camcorders), filters (red, green, and blue, or cyan, magenta, and yellow are used) are applied photolithographically directly to the sensor areas using dyed photoresist (an older, dye transfer approach that may be of interest for micromachining applications such as optical and chemical sensors is discussed in Dillon, et al. (1978b). More "green" pixels are typically needed to compensate for the response of the sensor and the eye (see Dillon, et al. (1978a)). If one needs better UV and blue response, the substrate is often thinned and illuminated from the back.

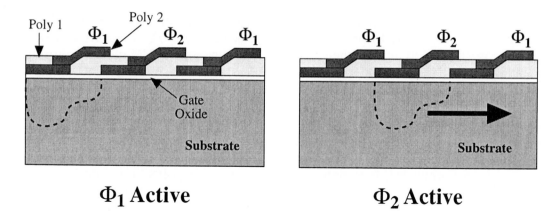

Illustration of the cross section of a two-phase CCD structure with overlapping polysilicon gates. Courtesy R. B. Darling.

A key parameter in CCD sensors is the charge-transfer efficiency, which is the fraction of the total charge in an MIS element that is successfully transferred on each clock cycle. While very large at low frequency, the efficiency falls off at high clock rates (this is the fundamental limit to CCD frame rates).

There are several possible CCD structures, and two common ones are discussed herein. Surface-channel CCD (SCCD) is an older technology in which the basic MIS structure is used, and charge is transferred along the surface channel. Charge-handling capacity varies linearly with bias voltage. Surface states and/or interface traps are the major limitations on charge-transfer efficiency for this type of design. Buried-channel CCD (BCCD) devices are based on majority carrier transport in a bulk layer buried below the surface. Charge-handling capacity varies nonlinearly with bias voltage and much higher charge-transfer rates can be achieved than SCCDs. This technology is available on a foundry basis through MOSIS (ISI, Inc., Marina del Rey, CA). CCDs also do not need to be silicon-based. For example, Kellner, et al. (1980) presented a two-phase, Schottky-barrier GaAs CCD imager, or Deyhimy, et al. (1980), whose GaAs CCD could be clocked at 500 MHz. In addition, CCDs do not need to be rectangular arrays of pixels, and "foveated" circular arrays that mimic the layout of the eye's retina have also been demonstrated (see, e.g., Debusschere, et al. (1990)).

MOS-CAPACITOR ULTRAVIOLET SENSORS

Researchers in biology and physics have long been interested in UV detectors, and due to increasing awareness of the negative health effects of UV overexposure, many people are interested in consumer applications. In the past, specialized processes were generally used to fabricate such sensors (often specially optimized photodiodes). Recently, some simple, direct UV detectors have been realized using conventional CMOS processes. The peak range of UV for DNA and skin damage is 280 to 320 nm, and hence a detector for consumer UV dose monitoring should be sensitive in that range.

Kerns (1993) demonstrated such structures fabricated using foundry CMOS (MOSIS). The basic principle of these devices is that incident UV photons can provide sufficient energy to allow electrons to cross an oxide layer between two polysilicon capacitor plates. This structure is basically a low-quality MOS capacitor, limited in performance due to defect levels at the oxide-polysilicon interface preventing MOS inversion or depletion. If a bias exists across such a capacitor, there will be a DC current that is proportional to the UV flux. This is similar to the Fowler-Nordheim tunneling mechanism by which EPROMs (erasable programmable ROM) are erased using UV light.

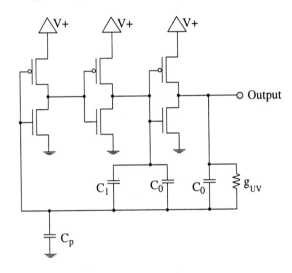

Simplified schematic of a UV-dependent relaxation oscillator. The frequency of oscillation depends on the UV-modulated conductance, g_{UV}. After Kerns (1993).

An important point raised by Kerns (1993) is that since the overlying polysilicon layer is thick enough (in a typical CMOS process) to be opaque to UV, it is the *edges* of the top poly layer where the transduction occurs (hence capacitors with large edge-to-area ratios were used). This is a common design approach for photo-detectors wherein these effects operate. The UV-sensitive capacitors, combined with shielded but matched counterparts (the second metal level was used to shield regions of the circuit where UV response was not desired), were integrated into an oscillator with a frequency proportional to the incident UV flux. The prototype device functioned over a three-decade range of UV intensities, 5 μW/cm^2 to 4 mW/cm^2.

2.3 INDIRECT OPTICAL SENSORS

As explained above, a class of optical sensors exists wherein there are two or more stages of transduction before electrical carriers (the final output signal) are generated. IR light, particularly the longer wavelengths, is difficult to detect directly with silicon, but silicon-based, micromachined, indirect optical sensors can do this very well, by converting the incident light into heat, which is then measured. The markets for such devices are varied and significant: surveillance, military, security, medicine, consumer, etc., with the potential for high-volume production due to the ability to deliver high performance, including a nominally flat wavelength response, at low cost. Generally speaking, all that is required for a good thermal optical detector is a temperature sensor with a low specific heat. Given the huge demand for such sensors, nearly every possible approach to this has been explored (e.g., thermistors, pyroelectrics, thermocouples, thermal expansion, etc.). This section covers some example devices that have resulted from this exploration.

2.3.1 PYROELECTRIC DETECTORS

Certain materials, referred to as *pyroelectrics*, with a non-centrosymmetric structure exhibit spontaneous electrical polarization that would give rise to a surface charge if it were not neutralized by external leakage current. Pyroelectric detectors are best thought of as capacitors whose charge can be altered by illumination or temperature change. The degree of polarization is temperature-dependent, so temperature changes give rise to surface charge changes. Looking at it another way, they are piezoelectric sensors where the signal-generating force arises from thermal expansion. This effect can be very pronounced. Pyroelectric detectors tend to have flat responses versus wavelength but lower sensitivity than direct detectors.

Since surface leakage cannot be made zero, these temperature-induced charge changes *do not provide a DC response*. Under AC illumination conditions, however, the temperature changes, and hence a voltage will appear across a piece of such material. Naturally, because these detectors are sensitive to temperature, they need to be temperature-compensated, and many include (for each sensing element) a reference element that is not exposed to the incident radiation.

Their physical construction is also very much like that of capacitors, that is, two plates on opposite sides of a slab of pyroelectric material. The surface charge induced by a temperature change is generally measured by a "charge-sensitive amplifier" or I-V converter (a transresistance amplifier).

Typically, materials used for pyroelectric materials are also piezoelectric and *ferroelectric* (having spontaneous electrical polarization that is adjustable via externally applied electric fields; i.e., their dielectric constants vary with applied voltage), and may include barium titanate, triglycine sulfate, polyvinylidene fluoride, polyvinyl fluoride, lithium tantalate ($LiTaO_4$), PLZT (lanthanum-doped lead zirconate tantalate), and several others. If one carries out an AC analysis of such a detector's performance, it turns out that for high responsivity, one needs to choose materials that have a large ratio of pyroelectric coefficient to dielectric constant. It is important to note that thermal (rather than electrical) time constants tend to determine the speed of such detectors, and thus the specific heat of the chosen material is critical, as is the overall thermal capacity of the detector. In addition, materials with large thermal expansion coefficients deliver the greatest sensitivities.

In some designs, an optical "chopper" can be used to produce an alternating temperature change, ΔT, (when DC temperature measurements are desired) at the sensor, this gives rise to an alternating charge on the electrodes,

$$\Delta Q = pqA\,\Delta T$$

where p is the pyroelectric coefficient in $\mu C/(cm^2 \bullet K)$, and A is the detector area over which radiation is absorbed, in cm^2.

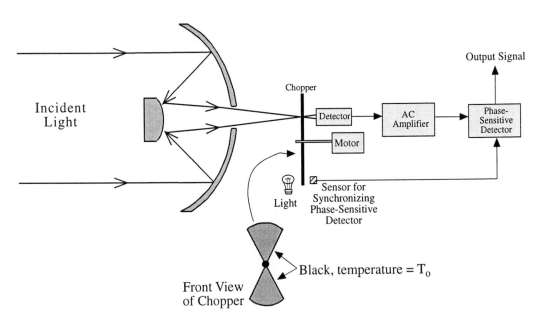

Example of an optical chopper system for using indirect optical sensors where temperature offsets need to be canceled. This ensures that the transducer sees the impinging radiation rapidly changing from ambient (reference) to the measured level. A lock-in, or synchronous detection scheme, is used to remove the background (reference) signal. After Cobbold (1974). This type of approach is used in many industrial and medical non-contact thermometers, such as those used to estimate temperature via IR emissions.

The resultant photocurrent is proportional to the rate of change of temperature,

$$i_p = pqA \frac{d(\Delta T)}{dt}$$

Some macroscopic pyroelectric detectors have excellent characteristics: response from far UV to far IR (often limited to IR by means of a wavelength-selective window [filter]), rise times that can be in the nanosecond range, and detectivities in the range of $D^* \approx 10^8$ cm•Hz$^{1/2}$/W. Classic references on conventionally fabricated pyroelectric detectors include Putley (1970, 1977) and Marshall (1978). Example values of pyroelectric coefficients (typically on the order of 0.001 to 0.07 μC/(cm^2•K) at 300 K) can be found in Stotlar (1994).

Several pyroelectric materials can readily be harnessed in micromachined devices, as seen in the examples discussed below. One example is the work of Polla, et al. (1986) wherein ZnO (1 μm thick) was deposited by RF-magnetron sputtering and etched with acetic acid:phosphoric acid:water (1:1:30). Since the substrate temperature can be kept below 250°C, it is relatively easy to integrate this type of pyroelectric film with MOS circuits (as Polla did).

Polla's process began with fabrication of poly-gate NMOS circuits processed to the step just before metallization is applied and encapsulated in 400 nm of phosphorous-doped SiO_2 (p-glass). This was followed by deposition and patterning of ZnO and encapsulation of the ZnO with more p-glass (400 nm) and etching of contact pads for transistor metallization. At this point, specific sensor metals (Pt, SnO, or Cr) were deposited using lift-off. Then EDP etching (from the back side) was carried out to form 25 μm thick membranes to thermally isolate the pyroelectric sensors.

Note that since ZnO is basically a temperature sensor (typical pyroelectric coefficients for ZnO are in the range of 0.001 to 0.002 μC/(cm^2•K) at 300 K), it can be used for many things other than for measuring IR light (Polla used it for an anemometer, an IR detector array, a chemical reaction sensor, and a CCD structure on the same chip). The piezoelectric properties of ZnO were also exploited to make a capacitive tactile sensor, a SAW chemical vapor sensor, and a microbeam accelerometer (also on the same chip). Using ZnO as a pyroelectric IR detector, Polla, et al. achieved $R_v = 4.3 \times 10^4$ V/W, a time constant of 0.76 s, and $D^* = 3.1 \times 10^7$ cm•Hz$^{1/2}$/W.

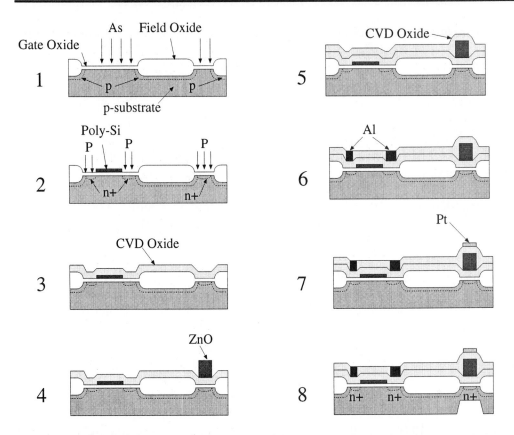

Illustration of the multisensor pyroelectric process used by Polla, et al. (1986): 1) initial oxidation, channel-stop implantation, LOCOS, and threshold implantations; 2) poly-Si deposition and patterning, self-aligned source, and drain ion implant; 3) CVD SiO$_2$ encapsulation; 4) ZnO deposition and patterning; 5) CVD SiO$_2$ encapsulation; 6) Al-Si sputtering; 7) sensor metallizations (Pt, SnO or Cr); and 8) EDP anisotropic backside etching. Adapted from Polla, et al. (1986).

ELECTRO-SPRAYED PVDF PYROELECTRIC SENSOR

As mentioned above, PVDF (polyvinylidene fluoride) is a very useful material for pyroelectric detectors, and can be patterned with oxygen RIE without reducing the pyroelectric coefficient of the remaining film. However, PVDF, like most piezoelectric/pyroelectric materials, requires "poling." This is carried out by heating or stretching the material while applying a high electric field to orient the dipoles within the film. Integrated pyroelectric sensor approaches have included gluing pre-poled sheets of PVDF to silicon substrates, spinning on copolymers such as P(VDF-TrFE) (see Lee, et al. (1989)), and then trying to pole it on the substrate. On-chip poling is often unsuccessful due to the relatively high voltages/temperatures required, tending to lead to dielectric breakdown of the materials being poled.

As an alternative, Asahi, et al. (1993) demonstrated an "electro-spray" method, wherein the PVDF solution is discharged through a needle held at 8 to 15 kV relative to a wafer. The droplets are transported 1 to 2.5 cm through a nitrogen flow that removes most of the solvent. The electro-sprayed PVDF droplets land on the substrate in the poled orientation, giving rise to a high pyroelectric constant without a poling step (as long as total current, monitored at the grounded chuck, stayed above a critical value). Further photolithography steps required baking photoresist at up to 120°C, potentially reversing some of the poling, but this only caused a 15% loss in pyroelectric coefficient.

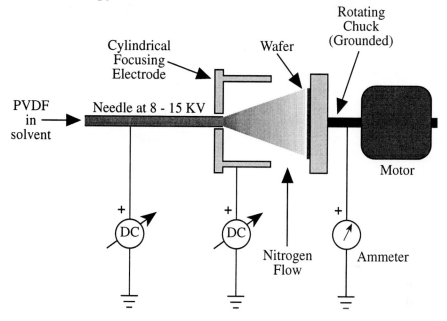

Illustration of the apparatus for depositing electro-sprayed PVDF that is effectively poled while it is being deposited. After Asahi, et al. (1993).

Unfortunately, many process details were not given in the paper, but a general summary is provided here. The PVDF (1 to 2 µm) was deposited atop large aluminum electrodes under which thermal isolation cavities were undercut (etchant was not disclosed; it probably was TMAH). Then an Au-black IR absorbing layer was deposited above the active regions of the PVDF. To prevent electrostatic-discharge damage to the input MOSFETS of the on-chip (nMOS) amplifiers during the electro-spray process, diode-connected MOSFETS (presumably tied to ground) were used to shunt potentials higher than roughly V_T. The reported performance included $R_v = 125$ V/W, $D^* = 1.4 \times 10^7$ cm•Hz$^{1/2}$/W, NEP = 2.9×10^9 W/Hz$^{1/2}$, and a thermal time constant of 1.3 ms.

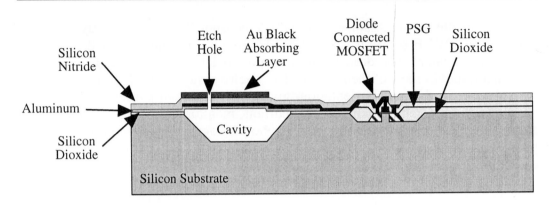

Illustration of a cross section of a completed PVDF-on-MOS optical sensor. After Asahi, et al. (1993).

The electro-sprayed pre-poled pyroelectric approach has many potential applications, including two-dimensional active IR imagers that are enabled by the relative ease of combining such detectors with active integrated circuitry.

2.3.2 BOLOMETERS

This class of devices is essentially a circuit consisting of two (or more) thermally sensitive resistors with one of them shielded from the incident radiation to serve as a reference. Typically, the (closely matched) sensors are connected in opposite arms of a Wheatstone bridge circuit, as shown below. A macroscopic design might use semiconductor- (e.g., Si or InSb) or thermistor- (e.g., oxide of Mn, Co, Ni, etc., with a negative temperature coefficient, or positive TC types) type detectors. A wavelength-selective filter window (e.g., a Ge window for 2 to 20 μm response) is usually included as part of the packaging, and responsivities on the order of $R_v \approx 300$ V/W are generally obtained (note that the actual response is a change in resistance, so R_v figures assume a particular circuit configuration). Without an input filter, bolometers have a virtually flat response versus wavelength (like all thermal optical sensors) and do have a DC response.

Bolometers exhibit voltage noise due to the Johnson (thermal) noise of the resistors given by (for a particular resistor),

$$v_n = \sqrt{4kTRB}$$

where,
T = temperature, in K
R = value of a particular resistor
B = bandwidth, in Hz

Energy fluctuation due to heat conduction through the support gives rise to a thermal "Johnson-like" noise, the power of which is given by,

$$P_n = \sqrt{4kT^2GB}$$

where G is the thermal conductivity through supports of the bolometer. Well-designed bolometers are limited by both of these noise sources.

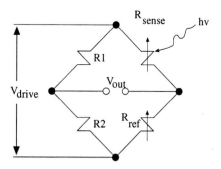

Bolometric Detection

Illustration of a typical Wheatstone bridge circuit used within a bolometer. Thermally sensitive resistors are shown on the right, with one exposed to incident optical energy and the other shielded from it.

A key specification for the bolometer design is the thermal coefficient of resistance for the sensor material, for a bolometer resistor value, R,

$$\alpha \equiv \frac{1}{R}\frac{dR}{dT} \text{ in K}^{-1}$$

which can be used to determine R_V for a given circuit configuration (generally R must be specified as well).

The responsivity of a bolometer is given by,

$$R_V = \frac{\eta I R \alpha}{G_{th}\sqrt{1+\omega^2\tau^2}}$$

where,
   η = optical absorptivity of the sensing region (not quantum efficiency)
   I = current used to interrogate the bolometer resistor, in A
G_{th} = thermal conductance to the supporting structure (i.e., substrate), in W/K

ω = angular frequency of input light modulation, in rad/s

τ = thermal time constant, in s, of the bolometer structure = C/G_{th}

UNCOOLED BOLOMETERS

Micromachined bolometers are readily achievable, and can offer extremely good performance due to the ability to fabricate very well thermally isolated microstructures. With the exception of superconducting bolometers, discussed below, the vast majority are uncooled devices. For example, Wood, et al. (1992) demonstrated uncooled, high-pixel-count, active IR focal plane arrays based on thin-film microbolometers. The thermoresistive elements were formed from a 50 nm semiconducting layer of non-stoichiometric vanadium oxide (VO_x) with a TCR of 20,000 ppm/°C, yet having low 1/f noise (a property usually found in metallic thin-films, with far lower TCR values).

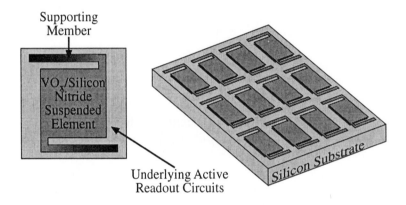

Conceptual illustration of micromachined bolometers developed by Wood, et al. (1992), showing a single pixel at the left and an array of bolometers at the right.

They thermally isolated the thermoresistive material by undercutting thin silicon nitride islands supported by narrow silicon nitride supporting members, and achieved a per pixel thermal conductance of 1×10^{-7} W/K and a thermal capacity of 10^{-9} J/K, corresponding to a thermal time constant of 10 ms. They reported a responsivity of 7×10^4 V/W (tested with 300 K blackbody radiation). A 336 × 240 pixel array was demonstrated with a 48% fill-factor for 50 μm pixels. Other examples of actively scanned, micromachined bolometer arrays have been presented by Oliver, et al. (1995) and Tanaka, et al. (1995).

Ringh, et al. (1995) demonstrated a bolometer array where the resistance of the individual thin-film amorphous silicon bolometer elements form part of an RC oscillator for each pixel, generating an output frequency proportional to optical power input. The operational principle is to count the number of oscillator output pulses since the beginning of an imaging frame and then very accurately (with a much higher frequency clock) determine the phase of the last rising edge in the imaging frame. This provided extremely high accuracy frequency/phase measurements. Their prototype used separate amorphous silicon thin-film bolometers (see Liddiard, et al. (1994)) and CMOS RC oscillators and reported an α of 0.02 K^{-1}, a resistance in the range of 1 to 2 MΩ, a D* of 1 to 3 $\times$ 10^8 cm•Hz$^{1/2}$/W, and a thermal time constant of 0.4 ms. Using this approach of optical-to-frequency conversion (a simple form of analog-to-digital conversion), it is anticipated that a high performance N $\times$ N array of CMOS bolometer IR sensors could be integrated that would require only very simple off-chip circuitry for readout.

Several other groups are actively pursuing similar uncooled micromachined bolometer devices that are now capable of offering performance comparable to cryogenically cooled HgCdTe, InSb, AlGaAs/GaAs, and PtSi detectors that are considerably more expensive and more difficult to integrate with active circuits.

SUPERCONDUCTING BOLOMETERS

A distinct class of bolometers makes use of high-temperature superconductive films to obtain extremely high sensitivities. The high-T_c superconductor film is formed into thermally isolated regions that are heated by incident radiation. By maintaining the film near its transition temperature, one sees that very small changes in temperature result in immense changes in resistance. For a superconductor near T_c, the temperature coefficient of resistance, α, can be on the order of 0.2 K^{-1}, as opposed to $\approx$ 0.002 K^{-1} for metals and $\approx$ 0.01 K^{-1} for semiconductors. Examples of such bolometers in the literature include Kruse (1990), Quelle (1990), Neff, et al. (1995), and Sánchez, et al. (1997).

Sánchez, et al. (1997) demonstrated a micromachined, superconducting bolometer making use of a GdBa$_2$Cu$_3$O$_{7-\delta}$ high-T_C superconductor film (0.64 mm^2 in area) thermally isolated using a 2 mm $\times$ 2 mm $\times$ 1 μm silicon nitride membrane. The process flow for such a superconducting device is sufficiently unusual to merit discussion here.

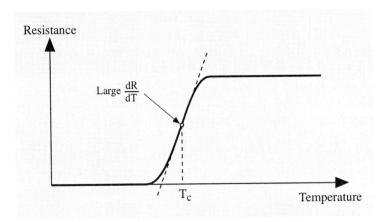

Illustration of a large temperature coefficient of resistance of superconducting materials near the transition temperature, T_c.

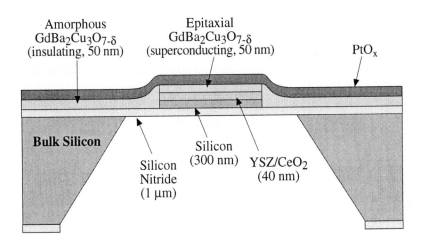

Illustration of a micromachined, superconducting bolometer element (electrical contacts to superconductor not shown, and not to scale). Adapted from Sánchez, et al. (1997).

Fusion bonding between an LPCVD silicon-nitride-coated silicon wafer and a bare silicon wafer was used to fabricate the starting "silicon-on-nitride" wafers. Following LPCVD nitride growth, the nitride-coated wafer was prepared using chemical-mechanical polishing (CMP) and fusion bonded to the second wafer. The second wafer's bonded surface contained either a thin p++ boron layer or a buried oxide layer to serve as an etch-stop for defining a thin silicon layer. Following bonding, the second wafer was etched using KOH and a 300 nm Si layer was formed (using one of the two etch-stop layers) above a 1 μm silicon nitride layer on the first silicon wafer. A 40 nm thick yttria stabilized ZrO_2 (YSZ), and CeO_2 layer was deposited on the silicon and patterned using ion-milling. (YSZ and CeO_2 are buffer layer materials typically used when depositing high-T_c superconductors on silicon and similar materials, and they not only compensate for lattice mismatch between the superconductor and substrate, but also prevent diffusion of silicon into the superconductor.) A 50 nm thick $GdBa_2Cu_3O_{7-\delta}$ high-T_C superconductor film was then deposited above the buffer layer via magnetron sputtering, and was coated with 200 nm PtO_x. The superconductor layer is only epitaxial (and thus functional) over the buffer layer; thus it did not need to be patterned, since amorphous $GdBa_2Cu_3O_{7-\delta}$ is an insulator. At the bond pads, the PtO_x layer was reduced to metallic Pt via laser heating. Finally, the membranes were released using a front-side protection chuck in a KOH bath.

Operating at 85 K, the bolometer's measured parameters were thermal conductance, $G = 3.3 \times 10^{-5}$ W/K; time constant, $\tau = 27$ ms; NEP at 5 Hz = 3.7×10^{-12} $W/Hz^{1/2}$ (close to the theoretical phonon noise limit of 3.6×10^{-12} $W/Hz^{1/2}$); maximum responsivity of 2,800 V/W at the temperature where dR/dT is maximal (≈ 85.3 K); and detectivity $D^* = 2.2 \times 10^{10}$ cm•$Hz^{1/2}$/W. At this temperature, the superconductor exhibited a TCR, $\alpha = 1.1$ K^{-1}, and a resistivity, $\rho = 26$ $\mu\Omega$•cm. While superconducting bolometers require cooling, for high-T_c materials, this is not a particularly difficult requirement, and the available performance is excellent.

2.3.3 THERMOPILES

Thermopiles use the *thermoelectric effect* as the transduction mechanism to generate electrical signals. If junctions are made between two electrical conductors (typically, but not necessarily metals) forming a thermocouple at each junction, and one junction is preferentially heated, a voltage is generated that is a function of the temperature difference. Many junctions can be placed in a "pile" (an obsolete term derived from series-connected piles of batteries) so that they are electrically in series and thermally in parallel, and are collectively referred to as a *thermopile*. This approach can be used to greatly increase the output voltage over that from a single thermocouple. Coating the preferentially heated junction(s) with an absorber (i.e., gold or bismuth black) converts incident radiation into heat, in turn generating a voltage signal.

These devices do not need an electrical bias to operate and have a DC response as long as the temperature difference across the junctions can be maintained (for micromachined versions, this may not really be possible in steady-state, so choppers may be necessary). Without an optical input filter, thermopiles have a virtually flat response versus wavelength.

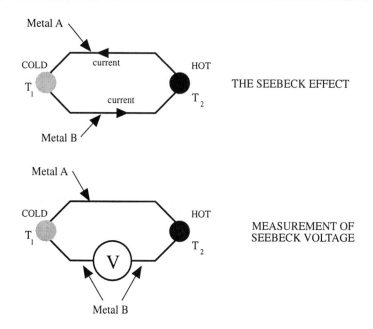

Illustration of the Seebeck (thermocouple) effect (top) and its use to measure temperature (bottom).

The basic thermocouple effect (*Seebeck effect*) depends on the types of electrical conductors used and their *thermoelectric power* (sensitivity). The Seebeck voltage is really a complex function of temperature and is best fit with a high-order polynomial, but it is often approximated as a lower-order function for simplicity. The Seebeck voltage is (approximately) related to the (two) junction temperatures as (first two terms of the polynomial),

$$V \approx \alpha(T_1 - T_2) + \gamma(T_1^2 - T_2^2)$$

where α and γ are constants for the thermocouple pair (see note below — often α is referred to as the thermoelectric power). Even though the Seebeck effect is nonlinear, it is well known for many metals, allowing for very accurate temperature measurements. The thermoelectric power, S, (good means of comparison between materials) is defined as the derivative of the above equation with respect to T_1, yet it is often given as α alone for small signals,

$$S = \frac{dV}{dT_1} = \alpha + 2\gamma T_1 \approx \alpha \quad \text{in V/°C}$$

Tabulated values of S for bimetallic thermocouples vary from 6.2 to 80 μV/°C in commonly used metal pairs (e.g., type T: copper/constantan

$[Cu_{1.00}/Cu_{0.57}Ni_{0.43}]$, type K: chromel/alumel $[Ni_{0.90}Cr_{0.10}/Ni_{0.94}Mn_{0.3}Al_{0.2}Si_{0.1}]$, etc.). More often, the thermoelectric properties are tabulated for a given material and the small-signal Seebeck voltage for a given temperature difference is given by,

$$\Delta V = N(\alpha_1 - \alpha_2)\Delta T$$

where,

N = number of thermocouple pairs in the thermopile

α_1, α_2 = thermoelectric powers of the two thermocouple materials, in V/K

ΔT = temperature difference across the thermocouples, in K

An example of a micromachined thermopile that is worth studying is that of Choi and Wise (1986). They presented a full analysis of the relationship of various design parameters and device performance. As illustrated below, the device was based on doped-polysilicon/gold thermocouples in series electrically and in parallel thermally. The thermocouple hot junctions were at the center of a thin oxide/nitride membrane that is "blackened" with an IR-absorbing material (bismuth black) in this region. The cold junctions were at the surrounding silicon support rim.

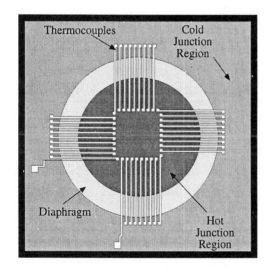
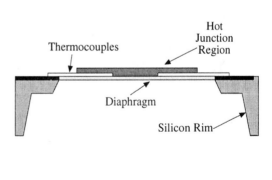

Illusration of a micromachined thermopile IR detector. After Choi and Wise (1986).

Their process began with the growth and patterning of 1 μm of thermal silicon dioxide on (100) silicon substrates, a 20 h solid-source boron diffusion at 1,150°C (to define the rim of the central membrane), followed by an oxide strip and regrowth. Using atmospheric pressure CVD, 200 nm of silicon nitride and 800 nm of silicon dioxide were deposited, followed by 600 nm of polysilicon deposited pyrolitically at 650°C, with a subsequent patterning step. At this point, 250 nm of

CVD silicon dioxide was deposited, with thermocouple contacts opened to the polysilicon below. A 300 nm Au/Cr layer was deposited and patterned to form thermocouple junctions and electrical interconnects, followed by EDP etching to form the thermally isolated membranes for the thermopiles.

To optimize such a structure and to obtain design insight, one needs to write equations that predict the performance of the device based on structural and materials parameters. One assumption made by Choi and Wise (1986) was that there was no radiative heat loss (reasonable, since objects have to become very hot to radiate much energy) and no convective heat flow (probably reasonable, and readily checked by vacuum operation). Thus only conductive heat flow was considered.

A key figure of merit is the responsivity, defined as,

$$R_V = \frac{\text{output voltage}}{\text{optical input power}} = \frac{\Delta V}{Q_{in} A_t} \quad \text{in V/W}$$

where Q_{in} is the power density of the incident radiation, in W/cm^2, and A_t is the total active area of the device (discussed below), in cm^2, and it is this parameter that their analysis sought to relate to design choices.

Typically, thermal conductance is used rather than resistance, and thus it is useful to write the equations in the form (analogous to Ohm's law),

$$V = IR = \frac{I}{G} \quad \Leftrightarrow \quad \Delta T = \frac{\text{heat flow}}{\text{thermal conductance}} = \frac{QA}{G}$$

One can write an expression giving ΔT in terms of the temperature of the heated central absorber plus the temperature drop across the membrane from the hot to cold junctions,

$$\Delta T = \frac{\eta_1 Q_{in} A_1}{G_t(r_1)} + \eta_2 Q_{in} \int_{r_1}^{r_2} \frac{2\pi r}{G_t(r)} dr$$

where,

η_1, η_2 = *absorptivities* of each region (at r_1 it is the η for the bismuth black), not quantum efficiencies

$G_t(r)$ = total thermal conductance at a distance r from the center of the device

Q_{in} = power density of the incident radiation

A_1 = area of the hot junction (blackened) region = πr_1^2

r_1 = outer radius of the blackened region

r_2 = radius at which the cold junction lies

(Note that polar coordinates were used for this analysis, despite the square shape of the silicon rim, since they simplify the analysis.)

There are two components to the conductance: $G_{lead}(r)$, the thermal conductance of the thermocouple leads and $G_{diap}(r)$, the thermal conductance of the diaphragm (in parallel). Adding the two conductances gives the total conductance,

$$G_{total} = G_{lead}(r) + G_{diap}(r) = N \sum_{i=Au, \, poly\text{-}Si}^{N} \frac{\sigma_i a_i}{(r_2 - r)} + \frac{2\pi t \sigma_{diaph}}{\ln\left(\frac{r_2}{r}\right)}$$

where,

N = total number of thermocouple leads

σ_i = thermal conductivity of each thermocouple lead (Au or poly-Si)

a_i = cross-sectional area of each thermocouple lead

$r_2 - r$ = length of lead from a radius r out to the cold junction

t = thickness of the diaphragm

σ_{diaph} = thermal conductivity of the diaphragm material

The equations can be combined to yield an expression for the responsivity,

$$R_V = \frac{\Delta V}{Q_{in} A_t} = \frac{N(\alpha_1 - \alpha_2)\Delta T}{Q_{in} A_t} = \frac{N(\alpha_1 - \alpha_2)}{A_t}\left(\frac{\eta_1 A_1}{G_t(r_1)} + \eta_2 \int_{r_1}^{r_2} \frac{2\pi r}{G_t(r)} dr\right)$$

which can be simplified assuming the blackened area is much more absorptive than the cold junction area ($\eta_1 \gg \eta_2$). The second term can be neglected and the total area taken to be the area of the blackened region, A_1,

$$R_V \approx \frac{\eta_1 N(\alpha_1 - \alpha_2)}{G_t(r_1)} = \frac{\eta_1 N(\alpha_1 - \alpha_2)}{N \sum_{i=Au, \, poly\text{-}Si}^{N} \frac{\sigma_i a_i}{(r_2 - r)} + \frac{2\pi t \sigma_{diaph}}{\ln\left(\frac{r_2}{r}\right)}}$$

Since one generally wants to maximize the responsivity, R_V, it can be seen from this analysis that 1) the hot region's absorptivity, η_1, should be maximized, 2) a large number of thermocouples, N, is helpful as long as the added thermal conductivity of their leads does not significantly reduce ΔT, 3) the difference between the thermoelectric powers for the two materials should be maximized, and 4) the diaphragm materials should have low thermal conductivities.

Items 1 and 2 are quite clear, with 1 being relatively simple and 2 being a lithographic resolution and transducer design issue. Item 3 presents a trade-off because for semiconductors, the thermoelectric power increases as doping concentration decreases, but this also causes the resistance to increase (and hence thermal noise increases as $R^{1/2}$), thus reducing the SNR at some point. Item 4 is pretty straightforward, but one needs diaphragm material that meets this goal and is strong enough and etchant resistant.

It is now useful to consider the detectivity of the device, which takes into account responsivity and detector noise to yield a specification analogous to SNR. Recalling that the DC detectivity is,

$$D^* = \frac{\sqrt{A}}{NEP} \quad \text{and} \quad NEP = \frac{\text{RMS noise voltage}}{R_V} \quad \text{or} \quad \frac{\text{RMS noise current}}{R_I}$$

one can write,

$$D^* = \frac{\eta_1(\alpha_1 - \alpha_2)}{G_t(r_1)}\left(\frac{NA_1}{4kTR_e}\right)^{\frac{1}{2}} \quad \text{in} \quad \frac{cm\sqrt{Hz}}{W}$$

where R_e is the electrical resistance of each thermocouple in the thermopile.

In addition to the equation for R_V, the D* equation provides insight into the trade-off of D* for increased R_V due to greater thermal noise from the thermocouples.

As for an electrical RC time constant, one can obtain a thermal time constant,

$$\tau = \frac{C_t}{G_t}$$

where,

$$C_t = \sum_i v_i \rho_i c_i \quad \text{in J/K}$$

where,
v_i = volume of material i, in m^3
ρ_i = mass density of material i, in kg/m^3
c_i = volume specific heat of material i, in J/(kg•K)

This simply illustrates the parameters that could potentially be in the designer's control to minimize the thermal capacitance of the structure, which (for all such micromachined thermal optical sensors is a key response-time-limiting factor).

It is worth reviewing the design issues involved in these devices. Since Choi and Wise (1986) chose to use EDP as the bulk silicon etchant, their material choices were contstrained (e.g., no Al). The boron diffusion etch stop made the active membrane size insensitive to wafer thickness, making the process less dependent on etch timing. Because they did not have low-stress dielectrics, stacked SiO_2 (compressive), Si_3N_4 (tensile), and SiO_2 (compressive) layers were used, illustrating an important concept: the use of *stress compensation* by stacking multiple layers to approximate neutral stress. In terms of device performance, responsivity was found to increase nearly linearly with diaphragm radius (r_1 fixed), but detectivity rose at a much lower rate (square root) due to increased resistance and thermal noise. This illustrates that for these devices, geometry is key.

A recent overview of thermopile-based infrared sensors fabricated by post-processing CMOS integrated circuits can be found in Lenggenhager (1994).

2.3.4 GOLAY CELLS

Invented around 1947 by Marcel Golay (Golay (1947, 1949)), the basic principle of operation of the Golay cell is that incident radiation heats a volume of trapped gas that expands (approximately) following the ideal gas law,

$$PV = nRT$$

where,
 P = pressure, in Pa = N/m^2
 V = volume, in m^3
 n = number of moles (mol)
 R = gas constant, 8.31451, in N•m/(mol•K)
 T = absolute temperature, in K

And the expansion of the gas is measured as a deflection of a membrane mirror that forms one wall of the gas cavity.

Without an input filter, Golay cells have a virtually flat response versus wavelength, but, in practice, they are often used with optical choppers to compensate for ambient temperature changes (which also affect gas expansion). Macroscopic Golay cells have been able to achieve detectivity values on the order of 10^9.

It turns out that Golay cells can readily be implemented in a micromachined form and retain excellent performance characteristics. The key transduction mechanism used in a micromachined version discussed herein is displacement sensing through tunneling, as explained below.

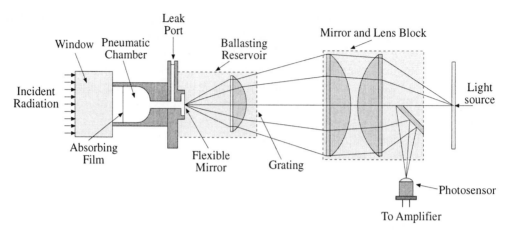

Illustration of a typical macroscopic Golay cell structure. The incident radiation is converted to heat on an absorbing film, heating gas trapped in a pneumatic chamber. The expanding gas deflects a flexible mirror, the movement of which is detected optically with high precision. After Norton (1989).

MICROMACHINED TUNNELING-BASED GOLAY CELL

Prior to discussing the micromachined Golay cells developed by Kenny, et al. (1991), it is necessary to introduce the concept of tunneling as a means of sensing displacement. The first major use of tunneling transducers (at IBM's Zurich research labs in 1982) was scanning tunneling microscopy. Using tunneling of electrons from an ultra-sharp tip through a very narrow gap as a means of measuring distance (moving the tip to keep the gap and tunneling current constant), the IBM researchers obtained atomic resolution images of surfaces using piezoelectric X-Y-Z scanners (see Binnig, et al. (1982) and Binnig and Rohrer (1987))

As described in the Mechanical Transducers chapter, the basic equation for the DC tunneling current through the tip, I, is,

$$I = I_o e^{(-\beta \sqrt{\phi} z)}$$

where,

I_o = scaling factor, a function of the materials, tip geometry, etc.

β = conversion factor, in eV$^{-1/2}$/nm, (typical value = 10.25 eV$^{-1/2}$/nm)

ϕ = tunnel barrier height, in eV (typical value = 0.5 eV)

z = tip/surface separation, in nm, (typical value = 1 nm)

Even with ideal electronics, over a bandwidth, Δf, the measurement of tunneling current is limited by shot noise,

$$I_n = \sqrt{2qI\,\Delta f}$$

and this gives a fundamental sensitivity to displacement limited by the shot noise (normalized for bandwidth),

$$z_n = \frac{\sqrt{2qI}}{\dfrac{\partial I}{\partial z}} = \frac{1}{\beta}\sqrt{\frac{2q}{I\phi}} \quad \text{in} \quad nm/\sqrt{Hz}$$

which is *far less than the radius of a single atom* for reasonable bandwidths.

For tunneling microscopes, one needs extremely sharp tips to avoid ambiguity as to which region of the surface being imaged is the tunneling source. For tunneling displacement transducers, as discussed below, this requirement is significantly relaxed.

A tunneling tip was employed as a displacement sensor in a micromachined Golay cell developed by Kenny, et al. (1991). The basic idea is to use electrostatic force with tunneling position feedback to keep the Golay cell membrane from deflecting, and measuring the force required to ensure this. As for macroscopic Golay cells, a gas-filled chamber with an absorptive surface converts incident radiation into a temperature rise, a subsequent pressure rise, and a membrane deflection (or force). The basic design is illustrated below, showing the closed-loop feedback circuit (implemented using conventional integrated circuits and components) that varies the electrostatic deflection voltage to keep the tunneling current constant.

A key difference between this application of tunneling in this sensor versus its use for microscopy is that the tunneling tip does not need to be very sharp. This is because the servo effect of the feedback circuitry brings the tip into close enough approach to the opposing electrode to start tunneling, inherently selecting the dominant (closest) tunneling site. This effect contributes to the high robustness of this design, which is much greater than that of conventional Golay cells.

The device reported by Kenny, et al. (1991) had a membrane area of 10^{-2} cm^2, thermal conductance between the membrane and surroundings, G, (dominated by air in the cavity) of 2×10^4 W/K, heat capacity, C, of the membrane and gas (dominated by membrane) of 8×10^{-7} J/K, and a time constant (C/G), τ, of 4×10^{-3} s. Measuring or calculating these numbers for a given design can be used to estimate the effects of design changes on temporal response characteristics.

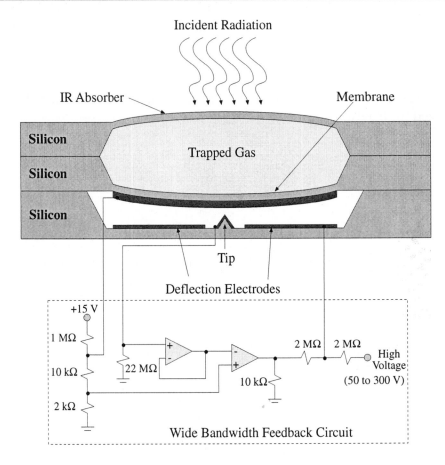

Illustration of micromachined Golay cell and external feedback circuitry. After Kenny, et al. (1991).

The frequency response of deflection (mechanical) was measured by feeding white noise into the reference channel of the feedback circuit and measuring the ratio of measured tunneling current versus reference current versus frequency. For the devices described above, the 3 dB point occurred at $\approx$ 50 KHz.

While predicted noise types included photon, phonon, electron, and amplifier noise, an "unexpected" noise source that turned out to be quite significant is that of accelerations of the transducer elements (taking advantage of this, very sensitive accelerometers have been built), often referrred to as microphonics when they are associated with acoustic events.

The responsivity, R_V, for these devices was $\approx$ 3 $\times$ 10^5 V/W (from 1 to 30 Hz, decreasing as 1/f above 30 Hz) and the detectivity, D*, was on the order of 7 $\times$ 10^8 cm•$Hz^{1/2}$/W. The NEP for the tunneling Golay cells was nearly an order of magnitude lower than, for example, a state-of-the-art pyroelectric sensor, in the range of 3 to 8 × 10^{-10} W/√Hz. The smaller end of this NEP range is near the state-of-the-art for uncooled IR detectors with active areas near 2 × 2 mm.

Modifications for improved performance have included adding a pinhole to the top membrane to prevent temperature/pressure drift from affecting performance and adding corrugations to the lower membrane to allow operation at larger separations and lower voltages. Also, Grade, et al. (1997) described a wafer-scale, production style process for fabricating the tunneling Golay cells. Their new design yielded devices with $R_V = 1.25 \times 10^5$ V/W (from 0.5 to 20 Hz) and NEP = 6 × 10^{-10} W/√Hz.

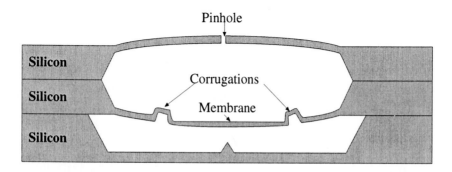

Illustration of an improved tunneling Golay cell design. After Kenny, et al. (1991).

MICROMACHINED CAPACITIVE GOLAY CELL

Chévrier, et al. (1995) described a micromachined Golay cell based on modulation of the inter-electrode spacing of an air-gap capacitor by the expanding gas in the cell. The basic design of the devices, as illustrated below, made use of a 200 nm LPCVD silicon nitride membrane as the deflectable part of a parallel-plate capacitor. If it was heated via IR light warming an absorber layer, the trapped gas expanded and deformed the cavity and narrowed the gap between two 15 nm Au electrodes (one on the membrane and the other, fixed electrode, on an anodically bonded Pyrex™ glass layer). The gap width was determined by an initial dry etch step prior to membrane deposition (this technique could readily be adapted to other capacitive membrane designs). Anodic bonding was also used to attach the optical input window, another Pyrex™ region with predefined 10 nm thick Au IR absorbers.

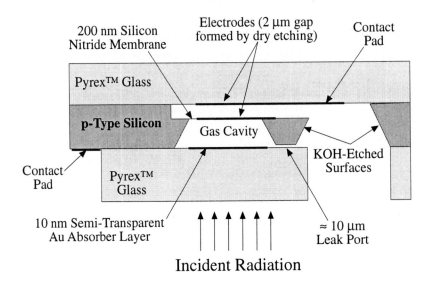

Illustration of capacitively sensed, micromachined Golay cell. After Chévrier, et al. (1995).

Their prototypes had a membrane area of 1 mm² and a cavity height of 0.5 mm (approximately standard 4 in. wafer thickness). An 0.5 μm × 10 μm × 1 mm pneumatic leak port was fabricated using dry etching to allow the low-frequency pressure offsets due to environmental temperature variations to equalize. It is worth noting that the fluidic time constant (analogous to an RC product) of the leak port is given by,

$$\tau_{leak} = \frac{128\mu_{air}Vd}{\pi D^4 P} \quad \text{in s}$$

where,
μ_{air} = viscosity of air at room temperature = 18×10^{-6} kg/(m•s)
 V = volume of the gas cavity, in m³
 d = length of the leak port, in m
 D = hydraulic diameter of the leak port (see the Microfluidic Devices chapter), in m
 P = static pressure, in Pa

This is an important design point for micromachined Golay cells, since this time constant should be designed to be much longer than that of the membrane/cavity system (i.e., the leak should have a much lower frequency response than the sensor), where the sensor time constant will be given by,

$$\tau_s = \frac{C}{G} \quad \text{in s}$$

where C is the thermal capacity of the gas in the cavity, in J/K, and G is the thermal conductance seen from the cavity, in W/K.

For a given design, knowing the values of C and G allow the design of an optimized leak port that does not compromise the desired frequency response of the sensor.

To date, the capacitive Golay cells have demonstrated lower performance than the tunneling-based devices with roughly the same complexity of construction. Arguments by Chévrier, et al. (1995) that capacitive transduction is more reliable than tunneling have yet to be proven in this case.

2.3.5 PHOSPHORS AND OTHER "INDIRECT" LIGHT SENSORS

Phosphors can be excited by electron bombardment (such as that used in a television or oscilloscope cathode-ray tube [CRT]) or photons ("light-to-light" transducers), depending on the type of phosphor. This phenomenon is referred to as *fluorescence*. Due to their use in a wide variety of industrial processes, many phosphors are well developed and commonly available. It should be noted that many phosphors are not only fluorescent, but also display *phosphorescence*, or slow emission of light after the removal of the initial exciting event (e.g., "glow-in-the-dark" paint).

Phosphors are sometimes used to make "indirect" photosensors to convert light of wavelengths invisible to an available sensor into wavelengths it can detect. One example of this is the scintillation detectors used to detect ionizing radiation events via the pulses of light that are generated by their interaction with the scintillator material. In general, the emitted wavelengths are longer than those exciting the phosphor.

A less common phosphor type with potential utility for micromachined devices are so-called *up-converting phosphors* that convert light to a *shorter* wavelength. An example is the IR-sensitive phosphors manufactured by Eastman Kodak (Rochester, NY), which can be used to re-emit invisible IR light at a wavelength detectable by simple detectors such as a silicon photodiode (or your eye). Sheets coated with this type of phosphor are very useful for locating and/or aligning low-power IR beams.

Another possible, but less developed "indirect" optical sensing approach involves direct optical-to-chemical conversion, with subsequent electronic detection.

An example of such optical-to-chemical conversion mechanism in nature is presented below in the discussion of animal visual systems.

It is quite likely that indirect light detection methods will find applications in the micromachined domain. For example, simply coating photodiodes with phosphors could produce simple but potentially quite efficient imaging scintillators.

2.3.6 COMPARISION OF INDIRECT OPTICAL MICROSENSORS

Device	Responsivity (R_V) (V/W)	Detectivity (D^*) $(cm \cdot Hz^{1/2} \cdot W^{-1})$
Choi and Wise (1986) Thermopile (n-type poly)	20 to 25	5 to 6×10^7
Choi and Wise (1986)Thermopile (p-type poly)	52 to 56	$\approx 7 \times 10^7$
Polla, et al. (1986) ZnO Pyroelectric	4.3×10^4	3.1×10^7
Asahi, et al. (1993) PVDF Pyroelectric	125	1.4×10^7
Kenny, et al. (1991, 1996), Golay cell	3×10^5	7×10^8
Liddiard, et al. (1994) Amorphous Si Bolometer	N/A	1 to 3×10^8
Wood, et al. (1992) Semiconductor Bolometer	7×10^4	$\approx 4 \times 10^8$
Sánchez, et al. (1997) Superconducting Bolometer	2.8×10^3	2.2×10^{10}

Table comparing the performance of example micromachined thermopile designs, two pyroelectric detectors, a tunneling-based Golay cell, two uncooled bolometers, and a superconducting bolometer (note that some specifications were not available). While not intended to be an exhaustive survey, this table does provide some general insight into the relative merits of each approach when applied to micromachined devices. Note that all but the last device listed are uncooled designs.

A brief comparison of the performance of the example micromachined IR detectors presented above are shown in the table above. It should be noted that non-micromachined pyroelectric detectors, thermopiles, and bolometers routinely achieve D* values of $> 5 \times 10^8$ for millimeter-sized devices. Micromachined versions should have advantages due to lower thermal capacity and conductance, but thermal sensor materials issues have limited their D* values to roughly an order of magnitude lower than their "conventional" counterparts.

2.4 BIOLOGICAL LIGHT SENSORS

Optical sensors are the rule rather the exception in higher animals. Clearly, the retina (or compound eyes in insects) is an amazingly sensitive and high-dynamic-range sensor (the human retina is capable of detecting single-photon events, although they are not perceived consciously). Animals have a wide variety of optical systems that are adapted to their environments. For example, pit vipers (*Crotalidae*) and boa constrictors (*Boidae*) use IR sensors in arrays under the eyes and along the jaw, respectively, to locate and target their prey. The human retina has two types of visual sensor cells: *rods* and *cones*, the properties of which are compared in the table below.

Rods	Cones
More photopigment	Less photopigment
Slow response: long integration time (can detect flickering light up to 12 Hz)	Fast response: short integration time (can detect flickering light up to 55 Hz)
High amplification: single quantum detection	Probably less amplification
Saturating response	Nonsaturating response
Not directionally sensitive	Directionally sensitive
Highly convergent retinal pathways	Less convergent retinal pathways
High sensitivity	Low sensitivity
Low acuity	High acuity
Achromatic: one type of pigment	Polychromatic: three types of pigment

Table comparing properties of rods and cones in the human retina. The two types of sensor are complementary, with rods serving in low-light vision and cones serving higher light level and higher acuity color vision. From Tessier-Lavigne (1991).

The plots below show the relative luminosity of human bright light color vision (*photopic*, mainly cones) and dim light monochrome vision (*scotopic*, rods only). The shift in sensitivity to shorter wavelengths for monochrome vision is referred to as *scototaxis*.

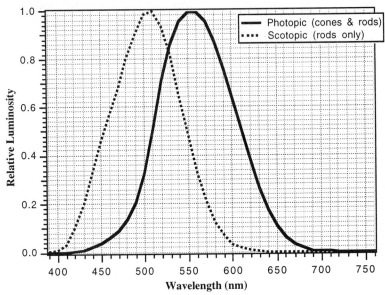

Plot showing relative luminosity of various wavelengths of light for monochrome and color vision. Plotted using data from Benson (1992).

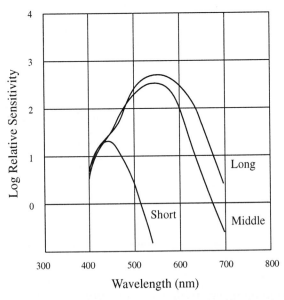

Logarithmic sensitivity plot of spectral sensitivities of the three types of human cone cells obtained from the study of individuals who lack one or more of them in their retinas (i.e., color blindness). After Bailey and Gouras (1985).

2.4.1 STRUCTURES OF HUMAN VISUAL TRANSDUCTION

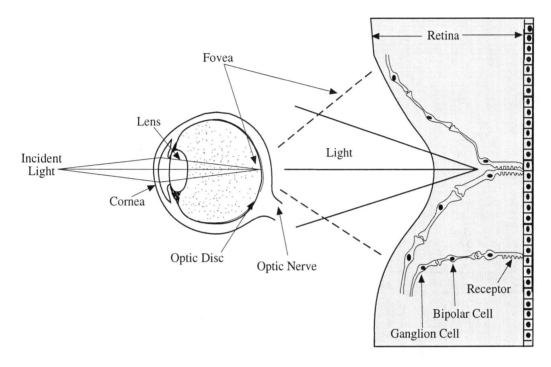

Illustration of the large-scale organization of the human eye, showing optical path and location of the retina. Note that the photosensors are at the "back" of the retina, furthest from the front of the eye. After Tessier-Lavigne (1991).

The retina, derived embryologically from cells in the *neural crest* (the region from which the nervous system arises), is a network of specialized neurons, as illustrated above. Like most of the sensory systems in animals, a key operating feature of the retina is its ability to compute spatial derivatives. In essence, it makes good use of edges and contrast information, as well as being able to detect rapid changes in visual information.

Higher-order processing takes place in the visual cortex, with both visual fields combined to form a three-dimensional perceived view. Neurons at this level carry out shape and object recognition functions. For further information on the visual system, the reader is referred to Tessier-Lavigne (1991).

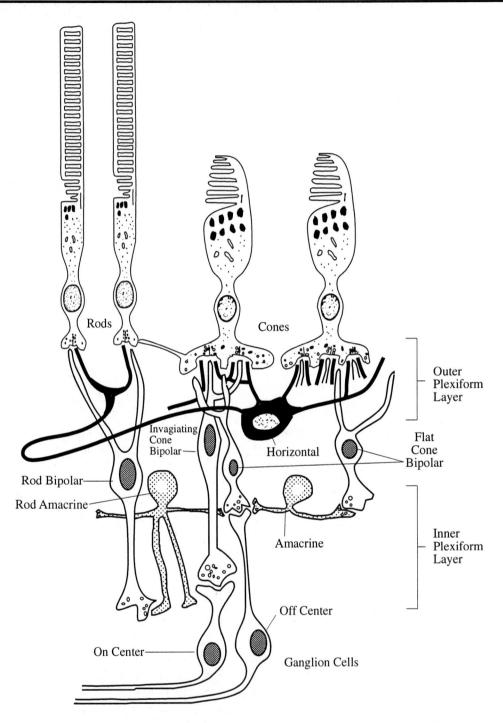

Illustration of a section of retina, showing a few cells of the network of neurons that carries out local signal processing. Note that light enters the eye from the bottom of the illustration. After Bailey and Gouras (1985).

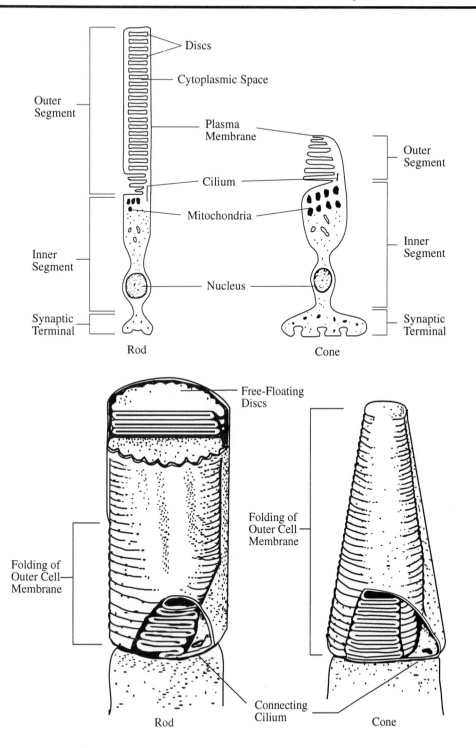

Idealized (top) and physical (bottom) structures of rods and cones. After Bailey and Gouras (1985).

2.4.2 BASIC VISUAL TRANSDUCTION MECHANISM

The actual mechanism of visual transduction is a chemical front-end with electrochemical output. A brief overview of the operation of these mechanisms within rods is presented below.

The basic transduction mechanism is the interaction of photons with the visual pigment, *rhodopsin.* Rhodopsin has two components, retinal (an aldehyde form of vitamin A) and the actual pigments, opsin(s), which are bound together in the dark when retinal is in the 11-cis form. (Note that the prolonged absence of vitamin A can lead to total blindness, since rhodopsin must, in part, be synthesized from it. Night blindness occurs first if vitamin A is deficient.) The absorption of photons changes 11-cis-retinal to trans-retinal, which cannot bind to opsin. (This is analogous to photogeneration in solid-state optical sensors: quantum efficiency is determined at this step.) If one is exposed too long to a bright light, much of the available rhodopsin is broken down, leading to difficulty in seeing when moving from brightly lit to darker areas.

The presence of rhodopsin through a "second messenger," cyclic GMP (guanosine monophosphate), keeps sodium channels in the cell membranes open in the dark. The absorption of a single photon by one rhodopsin molecule, results in the breakdown of the rhodopsin and a corresponding decrease of cyclic GMP, which in turn leads to the closure of several hundred sodium channels. This is the *gain* stage in the photodetection mechanism.

Illustration of the two states of the retinal component of rhodopsin.

3. OPTICAL ACTUATORS

3.1 LIGHT EMITTERS

3.1.1 LIGHT EMITTING DIODES

The light emitting diode (LED) is a p-n junction diode with a key difference (in comparison to "ordinary" diodes) that both the *p* and *n* regions are heavily doped (almost degenerate). The basic principle of an LED's light emission is simple: an electron makes a transition from the conduction band to the valence band, which is equivalent to saying an electron recombines with a hole and the energy is conserved by giving rise to a photon. This light emission is said to be "spontaneous" (i.e., nothing is needed to trigger it). The photon energy, or the frequency of emitted light is, by conservation of energy, almost identical to the energy difference between the conduction band and valence band (the energy gap, E_g, where $\lambda = 1.24/E_g$ in μm (E_g in eV).) The light emitted is *incoherent*, which means that there is no temporal or spatial relationship between the phases of different phase fronts of the emitted electromagnetic wave (since the emission is spontaneous).

The quantum efficiency, η, of a LED is determined primarily by the "ease" of an electron and a hole recombining, and as one expects, η may not exceed 1. This ease is a strong function of the band nature of the semiconductor. Direct band materials make good LED materials (e.g., GaAs, GaP), whereas indirect band materials are very inefficient in optical emission (e.g., Si, Ge, SiC). In a direct band material, an electron making an energy transition yields one photon (to conserve energy). The electron's momentum does not change in the process (momentum is thus conserved). This transition is a very probable one. The corresponding quantum efficiency is very large (> 80%). (The quantum efficiency for light emission is the ratio of the electron-hole pair recombination processes that are radiative to the total of all recombinations.)

In indirect materials, the energy transition requires a photon (to conserve energy) and a *phonon* (to conserve momentum) (a phonon is an interaction between the electron and the thermal oscillations of the atoms in a lattice). This two-particle process is *not* a probable process. The quantum efficiency is thus extremely low (e.g., in Si, it is $< 10^{-4}$). Another factor that determines the quantum efficiency of an LED is the I^2R losses. This is an important reason to heavily dope both sides of the diode.

The recombination that results in light generation occurs near the depletion region under the condition of forward bias as shown in the figure below. When a forward bias is applied to the p+/n+ diode, the electrons acquire enough energy to spill over beyond the n+ region into the depletion region and to propagate into the p+ region (minority carrier injection). The reverse happens to holes in the valence band.

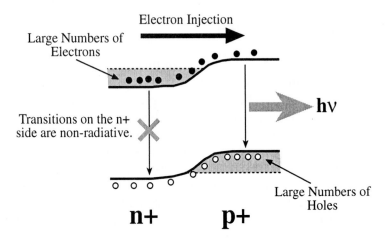

Band diagram of a forward biased LED. Under injection conditions, the carriers overcome the barrier, a current flows, and recombination occurs, resulting in the spontaneous emission of light. Note that emission only occurs from injected electrons on the p+ side.

The injection condition creates a spatial overlap of holes and electrons on both sides of the diode. Under sufficient injection by forward bias, band-to-band (conduction to valence band) transitions occur and light is emitted. Emission occurs only from electrons injected on the p+ side of the LED junction. The emitted photons have no phase relationship among each other.

It is important to realize that in addition to direct carrier recombination between the conduction band and valence band (via a photon whose energy equals the bandgap), it is possible to get recombination, radiative and non-radiative, by many other different mechanisms, such as transitions to acceptor and donor levels (remember the diode is heavily doped). These other mechanisms reduce the efficiency of the LED and result in a broad emission spectrum or no emission at all. As shown in the figure above, transitions of electrons from conduction to valence band in the n+ region are non-radiative.

GaP is the common material used for red LEDs (for red, one requires a bandgap on the order of 1.7 eV). Its quantum efficiency is very high, and this can be observed by noting that red LEDs can be extremely bright, despite the fact that GaP is an indirect semiconductor. Green LEDs can be made by reversing the principle used to fabricate extrinsic photoconductive detectors (see Groves, et al. (1971), Craford, et al. (1972) and Campbell, et al. (1974)). When GaP is doped with nitrogen (providing an extrinsic level in the band structure), one obtains a green emitting LED at 550 nm.

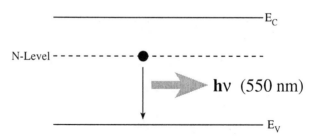

Band diagram of N-doped GaP showing the mechanism of green light emission.

Blue LEDs were initially fabricated using SiC. Their quantum efficiencies were very small because SiC is an indirect material, and this resulted in very dim LEDs. More recently, GaN has been used to fabricate very bright blue LEDs (see Nakamura, et al. (1991, 1993, 1994), placing the emphasis on developing more efficient blue-green devices for future color displays and other applications.

It is possible to improve the quantum efficiency of indirect material LEDs by creating states in the bandgap to facilitate the electron-hole recombination, as described above. In Si, this was achieved by bombarding single-crystal material with C atoms, or doping it with Er (Erbium) and Y (Yttrium). The result was a quantum efficiency an order of magnitude higher ($\approx 10^{-3}$) than without such doping, but still too small for any useful commercial application.

An extremely important non-display application of LEDs is in opto-isolators, which couple two circuits together through an insulating, but optically transmissive layer. Electrical input signals are sent through the layer by a modulated LED and are converted back into electrical signals on the other side using a phototransistor, photodiode, photo-SCR, etc. Opto-isolators are commonly used for power electronics, medical instruments, and other applications where such high-voltage-capable isolation is desirable.

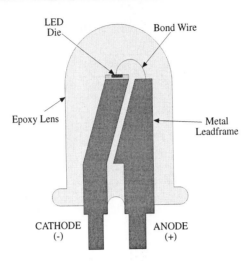

Illustration of a typical packaged LED. For low-cost applications, the LED die is mounted on a lead frame and, once the bond wire is attached, clear or tinted encapsulant is injected around the assembly.

3.1.2 SILICON LIGHT EMITTING DIODES

Since silicon is an indirect bandgap material, light emitting diodes fabricated from it will likely be very inefficient. In addition, as discussed by Haynes and Westphal (1956), emissions will have energies in a narrow band near the bandgap, making silicon detectors inefficient for detection. However, silicon LEDs have been fabricated using standard CMOS processes. Kramer, et al. (1993) demonstrated highly interdigitated pn junctions that, when forward biased at a forward voltage of ≈ 1.2 V, would emit a narrow-band IR light centered at 1,160 nm with an electrical-to-optical conversion efficiency of 10^{-4}. When operated in avalanche mode in reverse bias (6.5 to 52 V), the same junctions would emit visible light in the range of 450 to 800 nm with efficiencies of $\approx 10^{-8}$. While these LEDs are clearly extremely inefficient, their successful development could enable such devices as all-silicon opto-couplers and reflective sensors. As mentioned above, implantation with C atoms or doping with Er or Y can be used to increase η to $\approx 10^{-3}$. For further information on Si LEDs, the reader is directed to Kramer, et al. (1993) and Wolffenbuttel and van Drieënhuizen (1991), both of which have good lists of references in the area.

It has also been known for some time that porous silicon (methods for forming it are presented in the Micromachining Techniques chapter) would emit visible light if stimulated with green to blue, or UV light (see Canham (1990) and Jung, et al. (1993)). It is also possible to form Schottky diodes by depositing a thin (e.g., 25 nm), conductive electrode on the surface of a porous silicon region (e.g., indium tin oxide, or ITO, 90% In_2O_3 + 10% SnO_2 or a conductive polymer). As reviewed by Xu and Steckl (1994), a variety of electrically driven porous silicon LEDs have been demonstrated, with emission wavelengths between 480 and 700 nm. They demonstrated the use of pure chemical (as opposed to electrochemical) etching of boron-doped (100) silicon (6 to 16 Ω•cm) to form the porous silicon (using $HF:HNO_3:H_2O$, 1:3:5) and the use of ITO electrodes to obtain optical emission at $\approx$ 640 nm at current densities > 3 mA/cm^2. The important point regarding the use of purely chemical porous silicon etching is that it might be convenient to carry out such etching of silicon islands on (locally protected) glass substrates to fabricate large area displays. Considerable further work in this area seems to be required in order to better understand the mechanisms of emission, to investigate means of obtaining a variety of colors of emission, and to optimize quantum efficiency.

3.1.3 ORGANIC LIGHT EMITTING DIODES

There is considerable interest in developing *organic* LEDs, which might, if successful, enable thin, flexible displays. As described in Tang (1996), recombination electroluminescence in a molecular solid has a theoretical maximum quantum efficiency of 25%, but, due to loss mechanisms including optical coupling, an overall efficiency of 5% should be realistic with organic emitters. For comparison, direct bandgap semiconductor LEDs can have quantum efficiencies in the 80% range. A variety of such organic electroluminescent (EL) structures (sometimes referred to as organic LEDs) have been fabricated, and can produce a similar range of colors to CRT displays.

The basic structure of a two-layer organic EL device is a hole transport layer (generally an aromatic amine, typically in the benzidine family) in contact with an electron transport layer (typically a host material doped with a small concentration of fluorescent molecules), each with its respective carrier injection electrode. The mismatch in energy levels between the two layers creates a potential barrier at the interface, where recombination preferentially occurs. In some cases, a thin emission layer can be placed between the two main transport layers, and can be varied to tune the output wavelength. A typical hole injection electrode (anode) is ITO, and a typical electron injection electrode is a metallic alloy (examples of alternative electrode choices are discussed in Gustafsson, et al. (1992)).

Flexible, organic LEDs were demonstrated by Gustafsson, et al. (1992). Poly(ethylene terephthalate), or PET was used as the substrate, and soluble polyaniline was used as the hole-injecting electrode (anode). Substituted poly(1,4-phenylene-vinylene) was used as the electroluminescent layer (poly(2-methoxy, 5-(2'-ethyl-hexoxy)-1,4-phenylene-vinylene), or MEH-PPV). Metallic calcium was used as the electron-injecting top contact (cathode) (this requires that Ca be encapsulated to avoid oxidation by water vapor in the air). The LED's output was easily visible in room light with an external quantum efficiency of $\approx 1\%$ and a turn on voltage of ≈ 2 to 3 V. The completed LEDs are physically flexible enough to be bent more than 180° while operating.

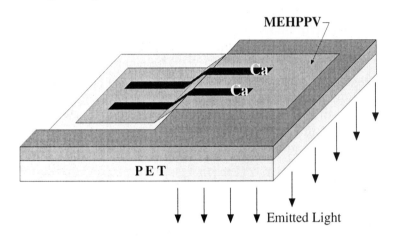

Illustration of an organic LED structure. After Gustafsson, et al. (1992).

A review of work in this area given by Tang (1996) shows that red, green, and blue emissions have been achieved with quantum efficiencies of (respectively) 1.2, 3, and 2.5%, at drive voltages on the order of 10 V. A major issue with such devices is their long-term stability. Most types have brightness half-lives of hundreds to thousands of hours, although some devices have yielded half-lives in excess of 6,000 hours. As the devices are refined, it is hoped that lifetimes greater than 50,000 hours (≈ 5.7 years of continuous operation) can be achieved. Considerable research effort continues in the development of such devices, and they have many potential uses, such as in thin, flexible displays and in "above CMOS" micro-displays, deposited above prefabricated CMOS driver circuits.

3.1.4 GAS AND SOLID-STATE LASERS

The acronym LASER stands for light amplification by stimulated emission of radiation. The laser was first proposed by Einstein around 1916, with the basic idea that if atoms with electrons in high-energy (excited) states were hit by photons at the wavelength of the high-energy-to-low-energy electron transition, the high-energy electrons would be stimulated to make that transition, releasing more photons at that wavelength. This chain reaction would continue until either the high-energy-state electrons were used up or some sort of energy input would be required to create more excited state electrons. The first actual demonstration of lasing was done by Theodore Maiman at Hughes Malibu Research Laboratories in 1960.

It is useful to compare LEDs and lasers. An LED can be thought of as an open-loop electrical-to-optical transducer, yet a laser inherently involves optical feedback in the form of a resonator, and is actually an electrically powered optical *oscillator*. LED emission is spontaneous: it is randomly initiated, and the photons created have no frequency or phase relation to those already present. Lasers utilize stimulated emission, in that the emission process is triggered by already existing photons so that newly emitted photons match in frequency and phase. This increases the amplitude of the existing optical wave, rather than adding a new wave of arbitrary frequency and phase. When the newly emitted photons match those that triggered the emission, spatial and temporal coherence is created in the optical wave and this is referred to as *optical gain.*

For laser operation, one must have an active region (a place where electrons and holes can easily recombine). This is the region in which the "emission of radiation" takes place. After its emission, a photon is reflected back to the active region by a partially reflective mirror. Under proper conditions, this photon will "stimulate" the recombination of an electron-hole pair (as opposed to the "spontaneous" emission of photons described above).

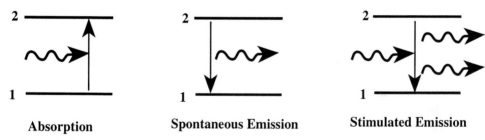

Absorption   **Spontaneous Emission**   **Stimulated Emission**

The three different ways light and electrons can interact. At the left, an incoming photon may be absorbed to excite an electron to a higher energy level (absorption). In the center, an electron drops in energy to a lower state (recombination) and a photon is emitted (spontaneous emission). At the right, an incoming photon triggers the emission of a second photon in matching frequency and phase (stimulated emission).

Stimulated emission, illustrated above, is the inverse process to optical absorption. In a semiconductor, an absorbed photon creates an electron-hole pair, while annihilation of an electron-hole pair will create a stimulated photon. Both processes are triggered by the presence of an existing photon, and their rates are proportional to the existing photon density (in spontaneous emission, the rate is independent of any existing photon densities). The maximum possible optical gain that can be developed is equal to the magnitude of the absorption coefficient, α, which, for optically active materials, will have a negative value. Producing optical activity requires *pumping* (optically or electrically) the medium to produce a population inversion in which there exists more electrons in an upper energy state than in a lower one. This is thermodynamically unfavored, and will never occur in equilibrium. In a semiconductor laser, for example, population inversion is achieved by forward-bias carrier injection in a pn junction.

The two reflective mirrors placed at opposite ends of the active luminescent region form what is known as a Fabry-Perot resonator. In essence, when the spacing between the mirrors is an integral multiple of a half-wavelengths ($n\lambda/2$), there is constructive interference and the light intensity at that particular wavelength λ increases dramatically. Under this condition, the stimulated emission of photons increases and "light amplification" occurs. The above description applies to all kinds of lasers, gas and solid-state lasers (including semiconductor lasers), although some types of pulsed lasers only use single-pass feedback.

There are many different types of lasers, and examples are provided below:

Ruby laser: Ruby consists of aluminum oxide (Al_2O_3) with about 0.03% chromium oxide (Cr_2O_3). The Cr^{3+} ions occupy locations otherwise filled by Al^{3+} and are responsible for the emission of light. The energy difference is 1.8 eV corresponding to a wavelength of 694 nm (red). Such lasers are usually optically pumped using an external xenon flash tube.

Neodymium laser: The active Nd^{3+} ion is incorporated into several hosts, in particular, YAG (yttrium aluminum garnet) or glass. The output wavelength is in the IR at 1.06 μm, but they are commonly doubled to 530 nm (green). These lasers can be operated with flashlamp pumping or in a continuous wave (CW) mode.

Organic dye laser: In this case, the active medium is an organic dye dissolved in a suitable solvent (the most common are ethylene glycol, dimethyl sulfoxide (DMSO), and ethanol). They suffer from the requirement of great pumping power to maintain a population inversion and the fact that the organic dyes (rhodamine, fluorescein, coumarin, etc.) can be extremely toxic. These lasers are usually pulsed and can achieve pulse durations well under 1 ps.

Helium-Neon (He-Ne) laser: Unlike the above lasers that are optically pumped, the He-Ne type is electrically pumped. The active medium is a mixture of five parts He to each part of Ne, at a pressure of 3 torr. Pumping takes place because of a plasma established in the tube. The main emission line occurs at 633 nm. These are very common lasers, used in most supermarket bar-code scanners.

Ion laser (Ar-ion and Kr-ion): The Ar laser has two main spectral lines in the blue (488 nm) and the green (514.5 nm), and typically are in the 1 to 20 W power range, despite very inefficient operation (they are popular nonetheless due to their very useful output wavelengths).

CO_2 laser: In this case, the oscillation is in the IR at 10.6 μm. The transitions occur between the different vibrational energy levels of the CO_2 molecule. It is the only molecular laser based on a vibrational/rotational energy transition. These are the most efficient of all lasers, with outputs of > 1 kW readily achievable.

Excimer laser: These designs use halides of the noble gases as the active media. Excimer lasers are pulsed and can provide very intense peak powers in the UV wavelengths (KrCl: 222 nm, KrF: 248 nm, XeCl: 308 nm, XeF: 351 nm), and are thus of interest for submicron lithography.

Semiconductor laser: The wavelength of semiconductor lasers depends on the energy levels, which may be tuned either thermally (tuning the bandgap) or by constructing "quantum wells" using heterojunctions. Semiconductor lasers are fabricated using III-V compounds (GaAs, AlGaAs, GaP, GaInAsP, InP, etc). Semiconductor lasers are usually low power (a few mW), but devices in the 1 to 10 W range are also available (often used for electromagnetic-interference (EMI) resistant detonators for explosives, as a replacement for such devices as blasting caps). These devices have certainly been enabled by micromachining technology, yet the approaches used are quite specific to III-V compounds. Such fabrication methods are discussed below.

MICROMACHINED SOLID-STATE LASERS

A variety of micromachining processes have long been employed in the fabrication of solid-state laser diodes. A complete review of these many approaches is beyond the scope of this book, but a representative example is presented below.

One method of forcing the optical wave in a solid-state laser to follow the active region is to limit the horizontal extent of the pumped area. Such optical waveguiding is easily created in the vertical (epitaxial growth) direction using the refractive indices of the materials, confinement in the horizontal axis is also needed. *Gain guiding* is the simplest means of achieving this, by creating the p-n junction along a stripe, which focuses the carrier injection, and hence population inversion, to a specified optical path (see Ripper, et al. (1971), Dyment, et al. (1972), and Yonezu, et al. (1973)). However, lateral spreading of the injected current makes the width of the stripe much larger than its physical dimension as fabricated. An attractive alternative is to use micromachining techniques to produce a horizontal refractive index pattern that confines the wave, and this is referred to as *transverse mode control*.

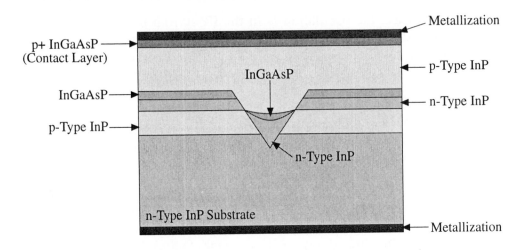

Illustration of a micromachined heterojunction laser diode with transverse mode control achieved using a buried crescent (BC) or a V-groove substrate buried heterostructure (VSB) design. Adapted from Hirano, et al. (1983).

In most cases, the fabrication process consists of the epitaxial growth of one or more layers of the device, followed by bulk anisotropic etching to form ridges, V-grooves, or channels, followed by further epitaxy. Often, the second epitaxy is carried out using liquid-phase epitaxy (LPE) methods (this is the oldest and least expensive method, versus molecular beam epitaxy (MBE) or metal-organic chemical vapor deposition (MOCVD)). As illustrated above, a so-called "buried crescent" structure can be formed by growing a first epitaxial layer (e.g., p-type InP) above a substrate such as n-type InP, anisotropically etching a V-groove, and carrying out a second epitaxial growth of, for example, n-type InP, InGaAsP, p-type InP, and a p+ InGaAsP contact layer. The growth of the epitaxial layers in the V-groove will form the crescent since LPE growth tends to produce such corner shapes. This example serves to illustrate the basic concept, and there are many additional papers in this area. For further information on transverse mode controlled designs, the reader is referred to Saito and Ito (1980), Mito, et al. (1983), Sugimoto, et al. (1984), and for general information to Kressel and Butler (1977), Casey and Panish (1978), Streifer and Ettenberg (1990) and Coleman (1994).

3.1.5 MICROMACHINED INCANDESCENT LAMPS

Incandescent light sources can be used to generate broad output spectra covering the long-wavelength IR through visible ranges. As for any hot-body radiator, the peak emission wavelength can be tuned by controlling the temperature of the radiator. For an incandescent source, this can be done by electrically dictating the dissipated power. Micromachined incandescent sources could bring these capabilities to potential integrated optical or chemical systems. For example, in IR absorption identification of organic compounds, relatively long IR wavelengths are needed, and incandescent sources are generally the only practical means by which to achieve this. However, very tiny "grain-of-wheat" bulbs are common and inexpensive, and thus a strong case for using micromachined devices would be required.

Mastrangelo and Muller demonstrated a micromachined incandescent lamp (Mastrangelo and Muller (1989), and Mastrangelo, et al. (1992)). The filaments were made from either bare or silicon-nitride-coated p+ polysilicon suspended above a silicon V-groove that helped to reflect some of the photons away from the underlying silicon. Silicon nitride was used to coat the filaments in some designs. The silicon nitride was between 300 and 500 nm thick and the polysilicon filaments were 900 nm thick. The filaments were varied in length between 110 and 510 µm. The filaments were sealed in a vacuum, protected from the atmosphere by a 2.5 to 2.8 µm silicon nitride window, which was thick enough to resist the pressure difference.

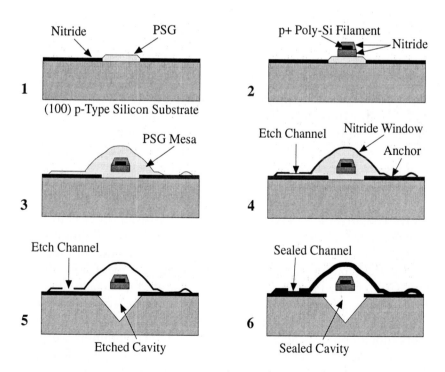

Fabrication sequence used to realize micromachined incandescent lamps. After Mastrangelo and Muller (1989).

As a useful example of sealed cavity formation, as well as detailing the incandescent filament fabrication, the process flow of Mastrangelo and Muller (1989) can be summarized as follows. The process began with the deposition of 0.5 μm of low-stress silicon nitride, patterned to define V-groove boundaries and 0.7 μm of phosphosilicate glass (PSG), patterned to form a spacer layer for the filament. This was followed by the deposition of 0.3 μm of low-stress nitride for the base of the filament and a quick HF dip to remove residual nitride and oxide. An 0.9 μm layer of undoped polysilicon was then deposited and ion implanted with boron for a resistivity of 4×10^{-3} Ω•cm. This polysilicon was then patterned using plasma etch (stopping at the nitride layer) with a quick HF dip to remove residual nitride and oxide. An 0.3 μm layer of low-stress silicon nitride was then deposited to form the top of the filament seal, and both filament nitride layers were then patterned. A 3 μm PSG layer was deposited, and the wafers were heated to 1,050°C for 30 min to activate dopants and to reflow the PSG. PSG mesas were then formed with an HF etch and 0.8 μm of PSG was deposited and patterned to form etch channel spacers. A 1 μm layer of low-stress silicon nitride was then deposited, patterned and etched to open etch channels from which the PSG was removed in HF in 2.5 min. After drying, the cavities thus formed were sealed in a vacuum with an additional silicon nitride layer, after which contacts to the polysilicon bond pads were opened and metal for bonding was deposited and sintered to complete the process.

Emitted power for these devices was in the microwatt range and was visible to the naked eye (the power required to reach visible incandescence was ≈ 5 mW for a $510 \times 5 \times 1$ μm filament). An interesting phenomenon was observed when testing these devices. The resistance of the polysilicon *decreased* (negative TCR) above a certain power level (not expected), leading to thermal runaway (it has been proposed that thermal breakdown of the polysilicon was occurring).

Parameswaran, et al. (1991) and Swart, et al. (1993) discussed the use of undercut dielectric/polysilicon layers in a standard CMOS process to fabricate an array of resistively heated IR emitters for use in projecting calibration images onto IR focal plane arrays. The authors used EDP etching to undercut dielectric plates consisting of the field oxide, polysilicon resistors, and an upper CVD silicon dioxide (SiO_2) layer. Cole, et al. (1995a, 1995b) demonstrated a surface micromachined 512×512 pixel IR scene projector with on-chip control electronics. Their arrays were fabricated by adding surface micromachined structures above prefabricated CMOS wafers (not all process details were provided). The emitters were serpentine TiN resistors on undercut silicon nitride plates, with underlying IR reflectors. A high level of thermal isolation (thermal resistances to the plates was on the order of 10^7 W/K) was achieved through the use of a pair of thin silicon nitride supports for each emitter plate.

3.1.6 PLASMA LIGHT SOURCES

Plasma light sources make use of ionization of gases at low pressure (generally "noble" gases such as neon, argon, etc.) to create the familiar plasma discharges of "neon signs." The emitted light spectra are characteristic for a given gas and ionization state, and phosphors are sometimes used to change the emitted wavelength (e.g., household fluorescent lamps, which use a mercury plasma to excite a phosphor on the inner surface of the tube, translating the largely UV plasma output to visible light).

A great deal of work has been done on plasma displays, despite their inevitable need for high voltages (> 60 V) to achieve gas ionization. As yet, there has been little use of plasma devices in micromachined systems, but it is worthwhile to consider for future applications. One effort in this area was made by Yeh (1991), who fabricated a variety of gas discharge lamps using polysilicon electrodes. Unfortunately, it was found that quite high voltages (300 V AC) were required to strike a Ne plasma, and that the 500 nm thick electrodes were sputtered away in $\approx$ 10 seconds. It is possible that such micromachined devices could be fabricated with lower operating voltages and longer lifetimes, but it seems that considerable further research would be required.

3.1.7 ELECTROLUMINESCENT LIGHT SOURCES

The operating principle of electroluminescence is the application of AC electric fields to phosphors between conductive electrodes, leading to light emission. Typically, the lower electrode is an opaque metal (generally reflective), while the top (light exit) electrode is a transparent conductor such as ITO. To prevent unwanted DC currents, the phosphor is typically covered on both sides with a thin dielectric layer. For example, starting with a glass substrate, one can deposit In_2O_3 as a bottom conductor, the phosphor layer, an insulating layer of Sm_2O_3 or Ta_2O_5, and aluminum metallization (electrodes) (see Kozawaguchi, et al. (1982)).

Today electroluminescence (EL) is most often used in "backlights" for LCD displays (see below), but his role is generally being taken over by conventional vacuum fluorescent tubes. Typically, the EL phosphor (i.e., ZnS doped with 0.5 wt% Mn or 3 wt% TbF_3) is driven by a high-voltage (100 to 400 V p-p) AC waveform in the range of 50 to 400 Hz. This approach to light generation also has tremendous potential for integration with electronics to form high resolution displays. One could start with a silicon substrate with Al bottom electrodes and a thin dielectric, add the phosphor locally, add a thin dielectric, and deposit top electrodes of something reasonably transparent (e.g., ITO) (see Kozawaguchi, et al. (1982)).

A consortium formed by Planar Systems (Beaverton, OR), the David Sarnoff Research Center (Princeton, NJ), Allied Signal Aerospace (Columbia, MD), and

Kopin (Taunton, MA) has resulted in the fabrication of an active matrix EL display on a single-crystal silicon-on-insulator (SOI) substrate. The display is based on a zinc sulfide:terbium (ZnS:Tb) green-emitting phosphor. The SOI substrate allows for monolithic integration of the high-voltage switching circuits required to drive the display. This approach shows great promise for head-mounted and other compact display applications.

3.1.8 FIELD EMISSION DISPLAYS

Field emission displays (FEDs) are of great interest and an application area for micromachining technologies. FEDs utilize arrays of ultra-sharp field emitter tips as cathodes to provide electron flux to illuminate phosphors (such as those used in color television tubes). The drive toward improved display quality and increased battery life in portable computers has motivated several companies to develop FEDs as a potential replacement for the currently dominant liquid crystal displays. LCDs generally require backlighting for computer applications, and this backlighting can account for 80 to 90% of the display's power consumption (it is also interesting to note that only $\approx 4\%$ of the backlight's output reaches the viewer). This combination leads to a very inefficient display. Meanwhile, the *cathodoluminescence* process used in FEDs is one of the most efficient means of converting electrical energy into light. FEDs should be at least two times more efficient than LCDs in terms of net power consumption.

Fields as high as 10^7 V/cm at the cathode tips (easy to generate) pull electrons off the tip where they are accelerated toward the anode(s) by a potential on the order of 200 V to 1 kV. Micromachined devices can operate at much lower voltages, such as 20 to 60 V. Because they only have to traverse a gap on the order of 200 μm, no electron focusing is required beyond that provided inherently by the control grid often placed near the tip (used to turn the current on and off).

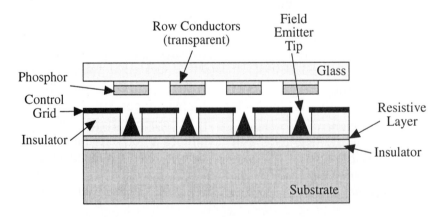

Example of a FED cross section. After Petrovich (1995).

Recent advances in micromachining FEDs have greatly improved performance. Nonuniformities in the tips, often resulting in excessive current (even "exploding" tips) were dealt with by connecting the tips together with an underlying resistive (ballasting) layer that limits current to each one. In some cases, individual pixels are driven by up to 2,000 individual field emitter tips, thus increasing reliability and improving uniformity.

There is a great deal of research going on in this area around the world. This is an example of "micromachining on a large scale," calling for the manufacture of as many as 2 billion field emitter tips, with 1 µm minimum features, in a single display.

Gated field emitter tips can also be used to illuminate solid-state light emitters such as ZnSe, AlGaN, GaN, etc., in which p-n junctions may be very difficult or impossible to fabricate. This is accomplished by gating the electron flow on and off, and gigahertz switching speeds for the optical output appear to be feasible. Busta, et al. (1994) demonstrated such an application of micromachined, gated field emitter tips.

3.1.9 BIOLUMINESCENCE

In nature, there are several chemical mechanisms that are capable of generating light by direct chemical means (i.e., without heat, hence the term "cold light"), referred to as *chemiluminescence* or *bioluminescence*. Fireflies are a common example, as are certain plankton and bacteria. These reactions are essentially oxidation/reduction systems. An example is luminol (5-Amino-2, 3-dihydro-1,4-phthalazinedione, $C_8H_7N_3O_2$), which produces a great deal of light when oxidized.

A general class of common biological chemiluminescent compounds are the *luciferins*, including those from the American firefly (*Photinus pyralis*), the limpet (*Latia neritoides*), and several other species. These compounds are activated by the enzyme luciferase. Light emission requires the involvement of magnesium ions, oxygen, adenosine triphosphate (ATP), luciferase, and the luciferin itself. It is interesting to note that if firefly luciferin (4,5-Dihydro-2-(6-hydroxy-2-benzothiazolyl)-4-thiazolecarboxylic acid, $C_{11}H_8N_2O_3S_2$) is extracted from fireflies, the yield is $\approx$ 9 mg from 15,000 active insects.

A variety of synthetic chemiluminescent systems have been developed. A common example is the so-called "light-stick," which is a plastic tube containing a mixture of chemicals and an inner glass vial filled with a catalyst and hydrogen peroxide. When the plastic tube is bent, the glass vial breaks, releasing its contents and causing the generation of a bright light (available in a variety of colors, including white) that can last several hours. Typical compositions of the contents include a fluorescer (e.g., 9,10-bis(4-methoxyphenyl)-2-chloroanthracene, which produces blue light), an oxalate compound (e.g., bis(2,4,5-trichloro-6-carbopentoxyphenyl)oxalate), an organic solvent (e.g., dibutyl phthalate), a perylene dye (to obtain a white output light by converting some of the blue light into a longer wavelength, such as red) (e.g., N,N'-bis(2,5-di-t-butylphenyl)-3,4,9,10-perylenedicarboximide), and hydrogen peroxide. A thorough discussion regarding the operation and variants of such mixtures can be found in the relevant patents of the American Cyanamid Company (Stamford, CT) such as Bollyky and Rauhut (1982), Dugliss (1987), and Koroscil (1988).

It is not clear whether or not these compounds can have any impact on micromachined systems, but it is worth considering that many redox reactions can be electrically controlled (see the discussion regarding electrochemiluminescence in the Chemical Transducers chapter).

3.2 LIGHT MODULATORS

This section covers light *modulators*, as opposed to light emitters. Light modulators of various forms are possible, including mechanical and non-mechanical means. To date, those that have been micromachined have nearly all been mechanical in nature, although without that restriction, the dominant optical modulator is certainly the liquid crystal display.

3.2.1 LIQUID CRYSTAL DISPLAYS

Liquid crystal displays are a simple, low-power and inexpensive type of optical "actuator" and have found widespread use in many types of electronic devices (industrial, consumer, military, etc.). The basic idea is to "gate" the passage of light through a layer of liquid crystal material that can be electrically controlled to modulate it. The term *liquid crystal* (LC) refers to liquid materials that are somewhat ordered (i.e., "anisotropic") as opposed to fully ordered (crystal) or fully disordered (ordinary liquids, i.e., "isotropic"). To fit into this category, the molecules within the liquid crystal must be "alignable" by various means, such as surface interactions or electrical fields. While liquid crystal materials had been known since the late 1800's, their application to displays did not occur until the mid-1960's, when researchers at RCA demonstrated the first working units, which required 80°C operating temperatures. This constraint was soon removed (Heilmeier, et al. (1968)), and ongoing development of this class of displays has been carried out ever since.

NEMATIC LIQUID CRYSTALS

Nematic liquid crystals (LCs) are rod-shaped molecules (typically 2×0.5 nm in diameter) that maintain a degree of parallel alignment despite the disruption from thermal energy. They are commonly used in consumer electronic devices as displays. They are fast enough for display applications, but generally are not fast enough for optical signal processing.

The key to the operation of nematic LC displays is the interactions between polarized light and the oriented molecules. The direction of polarization of the light can be made to follow that of the oriented molecules. Physically *buffing* the glass (orienting its surface by various means) can "program" the orientation of the nematic LC molecules that will end up in contact with the glass. They align with their long axes perpendicular to the direction of "buffing." In practice, manufacturers apply (by spin-coating or transfer printing) an "alignment layer," which is buffed, rather than the surface of the glass itself.

Typical nematic LCDs are constructed as shown in the illustration below, consisting of 1) an outer polarizer (analyzer) of buffed glass (often with an "alignment layer") with a transparent electrode evaporated on its inner surface (typically ITO), 2) a layer of nematic LC material, typically 6 µm thick, 3) an inner polarizer of buffed glass (with an alignment layer) with another ITO electrode on its inner surface (in contact with the LC), and 4) a reflective surface behind the assembly. To precisely form the LC gap, various spacer layers can be employed, and they are typically photolithographically patterned.

LCDs can be *reflective* (as shown below) or *transmissive* (without a bottom reflective layer, such that light either passes through or is blocked). Transmissive LCD displays are often used in projection television systems, "spatial light modulator arrays," etc.

In the case of reflective LCDs, light from crossed (90°) polarizers is blocked, but if the oriented LC molecules can rotate the light such that it is aligned to pass the outer polarizer (or *analyzer*) then light can be transmitted. The oriented nematic LC continuously twists from the direction imposed by the outer buffed glass plate and the inner one. The incident light polarized in the same direction as the outer polarizer (it is now coherent) is passed and then follows the twist of the LC and ends up aligned to the inner polarizer, which passes it on to the mirror. This light is reflected back through the inner polarizer, follows the twist of the LC on the way back out, and ends up aligned to the outer polarizer. One then can see that light as an "off" part of the LCD (bright). To turn "on" a segment (or pixel of the LCD), an electric field (usually AC) is applied across the two electrodes, forcing the LC molecules to take on an orientation orthogonal to the glass. This destroys the LCs ability to rotate the polarization of the incident light such that very little gets back out of the display. One can see this as an "on" part of the LCD (dark). The so-called "supertwist" displays use a 270° twisted nematic LC material to obtain a much greater viewing angle and four times the contrast of 90° twisted nematics.

Turning off the electric field allows the LC to return to its natural (twisted) orientation within 10 to 100 ms (typically). In general, the relaxation time is greater than the excitation time. AC signals are used, typically 3 to 10 V p-p (DC signals will cause irreversible and destructive chemical changes to occur).

Temperature effects are important in LCDs. If the temperature is too low, the orientation "freezes" and cannot be altered easily by external fields. With too high a temperature, the orientation changes from anisotropic to isotropic (random) at the transition temperature (loss of contrast). Typical LCD operation is from a few degrees Celsius (°C) and around room temperature, and can be extended by mixing materials of varying transition temperatures.

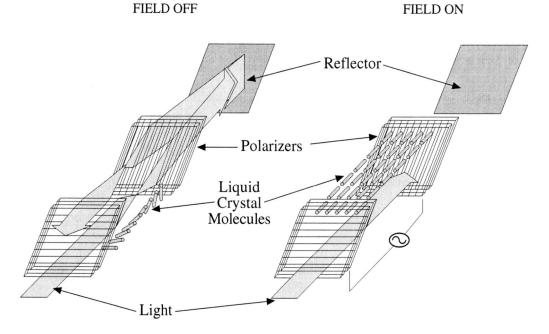

FIELD OFF   FIELD ON

Reflector

Polarizers

Liquid
Crystal
Molecules

Light

Diagram illustrating the principles behind the nematic LCD display. Note that the "buffing directions" (not shown) are orthogonal to the polarizer orientations. In the left-hand state (field off), the display appears transparent or bright, whereas in the right-hand state (field on), the display appears dark. After Cladis (1988).

In "passive" LCD displays a metal-insulator-metal (MIM) junction (made of tantalum pentoxide [Ta_2O_5]) is typically sandwiched between two conductive oxide layers, e.g., ITO). Another glass plate, coated with ITO is typically used to form the other side of the display. The MIM junction has a bidirectional transfer function like a Zener diode's reverse bias I-V characteristic. It blocks voltages lower than a threshold, allowing for *multiplexing* of the display. Unfortunately, MIM LCDs have only two states (on or off), and gray-scale modulation is very useful for computer or video applications.

Active matrix LCDs replace the MIM with a thin-film transistor to handle the switching, allowing for gray-scale modulation of light. The TFTs can be configured as flip-flops (inherent memory in the display) or as linear drivers for the pixels. A colored-pixel filter can also be inserted between the glass plates to make a color display.

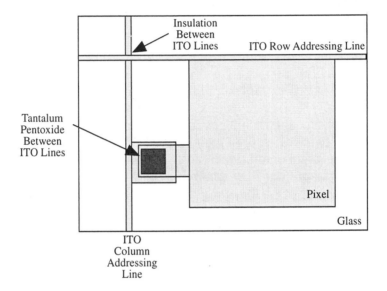

Illustration of a metal-insulator-metal LCD pixel.

DYNAMIC SCATTERING LIQUID CRYSTALS

Some of the first LC displays and later LCD televisions (circa 1969) were made by making use of the *dynamic scattering* mode of LC operation. In this mode, the LC material (specific types are designed for this type of operation) passes light when it is "off" (no applied field) because the molecules align themselves to the walls of the structure they are in (i.e., between sheets of ITO-coated glass). With an applied field, the molecules become disorganized (random) and scatter light, making a pixel appear cloudy, or "on." It is noteworthy that dynamic scattering operation can provide larger viewing angles than twisted nematic displays. Higher drive voltages may be needed (up to 15 V AC) but response times are roughly the same, on the order of 100 ms. The increased drive voltages may decrease the life of a display to about 30,000 hours (about 60% of the life of twisted nematics).

Dynamic scattering operation does not require polarizers. This makes it suitable for simple device structures, such as silicon substrate displays (see below). It is worth noting that one can pour most types of LC material directly onto silicon integrated circuits, power them up, and observe their operation under a microscope. The induced fields from the metallization and polysilicon conductors on the chip are sufficiently strong locally to affect the LC material's orientation. This technique is particularly suitable for static digital CMOS ICs where potentials are roughly either zero or V_{dd}, and the circuits can be clocked slowly enough for the eye to see.

TRANSFERRED SILICON ACTIVE LC DISPLAYS

A noteworthy technology in micromachined active LC displays has been developed by Kopin, Inc. (Taunton, MA). The display is a very high-density transmissive unit in the format of a 35 mm slide that can be inserted into a conventional slide projector to form a projection video system. Unfortunately, the technical details are proprietary at this time, but the essence of the fabrication process is to prefabricate all switching circuits on a single-crystal silicon-on-insulator substrate and then to transfer the single-crystal transistors to a glass substrate. Announced displays based on this technology provide 500 lines-per-inch resolution and the company claims that they can achieve 1,000. A key advantage of this approach is that the single-crystal transistors offer much higher switching speeds than amorphous or polysilicon devices typically used in active LC displays. The Kopin devices can run at video rates.

CHOLESTERIC LIQUID CRYSTALS

The main difference between *cholesteric* and nematic LCs is that the cholesteric types form helices spontaneously (i.e., without the need to be guided by the surfaces they interact with). The pitch of the helices can be comparable to the wavelengths of visible light. A layer of cholesteric LC will pass all incident light through except light that is circularly polarized at the wavelength of the pitch of the helix. This light is scattered back and is thus visible. In typical applications, the cholesteric LC is placed between an outer plastic cap and an optically absorptive (black) layer, and light at the wavelength of the LC helix pitch is scattered back. The pitch of the LC helix is *very sensitive to temperature* (the helix pitch varies with temperature), explaining why LC thermometers show different colors as they are heated, and can range from black (so cool that the pitch is in the IR), to red through blue. (This is the basic principle behind "mood rings" and LC thermometers.) One can glue sheets of cholesteric LC thermometer material to surfaces (i.e., heat sinks, operating silicon chips, etc.) or just spray on the material and directly visualize the temperature profiles. Cholesteric liquid crystals have not as yet been utilized with micromachined devices, but many applications are possible, such as directly visualizing temperature gradients (with a color microscope camera).

3.2.2 REFLECTIVE MICROMECHANICAL LIGHT MODULATORS

As mentioned above, the majority of micromachined light modulators have been mechanical. In this section, the evolution of such devices will be outlined. As for many micromachined devices, their origins go back decades before the terms "MEMS" or micromachining were coined (this should be noted by readers doing literature searches). There are several important issues with mechanical light modulators, including speed, sticking, fatigue, and long-term reliability. These areas are of concern for nearly all micromachined actuators, but are of particular importance for light modulators that are part of huge arrays used in displays.

Thanks to favorable scaling laws, the mechanical microdevices for light modulation have generally been extremely fast responding and have not used much power. From an actuation perspective, a light modulator is one of the few cases where an actuator only needs to move itself (as opposed to interacting with other objects). A down-side of this scaling is the increased possibility of failure due to sticking of the actuators. This latter issue, as well as that of long-term fatigue of flexures, is currently being addressed by several research teams.

Several different geometries of reflective mechanical light modulators have been micromachined, and the table above shows the four major categories or mechanical structures used. The bulk of the devices demonstrated to date have used electrostatic actuation, although magnetic, magnetic/electrostatic, and piezo-electric actuation schemes have also been demonstrated.

Modulator Type	Motion	Side and Top Views
Cantilever Beam	Bending	
Torsional Plate	Rotation about Torsion Axis	
Membrane	Drumhead	
Suspended Plate	Vertical	

Table comparing four basic surface micromachined reflective mechanical light modulator designs. Each design has different sensitivities to geometries, thin-film properties, squeeze-film damping, etc.

ELECTROSTATIC REFLECTIVE LIGHT MODULATORS

The simplest form of actuation for micromechanical light modulators is electrostatic, and thus it is logical that this would have been one of the earliest approaches used, despite its inherently nonlinear response to the control voltage.

WESTINGHOUSE MIRROR MATRIX TUBE

Apparently, the earliest micromachined mechanical light modulator was the "mirror matrix tube" projection display device developed at Westinghouse Research Labs (Pittsburgh, PA) in the mid-1970's (Nathanson and Davis (1973), Guldberg and Nathanson (1975), and Thomas, et al. (1975)). Previous projection TV systems used electrostatically dimpled oil films to modulate light in the projection beam (the "Eidophor"), a technology developed in the 1950's. The mirror matrix tube concept was to fabricate 300 nm aluminum-coated SiO_2 "flaps" (30 nm of aluminum), supported by 4 to 5 µm high silicon posts on a sapphire substrate. These flaps could be deflected relative to the substrate by charging them with an electron beam.

The entire assembly was sealed in a standard vidicon tube with normal electron beam deflection plates. When the electron beam was directed at one of the membranes, it would deflect up to 4°, which would modulate light on the screen through reflective Schlieren optics. Deflection would occur because the illuminating electron beam knocked more electrons (secondary) off the oxidized aluminum than those that hit it, rendering the flaps positively charged relative to the underlying aluminum grid. The patterns could be erased by holding the grid at a negative potential and flooding the array with low-energy electrons to equalize the different potentials on the flaps. In practice, images could be stored for several hours, demonstrating the inherent memory capability of the devices. Contrast ratios were on the order of 10:1.

Their process began with 1.5 in. silicon-on-sapphire wafer and a thermal silicon dioxide layer was grown to form the "flaps" (≈ 300 nm). The oxide was then patterned to form the four-flap structures. A second, "hinge" oxide was then grown and patterned to define the deflectable flaps. The silicon was then etched in HNA (with rotation of the substrate to increase mass transport) to define the posts. Finally, a 30 nm aluminum layer was deposited to form the reflective surfaces on the flaps and also the grid on the substrate.

It is noteworthy that, initially, the evaporated aluminum was too tensile and deflected the flaps up to 2°. If low partial pressures of oxygen were used during evaporation, the aluminum became compressive (a pressure at which the stress was nearly neutral was $\approx 10^{-6}$ torr).

Despite successful demonstration of the devices, the devices were never commercially viable. Difficulties with the HNA etch required to free the deflectable elements were very important (variations in post size across the arrays drastically affected yields). In addition, inability to achieve good enough electron-beam focusing prevented the use of individual pixels (each actual pixel was spread, on average, over four adjacent actuators). Since electron-beam resolution limited both cathode-ray tubes and the mirror matrix displays, there was no great advantage provided. Further details of the reasons these devices did not succeed commercially can be found in Petersen (1982).

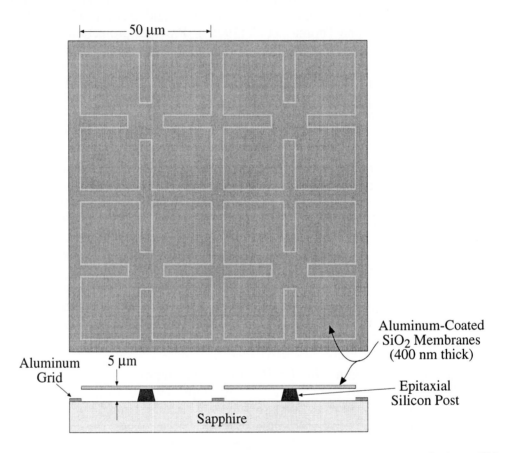

Illustration of top and side views of the Westinghouse mirror matrix device. SiO$_2$ membranes grown on epitaxial silicon on sapphire, released by undercutting the silicon and coated with aluminum formed the deflectable mirror surfaces. After Thomas, et al. (1975).

During the same time period, researchers at IBM also developed such electron-beam-addressed optical modulators. As described in Altman, et al. (1986), their designs used individual SiO$_2$ membranes on a silicon-on-sapphire substrate, with a single SiO$_2$ support post located at one corner of each membrane. The membranes were also aluminized to render them reflective, and the substrates were mounted in a vacuum tube with a scanned electron gun. Despite what were apparently substantial efforts by these two research groups, such display devices did not appear to have much commercial impact.

SILICON CANTILEVER LIGHT MODULATORS

Petersen demonstrated single-crystal cantilever micromechanical light modulators (Petersen (1977)), which are a useful example since the structure is relatively straightforward to model.

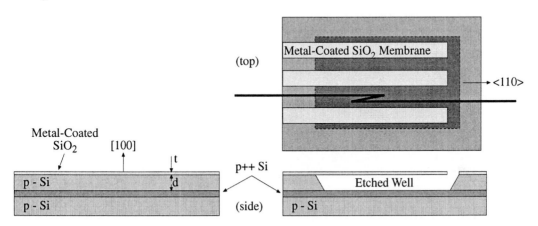

Illustration of undercut cantilever mechanical light modulators. After Petersen (1977).

Petersen's process began with a buried p+ epitaxial silicon layer under ≈ 12 μm of p-type epitaxial silicon. A silicon dioxide layer 500 nm thick was grown to form the cantilevers, followed by the deposition of a 50 nm Cr-Au film. The metal was then patterned to form the electrodes and bond pads, after which the silicon dioxide was etched using the patterned metal as a mask. Finally, EDP was used to form the cavities, with the etch stopping on the buried p+ epitaxial layer.

The cantilevers could be individually deflected by applying a voltage between the electrodes and the substrate p+ layer. Independent electrical connections were provided to each beam via the top metallization. Such electrostatic actuators are inherently nonlinear, and it was observed that once a cantilever was deflected past a threshold angle (i.e., at a threshold voltage), it spontaneously deflected all the way into the etched well.

A demonstration display was made using a 16-element modulator with beam-forming optics and a galvanometer to scan the modulated beam across a ground-glass target. In effect, each modulator element was responsible for forming a single horizontal line on the screen.

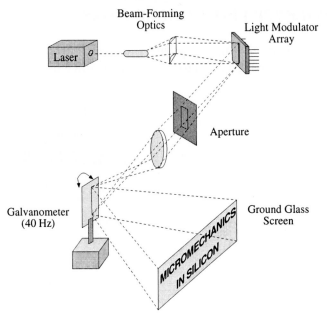

Illustration of experimental setup used with electrostatic cantilever light modulators. After Petersen (1977).

Three models of deflection versus voltage were considered (discussed in the Mechanical Transducers chapter): 1) parallel-plate capacitor, with uniform force along the cantilever (accurate for ≈ 10% deflection); 2) approximate numerical solution taking into account the fact that once the cantilever bends, the tip is closer to the other electrode and has a higher force on it (accurate to ≈ 30% deflection); and 3) assuming a curved cantilever (a still more accurate model that predicted the nonlinear deflection observed above the threshold voltage for full deflection). For modeling such electrostatically deflected devices, it is important to determine whether or not simple (small-signal) models will be suitable for the actual operating conditions or whether more complex, nonlinear models will be required.

The resonant frequency (a key factor for display update rate) was calculated using,

$$f_R = 0.162 \frac{t}{L^2} \sqrt{\frac{EK}{\rho}} \quad \text{in Hz}$$

where,
E = Young's modulus of Si = 190 GPa (19 MN/cm^2)
K = correction factor depending upon density (≈ 1)
 t = silicon cantilever thickness, in cm
L = cantilever beam length, in cm
ρ = density of Si = 2.32 g/cm^2

TORSIONAL SILICON ELECTROSTATIC LIGHT MODULATORS

As an alternative optical deflector for display or galvanometer applications, silicon torsion mirrors were made by Kurt Petersen at IBM based on a two-substrate construction (Petersen (1980, 1982)). A glass substrate was etched to form a well into which the torsion plate could deflect, a support ridge was used to hold the plate away from the well (i.e., to prevent it from being pulled down in parallel to the base of the well), and a pair of metal deflection electrodes were fabricated in the well. The silicon substrate was anisotropically etched to form the torsion plate and torsion bars. The two pieces were glued together and the silicon was grounded during the operation.

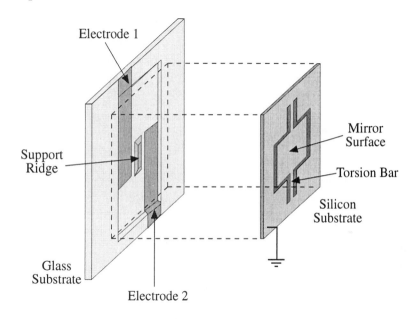

Illustration of silicon torsion mirror optical modulator. After Petersen (1980).

Based on previous analysis of such torsional structures (Higdon and Stiles (1961)), the resonant frequency was estimated from,

$$f_R = \frac{1}{2\pi} \sqrt{\frac{12 K E t^3}{\rho L b^4 (1 + \nu)}} \quad \text{in Hz}$$

where,
  t = silicon wafer thickness, in cm (132 μm in this case, or 0.0132 cm)
  ν = Poisson's ratio of silicon = 0.09
  L = length of each torsion bar, in cm

b = dimension of the square mirror, in cm

K = a dimensionless constant dependent on the cross-sectional shape of the torsion bar (K ≈ 0.24 in this case)

Both calculated and measured resonant frequencies were in the range of 15 kHz. In addition, an electrostatic equation predicting deflection versus voltage was determined and shown to be accurate to about 20% for angles smaller than 1°.

It is worth noting that the calulated stress on the torsion bars was $\approx 2.5 \times 10^9$ dyne/cm^2 = 250 MPa more than an order of magnitude below the fracture stress for single-crystal silicon (Pearson, et al. (1957)). Continuous operation of the torsional devices (near resonance) for months, followed by dislocation-sensitive Si etching showed virtually no damage. This highlights one of the important effects of using single-crystal materials and of scaling to these dimensions: fatigue mechanisms are not the same as those found in macroscopic materials due to the absence of grain boundaries. For example, in this case, over 10^{12} cycles were demonstrated with deflections on the order of ± 1°.

More recently, Dötzel, et al. (1997) demonstrated the fabrication of single-crystal silicon torsional light modulators as large as 3 × 3 mm using combinations of dry and wet etching. They also demonstrated the use of square modulator plates for 2-D light beam steering. These plates were designed with four underlying deflection electrodes, and were supported by compliant (folded) silicon members to allow semi-independent actuation in two orthogonal directions.

TORSIONAL ALUMINUM ELECTROSTATIC LIGHT MODULATORS

Texas Instruments, Inc. (Dallas, TX) has developed a high resolution, micro-mechanical reflective display device for the mass market arena (as projection TV and HDTV displays), as described in Hornbeck (1995). The electrostatically deflected micromirrors, referred to as digital micromirror devices, have been under development in various embodiments for more than a decade. The basic idea is to fabricate aluminum mirrors that are suspended above a silicon substrate by very thin (typically 60 nm in present versions) torsional hinges. When appropriate potentials are applied between the mirror plate and underlying deflection electrodes, the mirror rotates. Since the rotation of such electrostatic devices is highly nonlinear with respect to applied potential, it is often advantageous to use them in a "digital" manner (full deflection one way or another, with little or no time spent in the "undeflected" or flat state), hence the name "digital micromirror device," or DMD™.

Using organic sacrificial layers (e.g., positive photoresist and PMMA), torsional micromirrors are fabricated above conventional CMOS circuitry to form arrays of electro-optic modulators (thus all process steps must be MOS compatible, but the advantage is leveraging existing active circuit process development efforts).

After the underlying circuits are fabricated, the process begins with the fabrication of the deflection and "landing" electrodes (landing electrodes are kept at the same voltage as the mirror beams and are intended for the beam's tips to land on in order to prevent them from welding down in the "on" state).

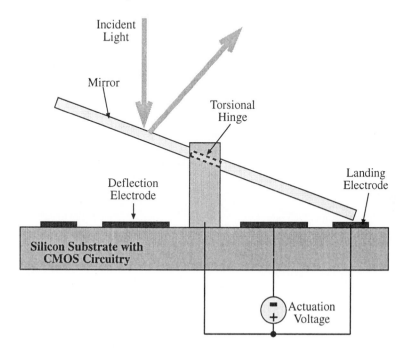

Conceptual illustration of the basic operation of a Texas Instruments torsional electrostatic micromirror. The tip of the torsional plate is allowed to land on an equipotential "landing electrode" to prevent sticking. Actual mirror designs have evolved considerably over time and are more complex.

It should be noted that the process described below has evolved over time, and this can be tracked through several published references (Hornbeck (1987, 1990, 1991a, 1991b, 1995) and Sampsell (1990)). Unfortunately, but understandably, none of the published process descriptions is complete. At present, an 0.8 μm CMOS process is first run to form the underlying static RAM address circuitry. As illustrated below, a thick oxide layer is deposited and planarized using CMP to provide an optically flat starting surface. A third-level aluminum metallization (above the two levels of the CMOS itself) is then deposited and patterned, followed by the deposition and patterning of the lower organic sacrificial layer (most likely photoresist or PMMA). A thin aluminum alloy layer ($\approx$ 60 nm) is then sputter-deposited and covered with plasma-deposited SiO_2, which is patterned to form as an etch mask for the hinges in a subsequent step. A thicker layer of aluminum is then sputter-deposited and patterned to form the yoke structures and the hinge support posts (by filling in holes in the lower organic sacrificial layer). A plasma etch is then used to pattern both the yoke metal and the underlying hinge metal (using the previously patterned silicon dioxide etch mask). Next, an upper organic sacrificial layer is spun on and patterned, followed by sputter deposition of the mirror layer, which is patterned with a plasma-deposited SiO_2 mask layer.

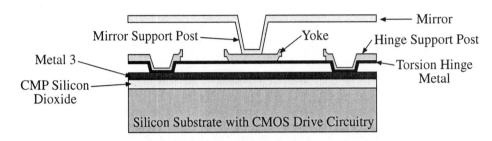

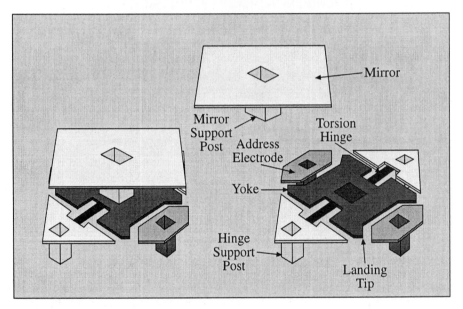

Illustration of current-generation torsional DMD devices, showing a "hidden hinge" design. After Hornbeck (1995) (not to scale). The top view shows a cross section and the bottom view shows the overall shape of the structure. The actual mirrors are 16 × 16 μm on 17 μm centers.

Critical to the processing of such surface micromachined devices is to leave the sacrificial layers in place during wafer sawing. The DMD™ wafers are partailly sawed and then plasma-etched (presumably in oxygen) to remove the sacrificial layers. At this point an anti-stiction self-assembled monolayer of perfluorodecanoic acid (PFDA) is deposited by vapor condensation (Henck (1997)) and the individual chips are tested after the wafers are mounted to sticky tape carriers. After testing, the tape backing is brought into contact with a dome that controllably breaks the wafers into dice, which are then mounted in packages and wire-bonded. Final plasma etch cleaning and repeat anti-stiction steps are done prior to the welding of an optical window on to the package.

It should be noted that careful selection of aluminum alloys for the different layers was carried out. For example, to prevent hillocking of the mirror surfaces, typical alloys include a small percentage of Cu and Si. Early publications also disclosed that the critically important hinge layer was 0.2% Ti, 1% Si, rest Al. A later patent (Webb, et al. (1996)) described a large variety of possible hinge materials (e.g., Al, Al alloys, Ti/W alloy, titanium nitride, titanium tungsten nitride, Cr, etc.) that were implanted with low-solubility implanted ions intended to harden the hinges and to reduce creep during torsional cycling. Interestingly, the authors noted that in typical applications, such as video projection, the hinges spend 75% of their time deflected in one direction versus 25% in the other, producing permanent offsets if creep occurs. The preferred implanted ions included O, N, B, C, a nitride, a carbide, or a metal such as Co. Alternatively, Tregilgas (1996) describes methods of fabricating the hinges from various aluminum compounds with other metals or non-metals to prevent creep (preferred alloys include Al_3Fe, Al_3Nb, Al_3Ni, Al_3Zr, Al_4Mo, Al_4W, AlAs, AlLi, AlN, etc., but no particular one is identified as the best). It is noteworthy that Tregilgas (1996) states that these aluminum alloys can be etched using conventional aluminum etchants. Such anti-creep concepts are likely to be useful, or even necessary in a variety of other micromachined devices.

A variety of different DMD™ arrays have been demonstrated, including an 848×600 array for projection television and computer graphics applications, a $1,280 \times 1,024$ array for high performance displays, and a $7,056 \times 64$ array for 600 dpi printer applications. For display applications, amplitude control is obtained by time division multiplexing (tyically, 8-bit amplitude resolution is obtained), and a rotating RGB color wheel is used to add an extra level of time multiplexing for color operation. Current devices operate through $\pm 10°$ angles with mechanical switching times on the order of 15 µs and optical switching times on the order of 2 µs. Total deflection voltages are on the order of 20 V and a distributed bias electrode structure is used with an off-chip voltage source to minimize the voltage that the CMOS control circuitry must switch.

A key observation from this work is that the fatigue characteristics of the thin aluminum alloy hinges are radically different from those of macroscopic hinges of the same materials. In work at Texas Instruments, no gross failures have been observed in such torsional hinges after one trillion operations (Hornbeck (1995) and also Henck (1997)). In similar work at Stanford by Storment, et al. (1994), no resonant frequency shifts (very sensitive to material property changes) were observed for over 40 billion cycles (Honer, et al. (1996)). It is thought that for such thin deflection members (often only two to three grains of aluminum thick), the lack of a significant grain structure in one direction may greatly increase the fatigue lifetime.

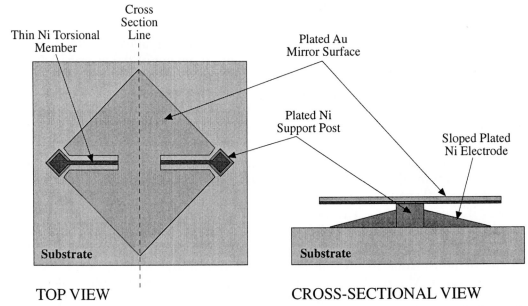

TOP VIEW   CROSS-SECTIONAL VIEW

Illustration of an all-electroplated, electrostatically deflected, reflective light modulator with sloped deflection electrodes. Adapted from Wagner, et al. (1997).

A metallic torsional mirror design was also demonstrated by Wagner, et al. (1997) consisting of electroplated nickel plates and torsional members (400 nm thick), with a 2 μm electroplated mirror layer and electroplated nickel support posts 13 μm tall. These torsional devices were arranged above sloped, electroplated electrodes realized using self-shorting, floating plating traces on either side of a single, electrically driven trace. This approach, discussed in the Micromachining Techniques chapter, resulted in the formation of symmetrical wedges of electroplated nickel beneath the mirrors (aligned with the peaks of the wedges parallel to the torsional axis), serving to reduce the threshold voltages. An electroplated copper sacrificial layer was deposited and mechanically planarized to form the gap beneath the mirrors. These devices operated over deflection angles of $\pm 15°$, and had 100×100 μm pixels and 35 V threshold voltages. Switching speeds were not reported.

POLYSILICON TORSIONAL OPTICAL MODULATORS

Another interesting example of torsional micromirror-type light modulators is the work of Jaecklin, et al. (1994). They demonstrated micromirrors using surface aluminized, phosphorus-doped polysilicon structures released by wet etching a sacrificial oxide using buffered HF and glycerine (NH_4F (from 40% solution):HF (from 48% solution):glycerine, 4:1:2). The added glycerine helped to prevent the etching of the aluminum. The 30×30 µm mirrors switched at ≈ 30 V, swinging through angles as large as 7.6° and landing on landing electrodes at the same potential (to prevent sticking). The measured resonant frequency of the micromirrors was 97 kHz, with a Q of 4.8. The fact that such devices are *bistable* (the mirrors can be held down by a voltage substantially lower then the pull-down voltage) was taken advantage of to fabricate an X-Y addressable array without on-board electronics (using row-column addressing with appropriate voltages).

DEFORMABLE GRATING LIGHT MODULATORS

Another interesting micromechanical light modulator is the deformable grating modulator (DGM) demonstrated by Solgaard, et al. (1992). Their modulators are based on arrays of taut aluminum beams that can be pulled down from planarity with the surrounding reflective surface to change from a fully reflective surface to a diffraction grating ("off" pixel), scattering light to the sides. Such modulators have been shown to switch in as little time as 20 ns, with contrast ratios as high as 200:1, and are bistable (useful for display applications). The 25×25 µm pixels could allow for high-density integration. A display with high color saturation can be built from a 2-D array of DGM devices. By using three different grating periods, each of the three color primaries can be generated. Such displays are under development by Silicon Light Machines, Inc., San Jose, CA.

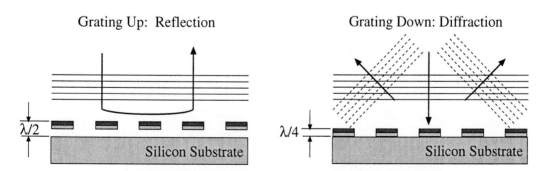

Illustration of the basic operating principle of a deformable grating light modulator. After Solgaard, et al. (1992). The actuators themselves consist of aluminum on silicon nitride.

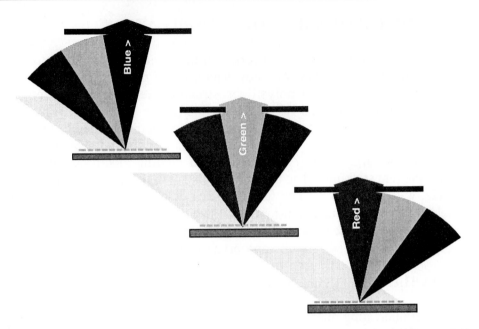

Illustration of the use of deformable grating light modulators with three different periods to control all three color primaries. After Solgaard, et al. (1992).

ELECTROSTATIC MEMBRANE LIGHT MODULATOR

Vdovin and Middelhoek (1995) described a physically simple membrane-based light modulator based on a thin, aluminum-coated, silicon nitride membrane that could be electrostatically deflected at multiple locations by underlying aluminum electrodes.

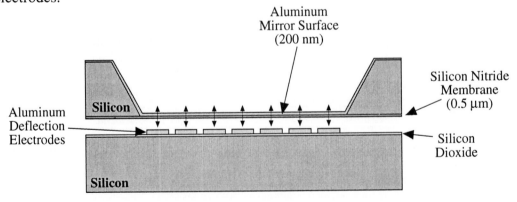

Illustration of an electrostatically deflected membrane light modulator. After Vdovin and Middelhoek (1995). The potentials applied to the deflection electrodes were individually controlled to deflect the membrane locally.

The membrane was formed by KOH etching an 0.5 μm thick silicon-nitride-coated wafer and then metallizing the back side with 200 nm of aluminum to form the mirror surface. A separate silicon die, containing independently addressable aluminum deflection electrodes on a SiO$_2$ insulating layer, was mounted a fixed distance (25 to 100 μm) from the membrane.

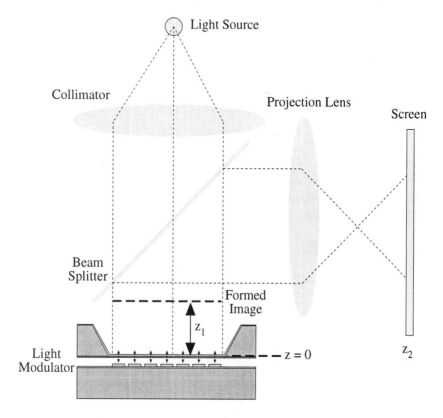

Illustration of the operating mode of an electrostatically deflected membrane light modulator. After Vdovin and Middelhoek (1995).

The intensity of the light intensity in the formed image plane (distance z_1 from the membrane) at a location (x, y) is given by,

$$I(x,y,z_i) \approx I_o\left(1 + \frac{2z_1\varepsilon_o\varepsilon_r V(x,y)^2}{d^2 T}\right)$$

where,

I_o = intensity in the plane of the deformable mirror, in W/cm^2

$V(x, y)$ = voltage applied to a deflection electrode at location (x, y) d = distance between the membrane and the electrodes

T = membrane tension, in Pa

ε_o = dielectric permittivity of free space = 8.8542×10^{-12} F/m^2

ε_r = relative dielectric permittivity of the medium beneath the membrane
       (dimensionless), which in this case was air, with $\varepsilon_r = 1$

Note that this equation does not contain any wavelength-dependent terms (assuming that the reflectivity of the mirror surface is fairly wavelength-independent for visible light), and thus the display can be used for color images, as can most of the reflective devices discussed herein.

By applying a sufficient voltage (not given in the paper) to a hard-wired pattern electrode array to deflect the membrane 10λ for the 633 nm light used, $10 \times$ 10 mm membranes were used to project images successfully. The measured time constants (80% of maximum brightness reached in 500 μs) indicated that the membrane would be capable of 100 Hz frame rates if simultaneous addressing of the electrodes was employed. Unfortunately, only relatively low-quality images have been demonstrated to date (possibly due to roughness of the aluminum surface).

A noteworthy advantage of membrane-based light modulators is that they can be used to achieve a 100% pixel fill-factor, which is not possible for the other architectures discussed. Another important point with such devices is their potential for low-cost manufacturing due to their simplicity. A significant negative aspect, however, is the difficulty in electrically and mechanically isolating individual pixels.

MAGNETICALLY DEFLECTED LIGHT MODULATORS

Due to the relatively large forces that can be obtained from electromagnetic actuators, they provide an alternative to electrostatic operation in optical modulators, although higher power levels are likely. In general, an external permanent magnet is used to provide a field against which micromachined coils can generate a force when energized.

Miller, et al. (1996) demonstrated an electromagnetically actuated cantilever-type light modulator developed for a holographic data storage application. The basic idea was to form an electromagnetic coil as well as a Permalloy magnetic layer on a thin cantilever beam (with serpentine, single-crystal silicon supports). The cantilever could then be actuated using an external magnetic field (alone, with the external field interacting directly with the Permalloy, or interacting with a field due to current in the electromagnetic coil on the cantilever). When the field was first applied with no current in the coil, the cantilever deflected to a "bias" angle, due to the Permalloy only. Current supplied (in either direction) in the coil allowed for movement about this bias point. Thus the external and on-board coil current field could be used for "coarse" and "fine" positioning, respectively.

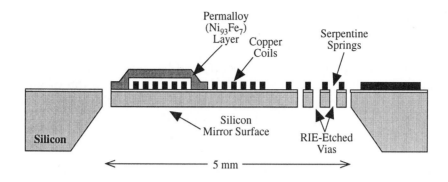

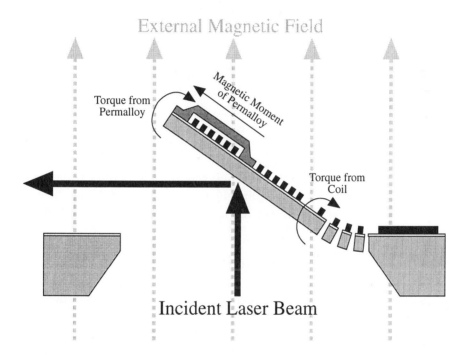

Illustration of (top) components of an electromagnetic optical modulator and (bottom) operation of the modulator showing torques generated by interaction of the external magnetic field with the Permalloy and with the electromagnetic field from the coil (when energized). Note that in the top drawing, the other dimension of the mirror membrane was 4 mm. Also, the restoring torques from both the serpentine spring and thermal bimorph (expansion coefficient mismatch) forces are not shown. After Miller, et al. (1996).

Their fabrication process used a 40 μm thick epitaxial layer to form a membrane, using a thin, buried p++ boron etch-stop layer. Back-side etching was carried out to form the rectangular membrane, the bottom of which would ultimately form the optical reflecting surface. The p++ layer was removed, and a SiO_2 layer was grown to serve as an electrical insulator. A Ti/Cu seed layer was evaporated onto the top surface, followed by the electroplating of a 30-turn, 9 μm thick, planar actuation coil from copper. This was followed by the deposition and hard baking of an insulating photoresist layer over the coil, the evaporation of another Ti/Cu seed layer, and the electroplating of an 11 μm thick Permalloy layer. The final step involved the selective removal of the SiO_2 and RIE etching of the serpentine hinges and cuts to free the 4×5 mm mirror membrane.

Using a high external field to saturate the Permalloy (79,122 A/m, or 994 Oe), the bias angle was 58°, with ±10° deflection with moderate linearity (apparently, suitable for their application). They typically used a magnetic field of ≈ 200 Oe at the mirror, generated using a permanent magnet located 2 mm above it. The first resonant mode of the structures was 65 Hz, with a step response time of 25 ms.

Miller, et al. (1997) demonstrated electromagnetically actuated cantilever and torsional fiber-optic light switches. These single-crystal silicon devices consisted of 30 μm thick silicon plates, each with a 70-turn electroplated copper coil in two layers (insulated with ITO). The reflecting surfaces were separately fabricated (110) silicon mirror, etched using TMAH. A permanent magnet was mounted beneath each device, which was deflected by passing current through the coils (30 mA produced 200 μm of deflection). They reported switching times of 10 ms at currents of 20 to 30 mA (corresponding to power levels of 17 to 38 mW). It is interesting to note that the cantilever devices could not survive a 20 g shock test, while the torsional designs could readily do so.

MAGNETIC/ELECTROSTATIC LIGHT MODULATORS

Judy and Muller (1996) demonstrated a combination magnetic/electrostatic torsional light modulator. By using magnetic fields from external or on-chip coils, nickel-plated polysilicon plates with thin torsional members at one end could be rotated out of the plane of the substrate. If a sufficient voltage was applied to some of the plate(s) (relative to the substrate) prior to application of the magnetic field, those plates would not deflect because they were held down by electrostatic clamping.

Their fabrication process consisted of coating a 10 Ω•cm n-type silicon wafer with 2 μm of LPCVD silicon nitride, followed by a 2 μm LPCVD PSG sacrificial layer. Vias were etched down to the silicon nitride (anchors) and directly to the silicon (ground contacts). A 2 μm thick LPCVD polysilicon structural layer was then deposited at 605°C, followed by an 0.5 μm layer of LPCVD PSG and a nitrogen anneal at 1,000°C for 1 h to dope the polysilicon with phosphorus from the PSG. The top PSG was patterned and used as a RIE etch mask to pattern the underlying polysilicon. The top PSG was then removed in an HF etch that also partially undercut the mechanical polysilicon structures. At this point, a 100 nm

NiFe seed layer was sputtered and a 4.0 μm photoresist plating mask was used to electroplate the nickel ferromagnetic layer, followed by a sputter etch step to remove excess seed layer. Finally, an HF release etch was used to free the polysilicon structures, released via a CO_2 critical point drying technique.

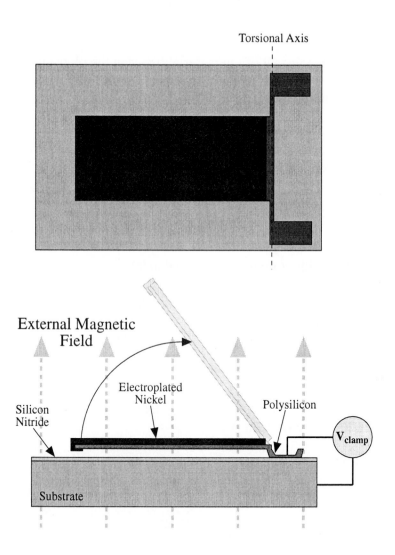

Simplified illustration of the design and basic operating principle of a combination magnetic/electrostatic torsional light modulator. After Judy and Muller (1996).

The clamping force acting on the torsional plate (assumed to be the same as for a simple parallel-plate capacitor) is,

$$F_c = \frac{\varepsilon_r \varepsilon_o A V^2}{2d^2}$$

where,
ε_r = relative dielectric constant of the material between the plate and ground plane
A = plate area
d = electrode separation when clamped
V = applied voltage

The resulting electrostatic torque can be obtained by integrating this force over the length of the plate, L,

$$T_c = \int_{x=0}^{1} F_c dx = \frac{\varepsilon_o \varepsilon_r w L^2 V^2}{4d^2}$$

where w is the plate width (along the torsional axis).

The magnetic torque when the plate is clamped (zero angle) is given by,

$$T_M (\text{clamped}) = V_{mag} MH$$

where,
V_{mag} = volume of magnet
M = magnetization of the film (calculated from the coercivity of the film, as discussed in Judy and Muller (1996)), in tesla (T)
H = external magnetic field, in A/m

By equating the electrostatic clamping force to the induced magnetic torque, one can obtain an equation for the minimum clamping voltage,

$$V_c > \sqrt{\frac{4MHtd^2}{\varepsilon_o \varepsilon_r L}}$$

where t is the thickness of the ferromagnetic material.

From this, a clamping voltage of 4.1 V was computed for their design and materials, which agreed well with experimental results ($\approx$ 5 V in a 10 kA/m magnetic field). The devices were capable of deflecting through 80°, with resonant frequencies on the order of 200 Hz (the response was, in fact, somewhat nonlinear and hysteretic, so that there was no resonant frequency in the true sense).

3.2.3 TRANSMISSIVE MICROMECHANICAL LIGHT MODULATORS

Another category of micromechanical light modulator is that of transmissive devices, wherein the mechanism interrupts the path of light passing through the substrate (or through a micromachined hole in the substrate). These devices generally involve surface micromachined structures that are electrostatically driven, such as motors or in-plane springs.

Kraus, et al. (1997) demonstrated the use of an electrostatic motor to serve as a rotating transmissive light modulator (or shutter) for pyroelectric IR detectors. They used a standard sacrificial oxide, polysilicon mechanism process, combined with the deposition of a gold layer on pie-slice-shaped regions of the center of the rotor as well as back-side KOH etching to form a path for light to travel through the substrate. Once these steps were completed, the surface microstructures were released using an HF etch followed by a p-dichlorobenzol sublimation step to prevent stiction. Rotor diameters of 800 μm and 1.2 mm were demonstrated, with drive voltages ranging from 35 to 60 V (starting voltages ranged from 60 to 100 V), and rotation rates of 0 to 7,000 RPM (two shutter events per rotation, since two of the rotor quadrants were gold-coated). The 50 nm gold layer used to block light showed a transmittance of < 0.1% over the wavelength range of 2.5 to 25 μm. It should be noted that for operation above a few microns of wavelength, the back-side via may not be necessary, since IR light of those wavelengths passes readily through silicon (although with a wavelength-dependent attenuation).

Perregaux, et al. (1997) demonstrated the use of an in-plane spring to move a polysilicon shutter blade along a circular path to modulate visible light (200 to 500 nm). They fabricated the lateral shutters using a sacrificial oxide, polysilicon mechanism process with a back-side etch (not specified) to form vias for light transmission. No metal coating was apparently used on the polysilicon shutters, since they are opaque to light at the specified wavelengths. Shutters with dimensions of 30×90 μm, on 190 μm long, 2.5 μm wide, and 2 μm thick polysilicon beams were used, yielding a resonant frequency of 9.5 kHz, a quality factor of 33, a switching voltage of ≈ 70 V, and a release voltage of 28 V. Apparently, switching times were close to the required 50 μs. Arrays of 28×21 elements on a 180×180 μm pitch were fabricated.

3.2.4 OTHER LIGHT MODULATORS

Several non-mechanical optical modulation effects exist, such as the *Faraday* (magneto-optical), *Kerr* (square-law electro-optical), and *Pockels* effects (linear electro-optical), of which a good overview is provided in Hecht (1987).

Faraday discovered that the application of a strong magnetic field in the direction of propagation of light through a physical medium would lead to a rotation of the plane of vibration of the light. If rotated polarizers were placed at both entry and exit regions of the medium in question, a so-called Faraday effect optical modulator could be formed. The rotation angle, β, is given by,

$$\beta = VBd$$

where,
V = the Verdet constant for a given material, in min of arc/(gauss•cm)
B = magnetic flux density, in gauss (G)
d = distance the light travels through the modulator medium, in cm

In practice, the effect is quite wavelength-dependent (the Verdet constant falls as the wavelength increases), and relatively small for typical materials. For visible or UV applications, paramagnetic materials (e.g., terbium-doped borosilicate glass) or diamagnetic materials (e.g., zinc selenide) are used (see the table below). Ferromagnetic materials such as rare earth garnets can have quite large Verdet constants and can be used in a variety of IR modulator applications (particularly in the 1.1 to 1.6 µm range commonly used for IR laser diodes in communcations systems). It is conceivable that suitable materials for Faraday effect electro-optic modulators could be monolithically integrated (or attached as a hybrid) with micro-machined coils and optical elements to yield extremely compact modulator or complete systems., particularly since on-chip magnets could be very closely coupled to the active material.

In order to discuss two other important effects, the Kerr and Pockels effects, the term *birefringence* must be introduced. Birefringent materials effectively have two different refractive indices, such that an image viewed along certain directions through such a material will appear "doubled." For example, the refractive indices in two orthogonal directions, referred to as n_o and n_e, can be quite different in some materials, such as calcite crystal ($n_o = 1.6584$ and $n_e = 1.4864$ at $\lambda = 589.3$ nm).

The Kerr effect occurs when an isotropic, transparent substance is placed in an electric field and thus becomes birefringent, with a refractive index difference given by,

$$\Delta n = \lambda_o KE^2$$

where K is the Kerr constant in peculiar units of 10^{-7} cm/statvolt^2, where a statvolt ≈ 300 V, and E is the field strength, in statvolts/cm.

Material	Conditions	Verdet Constant in min. of arc/(gauss•cm)
Yttrium-Iron-Garnet (YIG)	λ = 1,300 nm, T = N/A	10.5
Yttrium-Iron-Garnet (YIG)	λ = 1,550 nm, T = N/A	9.2
Light Flint Glass	λ = 578 nm, T = 18°C	0.0317
water	λ = 578 nm, T = 20°C	0.0131
NaCl	λ = 578 nm, T = 16°C	0.0359
Single-Crystal SiO_2 (quartz)	λ = 578 nm, T = 20°C	0.0166
Air	λ = 578 nm, T = 0°C, P = 1 atm	0.00000627
CO_2	λ = 578 nm, T = 0°C, P = 1 atm	0.00000939
$NH_4Fe(SO_4)_2 \cdot 12H_2O$	λ = 578 nm, T = 26°C	-0.00058
Bismuth-Substituted YIG (BIG)	λ = 1,300 nm, T = N/A	-600
Bismuth-Substituted YIG (BIG)	λ = 1,550 nm, T = N/A	-806

Table of Verdet constants for various materials. From Hecht (1987) and Young, et al. (1993).

Since the Kerr effect is proportional to the square of the field strength, it may not be ideal for some modulation applications (nonlinear), although frequency responses as high as 10 GHz are achievable. In addition, most efficient Kerr modulators (or Kerr cells) use liquids such as nitrobenzene or carbon disulfide, not readily adapted to micromachined applications. Some solid materials such as potassium tantalate niobate or barium titanate exhibit the Kerr effect and are potentially suitable in such cases. It is important to note that at the micromachined scale, the voltages (which can be multiple kilovolts for macroscale Kerr cell modulators) can be reduced greatly since it is the field strength that is of importance.

The Pockels effect also makes use of electric-field-induced birefringence, and occurs in certain crystals lacking a center of symmetry (these crystals, referred to as *non-centrosymmetric*, are also piezoelectric). Typical materials used for such modulators are KD_2PO_4 (potassium dideuterium phosphate) or CsD_2AsO_4 (cesium dideuterium arsenate) are used, fabricated by growing the crystals in a solution of heavy water (D_2O). As for the Kerr effect, Pockels modulators can operate at extremely high frequencies (as high as 30 GHz), and they are not particularly suited to micromachining applications. Unlike the Kerr effect (square law), the Pockels effect is linearly related to the applied electric field. For more information on the electro-optical effects, the reader is referred to Kaminow and Turner (1966), Kaminow (1974), Yariv (1976), and Yariv and Yeh (1984) (this last reference also discusses the magneto-optical effect). With particular attention to silicon, electro-optical effects are also discussed in Soref and Bennett (1987).

In addition, another family of non-mechanical electro-optical modulators, multiple quantum well (MQW) devices, have also been investigated. These devices, the details of which are beyond the scope of this book, have been fabricated from III-V compounds, co-integrated with photodetectors and flip-chip-bonded to silicon substrates with CMOS control circuitry. For example, Lentine, et al. (1996) present a discussion of such integrated optical modulators.

4. MICROMACHINED OPTICAL STRUCTURES

Micromachining various materials can yield a large variety of optical structures that do not directly participate in the transduction process. Examples of such structures include lenses, reflectors, gratings, couplers and switches for fiber optics, etc. Representative examples are presented below, and the reader is also referred to several reviews of the subject, such as Tien (1977), Peckerar, et al. (1994), Nishihara, et al. (1994), Peckerar, et al. (1994), and Deimel (1995).

4.1 MICROMACHINED FIBER-OPTIC COUPLERS

Micromachined grooves and channels have been used for some time as mechanical couplings for fiber-optic components due to the extremely high precision achievable. For example, the triangular cross section of long horizontal grooves wet etched anisotropically into (100) silicon are good choices for aligning cylindrical fiber-optic components. Deep reactive ion etching and other high-aspect-ratio etching approaches can also be used, but if reflective surfaces are also needed, these may compromise surface smoothness. Vertical fiber-optic couplers have also been demonstrated by Ade, et al. (1987). They used laser-driven wet etching with HF to fabricate 15 to 18 µm diameter, 20 to 25 µm deep cylindrical couplers in silicon (the etching process is described in the Micromachining Techniques chapter), leading directly to a monolithic photodiode. While the density of vertical couplers can be much greater than horizontal versions, the assembly task can be much more complex.

In general, there are many applications wherein the precision of micromachined mechanical components is extremely useful in optical systems, as discussed further in Section 4.5 below.

4.2 MICROMACHINED REFLECTIVE COMPONENTS

4.2.1 FIXED MICROMIRRORS

Since there are many anisotropic etching approaches that yield nearly perfect crystal plane surfaces, it is natural to expect these to be taken advantage of as mirror surfaces in optical systems.

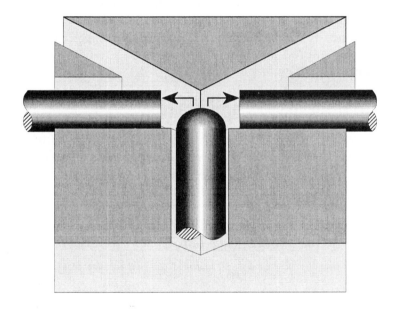

Corner-type, bulk micromachined silicon beam splitter for fiber-optic applications. After Rosengren, et al. (1994).

For example, Rosengren, et al. (1994) discuss the fabrication of optical planes and reflectors from silicon. They took advantage of the well-defined 45° angle between the (100) and (110) planes to make optical reflectors and fabricated V-grooves bounded by (111) planes to hold optical fibers. They used KOH etching (28 wt%, 80°C, 1 μm thermal SiO_2 mask) of (100) wafers to realize the structures, which were coated with evaporated aluminum after etching to increase their reflectivity. Beam splitters based on either vertical semitransparent walls or reflecting corners were successfully implemented. It is interesting to note that the vertical walls were obtained by etching to reveal the <100> planes perpendicular to the wafer surface. Etching does not stop on these planes, but rather continues etching laterally at the same rate as vertical etching. The point is that the planes must remain vertical.

In some cases, an optical transducer itself has built-in micromachined optical surfaces. For example, cleaved-coupled-cavity laser diodes are often fabricated by making use of the (110) cleavage planes of GaAs and its alloys to allow tuning of longitudinal lasing modes (see, e.g., Antreasyan, et al. (1986)).

High aspect ratio dry etching can be used to remove dependencies on crystal planes, as demonstrated by Marxer, et al. (1997). They reported sidewall angles of greater than 89.3° and surface roughnesses of ≈ 40 nm RMS. They etched 75 μm deep channels down to a buried oxide etch-stop, after which the surfaces were coated with various reflective metal films using electron-beam evaporation. In addition, electrostatically driven mechanical light switches, released by etching the oxide etch-stop layer, were fabricated on the same substrates (the switches could be actuated in under 200 μs with a drive voltage of 30 V).

4.2.2 POSITIONABLE MICROMIRRORS

Positionable micromirrors (or microreflectors), as opposed to moving light modulators, are generally moved at relatively low speeds and then fixed in position for optical alignment (however, there is not always a useful distinction between the two).

A very useful structure constructed from three square mirrors meeting at a single corner is known as a *corner-cube retroreflector* (CCR). A CCR can reflect light directly back down its incident path (this type of reflector is commonly employed in bicycle and road sign reflectors).

A micromachined polysilicon CCR with an aluminum reflecting surface was designed by Comtois and Bright (1996) and fabricated using the MCNC polysilicon foundry process. The mirrors were folded out of plane and held in place using hinges and latching elements of the type developed by Pister (1992). They used polysilicon thermal actuators to modulate the position of one of the CCR's mirrors. Similarly, Chu, et al. (1997) used electrostatic actuation of Au-coated polysilicon CCRs to modulate the reflected beam at rates as high as 1 kbps.

4.3 MICROMACHINED TRANSMISSIVE COMPONENTS

4.3.1 OPTICAL WAVEGUIDES

Optical waveguides are analogous to wires for electrical signals in the sense that they can be used in micromachined optical systems to route optical energy from one region to another. Typically, a high refractive index material is sandwiched between two layers of a lower-index material (or, in the case of a cylindrical optical fiber, the high-index core is clad with a lower-index coating). It should be noted that the low-index material can be, for example, air, but usually both materials are deposited solids. Light in the waveguide tends to follow the higher-index material if approaching the interface at a sufficiently glancing angle. The angle of incidence of light propagating down a waveguide is generally measured relative to the perpendicular direction to propagation, so that the angle needs to be greater than the critical angle θ_c for total internal reflection (theoretically, lossless propagation) to occur. The critical angle is given by,

$$\theta_c = \sin^{-1}\left(\frac{n_c}{n_w}\right)$$

where n_c is the refractive index of the cladding (outer) layer, and n_w is the refractive index of the waveguide (inner) layer.

Fullin, et al. (1994) demonstrated integration of photodiodes with CMOS as well as single-mode strip waveguides and waveguide-to-photodiode couplers. Their approach relied on adding three extra layers to a commercial 2 μm CMOS process, most notably a silicon oxynitride ($Si_xO_yN_z$, n = 1.51) layer in between to SiO_2 layers (n = 1.46) to act as the waveguide for 780 nm laser light. By etching pits with mirror surfaces (the details of the process are not given), reflective coupling between the waveguide and the photodiode was achieved.

Silicon nitride has a refractive index greater than silicon dioxide so it could theoretically be used to form a guided wave structure without the need to invoke oxynitrides. However, silicon nitride is generally too lossy for this to be practical.

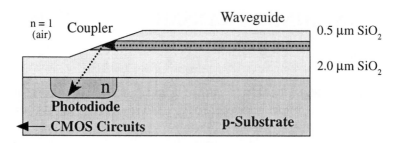

Illustration of a CMOS-integrated waveguide and reflective coupler (not to scale). After Fullin, et al. (1994).

4.3.2 LENSES

There are two types of lenses: *refractive* and *diffractive*. Refractive optics encompass the well-known optical lenses used in everyday applications such as cameras, and operate on bending, or refraction, of light passing through optical elements. Diffractive optics, used for gratings and other applications, operate on the basis of forming multiple wavefronts from incident light and using constructive and destructive interference to obtain desired optical functions (potentially equivalent to those of refractive optics). Both types of optics have focal length and other effects that are wavelength-dependent, leading to *chromatic aberrations*, but these effects are different in the two types and can sometimes be cancelled by combining them.

REFRACTIVE LENSES

The refractive index, n, is typically uniform in the refractive material (usually glass), and relates the speed of light in the refractive material, v, to its speed in vacuum, c,

$$n = \frac{c}{v}$$

It is also useful to note that the refractive index can be related to a material's dielectric constant and magnetic permeability by,

$$n = \sqrt{\varepsilon_r \mu_r}$$

where ε_r is the relative dielectric constant of a particular material, and μ_r is the relative magnetic permeability of a particular material. However, the above equation must be used with care in practice. While μ_r can generally be taken to be one for optical materials, ε_r at the frequencies of interest in optical systems is typically substantially less than the value calculated from electrical measurements (DC value of n). This is because ε_r values decrease with λ to a final value of 1 for very short wavelengths (this happens since the dielectric properties of materials arise due to polarizability at the level of domains, atoms or electrons, all of which can only track electromagnetic perturbations up to certain maximum frequencies). The variation of n with λ is an important effect, varying the angle through which light of a particular wavelength is bent when refracted. This phenomenon is referred to as *dispersion*, and is the basic mechanism by which conventional prisms produce spectrums from white light.

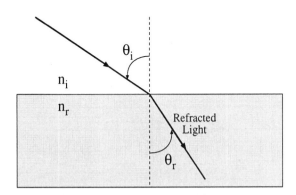

Illustration of refraction, with definitions of angles for Snell's law.

In refractive optics, light enters a refractive element through a smooth interface (specular) and is bent due to the difference in refractive indices between the incident (external) medium and the refractive material. The relationships between the angle of incidence, θ_i, the angle of refraction θ_r, and the refractive indicies in the incident side, n_i, and the refractive medium, n_r, is given by *Snell's law*,

$$n_i \sin(\theta_i) = n_r \sin(\theta_r)$$

Light rays entering a refractive medium wherein the speed of light is lower than the external medium will bend toward the normal (shown as a dashed line in the illustration) and will bend away from it if the speed of light is higher in the refractive medium. The table below lists some refractive indices of common materials and several that can be used in micromachining applications.

There are several different types of refractive lenses. While the three-dimensional (conventional) lens designs are often difficult to realize in micromachined devices, several methods have been devised to approximate them, as described below.

Material	Refractive Index (n = c/v)
Air (0°C, 1 atm)	1.0002926
Liquid Nitrogen	1.21
Ice	1.31
Methanol	1.326
Water (20°C)	1.333
Acetone	1.357
Ethanol	1.359
Fused Silica (SiO_2)	1.438
Hoya QZ Quartz Glass	1.46
Thermal Silicon Dioxide (TF)	1.46
PECVD Silicon Dioxide (TF)	≈ 1.47 (may vary)
Corning 7740 Pyrex™ Glass	1.473
Benzene	1.498
Plexiglass™ (PMMA)	1.5014
Borosilicate Glass	1.517
Phosphate Crown Glass	1.518
Hoya SL (70% SiO_2, 8% Na_2O, 9% K_2O, 13% Refractory Metal Oxides)	1.52
Sodium Chloride	1.53
Hoya LE (Aluminosilicate)	1.53
Light Flint Glass	1.581
Flint Glass	1.620
PECVD Silicon Nitride (TF)	1.8 to 2.5
LPCVD Silicon Nitride (TF)	2.01
Gallium Nitride (ordinary-ray)	2.35
Diamond	2.38
Gallium Arsenide	3.32
Silicon	3.415
Germanium	4.001

Table of refractive indices for various materials. Note that with the exception of the thin-film materials marked (TF), these are bulk values. It is also worth noting that many of the tabulated values in the literature do not include the wavelengths at which they were measured. From Sze (1981), Weast (1988), Waynant and Ediger (1994), Hoya Electronics, Inc., Corning Incorporated, and Giancoli (1989).

CONVEX   CONCAVE

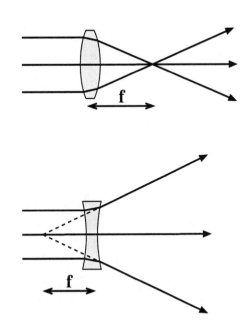

CONVEX	CONCAVE
$R_1 > 0$ $R_2 < 0$ Bi-Convex	$R_1 < 0$ $R_2 > 0$ Bi-Concave
$R_1 = \infty$ $R_2 < 0$ Planar Convex	$R_1 = \infty$ $R_2 > 0$ Planar Concave
$R_1 > 0$ $R_2 > 0$ Meniscus Convex	$R_1 > 0$ $R_2 > 0$ Meniscus Concave

Illustration of the different types of refractive lenses (left) and a comparison of ray diagrams for bi-convex (top right) and bi-concave (lower right). Note that R_1 is the entrance radius (first one encountered) and R_2 is the exit radius. After Hecht (1987).

For a physically thin lens, an approximation to its focal point can be made using the so-called "lens maker's equation,"

$$\frac{1}{f} \approx \left(n_1 - n_m \right) \left(\frac{1}{R_1} - \frac{1}{R_2} \right)$$

where,
n_1 = refractive index of the lens material
n_m = refractive index of the medium (for air, $n_m = 1$)
R_1 = entrance radius of lens
R_2 = exit radius of lens

Also, as mentioned above, variations for n with λ give rise to different focal lengths as the wavelength is changed, referred to as *chromatic aberration*.

One approach to allow refractive optics to be realized in micromachined devices is the surface relief lens (useful for slab-waveguide integrated optics), which takes advantage of the fact that the effective refractive index of thin films is a function of their thickness (this effective n can be estimated analytically but is generally determined empirically for a given device structure). A planar thin-film layer is patterned to form a shaped mesa structure wherein the effective refractive index is different from the rest of the thin-film layer. The mesa is generally shaped like one of the cross sections of typical 3-D lenses illustrated above. Surface relief lenses are their 2-D equivalents.

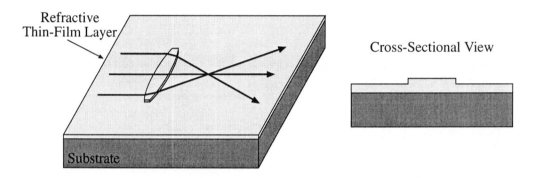

Illustration of a surface relief lens fabricated by forming a mesa in a thin-film refractive layer.

A related lens structure is the Luneberg lens, in which a thin-film layer is mechanically ground or chemically etched to form a spheroidal depression that acts as a lens in the plane of the substrate. Again, the basic principle is to make use of the fact that the effective refractive index of the thin-film layer is a function of its local thickness. The use of KOH etching to form such structures through the timed removal of an etch mask (allowing for the formation of the rounded features rather than pyramidal pits) was demonstrated by Kendall, et al. (1994). They noted that the surface of a 550 μm diameter structure, 9.6 μm deep, had undesirable surface features constrained to less than 5 nm in height.

Glass, a common transmissive optical material, can readily be etched using etchants such as HF, but the resulting surfaces are often unsuitable for optical uses (typically too rough on a microscale). In addition, it is difficult to precisely control cross sections and 3-D shapes with such isotropic etchants.

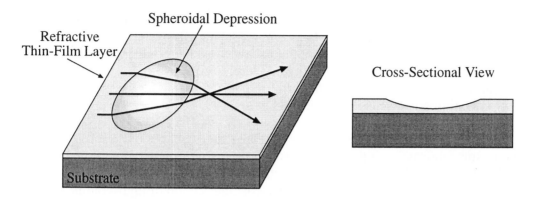

Illustration of a Luneberg lens structure fabricated by forming a spheroidal depression part way through a refractive thin-film layer.

Photosensitive glasses are available that allow direct photolithographic patterning to form lenses, etc. For example, Corning, Inc., (Corning, NY) has a proprietary process to fabricate so-called "SMILE" lenses by subjecting a light-sensitive glass material to a photothermal process.

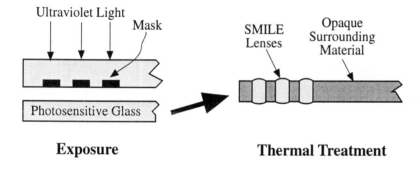

Illustration of the fabrication of microlens arrays using photosensitive glass. Adapted from Corning datasheet.

In the Corning process, the glass is exposed to UV light through a chrome photomask, as is typical for most microlithography. The glass is then thermally treated, causing the area exposed to UV light to undergo crystallization and densification. During crystallization, the exposed glass physically contracts in three dimensions, forcing the unexposed glass to erupt into spherical shapes. In addition to forming the lenses, the thermal process renders the area surrounding the lens optically opaque (in short-wavelength applications, this provides optical isolation and eliminates the need for apertures).

Another photosensitive glass product (also mentioned in the Micromachining chapter), Foturan® from Schott Glass, Inc. (Schott Glaswerke, Mainz, Germany) allows UV-exposed regions to be etched away. This could allow the resulting voids to be filled in with a material of different refractive index (or left empty to use the refractive index of air) to form optical devices. The use of Foturan® for micromachining is discussed in Dietrich, et al. (1993).

Illustrating yet another approach, 3-D (extruded shapes only) refractive lenses and prisms fabricated using LIGA molds to form them from plastic were demonstrated by Brenner, et al. (1993). This approach, while potentially having a high up-front cost (mold fabrication) allows for the low-cost fabrication of high-volume precision optical assemblies.

DIFFRACTIVE GRATINGS AND LENSES

As mentioned above, diffractive optics operate by generating phase or ampli-tude differences between different parts of a wavefront passing through them. Constructive and destructive interference occurs and can, for example, focus light as a refractive lens would. There is no physical difference between interference and diffraction, but interference generally refers to the interaction of a small number of wavefronts (from distinct apertures many λ apart) and diffraction to the interaction of a large number of wavefronts (also in diffraction, the aperture spacing is on the order of λ or smaller).

Diffractive Gratings

Diffractive gratings (discussed below in the Section 4.4.4) are arrays of equally spaced, parallel slits in a material opaque to the wavelengths of interest (note that gratings can be transmissive, as described, but also reflective). Light rays that pass through all of the slits without deviation (i.e., are oriented perpendicular to the grating, or $\theta = 0°$ referenced to the perpendicular direction) interfere con-structively to form a bright peak (a maximum). Light passing through the grating at angles also leads to maxima. This occurs when the light travels a distance equal to an integer number of wavelengths, $m\lambda$, before encountering light from an adjacent slit. These maxima occur at angles given by *Bragg's law,*

$$\sin(\theta) = \frac{m\lambda}{d}$$

where d is the distance between slits (grating pitch), and m is the (integer) order of the maximum (0, 1, 2, ...).

Thus, it can be seen that the location of the maxima are a function of slit width (narrower slits result in more spreading apart of the peaks) and the wavelength (longer wavelengths end up at larger angles). The ability of such gratings to separate light into its constituent wavelengths (like a conventional prism) is thereby apparent.

The resolution (sharpness of resolved wavelength peaks, otherwise referred to as *resolving power*) of a grating is a function of the number of slits, and is given by,

$$R \equiv \frac{\lambda}{\Delta\lambda} = Nm$$

where N is the number of grating lines illuminated.

As mentioned above, reflective gratings can also be realized. One approach is to pattern fine lines in a reflective surface such as aluminum on a nonreflective substrate.)

Fresnel Optics

If light from a point source is passed through a small, circular hole in an opaque plate, the emerging light will be modulated spatially such that there is a central peak with decreasing amplitude side rings (in fact, the amplitude of the far-field pattern is approximately a sin(r)/r pattern resulting from the frequency domain multiplication of the incoming light with the circular aperture). The individual rings (negative and positive in amplitude) are referred to as *Fresnel zones*. The radius of a zone, m, is given by (note that the second term is often negligible for small m),

$$R_m = \sqrt{mr_o\lambda + \frac{m^2\lambda^2}{4}}$$

where r_o is the distance of the plane of observation from the aperture, and m is the (integer) zone number (m = 1, 2, 3,...).

The number of zones within a given aperture of radius R is approximated by,

$$N_z = \frac{(\rho + r_o)R^2}{\rho r_o \lambda}$$

where ρ is the distance from the point source to the aperture (for a plane wave, $\rho = \infty$). Thus larger apertures have more zones, and the number of zones increases as the wavelength becomes shorter.

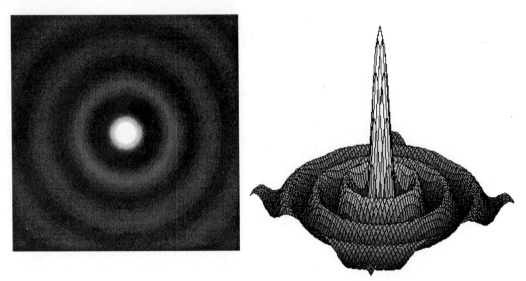

Simulated far-field intensity pattern from a circular aperture. A 2-D gray-scale intensity plot is shown at the left and a 3-D representation is shown at the right (intensity mappings were nonlinearly scaled to highlight detail).

Since the odd and even zones tend to cancel each other out, selectively removing one set will greatly increase the irradiance observed at the center (an experiment carried out by Lord Rayleigh in 1871). A plate that alters (in amplitude or phase) every other zone is referred to as a *Fresnel zone plate*. Diffracted light from the plate openings converges at a primary focal length f_1, given by,

$$f_1 = \frac{R_m^2}{m\lambda}$$

It is apparent from this equation that while operating in a manner similar to a refractive lens to form an image, the focal point will be much more wavelength-dependent (giving rise to extensive chromatic aberrations) than for refractive optics. Another important difference between Fresnel zone plates and refractive lenses is the existence of lower-intensity focal points at distances $f_1/3$, $f_1/5$, $f_1/7$, etc., from the plate.

The Fresnel-zone-plate approach works not only for visible light, but also for far-UV and x-ray light (critical for these shorter wavelengths where refractive optics do not exist).

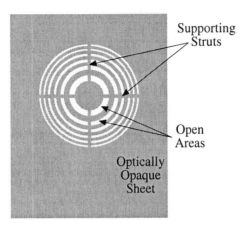

Illustration of a Fresnel zone plate fabricated from a thin-film sheet that is opaque to the wavelengths of interest.

Lin, et al. (1994, 1995), demonstrated Fresnel zone plates fabricated by etching patterns in polysilicon in a sacrificial oxide process. As described below, this approach, when combined with polysilicon hinge technology, allows the Fresnel zone plates to be raised into an out-of-plane position to form complete optical systems.

4.4 FILTERS AND SPECTROMETERS

4.4.1 INTERFERENCE FILTERS

A very common wavelength-selective structure is the interference filter. The basic principle is to stack layers of alternating high- and low-index thin-films on a transparent substrate (e.g., glass). The typical arrangement is to make the dielectric layers equal to one-quarter the wavelength of the cutoff (or center, depending on the type of filter) wavelength and arrange the stacks to obtain a band-pass response.

As discussed by Hecht (1987), these stacks are given designations based on starting with the glass layer (g) and naming each high-n quarter-wave layer "H," each low-n quarter-wave layer "L," and the air (or ambient), "a." For example, a stack of one high- and low-n layer would be referred to as, "gHLa," whereas a set of three repeating high/low-n layers would be referred to as "g(HL)³a." As more alternating layers are added, the band-pass response is sharpened (i.e., given steeper out-of-band roll-off without changing the bandwidth). These filters reflect light in their pass-band and pass light outside the pass-band. Similar dielectric stacks can be used as antireflection coatings and embedded laser diode mirrors.

To obtain high- and low-pass filters, eighth-wave layers and one extra quarter-wave layer can be added to the stacks. A high-pass (short wavelength passing) filter would have a designation of "g(0.5L)(HL)mH(0.5L)a," while a low-pass filter would have a designation of "g(0.5H)L(HL)m(0.5H)a."

Since this type of interference filter is basically a thin-film structure, it can generally be directly integrated with micromachined devices through the deposition of suitable films such as MgF_2 (n = 1.38) or CeF_3 (n = 1.63) for L layers and TiO_2 (n = 2.40) or ZnS (n = 2.32) as H layers. For example, for non-conductive layers, RF sputtering is a suitable deposition technique.

4.4.2 FABRY-PEROT FILTERS

The basic Fabry-Perot filter is a tuned cavity (resonator) consisting of two semi-transmissive mirrors (partially, or "half-silvered" glass mirrors are commonly used in macroscopic versions) aligned in parallel at a certain distance. Light entering through one mirror is multiply reflected within the cavity, and when the gap between mirrors is an integer multiple of a particular wavelength, the light is passed through the cavity (this technique is widely used to tune laser cavities). If the distance between mirrors is varied, the extremely narrow pass-band of the resonant cavity can be tuned. In terms of nomenclature, if the mirrors are fixed, the assembly is referred to as an *etalon*, and if they are movable, it is referred to as an *interferometer* (discussed below).

The ratio of the incident light intensity, I_i, to the light intensity that is transmitted, I_t, is given by,

$$\frac{I_t}{I_i} = \left(\frac{T}{1-R}\right)^2 \frac{1}{1 + \dfrac{4R}{(1-R)^2}\sin^2\left(\dfrac{2\pi nd\cos(\theta)}{\lambda} + \phi\right)}$$

where,
T = transmittance of the reflecting surfaces
R = reflectance of mirror surfaces
n = refractive index of gap
d = gap spacing
θ_i = angle of incidence of incoming light
ϕ = phase shift on reflection

While these devices can be used as spectrometers in some cases (see below), a fundamental limitation is that multiple wavelengths will pass through a given Fabry-Perot cavity (subject to the above limitation of integer wavelength multiples). In practice, this effect may be mitigated through the use of input band-pass filters (e.g., interference filters, as discussed above). For detectors with broad, flat responses with wavelength (e.g., indirect thermal optical sensors), this must be considered.

All-dielectric (no metallic mirrors) thin-film (fixed wavelength) Fabry-Perot interferometers can be fabricated by forming quarter-wave dielectric stacks with designations, such as "gHLHLLHLHa" or "gHLHLHHLHLHa" (see Section 4.4.1 above).

4.4.3 MECHANICALLY TUNABLE IR FILTERS

Ohnstein, et al. (1995, 1996) described mechanically tunable IR filters designed to operate over the 8 to 32 μm range. The filters are constructed as a set of metal, high aspect ratio plates supported by flexures that allow the inter-plate spacing to be varied simply by pulling on the assembly. A pair of such parallel-plate arrays, oriented orthogonally to each other in the light path, act as a 2-D waveguide array. Each plate array acts as a polarizer, and the net effect of these facts is that TE polarized radiation does not pass through the filter if it is a longer wavelength (lower frequency) than a cutoff given by, $\lambda_c = 2d$, where d is the plate spacing. The total energy admitted by the filter increases as the cutoff wavelength is increased, as expected for a low-pass filter.

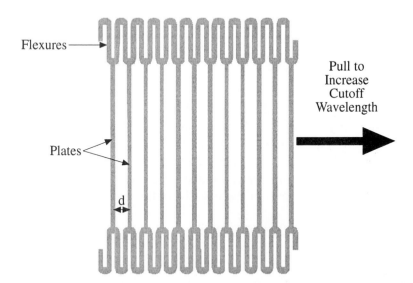

Simplified top view illustration of mechanically tunable IR low-pass filter as described by Ohnstein, et al. (1995, 1996), showing flexures at upper and lower extremes of the drawing and plates that are tall in the direction orthogonal to the drawing. In operation, two such arrays are stacked, with the plates oriented at ninety degrees to each other.

The devices were fabricated using the SLIGA (sacrificial LIGA) process to realize the 200 μm tall, 60:1 aspect ratio plates desired. While they are still in development, the basic components have all been demonstrated, including semi-monolithic micromachined stepper motors to elongate the plate arrays (the electromagnetic coils use LIGA cores, but are externally wound and assembled on to the stepper motor bases).

4.4.4 SPECTROMETERS

Spectrometers are devices used to examine the different wavelength components of light (if light intensity versus wavelength is quantitatively measured, the more proper term is *spectrophotometer*). This can be used to examine the chemical composition or state of a substance by measuring its emission or absorption spectrum, to measure the temperature of an object (optical pyrometry), to quantify color matching, etc. One way to achieve this function is to provide a tuned filter, such as a Fabry-Perot resonant cavity, with a variable pass-band wavelength (i.e., movable reflector(s)). This device can then be tuned to, or swept over, the wavelengths of interest and the light passing through is detected (analogous to a fixed or tunable band-pass filter for electrical signals).

Another spectrometer approach is to use a dispersive element to break up the incoming light into its constituent wavelengths by spreading them out in space (a common dispersive element is a simple refractive prism). This latter approach provides a continuous output representing a continuum of wavelengths simultaneously. Both types of spectrometers have been implemented in micromachined forms.

There are several terms that are commonly used in spectrometry that are worth defining. The *full-width half-maximum* (FWHM), or *half-width* (HW) is the width of a resolved peak at half-amplitude and is a measure of spectral resolution (resolution is defined as FWHM^{-1}). The *free spectral range* (FSR) is the frequency range over which the spectrometer can make accurate measurements. Finally, the *finesse, F*, is defined as the ratio of the FWHM to the specified wavelength λ_o.

FABRY-PEROT SPECTROMETERS

Several groups have fabricated micromachined Fabry-Perot spectrometers using bulk or thin-film techniques. For such devices, with a mirror spacing d, the transmission peaks are separated by a wavelength range, $\Delta\lambda$, defining the FSR as,

$$\Delta\lambda = \frac{\lambda^2}{2d}$$

and the FWHM is given by,

$$\text{FWHM} = \frac{\lambda(1-R)}{n\pi\sqrt{R}}$$

While developed as light-modulating devices, Aratani, et al. (1994) demonstrated an array of surface micromachined Fabry-Perot cavities designed for an incident wavelength of 780 nm (near infrared, just beyond the typical upper wavelength of visible light at $\approx$ 750 nm). By electrostatically varying the cavity spacing, the reflectance (note that these devices were not transmissive Fabry-Perot elements) could be adjusted from nearly one down to a small value, modulating the light.

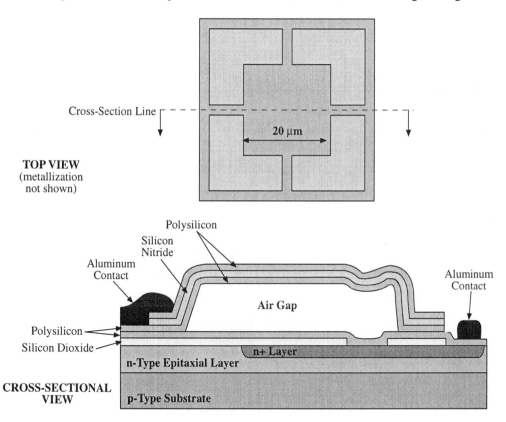

Illustration of a surface micromachined, electrostatically tunable Fabry-Perot interferometer. After Aratani, et al. (1994). The illustration (not to scale) shows a top view of a cavity as well as a cross section along one set of supporting beams.

The cavities were formed by removing a sacrificial PSG layer separating an electrostatically deflectable polysilicon/nitride/polysilicon top reflector from an underlying polysilicon/oxide/silicon (substrate) reflector. The deflectable reflectors were supported by four beams consisting of the polysilicon/nitride/polysilicon layer used to form the reflective surface itself. In addition, an underlying n-epi/p-substrate photodiode structure was integrated with the modulators. The active regions of the modulators were 20×20 µm, with a nominal cavity gap of 410 to 430 nm. Aratani, et al. (1994) demonstrated their modulators using a ± 5 V, 500 kHz sinusoidal drive signal. Since the membranes were attracted by both polarities of applied potential, the net drive frequency was 1 MHz. They reported a 25% modulation depth for the reflectance of the devices.

Jerman, et al. (1991) and Raley, et al. (1992) demonstrated bulk micromachined Fabry-Perot interferometers based on bonding two bulk-etched wafers together, with electrostatically controlled cavity gaps and peripherally corrugated membranes to linearize deflection. The two designs were for near IR and visible light, respectively. Jerman, et al. (1991) proposed the interferometers primarily as tunable, narrow-linewidth optical filters for wavelength division multiplexing in fiber-optic communications networks. By using silicon in the optical path, these devices were limited to operation at wavelengths longer than 1.15 µm, and were designed to operate around a center wavelength of 1.3 µm.

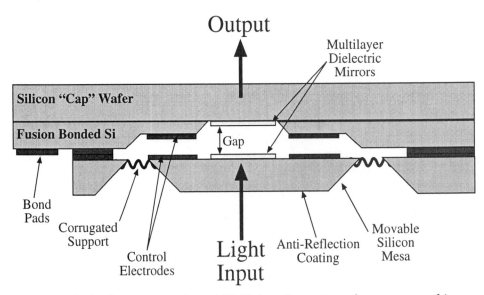

Illustration of a bulk micromachined IR Fabry-Perot interferometer making use of electrostatic tuning control. After Jerman, et al. (1991).

Unfortunately, not all of the process details are given in their paper. As shown above, a two-step anisotropic wet etch was carried out on the fusion-bonded "cap" wafer first to define the gap depth (≈ 25 µm) and then to define the spacing between the control electrodes (≈ 6 µm). A key process step was the subsequent brief etch that exposed the SiO_2 layer between the fusion-bonded regions of the cap wafer to provide an optically flat surface for the top dielectric mirror. The two wafers were then bonded using a metal-to-metal scheme (while not stated, Au/Sn or Au/Ti alloy systems are commonly used), which also provided electrical signal connections between them. On the lower wafer (containing the movable tuning mesa), an etch-stopped, corrugated support membrane was used to extend the linear operating range.

Jerman, et al. (1991) reported measured performance figures at 1.3 μm of FWHM = 0.9 nm, FSR = 38 nm, and finesse = 40. For the electrostatic tuning, a 0 to 70 V applied potential provided tuning over the entire FSR. In addition, they provide an interesting discussion of the use of such interferometers as highly accurate membrane position sensors, including their use as pressure sensors.

The design of Raley, et al. (1992), did not require that the incident or exiting radiation pass through silicon. Instead, they used CVD silicon nitride membranes, supporting 14-layer quarter-wave interference filters with a center frequency (for maximum reflectance) of 450 nm. The filter layers were alternating SiO_2 (n = 1.44, 78 nm thick) and HfO_2 (n = 1.80, 63 nm thick). The maximum reflectance achieved was ≈ 92% at the center frequency, with an overall bandwidth of ≈ 400 to 520 nm. Although the integrated devices were not fully functional, several interesting design ideas were presented.

DISPERSIVE SPECTROMETERS

For dispersive spectrometers, gratings are often used (particularly attractive for micromachined implementations due to their relative simplicity), and a key parameter is the dispersion, D, which is the amount of lateral distance change with respect to wavelength, given by,

$$D \equiv \left. \frac{dx}{d\lambda} \right|_{\theta=\theta_o} = \frac{m\,h}{d\cos^3(\theta_o)} \quad \text{(dimensionless, but often given in nm/nm)}$$

where,
m = spectral order (typically the first spectral order is used, where m = 1)
 h = optical distance between the dispersive element and the detector
θ_o = angle of dispersion

Three basic grating types are illustrated in the figure below: *amplitude*, *linear blazed,* and *binary phase*. An amplitude grating consists of an array of slits that give relatively low efficiency in terms of the fraction of the incoming light that ends up in the first spectral order (multiple "copies" of the light's spectrum appear at successively shallower angles to the grating), which is typically where the detector is located. Linear blazed gratings theoretically have an efficiency (defined as the fraction of the incident light that yields a useful signal) of 100%, and can be approximated by breaking them up into binary steps (an infinite number of steps results in a linear blaze). The amplitude and binary phase gratings are the two types most readily achieved with current micromachining techniques.

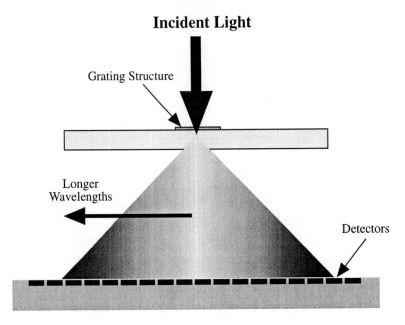

Illustration of a basic grating-type spectrometer. The grating could be replaced with a prism or other dispersive element.

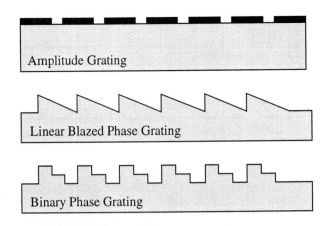

Illustration of three basic grating types: (top) amplitude grating, (center) linear blazed phase grating, and (bottom) binary phase grating.

As described by Goodman (1996), the diffraction efficiency, η, of a binary phase grating is given by,

$$\eta = \left| \, sinc\left[\frac{m}{I}\right] \frac{sinc\left[m - \dfrac{\phi_o}{2\pi}\right]}{sinc\left[\dfrac{m - \dfrac{\phi_o}{2\pi}}{I}\right]} \, \right|$$

where,
m = diffraction order
I = number of binary steps in the grating structure
ϕ_o = peak-to-peak phase difference (maximum) introduced by the grating

with,
$$\phi_o = 2\pi \frac{\Delta_o \left(n_g - n_o\right)}{\lambda_o}$$

where,
Δ_o = total distance from top surface of grating to the deepest etched level
n_o = refractive index of ambient medium
n_g = refractive index of grating substrate material
λ_o = wavelength in ambient medium

Cremer, et al. (1992) fabricated a monolithic reflective grating spectrometer with integrated photodiodes on a 4×7 mm InP substrate. The spectrometer was designed to operate in the 1.5 μm range for wavelength division multiplexing of fiber-optic communications (the goal was to demonstrate a fully integrated device, potentially suited to mass production). Their design had 7 input waveguides and 42 output photodiodes. Waveguides were fabricated using (from the substrate up) an 0.5 μm InP buffer layer, an 0.6 μm InGaAsP waveguide layer, an 0.2 μm InP intermediate layer, a 1 μm GaAs absorption layer, and an 0.5 μm InP cladding layer all deposited using molecular vapor phase epitaxy (MOVPE). A silicon nitride passivation layer was also deposited.

Photodiodes were formed using Zn diffusion through the (patterned) silicon nitride layer, and the reflective grating and output waveguides were RIE etched, and the grating was metallized with Au to increase its reflectivity.

In operation, the light input through a given input waveguide diverges freely within the slab waveguide and is diffracted and focused onto the appropriate output waveguides and transmitted to their individual photodiodes. The waveguides were separated by 7 μm, which corresponded to a spectral separation of 7 nm, with cross-talk of -15 dB between channels.

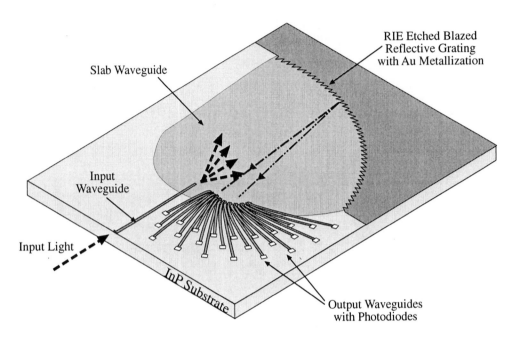

Illustration of a fully integrated InGaAsP/InGaAs/InP spectrometer with integrated detectors. After Cremer, et al. (1992) (not to scale).

Kwa and Wolffenbuttel (1992) presented a bulk micromachined, integrated grating spectrometer with a detector array. Incoming light entered through a binary grating and was reflected three times from silicon surfaces prior to impinging on an integral photodiode array. They used n-type epitaxial silicon layers (with diffused n+ contacts for electrochemical etch bias) on p-type substrates to form thin, etch-stop layers. The device was constructed through the fusion bonding of two bulk micro-machined silicon wafers. In the lower wafer, the n-type silicon exposed via anisotropic etching served as one of the reflecting surfaces, as did the exposed (111) planes of the p-type silicon. The upper wafer received a p-type implant to form 2 μm wide photodiodes on a 4 μm spacing. A silicon nitride layer was deposited above the silicon, followed after patterning, with an aluminum layer that was also patterned to form the grating as well as interconnects and the photodiode light shield.

With a 3.4×2 mm channel in the reflecting wafer, a 32-slit grating, and a 200-element photodiode array, they demonstrated a 380 to 720 nm FSR with a FWHM of 3.4 nm. Due to alignment variations and silicon wafer thickness tolerances, this design requires a single-wavelength calibration. In a later paper, Kwa, et al. (1997) describe the fabrication of the back-illuminated photodiodes in detail.

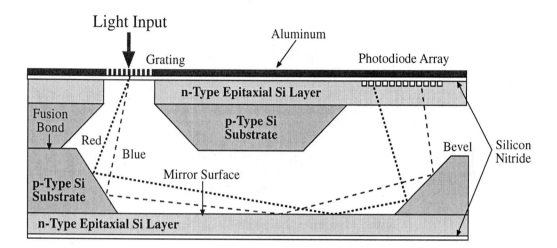

Illustration of a bulk micromachined grating-type spectrometer. After Kwa and Wolffenbuttel (1992) (not to scale).

Yee, et al. (1996) demonstrated a simple spectrometer structure based on the combination of a micromachined grating and a silicon CCD detector. Their approach was to fabricate the arrays using electron-beam lithography to expose a polymethyl methacrylate (PMMA) layer on a quartz lithographic mask plate with a 105 nm Cr layer beneath the PMMA. Binary phase gratings were fabricated by patterning the Cr and using RIE etching with oxygen to etch the quartz. This process was repeated for the necessary number of steps. The grating stuctures thus fabricated were bonded to CCD detector arrays, and a diffraction efficiency of 63% was achieved, with spectral resolution of 1.7 nm/pixel and 2.55 nm/pixel for two different types of CCD (specialized, 512×64 element detector versus consumer-grade 320×240 element device). The free spectral range of these devices was 400 to 800 nm, limited by the CCD's response spectrum.

In addition to the basic linear binary phase gratings, a Fresnel zone plate was added in the direction perpendicular to that of the dispersion, to help focus more light into the detector, nearly doubling the signal amplitude. Computed gratings were also demonstrated, wherein numerical techniques were used to generate more complex grating patterns than the simple slit patterns described above. The continuous tone grating functions computed were implemented using dithering algorithms (as used in laser printers) to simulate them. Using this computational approach, the diffracted spectra could be compressed in certain regions to discard areas of little interest (such as broad, nonspecific absorption regions for chemicals) and thus utilize a smaller detector. These nonlinear gratings can also be computed such that they form filters matched to a particular absorption or emission spectrum.

Goldman, et al. (1990) fabricated and tested a glass substrate, micromachined guided-wave spectrometer. While no detectors were integrated on to the devices, the fabrication process is very simple, making this a potentially attractive approach. They used a phase grating etched into the glass (n = 1.52) to provide light at an appropriate angle (i.e., wavelength) to propagate down a submicron waveguide layer on the glass (polystyrene, n = 1.59 or Ta_2O_5, n = 2.12) in a way that it could interact with an external sample, and a second phase grating to act as a spectrometer for the light exiting the waveguide.

The binary phase gratings were HF etched into glass microscope slides using photoresist masks (exposed with an interference pattern to produce the gratings, with 0.8 or 0.4 μm periods). A 15 s etch in 1:10 buffered $HF:H_2O$ yielded the desired depth of ≈ 0.1 μm. Polystyrene was spun on from an isobutyl ketone solvent or Ta_2O_5 was sputtered from a Ta target in an oxygen-argon ambient gas mixture (the stoichiometry was controlled via the total pressure and oxygen content).

They reported resolution as high as 0.15 nm/channel with a 1024-element photodiode array. Unfortunately, due to the low coupling efficiency of light into the waveguide, the overall efficiency of the system was only 0.01%. However, in theory, the efficiency could be increased by several orders of magnitude if the coupling of the gratings to the waveguide could be increased (e.g., by increasing the depth of the gratings).

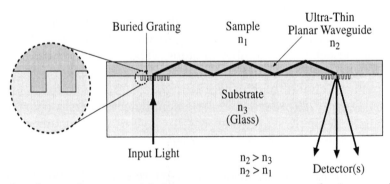

Illustration of a glass substrate guided-wave spectrometer employing a submicron waveguide layer to promote interaction of the guided light with an external sample. After Goldman, et al. (1990) (not to scale). Note that the actual grating cross section would be much more isotropic with HF etching.

4.5 INTEGRATED OPTICAL SYSTEMS

The idea of integrated optical systems is certainly not new (see, for example, Tien (1977)), but the advent of micromachined structures that access the space above the substrate has certainly enabled a variety of new configurations. Early efforts to develop monolithic optical devices relied entirely upon thin-film technologies to fabricate optical switches, modulators, waveguides, gratings, etc., not all of which were easy to realize using such methods. With modern micromachining techniques, one can realize systems that combine the best of in-plane and out-of-plane optical actuation, transmission, processing, and sensing modes. While it remains to be seen if such systems, often requiring "assembly" to access the third dimension, will be commercially viable, several exciting demonstrations have occurred.

4.5.1 INTEGRATED FREE-SPACE OPTICAL SYSTEMS

Completely integrated, free-space, micro-optical systems are being developed using micromachining technology. An important advantage of the free-space approach (as opposed to guided-wave designs) is that the region above the substrate can be utilized. Most such systems to date have been designed to optimize laser coupling to and from optical fibers, but many other potential applications exist.

An approach used by Lin, et al. (1994, 1995) and Daneman, et al. (1995, 1996a, 1996b) is to fabricate the necessary optical elements (reflectors, gratings, Fresnel optics, electrostatic positioning motors, etc.) using a polysilicon/sacrificial oxide technology that is well established. In order to achieve the necessary out-of-plane orientation for the optical elements, the polysilicon hinge designs of Pister, et al. (1992) were adapted.

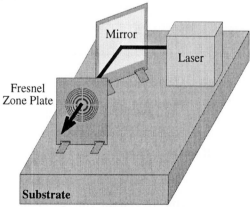

Conceptual illustration of a surface micromachined, free-space optical system. Optical devices that can be implemented in two-dimensional structures (i.e., micromachined polysilicon) can be rotated out-of-plane using hinges to form systems wherein the optical path is not confined to the substrate plane.

Lee, et al. (1995) demonstrated the use of Fresnel lenses to collimate the light from a 670 nm solid-state laser and gratings to produce multiple output beams from the laser as a demonstration vehicle for these concepts as they could be applied to future optical interconnects.

Daneman, et al. (1996b) demonstrated this approach in the form of impact drive vibromotors (as described in the Mechanical Transducers chapter) to push on sliders that in turn changed the angle of a mirror with two degrees of freedom. The mirror was front-coated with 40 nm of gold to increase its reflectivity. They demonstrated an angular range of over 90°, a translational travel range of 60 μm and a maximum slider velocity of > 1 mm/s. Reported beam positioning resolution (using a 1.3 μm laser and single-mode optical fiber) was 0.81 μm, and maximum achieved coupling efficiency was 32%. Once positioned, static friction holds the devices in place, and they can apparently withstand accelerations up to 500 g without moving. This approach could be used to align laser beams with other optical components that were not monolithically fabricated (hence would need to be aligned), such as fiber optics, etc.

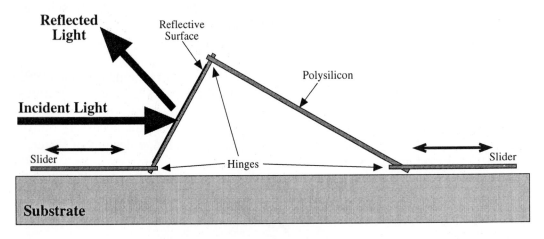

Illustration of a two-degree-of-freedom, vibromotor-actuated mirror as described by Daneman, et al. (1996b). Vibromotors (not shown) could push or pull on the sliders at either end of the structure. By coordinating the operation of the two vibromotors, the mirror's angle and position (relative to incident light) could be adjusted. Static friction held the assembly in place when it was not being moved.

A fold-up polysilicon scanning mirror for raster displays was described by Kiang, et al. (1997a). Their design used a resonance-driven mirror (2 to 3 kHz, depending on the design) to deflect a laser beam in an experimental 7 × 10 pixel display. If sufficient bandwidth, as well as two-axis scanning can be achieved, it is conceivable that a single laser or LED could be used to form images directly on the retina with this approach. With a very similar design, Kiang, et al. (1997b) added a micromachined diffraction grating to the deflectable mirror with the potential to form a scanning spectrophotometer.

Lin, et al. (1997) presented a full x-y-z reflective beam positioning control using a pair of independent fold-up mirrors, movable in orthogonal directions. By using electrostatic "scratch-drive" actuators, they were able to achieve beam motion in all three directions with a range of > 30 μm and a step size of 11 nm.

A fold-up x-y-z reflective beam positioning system was demonstrated by Fan, et al. (1997), also based on polysilicon scratch-drive actuators, but used an elevated polysilicon plate containing four Fresnel lenses as focusing elements. They demonstrated x- and y-axis travels of 120 μm, vertical travel of 250 μm, and 27 nm step size for ± 90V drive pulses. It is noteworthy that the structures did not require intervention other than electrical drive for their own assembly, eliminating the hand assembly that is common with fold-up polysilicon optical systems (not all geometries, however, are suitable for such self-driven assembly).

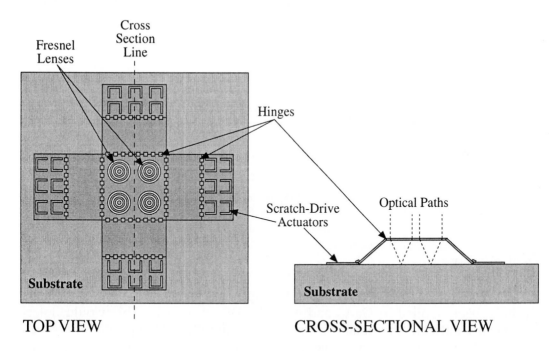

Illustration of a fold-up polysilicon, electrostatically driven x-y-z beam positioning system with integrated Fresnel lenses for focusing. Adapted from Fan, et al. (1997).

In addition, Lee, et al. (1997) demonstrated a polysilicon fold-up optical system for switching light between four optical fibers. The switch was operated by using electrostatic scratch-drive actuators to move a mirror out of the convergence of four optical fibers, thus controlling the path of the light. To move from reflecting to transmitting light across its center, the mirror was moved until a mechanical latch was engaged, which would hold the mirror in position without further power input. By briefly energizing a thermal actuator, the latch could be released, allowing a polysilicon pull-back spring to return the mirror to its centered position. By driving the scratch-drive actuators with an 80 V sinusoidal signal at 50 kHz, they reported linear velocities of 2.5 mm/s, and rise/fall times of 6 and 15 ms, respectively. These authors addressed a concern about vibrational effects on such micro-optical systems by subjecting an operating switch to 200 Hz to 10 kHz vibrations at up to 89 g amplitudes, and showing that there was no observable effect on its operation.

The beam steering examples above all are designed for relatively small-width light beams, which are typically appropriate for fiber-optic applications. An approach to larger beam (> 1 mm) steering was demonstrated through the use of an array of rectangular, metallized, tiltable polysilicon structures by Burns and Bright (1997). The array, analogous to the angled slats of window blinds, collectively formed a reflective blaze grating, where the blaze angle of each slat could be controlled (via electrostatic or thermal actuation). Through the appropriate choice of the slat angle, chosen diffraction orders could be steered as needed.

While only a few degrees of steering is generally possible with the variable blaze grating approach, no "folding" or "assembly" of the components was required. A potential drawback to the approaches that do require such final assembly, when applied to mass production, is that the final configuration of the optical system requires a lot of precision positioning. Thus, the mass-production benefits of lithographic fabrication may not confer the cost reductions often associated with them. As mentioned above, however, some self-powered assembly methods have been demonstrated and have considerable promise (for example, see Fan, et al. (1997) or, for a paper focused on automated assembly of such structures, see Reid, et al. (1997)).

It is interesting to note that such free-space optical devices appear not as yet to have been applied to GaAs and other III-IV substrates. This would not be technically difficult, particularly if low-temperature PECVD-deposited films were used for the structural regions of the devices. The advantage of taking this approach would be that high-efficiency optical emitters (and detectors) could be integrated directly into the substrate.

OPTICAL TRANSDUCERS REFERENCES

GENERAL REFERENCES

Benson, K. B., "Television Engineering Handbook," McGraw-Hill, New York, NY, 1992.

Bube, R. H., "Photoconductivity of Solids," John Wiley and Sons, New York, NY, 1962.

Cobbold, R. S. C., "Transducers for Biomedical Applications," John Wiley and Sons, New York, NY, 1974.

Giancoli, D. C., "Physics for Scientists and Engineers with Modern Physics," Prentice-Hall, Inc., Englewood Cliffs, NJ, 1989.

Goodman, J. W., "Introduction to Fourier Optics," McGraw-Hill, New York, NY, 1996.

Hecht, E., "Optics," Addison-Wesley Publishing, Inc., Reading, MA, 1987.

Khazan, A. D., "Transducers and Elements," Prentice-Hall, Inc., Englewood Cliffs, NJ, 1994.

Norton, A. N., "Handbook of Transducers," Prentice-Hall, Inc., Englewood Cliffs, NJ, 1989.

Sze, S. M., "Physics of Semiconductor Devices," John Wiley and Sons, New York, NY, 1981.

Waynant, R. W., and Ediger, M. N. [eds.], "Electro-Optics Handbook," McGraw-Hill, New York, NY, 1994.

SPECIFIC REFERENCES

Ade, R. W., Harstead, E. E., Amirfazli, A. H., Cacouris, T., Fossum, E. R., Prucnal, P., and Osgood, R. M., Jr., "Silicon Photodetector Structure for Direct Coupling of Optical Fibers to Integrated Circuits," IEEE Transactions on Electron Devices, vol. ED-34, no. 6, June 1987, pp. 1283 - 1289.

Altman, C., Bassous, E., Osburn, C. M., Pleshko, P., Reisman, A., and Skolnik, M. B., "Mirror Array Light Valve," U.S. Patent No. 4,592,628, issued June 3, 1986.

Antreasyan, A., Napholtz, S. G., Wilt, D. P., and Garbinski, P. A., "Monolithic Integration of InGaAsP/InP Semiconductor Lasers Using the Stop-Cleaving Technique," IEEE Journal of Quantum Electronics, vol. QE-22, no. 7, July 1986, pp. 1064 - 1072.

Aratani, K., French, P. J., Sarro, P. M., Poenar, D., Wolffenbuttel, R. F., and Middelhoek, S., "Surface Micromachined Tuneable Interferometer Array," Sensors and Actuators, vol. A43, nos. 1 - 3, May 1994, pp. 17 - 23.

Asahi, R., Sakata, J., Tabata, O., Mochizuki, M., Sugiyama, S., and Taga, Y., "Integrated Pyro-electric Infrared Sensor Using PVDF Thin-Film Deposited by Electro-Spray Method," Proceedings of Transducers '93, the 7th International Conference on Solid-State Sensors and Actuators, Yokohama, Japan, June 7 - 10, 1993, Institute of Electrical Engineers, Japan, pp. 656 - 659.

Auston, D. H., "Impulse Response of Photoconductors in Transmission Lines," IEEE Journal of Quantum Electronics, vol. QE-19, no. 4, Apr. 1983, pp. 639 - 648.

Auston, D. H., "Picosecond Optoelectronic Switching and Gating in Silicon," Applied Physics Letters, vol. 26, no. 3, Feb. 1, 1975, pp. 101 - 103.

Auston, D. H., Lavallard, P., Sol, N., and Kaplan, D., "An Amorphous Silicon Photodetector for Picosecond Pulses," Applied Physics Letters, vol. 36, no. 1, Jan. 1, 1980, pp. 66 - 68.

Bailey, C. H., and Gouras, P., "The Retina and Phototransduction," Chapter 27 in "Principles of Neuroscience," Second Edition, Kandel, E. R., and Schwartz, J. H. [eds.], Elsevier, New York, NY, 1985.

Bandy, S. G., and Linvill, J. G., "The Design, Fabrication, and Evaluation of a Silicon Junction Field-Effect Photodetector," IEEE Transactions on Electron Devices, vol. ED-20, no. 9, Sept. 1973, pp. 793 - 801.

Beneking, H., "On the Response Behavior of Fast Photoconductive Optical Planar and Coaxial Semiconductor Detectors," IEEE Transactions on Electron Devices, vol. ED-29, no. 9, Sept. 1982, pp. 1431 - 1441.

Bethea, C. G., Chen, C. Y., Cho, A. Y., Garbinski, P. A., and Levine, B. F., "Picosecond $Al_xGa_{1-x}As$ Modulation-Doped Optical Field-Effect Transistor Sampling Gate," Applied Physics Letters, vol. 42, no. 8, Apr. 15, 1983, pp. 682 - 684.

Binnig, G., and Rohrer, H., "Scanning Tunneling Microscopy - from Birth to Adolescence," Reviews of Modern Physics, vol. 59, no. 3, pt. 1, July 1987, pp. 615 - 625.

Binnig, G., Rohrer, H., Gerber, C., and Weibel, E., "Tunneling Through a Controllable Vacuum Gap," Applied Physics Letters, vol. 40, no. 2, Jan. 15, 1982, pp. 178 - 180.

Bolliger, D., "Integration of an Ultraviolet Sensitive Flame Detector," Dissertation for the Doctor of Natural Sciences Degree, Swiss Federal Institute of Technology, Zurich, Switzerland, DISS. ETH No. 11359, 1995.

Bolliger, D., Popovic, R. S., and Baltes, H., "Integration of a Smart Selective UV Detector," Proceedings of Transducers '95, the 8th International Conference on Solid-State Sensors and Actuators, Stockholm, Sweden, June 25 - 29, 1995, vol. 2, pp. 144 - 147.

Bollyky, L. J., and Rauhut, M., "Superior Oxalate Ester Chemical Lighting System," U.S. Patent No. 4,313,843, issued Feb. 2, 1982.

Brenner, K.-H., Kufner, M., Kufner, S., Moisel, J., Müller, A., Sinzinger, S., Testorf, M., Göttert, J., and Mohr, J., "Application of Three-Dimensional Micro-Optical Components Formed by Lithography, Electroforming, and Plastic Molding," Applied Optics, vol. 32, no. 32, Nov. 10, 1993, pp. 6464 - 6469.

Burns, D. M., and Bright, V. M., "Micro-Electro-Mechanical Variable Blaze Gratings," Proceedings of the Tenth Annual Workshop of Micro Electro Mechanical Systems, Nagoya, Japan, Jan. 26 - 30, 1997, pp. 55 - 60.

Busta, H., Dallesasse, J., Smith, S., Pogemiller, J., Zimmerman, B., and Mathius, R., "Light Emission from an AlGaAs Single-Quantum-Well Heterostructure by Electron Excitation from a Micromachined Field Emitter Source," Journal of Micromechanics and Microengineering, vol. 4, no. 2, 1994, pp. 55 - 59.

Campbell, J. C., Holonyak, N., Jr., Craford, M. G., and Keune, D. L., "Band Structure Enhancement and Optimization of Radiative Recombination in GaAs$_{1-x}$P$_x$:N (and In$_{1-x}$Ga$_x$P:N)," Journal of Applied Physics, vol. 45, no. 10, Oct. 1974, pp. 4543 - 4553.

Canham, L. T., "Silicon Quantum Wire Array Fabrication by Electrochemical and Chemical Dissolution of Wafers," Applied Physics Letters, vol. 57, no. 10, Sept. 3, 1990, pp. 1046 - 1050.

Capon, J., de Baets, J., de Rycke, I., de Smet, H., Doutreloigne, J., van Calster, A., and Vanfleteren, J., "A Lensless Contact-Type Image Sensor Based on a CdSe Photoconductive Array," Sensors and Actuators, vols. A37 - A38, June - Aug. 1993, pp. 546 - 551.

Carruthers, G. R., "Ultraviolet and X-Ray Detectors," Chapter 15 in "Electro-Optics Handbook," Waynant, R. W., and Ediger, M. N. [eds.], McGraw-Hill, New York, NY, 1994, pp. 15.1 - 15.39.

Casey, H. C., and Panish, M., "Heterostructure Lasers: Parts A and B," Academic Press, Inc., New York, NY, 1978.

Chévrier, J.-B., Baert, K., and Slater, T., "An Infrared Pneumatic Detector Made by Micromachining Technology," Journal of Micromachining and Microengineering, vol. 5, no. 2, June 1995, pp. 193 - 195.

Choi, I. H., and Wise, K. D., "A Silicon-Thermopile-Based Infrared Sensing Array for Use in Automated Manufacturing," IEEE Transactions on Electron Devices, vol. ED-33, no. 1, Jan. 1986, pp. 72 - 79.

Chu, P. B., Lo, N. R., Berg, E. C., and Pister, K. S. J., "Optical Communication Using Micro Corner Cube Reflectors," Proceedings of the Tenth Annual Workshop of Micro Electro Mechanical Systems, Nagoya, Japan, Jan. 26 - 30, 1997, pp. 350 - 355.

Cladis, P. E., "The Liquid Crystal State of Materials or How Does the Liquid Crystal Display (LCD) in My Wristwatch Work?," Essay #20 in "Fundamentals of Physics," Extended Third Edition, Halliday, D., and Resnick, R., John Wiley and Sons, New York, NY, 1988, pp. E20-1 - E20-8.

Cobbold, R. S. C., "Transducers for Biomedical Applications," John Wiley and Sons, New York, NY, 1974.

Cole, B. E., Han, C. J., Higashi, R. E., and Ridley, J., "512 X 512 Infrared Cryogenic Scene Projector Arrays," Sensors and Actuators, vol. A48, no. 3, May 1995, pp. 193 - 202.

Cole, B. E., Han, C. J., Higashi, R. E., Ridley, J., and Holmen, J., "Monolithic 512 x 512 CMOS-Microbridge Arrays for Infrared Scene Projection," Proceedings of Transducers '95, the 8th International Conference on Solid-State Sensors and Actuators, Stockholm, Sweden, June 25 - 29, 1995, vol. 2, pp. 628 - 631.

Coleman, J. J., "Semiconductor Lasers," Chapter 6 in "Electro-Optics Handbook," Waynant, R. W., and Ediger, M. N. [eds.], McGraw-Hill, New York, NY, 1994, pp. 6.1 - 6.26.

Collet, M. G., "Solid-State Imaging Sensors," Sensors and Actuators, vol. 10, nos. 3 - 4, Nov. - Dec. 1986, pp. 287 - 302.

Comtois, J. H., and Bright, V. M., "Surface Micromachined Polysilicon Thermal Actuator Arrays and Applications," Proceedings of the 1996 Solid-State Sensor and Actuator Workshop, Hilton Head Island, SC, June 3 - 6, 1996, pp. 174 - 177.

Constant, M., Boussekey, L., Decoster, D., and Vilcot, J. P., "Use of GaAs High-Gain Photocon-ductors as New Detectors in Spectroscopic Systems," Electronics Letters, vol. 24, no. 3, Feb. 4, 1988, pp. 141 - 147.

Constant, M., Lefebvre, D., Boussekey, L., Decoster, D., and Vilcot, J. P., "Detectivity of High-Gain GaAs Photoconductive Detectors," Electronics Letters, vol. 24, no. 16, Aug. 4, 1988, pp. 1019 -1021.

Craford, M. G., Shaw, R. W., Herzog, A. H., and Groves, W. O., "Radiative Recombination Mechanisms in GaAsP Diodes with and without Nitrogen Doping," Journal of Applied Physics, vol. 43, no. 10, Oct. 1972, pp. 4075 - 4083.

Cremer, C., Emeis, N., Schier, M., Heise, G., Ebbinghaus, G., and Stoll, L., "Grating Spectrograph Integrated with Photodiode Array in InGaAsP/InGaAs/InP," IEEE Photonics Technology Letters, vol. 4, no. 1, Jan. 1992, pp. 108 - 110.

Cuevas, A., Sinton, R. A., Midkiff, N. E. and Swainish, R. M., "26-Percent Efficient Point-Junction Concentrator Solar Cells with a Front Metal Grid," IEEE Electron Device Letters, vol. 11, no. 1, Jan. 1990, pp. 6 - 8.

Daneman, M. J., Solgaard, O., Tien, N. C., Lau, K. Y., and Muller, R. S., "Integrated Laser-to-Fiber Coupling Module Using a Micromachined Alignment Mirror," IEEE Photonics Technology Letters, vol. 8, no. 3, Mar. 1996, pp. 396 - 398.

Daneman, M. J., Tien, N. C., Solgaard, O., Lau, K. Y., and Muller, R. S., "Linear Vibromotor-Actuated Micromachined Microreflector for Integrated Optical Systems," Proceedings of the 1996 Solid-State Sensor and Actuator Workshop, Hilton Head Island, SC, June 3 - 6, 1996, pp. 109 - 112.

Daneman, M. J., Tien, N. C., Solgaard, O., Pisano, A. P., Lau, K. Y., and Muller, R. S., "Linear Microvibromotor for Positioning Optical Components," Proceedings of the IEEE Micro Electro Mechanical Systems Conference, Amsterdam, Netherlands, Jan. 29 - Feb. 2, 1995, pp. 55 - 60.

Darling, R. B., "Surface Sensitivity and Bias Dependence of Narrow-Gap Metal-Semiconductor-Metal Photodetectors," Journal of Applied Physics, vol. 67, no. 6, Mar. 15, 1990, pp. 3152 - 3162.

Debusschere, I., Bronckaers, E., Claeys, C., Kreider, G., Van Der Spiegel, J., Sandini, G., Dario, P., Fantini, F., Bellutti, P., and Soncini, G., "A Retinal CCD Sensor for Fast 2D Shape Recognition and Tracking," Sensors and Actuators, vol. A22, nos. 1 - 3, Mar. 1990, pp. 456 - 460.

DeFonzo, A. P., "Picosecond Photoconductivity in Germanium Films," Applied Physics Letters, vol. 39, no. 6, Sept. 15, 1981, pp. 480 - 483.

Degani, J., Leheny, R. F., Nahory, R. E., Pollak, M. A., Heritage, J. P., and DeWinter, J. C., Applied Physics Letters, vol. 38, no. 1, Jan. 1, 1981, pp. 27 - 29.

Deimel, P. P., "Micromachined Devices for Optical Applications," Proceedings of Transducers '95, the 8th International Conference on Solid-State Sensors and Actuators, Stockholm, Sweden, June 25 - 29, 1995, vol. 1, pp. 340 - 343.

Denyer, P. B., "Intelligent CMOS Imaging," Proceedings of the SPIE Conference on Charge-Coupled Devices and Solid State Optical Sensors V, San Jose, CA, Feb. 6 - 7, 1995, Proceedings of the SPIE, vol. 2415, 1995, pp. 285 - 291.

Denyer, P. B., Renshaw, D., Guoyu, W., and Mingying, L., "CMOS Image Sensors for Multimedia Applications," Proceedings of the Custom Integrated Circuits Conference, San Diego, CA, May 9 - 12, 1993, pp. 11.5.1-11.5.4.

Deyhimy, I., Eden, R. C., Anderson, R. J., and Harris, I. S., Jr., "A 500-MHz GaAs Charge-Coupled Device," Applied Physics Letters, vol. 36, no. 2, Jan. 15, 1980, pp. 151 - 153.

Dichter, B. K., "Fluorescent-Phosphor-Based Broadband UV Light Sensors," Journal of Machine Perception, vol. 10, no. 4, Apr. 1, 1993, p. 19.

Dietrich, T. R., Abraham, M., Diebel, J., Lacher, A., and Ruf, J., "Photoetchable Glass for Microsystems: Tips for Atomic Force Microscopy," Journal of Micromechanics and Microengineering, vol. 3, no. 4, Dec. 1993, pp. 187 - 189.

Dillon, P. L. P., Brault, A. T., Horak, J. R., Garcia, E., Martin, T. W., and Light, W. A., "Fabrication and Performance of Color Filter Arrays for Solid-State Imagers," IEEE Transactions on Electron Devices, vol. ED-25, no. 2, Feb. 1978, pp. 97 - 101.

Dillon, P. L., Lewis, D. M., and Kaspar, F. G., "Color Imaging System Using a Single CCD Area Array," IEEE Transactions on Electron Devices, vol. ED-25, no. 2, Feb. 1978, pp. 102 - 107.

Dötzel, W., Gessner, T., Hahn, R., Kaufmann, C., Kehr, K., Kurth, S., and Mehner, J., "Silicon Mirrors and Micromirror Arrays for Spatial Laser Beam Modulation," Proceedings of Transducers '97, the 1997 International Conference on Solid-State Sensors and Actuators, Chicago, IL, June 16 - 19, 1997, vol. 1, pp. 81 - 84.

Dugliss, C. H., "Chemiluminescent Composition," U.S. Patent No. 4,678,608, issued July 7, 1987.

Dyment, J. C., d'Asaro, L. A., North, T. C., Miller, B. I., and Ripper, J. E., "Proton Bombardment Formation of Stripe-Geometry Heterostructure Lasers for 300 K CW Operation," Proceedings of the IEEE, vol. 60, no. 6, June 1972, pp. 726 - 728.

Edwards, W. D., "Two and Three Terminal Gallium Arsenide FET Optical Detectors," IEEE Electron Device Letters, vol. EDL-1, no. 8, Aug. 1980, pp. 149 - 150.

Escher, J. S., "NEA Semiconductor Photoemitters," Chapter 3 in "Semiconductors and Semimetals - Volume 15: Contacts, Junctions and Emitters," Willardson, R. K., and Beer, A. C. [eds.], Academic Press, Inc., New York, NY, 1981, pp. 195 - 301.

Fan, L., Wu, M. C., Choquette, K. D., and Crawford, M. H., "Self-Assembled Microactuated XYZ Stages for Optical Scanning and Alignment," Proceedings of Transducers '97, the 1997 International Conference on Solid-State Sensors and Actuators, Chicago, IL, June 16 - 19, 1997, vol. 1, pp. 319 - 322.

Figueroa, L., and Slayman, C. W., "A Novel Heterostructure Interdigital Photodetector (HIP) with Picosecond Optical Response," IEEE Electron Device Letters, vol. 2, no. 8, Aug. 1981, pp. 208 - 210.

Forrest, S. R., "The Sensitivity of Photoconductor Receivers for Long-Wavelength Optical Communications," Journal of Lightwave Technology, vol. LT-3, no. 2, Apr. 1985, pp. 347 - 360.

Fossum, E. R., "CMOS Image Sensors: Electronic Camera on a Chip," Technical Digest - International Electron Devices Meeting, IEDM '96, Washington, DC, Dec. 10 - 13, 1995, pp. 17 - 25.

Fullin, E., Voirin, G., Chevroulet, M., Lagos, A., and Moret, J.-M., "CMOS-Based Technology for Integrated Optoelectronics: A Modular Approach," Technical Digest of the International Electron Devices Meeting, San Francisco, CA, Dec. 11 - 14, 1994, pp. 527 - 530.

Golay, M. J. E., "A Pneumatic Infrared Detector," Review of Scientific Instruments, vol. 18, no. 5, 1947, p. 347.

Golay, M. J. E., "The Theoretical and Practical Sensitivity of the Pneumatic Infra-Red Detector," The Review of Scientific Instruments," vol. 20, no. 11, Nov. 1949, pp. 816 - 820.

Goldman, D. S., White, P. L., and Anheier, N. C., "Miniaturized Spectrometer Employing Planar Waveguides and Grating Couplers for Chemical Analysis," Applied Optics, vol. 29, no. 31, Nov. 1, 1990, pp. 4583 - 4589.

Grade, J., Barzilai, A., Reynolds, J. K., Liu, C.-H., Partridge, A., Jerman, H., and Kenny, T., "Wafer-Scale Processing, Assembly and Testing of Tunneling Infrared Detectors," Proceedings of Transducers '97, the 1997 International Conference on Solid-State Sensors and Actuators, Chicago, IL, June 16 - 19, 1997, vol. 2, pp. 1241 - 1244.

Groves, W. O., Herzog, A. H., and Craford, M. G., "The Effect of Nitrogen Doping on $GaAs_{1-x}P_x$ Electroluminescent Devices," Applied Physics Letters, vol. 19, no. 6, Sept. 15, 1971, pp. 184 - 186.

Guldberg, J., and Nathanson, H. C., "Electrostatically Deflectable Light Valve with Improved Diffraction Properties," U.S. Patent No. 3,886,310, issued May 27, 1975.

Gustafsson, G., Cao, Y., Treacy, G. M., Klavetter, F., Colaneri, N., and Heeger, A. J., "Flexible Light-Emitting Diodes Made from Soluble Organic Polymers," Nature, vol. 357, June 11, 1992, pp. 477 - 479.

Hamaguchi, H., Makiuchi, M., Kumai, T., and Wada, O., "GaAs Optoelectronic Integrated Receiver with High-Output Fast-Response Characteristics," IEEE Electron Device Letters, vol. EDL-8, no. 1, Jan. 1987, pp. 39 - 41.

Harris, J. F., and Gamow, R. I, "Snakes," Science, vol. 172, 1971, pp. 1252 - 1253.

Haynes, J. R., and Westphal, W. C., "Radiation Resulting from Recombination of Holes and Electrons in Silicon," Physics Review, vol. 101, 1956, pp. 1676 - 1678.

Heilmeier, G. H., Zanoni, L. A., and Barton, L. A., "Dynamic Scattering: A New Electro-Optic Effect in Certain Classes of Nematic Liquid Crystals," Proceedings of the IEEE, vol. 56, 1968, pp. 1162 - 1171.

Henck, S. A., "Lubrication of Digital Micromirror Devices™," Tribology Letters, vol. 3, no. 3, June 1997, pp. 239 - 247.

Higdon, A., and Stiles, W. D., "Engineering Mechanics, Volume II: Dynamics," Prentice-Hall, Inc., Englewood Cliffs, NJ, 1961, p. 555.

Hirano, R., Oomura, E., Higuchi, H., Sakakibara, Y., and Suzaki, Y., "Low Threshold Current 1.3 μm InGaAsP Buried Crescent Lasers," Japanese Journal of Applied Physics (Supplement), vol. 22, 1983, pp. 231 - 234.

Honer, K. A., Maluf, N. I., Martinez, E., and Kovacs, G. T. A., "Characterizing Deflectable Microstructures via a High-Resolution Laser-Based Measurement System," Sensors and Actuators, vol. A52, nos. 1 - 3, Mar. - Apr. 1996, pp. 12 - 17.

Hornbeck, L. J., "Deformable-Mirror Spatial Light Modulators" Proceedings of the SPIE Workshop on Spatial Light Modulators and Applications III, Proceedings Of The SPIE - The International Society for Optical Engineering, San Diego, CA, Aug. 7 - 8, 1989, vol. 1150 (published 1990), pp. 86 - 102.

Hornbeck, L. J., "Digital Light Processing and MEMS: Timely Convergence for a Bright Future," Proceedings of the SPIE Workshop on Micromachining and Microfabrication Processes, Proceedings Of The SPIE - The International Society for Optical Engineering, Austin, TX, Oct. 23 - 24, 1995, vol. 2639, pp. 3 - 21.

Hornbeck, L. J., "Spatial Light Modulator and Method," U.S. Patent No. 4,710,732, issued Dec. 1, 1987.

Hornbeck, L. J., "Spatial Light Modulator and Method," U.S. Patent No. 5,061,149, issued Oct. 29, 1991.

Hornbeck, L. J., "Spatial Light Modulator System," U.S. Patent No. 5,028,939, issued July 2, 1991.

Hornbeck, L. J., "Spatial Light Modulator," U.S. Patent No. 4,956,619, issued Sept. 11, 1990.

Horton, J. R., Tasker, G. W., Fijol, J. J., "Characteristics and Applications of Advanced Technology Microchannel Plates," Proceedings of the SPIE Workshop on Sensor Fusion III Workshop, Proceedings Of The SPIE - The International Society for Optical Engineering, Orlando, FL, Apr. 4 - 20, 1990, vol. 1306, pp. 169 - 178.

Huang, Y., Zhang, H., Kim, E. S., Kim, S. G., and Jeon, Y. B., "Piezoelectrically Actuated Microcantilever for Actuated Mirror Array Application," Proceedings of the 1996 Solid-State Sensor and Actuator Workshop, Hilton Head Island, SC, June 3 - 6, 1996, pp. 191 - 195.

Jaecklin, V. P., Linder, C., de Rooij, N. F., Moret, J.-M., and Vuilleumier, R., "Line-Addressable Torsional Micromirrors for Light Modulator Arrays," Sensors and Actuators, vol. A41, nos. 1 - 3, Apr. 1994, pp. 324 - 329.

Jerman, J. H., Clift, D. J., and Mallinson, S. R., "A Miniature Fabry-Perot Interferometer with a Corrugated Silicon Diaphragm Support," Sensors and Actuators, vol. A29, no. 2, Nov. 1991, pp. 151 - 158.

Jones, R. C., "The General Theory of Bolometer Performance," Journal of the Optical Society of America, vol. 43, 1953, p. 1.

Jones, R. C., "The Ultimate Sensitivity of Radiation Detectors," Journal of the Optical Society of America, vol. 37, 1947, p. 879.

Judy, J. W., and Muller, R. S., "Batch-Fabricated, Addressable, Magnetically Actuated Microstructures," Proceedings of the 1996 Solid-State Sensor and Actuator Workshop, Hilton Head Island, SC, June 3 - 6, 1996, pp. 187 - 190.

Jung, K. H., Shih, S., and Kwong, D. L., "Developments in Luminescent Porous Si," Journal of the Electrochemical Society, vol. 140, no. 10, Oct. 1993, pp. 3046 - 3064.

Kaminow, I. P., "An Introduction to Electrooptic Devices," Academic Press, Inc., New York, NY, 1974.

Kaminow, I. P., and Turner, E. H., "Electrooptic Light Modulators," Proceedings of the IEEE, vol. 54, no. 10, Oct. 1966, pp. 1374 - 1390.

Kaneda, T., "Silicon and Germanium Avalanche Photodiodes," Chapter 3 in "Semiconductors and Semimetals - Volume 22, Part D: Lightwave Communications Technology," Willardson, R. K., and Beer, A. C. [eds.], Tsang, W. S. [volume ed.], Academic Press, Inc., New York, NY, 1985, pp. 247 - 328.

Kasper, E., and Schäffler, F., "Group-IV Compounds," Chapter 4 in "Semiconductors and Semimetals - Volume 33: Strained-Layer Superlattices: Materials Science and Technology," Willardson, R. K., and Beer, A. C. [eds.], Academic Press, Inc., New York, NY, 1991, pp. 223 - 309.

Kellner, W., Ablassmeier, U., and Kniepkamp, H., "A Two-Phase CCD on GaAs with 0.3-μm-Wide Electrode Gaps," IEEE Transactions on Electron Devices, vol. ED-27, no. 6, June 1980, pp. 1195 - 1197.

Kendall, D. L., Eaton, W. P., Manginell, R., and Digges, T. G., Jr., "Micromirror Arrays Using KOH:H_2O Micromachining of Silicon for Lens Templates, Geodesic Lenses and Other Applications," Optical Engineering, vol. 33, no. 11, Nov. 1994, pp. 3578 - 3587.

Kenny, T. W., Kaiser, W. J., Waltman, S. B., and Reynolds, J. K., "Novel Infrared Detector Based on a Tunneling Displacement Transducer," Applied Physics Letters, vol. 59, no. 19, Oct. 7, 1991, pp. 1820-1822.

Kerns, D., "A Monolithic Si UV Detector-Dosimeter," Sensors and Actuators, vol. A39, no. 3, Dec. 1993, pp. 225 - 229.

Kiang, M.-H., Francis, D. A., Chang-Hasnain, C. J., Solgaard, O., Lau, K. Y., and Muller, R. S., "Actuated Polysilicon Mirrors for Raster-Scanning Displays," Proceedings of Transducers '97, the 1997 International Conference on Solid-State Sensors and Actuators, Chicago, IL, June 16 - 19, 1997, vol. 1, pp. 323 - 326.

Kiang, M.-H., Nee, J. T., Lau, K. Y., and Muller, R. S., "Surface-Micromachined Diffraction Gratings for Scanning Spectroscopic Applications," Proceedings of Transducers '97, the 1997 International Conference on Solid-State Sensors and Actuators, Chicago, IL, June 16 - 19, 1997, vol. 1, pp. 343 - 345.

Koroscil, A., "Chemiluminescent Composition," U.S. Patent No. 4,717,511, issued Jan. 5, 1988.

Kosonocky, W. F., Shallcross, F. V., Villani, T. S., and Groppe, J. V., "160 X 244 Element PtSi Schottky-Barrier IR-CCD Image Sensor," IEEE Transactions on Electron Devices, vol. ED-32, no. 8, Aug. 1985, pp. 1564 - 1573.

Kozawaguchi, H., Ohwaki, J., Tsujiyama, B., and Murase, K., "Low-Voltage AC Thin-Film Electroluminescent Devices," Proceedings of the 1982 SID International Symposium, Cherry Hill, NJ, Oct. 19 - 21, 1982, Proceedings of the Society for Information Display, vol. 23, no. 3, 1982, pp. 181 - 186.

Kramer, J., Seitz, P., and Baltes, H., "Industrial CMOS Technology for the Integration of Optical Metrology Systems (Photo-ASICS)," Sensors and Actuators, vol. A34, no. 1, July 1992, pp. 21 - 30.

Kramer, J., Seitz, P., Steigmeier, E. F., Auderset, H., and Delley, B., "Light-Emitting Devices in Industrial CMOS Technology," Sensors and Actuators, vols. A37 - A38, no. 2, June - Aug. 1993, pp. 527 - 533.

Kraus, Th., Baltzer, M., and Obermeier, E., "A Micro Shutter for Applications in Optical and Thermal Detectors," Proceedings of Transducers '97, the 1997 International Conference on Solid-State Sensors and Actuators, Chicago, IL, June 16 - 19, 1997, vol. 1, pp. 67 - 70.

Kressel, H., and Butler, J. K., "Semiconductor Lasers and Heterojunction LEDs," Academic Press, Inc., New York, NY, 1977.

Kruse, P. W., "High T_c Superconducting IR Detectors," Proceedings of the SPIE Conference on Superconductivity Applications for Infrared and Microwave Devices, Orlando, FL, Apr. 19 - 20, 1990, Proceedings of the SPIE, vol. 1292, 1990, pp. 108 - 177.

Kwa, T. A., and Wolffenbuttel, R. F., "Integrated Grating/Detector Array Fabricated in Silicon Using Micromachining Techniques," Sensors and Actuators, vol. A31, nos. 1 - 3, Apr. 1992, pp. 259 - 266.

Kwa, T. A., Sarro, P. M., and Wolffenbuttel, R. F., "Backside-Illuminated Silicon Photodiode Array for an Integrated Spectrometer," IEEE Transactions on Electron Devices, vol. 44, no. 5, May 1997, pp. 761 - 765.

Lee, A., Fiorillo, A. S., van der Spiegel, J., Bloomfield, P. E., Dao, J., and Dario, P., "Design and Fabrication of a Silicon-P(VDF-TrFE) Piezoelectric Sensor," Thin Solid Films, vol. 181, Dec. 1989, pp. 245 - 250.

Lee, C. H., "Picosecond Optoelectronic Switching in GaAs," Applied Physics Letters, vol. 30, no. 2, Jan. 15, 1977, pp. 84 - 86.

Lee, S. S., Lin., L. Y., and Wu, M. C., "Surface-Micromachined Free-Space Micro-Optical Systems Containing Three-Dimensional Microgratings," Applied Physics Letters, vol. 67, no. 15, Oct. 9, 1995, pp. 2135 - 2137.

Lee, S.-S., Motamedi, E., and Wu, M. C., "Surface-Micromachined Free-Space Fiber Optic Switches with Integrated Microactuators for Optical Fiber Communication Systems," Proceedings of Transducers '97, the 1997 International Conference on Solid-State Sensors and Actuators, Chicago, IL, June 16 - 19, 1997, vol. 1, pp. 85 - 88.

Lenggenhager, R., "CMOS Thermoelectric Infrared Sensors," Dissertation for the Doctor of Natural Sciences Degree, Swiss Federal Institute of Technology, Zurich, Switzerland, DISS. ETH No. 10744, 1994.

Lentine, A. L., Goossen, K. W., Walker, J. A., Chirovsky, L. M. F., D'Asaro, L. A., Hui, S. P., Tseng, B. T., Leibenguth, R. E., Kossives, D. P., Dahringer, D. W., Bacon, D. D., Woodward, T. K., and Miller, D. A. B., "Arrays of Optoelectronic Switching Nodes Comprised of Flip-Chip-Bonded MQW Modulators and Detectors on Silicon CMOS Circuitry," IEEE Photonics Technology Letters, vol. 8, no. 2, Feb. 1996, pp. 221 - 223.

Leonberger, F. J., and Moulton, P. F., "High-Speed InP Optoelectronic Switch," Applied Physics Letters, vol. 35, no. 9, Nov. 1, 1979, pp. 712 - 714.

Liddiard, K. C., Unewisse, M. H., and Reinhold, O., "Design and Fabrication of Thin-Film Monolithic Bolometer Infrared Detector Arrays," Proceedings of the SPIE Conference on Infrared Detectors and Focal Plane Arrays III, Orlando, FL, Apr. 5 - 6, 1994, Proceedings of the SPIE, vol. 2225, 1994, pp. 62 - 71.

Lin, L. Y., Lee, S. S., Pister, K. S. J., and Wu, M.C., "Micro-Machined Three-Dimensional Micro-Optics for Free-Space Optical Systems," IEEE Photonics Technology Letters, vol. 6, no. 12, Dec. 1994, pp. 1445 - 1447.

Lin, L. Y., Lee, S. S., Wu, M. C., and Pister, K. S. J., "Micromachined Integrated Optics for Free-Space Interconnections," Proceedings of the IEEE Micro Electro Mechanical Systems Conference, Amsterdam, Netherlands, Jan. 29 - Feb. 2, 1995, pp. 77 - 82.

Lin, L. Y., Shen, J. L., Lee, S. S., Su, G. D., and Wu, M.C., "Microactuated Micro-XYZ Stages for Free-Space Micro-Optical Bench," Proceedings of the Tenth Annual Workshop of Micro Electro Mechanical Systems, Nagoya, Japan, Jan. 26 - 30, 1997, pp. 43 - 48.

Liu, R. H., Vasile, M. J., Goettert, J., and Beebe, D. J., "Investigation of the LIGA Process to Fabricate Microchannel Plates," Proceedings of Transducers '97, the 1997 International Conference on Solid-State Sensors and Actuators, Chicago, IL, June 16 - 19, 1997, vol. 1, pp. 645 - 648.

Marshall, D. E., "A Review of Pyroelectric Detector Technology," Proceedings of the SPIE Conference, Los Angeles, CA, Jan. 16 - 18, 1978, Proceedings of the SPIE, vol. 132, 1978, pp. 100 - 117.

Marxer, C., Grétillat, M.-A., de Rooij, N. F., Bättig, R., Anthamatten, O., Valk, B., and Vogel, P., "Vertical Mirrors Fabricated by Reactive Ion Etching for Fiber Optical Switching Applications," Proceedings of the Tenth Annual Workshop of Micro Electro Mechanical Systems, Nagoya, Japan, Jan. 26 - 30, 1997, pp. 49 - 54.

Mastrangelo, C. H., and Muller, R. S., "Vacuum-Sealed Silicon Micromachined Incandescent Light Source," Technical Digest of the International Electron Devices Meeting, Washington, DC, Dec. 3 - 6, 1989, pp. 503 - 506.

Mastrangelo, C. H., Yeh, J. H.-J., and Muller, R. S., "Electrical and Optical Characteristics of Vacuum-sealed Polysilicon Microlamps," IEEE Transactions on Electron Devices, vol. 39, no. 6, June 1992, pp. 1363 - 1375.

Mendis, S., Kemeny, S. E., and Fossum, E. R., "CMOS Active Pixel Image Sensor," IEEE Transactions on Electron Devices, vol. 41, no. 3, Mar. 1994, pp. 452 - 453.

Miller, R. A., Burr, G. W., Tai, Y.-C., Psaltis, D., Ho., C.-H., and Katti, R. R., "Electromagnetic MEMS Scanning Mirrors for Holographic Data Storage," Proceedings of the 1996 Solid-State Sensor and Actuator Workshop, Hilton Head Island, SC, June 3 - 6, 1996, pp. 183 - 186.

Miller, R. A., Tai, Y.-C., Xu, G., Bartha, J., and Lin, F., "An Electromagnetic MEMS 2 X 2 Fiber Optic Bypass Switch," Proceedings of Transducers '97, the 1997 International Conference on Solid-State Sensors and Actuators, Chicago, IL, June 16 - 19, 1997, vol. 1, pp. 89 - 92.

Millman, J., "Microelectronics: Digital and Analog Circuits and Systems," McGraw-Hill, New York, NY, 1979, pp. 298 - 312.

Mito, I., Kitamura, M., Kobayashi, K., Murata, S., Seki, M., Odagiri, Y., Nishimoto, H., Yamaguchi, M., and Kobayashi, K,. "InGaAsP Double-Channel-Planar-Buried-Heterostructure Laser Diodes (DC-PBH LD) with Effective Current Confinement," IEEE Journal of Lightwave Technology, vol. LT-1, no. 1, Mar. 1983, pp. 195 - 202.

Miyajima, H., Yamamoto, E., Ito, M., Hashimoto, S., Komazaki, I., Shinohara, S., and Yanagisawa, K., "Optical Micro Encoder Using Surface-Emitting Laser," Proceedings of IEEE International Workshop on Micro Electro Mechanical Systems, San Diego, CA, Feb. 11 - 15, 1996, pp. 412 - 417.

Nakamura, J.-I., Kemeny, S. E., and Fossum, E. R., "CMOS Active Pixel Image Sensor with Simple Floating Gate Pixels," IEEE Transactions on Electron Devices, vol. 42, no. 9, Sept. 1995, pp. 1693 - 1694.

Nakamura, S., Mukai, T., and Senoh, M., "Candela-Class High-Brightness InGaN/AlGaN Double-Heterostructure Blue-Light-Emitting Diodes," Applied Physics Letters, vol. 64, no. 13, Mar. 28, 1994, pp. 1687 - 1689.

Nakamura, S., Mukai, T., and Senoh, M., "High-Power GaN P-N Junction Blue-Light-Emitting Diodes," Japanese Journal of Applied Physics, Part 2 (Letters), vol. 30, no. 12A, Dec. 1, 1991, pp. L1998 - L2001.

Nakamura, S., Senoh, M., and Mukai, T., "p-GaN/N-InGaN/N-GaN Double Heterostructure Blue-Light-Emitting Diodes," Japanese Journal of Applied Physics, Part 2 (Letters), vol. 32, no. 1A-B, Jan. 15, 1993, pp. L8 - L11.

Nathanson, H. C., and Davis, J. R., Jr., "Electrostatically Deflectable Light Valves for Projection Displays," U.S. Patent No. 3,746,911, issued July 17, 1973.

Neff, H., Laukemper, J., Khrebtov, I. A., Tkachenko, A. D., Steinbeiß, E., Michalke, W., Burnus, B., Heidenblut, T., Hefle, G., and Schwierzi, B., "Sensitive High-Tc Transition Edge Bolometer on a Micromachined Silicon Membrane," Applied Physics Letters, vol. 66, no. 18, 1995, pp. 2421 - 2423.

Nishihara, H., Haruna, M., and Suhara, T., "Optical Integrated Circuits," Chapter 26 in "Electro-Optics Handbook," Waynant, R. W., and Ediger, M. N. [eds.], McGraw-Hill, New York, NY, 1994, pp. 26.1 - 26.39.

Nixon, R. H., Kemeny, S. E., Staller, C. O., and Fossum, E. R., "256 X 256 CMOS Active Pixel Sensor Camera-on-a-Chip," Digest of Technical Papers - IEEE International Solid-State Circuits Conference, San Francisco, CA, Feb. 8 - 10, 1996, pp. 178 - 179.

Ohnstein, T. R., Zook, J. D., Cox, J. A., Speldrich, B. D., Wagener, T. J., Guckel, H., Christenson, T. R., Klein, J., Earles, T., and Glasgow, I., "Tunable IR Filters Using Flexible Metallic Microstructures," Proceedings of the IEEE Micro Electro Mechanical Systems Conference, Amsterdam, Netherlands, Jan. 29 - Feb. 2, 1995, pp. 170 - 174.

Ohnstein, T. R., Zook, J. D., French, H. B., Guckel, H., Earles, T., Klein, J., and Mangat, P., "Tunable IR Filters with Integral Electromagnetic Actuators," Proceedings of the 1996 Solid-State Sensor and Actuator Workshop, Hilton Head Island, SC, June 3 - 6, 1996, pp. 196 - 199.

Oliver, A. D., Baer, W. G., and Wise, K. D., "A Bulk-Micromachined 1024-Element Uncooled Infrared Imager," Proceedings of Transducers '95, the 8th International Conference on Solid-State Sensors and Actuators, Stockholm, Sweden, June 25 - 29, 1995, vol. 2, pp. 636 - 639.

Panicacci, R., Kemeny, S. E., Matthies, L. H., Pain, B., and Fossum, E. R., "Programmable Multiresolution CMOS Active Pixel Sensor," Proceedings of the SPIE Conference on Solid State Sensor Arrays and CCD Cameras, San Jose, CA Jan. 31 - Feb. 2, 1996, Proceedings of the SPIE, vol. 2654, 1996, pp. 72 - 79.

Parameswaran, M., Robinson, A. M., Blackburn, D. L., Gaitan, M., and Geist, J., "Micromachined Thermal Radiation Emitter from a Commercial CMOS Process," IEEE Electron Device Letters, vol. 12, no. 2, Feb. 1991, pp. 57 - 59.

Pearson, G. L., Reed, W. T., Jr., and Feldman, W. L., "Deformation and Fracture of Small Silicon Crystals," Acta Mettalurgica, vol. 5, 1957, p. 181.

Peckerar, M., Ho, P.-T., and Chen, Y. J., "High-Resolution Lithography for Optoelectronics," Chapter 22 in, "Electro-Optics Handbook," Waynant, R. W., and Ediger, M. N. [eds.], McGraw-Hill, New York, NY, 1994, pp. 22.1 - 22.40.

Perregaux, G., Weiss, P., Kloek, B., Vuilliomenet, H., and Thiéaud, J.-P., "High-Speed Micro-Electromechanical Light Modulation Arrays," Proceedings of Transducers '97, the 1997 International Conference on Solid-State Sensors and Actuators, Chicago, IL, June 16 - 19, 1997, vol. 1, pp. 71 - 74.

Petersen, K. E., "Dynamic Micromechanics on Silicon: Techniques and Devices," IEEE Transactions on Electron Devices, vol. ED-25, no. 10, Oct. 1978, p. 1241

Petersen, K. E., "Micromechanical Light Modulator Array Fabricated on Silicon," Applied Physics Letters, vol. 31, no. 8, Oct. 1977, pp. 521 - 523.

Petersen, K. E., "Silicon as a Mechanical Material," Proceedings of the IEEE, vol. 70, no. 5, May 1982, pp. 420 - 457.

Petersen, K. E., "Silicon Torsional Scanning Mirror," IBM Journal of Research and Development, vol. 24, no. 5, Sept. 1980, pp. 631 - 637.

Petrovich, T., "Industry Trends Favor Low-Power Displays," Electronic Design, Jan. 9, 1995, pp. 95 - 102.

Pister, K. S. J., Judy, M. W., Burgett, S. R., and Fearing R. S., "Microfabricated Hinges," Sensors and Actuators, vol. A33, no. 3, June 1992, pp. 249 - 256.

Pohjonen, H., and Andersson, M., "Integrated Optoelectronic Circuits Using Metal-Semiconductor-Metal Photodetectors," Sensors and Actuators, vol. A23, nos. 1 - 3, Apr. 1990, pp. 1124 - 1127.

Polla, D. L., Muller, R. S., and White, R. M., "Integrated Multisensor Chip," IEEE Electron Device Letters, vol. EDL-7, no. 4, Apr. 1986, pp. 254 - 256.

Polla, D. L., Muller, R. S., and White, R. M., "Pyroelectric Properties and Applications of Sputtered Zinc-Oxide Thin Films," Record of the IEEE Ultrasonics Symposium, San Francisco, CA, Oct. 16 - 18, 1985, pp. 495 - 498.

Putley, E. H., "The Pyroelectric Detector - An Update," in "Semiconductors and Semimetals," Willardson, R. K., and Beer, A. C. [eds.], Academic Press, Inc., New York, NY, 1977, pp. 441 - 449.

Putley, E. H., "The Pyroelectric Detector," in "Semiconductors and Semimetals," Willardson, R. K., and Beer, A. C. [eds.], Academic Press, Inc., New York, NY, 1970, pp. 259 - 285.

Quelle, F. W., "Superconducting IR Focal Plane Arrays," Proceedings of the SPIE Conference on Electron Image Tubes and Image Intensifiers, Santa Clara, CA, Feb. 15 - 16, 1990, Proceedings of the SPIE, vol. 1243, 1990, pp. 206 - 213.

Raley, N. F., Ciarlo, D. R., Koo, J. C., Beiriger, B., Trujillo, J., Yu, C., Loomis, G., and Chow, R., "A Fabry-Perot Microinterferometer for Visible Wavelengths," Proceedings of the 1992 Solid-State Sensor and Actuator Workshop, Hilton Head Island, SC, June 22 - 25, 1992, pp. 170 - 173.

Redington, R. W., "Gain Band-Width Product of Photoconductors," Physical Review, vol. 115, no. 4, Aug. 15, 1959, pp. 894 - 896.

Reid, R. J., Bright, V. M., and Comtois, J. H., "Automated Assembly of Flip-Up Micromirrors," Proceedings of Transducers '97, the 1997 International Conference on Solid-State Sensors and Actuators, Chicago, IL, June 16 - 19, 1997, vol. 1, pp. 347 - 350.

Renshaw, D., Denyer, P. B., Wang, G., and Lu, M., "ASIC Vision," Proceedings of the IEEE 1990 Custom Integrated Circuits Conference, Boston, MA, May 13 - 16, 1990, pp. 7.3.1 - 7.3.4.

Ricquier, N., and Dierickx, B., "Random Addressable CMOS Image Sensor for Industrial Applications," Sensors and Actuators, vol. A44, no. 1, July 1994, pp. 29 - 35.

Ringh, U., Jansson, C., Svensson, C., and Liddiard, K., "CMOS RC-Oscillator Technique for Digital Readout from an IR Bolometer Array," Proceedings of Transducers '95/Eurosensors IX, Stockholm, Sweden, June 25 - 29, 1995, vol. 1, pp. 138 - 1412.

Ripper, J. E., Dyment, J. C., d'Asaro, L. A., and Paoli, T. L., "Stripe-Geometry Double-Heterostructure Junction Lasers: Mode Structure and CW Operation above Room Temperature," Applied Physics Letters, vol. 18, no. 4, Feb. 15, 1971, pp. 155 - 157.

Rogalski, A., "New Trends in Infrared Detector Technology," Infrared Physics and Technology, vol. 35, no. 1, Feb. 1994, pp. 1 - 21.

Rosengren, L., Smith, L., and Bäcklund, Y., "Micromachined Optical Planes and Reflectors in Silicon," Sensors and Actuators, vol. A41, nos. 1 - 3, Apr. 1994, pp. 330 - 333.

Saito, K., and Ito, R., "Buried Heterostructure AlGaAs Lasers," IEEE Journal of Quantum Electronics, vol. QE-16, no. 2, Feb. 1980, pp. 205 - 215.

Sampsell, J. B., "Spatial Light Modulator," U.S. Patent No. 4,954,789, issued Sept. 4, 1990.

Sánchez, S., Elwenspoek, M., Gui, C., de Nivelle, M. J. M. E., de Vries, R., de Korte, P. A. J., Bruijn, M. P., Wijnbergen, J. J., Michalke, W., Steinbeiß, E., Heidenblut, T., and Schwierzi, B., "A High-T_c Superconducting Bolometer on a Silicon Nitride Membrane," Proceedings of the Tenth Annual Workshop of Micro Electro Mechanical Systems, Nagoya, Japan, Jan. 26 - 30, 1997, pp. 506 - 511.

Slayman, C. W., and Figueroa, L., "Frequency and Pulse Response of a Novel High-Speed Interdigitated Surface Photoconductor (IDPC)," IEEE Electron Device Letters, vol. 2, no. 5, May 1981, pp. 112 - 114.

Solgaard, O., Sandejas, F. S. A., and Bloom, D. M., "Deformable Grating Optical Modulator," Optics Letters, vol. 17, no. 9, May 1, 1992, pp. 688 - 690.

Soref, R. A., and Bennett, B. R., "Electrooptical Effects in Silicon," IEEE Journal of Quantum Electronics, vol. QE-23, no. 1, Jan. 1987, pp. 123 - 129.

Stillman, G. E., and Wolfe, C. M., "Avalanche Photodiodes," Chapter 5 in "Semiconductors and Semimetals - Volume 12: Infrared Detectors II," Willardson, R. K., and Beer, A. C. [eds.], Academic Press, Inc., New York, NY, 1977, pp. 291 - 393.

Storment, C. W., Borkholder, D. A., Westerlind, V., Suh, J. W., Maluf, N. I., and Kovacs, G. T. A., "Flexible, Dry-Released Process for Aluminum Electrostatic Actuators," IEEE/ASME Journal of Microelectromechanical Systems, Sept. 1994, vol. 3, no. 3, pp. 90 - 96.

Stotlar, S. C., "Infrared Detectors," Chapter 17 in "Electro-Optics Handbook," Waynant, R. W., and Ediger, M. N. [eds.], McGraw-Hill, New York, NY, 1994, pp. 17.1 - 17.25.

Strandman, C., Rosengren, L., and Bäcklund, Y., "Fabrication of 45° Optical Mirrors on (100) Si Using Wet Anisotropic Etching," Proceedings of the IEEE Micro Electro Mechanical Systems Conference, Amsterdam, Netherlands, Jan. 29 - Feb. 2, 1995, pp. 244 - 249.

Streifer, W., and Ettenberg, M. [eds.], "Semiconductor Diode Lasers," vol. 1, IEEE Press, New York, NY, 1990.

Sugimoto, M., Suzuki, A., Nomura, H., and Lang, R., "InGaAsP/InP Current Confinement Mesa Substrate Buried Heterostructure Laser Diode Fabricated by One-Step Liquid-Phase Epitaxy," IEEE Journal of Lightwave Technology, vol. LT-2, no. 4, Aug. 1984, pp. 496 - 503.

Swart, N. R., Parameswaran, M., and Nathan, A., "Optimisation of the Dynamic Response of an Integrated Silicon Thermal Scene Simulator," Proceedings of Transducers '93, the 7th International Conference on Solid-State Sensors and Actuators, Yokohama, Japan, June 7 - 10, 1993, Institute of Electrical Engineers, Japan, pp. 750 - 753.

Tanaka, A., Matsumoto, N., Itoh, S., Endoh, T., Nakazato, A, Kumazawa, Y., Hijikawa, M., Gotoh, H., Tanaka, T., and Teranishi, N., "Silicon IC Process Compatible Bolometer Infrared Focal Plane Array," Proceedings of Transducers '95, the 8th International Conference on Solid-State Sensors and Actuators, Stockholm, Sweden, June 25 - 29, 1995, vol. 2, pp. 632 - 635.

Tang, C. W., "Organic Electroluminescent Materials and Devices," Information Display, vol. 12, no. 10, Oct. 1996, pp. 16 - 19.

Tessier-Lavigne, M., "Phototransduction and Information Processing in the Retina," Chapter 28 in "Principles of Neuroscience," Kandel, E. R., Schwartz, J. H. and Jessel, T. M. [eds.], Third Edition, Elsevier Science Publishing Co., Inc., New York, NY, 1991.

Thomas, R. N., Guldberg, J., Nathanson, H. C., and Malmberg, P. R., "The Mirror-Matrix Tube: A Novel Light Valve for Projection Displays," IEEE Transactions on Electron Devices, vol. ED-22, no. 9, Sept. 1975, pp. 765 - 775.

Tien, P. K., "Integrated Optics and New Wave Phenomena in Optical Waveguides," Reviews of Modern Physics, vol. 49, no. 2, Apr. 1977, pp. 361 - 420.

Tregilgas, J. H., "Micromechanical Device Having an Improved Beam," U.S. Patent No. 5,552,924, issued Sept. 3, 1996.

Vdovin, G., and Middelhoek, S., "Deformable Mirror Display with Continuous Reflecting Surface Micromachined in Silicon," Proceedings of the IEEE Micro Electro Mechanical Systems Conference, Amsterdam, Netherlands, Jan. 29 - Feb. 2, 1995, pp. 61 - 65.

Wada, O., Hamaguchi, H., Makiuchi, M., Kumai, T., Ito, M., Nakai, K., Horimatsu, T., and Sakurai, T., "Monolithic Four-Channel Photodiode/Amplifier Receiver Array Integrated on a GaAs Substrate," Journal of Lightwave Technology, vol. LT-4, no. 11, Nov. 1986, pp. 1694 - 1703.

Wagner, B., Reimer, K., Maciossek, A., and Hofmann, U., "Infrared Micromirror Array with Large Pixel Size and Large Deflection Angle," Proceedings of Transducers '97, the 1997 International Conference on Solid-State Sensors and Actuators, Chicago, IL, June 16 - 19, 1997, vol. 1, pp. 75 - 78.

Wang, S. Y., and Bloom, D. M., "100 GHz Bandwidth Planar GaAs Schottky Photodiode," Electronics Letters, vol. 19, no. 14, July 7, 1983, pp. 554 - 555.

Wang, S. Y., Bloom, D. M., and Collins, D. M., "20-GHz Bandwidth GaAs Photodiode," Applied Physics Letters, vol. 42, no. 2, Jan. 15, 1983, pp. 190 - 192.

Weast, R. C., [ed.], "CRC Handbook of Chemistry and Physics," CRC Press, Inc., Boca Raton, FL, 1988.

Webb, D. A., and Gnade, B., "Method for Fabricating a DMD Spatial Light Modulator with a Hardened Hinge," U.S. Patent No. 5,504,614, issued Apr. 2, 1996.

Wolffenbuttel, R. F., and van Drieënhuizen, B. P., "Direct Electro-Optical Actuation in Silicon," Proceedings of Transducers '91, the 1991 International Conference on Solid-State Sensors and Actuators, San Francisco, CA, June 24 - 27, 1991, pp. 286 - 288.

Wood, R. A., Han, C. J., and Kruse, P. W., "Integrated Uncooled Infrared Detector Imaging Arrays," Proceedings of the 1992 Solid-State Sensor and Actuator Workshop, Hilton Head Island, SC, June 22 - 25, 1992, pp. 132 - 135.

Wu, S.-Y., "A Hybrid Mass-Interconnection Method by Electroplating," IEEE Transactions on Electron Devices, vol. ED-25, no. 10, Oct. 1978, pp. 1201 - 1203.

Xu, J., and Steckl, A. J., "Visible Electroluminescence from Stain-Etched Porous Si Diodes," IEEE Electron Device Letters, vol. 15, no. 12, Dec. 1994, pp. 507 - 509.

Yariv, A., "Introduction to Optical Electronics," Second Edition, Holt, Rinehart and Winston, New York, NY 1976.

Yariv, A., and Yeh, P., "Optical Waves in Crystals," Wiley Interscience, New York, NY, 1984.

Yee, G. M., Hing, P. A., Maluf, N. I., and Kovacs, G. T. A., "Miniaturized Spectrometers for Biochemical Analysis," Proceedings of the 1996 Solid-State Sensor and Actuator Workshop, Hilton Head Island, SC, June 3 - 6, 1996, pp. 64 - 67.

Yeh, H.-J., "Miniature Gaseous Light Sources," Master of Science Research Project Report, Department of Electrical Engineering and Computer Science, University of California, Berkeley, CA, May 1991.

Yonezu, H., Sakuma, I., Kobayashi, K., Kamejima, T., Ueno, M., and Nannichi, Y., "A GaAs-$Al_xGa_{1-x}As$ Double Heterostructure Planar Stripe Laser," Japanese Journal of Applied Physics, vol. 12, no. 10, Oct. 1973, pp. 1485 - 1492.

Young, D., Wang, C.-L., and Pu, Y, "Magnetooptics," Chapter 54 in "The Electrical Engineering Handbook," Dorf, R. C. [ed.], CRC Press, Inc., Boca Raton, FL, 1993, pp. 1162 - 1172.

CITED INDUSTRY REFERENCES AND SUPPLIERS

CCD Information

"Kodak CCD Primer, #KCP-001, Charge-Coupled Device (CCD) Image Sensors," Eastman Kodak Company - Microelectronics Technology Division, B81, Floor 4, RL, Rochester, NY, 14650-2010.

"Kodak CCD Primer, #KCP-002, Conversion of Light (Photons) to Electronic Charge," Eastman Kodak Company - Microelectronics Technology Division, B81, Floor 4, RL, Rochester, NY, 14650-2010.

Photoformed Glass

Corning Incorporated, Materials Business, MP 21-3-2, Corning, NY 14831, Phone: (607) 974-7964.

Liquid Crystal Materials

EM Industries, 5 Skyline Drive, Hawthorne, NY 10532, Phone: (914) 592-4660.

Texas Instruments Digital Imaging Group (Web Site)

http://www.ti.com/dlp

Chapter 5:

IONIZING RADIATION TRANSDUCERS

1. INTRODUCTION

Ionizing radiation refers to high-energy particles and x-rays that are capable of ionizing materials that they penetrate or pass through. They are treated separately from more common IR, visible, and UV photons primarily due to their different behavior, and the different transducer structures needed to detect them.

There are many uses for such detectors. A common application is in dosimetry (monitoring radiation doses to personnel) and in instruments to locate radioactive sources (e.g., "Geiger counters"). These devices are also used in x-ray spectrometry (capable of identifying the chemical constituents of a sample through the spectra of x-rays emitted during electron bombardment). Another routine application is imaging x-rays and other forms of radiation in medical applications. In this case, radiation may be passed through a person to visualize internal features or radioactive chemicals may be administered that are typically sequestered by certain organs or defects that are of medical interest. Clearly in these cases, it is of interest to use the most sensitive detectors possible so as to minimize human exposure to radiation.

The major types of detectors are Geiger-Müller tubes (gas ionization by radiation), scintillators (light generation following excitation by radiation), and semiconductor detectors (electron-hole pair generation). A relative comparison of their performance levels is provided below.

Sensor Type	Energy Required to Form e/h or e/ion Pair	Response Time
Geiger-Müller (gas-filled tube)	≈ 25 to 40 eV	≈ 0.1 to 1.0 ms
Scintillator (example NaI(Tl))	≈ 50 eV for pair, ≈ 1 keV for photoelectron at cathode of photomultiplier	≈ 0.01 to 1.0 µs
Semiconductor	1.12 eV (Si), 0.66 eV (Ge)	≈ 1 ns

Table showing approximate sensitivities and response times of the three major ionizing radiation sensor types.

In general, as described by Goulding and Stone (1970), if the energy of an incident particle is totally absorbed within the sensor (i.e., the particle stops within it), the amount of ionization, light emission, or electron-hole pair generation will be proportional to the particle's energy, allowing for spectroscopic analysis. Since solids are roughly three orders of magnitude denser than in gases, the detector thicknesses required to stop particles are much smaller in the former case. As a consequence, gas ionization detectors are best suited to detection of slow, charged particles, while semiconductor detectors are able to sense a much broader spectrum. Secondly, the energy required to produce ionization is much lower in solids than gases, such that approximately an order of magnitude more ionization is produced in the former (with correspondingly larger signals). This effect is also important in scintillators, which in some cases have higher absorption efficiencies than semiconductor detectors. However, the relatively low efficiencies of conversion of the absorbed energy into light limits their SNR and the energy resolution.

1.1 UNITS OF MEASURE FOR RADIATION

An older unit of measure for absorbed radiation dose is the Roentgen, R, which was defined as the amount of radiation that produces one electrostatic unit (esu, defined as the charge on each of two points 1 cm apart that gives rise to 1 dyne of force) of charge in 1 cm^3 of air at standard temperature and pressure (STP). This corresponds to 87 ergs of energy per gram of air (1 J = 10^7 erg), or 1.6×10^{12} ion pairs. At present, the Roentgen is defined as the amount of x-ray or gamma radiation that delivers 0.878×10^{-2} J of energy per kg of air. A more modern unit of absorbed dose is the rad, where 1 rad is the amount of radiation that delivers 1.00×10^{-2} J/kg in *any* absorbing material (1 rad is also equal to 100 ergs/g).

When ionization is not the main issue, the basic SI unit for an absorbed dose is the Gray (Gy), where,

$$1 \text{ Gray} \equiv \frac{1 \text{ Joule of absorbed energy}}{1 \text{ kg mass of absorber}} = 10^4 \frac{\text{ergs}}{\text{g}} = 100 \text{ rads}$$

Of critical importance in radiation interactions with biological tissues is that more of the energy from incident radiation is deposited in denser tissues (e.g., bone). In addition, the biological damage caused by radiation is strongly dependent on the nature of the radiation. For example, a given dose (e.g., in rads) of alpha radiation can be more than an order of magnitude more damaging than an equivalent dose of gamma or beta radiation since the alpha radiation moves more slowly than equivalent energy gamma or beta radiation due to its larger mass. This results in more closely spaced ionizing collisions (their close spacing means that biological molecules can be more focally, and hence seriously damaged). In order to relate the different types of radiation in these terms, the relative biological effectiveness (RBE) is defined as the (approximate) relative amount of x-ray or gamma radiation that causes biological damage equal to that from 1 rad of the radiation in question.

Type of Radiation	RBE
β^-, β^+, x, γ (E < 0.03 MeV)	≈ 1
β^-, β^+, x, γ (E > 0.03 MeV)	≈ 1.7
Slow (Thermal) Neutrons	≈ 3
Fast Neutrons, Protons	$\approx$ 10 to 20
α-Particles	$\approx$ 10 to 20
Recoil Ions (heavy ions)	up to 30

Table of approximate relative biological effectivenesses (RBEs) of different types of radiation.

Using the RBE, one can define the *rem*, which stands for *rad equivalent man* and represents the effective dose,

$$\text{rem} = (\text{dose in rad})(\text{RBE})$$

and the newer SI unit, the *sievert* (Sv), is defined as,

$$\text{Sv} = (\text{dose in Gy})(\text{RBE})$$

In the United States, the government-specified maximum allowed annual radiation dose is 0.5 rem, of which the natural background radioactivity provides on the order of 0.1 rem (varies with geographic location and altitude). The maximum occupational exposure is 5 rem/year. For further information, see Giancoli (1989).

1.2 TYPES OF RADIATION

The term "ionizing radiation" covers a wide spectrum of energies, from alpha particles that can be stopped by a sheet of ordinary paper, through gamma rays, which can penetrate a meter or more of concrete or metal. It is important to note the symbols used to refer to isotopes of given elements. The notation used is, $^A_Z X$, where A is the atomic mass number (sum of the number of protons and neutrons), Z is the atomic number (number of protons, or electrons if the atom is not ionized).

1.2.1 ALPHA RADIATION

Alpha radiation (or alpha particles) are doubly ionized helium nuclei produced by the decay of radioactive parent atoms. They are low energy, and can be stopped by a single sheet of ordinary paper (their low energy makes them difficult to detect, since they are not always capable of entering a sensors active region). An example of a decay reaction for the formation of an alpha particle is,

$$^{238}_{92}U \rightarrow {}^{234}_{90}Th + {}^4_2He$$

where ^4_2He is α. Since the alpha particle consists of two neutrons and two protons, it has a net positive charge and can be deflected by a magnetic field.

1.2.2 BETA RADIATION

Beta particles are high-energy electrons or their antiparticles, positrons. The notation used is "β^-" for electrons and "β^+" for positrons (combining the two releases pure energy). While the electrons that typically make up beta particles are no different from orbital electrons, they originate in the nucleus of the decaying atom. This type of radiation can generally be stopped by sheets of metal (e.g., 1 cm of aluminum).

Example decay reactions that give rise to β^- particles are,

$$^3_1H \rightarrow {}^3_2He + \beta^- \quad \text{and} \quad {}^{22}_{11}Na \rightarrow {}^{22}_{10}Ne + \beta^-$$

Similarly, β^+ are generated by other decay reactions, such as,

$$^{19}_{10}Ne \rightarrow {}^{19}_9F + \beta^+$$

Neutrinos or antineutrinos are also emitted during such decays (they are very small particles with tiny masses that travel at the speed of light). When β^- particles are released, antineutrinos are as well, and for β^+ decays, neutrinos are produced.

Since beta particles are charged, they can be deflected by a magnetic field in a manner similar to alpha particles, but in a direction depending on their sign.

1.2.3 GAMMA RADIATION AND X-RAYS

Gamma radiation (or gamma rays) and x-rays are high-energy electromagnetic radiation (photons), typically in the MeV range and keV range, respectively. If they are generated by an atomic nucleus transitioning from a higher- to a lower-energy state, they are referred to as gamma rays. If they are generated through the interaction of an energetic electron and an atom, they are referred to as x-rays. For the higher-energy gamma rays, more than one meter of concrete or metal may be required to stop them. Being photons, gamma and x-rays are not deflected in magnetic fields. A typical decay reaction that liberates gamma rays is,

$$^{60}Co \; (5.26 \; yr \; half\text{-}life) \rightarrow {}^{60}Ni \; (stable) + \; \beta^- + \gamma_1 \; (1.173 \; MeV) + \gamma_2 \; (1.332 \; MeV)$$

where the above decay is a sequence.

This form of radiation is often involved in damage to biological tissues, as discussed below.

1.2.4 NEUTRON RADIATION

Neutrons are massive particles without charge. Their mass is 1838.7 times the mass of an electron, the latter being 9.109×10^{-31} kg, for a neutron mass of 1.675×10^{-27} kg (for completeness, the mass of the proton is 1836.2 times the mass of the electron). They are broadly categorized as "thermal" or slow neutrons and fast neutrons. These particles have limited ionizing potential (and are thus more difficult to detect than other forms of ionizing radiation).

2. IONIZATION-BASED DETECTORS

Perhaps the simplest ionizing radiation sensor with direct electrical output is the *Geiger-Müller tube*. This device is a low-pressure chamber filled with an ionizable (usually inert) gas, and containing two electrodes biased with a moderately high ($\approx$ 100 V) DC voltage. A key advantage of this type of detector over the other types below is that it can survive in extreme radiation enviroments.

Ionizing radiation of sufficient energy passing through the gas ionizes it. The ions are attracted to the cathode and the electrons to the anode, and this results in a current that can be detected (typically routed through a transresistance amplifier or a resistor to convert it to a voltage).

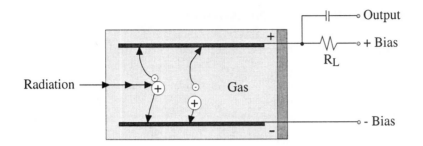

Illustration of the basic concept of a Geiger-Müller tube. After Norton (1989).

Physically, most macroscopic Geiger-Müller (G-M) tubes consist of a cylindrical (or disk-shaped "pancake") case with a central anode and a thin window for admitting the incident radiation. Implementation of the window presents a trade-off since making it thinner increases the sensitivity, but also increases its fragility.

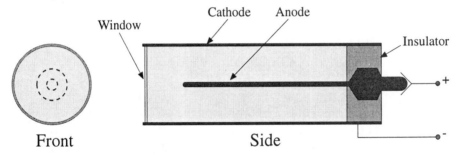

Illustration of the physical construction of a typical "end-window" Geiger-Müller tube design. Adapted from Norton (1989).

A "side-window" G-M tube design is shown below. This type of device is an alternative to designs with the window on the end of the cylinder and more closely resembles a structure that could be micromachined.

Although no publications appear to exist regarding micromachined G-M devices, there is no reason that they could not be realized, singly or in arrays for spatial localization of radiation sources. It should be quite possible to form the necessary windows from thin dielectrics and to deposit the electrodes using standard thin-film techniques. Low-pressure sealing of chambers with the necessary ionizable gas(es) would also be technically feasible.

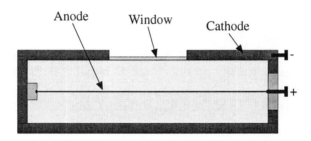

Illustration of a "side-window" Geiger-Müller tube. After Norton (1989). This device is shown in a cross-sectional view, and from the top view it may be virtually any shape (often circular).

3. SCINTILLATION DETECTORS

The operating principle of scintillators is to convert the incident radiation into light using an intermediate phosphor. The light is then detected using a conventional optical detector (the assembly is an "indirect" transducer). Simple materials such as doped zinc sulfide (as used in electroluminescent displays and cathode-ray tubes) can be used, or more exotic materials such as rare-earth phosphors or scintillation crystals may offer advantages in some cases. A variety of optical detectors can be used, such as CCDs, avalanche photodiodes, and photomultipliers.

One can easily imagine some of the structures that could be built in a micromachined format (i.e., phosphor plate imaged by a separate CCD array, a fully integrated "phosphor-on-detector" system, etc.). Naturally, the electronics would have to be sufficiently radiation hardened in many applications.

Important criteria for the efficiency of a scintillator material are that it be highly efficient in converting incoming photons into light, and that it be sufficiently transparent to the generated light so that it can effectively reach the detector. As for optical detectors, there is a "quantum efficiency" term, η_p, which represents the efficiency of down-conversion of photon wavelength.

In typical applications, the scintillator material is machined to the desired shape and coated with an optically reflective material on all surfaces except that from which light is coupled to the detector. Often, a collimator (see below) is used to provide some directionality to the detector.

For neutron detection, lithium iodide with europium activator, LiI(Eu), is typically used, and for nuclear medicine, the most popular scintillator materials are NaI, CsI, CaF_2, and $Bi_4Ge_3O_{12}$ (BGO). BGO was developed specifically for positron emission (e.g., PET [positron emission tomography] scanners) to detect the 170 keV x-rays produced by β^-/β^+ annihilation. NaI(Tl) is the most common scintillator for x-rays and gamma rays, but it is extremely hygroscopic. CsI(Tl) can be used in thin-film detectors, and has been used for space vehicles and satellites. It should also be noted that some organic materials, such as napthalene or anthracene crystals, are also good scintillators (such materials are discussed in Kelly (1995)). A review of position-sensitive (pixel-based) scintillation detectors is Akimov (1994).

Material	Peak Emission	Cutoff Emission	Time Constant	Refractive Index	Density	γ Scintillation Efficiency
* NaI(Tl)	410 nm	320 nm	230 ns	1.85	3.67	100%
CaF_2(Eu)	435 nm	405 nm	940 ns	1.44	3.18	50%
* CsI(Na)	420 nm	300 nm	630 ns	1.84	4.51	85%
CsI(Tl)	565 nm	330 nm	1.0 μs	1.80	4.51	45%
* LiI(Eu)	470 to 485 nm	450 nm	1.4 μs	1.96	4.08	35%
* CsF	390 nm	220 nm	5 ns	1.48	4.11	5%
$Bi_4Ge_3O_{12}$	480nm	350 nm	300 ns	2.15	7.13	8%
* KI(Tl)	426 nm	325 nm	≈ 1 μs	1.71	3.13	24%

Table of scintillation crystal materials. Note that hygroscopic materials are prefaced with an asterisk and that scintillation efficiencies are referenced to NaI(Tl). Activating dopants are shown in parentheses. From Padikal (1989).

While the sensitivity of direct semiconductor detectors is often adequate, there may be applications in which indirect radiation detection may be superior, and the possibility of micromachining detectors with integral scintillators would then be attractive.

4. DIRECT SOLID-STATE RADIATION DETECTORS

A common form of solid-state radiation detector is a crystal material, such as silicon or germanium, in which electron-hole pairs can be generated by incident radiation. Initially, large bulk crystals were used, but these are increasingly being replaced by microfabricated devices wherein the electron-hole pairs are generated in a semiconductor junction.

As described by Goulding and Stone (1970), in order to realize high-quality solid-state detectors, the free charges generated by the passage of radiation must have lifetimes that are long compared to the transit time across the active region of the detector (generally 10^{-6} to 10^{-9} s), allowing them to be collected before they recombine or are trapped. In order to accomplish this, the carrier velocity should be high in the material used.

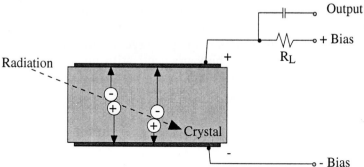

Illustration of electron-hole pair generation in a bulk semiconductor crystal. Adapted from Norton (1989).

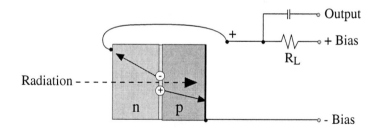

Illustration of electron-hole pair generation in a semiconductor device with a junction. Adapted from Norton (1989). Electron-hole pairs are generated throughout the material, but those generated in the depletion region are most rapidly and efficiently collected.

Electron-hole pair generation occurs at an average of 3 eV versus 30 eV for gaseous ionization. In some cases, semiconductor detectors have higher sensitivities and SNRs (although a scintillator and photomultiplier can generally provide a SNR one to two orders of magnitude higher than that of a silicon junction device).

While photomultipliers can operate with nanosecond response times, semiconductor devices also offer comparable speed with greatly reduced cost. Their lower cost is a key advantage over competing technologies. For example, a scintillator/photomultiplier combination can cost as much as one thousand times more than a comparable semiconductor diode detector. Another important advantage of semiconductor detectors is their small size (higher spatial resolution) and their inherent ability to be arrayed using integrated circuit technology.

A common semiconductor detector configuration is the p-i-n structure (p-type, intrinsic, n-type) as illustrated below, containing an active "intrinsic" region with as few free carriers as possible, and thin, heavily doped n- and p- regions for carrier collection. These detectors can be fabricated at low cost (in fact, the typical p-i-n detectors used for optical applications serve quite well as radiation detectors if incident light is blocked with an opaque layer) and can be integrated with circuitry (as discussed below). They are used with an applied reverse bias to prevent injection of majority carriers and move the generated carriers through the active region at relatively high velocities to reduce their transit times.

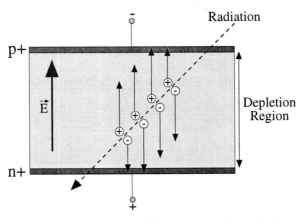

Illustration of a p-i-n diode radiation detector. Adapted from Middelhoek and Audet (1989). The typical reverse-biased operating mode is shown.

Depending on the type of radiation, many electron-hole pairs are generated per micron of linear ionization track distance (the path of the incident particle in the semiconductor). In some cases, this number may approach one hundred. The charge generated within the depletion region (or I-region widened by reverse bias in a non-p-i-n structure) is the signal component that is most rapidly and efficiently collected, thus the reverse bias voltages are often fairly high to reduce the carrier transit times.

Typically, intrinsic silicon is used as the substrate to maximize this effect (see Audet, et al. (1990) for an example of a low-energy [8 to 30 keV] detector array of this type), but lower-resistivity silicon can be used, with much worse performance due to a higher number of traps, scattering centers, and the density of donors or acceptors ionized at the operating temperature. For the best performance, the net doping of the silicon is typically on the order of 10^{12} cm^{-3}, although this is dependent on the methods used to manufacture the "intrinsic" silicon. It should be noted that if a detector that stops high-energy incident radiation as well as detecting it is desired, higher atomic weight materials such as germanium, cadmium telluride, or mercuric iodide must be considered.

The major sources of leakage current in p-i-n radiation detectors are surface conduction at the sides of the devices and thermal carrier generation. The latter effect can be mitigated by cooling the detector (often using liquid nitrogen [77 K]).

Other junction-based designs have been demonstrated that make use of heterojunctions. An example of this was presented by Yabe and Sato (1990), who fabricated such detectors using plasma-deposited amorphous Si:H on single-crystal silicon. They used a SiH_4/H_2 (9:1) in a plasma reactor to deposit 0.6 to 0.8 μm Si:H films on 10 to 30 kΩ•cm p-type silicon substrates. The resulting devices were capable of a resolution (full-width half-maximum) of 8.3 keV for 59.5 keV gamma rays (from [241]Am).

4.1 INTEGRATED DETECTOR ARRAYS

For high-energy physics applications, one is often interested in tracking the paths of individual particles and their decay products, if any. Multielement detectors can be made by selectively forming multiple n+ and/or p+ collectors on opposite sides of an I wafer. This allows the position of a particle to be localized to one or two dimensions. Multielement detectors such as these can be stacked to address the third dimension. Since many of the arrays are fabricated on silicon substrates, it would seem natural to integrate multiplexing and even signal-processing circuitry into them as well.

Tsoi, et al. (1985) demonstrated a deep depletion CCD imager for soft x-ray (1 to 10 keV) detection. They fabricated 512×96 pixel arrays in high-resistivity (50 kΩ•cm) silicon substrates. Since the depletion layer width of junctions increases with resistivity, large depletion regions (> 100 μm) could be obtained. This greatly reduced the pixel-to-pixel cross-talk problem seen with CCD detectors as x-ray detectors due to the fact that the x-ray photons penetrate tens of microns into the silicon, generating minority carriers that can diffuse laterally (this problem also affects CCDs with optical input wavelengths longer than ≈ 750 nm because of the lower absorption coefficients at longer wavelengths). Most of the incident x-ray photons are absorbed in the deep depletion layers, instead of the underlying substrate, and the field in the depletion layers pulls minority carriers toward the surface, greatly reducing lateral diffusion. Great care was required during processing to maintain the high resistivity of the substrates. The authors showed that single 5.9 keV x-ray photons could be resolved within single 26×26 μm pixels.

Snoeys (1992) (see also Snoeys, et al. (1991)) fabricated an integrated p-i-n diode detector array using a modified BiCMOS process flow (total of 16 masking steps) using high-resistivity p-type silicon doped at $\approx 10^{12}$ cm^{-3}. This substrate type provided a minority carrier lifetime in the range of 1 to 2 ms. A central n-well was formed, containing PMOS preamplification and switching circuitry, surrounding p+ collection electrodes that were in contact with the p-- substrate. The n-well not only provided Faraday cage-type shielding to isolate the collection field from transients generated by the circuitry, but also served to provide an electric field to guide the generated carriers to the collection electrodes. Three of the mask steps were back-side alignments carried out using infrared alignment to allow for accurately positioned back-side structures (including a large n-type diffusion for the back-side contact) to be fabricated. Fully integrated preamplifier, sample-and-hold, scan, and readout circuitry was incorporated into the devices. The array was 30×10 pixels, each of 34×125 μm in size, with an overall die size of $1.8 \times 2.2 \times 0.3$ mm. The arrays were successfully tested.

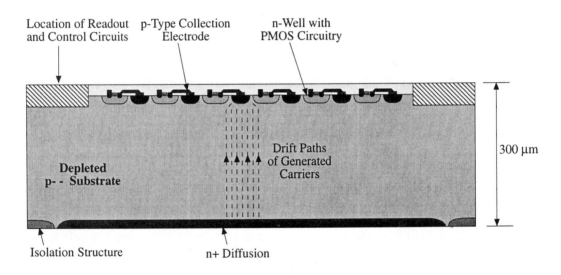

Illustration of the cross section of an integrated p-i-n diode detector array, including on-chip preamplification and multiplexing circuitry. Adapted from Snoeys (1992). Not to scale.

Streil, et al. (1994) reported on the use of modified dynamic RAM cells as the detectors. It has long been known that alpha particles can induce errors in dynamic random-access memories (DRAMs), and this group capitalized on that effect. They utilized a 1.0 μm CMOS DRAM process to fabricate 100×100 μm PMOS transistors in floating n-wells as sensors (unfortunately, this paper does not discuss the design of the PMOS transistors used, yet implies that it was somewhat "special"). Their entire array of 1,024 sensors/preamplifiers, a 4-bit A/D converter, an analog comparator with programmable threshold, and all necessary addressing circuitry were integrated into a single chip of 56 mm^2 total area. They reported detection of alpha particles with energies in the range of 5.5 to 7.9 MeV, with an energy resolution on the order of 300 keV. The use of such devices was demonstrated in a home tap water monitoring system to detect levels of radon.

5. OTHER DETECTOR TYPES

5.1 TRANSMUTATION-BASED SENSORS

Thin-film sensors have been demonstrated wherein the incident particles cause dose-related changes in their resistances by conversion of one type of atom into another. An important advantage of such sensors is that these processes are irreversible and thus provide an inherent "dose memory." Kervalishvili, et al. (1993) discussed the use of ^{10}B enriched boron-carbide films in which neutrons can be captured to form lithium atoms, which modulate the film's conductivity,

$$^{10}B + n \rightarrow ^7Li + ^4He$$

They grew the films epitaxially onto (insulating) sapphire substrates heated with a pulsed laser. The films were then irradiated with thermal neutrons in a nuclear reactor with neutron flows of 2×10^{11} cm$^{-2} \cdot$s^{-1} and their resistivities were measured. It was found that the normalized variations in resistivities were nearly linearly related to the neutron dose up to 3×10^{16} n/cm^2.

Similarly, silicon or germanium neutron sensors could, in principle, be fabricated by doping those crystals with ^{10}B. In this case, the concentration of holes decreases with neutron radiation due to the decrease in boron acceptor sites and their replacement with lithium donor atoms. Another approach would be to monitor the conversion of silicon atoms into phosphorus impurities by neutron irradiation.

5.2 CERENKOV EFFECT DETECTORS

The Cerenkov effect is radiation-induced light emission that occurs in some pure, transparent materials (e.g., certain glasses) and water. These materials can be used in conjunction with some type of photon detector such as a solid-state device or a photomultiplier. In these materials, the velocity of the radiation is greater than the velocity of light (in that particular material), and this gives rise to the emission of light (the optical equivalent of a "sonic boom"). For Cerenkov emissions to be produced, the product of the relative velocity of the radiation in the medium, β, and the refractive index of the medium, n, must be greater than one (β is defined as v/c, where v is the radiation velocity and c is the speed of light).

The blue glow seen in reactor fuel storage ponds is caused by this effect. It is unlikely that this effect would have much relevance to micromachined devices and is presented here only for completeness.

5.3 THERMAL DETECTORS

In some cases, it is possible to use miniature bolometers (as discussed in the Optical Transducers chapter) to detect incident ionizing radiation via the heat liberated during its interaction with the detector. In general, these devices are operated at temperatures near absolute zero, typically < 4 K.

Such detectors have been micromachined, and an example is presented in Racine, et al. (1990). Using bulk micromachining techniques, they fabricated suspended 2×2 mm single-crystal silicon islands containing thermistors. Each of the islands was thermally isolated from the substrate using four thin silicon support beams (3.5 mm long, 120 μm wide, and 5 to 40 μm thick). Operating at temperatures of 1 to 4 K, they demonstrated a sensitivity of 530 kV/W and a resolution of 33 keV for 5.5 MeV alpha particles (from ^{241}Am). Given the high degree of thermal isolation possible with micromachining approaches, it is likely that this approach could readily be adapted to large arrays of such detectors with on-chip scanning and amplification circuitry.

5.4 MOS THRESHOLD VOLTAGE SHIFT DETECTORS

Another category of solid-state detector, with inherent electrical readout capability is the MOS radiation-sensing transistor. The basic concept is that the threshold voltages of MOSFETs are sensitive to radiation-induced trapped charges in the oxide and at the oxide/semiconductor interface. The net shift in V_T is related to the accumulated dose, providing an inherent "memory." As discussed in Ristic, et al. (1997), the threshold voltage shift as a function of the absorbed dose, $\Delta V_T(D)$, is given by,

$$\Delta V_T(D) \approx \frac{q t_{ox}}{\varepsilon_o \varepsilon_{ox}} \Delta N_{ot}(D)$$

where,
        q = charge on the electron = 1.60218×10^{-19} C

t_{ox} = oxide thickness, in cm

ε_o = dielectric constant of free space = 8.854188×10^{-14} F/cm

ε_{ox} = relative dielectric constant of SiO_2 (≈ 3.9)

$\Delta N_{ot}(D)$ = areal density of trapped charges (in cm^{-2}) in the SiO_2 as a function of dose, D

It is assumed in the above equation that trapped charges in the SiO_2 dominate over those at the interface, and this was demonstrated by Ristic, et al. (1997).

This effect was harnessed in a dosimeter design many years ago (no author given (1971)) where R_{ds} was measured (with the transistor in a diode connection) as an indicator of accumulated charge. It was noted that the radiation-induced shift in the measured resistance was linear up to doses of 10^4 rads, and with simple circuitry it could be linearized up to a dose of 1.5×10^5 rads. Other useful references in this area are Holmes-Siedle (1974) and Ensell, et al. (1988). More recent work (e.g., Ristic, et al. (1997)) has provided insight into the mechanisms of device operation and their performance as dosimeters.

MOSFET-based detectors may prove quite useful in dosimetry due to their simple construction, potential for inherent storage of dose information, and ease of signal readout.

5.5 THERMOLUMINESCENT DETECTORS

In certain materials, such as LiF with an Mn activator a small fraction of the energy from absorbed radiation is stored in metastable states and can be recovered as emitted photons if the material is heated. This phenomenon is referred to as *thermoluminescence*. These materials (typically LiF(Mn) or CaF_2(Mn)) have wide energy responses from 30 keV to > 2 MeV (they respond to most forms of radiation) and can accumulate doses from mrads to 10^5 rads without errors due to the rate at which the dose is delivered. These materials also have no significant response nonlinearities and can be reset and reused repeatedly. While no micromachined thermoluminescent detectors have apparently been fabricated, such materials on thermally isolated structures with optical detection capabilities could theoretically be used to realize very versatile radiation monitors. A key feature of this approach would be that the monitors would continuously accumulate radiation without needing to be powered and the data could be accessed at any time.

5.6 COLLIMATORS

It is often desirable to measure gamma and x-ray (photon) radiation in a directional fashion. Devices referred to as collimators are typically used to make a detector more directional, although they do not operate as optical lenses. The principle of operation of radiation collimators is like that of pinhole cameras, diffractive rather than refractive optics. A dense material is used to absorb radiation incident in any directions other than those accepted by the collimator. A useful, but dated review of collimators in gamma-ray cameras is Moody, et al. (1970).

As for other photons, their absorption as a function of depth, x, into the absorber is given by *Beer's law*,

$$\Phi(x) = \Phi_o e^{-\alpha x}$$

where Φ_o is the photon flux at the surface, in photons/cm²•s, and α is the absorption coefficient, in cm⁻¹, which is a function of wavelength.

For lead, a commonly used absorber, the collimator thicknesses begin to become impractical for energies approaching 500 keV, at which point ≈ 5 cm of lead is required to absorb 99.9% of incident gamma radiation.

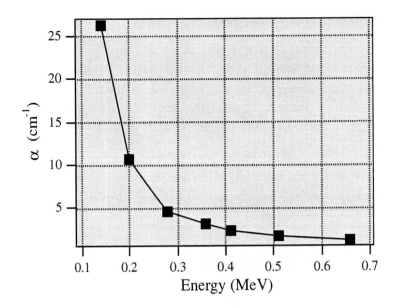

Plot of absorption coefficient versus energy for lead as an absorber. Note that the thickness of absorber required for 99.9% absorption is obtained from Beer's law and is ≈ 9.2 α^{-1}. Plotted from tabulated data in Moody, et al. (1970).

While this type of optical device is quite inefficient, they are nonetheless useful due to the highly efficient detectors that are available. In addition, micromachining techniques (e.g., electroplating of dense metals such as Ni, Pb, etc.) can be used to fabricate very precise collimating optics for such applications, potentially co-fabricated with solid-state detectors.

6. RADIATION EFFECTS ON ELECTRONICS

The transfer of mass, charge, or energy from ionizing radiation to electronic circuitry is determined by the means by which the radiation is stopped or absorbed within the circuitry (if the radiation passes through the circuitry, it is typically not a problem in terms of permanent damage). One way to achieve "radiation hardness" is to reduce the sensitive volume of the circuitry by using shallow junction devices (bulk devices tend to be more susceptible to radiation effects).

There are two basic radiation stopping effects: nuclear stopping (mass based, involving direct collision with nuclei, and dominant at low energies) and electronic stopping (electrostatic in nature, also referred to as coulombic scattering, and dominant at high energies). There are three basic classifications of device effects: prompt ionizing effects, long-term degeneration, and single-event upsets.

Prompt ionizing effects involve the creation of electron-hole pairs in semiconductors or photoemission of electrons in vacuum devices. Typically, these effects are caused by x-rays and gamma rays, but some beta particles can also cause them. This carrier generation ceases when the radiation is removed and underlies the operation of most radiation sensors. This type of effect typically does not cause long-term damage.

Long-term degradation is irreversible, cumulative structural damage to devices due mainly to nuclear stopping of heavy particles such as alpha particles and neutrons. Damage is usually in the form of dislocations in the crystal lattices and so-called Frenkel defects associated with the trajectories of particles and its secondaries. In polysilicon, this produces smaller grain sizes and drives the structure toward that of amorphous silicon. In single-crystal electronic devices, there is a great reduction in carrier lifetimes (affecting bipolar devices much more than MOS) and the introduction of trapping and deep level defect centers.

Single-event upsets are caused by individual impinging particles and manifest (usually in digital circuits) as a change in state. Floating-gate devices and small MOS capacitances (e.g., CCDs and dynamic RAMs) are most susceptible, and these events are the sources of classic "soft" (or random) errors in MOS memories, due to alpha particles (in some cases, polyimide shield layers above the passivation layer eliminates the effect, with alpha particles typically originating from the packaging materials. Also, as mentioned in Section 5.4 above as a sensing mechanism, ionizing radiation can give rise to trapped charges in MOS gate regions, leading to shifts in threshold voltages.

7. BIOLOGICAL INTERACTIONS WITH RADIATION

Ionizing radiation can cause harmful effects in biological systems due to the creation of free radicals and through direct damage to DNA. It is worth noting that while there are many well-documented effects of ionizing radiation on biological tissues, there are no known biological sensors for it (i.e., sensors that might have evolved to aid an organism in avoiding high-radiation areas).

An example of the types of radiation effects leading to damage of biomolecules is the creation of hydroxyl radicals, and in turn peroxides, which are extremely strong oxidixers (capable of damaging proteins and nucleic acids),

$$1) \quad H_2O + \text{radiation} \rightarrow (H_2O)^+ + e^-$$

$$2) \quad (H_2O)^+ + H_2O \rightarrow H_3O^+ + OH^\bullet$$

$$3) \quad OH^\bullet + OH^\bullet \rightarrow H_2O_2$$

Antioxidants should theoretically aid in preventing such damage to some extent.

RADIATION TRANSDUCERS REFERENCES

GENERAL REFERENCES

Audet, S. A. and Wouters, S. E., "Monolithic Integration of a Nuclear Radiation Sensor and Transistors on High-Purity Silicon," IEEE Transactions on Nuclear Science, vol. 37, no. 1., Feb. 1990, pp. 15 - 20.

Giancoli, D. C., "Physics for Scientists and Engineers," second edition, Chapter 46, Prentice-Hall, Inc., Englewood Cliffs, NJ, 1989.

Kelly, D. A., "Measurements Employing Nuclear Techniques," Chapter 4 in "Instrumentation Reference Book," Noltingk, B. E. [ed.], Butterworth-Heinemann, Ltd., Oxford, UK, 1995, pp. 3/124 - 3/143.

Middelhoek, S., and Audet, S. A., "Silicon Sensors," Academic Press, Inc., London, UK, 1989.

Norton, H. N., "Handbook of Transducers," Prentice-Hall, Inc., 1989.

Tove, P. A., "Review of Semiconductor Detectors for Nuclear Radiation," Sensors and Actuators, vol. 5, 1984, pp. 103 - 117.

SPECIFIC REFERENCES

Akimov, Y. K., "Position-Sensitive Scintillation Detectors of Nuclear Radiation (Review)," Instruments and Experimental Techniques, vol. 37, no. 6, Part 1, 1994, pp. 667 - 699.

Audet, S. A., Schooneveld, E. M., Wouters, S. E., and Kim, M. H., "High-Purity Silicon Soft X-Ray Imaging Sensor Array," Sensors and Actuators, vol. A22, nos. 1 - 3, Mar. 1990, pp. 482 - 486.

Ensell, G., Holmes-Siedle, A., and Adams, L., "Thick Oxide pMOSFET Dosimeters for High Energy Radiation," Nuclear Instruments and Methods in Physics Research, vol. A269, no. 3, June 20, 1988, pp. 655 - 658.

Goulding, F. S., and Stone, Y., "Semiconductor Radiation Detectors," Science, vol. 170, Oct. 16, 1970, pp. 280 - 289.

Holmes-Siedle, A., "The Space-Charge Dosimeter: General Principles of a New Method of Radiation Detection," Nuclear Instruments and Methods, vol. 121, no. 1, Oct. 1, 1974, pp. 169 - 179.

Kelly, D. A., "Nucleonic Instrumentation Technology," Chapter 3 in "Instrumentation Reference Book," Noltingk, B. E. [ed.], Butterworth-Heinemann, Ltd., Oxford, UK, 1995, pp. 3/95 - 3/123.

Kervalishvili, P. J., Karumidze, G. S., Shavelashvili, Sh., Kalandadze, G. I., and Shalamberidze, S. O., "Semiconductor Sensor for Neutrons," Sensors and Actuators, vol. A36, no. 1, Mar. 1993, pp. 43 - 45.

Moody, N. F., Paul, W., and Joy, M. L. G., "A Survey of Medical Gammay-Ray Cameras," Proceedings of the IEEE, vol. 58, no. 2, Feb. 1970, pp. 217 - 242.

(No author listed), "A New Use for MOS Transistors: To Detect Accumulated Radiation," Electronic Design, vol. 11, May 27, 1971, p. 32.

Padikal, T. N., "Medical Physics," Section 14 in "A Physicist's Desk Reference: The Second Edition of Physics Vade Mecum," Anderson, H. L. [ed.], American Institute of Physics, New York, NY, 1989, pp. 226 - 237.

Racine, G.-A., Buser, R. A., de Rooij, N. F., Stucki, G., Stucki, R., and Pretzl, K., "Low-Temperature Operating Silicon Bolometers for Nuclear Radiation Detection," Sensors and Actuators, vol. A22, nos. 1 - 3, Mar. 1990, pp. 478 - 481.

Ristic, G., Jaksic, A., and Pejovic, M., "pMOS Dosimetric Transistors with Two-Layer Gate Oxide," Sensors and Actuators, vol. A63, no. 2, Oct. 1997, pp. 129 - 134.

Snoeys, W. J., "A New Integrated Pixel Detector for High Energy Physics," Doctoral Dissertation in Electrical Engineering, Stanford University, Aug. 1992, Technical Report No. 92-017.

Snoeys, W., Plummer, J., Parker, S., and Kenny, C., "A New Device Structure and Process Flow for a Low-Leakage P-I-N Diode-Based Integrated Detector Array," summary of conference paper, in IEEE Transactions on Electron Devices, vol. 38, no. 12, Dec. 1991, pp. 2696 - 2697.

Streil, T., Klinke, R., Erlebach, A., Hübler, P., Kluge, W., Kück, H., and Zimmer, G., "New Alpha Radiation Detection Systems for Environmental Surveys," Sensors and Actuators, vol. A41, nos. 1 - 3, Apr. 1994, pp. 85 - 87.

Tsoi, H.-Y., Ellul, J. P., King, M. I., White, J. J., and Bradley, W. C., "A Deep-Depletion CCD Imager for Soft X-Ray, Visible and Near-Infrared Sensing," IEEE Transactions on Electron Devices, vol. ED-32, no. 8, Aug. 1985, pp. 1525 - 1530.

Yabe, M., and Sato, N, "Silicon Nuclear Radiation Detectors of a-Si:H/c-Si Heterojunction Structure with Sensitive Large Area or Array Area," Sensors and Actuators, vol. A22, nos. 1 - 3, Mar. 1990, pp. 487 - 493.

Chapter 6:

THERMAL TRANSDUCERS

1. INTRODUCTION

The measurement and/or regulation of temperature is an important everyday application of transducers. Accurate temperature-sensing mechanisms have evolved in most living organisms (certainly those that are warm-blooded, or *homoiothermal*, with body temperatures in the range of 37 to 44°C) because of the need to maintain important chemical reactions in their optimum temperature ranges. Thus the regulation of an individual's temperature can be imperative to survival, even in cold-blooded organisms (*poikilothermal*).

Temperature	K	°C	°F
Boiling point of copper	2,868.0	2,594.9	4,702.7
Boiling point of lead	2,017.0	1,743.9	3,170.9
Melting point of copper	1,356.0	1,082.9	1,981.1
Boiling point of mercury	630.0	356.9	674.3
Melting point of lead	601.0	327.9	622.1
Boiling point of water	373.15	100.0	212.0
Normal human body temperature	310.2	37.0	98.6
Comfortable room temperature	293.2	20.0	68.0
Freezing point of water	273.15	0.0	32.0
Zero of Farenheit scale	255.4	-17.8	0.0
Melting point of mercury	234.0	-39.1	-38.5
Coincidence of °C and °F scales	233.2	-40.0	-40.0
Boiling point of oxygen	90.2	-183.0	-297.3
Boiling point of nitrogen	77.4	-195.8	-320.4
Melting point of nitrogen	63.3	-209.9	-345.8
Melting point of oxygen	54.8	-218.3	-361.0
Boiling point of hydrogen	20.3	-252.8	-423.1
Melting point of hydrogen	14.0	-259.2	-434.5
Absolute zero	0.0	-273.15	-459.4

Table of selected important temperatures (phase transition temperatures are for a pressure of one atmosphere).

Measurement and regulation of temperature are important functions in a wide variety of consumer, industrial, and medical applications, covering a wide range of temperatures of interest. Since these application areas have generally been around for many years, this has led to the development of a wide variety of temperature-sensing approaches. The examples given below are drawn from them, with a focus on relevance to micromachined transducers.

2. TEMPERATURE MEASUREMENTS

2.1 BASIC TERMS

Some useful terms and definitions include are given below.

Heat flux, which is analogous to the "current" of heat flowing across a unit area per unit time.

Thermal conductivity, the thermal analog of electrical conductivity.

Heat capacity, the ability of a substance to "hold" thermal energy, in a manner analogous to electrical capacitance. Heat capacity is defined as the quantity of heat required to raise the temperature of something one degree (K or °C unless otherwise specified).

Temperatures are generally given in degrees Kelvin (K), the base SI unit, defined as the fraction 1/273.15 of the thermodynamic temperature of the triple point of water. For conversion from Celsius degrees, one can use $K = °C + 273.15$. (For reference, $°F = 1.8(°C) + 32$.)

The calorie (cal) is defined as the amount of heat required to raise the temperature of one gram of water from 14.5 to 15.5°C. (For reference, the BTU, or British Thermal Unit, is defined as the amount of heat to raise the temperature of one pound of water from 63 to 64°F.)

Some other common units of thermal energy are:

$$1 \text{ J} = 0.2389 \text{ cal} = 9.481 \times 10^{-4} \text{ BTU}$$

$$1 \text{ kcal (dietary calorie)} = 10^3 \text{ cal}$$

2.2 MODES OF HEAT TRANSFER

There are three modes of heat transfer to consider: conduction, convection, and radiation. In examining the relative contributions of these three components, one must take into account the composition and scale of the structures in question, particularly for micromachined devices. A brief summary is provided below, and there are many useful texts in the area of heat transfer to which the reader is referred for deeper coverage, for example, Incropera and DeWitt (1990).

2.2.1 CONDUCTION

Conduction refers to heat transfer by diffusion through solid material or non-moving fluid, and is analogous to electrical conduction. There is a wide range of thermal conductivities that must be considered in both macroscopic and micro-machining applications, as illustrated below.

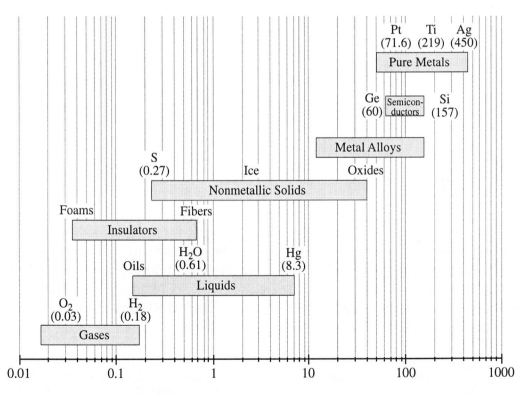

Illustration showing the relative thermal conductivities of various materials at normal temperatures and pressures. Adapted from Incropera and DeWitt (1990) with additional data from Weast (1988).

Fourier's law gives the heat flux or energy transfer rate (in one dimension), as a function of the temperature gradient across an object, in units of watts (W), as,

$$q = \frac{dQ}{dt} = -\kappa A \frac{dT}{dx}$$

where
Q = quantity of heat transferred, in joules (J)
κ = thermal conductivity, in W/(m•K) or J/(s•m•K)
A = cross-sectional area, in m^2
T = temperature, in K
x = distance in the direction of heat flow (normal to A), in m

(Note that quite often the dimensional parameter used in tabulated values is the centimeter rather than the meter, as shown above.)

The more general form of this equation is an expression of the heat flux in W/m^2, as,

$$\vec{q} = -\kappa \, \nabla T$$

where ∇ is the three-dimensional del operator.

The diffusion equation represents the temporal diffusion of thermal energy, and is,

$$\frac{dT}{dt} = \frac{\kappa}{\rho c} \, \nabla^2 T$$

where ρ is the mass density, in kg/m^3, and c is the heat capacity, in J/(kg•K). Note that $\kappa/(\rho c)$, or α, is often referred to as the *thermal diffusivity* of a material (in m^2/s). This parameter reflects the thermal conductivity of a material relative to its ability to store thermal energy (analogous to an electrical RC time constant). High α materials respond more quickly to thermal transients.

For most materials, the thermal conductivity is a function of temperature. For example, in the case of aluminum oxide, its thermal conductivity varies roughly twenty times over the temperature range of 100 to 2,000 K. However, for the temperature ranges encountered by most micromachined structures, κ is generally assumed to be constant.

The thermal properties of thin-film materials often differ significantly from those published for bulk samples. As discussed by Paul, et al. (1995), variations between standard CMOS processes and between runs of a single process can be significant enough to require the use of dedicated test structures for characterization on a per-run basis. For example, the thermal conductivities of n+ polysilicon layers from three different CMOS processes ranged from 16 to 24 W/m•K. It does appear that great care must be exercised when using published values for design, and that each process or deposition technique needs thorough characterization.

2.2.2 CONVECTION

Convection refers to heat transfer by the movement of fluid or gas. In general, convective processes are difficult to model numerically, but several useful empirical methods exist for the estimation of convective heat flows.

There are essentially two types of convection: free and forced. Free convection is the case wherein a fluid will move and transfer heat due to thermal gradients (e.g., a typical heat sink for power semiconductors). Forced convection refers to cases where the fluid transferring heat is actively pumped (e.g., a heat sink with a fan blowing air onto it). In general, one can estimate a convection coefficient that models the heat transfer by this mechanism. Unfortunately, this is often a complicated task, and for a thorough treatment of the relevant methods, a useful reference is Incropera and DeWitt (1990).

Of particular importance to micromachined structures, convective heat flow can be completely interrupted if the structure is operated in a vacuum.

2.2.3 RADIATION

Radiation refers to heat transfer via the emission of electromagnetic waves. The *Stefan-Boltzmann law* relates the total amount of heat radiated (in W) to the temperature, T (in K), of the emitting surface (for a blackbody [nonreflective] radiator),

$$\frac{dQ}{dt} = \varepsilon_T \sigma A T^4$$

where,
Q = total heat radiated from the surface of an ideal blackbody, in J
ε_T = emissivity (dimensionless correction factor for how closely a given material
    approximates the characteristics of an ideal blackbody radiator)
σ = Stefan-Boltzmann constant = 5.67032×10^{-8}, in W/m^2•K^4
A = area of emitting surface, in m

For reflective surfaces, there is a similar form of equation,

$$\frac{dQ}{dt} = CT^m$$

where C is a scaling constant, and m is a material dependent factor ($4 \leq m \leq 5.5$) (e.g., $m \approx 5$ for Pt).

The wavelength of maximum radiance from the emitting body can be determined using *Wien's displacement law* (discussed in the Optical Transducers chapter),

$$\lambda_{MAX} = \frac{2.8978 \times 10^{-3}}{T} \quad \text{in m}$$

where Wien's constant is 2.8978×10^{-3} m•K.

2.3 NON-CONTACT TEMPERATURE MEASUREMENTS

The concept of optical, non-contact temperature measurement, referred to in general as "pyrometry," is discussed in the Optical Transducers section and makes use of Wien's displacement law to determine the temperature of an object. Optical pyrometry is the only practical means for measuring temperatures above $\approx 2,000°C$. Using a spectrophotometer or one or more known filters, the temperature of an object can be accurately estimated. A common example is the infrared thermometers that measure the temperature of the eardrum by pyrometry.

2.4 THERMO-MECHANICAL SENSORS

Thermo-mechanical transduction is one of the most common methods of sensing and regulating temperature in household and automotive thermostats. It is also an actuation method, and this is discussed separately in the Mechanical Transducers chapter. The basic idea is that all materials have a *coefficient of* (linear) *thermal expansion*, α_L, which quantifies the relative dimensional change of an object that occurs for a change in temperature, ΔT,

$$\alpha_L = \frac{\Delta L}{L} \frac{1}{\Delta T} \quad \text{or} \quad \Delta L = L \alpha_L \Delta T \quad \text{in } °C^{-1}$$

As discussed in the Mechanical Transducers chapter, one can use the effect of two different thermal expansions of sandwiched materials to obtain a desired movement when the temperature of the assembly is changed. This is the basis for the common bimetallic (or *thermal bimorph*) thermostat switch, illustrated below.

The radius of curvature of a bimetallic strip is given (assuming that neither material has any residual strain) by,

$$R = \frac{(t_1 + t_2)^2}{6(\alpha_{L1} - \alpha_{L2})(T_f - T_o)t_1 t_2}$$

where,

t_1, t_2 = thicknesses of each material
T_o = resting temperature, in °C
T_f = final temperature, in °C
α_{L1}, α_{L2} = thermal coefficient of expansion of each material, in °C^{-1}

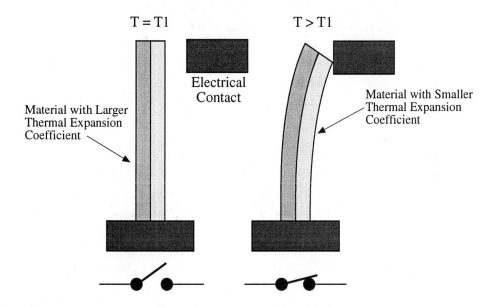

Illustration of the basic operating principle of a thermal bimorph switch.

Goldman and Mehregany (1995) demonstrated a thermal bimorph switch that was intended to provide an electrically sensed mechanical memory that would indicate if the structure had experienced hot or cold temperature excursions beyond designed values. As illustrated below, two bimorph beams could be manually "pre-latched" (using a probe tip) such that they would be released when cooling beyond a specified temperature, causing the beams to become "unlatched." This approach also works in the reverse direction to sense high-temperature excursions.

They began their fabrication process by growing 500 nm of thermal silicon dioxide on double-side polished (100) silicon wafers, patterning the oxide layer and using it as a mask to diffuse boron (from a solid source) into regions to form one of the two beams. The oxide was then stripped, 120 nm of LPCVD silicon nitride was deposited, and it was patterned to form a region where 180 nm of thermal silicon dioxide was grown to form a sacrificial layer between the beam tips. Phosphorus-doped LPCVD polysilicon was then deposited, followed by 120 nm of LPCVD silicon nitride. The thin-film layers on the front and back sides of the wafers were patterned, and 500 nm of Au/Cr was sputter-deposited and patterned to form the contacts. The back-side bulk etching was carried out with KOH and TMAH, and the sacrificial silicon dioxide holding the beams together was then released in HF.

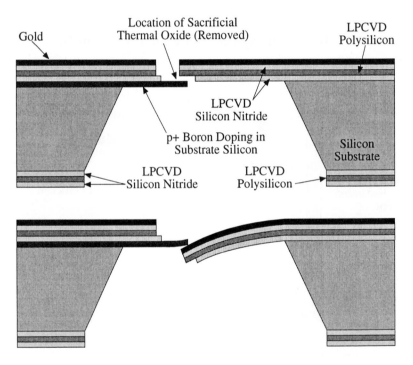

Illustration of a latching thermal bimorph switch, adapted from Goldman and Mehregany (1995). In the lower half of the illustration, the switch is shown in the "latched" state, prior to release through cooling beyond a specified temperature.

Each device was manually pre-latched and tested in liquid nitrogen to verify operation. Those that operated were tested successfully for 100 operations. Operation at a variety of geometrically determined operating temperatures was demonstrated, ranging from -105 to -20°C.

There are a large number of potential applications of thermal bimorph constructs for sensing and actuation in micromachined structures. The thermal conductivities and expansion coefficients from some important thin-film materials are given in the table below.

For liquids, the logical thermal expansion parameter is the volume coefficient of expansion, β, which gives the change in volume with temperature,

$$\Delta V = V \beta \, \Delta T$$

where $\beta = 3\alpha_L$, in $°C^{-1}$. (Note that this is the same α_L for linear expansion if the material is homogeneous and α_L is not orientation-dependent.)

And the above equation is obtained from the binomial expansion,

$$V + \Delta V = (L + \Delta L)^3 = L^3 + 3L^2 \Delta L + \cdots \approx L^3 + 3L^2 L \alpha_T \, \Delta T$$

Most materials expand when heated (see the table at the beginning of this chapter), although water ice initially *contracts* $\approx 0.2\%$ when heated above freezing (the fact that frozen water is less dense than liquid explains why bodies of water freeze from the top down), but at higher temperatures it expands. Silicon also expands upon freezing. A "classic" device that makes use of volume expansion for temperature sensing is the fluid-filled thermometer.

Single-crystal materials have coefficients of linear thermal expansion that are a function of direction with respect to crystal planes (e.g., [100], [110], and [111] for silicon).

Phase-change-based thermo-mechanical transducers (such as TiNi alloy) can also be used in such sensor applications, and have unusual heat-to-length or heat-to-force transfer functions (a very large length change with temperature). These materials are discussed in the Mechanical Transducers chapter. In addition, thermal expansion of a trapped fluid can also be used as a means of generating mechanical force in a sensor (e.g., the *Golay cell* discussed in the Optical Transducer chapter).

It is also worth noting that single-use micromachined thermal sensors could be constructed based on the use of the known melting points of various materials (e.g., a layer of such a material melting at a specific temperature and allowing two electrical contacts to close). Such devices would make use of stored mechanical energy in the structure, for example, using stress gradients in thin films that are normally a problem.

Material	Thermal Conductivity W/cm•K (@300K)	Temperature Coefficient of Expansion, ppm/K
Aluminum	2.37	25.0
Aluminum Oxide (polycrystalline	0.36	8.7
Aluminum Oxide (sapphire)	0.46	-
Carbon, Amorphous	0.016	-
Carbon, Diamond	23	-
Chromium	0.94	6.00
Copper	4.01	16.5
Gallium Arsenide	0.56	5.4
Germanium	0.60	6.1
Gold	3.18	14.2
Iridium	1.47	6.40
Iron	0.80	11.8
Molybdenum	1.38	5.00
Nickel	0.91	13.0
Platinum	0.716	8.8
Polyimide, Amoco Ultradel 1414	-	191
Polyimide, Dupont PI2611D	-	3.00
Polyimide, Hitachi PIQ-3200	-	50.0
Polysilicon	0.34	2.33
Silicon	1.49	2.60
Silicon Carbide	4.90	-
Silicon Dioxide (fused silica)	0.0138	0.4
Silicon Dioxide (thermal)	0.0138	0.35
Silicon Nitride	0.16	1.6
Silver	4.29	18.9
Teflon™ (PTFE)	0.0225	-
Tin	0.67	22
Titanium	0.219	8.6
Tungsten	1.73	4.50

Table of thermal conductivities and expansion coefficients for useful thin-film materials. (Sources: Riethmüller and Benecke (1988), Weast (1988), Incropera and DeWitt (1990), Lide (1996), Madou (1997), and Amoco/Hitachi/Dupont data from respective specification sheets.)

Thermal means can also be used for actuation, rather than sensing, in a large number of contexts. Thermally driven pumps, bubble-powered (phase-change) actuators, thermal bimorph actuators, and a variety of other thermal devices are discussed in the Microfluidic Devices and Mechanical Transducers chapters.

2.5 THERMORESISTIVE TRANSDUCERS

2.5.1 THERMORESISTIVE EFFECTS

Thermoresistive temperature transducers are based on the fact that the resistance (or resistivity) of most materials changes with temperature. This can be seen from the basic equation for the resistance of a bulk material,

$$R = \frac{\rho L}{A}$$

where
ρ = resistivity, in $\Omega \cdot cm$
L = length of conductor, in cm
A = cross-sectional area of conductor, in cm^2

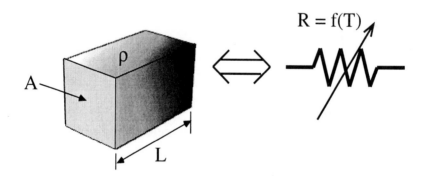

Illustration of the use of a slab of material as a thermoresistor.

Over temperature, dimensional variables (A and L) will vary, but here it is assumed that only ρ varies significantly. For most common materials, their resistance increases with temperature (positive temperature coefficient). Some materials (e.g., ceramics used in certain "thermistors," and most semiconductors) exhibit negative temperature coefficients (the thermistors with this characteristic are called "NTC," or negative temperature coefficient type). Examples are carbon and certain ceramics.

2.5.2 SIMPLE THIN-FILM THERMORESISTORS

For moderate temperature excursions (between 0 and 100°C), one can approximate the electrical resistance of a material using,

$$R_T = R_o \left(1 + \alpha_R [T - T_o]\right)$$

where R_o is the resistance at temperature T_o, and α_R is the temperature coefficient of resistance, in °C^{-1}.

If T_o is 0°C and R_o is measured at 0°C (common in published tables), then one can use,

$$R_T = R_o \left(1 + \alpha_R T\right)$$

For larger temperature excursions, material-dependent nonlinearities necessitate the use of third-order equations such as the *Callendar-Van Dusen equation* (exclusively for platinum wires, an international standard),

$$R_T = R_o + R_o \alpha \left[T - \delta(0.01T - 1)(0.01T) - \beta(0.01T - 1)(0.01T)^3\right]$$

or, more clearly showing the relative magnitudes of each coefficient,

$$R_T = R_o + R_o \alpha \left([1 + 0.01\delta]T - 0.0001\delta T^2 + 1 \times 10^{-6}\beta T^3 - 1 \times 10^{-8}\beta T^4\right)$$

where,
R_o = resistance at 0°C, in Ω
 α = 0.00392 for Pt (determined at 100°C)
 β = 0 if T is positive and 0.11 if T is negative (determined well below 0°C, typically -183.96°C)
 δ = 1.49 (determined well above 100°C, typically 444.7°C)

One can continue refining the correction factors *ad nauseum* (20th order polynomial fits are available), although the overall accuracy can be extremely high with even a few terms.

As mentioned above, platinum resistance thermometers are often used for calibration standards, but they are generally not low cost. However, Pt thin films are not particularly expensive and can be used to make extremely accurate micromachined thermometers that can be calibrated to achieve extremely high performance (most platinum thermometers can resolve 0.01°C). An overview of Pt thermometry, including linearization and instrument design can be found in Foster (1974).

In general, thermoresistive sensors directly make use of the temperature coefficient of resistance of a thin-film material to measure temperature and are easy to integrate on silicon or other substrates. The table below contains a list of useful materials for micromachining applications (silicon is dealt with separately below). From these tabulated values, it is apparent that for most commonly used thin-film materials, the value of α_R is relatively small (typically < 1% per °C). A potential problem associated with this is that the signal-to-noise ratio may be relatively poor. First, there is only a small change in value with temperature, which leads to a small signal amplitude, and second, the interconnections to the sensor can also exhibit a comparable α_R, thus careful design of the entire system is necessary.

In addition to sensing applications, the use of the resistance of a thin-film heater as a feedback mechanism for closed-loop temperature control has been exploited commercially with micromachined devices.

Material	Resistivity $\mu\Omega \cdot cm$	Temperature Coefficient of Resistance, ppm/°C
Carbon (graphite)	1,390	-500
Manganin (alloy)	48.2	2
Nichrome	101	1,700
Chromium	12.9	3,000
Aluminum	2.83	3,600
Silver	1.63	3,800
Copper	1.72	3,900
Platinum	10.6	3,927
Tungsten	4.20	4,500
Iron	9.71	6,510
Nickel	6.84	6,900
Gold	2.40	8,300

Table of resistivities and temperature coefficients of resistance for several materials relevant to micromachining (most resistivities are given for 0 or 20°C). From Weast (1988). Note that resistivities of conductors (generally metals) are sometimes given in units of $\Omega \cdot cm$ in the literature.

2.5.3 THERMISTORS

Thermistors are temperature sensors, generally made from semiconducting oxides, that exhibit large and well-characterized thermal coefficients of resistance (usually negative, hence the designation NTC, or negative temperature coefficient). Typical materials are sintered mixtures of metal oxides, selenides, sulfides, etc. Some example metals are Li, Cu, Co, Ti, Mn, Fe, Ni, and U. The metallic oxides are mixed in specific proportions to obtain the desired R versus T transfer curve. A slurry of the mixed oxides is formed and either pressed into individual "pills" or formed into wafers that are metallized and diced.

In general, the signals obtained from thermistors are quite a bit larger than those from simple thermoresistors (around 4 to 6% per K [40,000 to 60,000 ppm/K] versus < 1%). This increases the SNR and minimizes the effect of the thermal resistance change of long interconnect leads. However, this also reduces the temperature range over which thermistors can be used, generally limited to ≈ 100°C full-scale. In addition, modern thermistors are typically of such high repeatability that they can deliver an (interchangeable) temperature tolerance of ± 0.1°C.

The resistance of an NTC thermistor (typically, this is the type used, rather than PTC, which is the way nearly any ordinary material changes its resistance with temperature) can be approximated using the *Steinhart-Hart equation*. Here it is given in terms of T (measurand),

$$T = \left\{ a + b\ln(R) + c\left[\ln(R)\right]^3 \right\}^{-1}$$

where T is the temperature, in K, and a, b, and c are calibration constants determined for a given sensor.

The calibration coefficients are derived from measurements of resistance (R_1, R_2, and R_3) at three different temperatures, T_1, T_2, and T_3 which are used to generate a set of three simultaneous equations using the Steinhart-Hart equation (generally in terms of 1/T). A good description of thermistor operation in practical applications can be found in McGillicuddy (1993).

Thermistor materials could readily be integrated with silicon or other micromachined structures. For example, they could be applied to a substrate and fired in place, or plasma-sprayed on.

2.5.4 SEMICONDUCTOR THERMORESISTORS/THERMISTORS

Since the bulk of the micromachined transducer work today is done using silicon, it makes sense to consider how well "intrinsic" or doped silicon works as a

thermal sensor. It should be noted that the term intrinsic can be used only loosely with silicon, since, at present, "good" low-doped silicon has impurity levels on the order of 10^{12} cm^{-3}, which is still two decades from truly intrinsic silicon. Thus it would be more correct to refer to the typical intrinsic silicon as "unintentionally doped" instead.

For truly intrinsic semiconductors, one can express the conductivity (typically in $\Omega^{-1} \cdot$cm^{-1} or S/cm) as,

$$\sigma = \frac{1}{\rho} = n_i q(\mu_n + \mu_p)$$

where,
n_i = intrinsic carrier concentration, in cm^{-3}
μ_n = electron mobility, in cm^2/(V$\cdot$s)
μ_p = hole mobility, in cm^2/(V$\cdot$s)

As mentioned above, the impurity concentration in "good" unintentionally doped silicon is on the order of 10^{12} cm^{-3}, giving a resistivity of $\approx$ 5 k$\Omega \cdot$cm and a negative temperature coefficient of resistivity (reduction in carrier mobility with increasing temperature). The temperature-dependent components are the carrier mobilities and the intrinsic carrier concentration, n_i. The intrinsic carrier concentration varies with temperature as,

$$n_i(T) = C_1 T^{\frac{3}{2}} e^{-\left(\frac{E_g}{2kT}\right)}$$

where E_g is the bandgap of the semiconductor, in eV, and C_1 is a constant. The bandgap in the above equation varies as given by *Thurmond's formula* (for Si) (Thurmond (1975)),

$$E_g(T) = E_g(0) - \frac{\left(4.73 \times 10^{-4}\right)T^2}{T + 636}$$

where $E_g(0)$ is the bandgap of silicon at absolute zero = 1.17 eV.

Thermal effects on mobilities are more difficult to determine, but one can obtain equations such as (for silicon at light doping),

$$\mu(T) = C_2 T^{-\frac{5}{2}}$$

where C_2 is a constant. Thus the conductivity versus temperature relationship is extremely nonlinear, and indicates a negative temperature coefficient of resistance (from the reciprocal of the equation below),

$$\sigma(T) = n_i(T)q\left[\mu_n(T) + \mu_p(T)\right] \approx Te^{-\left(\dfrac{E_g(0) - \dfrac{\left(4.73 \times 10^{-4}\right)T^2}{T+636}}{2kT}\right)}$$

Intentionally doped silicon also has a very nonlinear dependence of conductivity on temperature (this discussion assumes an n-type dopant), as illustrated below. For low temperatures (typically below $\approx$ 50 K), the thermal energy is not sufficient to excite electrons from the dopant atoms into the conduction band. This is referred to as the "freeze-out" range. Thus, for temperatures low enough for freeze-out to occur, the conductivity decreases with further cooling. In this range, bulk Ge and InSb (and other narrow bandgap semiconductors are often used for cryogenic temperature sensing).

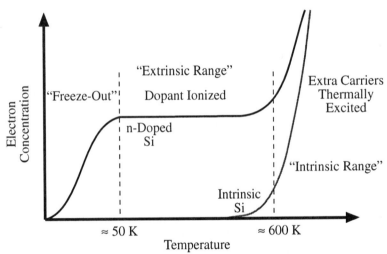

Illustration of electron concentration versus temperature. Adapted from Muller and Kamins (1986).

Above this temperature range there is a broad range of temperatures up to $\approx$ 600 K, wherein the dopant atoms are nearly all ionized and the carrier concentration in the doped silicon remains roughly constant. However, the conductivity of the silicon does vary due to the temperature dependence of mobility. Above $\approx$ 600 K, all dopants are ionized, and additional increases in conductivity come about through the thermal ionization of silicon atoms.

An example of the use of arrays of thin-film, amorphous Ge temperature sensors on glass, alumina, and polyimide substrates can be found in Urban, et al. (1990). These authors deposited the amorphous Ge by evaporation, and noted a TCR at 300 K of 2%/K.

2.5.5 SELF-HEATING EFFECTS IN RESISTIVE SENSORS

To measure any resistive thermal sensor, a current must be passed through it, and there is always a possibility that the power dissipated within the sensor will heat it and thus alter the measurements. This effect is referred to as "self-heating" and must be avoided by limiting the internal dissipation of the sensor. Unfortunately, as sense currents are reduced, thermal noise or circuit constraints begin to limit the SNR (in theory, with large enough resistors, their temperature could be measured via the Johnson noise, but this is not a practical technique).

If the thermal time constants of a thermoresistive sensor system are known, pulsed measurements may be used to increase accuracy by limiting self-heating if the electrical time constants are sufficiently short to allow measurement times substantially faster than the sensor's thermal time constant.

It should also be noted that self-heating with a negative temperature coefficient material can lead to thermal instability or even thermal runaway (destruction of the sensor through increased power dissipation with positive feedback), and this is the fundamental cause of several secondary breakdown modes in semiconductor devices.

2.5.6 INTERFACE CIRCUITS

Clearly, there are many circuits one can build to measure resistance, bearing in mind the self-heating effects mentioned above, and they are essentially either current- or voltage-driven ohmmeters (with explicit or indirect calculation of resistance), perhaps with a digital correction table for nonlinearities. An example of a commercial CMOS integrated interface circuit is the ATMOS (San Jose, CA) chip. Packaged in a 20-pin package, the chip includes an on-chip reference voltage generation, sensor drive and sense circuits, a 12-bit A/D and an internal calibration-curve EEPROM (16 bit × 4). The ATMOS chip can be preprogrammed to take into account the calibration curves of standard or nonstandard sensors.

As an aside, it should be noted that a classic example of thermal circuit stabilization (using the temperature coefficient of resistance of a light bulb to stabilize the amplitude of a Wein bridge sine wave oscillator, rather than to measure a temperature) was used in the first product produced by the Hewlett-Packard company, described in Hewlett (1939).

2.6 THERMOCOUPLES

Thermocouples are also discussed in the Optical Transducers chapter in the context of infrared-to-electrical transduction by simply absorbing the incident radiation, thus converting it into heat and measuring the temperature. The arrays of thermocouples often used to accomplish this are referred to as *thermopiles*. They have been used in a wide variety of micromachined transducers, and the reader is referred to van Herwaarden (1990) for a useful overview.

As a basic temperature transducer, the thermocouple is probably the most common because it is inexpensive, reliable, interchangeable, and covers much broader temperature ranges than other types (except non-contact methods such as optical pyrometry). The concept is to form a junction between two different materials and to measure the temperature-dependent voltage that arises across the junction. Typical thermocouples have output signals on the order of 50 μV/°C and some can span temperature ranges from -270 to +2,700°C with (calibrated) accuracies on the order of 0.5 to 2°C (e.g., K-type thermocouples).

The *Seebeck effect* (Seebeck (1822)) occurs when a temperature gradient (generally in a long, thin bar to reduce thermal conductivity) sets up an electric field that acts in the opposite direction to the temperature gradient. The measurement of the potential that arises is the basis for thermocouples. The physical basis of the Seebeck and Peltier effects can be explained using the fact that electric charge carriers (electrons and holes, the former being present alone in metals) not only carry a charge but also carry energy. Whenever there is a flux of carriers, there is, of course, a flux of charge (i.e., current), and also a flux of energy: heat flux. In the case of the Seebeck effect in metals, the "hot" electrons on the hot side migrate or diffuse to the cold side, setting up an electric field that will oppose the diffusion of any additional electrons (provided that current is allowed to flow). This is much like the built-in electric field in a diode that comes about due to the diffusion gradient across the junction (instead of a temperature gradient).

The constant that relates the amplitude of the electric field across a piece of material and the temperature is called the Seebeck coefficient and is usually denoted as α (in physics, it is usually defined in terms of an electric field, i.e., V/m•K; but in engineering, it is often given as V/K). The electric field (or voltage) that arises across such an object can be given as a simple function of the temperature gradient and the Seebeck coefficient, α,

$$V = \alpha \, \Delta T$$

The Seebeck voltage at a two-metal thermocouple junction is a complex function of temperature and is best fit with a high-order polynomial, but is often approximated as a lower-order function for simplicity. The Seebeck voltage is approximately related to the two junction temperatures, T_1 and T_2, as,

$$V \approx \alpha \left(T_1 - T_2 \right) + \gamma \left(T_1^2 - T_2^2 \right)$$

where α and γ are constants for the thermocouple pair (often α is referred to as the thermoelectric power, or the Seebeck coefficient, as mentioned above).

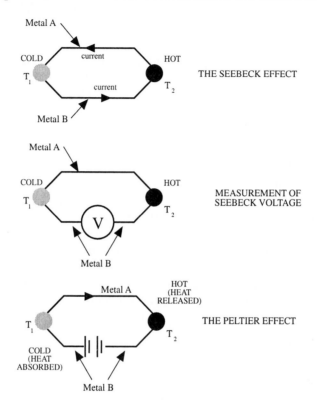

Illustration showing the basic thermoelectric effects.

Most metals have small Seebeck coefficients, on the order of 10^{-8} V/K, while for semiconductors, much higher values can be obtained. For an n-type semiconductor, α_n is given by,

$$\alpha_n = -\frac{(E_C - E_F) + 2kT}{qT}$$

where, E_C is the conduction band level, in eV, and E_F is the Fermi energy, in eV. For a p-type material, α_n is given by,

$$\alpha_p = -\frac{(E_F - E_V) + 2kT}{qT}$$

where E_V is the valence band level, in eV.

For moderately doped semiconductors, Seebeck coefficients on the order of 10^{-4} to 10^{-3} V/K are achievable, making integrated circuit thermocouples (discussed below) quite attractive, since polysilicon interconnects are available in a variety of processes.

For macroscopic thermocouples (and sometimes microscopic ones), there are unintended "parasitic" thermocouple junctions that are formed by the simple act of making electrical connections to the device (such as solder-to-wire or wire-to-screw-terminal junctions). For example, a solder-to-copper junction will produce a 3 μV/°C thermocouple, which can be quite significant compared to the ≈ 50 μV/°C output of the desired thermocouple junction. These parasitic junctions are in series with the intended junction(s), and can cause problems. One way to compensate for these unwanted effects is to put a second deliberate thermocouple junction (the "reference" or "cold" junction) at the temperature of the parasitic junctions in series with the first junction so that its voltage subtracts from that of the primary junction. Since the signal is proportional to the difference between the two junctions, this allows for a much more accurate measurement (e.g., if the reference junction is at 0°C, the offset can be theoretically eliminated with measurements relative to that reference temperature). If done correctly, with a known reference temperature, this replaces the original methods of keeping a reference junction in an ice bath (hence the name "cold" junction), and provides an *absolute* temperature reference.

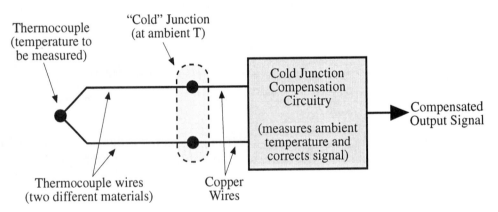

Illustration of the principle of "cold" junction compensation for thermocouple-based temperature measurements.

The idea is to subtract a known reference voltage from the total voltage, which includes not only the desired signal from the measurement junction, but also the voltages from the parasitic junctions. This can be done without the need for a physical cold junction, and it is typically done electronically. An example of a "cold-junction compensation" chip is the LT1025 from Linear Technology, Corp., shown below in use with a "J-type" thermocouple (from Williams (1988), which is an excellent practical overview of thermocouples and their applications). The operational amplifier (LT1001) generates an output signal that is the amplified difference between the thermocouple's output and that of the cold-junction compensator. The cold-junction compensator is basically a solid-state temperature sensor that measures the ambient temperature at the "cold junction" and generates a properly scaled voltage that simulates another thermocouple measuring the cold-junction temperature (i.e., has the same temperature-to-voltage slope).

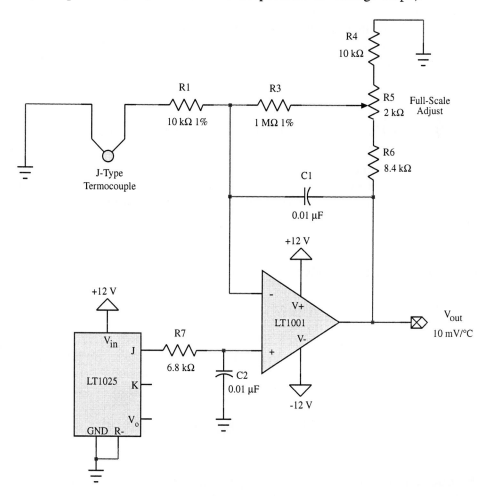

Schematic of a cold-junction compensation circuit in a signal conditioning amplifier for a J-type thermocouple. After Williams (1988).

In terms of micromachined thermocouple-based devices, the cold-junction compensation technique may still be important, and could in principle be accomplished directly or electronically. The effect of different metallizations in series must be considered (including bond wires and packaging), since these form additional parasitic junctions. The importance of having such compensation carried out locally may be decreased if one can control or eliminate the parasitic junctions. In addition, the need for compensation is eliminated if one is only interested in relative measurements (i.e., ∆T) such as those used with thermopile optical sensors discussed in the Optical Transducers chapter.

The above discussion has not addressed the problem of thermocouple nonlinearity, which can become significant over large temperature ranges (e.g., 500°C, which are not likely in micromachined devices, particularly if circuitry is to be included). Several techniques and even complete chips for doing this are discussed in the catalogs of manufacturers such as Analog Devices, Inc. (Norwood, MA) and an article by Williams (1988).

2.7 JUNCTION-BASED THERMAL SENSORS

For a vast number of day-to-day non-extreme temperature measurements, silicon-junction-based thermal sensors are ubiquitous. While the discussion below applies to any semiconductor junction, silicon is the most common due to its ready availability and the ease with which other circuitry can be combined with the sensors.

2.7.1 DIODE TEMPERATURE SENSORS

The temperature dependence of the diode I-V characteristics arises from a number of factors beyond the obvious appearance of temperature in the equation below (the intrinsic carrier density [the major contributor], the diffusion coefficients, the diffusion lengths, and the bandgap). For a forward-biased diode, the current flowing for a given applied voltage is given by the Shockley equation,

$$I_D = I_S \left[e^{\left(\frac{qV_D}{nkT} \right)} - 1 \right] \approx I_S e^{\left(\frac{qV_D}{nkT} \right)}$$

where,
I_D = diode current (DC)
q = charge on the electron = 1.60218×10^{-19} C
V_D = voltage across diode, in V
n = ideality factor (≈ 1 for integrated diodes and ≈ 2 for discrete diodes)
k = Boltzmann's constant = 1.38066×10^{-23} J/K
T = temperature, in K (or 273.15 + temperature in °C)

and I_S is the reverse saturation current (or scale current, proportional to the diode area and doping profile), given by,

$$I_S = qA\left(\frac{D_n}{L_n}\frac{n_i^2}{N_a} + \frac{D_p}{L_p}\frac{n_i^2}{N_d}\right)$$

where the diffusion coefficients (in cm²/s),

$$D_n \equiv \frac{kT}{q}\mu_n \quad \text{and} \quad D_p \equiv \frac{kT}{q}\mu_p$$

are proportional to $T^{-1/2}$, the diffusion lengths (in cm), $L_n \equiv \sqrt{D_n\tau_n}$ and $L_p \equiv \sqrt{D_p\tau_p}$ are proportional to $T^{-1/4}$, and the intrinsic carrier density, n_i, is the major source of temperature dependence of I_S,

$$n_i^2 = N_C N_V e^{-\frac{E_g}{kT}}$$

where N_C is the effective density of states in the conduction band, in cm⁻³, N_V is the effective density of states in the valence band, in cm⁻³, and the product $N_C N_V$ is proportional to T^3. In practice, either electron or hole current will dominate. It should be noted that the above expression for I_S is valid only for cases where the ideality factor for the diode, n, is one (as in integrated diodes). If the ideality factor is not one, the current flow is not due to diffusion, and this expression cannot be employed.

Relative to some reference temperature, T_o, the temperature dependence of n_i can be seen in,

$$n_i^2 = n_{io}^2 \left(\frac{T}{T_o}\right)^3 e^{-\left(\frac{E_g(T)}{kT}\right)} e^{+\left(\frac{E_g(T_o)}{kT_o}\right)}$$

Similarly, one can write an expression for I_S relative to $I_S(T_o)$, the reverse saturation current,

$$I_S = I_S(T_o)\left(\frac{T}{T_o}\right)^3 e^{-\left(\frac{E_g(T)}{kT}\right)} e^{+\left(\frac{E_g(T_o)}{kT_o}\right)}$$

This latter expression is extremely useful to evaluate temperature effects on semiconductor junctions and is accurate over the range of -200 to +500°C. Typical approximations near 300 K that are often cited as designer's "rules of thumb," make use of the fact that the predominant effect of temperature on I_S arises from the first exponential term. If one neglects the temperature cubed factor, as well as the temperature dependence of the bandgap, the equation simplifies to standard Arrhenius form,

$$I_S \approx I_S(T_o)e^{-E_g\left(\frac{1}{kT}-\frac{1}{kT_o}\right)}$$

and from which, using an Arrhenius plot, the bandgap of a material can be obtained (when $\ln(I_S/I_S(T_o))$ is plotted versus $1/T$, the slope is $-E_g/k$). Near 300 K, one can approximate (using a Taylor series about $T = T_o$),

$$\ln\left(\frac{I_S}{I_S(T_o)}\right) = \frac{E_g}{k}\left(\frac{1}{T_o}-\frac{1}{T}\right) \approx \frac{E_g}{kT_o^2}(T-T_o)$$

and it is from this equation that the often cited rule of thumb stating that I_S doubles per 5°C temperature rise originates, by solving for the required temperature rise for a doubling,

$$\Delta T = T - T_o = \frac{kT_o^2}{E_g}\ln(2) \approx \left(\frac{26\text{ mV}}{1.12\text{ V}}\right)(300\text{ K})\ln(2) \approx 4.83\text{ K}$$

It turns out that a practical circuit configuration for measuring the temperature of a diode junction is to measure the voltage across it, V_D, when operated at a fixed forward bias current. From the Shockley equation, it can be seen that,

$$V_D = \frac{nkT}{q}\ln\left(\frac{I_D}{I_S}\right)$$

At first glance, this equation implies that V_D is proportional to T, but, in fact, the variation of V_D with temperature is negative and is caused primarily by the temperature dependence of I_S. (It is worth noting that taking a derivative of the above equation with respect to T, incorrectly assuming I_D and I_S are not temperature-dependent, by chance gives the correct magnitude but the wrong slope for the diode temperature dependence.) Using the "full" expression for $I_S(T)$, one obtains,

$$V_D = \frac{kT}{q}\ln\left[\frac{I_D}{I_S(T_o)}\left(\frac{T_o}{T}\right)^3 e^{+\left(\frac{E_g(T)}{kT}\right)}e^{-\left(\frac{E_g(T_o)}{kT_o}\right)}\right]$$

which can be rearranged to show the contributions of each temperature-dependent term,

$$V_D = \frac{kT}{q} \ln\left(\frac{I_D}{I_s(T_o)}\right) - \frac{3kT}{q} \ln\left(\frac{T}{T_o}\right) + \frac{E_g(T)}{kT} - \frac{E_g(T_o)}{kT_o}\left(\frac{T}{T_o}\right)$$

This expression can be differentiated with respect to T to determine the contribution of each term (note that by differentiating the Thurmond's formula given above, for Si at 300 K, the derivative of E_g with respect to T at T = 300 K is -0.255 meV/K) and evaluated at $T = T_o = 300$ K,

$$\frac{dV_D}{dT} = \frac{V_{Do}}{T_o} - \frac{3k}{q} + \frac{dE_g(T)}{qT_o} - \frac{E_g(T_o)}{qT_o} = +2.000 - 0.260 - 0.255 - 3.750 \text{ mV}/\text{K}$$

Thus V_D decreases at $\approx$ -2.27 mV/K (varies from 2.0 to 2.4 mV/K in practice). (It should be noted that the first and fourth terms are nearly linear, with nonlinearities arising mainly from the second and third terms. The first term produces a slight bias-dependent effect.) This well-defined voltage variation with temperature has been used in numerous cases as a low-cost electronic thermometer (0.1° accuracy is possible with inexpensive silicon diodes, such as 1N4148 devices, once the circuit is calibrated).

An interesting example of the use of micromachined diode arrays for thermometry is Barth and Angell's (1982) micromachined linear diode arrays for temperature feedback in a medical application. *Hyperthermia* is a form of treatment for cancer in which the tissue to be destroyed is heated locally using ultrasound or RF energy (including microwave), sometimes in combination with chemo- or radiation therapy. A problem with this approach is to minimize the heating of healthy tissue while delivering a lethal (at the cellular level) dose to the cancer cells. In order to accomplish closed-loop heating control, temperature must be sensed within the tissue. This can be done by fabricating a linear array of temperature sensors that can be pushed into the tissue.

Diodes can be "multiplexed" by taking advantage of their rectifying transfer function, and thus fairly large numbers of them can be interrogated for their temperature-dependent forward I-V characteristics using relatively few wires, as illustrated below. Using such a scheme, N wires can be used to interrogate N × (N - 1) diodes.

Barth and Angell (1982) fabricated two types of such linear diode temperature probes, one based on manually attaching stainless steel wires to chemically etched (grooved) discrete diodes and one using a monolithic array, interconnected using polyimide supports. The latter devices were made by etching away the wafer, leaving the silicon diode "islands" attached to the polyimide at various points.

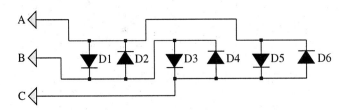

Schematic showing a diode multiplexing scheme for a thermal sensor array. After Barth and Angell (1982).

2.7.2 TRANSISTOR TEMPERATURE SENSORS

As seen above, any structures with a p-n junction will work as a temperature sensor, and V_{BE} of bipolar transistors is often used (examples are presented in Ohte and Yamaguta (1977) and Meijer (1986)). In a similar approach as that used for diode-based temperature sensors, one can operate a bipolar transistor at a constant collector current and measure the base-emitter voltage. The collector current versus the base-emitter voltage, V_{BE}, is given by,

$$I_C = I_S \, e^{\left(\frac{qV_{BE}}{kT}\right)} = \alpha \, I_E$$

where I_C is the collector current (DC), in A, and I_S is the saturation current (or scale current, proportional to the emitter area and doping profile, as for the diode case), in A. Rearranging to solve for V_{BE}, one obtains,

$$V_{BE} = \frac{kT}{q} \, \ln\left(\frac{I_C}{I_S}\right)$$

As for the diode, I_S is very strongly temperature dependent. A logical approach to developing a transistor-based temperature sensor would be to remove the I_S term entirely, and end up with a sensor with an output proportional to absolute temperature. One way to accomplish that is to use a pair of transistors of different emitter areas and to measure the difference in V_{BE}'s between them, as discussed in Timko (1976) (it should be noted that Timko normalizes for the emitter area and uses current density, J, instead of current),

$$\Delta V_{BE} = V_{BE1} - V_{BE2} = \frac{kT}{q} \, \ln\left(\frac{I_{C1}}{I_{C2}} \frac{I_{S2}}{I_{S1}}\right)$$

This means that one can use either equal collector currents and different emitter areas or different collector currents and equal emitter areas to set the scaling factor (assuming identical current densities in both devices) within the logarithm. In any case, the above approach can result in temperature sensors that (to first order) have an output of the form,

$$\Delta V_{BE} = (constant) \times T$$

Such a circuit that generates a voltage that is "proportional to absolute temperature" in this way is often referred to as a "PTAT" circuit, as discussed in a temperature sensor application by Dobkin (1974).

BASIC PTAT CIRCUIT

A basic PTAT circuit (Timko (1976)) is shown below. The base-emitter voltage of Q2 mirrors V_{BE} of diode-connected Q1, constraining the collector currents of Q3 and Q4 to be equal. Since Q4 is diode-connected and tied to V-, the difference in V_{BE}'s appears across the resistor, R1.

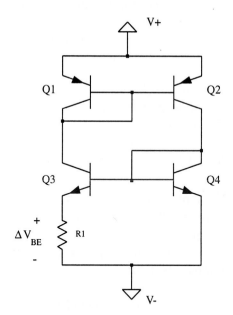

Schematic of a basic PTAT circuit. After Timko (1976).

By scaling the ratio of emitter areas, one can adjust the scaling factor for the output voltage. If the emitter of Q3 is "r" times larger than that of Q4, the voltage is simply,

$$\Delta V_{BE} = \frac{kT}{q} \ln(r) \approx 26 \ln(r) \quad \text{in mV (for T = 300 K)}$$

In practice, there are some problems with this circuit (start-up, high output impedance, potential for oscillation, etc.), as discussed in Timko (1976), but these have generally been addressed in commercially available devices with both voltage and current outputs (to make a current-output device, one starts by observing that if R1 had a zero temperature coefficient, the total current drawn by the circuit will also be PTAT).

An example of a current-output PTAT device is the Analog Devices AD592, which draws 1 µA/K and has a maximum nonlinearity of 1.8°C over the range of -55 to +125°C. This is achieved through a more complex circuit than that shown above, and uses laser trimming for one-time calibration at the wafer level. Current mode operation allows measurements over long cables without the need to compensate for voltage drops (the current can readily be converted back into a voltage at the receiving end). In addition, an interesting example of a Fahrenheit-scaled temperature sensor is presented in Pease (1984).

Newer commercial devices incorporate some form of on-chip analog-to-digital conversion, such as a serial A/D converter, frequency output proportional to temperature, or a D/A converter combined with a comparator to provide a programmable set point. An example of a fully integrated CMOS temperature sensor with an on-chip sigma-delta A/D converter, a ± 1°C accuracy (after two-point calibration) over -40 to 120°C, and a 7 µW power dissipation is presented in Bakker and Huijsing (1996). A fully integrated PTAT temperature sensor with a set-point output switch (one-bit A/D) is described in Brokaw (1996).

SWITCHED PTAT CIRCUIT

An alternative implementation is the *switched* PTAT circuit (as discussed in Williams (1991)) that achieves ± 1°C accuracy for *randomly selected* NPN transistors (shown as Q2 in the schematic). In this circuit, instead of looking at the ratio of V_{BE}'s between two transistors with different I_C's, just one transistor is used with two different I_C values switched over time.

The idea is to change the current periodically through diode-connected transistor Q2 from 10 to 100 µA (it is the *ratio* of the currents that must remain constant, not their absolute values). Since V_{BE} increases 59.16 mV per decade of current at room temperature (this number is simply kT/q times the logarithm of ten), and varies at -198 µV/°C from that temperature (dividing the former factor by 300 K). In the circuit shown below, capacitor C1 removes the DC bias from Q2 and allows for the extraction of the ΔV_{BE} signal. The remainder of the circuit (noninverting op-amp configured with a deliberate offset voltage) provides amplification and offset (so that 0°C = 0 V) to scale the signal such that the full-scale range of 0 to 10 V corresponds to 0 to 100°C.

This approach proved important in instruments built using discrete bipolar transistors as temperature probes. It may have applications in integrated temperature sensors in micromachined devices, although typical errors of matching transistors are on the order of 0.1%. In any event, this approach could readily be integrated and could, for example, address multiple transistors in an array with only simple multiplexing required and virtually no offset signal problems.

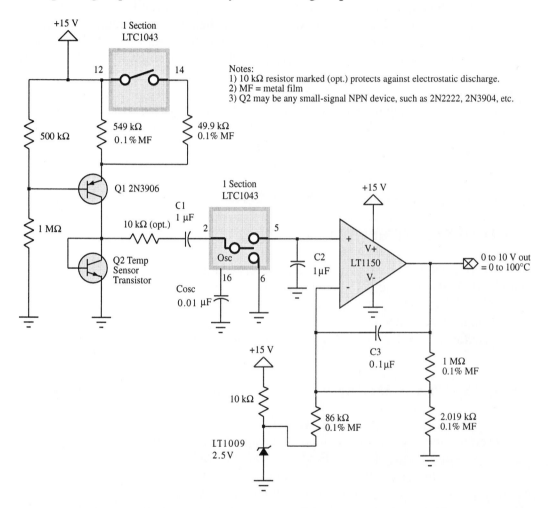

Schematic of a switched PTAT circuit. From Williams (1991).

DESENSITIZING CIRCUITS TO TEMPERATURE

Having shown above that p-n junctions are temperature-sensitive, it is worth mentioning that *desensitizing* circuits to temperature variations is an important area, especially for high-performance analog circuits. One basic idea is to make a temperature sensor and to use it to correct for the unwanted temperature sensitivities of the rest of the circuitry. This subject is not within the scope of this book, but the interested reader is referred to Gray and Meyer (1993) for in-depth coverage of this area in the context of bandgap references.

Another temperature desensitization approach is to deliberately heat the entire circuit above the expected ambient temperature range, thus rendering it insensitive to ambient variations. Naturally, the stability of the heater servo circuit is important to avoid drift. This approach has been used for many years to stabilize quartz crystals in oscillators, voltage references, and other circuits, as well as early navigation-grade accelerometers for ballistic missiles. The only serious drawback to this approach is the relatively high power consumption of the heaters. Through micromachining, one can deliberately thermally isolate a section of a microcircuit and perform this type of thermal stabilization using very little power, as discussed below.

2.8 OTHER THERMAL SENSORS

2.8.1 ACOUSTIC TEMPERATURE SENSORS

Another temperature-sensing mechanism that is potentially relevant in micromachined systems is the use of the variation in acoustic propagation velocities in materials. For example, one can construct a surface acoustic wave oscillator whose self-resonant frequency varies with temperature (due to variations in propagation velocity as well as substrate expansion), potentially providing very high resolution. An example of such a sensor can be found in Neumeister, et al. (1990).

2.8.2 QUARTZ AND OTHER RESONANT TEMPERATURE SENSORS

One of the most sensitive thermal measurement methods (which has not been used to fabricate micromachined devices) is the use of special quartz oscillators (see Benson and Krause (1974), Fraden (1997) and Schäfer (1989)). Single-crystal SiO_2 is cut at an orientation that maximizes the linearity of the sensitivity of resonant frequency to temperature via the temperature dependence of the piezoelectric effect. In general, resonant frequency shift, Δf, can be approximated for temperature shift, ΔT, by,

$$\frac{\Delta f}{f_o} \approx 1 + a_1 \, \Delta T + a_2 \, \Delta T^2 + a_3 \, \Delta T^3$$

where a_o, a_1, a_2 and a_3 are calibration coefficients, and f_o is the resonant frequency at the calibration temperature. With a careful choice of crystal cut (e.g., doubly rotated Y-cut), the second and third coefficients can be made negligibly small, and a_1 can be on the order of 35 ppm/K (as discussed in Norton (1989), such a crystal can be excited at its third harmonic resonance and can provide sensitivities on the order of 1 kHz/K). A frequency counter can be used to obtain temperature information. Since frequency is easily measurable to 0.1 Hz resolution (with a 10 s count time), extremely high-resolution temperature measurements are possible (improved with ROM-based digital corrections, which also allow more sensitive but less linear crystal cuts to be used). An instrument based on this principle, with 100 μK resolution, is the Hewlett-Packard 2804A quartz thermometer. If nonlinearities can be externally compensated, more sensitive cuts can be used, such as a singly rotated AT-cut (θ = -20°) with $a_1 \approx$ 67 ppm/K, $a_2 \approx$ 0.02 ppm/K^2, and $a_3 \approx$ 0 ppm/K^3 (see Ziegler (1984).

An example of a bulk micromachined quartz thermometer was presented by Ueda, et al. (1986). They used an NH_4F/HF bulk etch with a Cr/Au mask to pattern tuning fork-shaped resonators from singly rotated Y-cut quartz (θ = -40.23°), for theoretical sensitivity coefficients of a_1 = -58 ppm/K, a_2 = -0.027 ppm/K^2, and a_3 = -2.8 × 10^{-6} ppm/K^3. Their measured a_1 coefficient was -54.1 ppm/K. Other micromachined versions are possible, such as thermally isolated arrays of quartz resonators for an uncooled IR imaging array, as proposed by Vig, et al. (1996).

A potential limitation of this approach is that the resonators require environmental protection (sealed packaging), and thus often take considerable time to equilibrate to an external temperature change. However, if the devices are scaled, and suitably scaled packaging is found, these time constants can probably be drastically reduced.

2.8.3 TUNNELING TEMPERATURE SENSORS

The use of thermal, or Johnson, noise (electrical or mechanical) as a means of determining temperature has been considered for thermometers that theoretically do not need absolute calibration. Chmielowski and Witek (1994) proposed fabricating such a device based on electron tunneling from a sharp emitter tip to a small quartz resonator beam. It remains to be seen if the sensor's temperature-dependent thermal noise can be adequately separated from other noise sources such as those associated with tunneling, electronics, and external acoustic disturbances.

2.9 BIOLOGICAL THERMAL SENSORS

Thermal sensation in living things is relatively poorly understood at the present time. Thermal sensors in human skin are localized to $\approx$ 1 mm diameter regions, but can respond outside these regions due to thermal diffusion. These sensors are specialized in "hot" or "cold" functions. Cold receptors are activated within the range of 1 to 20°C below normal skin temperature (≈ 34°C) and produce trains of electrical discharges (*action potentials*) at frequencies related to the degree of cooling (most of the nervous system uses a similar form of amplitude-to-frequency A/D conversion for information transfer between the sensor and signal processing regions). Paradoxical sensations of cold can be elicited by touching a hot probe (≈ 45°C) to a known cold-sensitive region. In such a case, a person will report feeling cold because the receptors are inappropriately activated by heat, but are "wired" to the brain as sensing cold only. Heat receptors are selectively activated within the range of 32 to 45°C and generate action potentials at a rate that increases with temperature. Temperatures above 45°C are perceived as pain rather than as heat.

Interestingly, decreases in a diabetic person's sensitivity to temperature is a sensitive method for detecting the peripheral nerve damage (*neuropathy*) that is common in the extremities. A miniature, but not micromachined, Peltier effect thermal signal for use in such sensitivity testing and other biological thermal sensor research was described by Maluf, et al. (1994).

3. THERMAL ACTUATORS

Thermal actuators, or simply "heaters," are useful in a large variety of micromachined devices, and are thus quite common. A simple resistor can be used to heat a structure, such as an entire die, a floating island of dielectric, a floating island of single-crystal silicon, a bilayer actuator, etc. Similarly, other approaches to generate heat from electricity can be employed, such as dissipated power in an active device (such as a MOSFET) or dielectric loss heating through the application of radio frequency power to lossy dielectrics. These heating methods are discussed below and in other chapters, as appropriate to the type of sensors or actuators in which they are used. The discussion of thermal actuators below is primarily limited to the basic approaches used to convert electrical or other energy into thermal energy or to transfer thermally energy from place to place. Thermal actuators with mechanical motion or force as output signals are discussed in the Mechanical Transducers chapter, and heater arrays for generating IR imagery are discussed in the Optical Transducers chapter.

The reader should note that there are many as yet unexplored approaches to thermal actuation that could be applied to micromachining, such as the use of chemical energy for heating or cooling (common in macroscopic devices). As micromachined structures increasingly are made for disposable applications, these potentially "one-shot" actuation mechanisms may become more attractive.

A device that transfers heat from a cold to a hot reservoir is referred to as a refrigerator, and has an efficiency (referred to as the *coefficient of performance*), K, defined as,

$$K \equiv \frac{|Q_C|}{|Q_H| - |Q_C|}$$

where Q_C is the quantity of heat extracted from a cold reservoir, in J, and Q_H is the quantity of heat that is discharged to a hot reservoir (includes extracted and waste heat), in J. It should be noted that this equation relates the amount of heat transferred to the amount of waste energy expended to accomplish that, so K values are often much greater than one, and an ideal refrigerator would have a K value of infinity. A key limitation to efficiency is the unavoidable thermal leakage between the cold and hot reservoirs. In micromachined devices, this becomes even more critical.

3.1 JOULE-THOMPSON REFRIGERATORS

The expansion of an ideal gas through a nozzle is an *isenthalpic* process (i.e., the internal energy plus the pressure-volume product, U + PV, remains constant for the gas). In ideal gases, both terms in this sum are proportional to absolute temperature, and thus the temperature of the exiting gas remains constant. For real gases, however, intermolecular forces cause work to be done as the result of such a pressure change, causing a temperature change in the gas. This is known as the *Joule-Thompson effect*.

Every gas has a Joule-Thompson inversion temperature, T_{inv}, for which $T < T_{inv}$ causes cooling and $T > T_{inv}$ causes heating (the latter generation of extra heat becomes an important issue in many jet engines and rocket motors). Joule-Thompson cooling is only feasible at temperatures where the gas temperature remains above its critical temperature to avoid condensation.

Thus, for input and output gas temperatures T_{in} and T_{out}, respectively, the practical limits on the Joule-Thompson effect (for cooling) are,

$$T_{crit} < T_{out} < T_{in} < T_{inv}$$

For N_2, these values give 126 K $< T_{out} < T_{in} <$ 621 K, which provides a broad working temperature range. Similarly, O_2 has values of $T_{crit} =$ 155 K and $T_{inv} =$ 893 K. Not all gases (e.g., CO_2, CH_4, or NH_3), provide such useful operation, but this is of little consequence, since N_2, for example, is so readily available. It should be noted that using high gas pressures is not necessary, but increases the amount of energy transfer.

This approach to cooling is used in most air-conditioning plants, and can also be applied to micromachined refrigerators, as demonstrated by Little (1984). His micromachined Joule-Thompson refrigeration devices could cool very quickly from room temperature to 80 K using 1,800 psi nitrogen. (In the published examples, the energy used for cooling came from compressed N_2 from a gas cylinder, but if some form of compressor were used at the same time as the cooler, its efficiency could be computed more readily.) In the micromachined devices, as for their macroscopic counterparts, the gas was routed through a fluidic channel until it reached either a porous plug or an expansion nozzle, at which point cooling occurred.

Additional cooling can be obtained through "pre-cooling" the incoming high-pressure gas using a countercurrent heat exchanger principle wherein the incoming gas passes by, and comes into thermal "contact" with the outgoing, cooler gas. This approach was invented by Carl von Linde in 1895, and is known as the Linde cycle (it is the standard process for liquifying air, and is used to produce liquid nitrogen and oxygen in bulk for industrial use).

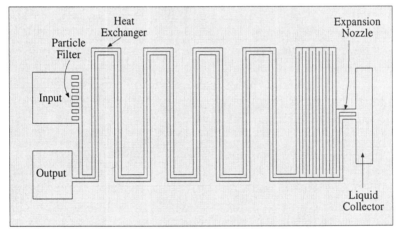

Diagram illustrating the fluidic channel design of a micromachined Joule-Thompson refrigerator. Adapted from Little (1984).

The key requirement for Joule-Thompson refrigerators is relatively long, closed channels that can withstand the pressures of the incoming gas. At first, silicon was used as a substrate for the channels (using SiO_2 masks with EDP etchant) and a Pyrex™ top-plate was anodically bonded on to seal the channels. The high thermal conductivity of the silicon provided a large thermal leakage between the ambient and cooled ends of the devices, and efforts to reduce this effect by using thinner silicon substrates failed because adequately thin silicon could no longer withstand the pressures. (Some success was achieved with silicon-based Joule-Thompson coolers using CO_2 and ethylene at lower pressures.)

Present designs rely on the deep etching of glass plates using extremely resilient gelatin-based photoresist (described in the Micromachining Techniques chapter) and "sand-blasting" the glass using 27 μm Al_2O_3 (alumina) grit in a high-pressure gas stream (forced through a small nozzle that is scanned across the glass). Precise channels from 2 to 100 μm in depth can be etched in this manner (this approach should work with almost any substrate material) with nearly vertical sidewalls (unfortunately, quantitative aspect ratio information was not included in the cited publication). To close the channels, a glass top cover is bonded onto the substrate using ultraviolet-cured adhesive, "solder glass" (low-temperature glass frit), or methods proprietary to the manufacturer (MMR, Inc., Mountain View, CA).

Commercial devices fabricated in this manner can reach temperatures on the order of 70 K and "fast cool-down" versions can reach 90 K in 2 seconds using 3,500 psi N_2. They are typically used to cool optical detectors and for bench-top cryogenic experiments.

Such Joule-Thompson refrigerator fluidic channels could readily be micro-machined as part of high-power microelectronic circuits (for local cooling), combined with micromachined valves for proportional temperature control, or simply fabricated using more thermally isolated cooled regions to achieve faster response times and greater efficiencies. In addition, it is likely that the construction of mesoscopic refrigerators from arrays of such devices will prove to be quite effective.

3.2 OTHER FLUIDIC COOLERS

As discussed in the Micromachining Techniques chapter, Tuckerman and Pease (1981) etched deep channels in (110) silicon to fabricate liquid-filled cooling channels intended for use as a heat exchanger with high-power integrated circuits. The basic concept was to use a pumped fluid flow to carry away excess heat, as is done on a macroscopic scale in some mainframe computers. Another useful reference on this work is Tuckerman (1984). While not yet broadly applied, several more recent efforts are described in Weber (1991) and Joo, et al. (1995). A variety of fluid channel fabrication methods that could be used for such applications are discussed in the Microfluidic Devices chapter.

3.3 PELTIER EFFECT HEAT PUMPS

The *Peltier effect* (also referred to as the *thermoelectric effect* in semiconductors) makes use of an electric current to generate an intentional heat flux, cooling one region and heating another. Electrons and holes both carry positive kinetic energy (the energy required to remove a deep-lying electron from the valence band is the kinetic energy of the hole, and is thus positive). An electron current will transport energy in the direction opposite to that of the current, while a hole current will transport energy in the same direction as the current. By adjusting the polarity and type of current, heat can be transported to or from one side of a junction. The coefficient that relates the current flux to the heat flux per junction is called the *Peltier coefficient* and is denoted by the symbol π_{AB} (for two materials A and B forming the junction), and is approximated for narrow temperature ranges by,

$$\pi_{AB} \approx (\alpha_A - \alpha_B)T$$

It is clear that for Peltier heat pumps, one would want to eliminate any type of ohmic heating (I^2R loss), therefore, a material with a high electrical conductivity is very desirable. Once a temperature gradient is set up (due to the heat transfer) across both sides of the junction, thermal conductivity acts to reduce that temperature gradient and thus in a direction opposite to the heat flux setup by the Peltier effect. This results in a reduction in the efficiency of Peltier devices. Thus the ideal material for such refrigerators is one with a high electrical and low thermal conductivity. Major programs were established in the 1950's to build high-performance miniature refrigerators based on the Peltier effect, but they were ultimately abandoned because it is generally difficult to find pure materials that meet both of these criteria simultaneously (this violates the Wiedemann-Franz law). Alloying two semiconductors is one method that has been used to increase the electrical-to-thermal conductivity ratio (approximately a factor of ten can be gained in this way). Examples include bismuth telluride and silicon-germanium.

Commercial devices exist that use Peltier effect heat pumps to transfer heat from a hot region (e.g., a microprocessor) to a cold region (e.g., a heat sink). For optimal efficiency, these devices typically use bismuth or bismuth telluride blocks. Typically, alternating n- and p-type bismuth telluride elements are connected in series electrically by tin interconnections and thermally in parallel between two ceramic substrates (as illustrated below). Flowing electrons carry kinetic energy (and hence heat) downward in the n-type blocks, while holes carry heat downward in the p-type blocks.

Since bismuth is a poor thermal conductor, yet still reasonably conductive electrically, such devices work quite well. Since no physical p-n junctions are used, this design can be operated in either direction, so reversing the polarity of the applied voltage reverses which substrate is cooled and which is warmed.   At present, these devices are used to cool high-performance microprocessors and other circuit elements, as well as infrared sensors.   In addition, they can be used to generate electricity if a suitable temperature gradient is applied across them.   For example, radioactive isotopes have been used as heat sources for this purpose to power deep-space probes and remotely located radio transmitters.

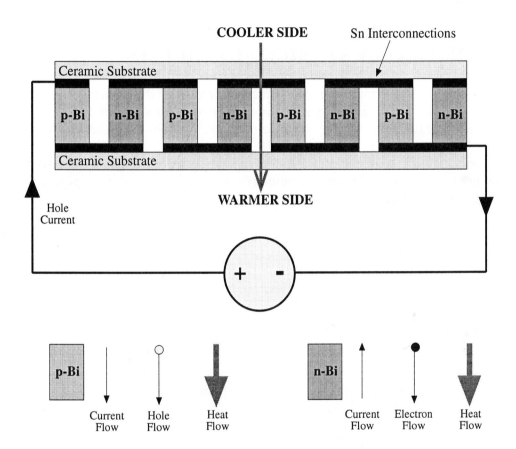

Illustration of (top) the physical structure of a conventional (macroscopic) Peltier effect refrigerator constructed from discrete p- and n-type bismuth blocks and (bottom) the polarity of energy and charge flows in each block.   Courtesy R. B. Darling.

It is at least theoretically possible to micromachine such devices, but since their macroscopic equivalents are inexpensive and highly effective, there may be no benefit from this.

4. THERMAL SENSOR/ACTUATOR COMBINATIONS

As mentioned above, thermal actuators are used in a large variety of micromachined devices. There is a multitude of possible transducer structures that combine both heaters and temperature sensors to accomplish a specific purpose, such as to allow for localized closed-loop temperature control or to sense the thermal properties of the surrounding medium.

4.1 THERMALLY STABILIZED CIRCUITS

Many electronic circuits exhibit undesirable drift as the ambient temperature changes and, as mentioned above, a potential method to stabilize this drift is to deliberately heat the circuits to a temperature well above the expected range of ambient temperatures. If the thermal mass of the circuit to be stabilized were made very small and well isolated thermally, this type of approach could be used efficiently. Since integrated circuit devices are already fabricated lithographically, micromachining is a logical means of achieving this.

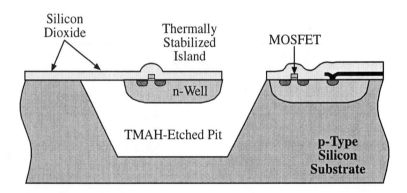

Illustration of isolated n-well and its relationship to other bandgap circuitry on the substrate. Courtesy of R. Reay.

Reay, et al. (1994) demonstrated an electrochemical technique for use with standard foundry CMOS for forming n-type single-crystal silicon islands suspended from silicon dioxide beams (discussed in the Micromaching Techniques chapter). Reay, et al. (1995) applied that approach to the fabrication of a closed-loop, thermally stabilized bandgap voltage reference. Key components of the bandgap circuit were isolated on a silicon island heated to 80°C using a p-channel MOSFET as the heater with only 1 mW of power (thermal isolation ≈ 60,000 K/W).

As illustrated below, the core of the reference was a Brokaw bandgap cell, with the two bipolar transistors scaled so that the emitter area of Q_2 was twice that of Q_1. The upper servo amplifier holds the reference voltage at a voltage where the currents through both transistors are equal, generating the bandgap reference voltage at a value given by,

$$V_{out} = V_{be1} + 2\frac{R_1}{R_2}\frac{kT}{q}\ \ln(2)$$

To implement closed-loop control of the thermally isolated island heater, the temperature-dependent PTAT voltage between R_1 and R_2 was compared to a scaled copy of the output voltage so that the PTAT voltage (and hence the temperature of the isolated silicon island) was regulated at a value set by the scaling resistors, R_3 and R_4 (typically the island was heated to 90°C to be above the upper limit of the commercial temperature range of 0 to 80°C). A PMOS transistor was used directly as the heater, rather than a resistor driven by a transistor to avoid the excess power dissipation of the drive transistor in the latter case. While the bandgap circuit itself was far from optimized, its total drift was reduced from $\approx$ 400 to 9 ppm/°C. Optimized bandgap references, with far lower open-loop drift are likely to exhibit even greater improvements in stability.

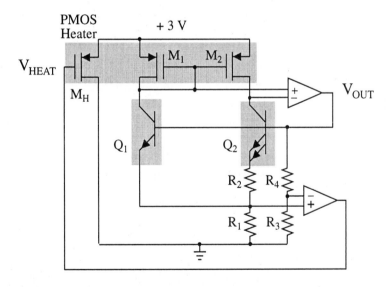

Simplified schematic of a thermally isolated bandgap circuit indicating components on the isolated silicon island with shading. Courtesy of R. Reay.

This technique can also be applied to other structures that might benefit from thermal stabilization, such as micromachined resonators and other electronic circuits.

4.2 THERMAL AC/RMS CONVERTERS

One of the most accurate methods of computing the root-mean-square (RMS) value of an unknown AC signal is to utilize the basic idea behind RMS and to directly measure the DC equivalent power of the signal. The idea is to use the signal to generate heat and to measure the heat. Somewhat like thermal infrared sensors, one can obtain "flat" frequency responses over seven or more decades of frequency. For many years, thermal AC/RMS converters have been designed and built using discrete components (see Hermach and Williams (1966), Cox and Kusters (1974), Inglis (1985), and Tejwani and Moore (1986)), but recently it has been possible to micromachine them. This approach takes advantage of the extremely high degree of thermal isolation, and low thermal masses that can be achieved.

Micromachined thermal AC/RMS converters provide examples of where physical scaling clearly has major advantages. The time constants of the converter can be decreased from minutes to milliseconds. The high-frequency cutoff of such transducers occurs when parasitics attenuate the input signal (capacitances tending to short it and inductances, such as bond wires, tending to impede it). The low-frequency cutoff is limited by the thermal time constants of the converter and occurs when the temperature of the converter begins to track the time course of the signal rather than its RMS value. A useful approach is to micromachine a thermally isolated "island" with a minimal thermal mass and to locate the power-to-heat and temperature-to-voltage transducers on it.

Prior to the broad use of micromachining technologies, Ott (1974) demonstrated the use of a 610×610 µm silicon chip, mounted on an insulating material, as a thermal AC/RMS converter. A diffused 50 Ω resistor accomplished the power-to-heat conversion, and the resulting die temperature was sensed using the base-emitter voltage of a monolithic bipolar junction transistor. While the entire die had to be heated, the devices still demonstrated good performance, with a ± 2 % bandwidth of 100 MHz, a dynamic range of 30 dB, and a thermal time constant of 65 ms (≈ 1 s to settle to $\pm 0.05\%$ up to 10 MHz). It should be noted that there is a commercial product on the market (LT1088, from Linear Technology, Inc., Milpitas, CA) that uses an air-impregnated polymer die attach technique to provide 300 K/W thermal isolation (thermal resistance) for an entire silicon AC/RMS converter die (see Williams (1987)). In contrast, however, thermal isolation resistances on the order of 10,000 K/W are not uncommon in micromachined thermal structures.

The first examples of micromachined versions appear to be those developed by Jackson (1974) (discussed below among other microwave power sensors) and O'Neill (1980), for use in Hewlett Packard test instruments. In this case of O'Neill (1980), a pair of isolated silicon islands was formed using wet etching, each containing a heater resistor (tantalum nitride) and a silicon diode for use as the sensor. These devices were used to stabilize the output amplitude of the Model 3336 family of frequency synthesizers. One heater resistor was driven with the output signal and the other was DC driven using feedback so that the temperatures of the two isolated structures matched. The DC drive signal thus corresponded to the RMS value of the AC drive signal.

Goyal and Brodie (1984) also presented a commercially applied (in Fluke test instruments, such as the Model 8506A digital multimeter) micromachined thermal AC/RMS converter. Pairs of thermally isolated silicon islands were formed by bulk anisotropic etching (apparently from the front side of the substrate) of 75 μm (100) silicon wafers. Thin-film resistors and diffused bipolar transistors were formed prior to etching, and served as the thermal sources and sensors, respectively. As for O'Neill (1980), the pair of converters was used differentially so as to null out ambient temperature effects. The reported frequency response of the Goyal and Brodie design was 40 to 20 kHz, with an achieved thermal isolation of 8,400 K/W. Many of the fabrication details were not provided for these two commercial devices. However, patents can often serve to fill in the details, since they must enable someone of ordinary skill in the art to fabricate the devices. For example, the Fluke device is explained in detail in the patent to Chapel and Gurol (1981), where it is disclosed that low thermal conductivity 304 stainless steel (with NiCr adhesion layer) was used to form the leads to the undercut structures. In any case, several more recent examples of micromachined AC/RMS converters have been published in detail and are discussed below.

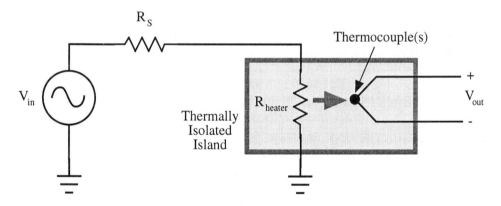

Basic concept of a thermoelectric AC/RMS converter. Adapted from Jaeggi, et al. (1992).

Jaeggi, et al. (1992) developed a thermoelectric AC/RMS converter made by post-processing devices fabricated using a conventional CMOS IC process (described below and in Moser (1993)). Thermal conversion was accomplished by using aluminum/polysilicon thermopiles as the power sensors. Since the output voltage of thermocouples tends to be quite small, thermopiles are often fabricated with a number of thermocouples in series electrically and in parallel thermally (the trade-offs in terms of signal, noise, and sensitivity are discussed in the Optical Transducers chapter). The output voltage of a thermopile of N elements, V_{out}, is given by,

$$V_{out} = N\alpha \Delta T$$

where α is the Seebeck coefficient (≈ 50 μV/°C for CMOS aluminum/polysilicon junctions), and ΔT is the temperature difference between "hot" and "cold" junctions, in K. ΔT as a function of input power, P_{heater}, and thermal conductance of the thermally isolated island, κ, is,

$$\Delta T = \frac{P_{heater}}{\kappa} = \frac{\left(\frac{V_{in}^2}{R_{heater}}\right)}{\kappa}$$

And the overall equation for signal out versus voltage in (neglecting the series resistance R_s) is,

$$V_{out} = \frac{N\alpha}{\kappa} \frac{V_{in}^2}{R_{heater}}$$

A very interesting aspect of their work was the use of a standard CMOS process to fabricate the devices, with predefined etching windows through the field oxide. The etching windows allowed them to carry out a simple post-fabrication etch using EDP, thus forming the thermally isolated silicon dioxide membranes on which the thermal circuits were located. The etching was done using an EDP solution of the following composition: 1,000 ml ethylenediamine, 160 g pyrochate-chol, 133 ml deionized (DI) water, and 6 g pyrazine. The actual post-process etching began with a 15 s buffered oxide etch dip (the authors note that this process begins to attack the aluminum metallization, so that NH_4F-based etchants are pre-ferred, although they are not necessarily much better) followed by a 60 s rinse in DI water. After etching in EDP (with an etch rate of ≈ 30 μm/min) and a 20 s rinse in DI water, the surfaces of the silicon pits were encrusted with polymerized $Si(OH)_4$, requiring a specific cleaning step (without allowing the devices to dry): a 60 s dip in cold ethylenediamine followed by a 20 s rinse in DI water.

This was followed by a sequence of steps to prevent the formation of aluminum hydroxides on the bond pads (which would make it nearly impossible to wire bond to them) and to help in the drying process. A 120 s dip in 5% vitamin C (ascorbic acid, $C_6H_8O_6$) solution (followed by a 120 s rinse in DI water) was used to remove any excess OH⁻ ions and a 60 s dip in hexane (C_6H_{14}) provided a low surface-tension liquid under the membranes so that they could dry without distortion. This approach, or a similar one using safer TMAH etching solutions, can be carried out on virtually any foundry-fabricated CMOS devices.

The AC/RMS converter results reported by Jaeggi, et al. (1992) were: sensitivity = 9.9 V/W, time constant = 1.85 ms, signal-to-noise ratio (SNR) = 8×10^9, linearity error = 0.1% for < 400 MHz, and 1% for < 1.2 GHz.

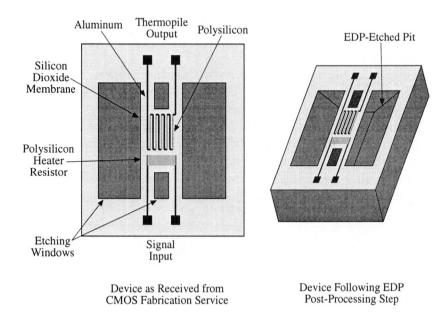

Device as Received from
CMOS Fabrication Service

Device Following EDP
Post-Processing Step

Illustration of CMOS post-processing used to fabricate a thermopile-based AC/RMS converter. Adapted from Jaeggi, et al. (1992).

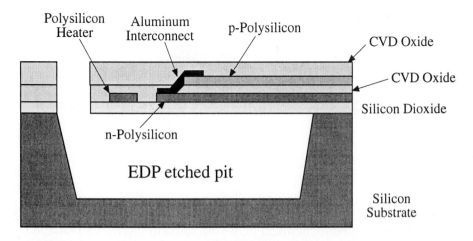

Illustration of the cross section of a thermopile fabricated using n- and p-type polysilicon in a CMOS process. Adapted from Baltes and Moser (1993).

A newer EDP-based membrane fabrication process (Baltes and Moser (1993)) used both n- and p-doped polysilicon as well as aluminum interconnects to form n-to-p thermocouple junctions (indirectly through aluminum bridges to avoid forming p-n diodes), which have much higher Seebeck coefficients: 190 to 320 μV/°C. This approach could increase the sensitivity of such an AC/RMS converter.

Yoon and Wise (1989, 1994) presented a micromachined thermal AC/RMS converter based on a modified 3 μm CMOS process. Back-side EDP etching was used to form a low thermal conductivity membrane using a p++ etch-stop ring. Two matched polysilicon heaters and Au/Cr thin-film thermal sensors were servo-controlled to operate at a constant temperature using on-chip circuitry. The devices had nonlinearities on the order of 1% and -3 dB bandwidths on the order of 20 MHz, with thermal isolation of 7,000 K/W for the transducer structures on the membrane. The overall size of the converter was 3 × 3.5 mm (when back-side bulk micromachining processes are employed for thermal isolation, die sizes are typically larger than when front-side undercutting techniques are used).

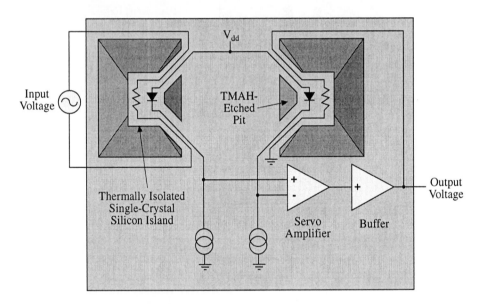

Illustration of servo-controlled, diode-based thermal AC/RMS converter. Adapted from Klaassen, et al. (1996).

Klaassen, et al. (1996), described a thermal AC/RMS converter based on the use of a TMAH-based electrochemical etch-stop to form thermally isolated n-wells from standard CMOS integrated circuits. (The process used is described in the previously referenced Reay, et al. (1994) paper and made use of the MOSIS 2.0 μm analog CMOS technology.) The design made use of two matched, isolated n-wells, suspended by silicon dioxide beams (containing aluminum interconnects) for a thermal resistance up to 37,000 K/W. The thermal response of each n-well island was modeled almost exactly by a first-order low-pass filter with a -3 dB frequency of 75 Hz (below which the AC/RMS converter would be increasingly inaccurate).

As illustrated above, the basic design of this AC/RMS topology was to use two matched thermally isolated structures, each complete with a heater resistor and a silicon diode. The input power is applied to one of the (50 Ω) resistors and the resulting temperature is measured via the forward voltage of the diode on that island (driven by a current source). An op-amp servo circuit applies an equivalent DC voltage to the heater resistor on the other island to bring it to a matching temperature. For this topology, one can write the voltages at the positive and negative terminals of the servo op-amp as,

$$V_+ = V_{dd} - \frac{\lambda_1 R_{TH1} \overline{V_{in}^2}}{R_{E1}} = V_{dd} - K_1 \overline{V_{in}^2}$$

and,

$$V_- = V_{dd} - \frac{\lambda_2 R_{TH2} \overline{V_{out}^2}}{R_{E2}} = V_{dd} - K_2 \overline{V_{out}^2}$$

where,
$\quad V_{dd}$ = positive supply voltage
$\quad \lambda_1, \lambda_2$ = diode temperature coefficients, in V/K
R_{TH1}, R_{TH2} = thermal resistance of each isolated element, in K/W
$\quad R_{E1}, R_{E2}$ = electrical resistance of each heating resistor, in Ω
$\quad K_1, K_2$ = lumped constants

Assuming an ideal op-amp with infinite gain, the voltages at the two input terminals are assumed equal due to the negative feedback, allowing the above two equations to be combined into,

$$K_2 \overline{V_{out}^2} = K_1 \overline{V_{in}^2}$$

which can be solved to show that V_{out} is the (scaled) RMS equivalent of the input voltage,

$$V_{out} = \sqrt{\frac{K_1}{K_2}} \sqrt{\overline{V_{in}^2}}$$

The ICs included all of the necessary feedback and control circuitry. The resulting specifications were a packaging-limited -3 dB bandwidth of 415 MHz, a dynamic range of 60 dB, nonlinearity of 1%, a quiescent power dissipation of 1 mW, and an area of 400×400 µm. A detailed analysis of the design issues and fundamental limits of this approach, as well as a review of several other micromachined thermal AC/RMS converters is presented in Klaassen (1996a).

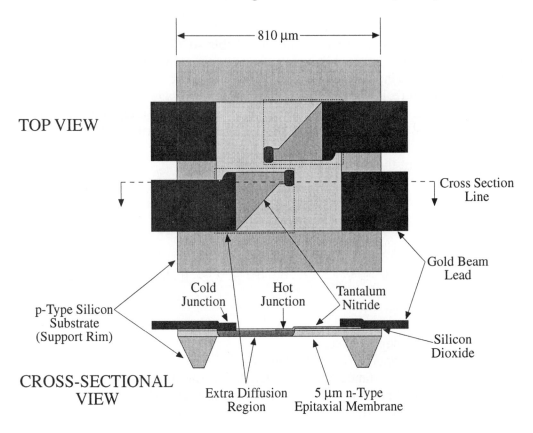

Illustration of a micromachined, microwave-frequency thermal RMS converter. Adapted from Jackson (1974).

Microwave frequency thermal AC/RMS converters, with careful attention to parasitics and controlled impedance transmission lines have also been developed, such as the devices implemented at Hewlett Packard (Jackson (1974)) for use in the Model 8481A microwave power sensor. This design used tantalum nitride/silicon and gold/silicon thermocouples situated on a 5 µm thick n-type silicon membrane on a square p-type frame (presumably electrochemical etching was used to form the membrane) with dimensions of 810×810 µm. A key design concept in these devices was that the tantalum nitride films used as part of the thermocouples were also used as the (100 Ω) heat dissipating resistive elements, so that the thermocouples were essentially self-heating. (It is also worth noting that the the two resistor/thermocouples on the chip were capacitively coupled so that they were in parallel for incident power [for a 50 Ω input resistance] and in series to double the thermocouple output voltage, as described by Lamy (1974).) These devices could operate at frequencies up to 18 GHz, with a reported sensitivity of 160 µV/mW and a thermal time constant of 120 ms. A similar, membrane-based microwave thermal AC/RMS converter (with monolithic gold beam-lead interconnects) was described by Christel and Petersen (1992) using a dissolved wafer process for fabrication. Their device had a reported maximum frequency of 20 GHz.

Using a laser-trimmed 50 Ω NiCr input resistor and metal/silicon thermocouples on a silicon membrane, Kopystynski, et al. (1991) developed another such device. Membranes were formed using KOH etching with no etch-stop (timed etching). They reported sensitivities in the range of 0.2 to 0.27 V/W and a bandwidth of 26 GHz (no time constant was reported). A more recent microwave thermal AC/RMS converter design was described by Kodato, et al. (1997), fabricated on either sapphire or glass, and utilized thermopiles formed from platinum and microcrystalline SiGe films. For the sapphire-substrate devices, they reported a frequency response up to 65 GHz, a sensitivity of 0.25 V/W (2.10 V/W for the glass-substrate devices), and symmetrical rise/fall times of $\approx$ 3 ms.

Milanovic, et al. (1997) reported a 20 GHz thermal AC/RMS converter fabricated using simple post-processing of standard CMOS dice. Two arrays of Al/polysilicon thermocouples (26 in total) were designed using standard CMOS layers and designed so that the hot junctions could be heated on a thermally isolated region while the cold junctions were located above un-etched substrate. Using a combination of XeF_2 dry, and EDP wet etching, the thermally isolated region was formed by etching through bare silicon regions (formed using the superposition of cut layers for all levels of dielectric during the CMOS processing). The combination etch allowed separate etching regions to be connected via the rapid lateral etching of the XeF_2, then deepened using EDP. Sixteen 30 s pulses of XeF_2 etching were used, followed by 45 min in EDP at 92°C to fully undercut the isolated area. They reported a sensitivity of 5.32 V/W, a dynamic range of 40 dB, linearity of ± 0.16%, and a frequency response of at least 20 GHz.

Using a notably different thermal AC/RMS conversion approach, Dehé, et al. (1995, 1996) described a microwave thermal RMS converter combining an electroplated gold coplanar waveguide (50 Ω) on a 1 µm AlGaAs membrane on a GaAs substrate. Rather than directly converting the incoming RF power entirely to heat using a resistor, the inherent ohmic losses of the coplanar waveguide's center conductor were used to convert only a fraction of the power to heat (providing for an in-line RF power monitor between stages of an RF system or in similar applications). The high-resistivity epitaxial $Al_{0.48}Ga_{0.52}As$ membrane was formed by isotropic bulk etching from the back side of the GaAs wafer using sprayed NH_3OH/H_2O_2. Thermal sensing was carried out using two arrays of series-connected $Au/Al_{0.40}Ga_{0.60}As$ thermocouples, which provided a sensitivity of 1.1 V/W for frequencies up to 14 GHz. The thermal time constant of the sensor was 1.2 ms.

5. THERMAL GAS PRESSURE SENSORS

The basic operating principle of *Pirani-type* gas pressure sensors is that an electrically heated wire (or other thermally isolated structure) will lose heat to the external gas (as well as the surrounding structures, which is a parasitic thermal conductance that should be minimized through the design of the sensor) as a function of the gas pressure. By monitoring the resistance of the wire and knowing its temperature coefficient of resistance, one can determine its temperature. Then, knowing the electrical input power and the temperature, one can determine the losses of heat to the ambient (by calibrating the sensor in a vacuum, one can measure and later null the parasitic conductances).

As described in Klaassen (1996b) and Klaassen and Kovacs (1997), the thermal conductivity, κ, of a gas can be approximated as,

$$\kappa \approx \frac{1}{3}\overline{m}\,\lambda n \overline{v} c_v$$

where,
$\overline{m}$ = average molecular mass
  n = number of molecules per unit volume (for air, $n \approx 2.7 \times 10^{19}$ cm^{-3})
  $\overline{v}$ = mean velocity of the gas molecules, in m/s
  c_v = specific heat, at constant volume, of the gas, in K/(J•kg)

and where, λ, the mean free path of gas molecules is given by,

$$\lambda = \frac{kT}{\sqrt{2}\pi P \delta^2}$$

where,
k = Boltzmann's constant = 1.38066×10^{-23} J/K
P = pressure, in N/m^2
δ = the molecular diameter of the gas molecules, in m

Thus it can be seen that the thermal conductivity of a gas is theoretically a property of the gas itself, and not a function of pressure, since from the ideal gas law, $P = nkT$ and thus the pressure-dependent terms cancel. The key operating principle of thermal pressure sensors is that their geometry must somehow *limit the mean free path* of the gas molecules. This is usually accomplished by spacing the heating element and the substrate sufficiently close so that the spacing is $< \lambda$. In such a regime, the gas molecules collide much more frequently with the structures of the sensor than with each other (referred to as *viscous heat conduction*, as opposed to free molecular conduction when this is not the case). At a sufficiently small spacing, the thermal conductance through the gas from the heater to the remainder of the structure can be written as,

$$G = G_o \frac{P\, P_t}{P + P_t}$$

where G_o is a reference conductance at low pressure (e.g., 1 Pa), and P_t is the transition pressure at which the mean free path is on the same order as the geometrical spacing between the heater and substrate.

Ideally, gas pressure sensors measure pressure independently of the composition of ambient gas(es). This is true for mechanical-type (e.g., membrane) pressure sensors. Unfortunately, this is not true for Pirani-type sensors, but once corrected for a given gas, they can be extremely accurate and cover large dynamic ranges.

For a Pirani-type sensor, there will be a low-pressure range wherein the heat loss is essentially only through structural conduction, a fairly linear range wherein the heat loss is proportional to pressure, and a high-pressure range wherein the heat loss is "saturated."

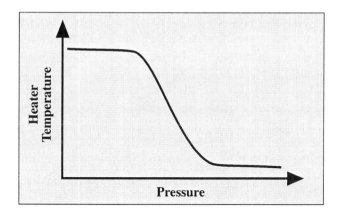

Typical heater temperature versus pressure curve for a Pirani-type pressure sensor operated at constant power.

Mastrangelo and Muller (1991a, 1991b) and Mastrangelo, et al. (1992) demonstrated a micromachined Pirani vacuum sensor, with on-board NMOS processing circuitry. The same basic fabrication sequence used to make the "on-chip incandescent lamps" discussed in the Optical Transducers chapter was employed, except that the filament was exposed in the case of the Pirani sensor. The sensor filament was a 1 μm thick n+ polysilicon layer doped to a resistivity of $10^{-3}\ \Omega{\cdot}cm$ (TCR = 900 ppm/°C), encased in silicon nitride and released in EDP. The sensor process was integrated with a 14-mask, 4 μm feature size NMOS process. On-chip sensor drive, signal conditioning, and an 8-bit analog-to-digital conversion (successive approximation type) circuitry was integrated. The output of the device is a serial, logic-compatible output corresponding to a temperature range of 10 to 10^5 Pa, with good linearity over the range of $\approx 5 \times 10^2$ to 5×10^3 Pa.

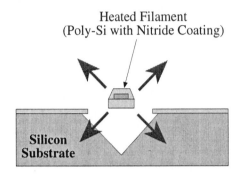

Simplified diagram of a micro-Pirani pressure sensor. Adapted from Mastrangelo and Muller (1991a).

Klaassen (1996b and Klaassen and Kovacs (1997) demonstrated the use of electrochemically etched n-wells in CMOS (the MOSIS 2.0 μm analog CMOS technology was used) to realize a Pirani-type pressure sensor with on-chip thermal servo circuitry and a temperature setting D/A converter. The sensor operated over a pressure range of 0.8 to 9.2×10^4 Pa, with a quiescent power dissipation of 1 mW, and a sensing element thermal resistance of 180,000 K/W at 0.1 Pa.

As the gap between the heating element and the substrate is reduced, the upper pressure limit of the sensor is increased (by raising the transition pressure). A process used by Hierold, et al. (1996) to fabricate an accelerometer by undercutting a standard CMOS polysilicon has also been used by that group to fabricate high-pressure Pirani sensors (Hierold (1996)). The sacrificial layer was a 600 nm field oxide, resulting in a very narrow gap. This approach can be used to increase the pressure range up to that typically encountered in automotive manifold air pressure and industrial ventilation and heating applications, for example.

The nonlinearities of Pirani-type sensors may not be an issue in an actual system implementation since it is relatively simple to digitally compensate such a sensor, although typically more A/D converter resolution is necessary than that ultimately required for the sensor if it were linear.

6. THERMAL FLOW SENSORS

One can deliberately allow a thermal microsensor to be exposed to a flowing gas or liquid and use the cooling from forced convection to measure the flow. The simplest implementation of this is the well-known "hot-wire" anemometer, which is simply a heated wire, the temperature of which varies proportionally to the flow (the temperature can be determined through knowledge of the temperature coefficient of resistance of the wire and simple electrical measurement of its resistance).

Stemme (1986) presented a bulk micromachined hot-wire-type anemometer based on a pair of diodes and a heating resistor on a polyimide-suspended silicon island. The dimensions of the suspended island were $400 \times 300 \times 30$ µm, resulting in a thermal time constant of 50 ms. The device was successfully tested over flow velocities of ≈ 0.6 to 6 m/s. Johnson and Higashi (1987) demonstrated another thermal flow sensor based on platinum or permalloy (Ni-Fe alloy) resistors and heaters on silicon nitride thermally isolated membranes. Their device was operated in a Wheatstone bridge configuration and could measure flows in both directions with a range of $\pm$ 1,000 sccm. These sensors are commercially available from Honeywell, Inc. (Minneapolis, MN). A more recent example of such a membrane-based anemometer was presented by Lammerink, et al. (1993), who built a model for the sensor's operation and demonstrated it using gases and liquids. In addition, Robadey, et al. (1995) demonstrated a thermopile-based thermal gas flow sensor fabricated by post-processing structures fabricated using a standard CMOS process.

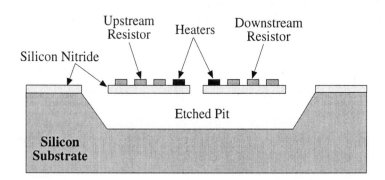

Illustration of a micromachined "hot-wire" anemometer flow sensor. Adapted from Johnson and Higashi (1987).

An alternative technique that can be applied to micromachined devices is thermal tracing, in which a heater is pulsed and the "time of flight" of the heated gas or fluid is determined using a downstream temperature sensor.

As for Pirani-type gas pressure sensors, gas composition can have an important effect on the accuracy of thermal flow sensors. A recent paper discussing a wide range Pt-on-nitride, serpentine thermal pressure sensor that highlights some of the practical issues involved with gas composition-dependent variables is Bonne and Kubisiak (1994). For further information on micromachined thermal flow sensors, an interesting patent on the subject is Higashi, et al. (1985) and an excellent overview of this area is provided by Moser (1993).

Such sensors can readily be incorporated into more complex arrays, and can certainly be fabricated with on-chip electronics if it were justified on a cost-performance basis. An interesting example of this was presented by Kersjes, et al. (1995), who demonstrated a thermal flow sensor on a 1×5 mm die, complete with on-chip CMOS circuitry to drive the sensor and to convert the measured flow signal into a pulse-width-modulated digital output. The sensor was mounted in a 2 mm catheter (it was intended for blood flow measurements) and tested to show that it could provide > 7 bit resolution over a flow range of 0 to 0.5 m/s, with a power dissipation of 5 mW. If multi-directional flow sensing is needed, one can arrange sensors at various locations and correlate their outputs. They can also be combined with other types of sensors, as demonstrated by Kälvesten, et al. (1995), who fabricated a combination thermal flow sensor and pressure sensor for studying turbulence phenomena.

Resonant flow sensor mechanisms using thermal actuation are also feasible. For example, Bouwstra, et al. (1989) fabricated a thermally driven silicon nitride membrane (with a polysilicon resistor) whose resonant frequency was proportional to the flow rate. In addition, Joshi (1994) presented a surface acoustic wave resonator, with an on-board heater, as a flow sensor. The operating principle in this case was that cooling of the SAW device by the flowing fluid changed the oscillation frequency, and they reported a frequency shift of 140 kHz (for a 73 MHz sensor) over a flow range of 0 to 1,000 ml/min for nitrogen gas.

It also should be noted that a variety of non-thermal flow sensors are discussed in the Microfluidic Devices chapter, as heating of the fluids to be measured is sometimes not desirable.

7. OTHER THERMAL SENSORS

There are several other types of sensors wherein temperature measurement is the basic transduction mechanism, and there is considerable overlap with the material in other chapters, particularly regarding chemical sensing. However, some examples are presented here.

7.1 MICROMACHINED CALORIMETERS

Calorimeters are devices where the energy of exothermic chemical reactions is measured to characterize them in some way (e.g., to identify them or to examine their purity). Essentially what is required is a low thermal mass structure onto (or into) which the reactants can be placed and a means for determining the temperature of that structure. Micromachining approaches should readily improve performance of such devices since they allow for much shorter time constants than macroscopic devices as well as for greatly decreased sample size, easy differential measurements (by fabricating arrays of calorimeter devices on a single chip) and disposability. For example, Bataillard, et al. (1993) describe the use of a 4 µm-thick silicon membrane and a Si/Al thermopile as a microcalorimeter.

7.2 DEW-POINT (THERMAL) HUMIDITY SENSORS

Dew-point humidity sensors are used to measure the humidity of a gas (usually air). They require a refrigerator and a sensor that can detect condensation. Their basic operating principle is to cool the condensation sensor and to measure the temperature at which dew forms. The underlying mechanism is that as a gas bearing a vapor (such as water) is cooled, the amount of vapor it can contain decreases. Thus, as a sensor in contact with such a gas is cooled, the gas near it eventually becomes saturated and the liquid form of the vapor condenses onto the sensor.

The dew can be detected by optical or capacitive means, the latter perhaps being more interesting for micromachining since a simple interdigitated capacitor will do the job (the relative dielectric constant of water is roughly 80 times that of air). Micromachined refrigerators and conventional "macroscopic" Peltier effect coolers are described elsewhere, and completely micromachined thermal dew-point sensors are possible. However, most micromachined humidity sensors available at present make use of the change in capacitance of a water-permeable dielectric (such as polyimide) as the sensing mechanism. This type of humidity sensor is described in the Chemical Transducers chapter.

7.3 THERMAL ELECTROMAGNETIC MIXERS

Since resistive heaters are square-law devices, they can be used as total power detectors for electromagnetic energy, as discussed above in Section 4.2. Such devices do not provide an instantaneous response to the input signal, as do other nonlinear devices used as mixers to heterodyne down high-frequency signals. However, if appropriate bolometer structures are scaled down, they can be used in such applications. Skalare, et al. (1996) demonstrated a superconducting terahertz mixer based on this principle (proposed by Prober (1993)). This design employed a Nb microbridge (270 nm long, 140 nm wide, 10 nm thick, with critical temperature $T_C = 5$ K and transition width ≈ 1 K) suspended between large Au contacts. In operation, the Nb is kept at the edge of the transition temperature between superconductivity and its cessation (≈ 4.3 K). At this temperature, the temperature coefficient of resistance is extremely high. The time constant of the microbridge is extremely small, primarily because its scale is small enough that direct diffusion of hot electrons into the Au supports dominates, leading to thermal time constants on the order of *tens of picoseconds*. When two RF signals were coupled into it, its temperature is able to respond quickly enough to track the power variation at the lower intermediate frequency (difference). They demonstrated that with a local oscillator at 533 GHz, heterodyne detection was feasible and that the bolometer's performance would likely be competitive with conventional Schottky mixers currently used in the THz frequency range.

THERMAL TRANSDUCERS REFERENCES

GENERAL REFERENCES

Carr, J. J., "Sensors and Circuits," PTR Prentice-Hall, Inc., Englewood Cliffs, NJ, 1993.

Halliday, D., Resnick, R., and Walker, J., "Fundamentals of Physics," Fourth Edition, John Wiley and Sons, New York, NY, 1993.

Holman, J. P., "Heat Transfer," Seventh Edition, McGraw-Hill, New York, NY, 1990.

Incroprera, F. P., and DeWitt, D. P., "Fundamentals of Heat and Mass Transfer," Third Edition, John Wiley and Sons, New York, NY, 1990.

Lide, D. R. [ed.], "CRC Handbook of Chemistry and Physics," 77th Edition, CRC Press, Inc., Boca Raton, FL, 1996.

Weast, R. C. [ed.], "CRC Handbook of Chemistry and Physics," CRC Press, Inc., Boca Raton, FL, 1988.

SPECIFIC REFERENCES

Analog Devices 1990/91 Linear Products Databook, "AD590 Two-Terminal IC Temperature Transducer," pp. 12-7 to 12-15.

Analog Devices 1990/91 Linear Products Databook, "AD592 Low Cost, Precision IC Temperature Transducer," pp. 12-17 to 12-24.

Bakker, A. H., and Huijsing, J. H., "Micropower CMOS Temperature Sensor with Digital Output," Proceedings of the 1995 21st European Solid-State Circuits Conference, ESSCIRC, Lille, France, in IEEE Journal of Solid-State Circuits, vol. 31, no. 7, July 1996, pp. 933 - 937.

Baltes, H., and Moser, D., "CMOS Vacuum Sensors and Other Applications of CMOS Thermopiles," Proceedings of Transducers '93, the 7th International Conference on Solid-State Sensors and Actuators, Yokohama, Japan, June 7 - 10, 1993, pp. 736 - 741.

Barth, P. W., and Angell, J. B., "Thin Linear Thermometer Arrays for Use in Localized Cancer Hyperthermia," IEEE Transactions on Electron Devices, vol. ED-29, no. 1, Jan. 1982, pp. 144 - 150.

Bataillard, P., Steffgen, E., Haemmerli, S., Manz, A., and Widmer, H. M., "An Integrated Silicon Thermopile as Biosensor for the Thermal Monitoring of Glucose, Urea and Penicillin," Biosensors and Bioelectronics, vol. 8, no. 2, 1993, pp. 89 - 98.

Benson, B. B., and Krause, D., "Use of the Quartz Crystal Thermometer for Absolute Temperature Measurements," Review of Scientific Instruments, vol. 45, no. 12, Dec. 1974, pp. 1499 - 1501.

Bonne, U., and Kubisiak, D., "Burstproof, Thermal Pressure Sensor for Gases," Proceedings of the Solid-State Sensor and Actuator Workshop, Hilton Head Island, SC, June 13 - 16, 1994, pp. 78 - 81.

Bouwstra, S., Kemna, P., and Legtenberg, R., "Thermally Excited Resonating Membrane Mass Flow Sensor," Sensors and Actuators, vol. 20, no. 3, Dec. 1989, pp. 213 - 223.

Brokaw, A. P., "Temperature Sensor with Single Resistor Set-Point Programming," Digest of Technical Papers - IEEE International Solid-State Circuits Conference, Feb. 1996, pp. 334 - 335.

Chapel, R. W., and Gurol, M., "Thermally Isolated Monolithic Semiconductor Die," U.S. Patent No. 4,257,061, issued Mar. 17, 1981.

Chmielowski, M., and Witek, A., "Tunneling Thermometer," Sensors and Actuators, vol. A45, no. 2, Nov. 1994, pp. 145 - 151.

Christel, L. A., and Petersen, K., "A Miniature Microwave Detector Using Advanced Micromachining," Proceedings of the Solid-State Sensor and Actuator Workshop, Hilton Head Island, SC, June 22 - 25, 1992, pp. 144 - 147.

Cox, L. G., and Kusters, N. L., "An Automatic RMS/DC Comparator," IEEE Transactions on Instrumentation and Measurement, vol. IM-23, no. 4, Dec. 1974, pp. 322 - 325.

Dehé, A., Klingbeil, H., Krozer, V., Fricke, K., Beilenhoff, K., and Hartnagel, H. L., "GaAs Monolithic Integrated Microwave Power Sensor in Coplanar Waveguide Technology," Digest of the 1996 IEEE MTT-S International Microwave Symposium, San Francisco, CA, June 17 - 21, 1996, vol. 1, pp. 161 - 164.

Dehé, A., Krozer, V., Fricke, K., Klingbeil, H., Beilenhoff, K., and Hartnagel, H.L., "Integrated Microwave Power Sensor," Electronics Letters, Dec. 7, 1995, vol. 31, no. 25, pp. 2187 - 2188.

Dobkin, R. C., "Monolithic Temperature Transducer," Digest of Technical Papers, International Solid-State Circuits Conference, Feb. 1974, pp. 126 - 127.

Foster, T. E., "An Easily Calibrated, Versatile Platinum Resistance Thermometer," Hewlett Packard Journal, Apr. 1974, pp. 13 - 17.

Fraden, J., "Handbook of Modern Sensors: Physics, Designs and Applications," American Institute of Physics Press, Woodbury, NY, 1997.

Goldman, K., and Mehregany, M., "A Novel Micromechanical Temperature Memory Sensor," Proceedings of Transducers '95, the 8th International Conference on Solid-State Sensors and Actuators, Stockholm, Sweden, June 25 - 29, 1995, vol. 2, pp. 132 - 135.

Goyal, R., and Brodie, B. T., "Recent Advances in Precision AC Measurements," IEEE Transactions on Instrumentation and Measurement," vol. IM-33, no. 3, Sept. 1984, pp. 164 - 167.

Gray, P. R., and Meyer, R. G., "Analysis and Design of Analog Integrated Circuits," John Wiley and Sons, New York, NY, 1993, pp. 338 - 346.

Hermach, F. L., and Williams, E. S., "Thermal Converters for Audio-Frequency Voltage Measurements of High Accuracy," IEEE Transactions on Instrumentation and Measurement, vol. IM-15, no. 4, Dec. 1966, pp. 260 - 268.

Hewlett, W. R., "A New Type Resistance-Capacity Oscillator," M.S. Thesis, Stanford University, Stanford, CA, 1939.

Hierold, C., Hildebrandt, A., Näher, U., Scheiter, T., Mensching, B., Steger, M., and Tielert, R., "A Pure CMOS Surface Micromachined Integrated Accelerometer," Proceedings of the 9th Annual IEEE 1996 Workshop on Micro Electro Mechanical Systems (MEMS '96), San Diego, CA, Feb. 11 - 15, 1996, pp. 174 - 179.

Hierold, C., Siemens Corporate Research and Development Center, Munich, Germany, personal communication, 1996.

Higashi, R. E., Johnson, R. G., and Bohrer, P. J., "Flow Sensor," U.S. Patent No. 4,501,144, issued Feb. 26, 1985.

Huijsing, J. H., Schuddemat, J. P., and Verhoef, W., "Monolithic Integrated Direction-Sensitive Flow Sensor," IEEE Transactions on Electron Devices, vol. ED-29, no. 1, Jan. 1982, pp. 133 - 136.

Inglis, B. D., "AC-DC Transfer Standards - Present Status and Future Directions," IEEE Transactions on Instrumentation and Measurement, vol. IM-34, no. 2, June 1985, pp. 285 - 290.

Jackson, W. H., "A Thin-Film/Semiconductor Thermocouple for Microwave Power Measurements," Hewlett Packard Journal, Sept. 1974, pp. 16 - 18.

Jaeggi, D., Baltes, H., and Moser, D., "Thermoelectric AC Power Sensor by CMOS Technology," IEEE Electron Device Letters, vol. 13, no. 7, July 1992, pp. 366-368.

Johnson, R. G., and Higashi, R. E., "A Highly Sensitive Silicon Chip Microtransducer for Air Flow and Differential Pressure Sensing Applications," Sensors and Actuators, vol. 11, no. 1, Jan. 1987, pp. 63 - 72.

Joo, Y., Dieu, K., and Kim, C., "Fabrication of Monolithic Microchannels for IC Chip Cooling," Proceedings of the IEEE 1995 Micro Electro Mechanical Systems Workshop (MEMS '95), Amsterdam, Netherlands, Jan. 29 - Feb. 2, 1995, pp. 362 - 367.

Joshi, S. G., "Flow Sensors Based on Surface Acoustic Waves," Sensors and Actuators, vol. A44, no. 3, Sept. 1994, pp. 191 - 197.

Kälvesten, E., Vieider, C., Löfdahl, L., and Stemme, G., "An Integrated Pressure-Flow Sensor for Correlation Measurements in Turbulent Gas Flows," Proceedings of Transducers '95, the 8th International Conference on Solid-State Sensors and Actuators, Stockholm, Sweden, June 25 - 29, 1995, vol. 2, pp. 428 - 431.

Kersjes, R., Liebscher, F., Spiegel, E., Manoli, Y., and Mokwa, W., "An Invasive Catheter Flow Sensor with On-Chip CMOS Read-Out Electronics for the Online Determination of Blood Flow," Proceedings of Transducers '95, the 8th International Conference on Solid-State Sensors and Actuators, Stockholm, Sweden, June 25 - 29, 1995, vol. 2, pp. 432 - 435.

Kittel, C., and Kroemer, H., "Thermal Physics," W. H. Freeman, San Francisco, CA, 1980.

Klaassen, E. H., "Thermal AC to RMS Converter," Chapter 4 in, "Micromachined Instrumentation Systems," Doctoral Dissertation in Electrical Engineering, Stanford University, Stanford, CA, May 1996, pp. 71 - 113.

Klaassen, E. H., "Thermal Conductivity Vacuum Sensor," Chapter 5 in, "Micromachined Instrumentation Systems," Doctoral Dissertation in Electrical Engineering, Stanford University, Stanford, CA, May 1996, pp. 114 - 142.

Klaassen, E. H., and Kovacs, G. T. A., "Integrated Thermal Conductivity Vacuum Sensor," Sensors and Actuators, vol. A58, no. 1, Jan. 1997, pp. 37 - 42.

Klaassen, E. H., Reay, R. J., and Kovacs, G. T. A., "Diode-Based Thermal R.M.S. Converter with On-Chip Circuitry Fabricated Using CMOS Technology," Sensors and Actuators, vol. A52, nos. 1 - 3, Mar. - Apr. 1996, pp. 33 - 40.

Kodato, S., Wakabayashi, T., Zhuang, Q., and Uchida, S., "New Structure for DC-65 GHz Thermal Power Sensor," Proceedings of Transducers '97, the 1997 International Conference on Solid-State Sensors and Actuators, Chicago, IL, June 16 - 19, 1997, vol. 2, pp. 1279 - 1282.

Kopystynski, P., Obermeier, E., Delfs, H., and Löser, A., Silicon Power Microsensor with Frequency Range from DC to Microwave," Proceedings of Transducers '91, the 1991 International Conference on Solid-State Sensors and Actuators, San Francisco, CA, June 24 - 27, 1991, pp. 623 - 626.

Lammerink, T. S. J., Tas, N. R., Elwenspoek, M., and Fluitman, J. H. J., "Micro-Liquid Flow Sensor," Sensors and Actuators, vols. A37 - A38, June - Aug. 1993, pp. 45 - 50.

Lamy, J. C., "Microelectronics Enhances Thermocouple Power Measurements," Hewlett-Packard Journal, Sept. 1974, pp. 19 - 23.

Legtenberg, R., Bouwstra, S., and Fluitman, H. J., "Resonating Microbridge Mass Flow Sensor with Low-Temperature Glass-Bonded Cap Wafer," Sensors and Actuators, A27, nos. 1 - 3, May 1991, pp. 723 - 727.

Little, W. A., "Microminiature Refrigeration," Review of Scientific Instruments, vol. 55, no. 5, May 1984, pp. 661 - 680.

Madou, M., "Fundamentals of Microfabrication," CRC Press, Inc., Boca Raton, FL, 1997.

Maluf, N. I., McNutt, E. L., Monroe, S., Tanelian, D. L., and Kovacs, G. T. A., "A Thermal Signal Generator Probe for the Study of Neural Thermal Transduction," IEEE Transactions on Biomedical Engineering, vol. 41, no. 7, July 1994, pp. 649 - 655.

Martin, J. H., and Jessell, T. M., "Modality Coding in the Somatic Sensory System," Chapter 24 in "Principles of Neural Science," Third Edition, Kandel, E. R., Schwartz, J. H., and Jessell, T. M. [eds.], Elsevier, New York, NY, 1991, pp. 341 - 352.

Mastrangelo, C. H., and Muller, R. S., "Fabrication and Performance of a Fully Integrated μ-Pirani Pressure Gauge with Digital Readout," Proceedings of Transducers '91, the 1991 International Conference on Solid-State Sensors and Actuators, San Francisco, CA, June 24 - 27, 1991, pp. 245 - 248.

Mastrangelo, C. H., and Muller, R. S., "Thermal Absolute-Pressure Sensor with On-Chip Digital Front-End Processor," IEEE Journal of Solid-State Circuits, vol. 26, no. 12, Dec. 1991, pp. 1998 - 2007.

Mastrangelo, C. H., Yeh, J. H.-J., and Muller, R. S., "Electrical and Optical Characteristics of Vacuum-Sealed Polysilicon Microlamps," IEEE Transactions on Electron Devices, vol. 39, no. 6, June 1992, pp. 1363 - 1375.

McGillicuddy III, D. J., "NTC Thermistor Basics and Principles of Operation," Sensors, vol. 10, no. 12, Dec. 1993, pp. 40 - 44.

Meijer, G. C. M., "Thermal Sensors Based on Transistors," Sensors and Actuators, vol. 10, nos. 1 - 2, Sept. - Oct. 1986, pp. 103 - 125.

Middelhoek, S., and Audet, S. A., "Silicon Sensors," Academic Press Ltd., London, UK, 1989.

Milanovic, V., Gaitan, M., Bowen, E. D., Tea, N. H., and Zaghloul, M. E., "Thermoelectric Power Sensor for Microwave Applications by Commercial CMOS Fabrication," IEEE Electron Device Letters, vol. 18, no. 9, Sept. 1997, pp. 450 - 452.

Moser, D., "CMOS Flow Sensors," Doctoral Dissertation, Swiss Federal Institute of Technology (ETH), Zurich, Switzerland, No. 10059, 1993.

Muller, R. S., and Kamins, T. I., "Device Electronics for Integrated Circuits," Second Edition, John Wiley and Sons, New York, NY, 1986.

Neumeister, J., Thum, R., and Lüder, E., "A SAW Delay-Line Oscillator as a High-Resolution Temperature Sensor," Sensors and Actuators, vol. A22, nos. 1 - 3, Mar. 1990, pp. 670 - 672.

Norton, H. N., "Handbook of Transducers," Prentice-Hall, Inc., Englewood Cliffs, NJ, 1989, pp. 388 - 389.

O'Neill, P., "A Monolithic Thermal Converter," Hewlett Packard Journal, May 1980, p. 12.

Ohte, A., and Yamaguta, M., "A Precision Silicon Transistor Thermometer," IEEE Transactions on Instrumentation and Measurement, vol. 26, 1977, pp. 335 - 341.

Ott, W. E., "A New Technique of Thermal RMS Measurement," IEEE Journal of Solid-State Circuits, vol. SC-9, no. 6, Dec. 1974, pp. 374 - 380.

Paul, O., von Arx, M., and Baltes, H., "Process-Dependent Thermophysical Properties of CMOS IC Thin Films," Proceedings of Transducers '95, the 8th International Conference on Solid-State Sensors and Actuators, Stockholm, Sweden, June 25 - 29, 1995, vol. 1, pp. 178 - 181.

Pease, R. A., "A New Fahrenheit Temperature Sensor," IEEE Journal of Solid-State Circuits, vol. SC-19, no. 6, Dec. 1984, pp. 971 - 977.

Prober, D. E., "Superconducting Terahertz Mixer Using a Transition-Edge Microbolometer," Applied Physics Letters, vol. 62, no. 17, Apr. 26, 1993, pp. 2119 - 2121.

Reay, R. J., Klaassen, E. H., and Kovacs, G. T. A., "A Micromachined Low-Power Temperature-Regulated Bandgap Voltage Reference," IEEE Journal of Solid-State Circuits, vol. 30, no. 12, Dec. 1995, pp. 1374 - 1381.

Reay, R. J., Klaassen, E. H., and Kovacs, G. T. A., "Thermally and Electrically Isolated Single Crystal Silicon Structures in CMOS Technology," IEEE Electron Device Letters, vol. 15, no. 10, Oct. 1994, pp. 399 - 401.

Riethmüller, W., and Benecke, W., "Thermally Excited Silicon Microactuators," IEEE Transactions on Electron Devices, vol. 35, no. 6, June 1988, pp. 758 - 763.

Robadey, J., Paul, O., and Baltes, H., "Two-Dimensional Integrated Gas Flow Sensors by CMOS IC Technology," Journal of Micromechanics and Microengineering, vol. 5, no. 3, Sept. 1995, pp. 243 - 250.

Schäfer, W., "Temperature Sensors: New Technologies on Their Way to Industrial Applications," Sensors and Actuators, vol. 17, nos. 1 - 2, May 1989, pp. 27 - 37.

Seebeck, T., "Magnetische Polarisation der Metalle und Erze durch Temperatur-Differenz," Abhaandlungen der Preussischen Akademic der Wissenschaften (1822 - 1823), pp. 265 - 373.

Skalare, A., McGrath, W. R., Bumble, B., LeDuc, H. G., Burke, P. J., Verheijen, A. A., Schoelkopf, R. J., and Prober, D. E., "Large Bandwidth and Low Noise in a Diffusion-Cooled Hot-Electron Bolometer Mixer," Applied Physics Letters, vol. 68, no. 11, Mar. 1996, pp. 1558 - 1560.

Stemme, Göran, N., "A Monolithic Gas Flow Sensor with Polyimide as Thermal Insulator," IEEE Transactions on Electron Devices, vol. ED-33, no. 10, Oct. 1986, pp. 1470 - 1474.

Tejwani, P., and Moore, T., "A Fully Automated Digital AC/DC Transfer Standard," Digest of the 1986 Conference on Precision Electromagnetic Measurements, Gaithersburg, MD, June 23 - 27, 1986, pp. 220 - 222.

Thurmond, C. D., "The Standard Thermodynamic Function of the Formation of Electrons and Holes in Ge, Si, GaAs, and GaP," Journal of the Electrochemical Society, vol. 122, no. 8, Aug. 1975, p. 1133.

Timko, M. P., "A Two-Terminal IC Temperature Transducer," IEEE Journal of Solid State Circuits, vol. SC-11, no. 6, Dec. 1976, pp. 784 - 788.

Tuckerman, D. B., "Heat-Transfer Microstructures for Integrated Circuits," Ph.D. Thesis, Stanford University, Stanford, CA, 1984.

Tuckerman, D. B., and Pease, R. F. W., "High-Performance Heat Sinking for VLSI," IEEE Electron Device Letters, vol. EDL-2, 1981, pp. 126 - 129.

Ueda, T., Kohsaka, F., Iino, T., and Yamazaki, D., "Temperature Sensor Utilizing Quartz Tuning Fork Resonator," Proceedings of the 40th Annual Frequency Control Symposium, Philadelphia, PA, May 28 - 30, 1986, pp. 224 - 229.

Urban, G., Jachimowicz, A., Kohl, F., Kuttner, H., Olcaytug, F., Kamper, H., Pittner, F., Mann-Buxbaum, E., Schalkhammer, T., Prohaska, O., and Schönauer, M., "High-Resolution Thin-Film Temperature Sensor Arrays for Medical Applications," Sensors and Actuators, vols. A21 - A23, Feb. - Apr. 1990, pp. 650 - 654.

Van Herwaarden, A. W., Van Duyn, D. C., Van Oudheusden, B. W., and Sarro, P. M., "Integrated Thermopile Sensors," Sensors and Actuators, vols. A21 - A23, Feb. - Apr. 1990, pp. 621 - 630.

Vig, J. R., Filler, R. L., and Kim, Y., "Uncooled IR Imaging Array Based on Quartz Microresonators," Journal of Microelectromechanical Systems, vol. 5, no. 2, June 1996, pp. 131 - 137.

Weber, R. J., "High Power Semiconductor Devices with Integral Heat Sink," U.S. Patent No. 5,057,908, issued Oct. 15, 1991

Williams, J. "A Monolithic IC for 100 MHz RMS-DC Conversion," Linear Technology Corporation Application Note 22, Linear Technology Corporation, Milpitas, CA, Sept. 1987.

Williams, J. "Measurement and Control Circuit Circuit Collection," Linear Technology Corporation Application Note 45, Linear Technology Corporation, Milpitas, CA, June 1991.

Williams, J., "Thermal Techniques in Measurement and Control Circuitry," Linear Technology Corporation Application Note 5, Linear Technology Corporation, Milpitas, CA, Dec. 1984.

Williams, J., "Thermocouple Measurement," Linear Technology Corporation, Application Note No. 28, Linear Technology Corporation, Milpitas, CA, Feb. 1988.

Yoon, E., and Wise, K. D., "A Monolithic RMS-DC Converter Using Planar Diaphragm Structures," Proceedings of the International Electron Devices Meeting, Washington, DC, 1989, pp. 491 - 494.

Yoon, E., and Wise, K. D., "Wideband Monolithic RMS-DC Converter Using Micromachined Diaphragm Structures," IEEE Transactions on Electron Devices, vol. 41, no. 9, Sept 1994, pp. 1666 - 1668.

Ziegler, H., "A Low-Cost Digital Sensor System," Sensors and Actuators, vol. 5, no. 2, Feb. 1984, pp. 169 - 178.

CITED INDUSTRY REFERENCES AND SUPPLIERS

Platinum Resistance Thermal Sensors

Sensing Devices, Inc., 1809 Olde Homestead Lane, Lancaster, PA 17601, Phone: (717) 295-2311, Fax: (717) 295-2314.

Thermistors

BetaTherm Corporation, 910 Turnpike Road, Shrewsbury, MA 01545, Phone: (508) 842-0516, Fax: (508) 842-0748.

Thermometrics, 808 U.S. Highway 1, Edison, NJ, 08817, Phone: (201) 287-2870.

Yellow Springs Instruments, Inc., P.O. Box 279, Yellow Springs, OH 45387, Phone: (513) 767-7241.

Thermistor Interface Chip

ATMOS Technology, Inc., 1060 Lincoln Avenue, San Jose, CA 95125, Phone: (408) 292-8066, Fax: (408) 292-8241.

Silicon Thermal Sensors and Thermocouple Interface Devices

Analog Devices, Inc., One Technology Way, P.O. Box 9106, Norwood, MA 02062-9106, Phone: (617) 329-4700, Fax: (617) 326-8703.

Linear Technology Corp., 1630 McCarthy Blvd., Milpitas, CA 95035, Phone: (408) 432-1900, Fax: (408) 434-0507.

Thin-Film, Macroscopic Heaters on Kapton™, Polyester, Etc.

Minco Products, Inc., 7300 Commerce Lane, Minneapolis, MN 55432-3177, Phone: (612) 571-3121, Fax: (612) 687-9025.

Macroscopic Bismuth Telluride Peltier-Effect Heat Pumps

Melcor (Materials Electronic Products Corporation), 990 Spruce Street, Trenton, NJ 08648, Phone: (609) 393-4178, Fax: (609) 393-9461.

Chapter 7:

MAGNETIC AND ELECTROMAGNETIC TRANSDUCERS

1. INTRODUCTION

Magnetic sensors have an extremely large number of applications in consumer, industrial, and military circuits. Some magnetic sensors have digital outputs (e.g., sealed switches, keyboards, tachometer timing sensors, etc.) and others have analog outputs proportional to the magnetic field (e.g., position sensors, current sensors, "magnetic potentiometers," magnetic compasses, magnetic tape heads, etc.). In terms of micromachined magnetic sensors being shipped in volume, Hall effect devices are produced by the millions and are very common in automotive and other applications. Hall sensors are also easily integrated with active circuitry on silicon substrates, making them particularly attractive for a variety of applications. On the horizon for micromachined sensors, several other magnetic transduction mechanisms are available. For these new sensing techniques to displace ubiquitous Hall sensors, they must provide some form of performance (including cost) beyond that achievable with them — not an easy feat for most day-to-day applications.

Magnetic actuators are also very common on the macroscopic scale (motors, solenoids, relays, etc.), and increasingly so in micromachined devices. In terms of commercial availability, however, they are in their early stages of development. One key exception is that magnetic read/write heads for disk drives have long been fabricated using lithographic techniques we would now refer to as micromachining. These devices are currently shipped in enormous volumes, and new fabrication techniques will likely increase both those volumes and the storage densities of the resulting disk drives. Examples of emerging micromachined magnetic actuators include optical modulators (discussed in the Optical Transducers chapter) and electromagnetic relays and motors.

A common macroscopic electromagnetic circuit component, the inductor, has also been micromachined. In the low-frequency domain, it is not clear that micromachined inductors will have an impact in the high-volume applications of noise and signal filtering, resonators, and energy storage in switching power supplies. For the bulk of these uses, their characteristics would have to compete with those of conventionally wound coils, whose geometries are not so highly constrained. Meanwhile, at radio frequencies (particularly microwave), micromachined inductors (and related components) have been used for many years on non-silicon substrates (e.g., GaAs), and recent applications of micromachining have led to the demonstration of good quality RF inductors fabricated on standard CMOS silicon substates. As RF devices use higher frequencies, scaling of circuit components will undoubtedly make micromachining more attractive, as well as enable the application of well-established techniques like mechanical resonators to previously impossible frequencies (this is discussed in the Mechanical Transducers chapter).

This chapter begins with a review of some of the magnetic phenomena that underly the operation of the transducers discussed thereafter. Examples from a wide variety of these diverse application areas are discussed, with the focus on how micromachining was used, or could be used, to realize them. The chapter covers, in order, sensors, actuators, and electromagnetic devices as applied to electronic circuits.

1.1 TERMS AND DEFINITIONS

Magnetic flux density (or magnetic induction), B, is expressed in teslas (T = (N•s)/(C•m), N/(A•m), or 1 Weber/m^2 [Wb/m^2]). The gauss, a non-SI unit that is still commonly used, is simply 10^{-4} T. Some example magnetic flux densities, useful for gaining a sense for the magnitudes of typical values, are given in the table below.

Magnetic field strength (or magnetic field intensity), H, has units of A/m. An older unit is the Oersted (Oe) where one A/m = 1.257×10^{-2} Oe.

Calculated magnetic flux at the surface of a neutron star.	100 MT
Magnetic flux produced by the strongest superconducting electromagnets.	40 to 60 T
Magnetic flux produced by the strongest conventional electromagnets.	4 to 6 T
Magnetic flux near a small bar magnet.	10 mT
Magnetic storage media.	≈ 1 mT
Earth's magnetic flux at near equatorial latitude.	≈ 100 µT
Limit in magnetic flux below which superconducting quantum interference detector (SQUID) or specialized flux-gate magnetometers are typically required.	10 to 100 nT
Magnetic flux produced by electrical currents in the human heart.	10 nT
Magnetic flux in interstellar space.	100 pT
Magnetic flux produced by our galaxy.	1 pT
Lowest magnetic flux achievable in carefully shielded room.	10 fT

Table of some example magnetic flux densities.

The *permeability* of a material refers to its conductivity to magnetic fields, and is usually given as the dimensionless *relative permeability*, μ_r, in relation to the permeability of free space ($\mu_o = 4\pi \times 10^{-7}$ T•m/A or H/m). The permeabilty of a material is its ratio of magnetic flux density to magnetic field strength,

$$\mu = \mu_o \mu_r = \frac{B}{H}$$

Some example permeabilities (from Bate (1993)) are given in the table below.

The *coercivity* of a material is a relative measure of the ease with which a material can be magnetized or demagnetized, and has units of A/m or Oe.

The *Curie temperature*, Φ_c, is the temperature at which the spontaneous magnetization of a magnetic material becomes zero, and is given in °C or K.

Material	Relative Permeability (μ_r)
Mercury	0.999968
Silver	0.9999736
Copper	0.9999906
Water	0.9999912
Air	1.00000037
Tungsten	1.00008
Platinum	1.0003
Nickel-Zinc Ferrite	650
Manganese-Zinc Ferrite	1,200
Permalloy (78.5% Ni, 21.5% Fe)	70,000
Iron (99.96% pure)	280,000
"Supermalloy" (79% Ni, 15% Fe, 5% Mo, 0.5% Mn)	1,000,000

Table of relative permeabilties of some materials with potential relevance to micro-machining applications (from Bate (1993)).

2. MAGNETIC PHENOMENA

2.1 THE HALL EFFECT

The *Hall effect* was discovered in 1879 by Edwin Hall (Hall (1879)) while he was a graduate student at Johns Hopkins University (this paper is wonderful to read, and is highly recommended). The basic effect is quite simple conceptually: charge carriers traveling through a perpendicular magnetic field are subject to a deflection by a force known as the *Lorentz force*. If the carriers are flowing in a slab of metal or semiconductor, they are deflected preferentially to one side of the slab, producing a voltage as illustrated below.

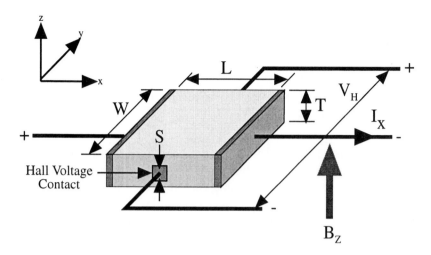

Illustration of the Hall effect. Adapted from Middelhoek and Audet (1989). Note that hole current is shown.

Referring to the above figure, the Hall voltage is given by,

$$V_H = \frac{R_H I_X B_Z}{T} = R_H J_X W B_Z$$

where,

R_H = is defined as the Hall coefficient (material-dependent), in m^3/C
T = plate thickness, in m
L = plate length, in m (not used in this equation, but shown in the figure)
W = plate width, in m
I_x = current (in the x-direction) through the plate, in A (or C/s)
B_z = magnetic flux density (z-direction), in N/(A•m) or (N•s)/(C•m) or teslas

The above equation is valid for L >> W >> T, approximate where L is not much greater than W and W >> T, and assumes small Hall voltage contacts (dimension shown as S in the above figure. It should also be noted that the Hall coefficient is actually weakly dependent on B, and thus introduces a small nonlinearity. For cases where the equation is valid, and the nonlinearity in Hall coefficient can be neglected, the Hall voltage is *directly* proportional to the current and the magnetic flux density and *inversely* proportional to the thickness of the plate.

In a Hall plate, electrons moving with average drift velocity $\vec{v}$ are subject to the Lorentz force (first term) and the force due to the electric field in the plate (second term),

$$\vec{F} = -q(\vec{v} \times \vec{B} + \vec{E}) \quad \text{in N}$$

where,

q = charge of the electron = 1.602177×10^{-19} C

$\vec{v}$ = average drift velocity, in m/s

$\vec{B}$ = magnetic flux density, in N/(A•m)

$\bar{E}$ = electric field strength, in V/m

The electrons are deflected in the (negative) y-direction building up charge near the nearby Hall contact. This built-up charge produces an electric field that counterbalances the Lorentz force, causing the electrons traversing the Hall plate to flow in the original direction (straight across the plate). Note that this discussion assumes that L >> W >> T, and W >> S, and is approximate for L about four times or more larger than W. This equilibrium is reached extremely quickly ($\approx 10^{-14}$ s, the *scattering relaxation time*).

For the above example, with the average drift velocity of the electrons in the x-direction, the Hall field that counterbalances the Lorentz force is given by,

$$E_H = +v_X B_Z \quad \text{in V/m}$$

The current density (for electrons) is,

$$J_X = -nqv_X \quad \text{in A/m}^2$$

Thus the Hall voltage (for electrons) is,

$$V_H = E_H W = -\frac{I_X B_Z}{WTnq} W = -\frac{R_H I_X B_Z}{T}$$

And thus the Hall coefficient, R_H, is given in terms of the electron density, n (in m^{-3}), by,

$$R_{H(electrons)} = -\frac{1}{nq} \quad \text{in m}^3/\text{C}$$

or, for hole current, in terms of the hole density, p (in m^{-3}), by,

$$R_{H(holes)} = +\frac{1}{pq} \quad \text{in m}^3/\text{C}$$

For metallic Hall plates (e.g., Cu, with $n = 8.47 \times 10^{28}$ electrons/m³, or 8.47 $\times 10^{22}$ cm⁻³), the Hall coefficients are quite small, on the order of -0.5×10^{-10} m³/C.

2.1.1 HALL EFFECT IN SEMICONDUCTORS

R_H is between four and five orders of magnitude larger for semiconductors than for most metals due to the fact that the carrier density is generally that much smaller in semiconductors. This translates directly into increased sensitivity. Hall effect devices are relatively easy to fabricate (although special care must be taken to reduce their offset voltages), and can be made as an intrinsic part of nearly any bipolar or MOS process. A great deal of thought has been given to how to realize high-quality Hall devices from available active circuit processes.

Again referring to the above illustration, for n-type material, electrons would be deflected toward the lower Hall contact (shown with a negative developed potential), leading to a Hall voltage as shown (and a negative Hall coefficient). For p-type material, the Hall voltage would be reversed, and this is often used to differentiate between the two types of doping in semiconductors. Since hole conduction is really the movement of electrons in the valence band (i.e., electrons flowing the same direction as discussed above), a common question is why the Hall voltage is not also negative for holes. It turns out that the valence band electrons have a negative effective mass, meaning that they will be accelerated in a direction *opposite* to that given by the Lorentz force equation.

In the above equations for R_H, it was assumed that either p >> n or n >> p. A general purpose equation for R_H when both electrons and holes contribute is,

$$R_H = \frac{\left(p\mu_p^2 - n\mu_n^2\right)}{q\left(p\mu_p - n\mu_n\right)^2} \quad \text{in cm}^3/\text{C}$$

where μ_p is the hole mobility, in cm²/(V•s), and μ_n is the electron mobility, in cm²/(V•s). (Note that the units are cm³/C to be consistent with semiconductor literature conventions.)

Examining the above equation and realizing that the mobilities of electrons and holes are different in silicon ($\mu_n \approx 1{,}400$ cm²/(V•s) and $\mu_p \approx 500$ cm²/(V•s) for low doping and room temperature, and, for comparison, $\mu_n \approx 3{,}800$ cm²/(V•s) and $\mu_p \approx 1{,}800$ cm²/(V•s) for germanium), there is a value of p-type doping at which R_H becomes zero (i.e., one can theoretically make regions of the silicon completely insensitive to magnetic fields). By solving for the case wherein the numerator of the above equation equals zero, one sees that this occurs for p ≈ 8n.

Corrections for real Hall constants are required because of the velocity distribution of the electrons (i.e., not all of the charge carriers have the same velocity), which is affected by lattice and impurity scattering. The corrected equations are,

$$R_{H(electrons)} = -\frac{r}{nq} \quad \text{in cm}^3/\text{C}$$

and,

$$R_{H(holes)} = +\frac{r}{pq} \quad \text{in cm}^3/\text{C}$$

where, r is the Hall scattering factor (dimensionless), between 0.8 and 2 depending on the type of semiconductor and the temperature.

The temperature coefficient of the Hall effect in silicon is determined mainly by the temperature coefficient of the Hall scattering factor and is $\approx -0.6\%/°C$. Nonlinearity is typically $< 1\%$ in realistic Hall devices.

In terms of a design summary for Hall plates, the following points are important to consider:

1) The *Hall voltage is proportional to the current*, so if high sensitivity without high power is important, other parameters must be optimized.

2) The *Hall voltage is linearly related to the magnetic flux density* (R_H is somewhat modulated by B).

3) The *Hall voltage is inversely proportional to the plate thickness*, so thin plates are highly desirable.

4) The *Hall voltage is inversely proportional to the carrier density*, which is why the Hall effect is greater in semiconductors than in metals.

2.2 PHYSICAL MAGNETORESISTIVE EFFECT

Since the charge carriers in a Hall plate do not all have the same velocity, the Lorentz force is different for each one. For some carriers, the Lorentz force is larger than the (net) counterbalancing force of the Hall field, and for others, the Lorentz force is smaller. Thus some carriers will deflect up or down relative to the straight path across the Hall plate.

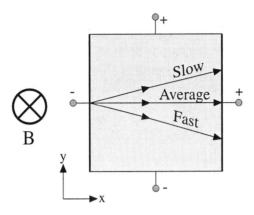

Illustration of the different path lengths of carriers of different velocities in a Hall sensor, giving rise to the physical magnetoresistive effect. After Middelhoek and Audet (1989).

The paths of these deflected carriers are *longer* than the straight one, so that there is a slight increase in resistance. This is the *physical magnetoresistive effect,* expressed (in terms of conductivity) as,

$$\sigma(B) = \sigma_o \left(1 - r^2 \mu^2 B_z^2\right) \quad \text{in } \Omega^{-1} \bullet \text{cm}^{-1}$$

where σ_o is the baseline conductivity (no magnetic flux density), in $\Omega^{-1} \bullet \text{cm}^{-1}$.

This effect is very slight in silicon, but it can be appreciable in other materials with higher mobilities, such as InSb, with an electron mobility of 80,000 cm^2/(V•s) (see Shibasaki, et al. (1991) and Partin, et al. (1992). InP is also a useful material in this regard. Ferroelectric materials also exhibit a large dependence of resistance on the magnetic field, and this has been exploited using thin-film techniques to build sensors with nanosecond response times (see Rottmann and Dettmann (1991)).

2.3 GEOMETRIC MAGNETORESISTIVE EFFECT

By shorting out the Hall voltage through the geometry of the sensor, the effective resistance between the current source and sink terminals in a Hall plate increases, as explained below. This allows a magnetic field measurement to be made with a two-terminal resistive device, rather than a four-terminal device as in the basic Hall plate.

As the name implies, the geometric magnetoresistive effect is a function of the geometry of the Hall plate. This effect is not significant as long as the length, L, of the Hall plate is greater than twice the width W. For short L Hall plates (in practice, W is approximately four times or more larger than L), the current contacts (up until now assumed to completely cover the W sides of the plate) *tend to short out the Hall voltage,* particularly near to the voltage contacts.

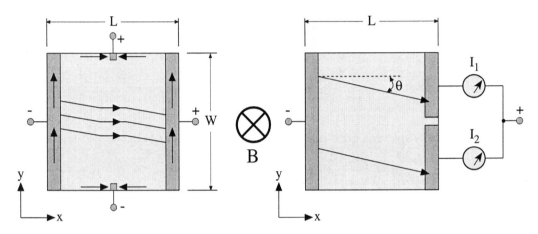

Illustration of the geometric magnetoresistive effect (left), adapted from Middelhoek and Audet (1989), and a practical implementation of a magnetoresistive sensor using two current-sensing electrodes to measure carrier deflection.

If the entire Hall voltage is short-circuited, then the current will flow at an angle to the direct path (the phenomenon is known as *Lorentz deflection*) given by,

$$\tan(\theta) = \mu B_Z$$

This extra path length corresponds to an increased resistance as seen from the two driven terminals. For small angles, the change in resistance is given by,

$$R(B) = R_o\left(1 + \tan^2(\theta)\right) = R_o\left(1 + a\mu^2 B_Z^2\right) \quad \text{in } \Omega$$

where R_o is the baseline resistance (no magnetic flux density), in Ω, and a is a correction factor, which is 1 for L << W and decreases to zero if L/W $\approx$ 4.

To make a high-performance magnetoresistor, one needs to design it so that the Hall voltage is nearly totally short-circuited. One approach is to make a disc (the so-called "Corbino disc") in which there can effectively be no Hall voltage. Another approach is to stack multiple thin Hall elements where L << W. However, for many materials, however, this effect is still quite weak.

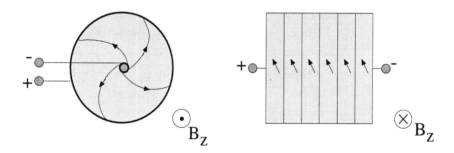

Illustration of configurations for magnetoresistors: (left) the Corbino disc and (right) a series connection of a multitude of Hall plates with a small L/W ratio. After Middelhoek and Audet (1989).

2.4 GIANT MAGNETORESISTIVE EFFECT

As discussed in White (1992, 1994) the *giant magnetoresistive (GMR) effect* was first reported by Baibich, et al. (1988). Compared to magnetoresistive materials that might exhibit a $\Delta R/R$ on the order of 6% at best, GMR materials can achieve values as high as 50%. While magnetoresistive phenomenon is based on the effect of the Lorentz force (moving carriers are deflected perpendicular to current flow, increasing path length and consequently resistance), the GMR effect is based on the behavior of electrons flowing in a magnetic material. In such a material, electrons with spin orientation opposed to the prevailing magnetization are scattered much more strongly than those with spins aligned with the field. It was found that by layering certain magnetic materials (e.g., Fe or Co) with thin non-magnetic materials (e.g., Cr, Cu, Ag, or Ag) at precisely the correct thicknesses, successive magnetic layers would couple antiferromagnetically (i.e., in opposite directions). Typically, nanometer-scale alternating layers of magnetic and non-magnetic layers are built up (e.g., 15 layers of Co(3 nm)/Cu(5nm)/NiFe(3nm)/Cu(5nm), as reported by Shijo and Yomamoto (1990)). As a result, electrons flowing through a GMR material experience increased scattering every other layer (the mean free path of electrons is longer than typical layer thicknesses). This antiferromagnetic coupling, however, can be overcome by a sufficiently high applied field. As the applied field causes a greater percentage of the GMR material to align in a unified direction, scattering is significantly reduced for half of the electrons. The net effect is a large decrease in resistance.

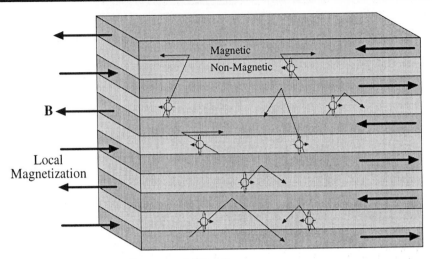

No External Field

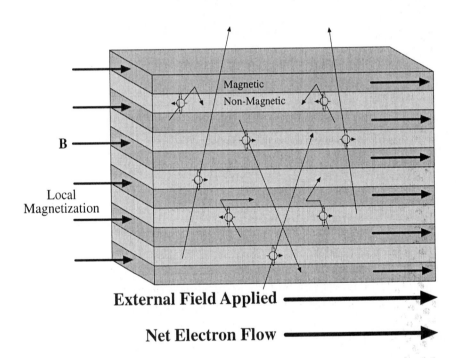

External Field Applied ➝

Net Electron Flow ➝

Illustration of the giant magnetoresistive effect in multilayer film showing the difference in scattering of electrons in antiparallel (top) and parallel (bottom) alignment. Electrons with oppositely aligned magnetic moments to the local magnetization are scattered more readily. Adapted from White (1994).

One drawback to the use of GMR sensors in the magnetic recording industry and elsewhere is the fact that while relatively large $\Delta R/R$ ratios can be obtained, they typically require high coercive fields (on the order of 0.1 T) to realize that level of response. This is a prohibitively high field strength. Typical field strengths from disk drive media, for example, are approximately two orders of magnitude lower. At these levels, ΔR/R is only a few percent above that of conventional Permalloy magnetoresistive heads.

2.5 MAGNETO-OPTICAL EFFECTS

The magneto-optical effect, or *Faraday effect* (optical modulation effects are also discussed in the Optical Transducers chapter), is based on the fact that the polarization of an electromagnetic plane wave propagating through a medium (e.g., metallic waveguide, optical fiber, or bulk crystal) is rotated (see Yariv and Yeh (1984)). The direction of rotation (clockwise or counterclockwise) is dependent on the orientation of the magnetization in the medium. Considering a plane wave as a linear combination of two circularly polarized components, these components have opposite orientation (left-handed versus right-handed). One consequence of this opposite orientation is that each component experiences a slightly different refractive index in a magnetic material. As these components enter the magnetic material, they are perfectly in phase. One component, however, experiences a delay relative to the other, due to the difference in its respective refractive index. The linear combination of these two components is thus rotated relative to the original orientation, the direction of rotation being dependent on the polarity of the magnetization. The rotation angle, β, is given by,

$$\beta = VBd$$

where,
V = the Verdet constant for a given material, in min of arc/(gauss•cm) (see the
    Optical Transducers chapter for a table of Verdet constants)
B = magnetic flux density, in gauss (G)
d = distance the light travels through the modulator medium, in cm

In terms of reflected light, the electro-optical effects (Kerr and Pockels, as described in the Optical Transducers chapter) are such that a reflected wave retracing its path into a material will have any induced rotation reversed. For Faraday rotation, this is not the case, and reflected light will be rotated twice. While this effect has apparently not yet been applied in micromachined devices, it is commonly used in magneto-optical data storage. As described in Mansuripur (1993), a low-power laser beam is reflected from individual storage (bit) locations on a magneto-optical disk's surface. Depending on the state of magnetization of a given location, the polarization of the reflected laser beam is modulated and decoded using a polarization-sensitive optical system. In order to change the state of a given bit, it is heated above its Curie temperature with the same laser beam (at higher power) and magnetized in the desired direction using a magnetic head. If the external magnetic field is maintained until the recording medium cools below the Curie temperature, the datum is retained. In general, the materials used for the medium are alloys of Co and Fe with transition metals (e.g., $(Tb_yGd_{1-y})_x(Fe_zCo_{1-z})_{1-x}$, and they are deposited by sputtering.

2.6 MAGNETIC ANISOTROPY

Magnetic anisotropy is the tendency of a material, crystal, or object to magnetize along a particular axis (called the *easy axis*) and to return to this axis in the absence of an external magnetic field. External fields cause the magnetization axis to rotate. The phenomenon of magnetic anisotropy is particularly important for such devices as magnetoresistive (MR) read heads for disk drives and bubble memories. For a thorough discussion of magnetic anisotropy, see Cullity (1972).

Of the factors that influence the easy axis of magnetization, three stand out as being particularly significant for micromachining applications, and are discussed below. These are shape, stress, and crystalline anisotropy. In designing magnetic transducers, one must understand how the energetics of these factors can interact. Such issues as device proportions, annealing temperatures, and operating temperature can be critical for creating the correct easy axis orientation.

2.6.1 SHAPE ANISOTROPY

Shape anisotropy has its origins in the *demagnetizing field*. The demagnetizing field is the field produced by the north and south poles of a magnet. This field opposes the existing (or induced) magnetization. Because the strength of this demagnetizing field varies roughly as $1/r^2$, it is much more effective at opposing short axis magnetization than long axis. A long, thin rod exhibits the maximum degree of shape anisotropy, preferring strongly to magnetize along the long axis. Shape anisotropy can have a major influence on the energetics of magnetization. In disk drives, for example, vertical recording (recording by magnetizing perpendicular to the plane of the media) is more difficult (less energetically favorable) than longitudinal recording (recording by magnetizing in the plane of the medium).

2.6.2 STRESS ANISOTROPY

Stress anisotropy arises as a complement to the magnetostrictive effect. Just as an applied magnetic field can induce small changes in the dimensions of a crystal, applied stress can cause dimensional changes, which create a favored axis of magnetization. At the root of these interrelated effects is the fact that some crystals are slightly longer or shorter along the axis of magnetization. Applying stress induces magnetic domain rotation to accommodate the induced strain.

2.6.3 CRYSTALLINE ANISOTROPY

The third type of anisotropy, *crystalline anisotropy*, stems from an interaction between the spin of the atom and the magnetic moment generated by the orbiting electron (spin-orbit coupling). This creates an easy axis aligned with a particular axis of the unit crystal. Iron, a body-centered cubic (bcc) crystal is more easily magnetized along the <100> axis. In nickel, a face-centered cubic (fcc) crystal, it is along the <111> axis. The hexagonal close-packed structure of cobalt has an easy axis along the [0001] orientation.

2.7 MAGNETIC DOMAINS

Within an object composed of a magnetic material (such as the core of an inductive read/write head), anisotropic factors influence the orientation of magnetic moments at the atomic level. Regions where all of the moments are aligned parallel to each other are called *magnetic domains*. These regions can be very localized or occupy an entire slab of material. Controlling the formation of magnetic domains is critical in many micromachining applications such as in inductive read heads for disk drives, magnetoresistive read heads, and bubble memory substrates. These applications are discussed below.

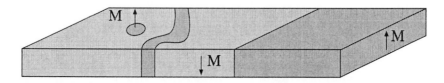

Illustration of magnetic domains. Shaded and unshaded regions are areas of oppositely aligned magnetic moments. Courtesy A. Flannery.

Magnetic domains can be visualized through *polarization microscopy*, via the magneto-optic effect (Faraday rotation). As described by Vishnevski, et al. (1997), thin-film magneto-optic garnet transducers can be used to visualize magnetic fields in a variety of magnetic media and materials. The external fields locally remagnetize the garnet films and alter their optical properties. They used films of $(Bi,Lu,Ca)_3(Fe,Ge)_5O_{12}$, $(Bi,Tm)_3(Fe,Ga)_5O_{12}$, and $(Bi,Lu,Sm)_3(Fe,Ga,Al)_5O_{12}$, prepared using liquid-phase epitaxial growth. With thin films of these materials, they were able to visualize magnetic field patterns of credit cards and an aircraft "black box" recorder (even after heating to 773 K).

3. MAGNETIC SENSORS

As discussed below, there is a wide variety of possible magnetic sensors, with wide ranges in sensitivity, noise, power consumption, temperature stability, etc. Due to the fact that the high-volume drivers for manufacturing such sensors demand low cost above all, the bulk of the devices in use today are silicon-based. This is the case not only because of the ability to use standard bipolar and CMOS processes to realize them (leveraging existing infrastructure), but also due to the relative ease with which circuitry can be integrated with them. These silicon devices offer extremely good performance-to-cost ratios, but generally they are limited to flux densities above 10 μT for current production devices. Several examples of existing and emerging micromachined magnetic sensors are presented below. It should be noted that excellent reviews of macroscopic and micromachined magnetic sensors include Baltes and Popovic (1986), Baltes and Nathan (1989), Middelhoek and Audet (1989), Lenz (1990), and Baltes and Castagnetti (1994).

3.1 HALL EFFECT SENSORS

3.1.1 PLATE DESIGN FOR HALL-VOLTAGE SENSORS

Many different designs for Hall plates are possible for micromachined sensors. Important considerations include the effects of geometry and the process(es) available for fabrication on sensitivity, offset, thermal drift, etc.

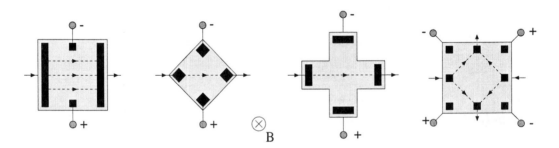

Illustration of several Hall plate designs. Adapted from Middelhoek and Audet (1989).

A great deal of thought has gone into the design of "optimal" Hall plates in terms of geometric effects. In practice, L ≈ W, and the "diamond" pattern shown above is often used.

3.1.2 OFFSET VOLTAGE ISSUES

Offset voltages (voltages present on the sensing terminals with no applied magnetic field) are a problem in commercial devices. Low-cost plastic packaging (as opposed to metal or ceramic hermetic packaging with an actual cavity above the chip) often imposes considerable mechanical forces on silicon dice during cooling and/or curing of the overcoated plastic. In addition, voltage offsets in Hall effect sensors can also be caused by misalignment of the contacts, which are generally lithographic/fabrication issues. These can often be corrected by chip-level trimming, but this approach cannot "pre-compensate" for packaging-induced stresses.

There are a number of electronic schemes for correcting offsets. One of the most interesting is to periodically reverse the sensing and driven terminals (in a symmetrical Hall plate design) and to average the output signals. This approach compensates for thermal or stress-induced (piezoresistive) offsets that may change over time (see Maupin and Geske (1980)).

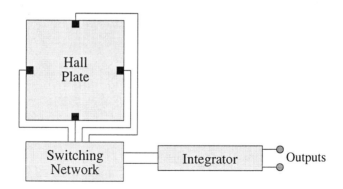

Illustration of an integrated Hall effect sensor with periodic switching of current and voltage contacts on the Hall plate. After Maupin and Geske (1980).

One can also design a round Hall plate with multiple pairs of plates and rotate the driven and sensing contacts (always perpendicular to each other for any rotation). Munter (1990, 1991) described a 16-contact round Hall plate wherein the current direction was made to spin in $\pi/8$ radian steps while all available Hall voltages were averaged over time. This approach greatly reduced the offset voltage (ten times). A more recent paper by Steiner, et al. (1997) describes the use of this approach in a CMOS Hall effect sensor to reduce offsets to below 10 μT using four pairs of opposed electrodes.

3.1.3 HALL PLATES IN BIPOLAR PROCESSES

The addition of Hall plates to a bipolar process such as the n-epitaxial layer process illustrated below requires only the layout of the desired structures (no special processing). The lower-doped n-type regions normally used for the collectors of transistors is used for the Hall plate because it has a lower carrier concentration and because electrons have a higher mobility (i.e., higher drift velocity) than holes, and hence a higher Lorentz force. The top view (not shown) of the Hall plate section would be a symmetrical structure with equally sized contacts.

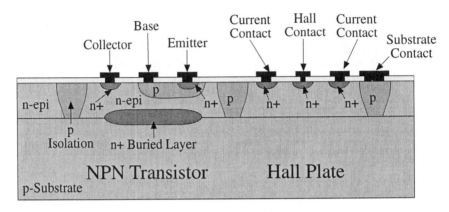

Cross section of an NPN transistor (left) and a Hall plate (right) implemented in a typical bipolar process. Adapted from Middelhoek and Audet (1989).

3.1.4 MOS HALL PLATES

Hall plates can readily be fabricated using MOS processes. In this case, the Hall plate can be the inversion region below an MOS gate and hence the carrier concentration, and thus the sensitivity can be deliberately modulated. All that is needed is two additional contacts to the well under the gate region, aligned along a line perpendicular to the source-drain axis.

Considering a p-channel enhancement MOSFET example, if the device is not saturated, the channel thickness, t, is essentially uniform and one can relate it to the previous expression for the Hall plate,

$$V_H = \frac{IB}{pqt} = \frac{IB}{\text{charge density}}$$

For a MOSFET, the charge density (per unit area, in C/cm^2) in the channel is given in terms of the gate-source voltage, V_{GS} and, the silicon dioxide thickness, C_{ox} and the threshold voltage, V_T, by,

$$Q_{ch} = C_{OX}(V_{GS} - V_T) \quad \text{in C/cm}^2$$

which yields the expression for V_H in a MOS Hall plate,

$$V_H = \frac{IB}{C_{OX}(V_{GS} - V_T)}$$

While the MOS Hall plate has significant drawbacks (including 1/f noise), the advantage it has over bipolar structures is the ability to modulate the sensitivity with the gate-source voltage. The design shown below can be used to provide Hall sensors along two axes parallel to the substrate surface.

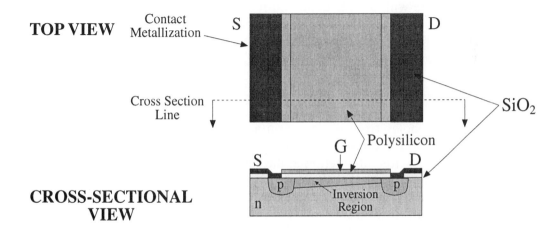

Illustration of a p-channel MOS Hall plate. Adapted from Middelhoek and Audet (1989).

An interesting example of making the MOSFETs on the sloped (111) sidewalls of an etched trench can be seen in Kawahito, et al. (1993a). While this approach does appear to produce very good sensitivity in the direction vertical to the substrate, there is a large added overhead associated with carrying out lithographic steps in the 100 μm deep trenches etched into the silicon substrates in their process.

3.1.5 NON-SILICON HALL PLATE MATERIALS

GaAs is quite useful in the fabrication of Hall effect magnetic sensors (see Mathieu, et al. 1991) and can provide higher temperature performance than silicon (GaAs devices are generally preferred for temperatures above ≈ 150°C, and can operate up to temperatures near 250°C due to the large bandgap of this material). AlGaAs/InGaAs/GaAs heterostructures have also been used to achieve high sensitivities (900 V/(A•T)) with low temperature coefficients (< 2% drift over a 120°C range) by Mosser, et al. (1993). The Hall effect can also be observed in conductive liquids (electrolytes, liquid metals, etc.), but in this case, AC drive signals are preferrable to avoid electrode reactions (see Ogita and Yasuda (1988)).

3.1.6 MAGNETODIODES

In an ordinary silicon diode, the forward current arises from injection and diffusion of carriers, as well as from generation and recombination in the depletion layer, the latter effects generally being negligible. As described by Middelhoek and Audet (1989), diode designs are possible wherein the recombination rates at certain surfaces of the diode are significantly different from those in the bulk of its depletion region (this is known as the *Suhl effect*). If two different surfaces can be arranged on a diode with a wide depletion region, the deflection of carriers by an external magnetic field will result in varying degrees of recombination depending on which surface the deflection is directed toward. This is the underlying principle of a *magnetodiode*.

A typical magnetodiode consists of a p-i-n or p+/n-/n+ structure, typically sandwiched between a sapphire substrate and a silicon dioxide layer. If the diode is forward-biased, holes from the p-region and electrons from the n-region flow through the intrinsic region where some of them recombine at both the sapphire and oxide interfaces. Recombination at the sapphire interface is much greater (on the order of one hundred times) than at the oxide interface. In the presence of a magnetic field, both holes and electrons are deflected in the same direction within the depletion region. If they are deflected toward the sapphire, more of them will recombine, the carrier concentration will decrease, and the resistance of the diode will increase. Conversely, if they are deflected toward the oxide, fewer will recombine, the carrier concentration will increase, and the resistance of the diode will decrease. These changes in resistance manifest as concomitant changes in the diode's forward voltage.

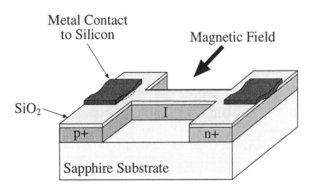

Cross-sectional illustration of a magnetodiode. After Lenz (1990).

An important advantage of a magnetodiode over a Hall sensor is that it is roughly ten times more sensitive; however, the requirement for an SOS or other SOI technology is potentially a drawback. It is also nonlinear and extremely temperature-dependent, as seen by the forward voltage sensitivity, S_{VF}, (Lutes, et al. (1980) and Middelhoek and Audet (1989)) given for fields below 20 mT by,

$$S_{VF} = \frac{dV_F}{dB} = \frac{q(\mu_n + \mu_p)\tau_{eff}(S_{sapphire} - S_{oxide})V_F^2}{8kTL}$$

where,

τ_{eff} = effective carrier lifetime, in s
$S_{sapphire}$ and S_{oxide} = complex functions of recombination rates at the sapphire and oxide interfaces, respectively
V_F = diode forward voltage, in V
k = Boltzmann's constant = 1.38066×10^{-23} J/K
T = temperature, in K
L = length of the intrinsic or n- region, in m
(Note that μ_n and μ_p should be in $m^2/(V\bullet s)$.)

The difficulties caused by this nonlinearity can be mitigated by taking differential measurements from magnetodiodes in opposite orientations, as described in Roumenin (1995), who obtained a sensitivity of 4 V/T with diodes biased at 2 mA at 20°C. Another difficulty that may not be as easy to deal with is the need to carefully control the recombination effects at the two surfaces if repeatable device characteristics are to be obtained in a manufacturing setting.

3.1.7 MAGNETOTRANSISTORS

Magnetotransistors are transistors that are specifically designed to take advantage of the fact that the Lorentz force affects carriers (both types). They can be made in bipolar or MOS technologies and consist of a single emitter (or source) and several collectors (or drains) to pick up the current, depending on where it is deflected by the magnetic field and/or on the modulation of emitter injection.

A huge number of device geometries are possible, such as lateral or vertical magnetotransistors, depending on which current component is primarily influenced by the magnetic field (bipolar devices can be lateral or vertical and MOS devices are generally lateral). An interesting approach is to couple the output of a Hall plate to both bases of a differential pair of bipolar transistors, providing built-in gain. For a detailed categorization and discussion of the various possible structures of magnetotransistors (and other Hall effect magnetic sensors), the reader is referred to Baltes and Castagnetti (1994). Another valuable reference covering lateral magnetotransistors is Ristic (1994).

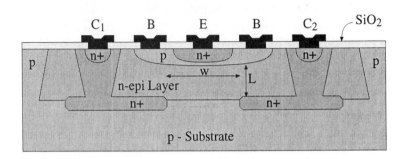

Illustration of a dual-collector vertical NPN magnetotransistor. Adapted from Middelhoek and Audet (1989) (original source: Takamiya and Fujikawa (1972)).

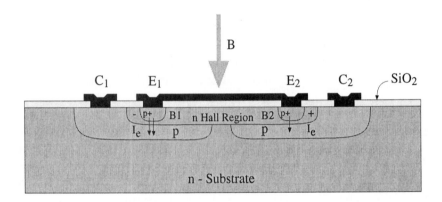

Illustration of a differential PNP magnetotransistor pair. Adapted from Middelhoek and Audet (1989).

In order to improve the sensitivity of CMOS magnetotransistors, Schneider, et al. (1995) demonstrated the use of electroplated Permalloy to form a flux concentrator above the integrated circuit structures. A photosensitive polyimide template was used over a metal/silicon seed layer (150 nm Cr, 500 nm Cu, 150 nm Cr, 200 nm Si). The silicon top layer served to improve photolithography by absorbing UV light, and the top Cr layer protected the Cu plating seed layer from oxidation. RIE etching in an SF_6 plasma was used to remove the top Si layer, followed by a wet etch to remove the top Cr layer prior to Permalloy electroplating, which is followed by polyimide stripping. For low fields (< 1 mT), the sensitivity of the device was greatly increased ($\approx$ 20 times), while at fields > 5 mT, the flux concentrator became saturated.

3.1.8 SPLIT-DRAIN MAGFETS

The *split-drain MAGFET* is essentially a conventional MOSFET with an extra drain. External magnetic fields (only those perpendicular to the substrate) induce a current difference between the two drains by preferentially steering carriers toward one or the other drain.

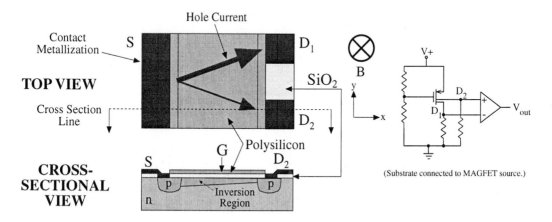

Illustration of a p-channel, split-drain MAGFET. Adapted from Middelhoek and Audet (1989). Note that this device is identical to the MOS Hall plate illustrated above, with the exception that here there are two drains in the same device. The effects of the Lorentz force on the hole current is illustrated. At the right, a simple application circuit for the p-channel MAGFET uses dual drain resistors and a differential amplifier to produce the output signal.

3.1.9 HALL DEVICES IN INTEGRATED CIRCUITS

Since they can be made from silicon, it is possible to integrate circuitry with Hall plates, as illustrated by a commercial example from Allegro Microsystems. Their three-terminal devices incorporate an on-chip differential amplifier and a Schmitt trigger to convert the amplified Hall voltage to a digital threshold output. In addition, the devices incorporate a voltage regulator. Further examples of integrated magnetic sensors can be found in Nakamura and Maenaka (1990).

Typical Hall sensors are housed in small (discrete transistor sized) packages and often can operate up to 150°C, making them suitable for automotive applications such as the sensing of engine timing for ignition control. In some early computer keyboards (and rarely today), Hall switches were used in place of metallic switches, and a magnet was placed in each spring-loaded key, brought into proximity of the underlying Hall switch by its depression. There are a number of other applications for integrated Hall devices, and one can combine multiple devices to get "3-D" magnetic sensors for more complex position-sensing applications.

It is noteworthy that as a means of reducing the effects of offset and drift in such integrated Hall sensors, Simon, et al. (1995) demonstrated the co-integration of a spiral coil with a Hall plate. The coil could be used to generate known calibration fields that can be employed to calibrate the sensor if the background field is constant during calibration. This generally requires a memory and a micro-processor to correct sensor outputs, but one is often available in typical applications.

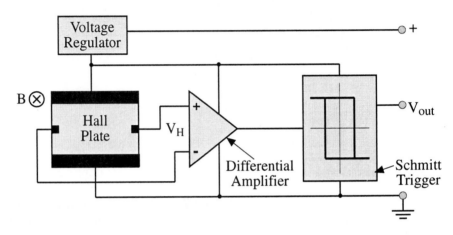

Illustration of a fully integrated Hall switch. Adapted from the Allegro Microsystems databook.

3.2 CARRIER DOMAIN MAGNETOMETERS

Carrier domain magnetometers were invented by Barrie Gilbert (Gilbert (1976) and Manley, et al. (1976)), and are discussed in Popovic and Baltes (1983) and Baltes and Castagnetti (1994); much of this information was drawn from the latter reference. Carrier domains are regions of a semiconductor containing a high density of carriers of equal densities (essentially an electron-hole plasma) confined via deliberate potential gradients or positive feedback. As for Hall effect devices, the Lorentz force from an external magnetic field acts on the moving carriers, shifting the entire domain (or affects its oscillatory movements), which leads to some form of measurable signals.

One form of the carrier domain magnetometer is the vertical, four-layer device demonstrated by Goicolea, et al. (1984). The device, which can readily be fabricated in a standard bipolar IC process, consists of an NPNP structure that can be regarded as an NPN and PNP transistor sharing a common base-collector junction.

Both transistors are maintained in their active regions by external current sources, and the base-collector junctions are kept in reverse biased states. Base currents to the two transistors are provided through dual contacts (labeled I_{BNPN1}, I_{BNPN2}, I_{BPNP1}, and I_{BPNP2} in the illustration). Electrons injected into the base of the NPN transistor exit its collector, which is also the base of the PNP transistor. Holes injected into the base of the PNP transistor exit its collector, which is also the base of the NPN transistor. Due to the lateral voltage drops to the base current contacts, the vertical base-emitter carrier injection occurs in a very small central region, forming the carrier domain. Under an external magnetic field, the movement of the carrier domain is detected via imbalances in the base currents between the left and right sets of contacts, and the domain is constrained from large movement by the built-in restoring forces.

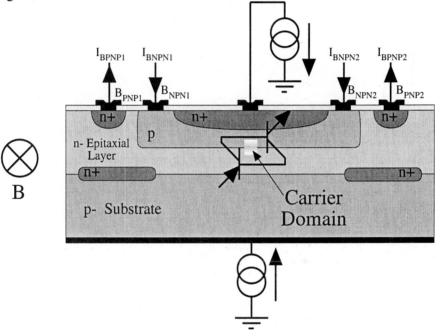

Illustration of a four-layer carrier domain magnetometer. Adapted from Baltes and Castagnetti (1994). The equivalent NPN and PNP transistor structures are shown schematically, and hole currents are illustrated. The carrier domain, shown at the center of the device, is a region of high densities of electrons and holes, moving vertically in opposite directions.

The circular carrier domain magnetometers proposed by Gilbert (1976) are also NPNP devices in which a current domain rotates under the influence of the external magnetic field perpendicular to the substrate (when the external magnetic field exceeds a threshold level, and is high enough to "dislodge" the domain from a preferred location). By monitoring currents at the segmented outer contacts, the position of the carrier domain can be determined. The frequency of rotation of the domain is proportional to the strength of the external magnetic field, and is generally in the range of 10 to 100 kHz/T, with the operating frequency in the range of 1 to 100 kHz. The direction of rotation depends on the sign of the field, and can be sensed using the segmented contacts.

It is not clear that there are many practical applications for carrier domain magnetometers where other magnetic sensors could typically perform equivalent functions. They do not necessarily offer greater sensitivity and often suffer from greater temperature coefficients.

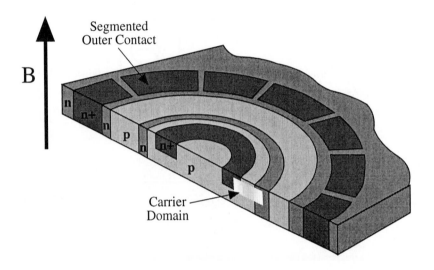

Illustration of a circular, horizontal four-layer carrier domain magnetometer. Adapted from Baltes and Castagnetti (1994). The carrier domain rotates due to an external magnetic field perpendicular to the substrate, causing currents at the segmented outer contacts to vary.

3.3 FLUX-GATE MAGNETOMETERS

The basic structure required to form a *flux-gate magnetometer* is a solenoid coil or transformer (as opposed, for example, to a toroidal transformer), and despite the need for helically wrapped conductors around the core, micromachined versions have been realized. The sensing principle of flux-gate magnetometers is based on the fact that the inductance of a solenoid is related to its magnetic permeability. The magnetic permeability (μ) of the solenoid core is in turn dependent on the magnetic field in a saturable fashion, as illustrated below.

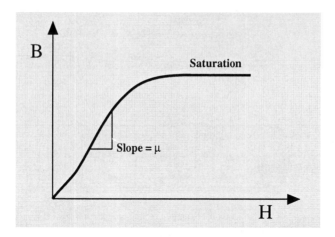

Illustration of the saturable nature of the magnetic flux density, B, with increasing magnetic field strength, H, and the fact that the magnetic permeability, μ, of a material is the slope of a local tangent to the B-H curvze.

For the B-H plot shown, μ is the slope of the curve at a given point. If the device is kept at a magnetic bias (i.e., a coil bias current) near the "knee" of the curve by a field applied by the sensor itself, slight changes in the external magnetic field will cause drastic changes in μ and hence inductance (the permeability is then a function of the sum of the applied and external magnetic fields). For a solenoid, the inductance, L, is given by,

$$L = \frac{\mu_o \mu_r N^2 A}{l} \quad \text{in H}$$

where,
μ_r = relative magnetic permeability of the solenoid core
N = number of turns of conductor in solenoid
A = cross-sectional area of solenoid, in m^2
l = length of solenoid over which conductor turns are arranged, in m

A simple measurement of L by various means (e.g., a tuned RLC circuit) will thus be extremely sensitive to magnetic field changes, and the overall sensitivity can be tuned by adjusting the bias.

Another way to take advantage of this effect is to make a solenoid transformer where the coupling between the primary and secondary coils is modulated by the externally applied magnetic field. If the core is driven away from the small-signal region near the origin by the external field (i.e., increasingly nonlinear), harmonics in the secondary's signal increase and can form the basis for the detection scheme.

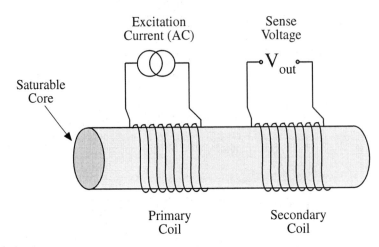

Illustration of the basic structure of an AC-driven flux-gate magnetometer. After Kawahito, et al. (1993b).

Kawahito, et al. (1993b) fabricated micromachined solenoids for flux-gate magnetometers. They described two processes, both involving evaporated aluminum traces under and over an electroplated Permalloy (NiFe) solenoid core. One process was subtractive and the other was additive with respect to the substrate. The subtractive process used a KOH etch to form a flat-bottomed groove in the silicon substrate, which was subsequently rounded off (for ease of photoresist application) using HF:HNO$_3$ (1:3). This was followed by fabrication of the magnetic core. This lithographic step within the groove required electron-beam lithography to pattern the bottom turns of the solenoid windings because the depth of focus must be so great (from the top of the substrate to the bottom of the grooves, 50 μm below). The additive process involved building the entire solenoid above the substrate, as illustrated below.

For both approaches, the lower aluminum turns of the winding were deposited on the thermal oxide above the substrate using lift-off, with PMMA as the electron-beam resist. For the groove structures, two microns of SiO$_2$ were RF-sputtered over the lower aluminum turns to insulate them from the core. For the above substrate version, polyimide was used as the insulator. In both cases, a copper seed layer was then deposited for electroplating the core. Permalloy (NiFe, ideally 81% Ni and 19% Fe) was then electroplated from a custom solution at a current density of 20 mA/cm^2. The Permalloy plating solution used had the following formulation: NiSO$_4$•7H$_2$O (300 g/l), NiCl$_2$•6H$_2$O (25 g/l), FeSO$_4$•6H$_2$O (5 g/l), H$_3$BO$_3$ (15 g/l), Na[C$_6$H$_4$CON]SO$_4$ (2 g/l), and Na[CH$_3$(CH$_2$)$_4$]SO$_4$ (2 g/l). (For further information on Permalloy plating, see the Micromachining Techniques chapter.) It is important to refer to Kawahito, et al. (1993b) to see the current density versus composition (Ni:Fe) curve they derived (20 mA/cm^2 gives roughly 80% Ni). Electroplating was followed by the application of a top polyimide layer, its patterning to form contacts to the underlying aluminum "windings," and the deposition/patterning of the top aluminum "turns."

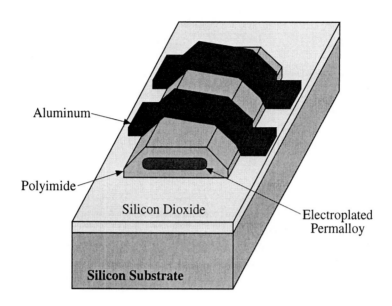

Micromachined, additive solenoid with a Permalloy core. Adapted from Kawahito, et al. (1993b). Note that the solenoid turns are completed by patterned aluminum traces (not shown) above the silicon dioxide and below the polyimide.

A 100 mA (peak) triangle wave was used to excite the 20-turn drive coil at 40 kHz and a sensitivity of 11.5 V/T was measured using the 30-turn pick-up coil (employing second harmonic detection). Extrapolating from the frequency dependence of the core permeability, they speculated that higher frequencies should greatly increase sensitivity (e.g., 1,000 V/T at 10 MHz). Such high frequencies may not be practical in "conventional" flux-gate magnetometers (due to inter-winding parasitics), potentially pointing to a true advantage of the micromachined approach (assuming that the parasitics could be better controlled in the latter case). It is also important to note that power dissipation increases with frequency, regardless of whether or not micromachining is employed.

Conventionally fabricated flux-gate magnetometers are being used in high-volume applications such as "solid-state" compasses (e.g., those found in many current production automobiles), and this might represent an application area for the micromachined devices as well. For the Earth's magnetic flux, on the order of 10^{-4} T, the Kawahito, et al. (1993b) designs could provide only $\approx$ 1 mV full-scale signals for compass applications (probably too small to provide adequate resolution).

In a more recent paper, Kawahito, et al. (1995) reported a two-level aluminum coil process using a polyimide insulator and a NiFeIn electroplated core. They noted that the addition of 0.2 to 1 g/l of $InCl_3$ to their previously reported Permalloy plating bath protects the core from degradation of its magnetic properties during subsequent thermal processing (notably the curing of the polyimide). At an excitation frequency of 3 MHz, they achieved a sensitivity of 2,700 V/T, which was, as they had predicted, greatly increased over their previous results.

A similar miniature flux-gate magnetometer system (the coil is illustrated below), but with complete on-chip CMOS processing electronics (part of a two-chip system, with a few discrete components used) was demonstrated by Gottfried-Gottfried, et al. (1995). As the magnetic core, they used a 460 nm thick electron-beam evaporated NiFe layer (81:19 ratio) coated with 20 nm of Ta above and below it (this deposition process is described in Sauer, et al. (1994). The patterning of the Ta/NiFe/Ta core was done using an aluminum mask based lift-off process. The coated core was located between the upper and lower metallization layers of a standard CMOS process, sandwiched between PECVD silicon dioxide layers. The reported coil resistance was on the order of 5.5 Ω/turn. The reported sensitivity for the system was 9.2 mV/μT over a magnetic field range of $\pm$ 90 μT (the noise level was not given), with a total power dissipation of 190 mW when operated with a 5 V supply (note that this power level is somewhat large for such a magnetometer).

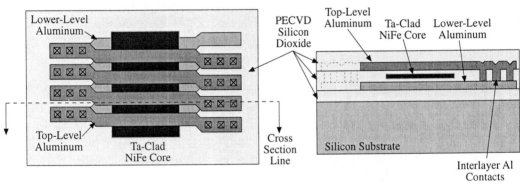

TOP VIEW   CROSS SECTION

Illustration of a process used to fabricate magnetic coils in a standard CMOS process with additional steps associated with the deposition of a Ta clad NiFe core between the two metallization layers. Adapted from Gottfried-Gottfried, et al. (1995).

In terms of non-micromachined flux-gate magnetometers, a good example is the APS533, three-axis flux-gate magnetometer manufactured by Applied Physics Systems, Inc. (Mountain View, CA) and sold for $\approx$ $3,000. This cylindrical device is 18 mm in diameter by 38 mm in length, weighs 18 g, and draws 20 mA from $\pm$ 5 V supply rails (200 mW). Its noise level is less than 100 pT/$\sqrt{Hz}$ and has a full-scale range ($\pm$ 4V) of $\pm$ 100 μT. It appears that micromachined flux-gate devices can be made with reasonable performance for a considerably lower cost if high-volume applications are addressed, but at present, they cannot approach the sensitivities of their conventional counterparts.

3.4 TUNNELING MAGNETOMETERS

As discussed in the Optical Transducers chapter, the use of tunneling trans-duction to measure small motions of micromachined devices is an exquisitely sensitive tool applicable to a variety of sensors. Miller, et al. (1996) used this principle to realize a micromachined tunneling magnetometer with nT resolution. The device was designed for spacecraft applications, where low power and low volume/mass are critical design criteria. In comparison to the more sensitive superconducting quantum interference devices (SQUIDs, discussed below), this design is favored due to its far lower power consumption (SQUIDs require consid-erable power for cooling to cryogenic temperatures).

The devices were fabricated from silicon using bulk micromachining tech-niques. A 2.5 $\times$ 2.5 mm low-stress silicon nitride membrane bearing a tunneling counter-electrode (also used to deflect the membrane) was electrostatically drawn toward the tunneling tip until the closed-loop control circuitry locked onto a pre-defined tunneling current. A three-loop thin-film coil fabricated on the opposite side of the membrane is excited by an oscillating current and generates a Lorentz force given by,

$$F = NL_w IB \sin(\theta)$$

where,

 N = number of loops of the coil

L_w = length of coil segments supported by the flexible membrane (again with care
        to avoid confusing this use of the symbol "L" with that for inductance), in m

 I = coil current, in A

 θ = angle between the magnetic field vector and the coil segments on the membrane

As discussed in the Optical Transducers chapter regarding tunneling Golay cells, the tunneling currents were on the order of 1 nA and the gap sensitivity was less than 100 fm, or 1 mÅ (1 mÅ = 10^{-13} m). With AC excitation (frequencies on the order of 200 Hz were used), the Lorentz force oscillates as well, allowing detection of static magnetic fields by synchronous demodulation (providing great immunity to low-frequency noise). The measured noise equivalent bandwidth

(NEB), corresponding to the minimum detectable magnetic field was 6.6 µT/√Hz. Excluding external circuitry, the tunneling magnetometers were 1 cm × 1 cm × 0.6 mm in size, consumed < 100 mW, had a dynamic range of > 100 dB, and had a bandwidth greater than 10 kHz). The theoretical resolution for such a design, with an increased number of turns and larger bias current, is on the order of 1 nT.

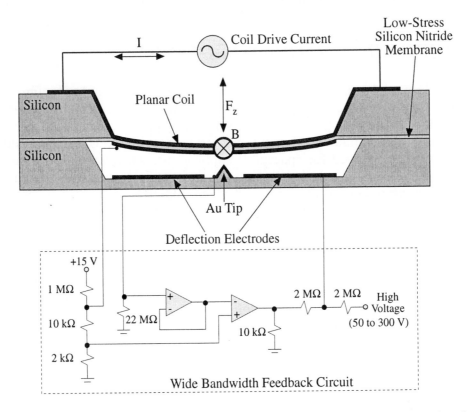

Illustration of a bulk micromachined tunneling magnetometer, showing the closed-loop tunneling current control circuit. Adapted from Miller, et al. (1996).

While few, if any, tunneling transducers have been used in commercial applications to date, this high level of performance implies that they can provide strong competition for such devices as SQUIDs in certain applications.

3.5 TUNNEL DIODE MAGNETIC SENSORS

Barjenbruch (1993) presented a novel magnetic sensor in which the oscillation frequency of a tunnel diode circuit was determined by the magnetic field acting upon its tuning inductor. The prototype was not micromachined, but used a thin strip of mu-metal to form the inductor core (mu-metal is an alloy of Fe, Ni, Cu, and Cr, and has a high magnetic permeability with a low resistivity, making it a popular choice for magnetic shielding). Reported sensitivities were as high as 100 GHz/T. The use of frequency readout can allow for extremely high-precision measurements with simple circuitry (essentially a simple frequency counter). It is conceivable that such an approach could be applied in a micromachined setting, and could provide a means for inherently wireless magnetic field measurement.

3.6 VACUUM ELECTRON MAGNETIC SENSORS

In solid-state Hall effect sensors, the sensitivity is proportional to the carrier mobility, and thus to velocity. In a vacuum, electrons can readily reach velocities of 10^6 m/s, which is more than one order of magnitude faster than in semiconductors. Thus, using the Lorentz force to steer currents in a vacuum device could lead to greatly enhanced sensitivities. A possible device structure is a single emitter (cathode), a control gate, and two symmetrical anodes (analogous to the split-drain MAGFET discussed above). By measuring the current received by each anode, one can detect deflection of the electrons (and hence a current imbalance) by the Lorentz force, which forces the electrons through circular arcs with radii given by,

$$r = \frac{m_e v}{qB}$$

where m_e is the mass of the electron $= 9.10939 \times 10^{-31}$ kg.

Using a small-angle approximation, the distance of lateral electron deflection, y, can be estimated using,

$$y = \frac{qBL^2}{2m_e v}$$

where L is the distance from anode to cathode (not inductance).

Sugiyama, et al. (1993) fabricated such devices, using micromachining techniques. Starting with quartz substrates, a 100 nm tungsten film was evaporated and etched to form the anode region, including a comb-shaped cathode emitter section. A niobium/aluminum bilayer was used to form the gate and anode regions in a shallow (few microns) depression etched into the substrate. The emitter array consisted of 150 tips over a 1.4 mm distance, with an overall die size of 2×2.5 mm. For their design, electron deflections as large as 15 µm were predicted for 0.1 T magnetic flux densities (with L = 100 µm and v = 6×10^6 cm/s).

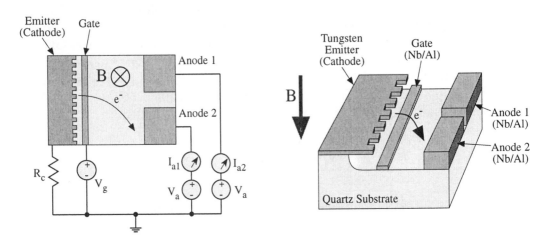

Illustration of the operation of a vacuum electron emission based magnetic sensor, showing the influence of the Lorentz force on the path of an electron (left) and the physical construction of such a device (right). Adapted from Sugiyama, et al. (1993).

They operated the devices with a gate voltage, V_g, of +90 to +105 V and an anode voltage, V_a, of +160V. In a packaged version, they used a top electrode, spaced 1 mm above the quartz substrate, consisting of an indium tin oxide (ITO) coated glass plate, biased at +100 V. They reported that the current (on the order of 20 nA total) could be fully differentially swung between anodes by a total magnetic flux density of 120 mT peak-to-peak. Estimated from published data, the current varied with a sensitivity of $\approx$ 100 nA/T.

Williams and Muller (1997) demonstrated a similar magnetic sensor that used a hot tungsten filament ($\approx$ 2,800 K) as the electron source. Since the hottest region of the filament was in the center, the emitted electrons were concentrated into more of a "beam" shape than those from the comb-shaped emitter of Sugiyama, et al. (1993). Reported sensitivity for the device was 65 nA/G (650 µA/T).

It is interesting to observe that this approach may have applications in areas where the extra sensitivity is important, but it does not warrant the use of more "exotic" sensors. The need for relatively high-voltage supplies (> 100 V) and vacuum packaging may well be balanced by the simple, low-cost fabrication methods that can be used. It should be noted, however, that some magnetotransistor designs, and certainly tunneling magnetometers, can and have reached comparable sensitivity levels. Also, and most importantly, a key performance parameter (often difficult to obtain from the sensor literature) is the signal-to-noise ratio (SNR).

3.7 SUPERCONDUCTING QUANTUM INTERFERENCE MAGNETOMETERS

Superconducting quantum interference devices, or SQUIDs, are by far the most sensitive of all magnetic sensors (reviewed in Clarke (1980)). At present, they are the only viable sensors for detecting the magnetic fields generated by minute ionic currents in the brain, those fields on the order of 10^{-9} T.

In small, superconducting rings, persistent currents can flow, but because of the periodic boundary conditions, only certain allowed modes can be sustained. A circulating current will produce a magnetic flux through the loop, and if the currents are quantized into specific modes, the magnetic field must also be. It is quantized into units of,

$$2\pi\Phi_{int} = 2\pi n\Phi_o = LI_{cc}$$

where,

n = quantum number, an integer

Φ_{int} = internal magnetic flux, in Wb (Webers)

$\Phi_o = \dfrac{h}{2q} = 2.07$ fWb, the magnetic flux quantum

h = Planck's constant = 6.626076×10^{-34} J•s

L = self-inductance of the loop, in H

I_{cc} = circulating loop current, in A

If an external magnetic flux passes through the ring, the circulating current will shift through the allowed modes (changes in the quantum number, n) so as to minimize the net flux $\Phi_{ext} + \Phi_{int}$, in accordance with Lenz's law (which states that when the flux through a circuit is changed, a current will be set up in a direction to oppose the change). Thus the shift in n is an indication of the external magnetic flux, measured in units of the magnetic flux quantum, Φ_o.

The sensitivity of a superconducting loop to a given flux density is thus determined by the size of the loop. For example, a loop of area 1 cm^2 = 10^{-4} m^2 has a quantum sensitivity limit of approximately,

$$\frac{\Phi_o}{A} = \frac{2.07 \times 10^{-15} \text{ Wb}}{10^{-4} \text{ m}} = 2.07 \times 10^{-11} \text{ T}$$

Still greater sensitivities can be obtained through the use of larger area rings, but the superconducting ring itself cannot be made much larger and still support quantum coherence. Thus flux is coupled into the superconducting ring using a low-resistance (but not necessarily superconducting) pick-up coil to form a transformer. This greatly increases the effective area of the superconducting ring and thus magnifies the effective flux density in it.

Given that this sort of device has a circulating loop current that jumps between modes in response to very small changes in magnetic fluxes, a very sensitive magnetic sensor can be realized if the small changes in the circulating ring current can be detected without breaking the loop (and thus removing the periodic boundary condition that gives rise to the quantized internal flux in the first place). In order to accomplish this, one or two *Josephson junctions* are introduced into the superconducting ring (see Fulton, et al. (1972)). Such junctions are created by separating the two superconducting paths with a thin, tunneling insulator, usually an oxide of 5 to 20 nm thickness. Electrons can easily tunnel through this layer, but their wave-function phase (wave-function definition of electrons from Schrödinger's equation) becomes perturbed and matched to the relative position of the junction in the loop and the existing value of loop current (i.e., the Josephson junction defines the phase of the electron wave function at its location).

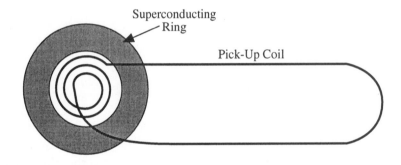

Illustration of the use of an external pick-up coil to increase the effective area of a superconducting loop in a SQUID. Courtesy Prof. R. B. Darling.

If a DC bias current is passed through a superconducting ring with two Josephson junctions, its value will jump with increases in the magnetic flux in order to satisfy the two conditions of flux minimization and resonance of the electron wave functions to the oxide insulator barriers (an integral number of wavelengths must fit into the closed path). The bias current, I_{bias}, splits into two components, I_1 and I_2, each of which must satisfy resonance in the ring, and their difference must be such so as to minimize the next flux through the ring. This results in a response of I_{bias} versus changing magnetic flux as illustrated below, with periodic interference peaks. By counting the interference peaks, one can determine the change of external magnetic flux in units of Φ_o. Because the output is multivalued, an absolute measurement of flux is not entirely simple, as for any interferometric sensor.

SQUIDs have been available for many years, as have Josephson junctions. They have always been made using microlithography since the rings must be very small to support quantum coherence. As for Josephson junctions, SQUIDs originally used the available type I and type II superconducting metals, requiring cooling to less than ≈ 20 K. Type I superconductors are usually pure elements, while type II are typically alloys, with type II materials exhibiting higher critical temperatures, T_C. An example of a type I material is Pb, with $T_C = 7.2$ K and an example type II material is Nb_3Sn, with $T_C = 18$ K. Most practical superconductors are type II.

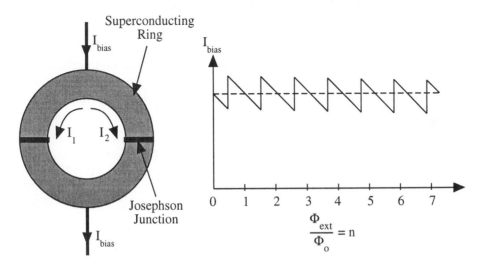

Illustration of the use of a split bias current and two Josephson junctions in a SQUID to enable readout of changes in external flux via changes in the bias current, I_{bias}. Courtesy Prof. R. B. Darling.

With the discovery of the superconducting rare-earth oxides, such as $YBa_2Cu_3O_{7-x}$ (type II), they must only be cooled to ≈ 100 K, allowing considerable growth in this area in the last decade or so. Films such as $YBa_2Cu_3O_{7-x}$ are usually sputtered, a technique that has enabled nearly all of the high-T_c superconductor applications (for further information, general references on superconductors are van Duzer and Turner (1981) and Orlando and Delin (1991)).

3.8 OTHER MAGNETIC SENSORS

3.8.1 WIEGAND WIRES

Wiegand wires (Wiegand (1975) and Hauptmann (1991)) are ferromagnetic wires, typically ≈ 250 μm in diameter, made by cold-forming an alloy called

Vicalloy, which is a combination of Co, Fe, and V. The magnetic domains in the core and the cladding (referred to as Weiss zones), can be reversed by an external magnetic field, causing a voltage pulse in a pick-up coil placed near or around the wire. These devices are used in rotary and linear position sensors and in coded card readers. While no micromachined Wiegand wires have been reported, there is certainly the potential to employ this principle in microscale devices.

3.9 BIOLOGICAL MAGNETIC SENSORS

In 1975, it was discovered (Blakemore (1975), Blakemore and Frankel (1981) as described in Bean (1993), and Moskowitz, et al. (1988)) that certain classes of *anaerobic* bacteria (growing in regions of low oxygen concentration) would orient toward to the Earth's magnetic field. These motile (capable of moving themselves) organisms, *Aquaspirillum magnetotacticum*, were seen to swim consistently north, or toward the north pole of a magnet. Using electron microscopy, it was discovered that each of them contains chains of ≈ 50 nm diameter particles of magnetite that is permanently magnetized (synthesized by the bacterium itself). Thus the entire 3×1 μm diameter organism acts like a compass needle. This tendency to swim toward a magnetic pole is referred to as *magnetotactic* behavior, and in this case, it is a passive magnetic torque orienting the organism.

It has been theorized (and there is some experimental evidence in support of this) that higher organisms such as bees, sea turtles, trout, salmon, and pigeons also possess magnetic senses. Magnetite has been found in all of these animals, but this does not prove that they use it for navigation. As explained in Wickelgren (1996), researchers have used strong external electromagnetic pulses to reorient the magnetization of magnetite in such organisms. Bees trained to fly north, once exposed to such pulses, flew south. Preliminary results indicate that certain neurons in rainbow trout have responses to changes in magnetic fields, and tracing their pathways led to magnetite-containing cells in the snout regions of the fish. Experiments with head-mounted magnetic field coils have successfully reversed the flight paths of homing pigeons. It has also been theorized that in some organisms, such as newts and fruit flies, certain visual pigments, when excited by appropriate wavelengths of light, become paramagnetic and hence magnetically sensitive. Developments in the area of animal magnetic senses are not only interesting, but may have impact on the design of man-made sensors as well.

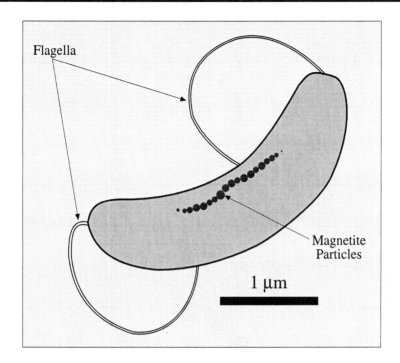

Illustration of a magnetotactic bacterium showing the internal chain of permanently magnetized magnetite particles (≈ 50 nm in diameter) and the organism's two flagella. The illustration was prepared from a transmission electron micrograph (after Bean (1993)) and is approximately drawn to scale. Note that the other internal details of the bacterium are not shown.

4. MAGNETIC ACTUATORS

4.1 MAGNETIC FIELD ACTUATORS

A current-carrying conductor will generate a magnetic field. Two such conductors, with parallel current directions are attracted (as shown below), and they are repelled with antiparallel currents. Similarly, a current-carrying wire (or coil) can generate a magnetic field that interacts with an external magnetic field from a magnet or distant coil to produce a mechanical force. Advantages of (electro)magnetic force generation include relatively high forces that are achievable and the ability both to attract and repel. Disadvantages include the potential for high power consumption and unintended interactions of the generated magnetic fields with nearby objects (such as moving charged particles or magnetic data storage media).

As described above, two lengths L of wire (being careful not to confuse the variable L with its other use for inductance) with parallel currents will experience a repulsive magnetic force given by Ampere's law,

$$F_{ba} = ILB = \frac{\mu_o \mu_r L I_a I_b}{2\pi d} \quad \text{in N}$$

where,

I_a and I_b = currents in wires a and b respectively, in A

    B = magnetic flux density, in T

    μ_r = relative permeability of the material between the wires

    d = distance between wires, in m

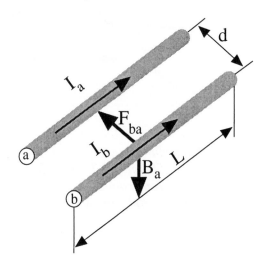

Illustration of parallel currents generating an attractive force between two wires.

It should be noted that a "U"-shaped piece of wire with a current flowing through it will thereby generate a repulsive force between the two straight wires. Similarly, a flexible wire (as illustrated below) or a current-carrying conductor and supporting platform (which has freedom to move) will deflect in a magnetic field.

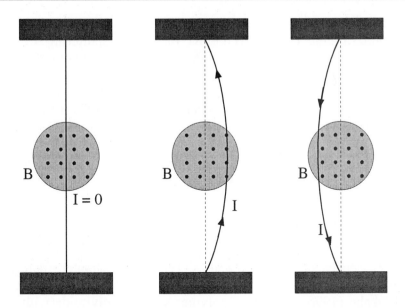

Illustration of a flexible wire unaffected by an external magnetic field (pointing out of the page) with no current flowing, and deflected right or left depending on the direction of current flow. Adapted from Halliday, et al. (1993).

4.2 MAGNETOSTRICTIVE ACTUATORS

As discussed in Hunter and Lafontaine (1992), Joule discovered in 1840 that a nickel rod would contract when a longitudinal magnetic field was applied. This is the *magnetostrictive* effect, and is due to magnetic domains in the material aligning themselves with the external magnetic field and causing a change in physical dimensions of the material.

Magnetostriction in nickel is capable of producing strains of -0.033×10^{-3}, and in some rare-earth metal alloys can generate much greater strains. For example, the terbium-iron and dysprosium-iron alloys, $TbFe_2$ (Terfenol) and $DyFe_2$ can generate peak magnetostrictive strains of -2.46×10^{-3} and 1.26×10^{-3}, respectively. Also, Terfenol-D, a mixed alloy of $Tb_{0.27}Dy_{0.73}Fe_{1.9}$ can generate peak strains of 2.0×10^{-3}. These strains are larger than those available with piezoelectrics, and if suitable processes can be developed to merge such magnetostrictive materials with micromachined structures, the generation of the magnetic fields required for actuation should not be difficult with monolithic or external coils.

Quandt, et al. (1997) discussed the fabrication of giant magnetostrictive multilayers where magnetostrictive materials (such as $Tb_{0.4}Fe_{0.6}$) are interleaved with materials with large magnetic polarizations (such as Fe or $Fe_{0.5}Co_{0.5}$) to enhance magnetic polarization at low field strengths. The multilayers were prepared by DC magnetron sputtering the TbFe films and RF magnetron sputtering the Fe or FeCo layers, followed by annealing at 300°C (un-annealed films were under 250 to 300 MPa compressive stress). The multilayers did produce an increase in magnetostriction at low fields, at the expense of the maximum magnetostriction available at the point where the effect saturates. Preliminary demonstrations of these multilayers was accomplished through the fabrication and testing of a diffuser-type fluidic pump and a thin-film linear motor, both operated using external magnetic fields. An additional reference with some relevance to micromachining applications of magnetostrictive actuators is Shearwood, et al. (1995).

4.3 MAGNET FABRICATION

As discussed in the Microfabrication Techniques chapter, there are several ways to fabricate thin-film permanent magnets, including multi-target sputtering (Araki and Okabe (1996), loading a polymer such as polyimide with a magnetic powder (Lagorce and Allen (1996)), and electroplating (Liakopoulos, et al. (1996)).

4.4 MAGNETIC ACTUATORS WITH DRIVE COILS

Various approaches have been taken to fabricate coils for magnetic actuators (or sensors, as discussed above in the flux-gate magnetometer example). Lithographic methods have been used to fabricate coils, although this approach severely restricts the number of turns possible per unit length, as compared to macroscopic coils in which multiple layers of insulated wires can be laid on top of each other. In addition, the typical use of thin-film interconnects for part of each turn (usually the pass-throughs beneath the core) limits the maximum current that can be used. Wire-bond loops have been used to form the tops of the turns, but these limitations still apply. If the generated forces scale appropriately, however, drive current limitations may not be a major issue.

In order to lessen the effects of these limitations but to retain some of the scaling advantages of micromachining, coils have been formed using conventional magnet wire wound onto microfabricated cores. While considerably more force can be generated with this approach, the resulting devices are not monolithic and require considerably more manual (or robotic) interaction in winding and assembly steps. Nonetheless, the long-established technology to produce miniaturized coils cannot be discounted as a serious competitor to micromachined devices. Fully monolithic and "hybrid" coils are discussed below, with the focus on specific methods for realizing monolithic versions.

4.4.1 ELECTROPLATED MAGNETIC COILS

Ahn and Allen (1992) described a process to fabricate coils for magnetic actuators based on polyimide mold and spacer layers, electroplated metal core and conductors, and metallic sacrificial layers. The "meander" coil structure described made use of a planar conductor in a serpentine path intertwined with a two-level magnetic core.

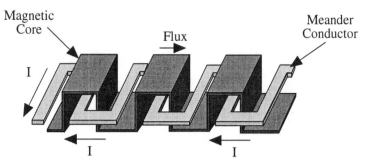

Illustration of a meander-type magnetic coil fabricated with a planar metal conductor and a two-level magnetic core. Adapted from Ahn and Allen (1992).

Their fabrication process for the meander coils began with the sputtering of a 200 nm Ti plating seed layer onto oxidized silicon wafers and multiple spin-coats of polyimide (DuPont PI-2600) with 10 min/120°C soft-bakes between coats and a final 1 h/350°C hard-bake. Following the deposition and patterning of an aluminum mask, the polyimide was etched using a CF_4/O_2 plasma (4% CF_4) to expose the underlying seed layer. Permalloy was electroplated to form the lower level of the magnetic core, using the plating formulation given in the Micromachining Techniques chapter, and plated up to the level of the polyimide template to achieve reasonable planarity. (The reader is also referred to the Permalloy plating formulation used by Kawahito, et al. (1993b) and described above.) Another polyimide layer was spun on to insulate the underlying core from the meander conductor to be deposited next. For this purpose, they used either a 5 μm sputtered/wet etched Al conductor or a 5 μm Cu layer electroplated above a 200 nm Cu seed layer using a 5 μm photoresist template (their Cu plating formulation is given in the Micromachining Techniques chapter).

Another layer of polyimide was then spun on, followed by the use of an Al mask and a 100% O_2 plasma to pattern the polyimide stack all the way down to the magnetic core base layer previously deposited. Due to exposure to the plasma etch, a thin oxide had formed on the Permalloy, and it was removed with a 30 s, 2% HF etch, and the vertical sections of the magnetic core were then electroplated within the newly etched openings (using Permalloy). A top layer of Permalloy was then electroplated (using a photoresist template) to complete the magnetic core. Bond-pad regions were formed using a 100% O_2 plasma to etch through the polyimide.

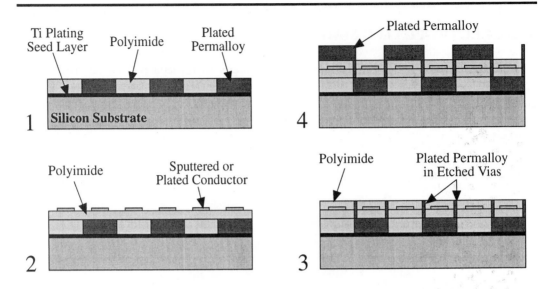

Illustration of a process for fabricating meander-type magnetic coils using multiple polyimide layers as both a plating template and an insulator/spacer. Adapted from Ahn and Allen (1992).

It is interesting to note that Ahn and Allen (1992) used such coils to drive cantilever magnetic actuators without removing the underlying Ti plating seed layer (this layer was not in contact with any current carrying conductors). They reported that if this were desired, it could be accomplished by O_2 plasma etching down to the seed layer and then etching it with a 90% CF_4/O_2 plasma. More recently, Ahn, et al. (1993) described another integrated inductor fabricated using a similar approach. In addition, a description of a low-frequency, two-layer spiral inductor/electromagnet (primarily intended for use in magnetic actuators), completely encased in a NiFe magnetic core, can be found in Ahn and Allen (1993).

Löchel, et al. (1995) demonstrated a thick-photoresist-based process that could achieve template thicknesses of up to 100 μm and aspect ratios up to 10:1. They described an assortment of coil structures based on electroplated NiFe cores, electroplated Au coils, and Cu sacrificial layers.

Such coils can be used directly as actuators. For example, the use of an array of two-layer electroplated Cu coils to form a two-dimensional micropositioning device was discussed by Nakazawa, et al. (1997). The principle employed was to use permanent magnet "carriers" to move small parts put on them. They fabricated an array of such coils over a 40 × 40 mm area of silicon, and were able to levitate and move NdFeB magnets and on-board objects with total masses up to 1.2 g with velocities up to 30 mm/s.

Using the fabrication techniques described above, or alternatives, it is possible to fabricate on-chip transformers and inductors, which may have circuit applications (e.g., filters and switching power supplies). At present, there does not appear to be an analysis in the literature comparing the relative costs and performance parameters for integrated and discrete inductors. The fabrication of on-chip inductors optimized for electronic applications is discussed below in Section 5.1.

4.4.2 WIRE-BONDED MAGNETIC COILS

Christenson, et al. (1992) reported the use of wire bonding to form the tops of the spiral turns of microfabricated magnetic coils with LIGA-fabricated cores and underlying conductors. They used conducting metal patterns to form the bottoms of the turns. These were covered with an insulating layer and their ends were exposed by patterning the insulator. Vertical "sides" of each turn, as well as the magnetic core were fabricated using LIGA (electroplated Ni). The tops of the turns of the coils were formed by wire bonding between the Ni "sides" of each turn. This approach was used to fabricate, among other structures, a toroidal transformer ≈ 5.25 mm in (outer) diameter. This approach was later employed to fabricate magnetically driven actuators capable of generating forces on the order of 10 mN. Despite a tendency to dismiss such processes as impractical for manufacturing, if the number of turns formed using wire bonding is small, the relative cost of fabrication (using high-speed wire bonding machines) may not be prohibitive.

It should be noted that wire-bonded magnetic coils have been used for some time in signal isolation applications (discussed in Section 5.1 below). For example, Burr-Brown Corporation (Tucson, AZ) manufactured the model 3656 analog isolation amplifier, which is a hybrid device. A toroidal core (≈ 1 cm in diameter) was mounted on a ceramic substrate and conventional wire bonding was used to form seven separate windings (for further information, see the Burr-Brown Handbook of Linear IC Applications, 1984). Such an approach is also applicable to the manufacture of pulse isolation transformers and other similar devices.

4.4.3 EXTERNALLY WOUND MAGNETIC COILS

In a paper arguing against micromachining electromagnetic coils for actuation applications, Guckel, et al. (1995) demonstrated coils that were manually wound around LIGA-fabricated cores. Their previously fabricated integrated coils had only 21 turns, with a resistance of ≈ 0.2 Ω/turn, and the low number of turns that were achievable greatly limited the maximum forces that could be generated. With their newer designs, the manually (or automatically) wound cores were fabricated with a Permalloy composition of 78% Ni and 22% Fe (coercivity < 0.3 Oe, saturation flux density ≈ 1 T, and $\mu_r \approx 2,000$). The cores were wound with 50 gauge magnet wire (≈ 25 μm in diameter), to achieve between 500 and 1,000 turns over their lengths of 3 mm. With a resistance of < 0.1 Ω/turn, this yielded coils that (limited by the maximum temperature sustainable without degradation of the magnet wire insulation) could withstand currents in the range of 50 to 100 mA.

With this approach, they demonstrated a LIGA-based linear actuator with inherent position feedback capabilities. The self-inductance of the moving plunger was shown to vary linearly with displacement with a factor of 1 μH/μm (readily measurable). The maximum travel achieved was 500 μm. A current on the order of 40 mA resulted in a force of 1 mN. In addition, this approach was used to fabricate electromagnetic coils for use in a microscale electromagnetic dynamometer, as described in Christenson, et al. (1995).

Extending the use of LIGA-based magnetic actuators with externally wound coils into centimeter-scaled mesoscopic machines, Garcia, et al. (1997) demonstrated a stepped, rotary motor with mN-level torque. Permalloy structures were manually assembled to form the completed motor assembly. How this device compares to conventional electromagnetic motors in this size range was unfortunately not addressed.

This approach may prove effective in other micromachined magnetic applications, and is an example of a case wherein lithographic fabrication of an entire transducer may not be the optimal approach. Given that the wristwatch industry has developed a wide variety of millimeter-scale coils for low-power actuation, it is likely that some potentially useful technologies for micromachined magnetics could be found in that field.

4.5 MAGNETIC ACTUATORS USING EXTERNAL FIELDS

Judy, et al. (1994) demonstrated large throw magnetic actuators using an electroplated NiFe alloy layer atop released polysilicon islands with flexure supports. While no specific function of the demonstrated actuators was given, the process is of considerable interest since it is an extension of a well-known polysilicon surface micromachining approach (see Tang, et al. (1990)).

The magnetic material part of their process began with prefabricated two-layer polysilicon microstructures with a PSG sacrificial layer. Adhesion and plating seed layers (10 nm Cr and 100 nm Cu, respectively) were deposited over the entire wafers by evaporation, followed by the deposition of a thick photoresist plating template layer. (It should be noted that the authors chose to deposit four layers of Olin Hunt 6512 resist for a 10 μm final thickness, although several resists such as Hoescht AZ4620 can be used to obtain the same thickness in one spin.) Permalloy was electroplated to a thickness of 7 μm, followed by the application and patterning of another photoresist layer to protect regions where the Permalloy would remain following a wet etch to pattern it (details not provided). After the Permalloy was patterned and the photoresist removed, the Cr adhesion layer was wet etched and the polysilicon structures were released in a concentrated HF etch and via electrically "blowing" polysilicon fuses (used to keep the actuators from moving during the release step).

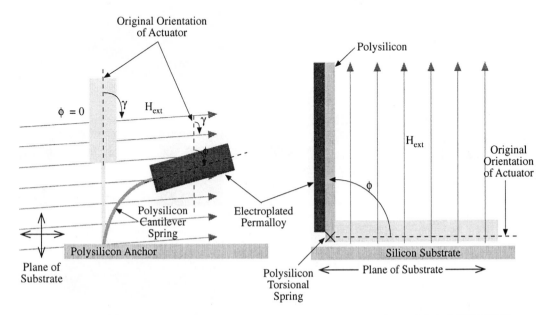

IN-PLANE ACTUATOR   **OUT-OF-PLANE ACTUATOR**

Illustration of polysilicon-based magnetic actuators, showing (left) an in-plane actuator (shown from above) and (right) an out-of-plane actuator (with a torsional spring, shown from the side). Adapted from Judy, et al. (1994) and Judy and Muller (1995), respectively.

The demonstration devices were used to show that movement of the magnetic region along a spiral path was possible with an external magnetic field whose angle was varied with respect to the original orientation of the actuator (with no field). The external magnetic field induced magnetic poles in the Permalloy region, which in turn interacted with the external field, resulting in reorientation of the actuator until it was aligned with the magnetic field (as for a compass needle). For a $400 \times (40 \text{ to } 47) \times 7$ μm Permalloy plate on a $400 \times (0.9 \text{ to } 1.4) \times 2.25$ μm polysilicon cantilever, 90° deflections were achieved for a magnetic field strength of $H_{ext} \approx 25$ kA/m. Deflections of > 180° were demonstrated prior to fracture of the cantilever springs.

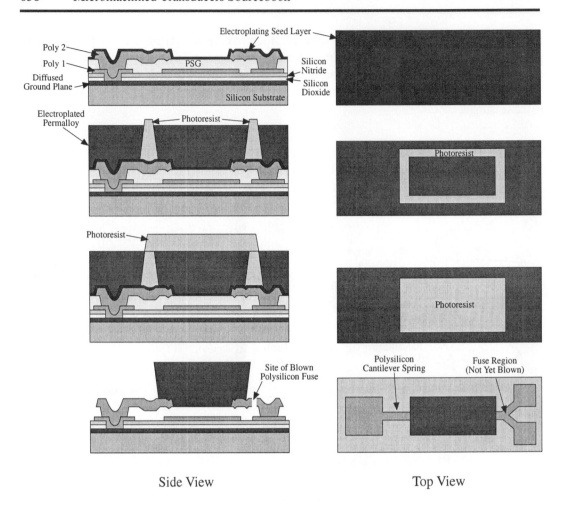

Side View   Top View

Illustration of a process for adding electroplated Permalloy to polysilicon surface micromachined structures to form magnetic actuators. Adapted from Judy, et al. (1994).

Judy and Muller (1995) demonstrated the use of a similar process to fabricate actuators with torsional supports, where the electroplated regions of the actuators could rotate out of the plane of the wafer. For a Permalloy plate with dimensions $430 \times 130 \times 15$ μm, with dual torsional beams $400 \times 2.2 \times 2.2$ μm in size, they achieved 90° deflection relative to the substrate plane for a magnetic field strength of $H_{ext} \approx 10$ kA/m, with a resulting torque > 3.0 nN•m. The out-of-plane devices were intended for use in microphotonic applications such as the chopping, scanning and steering of optical beams. Judy and Muller (1996) further reported the addition of on-chip spiral actuation coils and the use of electrostatic hold-down to allow individual torsional actuators in an array to be held down during the application of an external magnetic field.

For the in-plane and out-of-plane actuators, respectively, Judy, et al. (1994) and Judy and Muller (1995, 1996) provide very useful analyses of the magnetic forces generated and should be referred to for further information.

Another interesting demonstration of external magnetic fields to drive micro-actuators was the manually assembled "micro flying machine" demonstrated by Arai, et al. (1995). They reported a composite device with a pair of polyimide wings (10 mm × 10 mm × 7.5 µm), a pair of magnetic wings (10 mm × 2 mm × 8 µm of 4 µm thick Fe magnetic powder on a 7.5 µm organic substrate), polyimide hinges, and a 20 mm × 50 µm diameter Co-Fe-Si-B wire as a vertical stabilizer. With a total mass of 5.3 mg, the device was able to lift itself vertically using a 12 Hz-modulated external magnetic field of 400 Oe (0.04 T).

Liu, et al. (1995a) used an array out-of-plane "flap" actuators similar to those of Judy and Muller (1995) (with the notable difference that they used bending cantilevers, rather than torsional beams) to move small objects using ciliary motion. Their flaps consisted of 1 mm × 1 mm × 1.8 µm polysilicon plates, with 5 µm thick NiFe (80:20) electroplated over them, and with 100 or 400 µm long, 1.8 µm thick polysilicon flexural beams. With an external magnetic field strength of 20 kA/m, the maximum lifting force for these actuators was ≈ 87 µN each. They demonstrated fixed-pattern motion of silicon and glass chips over their 1 × 1 cm actuator array. This group also describes the use of such flaps to potentially modulate the vortices at the leading edge of a delta-wing aircraft (Liu, et al. (1995b)).

It is interesting to note that despite the large number of symmetrical torsional electrostatic actuators (i.e., capable of rotation in both directions about a torsional axis, such as the torsional micromirror devices discussed in the Optical Transducers chapter), it appears that as yet no magnetic actuators of this type have been micro-machined (e.g., a coil on the torsional platform, inducing rotation in an external magnetic field).

4.6 MAGNETIC MICROMOTORS

Magnetic motors have been aggressively scaled using "conventional" fabrica-tion technologies. Several efforts have also been made to micromachine them. Guckel, et al. (1991, 1993a, 1993b) reported the use of wire-bonded magnetic coils and LIGA fabrication to realize a magnetic micromotor with built-in photodiode-based rotation sensors in the substrate. A 0.6 µm layer of SiO_2 was thermally grown on silicon wafers, followed by an oxide etch, implant, and drive-in to define the photodiodes. Metallization for the coil feed-throughs and interconnects was then deposited (30 nm Ti and 4 µm Ni) and patterned, followed by the deposition and patterning of a 2 µm silicon dioxide layer, forming an insulating layer. A 30 nm Ti, 30 nm Ni layer was then sputtered as a plating base, and LIGA was carried out with Ni plating to form the coil frames, shafts, and bond pads. The stator poles and rotors were separately fabricated using LIGA and assembled into the final motors, after which wire bonding was used to close the stator coils. Rotor/shaft tolerances on the order of 0.5 μm and rotor/stator air gaps on the order of 3 μm were achieved.

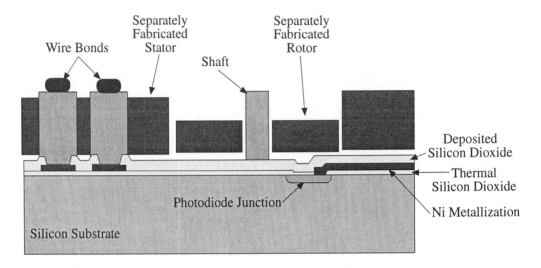

Illustration of the cross section of a magnetic micromotor, showing the separately fabricated and assemble stator and rotor in conjunction with the monolithic coil frame, rotor shaft, and integrated photodiode for rotation/position sensing. Adapted from Guckel, et al. (1993a).

Eighteen windings per pole were used, with two poles for each of the three phases. For a maximum current of 0.6 A per coil, this resulted in a maximum magnetomotive force of 10.8 A•turns. Torques on the order of 3 nN•m were achieved, and could theoretically be increased to over 100 nN•m if a Permalloy core were used (with its higher saturation flux density). Due to magnetic reluctance effects, the rotors were levitated above the substrate, minimizing friction. For 285 μm diameter rotors, stable rotation rates of 8,000 rpm were demonstrated, and could be pushed as high as 33,000 rpm. The motors were apparently tested for up to 5×10^7 rotation cycles without changes in characteristics (it should be noted that at 8,000 rpm, this corresponds to only 4.3 days). While not directly addressed in this paper, the rotors constitute a very small fraction of the overall dimensions of the motor ($\approx 3 \times 3$ mm), which is a consequence of this type of planar design.

Watanabe, et al. (1995) demonstrated a magnetic motor with overall dimensions of $\approx 2 \times 2 \times 2$ mm, with a 1 mm diameter rotor formed from 25 μm of sputtered NdFeB magnetic film on a 100 μm thick FeSi alloy substrate. The rotor was held over a polyimide template, electroplated Cu coil with three independent phases. The resulting motor was operated at 25,000 rpm with an input power of 60 mW (at 0.1 A current, for a coil resistance of 6 Ω). The output power developed by the motor (and hence its efficiency) was not reported.

At present, it remains to be seen if micromachined magnetic motors will have any large-scale commercial impact. They face stiff competition from millimeter-scale motors fabricated by non-lithographic manufacturing techniques, such as those used in electromechanical wristwatches.

4.7 EDDY-CURRENT DEFECT SENSORS

It is possible to locate defects in metallic structures such as pipes using eddy-current sensing. The basic principle is to apply an AC magnetic field to a coil in proximity to the structure being tested. Eddy currents are induced in the metal, and the magnetic fields that they in turn induce couple back into the coil. These reverse-coupled fields oppose those produced by the coil. The effect of this is that the coil's net inductance (typically measured as its impedance) varies with the eddy currents. The defect detection aspect of this is that the eddy currents are distorted by local defects that become apparent particularly when the probe coil is moved relative to the defect.

As described in Fraden (1993), the depth within the object under test to which eddy currents can be produced is given by the formula for electromagnetic skin depth,

$$\delta = \frac{1}{\sqrt{\pi f \mu_r \mu_o \sigma}}$$

where f is the excitation frequency, in Hz, and σ is the conductivity of object under test, in $\Omega^{-1} \cdot cm^{-1}$.

Thus if large depths are to be examined, low frequencies are advantageous. Typically, commercial sensors operate in the range of 50 kHz to 10 MHz and have impedance versus distance relationships that are quite nonlinear. To inspect small structures, such as narrow pipes, it would be desirable to have a micromachined eddy-current sensor that would not only fit, but would also be able to get fairly close to the surfaces being tested.

Hamasaki and Ide (1995) demonstrated a 3 mm $\times$ 3 mm $\times$ 80 μm micromachined eddy-current sensor based on a four-level spiral coil made of electroplated copper (cross-sectional dimensions of 10 μm $\times$ 10 μm) sandwiched between layers of polyimide built up on a silicon substrate. (It should be noted that conventional probes on the order of 2 to 3 mm in diameter already exist, but that the micromachined devices offer the possibility of being scaled down substantially below this.) Their device, with an inductance on the order of 90 µH, could be used to frequencies as low as 1 kHz, and was shown to detect defects as small as 1 mm in diameter (a resulting change in inductance of 8% was observed).

It is likely that this type of sensor can be scaled down much further, as well as potentially fabricated on flexible substrates so as to conform to the interiors of small pipes and other structures.

4.8 RESONANT MAGNETIC SENSORS

An interesting micromachined magnetic field sensor based on a magnetically deflected piezoresistively sensed resonator was presented by Donzier, et al. (1991). A single-crystal silicon beam was used as the resonant structure, with a 36-turn aluminum magnetic coil fabricated on its surface using lift-off methods. A p-type diffusion was used to form piezoelectric sensors on the beam, the shape of which was defined using a shallow top-side plasma etch and subsequent back-side thinning in the flexing region (method not given). An alternating frequency current was passed through the coil, and amplitude of the vibrations of the cantilever were proportional to the external magnetic field (against which the field of the coil would act via the Laplace force). They reported 10 nT sensitivities for a 10 Hz band, using a 2×4 mm resonant beam 30 µm thick.

A version of such a resonant magnetic field sensor was fabricated using XeF_2 etching to undercut a SiO_2 plate (with on-board current loops) one a die fabricated using a standard 2 µm CMOS process (Eyre and Pister (1997)). Piezoresistive sensing was used for detection. Considerable heating of such structures by the coil drive current is possible, since they are very well thermally isolated. However, the authors observed that the thermal response is at twice the frequency of the drive signal (due to the square-law response of resistive heating) and that it did not significantly interfere with the signal. Sensitivities were in the range of hundreds of mV per tesla when the current loops were operated at $\approx$ 10 mA.

Devices using this principle in reverse have also been fabricated. In this case, the resonator is driven mechanically and a voltage is generated by the coil in proportion to the strength of the external magnetic field through which it moves. Hetrick (1989) describes such a device, with a micromachined coil on a 50 µm thick single-crystal silicon cantilever, driven by a separate PZT piezoelectric driver. A ten-turn Al coil was lithographically fabricated on the cantilever, with conductor widths of 30 µm and a thickness of 0.8 µm. The reported sensitivities were as high as 180 mV/T when the coil was vibrated at $\approx$ 2 kHz at an amplitude of 1 mm.

4.9 MAGNETIC READ/WRITE HEADS

Magnetic recording heads, commonly used in disk drives, have long been fabricated using relatively high-resolution lithographic methods. This category of magnetic device is truly both a sensor and an actuator, since they must read from and write to magnetic media. It should be noted, however, that in some cases the mechanisms used for the two processes are different (e.g., magnetoresistive effect for reading and direct electromagnetic effects for writing). Excellent overviews of the evolution of these technologies can be found in Harker, et al. (1981), Stevens (1981), and Grochowski, et al. (1993).

4.9.1 MAGNETIC HEADS FOR STORAGE DEVICES

INDUCTIVE MAGNETIC HEADS

As described by Harker, et al. (1981), Stevens (1981), and Romankiw (1996), the first micromachined inductive thin-film head was used in the IBM 3370 hard disk drive, which was first shipped in 1979. Since then, although several key refinements have been made, the basic design has remained essentially the same.

Romankiw (1996) provided an example of a typical inductive read/write head fabrication process. The substrate is typically Al_2O_3-TiC, selected because of its hardness, small grain size, and resistance to chipping. In the process described, an undercoat of Al_2O_3 was sputter-deposited, followed by a NiFe seed layer. A photoresist electroforming mold was then applied and patterned. After plating 2 to 4 μm of the NiFe core, the photoresist and seed layers were removed using sputter etching. An Al_2O_3 magnetic gap was then sputter-deposited and patterned followed by a thin seed Cr/Cu plating seed layer. Photoresist was applied and patterned as before and copper coils were electroplated. The photoresist and seed layers were once again sputter-etched. This process could be repeated using baked photoresist as a dielectric between successive coil layers. The upper layer of the magnetic core was then electroformed using the same process as the lower layer, and a final overcoat of Al_2O_3 was sputtered. The heads were then diced and precision lapped to form an air bearing surface (ABS). This step is important because the floating height and frictional interactions with the recording surface must be tightly controlled to obtain consistent disk drive performance.

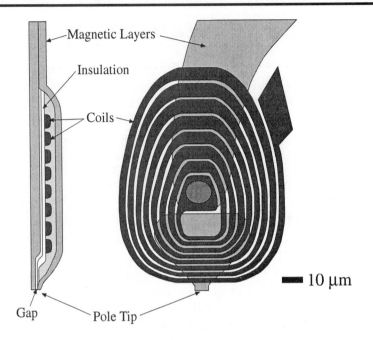

Illustration of the IBM 3370 inductive read/write head. Adapted from Romankiw (1996).

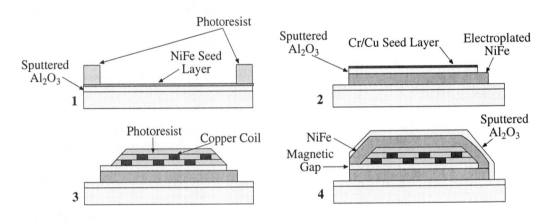

Typical fabrication process used to realize thin-film inductive heads. Courtesy A. Flannery, after description by Romankiw (1996).

Such thin-film heads are generally fabricated in the plane of the wafer, but Lazzari, et al. (1989) and Autino, et al. (1992) reported the development of a silicon planar head (SPH) as an alternative geometry. In these devices, the recording gap is fabricated so as to write perpendicular to the plane of the substrate. A two-layer magnetic coil structure was used, connected to bond pads on the opposite side of the substrate using metallized via holes. Performance demonstrated was slightly better than the conventional thin-film head used for comparison, and there are currently some efforts toward commericalization of this approach.

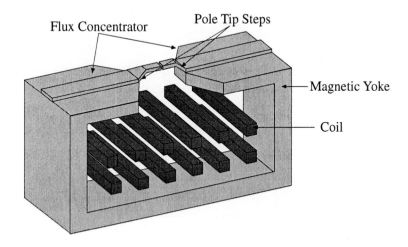

Conceptual illustration of the magnetic gap region of a silicon planar head with flux concentrator. Adapted from Lazarri and Cuchet (1995). Shown in the center of the magnetic yoke are cross sections through the double-layer coils used in this design.

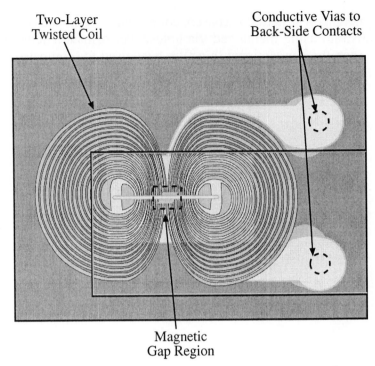

Two-Layer
Twisted Coil

Conductive Vias to
Back-Side Contacts

Magnetic
Gap Region

Illustration of the top view of a silicon planar head, showing the top layer of the two-layer coil (note that the coil shown is continuous, twisted in the center, and connected to the lower coil layer through via holes in the substrate). Adapted from Lazzari (1997).

MAGNETORESISTIVE READ HEADS

Magnetoresistive (MR) heads for magnetic recording were invented in 1970 (Hunt (1970, 1971)), and were originally used in tape drives, starting with the IBM 3480 in 1985, and first applied to hard disks in 1991. A useful overview of this evolution can be found in Shelledy and Nix (1992). As explained in Section 2.2, the magnetoresistive effect is due to changes in the path length of carriers in a region of material due to the Lorentz force, hence changes in the resistance measured across the region. This effect can be harnessed in magnetic sensors, and is most commonly used today in read heads for magnetic disk drives.

As illustrated below, a typical MR head (of the so-called "soft adjacent layer" type) is constructed of a trilayer consisting of an MR material (e.g., trilayers of NiFe, FeMn, and Mo), a non-magnetic spacer (e.g., SiO_2), and a soft magnetic film (e.g., NiFe). (The example compositions of the layers were taken from Guzman, et al. (1994).) Electrical current in the MR element induces a magnetic field in the magnetic film, which is in turn coupled back to the MR layer; the purpose of the magnetic film is to bias the MR element with a field at 45° to the surface of the magnetic recording medium. This biasing places the element in the middle of its linear range of response. Fields from the storage medium below influence this orientation and hence the resistance of the element. Several other MR biasing schemes exist, including permanent magnet, shunt, and dual stripe, as discussed in Shelledy and Nix (1992). In some cases, dual heads are fabricated and used in a differential mode (e.g., a Wheatstone bridge) to reduce sensitivity to common-mode signals (allowing the shields usually used with MR heads to be omitted), as discussed in Jones, et al. (1991) and Guzman, et al. (1994). Typical MR heads have sensitivities (expressed as $\Delta R/R$) on the order of 2% (the highest reported values are on the order of 6%), decreasing as they are scaled down with disks using increasing recording densities, as discussed in White (1994)

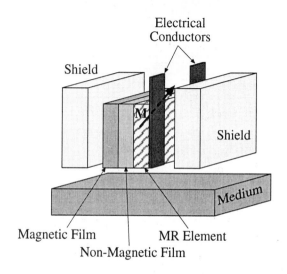

Illustration of the layers and orientation of a magnetoresistive read head. The MR element is magnetically biased at 45° (to improve linearity) relative to the substrate using an adjacent soft magnetic film (in which a magnetic field is induced by running current through the MR region, and this field is in turn coupled back to the MR region to bias it). Adapted from Shelledy and Nix (1992).

As discussed in Grochowski, et al. (1993), MR heads are several times more sensitive than inductive heads. As magnetic storage densities rise into the Gbit/in^2 and bit sizes become smaller, this sensitivity is critical for an adequate SNR. MR heads also have an advantage in that the magnitude of the signal is dependent on the absolute value of the magnetic field (H), and not on the rate of change of the magnetic field (dH/dT) as is the case with the inductive head. Consequently, the magnitude of the readback signal is practically constant over the entire radius of the disk platter. The MR head, however, can only read data, not write it. In order to be of practical use, it is often merged with an inductive write head, and the resulting device has several advantages. Because the inductive head only needs to perform the write function, it can be simpler and smaller than its read/write counterpart and thus requires fewer turns. The MR head's greater sensitivity also enables higher bit density. The result is a better performing hard disk head that is less expensive to manufacture.

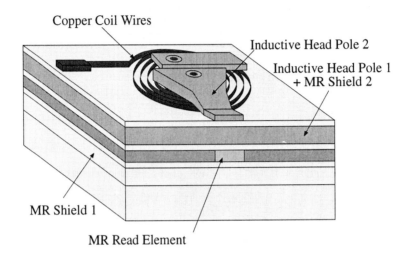

Illustration of a merged magnetoresistive-inductive head. Adapted from Grochowski, et al. (1993).

While far larger changes in resistance, $\Delta R/R$ for a given magnetic field are achievable with giant magnetoresistive sensors (see Section 2.4), they are relatively insensitive. Their low relative response, dR/dH, at small fields, has as yet prevented them from having much impact in the disk drive industry. However, it has been predicted (White (1994)) that with further effort, these sensors will represent a major advance for disk drive read channels. Recent work in this area, as well as a good explanation of the GMR effect, can be found in Lenssen, et al. (1997).

4.9.2 MAGNETIC INK PRINT HEADS

An interesting combination of active microcircuit fabrication, bulk micromachining and electroplating was used by Cardot, et al. (1993), to make large arrays of magnetic drivers for writing patterns on a rotating drum of a "magnetographic" printer. They fabricated single-level spiral coils using electroplated gold, with electroplated Fe/Ni alloy (50:50) cores, or "writing poles." These structures protruded ≈ 40 μm above the silicon substrate and were used in close proximity to a rotating magnetic drum onto which a magnetic "image" of the page to be printed was written. By transferring magnetic ink to the drum and the paper, images could be formed. Their array consisted of 6×80 "microelectromagnets" complete with 480 diodes for addressing. Each electromagnet was capable of producing a maximum magnetic flux density of ≈ 1.4 T.

4.10 MAGNETIC BUBBLE MEMORY

As described by Chang (1978), in the late 1960's, work on magnetic domains led to the realization that under appropriate conditions, discrete circular magnetic domains ("bubbles") within a substrate could be precisely manipulated with thin (electrically conductive) wire loops. These bubbles could be moved, replicated, or combined to perform information storage or logic functions.

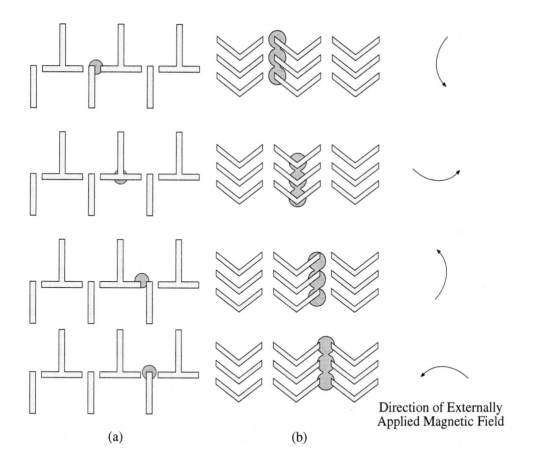

Direction of Externally
Applied Magnetic Field

(a)   (b)

Two variations of bubble memory propogation tracks are shown above: (a) the T-bar, and (b) the chevron. After Chang (1978).

In a bubble memory, propagation tracks are fabricated with specific geometries as shown above. The shape anisotropy is the dominant factor that determines the axis of magnetization. In a resting state with no power applied, the underlying magnetic bubbles are pinned to a particular location in the substrate underneath one of the poles of this axis. When an external rotating magnetic field is applied, the axis of magnetization is coerced into a new direction, and the bubble is pulled along with the poles as they move. Elaborate propagation schemes were developed to be able to read, write, and manipulate patterns of bubbles that could be used to store binary information. Such memory devices were nonvolatile as long as the bias field from an external permanent magnet remained present (the magnets were integrated into the packaging to ensure that this would be the case).

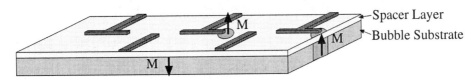

Cross-sectional illustration of a T-bar propagation track over a magnetic bubble layer. Courtesy A. Flannery.

In order for a bubble memory to function, anisotropic forces and domain wall energy must be precisely balanced to favor the formation of the magnetic bubbles. As mentioned above, magnetic bubble devices are typically biased with a permanent magnet layer to help achieve this. Early work was done with orthoferrites of the form $MFeO_3$ where M is a rare-earth element. Some that have been used previously include Tm, Ho, and Lu. Orthoferrites were later replaced with magnetic garnets such as $(YSmLuCa)_3(FeGe)_5O_{12}$. Minimum bubble diameter varies considerably for different materials and thicknesses. For $TmFeO_3$, it is $\approx 60 \ \mu m$. Some materials such as GdTbIG can support a submicron bubble diameter before collapse.

While they were used commercially for a short period (e.g., Intel and Fujitsu had families of commercial bubble memory devices on the market in the early 1980's), several inherent factors have prevented the long-term use of bubble memories. The need for an externally generated, rotating magnetic field requires large, expensive packaging and a prohibitive amount of power. Bubble memories are also slower than transistor-based memories and must be read serially (often byte-wise, however). Minimum domain sizes also limit the degree to which bubble memories can be scaled down in size. The high resistance of bubble memories to perturbation by ionizing radiation, however, has given them a special niche in aerospace applications. Unlike static and dynamic silicon RAM memories, bubble memories are not subject to errors by radiation such as cosmic rays and high-energy particles. They also do not dissipate any static power, making them an alternative for low-power applications where periodic memory access is separated by long stretches of inactivity.

5. MICROMACHINED ELECTROMAGNETIC DEVICES

This section covers some of the electromagnetic applications of micromachined devices, loosely defined here as their use as electronic circuit elements. These elements include passive components (resistors, inductors and capacitors), as well as relays and RF switches (these two categories are discussed in the Mechanical Transducers chapter).

Clearly, passive components (resistors and capacitors) are currently used in a wide variety of integrated circuits, and will not be discussed in depth here (useful references on this are Gray and Meyer (1984) and Laker and Sansen (1994)). However, inductors, which have remained an "off-chip" component for mainstream ICs, can certainly be microfabricated and, due to the limited size of inductances available, make sense mainly at high frequencies. Signal switching with mechanical relays, as discussed in the Mechanical Transducers chapter, may have some important advantages for precision instrumentation and perhaps power switching. Particularly in terms of RF applications, as higher and higher frequencies are used, the relevant scales of the components in the signal path become small enough so that micromachining makes sense.

5.1 PASSIVE COMPONENTS AND CIRCUITS

Thin-film techniques can readily be applied to fabricate microscale resistors, capacitors, and inductors. The former two are quite common in the mainstream silicon IC industry, but the latter is still a subject of active research in the micromachining community. Typically, resistors are diffused or implanted regions in the bulk material or polysilicon traces above an insulator. Alternative resistor structures can be used if necessary (not often), such as the use of an etched epitaxial layer formed into mesa resistors above the substrate containing the active devices (Müller, et al. (1996)).

Capacitors in conventional ICs are typically either MOS capacitors (using gate oxide as a dielectric, with one capacitor "plate" being the substrate) or a thin dielectric sandwiched between two conductors, often polysilicon or aluminum metallization. In some high-frequency applications, particularly with GaAs millimeter-wave devices, interdigitated metal fingers are used due to their simplicity (only requiring one metal level), as described in Chi and Rebeiz (1994, 1995) and Müller, et al. (1996). An interesting application of micromachining to capacitors is the formation of three-dimensional plates to increase the effective plate area without increasing the geometric area on the substrate. This is typically achieved by etching into the substrate and then fabricating the conductive plates and intervening dielectric layer. As long as the deposition techniques have appropriate step coverage, large surface area savings can be achieved, as discussed by Hirano, et al. (1991, 1993). (This approach has been used for several years to help shrink dynamic RAM memories.)

On-chip inductors for high-frequency applications are generally realized as spiral structures. (It should be noted that the often-cited equation of Gleason (1964) [apparently taken from Grover (1946)] will yield incorrect results (Darling (1997).) As described by Chi and Rebeiz (1995) (a similar expression can be found in Bahl and Bhartia (1988)), the inductance, L, of a planar circular spiral is estimated (without correction for conductor thickness, and converted to metric units) by,

$$ L = \left(0.00062 \ \frac{nH}{\mu m} \right) 2AN^2 \left[\ln\left(\frac{8A}{C} \right) + \frac{1}{24}\left(\frac{C}{A} \right)^2 \ln\left(\frac{8A}{C} + 3.583 \right) - \frac{1}{2} \right] $$

where,
 N = number of turns
 D_O = outer diameter of the spiral, in μm
 D_I = inner diameter of the spiral, in μm
 A = $(D_O + D_I)/4$, in μm
 C = $(D_O - D_I)/2$, in μm

For rectangular spiral coils, which are often more convenient to lay out with Manhattan-geometry CAD tools, Greenhouse (1974) gives a formula attributed to Terman (1943) for the inductance (in microhenries) of planar square coils (converted to dimensions in cm) of,

$$ L = 0.0184SN^2 \left[\log\left(\frac{0.7874\ S^2}{t+w} \right) - \log(0.9504S) \right] + 0.008SN^2 \left[0.914 + 0.2235\left(\frac{t+w}{S} \right) \right] $$

where,
 S = maximum side dimension, in cm
 t = conductor thickness, in cm
 w = conductor width, in cm
 N = number of turns

Another such approximation, but without correction for conductor thickness, can be found in Geen, et al. (1989). In addition, estimation methods for octagonal spiral coils were presented by Mondal (1990). All of the approximations referenced above are useful for initial designs, but more complex (and structure-specific) approximations or numerical simulations are often also useful. Naturally, the time taken to obtain and use such other tools must be weighed against that required to fabricate and test prototypes.

There is also interest in fabricating efficient on-chip inductors and transformers for use in switching power supplies. These inductors are generally for relatively high-frequency use, on the order of 1 to 10 MHz, and are optimized for higher currents while still seeking minimal losses. Daniel, et al. (1996) cited numerous previous efforts in this area and described specific techniques for optimizing the design of inductors for such applications. Two important design ideas they discuss are the use of a single-layer coil (to minimize interlayer contact resistance losses) and the use of two layers of magnetic material, one on either side of the coil to form a distributed gap between high permeability regions at the outer boundaries of the coil. It is also worth noting that micromachined transformers can be used in high-voltage circuit isolation applications as long as the inter-turn insulator has a sufficiently high breakdown voltage (thick-film lithography has been used for many years to fabricate such isolation transformers in hybrid analog signal isolators).

There is a long history of microfabricated inductors for use in GaAs and other III-IV integrated circuits. Typically, to increase the surface area available (since conduction at high frequencies is limited to a few skin depths), electroplating was used to fabricate conductors several microns tall. Brehm (1990) provides an interesting account of the evolution of monolithic GaAs RF subsystems in the microwave and millimeter wave frequency ranges (millimeter-wave ICs are often referred to as "MMIC" devices). Examples of micromachined single-layer, multilayer and three-dimensional designs can be found in Chi and Rebeiz (1994, 1995), Geen, et al. (1989), and Hirano, et al. (1991, 1993), respectively.

Although such inductors have been in common use since the early 1980's in GaAs and other III-V RF ICs, they had not been applied to silicon substrates until nearly a decade later. Spiral inductors can readily be fabricated from aluminum metallization in CMOS technology, but with increasing inductance (and size), increasing coupling capacitance to the substrate can decrease the self-resonant frequency. In addition, losses due to the spreading resistance of the substrate can create significant problems. Both of these factors are considerably less with semi-insulating GaAs, high-resistivity Si (see Reyes, et al. (1994, 1996)), or sapphire substrates (highly insulating).

The use of thick insulating layers beneath the inductors can help to reduce these losses, as reported by Kim, et al. (1995). They used a 10 μm thick polyimide to form insulating spacers beneath rectangular spiral inductors on a 20 Ω•cm boron-doped silicon substrate, and demonstrated 10 nH inductors with a self-resonant frequency of 6 GHz and a maximum Q of 5.5 at 1.2 GHz (comparable to results obtained on GaAs substrates). Another way to improve the performance of inductors on low-resistivity silicon is to electroplate them to increase their vertical dimension. This approach has been used commonly in III-V RF ICs. Nardin and Najafi (1995) described a 20 μm tall by 1.24 $\times$ 7.42 mm electroplated nickel rectangular spiral inductor on a low-resistivity, circuit-containing substrate. Their 1.2 μH inductor had a series resistance of 90 Ω, a self-resonance > 40 MHz, and a Q value of 2.8 at 33 MHz. This approach can be used to further decrease resistive losses in the coil if higher conductivity metals, such as Au or Ag, are electroplated.

The formation of cavities beneath such thin-film inductors can also greatly reduce these problems, even when inductors are fabricated on low-resistivity silicon. An example of such an approach was used by Chang, et al. (1993), who described spiral inductors formed by undercutting standard CMOS layers with EDP. They demonstrated the use of a 100 nH micromachined inductor (440×440 μm in size, and with a self-resonance at 3 GHz) in a tuned amplifier with a gain of 14 dB, centered at 770 MHz. They noted significant increases in the self-resonant frequency of the inductors when undercut, suggesting a large reduction in parasitic capacitance (previously, some authors such as Nguyen and Meyer (1990) had described substrate effects for such inductors as negligible).

Chi and Rebeiz (1994, 1995) fabricated both planar inductors and interdigitated capacitors on membranes formed using back-side bulk etching of GaAs and Si, and noted substantial decreases in parasitic coupling to ground. They fabricated these passive components on tensile silicon dioxide/silicon nitride/silicon dioxide trilayers, which became membranes after the back-side etch (KOH or EDP was used for silicon, and $H_2SO_4/H_2O_2/H_2O$ for GaAs). For example, they noted an increase in the resonant frequency of a 1.7 nH inductor from 17 GHz on high-resistivity Si to 54 GHz (the resonant frequency achieved on the undercut membrane is independent of the substrate type).

Milanovic, et al. (1997a, 1997b) demonstrated the use of a hybrid etching technique (isotropic dry etch with XeF_2 followed by anisotropic wet etching using EDP or TMAH) to realize large area suspended transmission lines and inductors using relatively small openings in all CMOS layers (defined by superimposing standard contact cut layers in the layout tool used). Beginning with an XeF_2 etch allows the large areas beneath the structures to be cleared, at which point deep etching could be accomplished using one of the wet etchants. They demonstrated low-loss 50 Ω transmission lines, 4.0 and 8.0 nH inductors, and LC resonators centered at 2 and 21 GHz (both resonators had measured Q values of 20).

It should be noted that a viable alternative to the post-processing of standard CMOS is the use of high-resistivity silicon or silicon-on-sapphire technology, and in some cases, the improved performance can outweigh the cost of developing specialized processes on such substrates. In addition, it is possible to fabricate a wide variety of RF inductor and capacitor structures in standard multilevel CMOS technology, without any post-processing, as described by Burghartz, et al. (1996a). For example, they noted Q factors as high as 10 and a self-resonance frequency of 20 GHz for 2 nH multilayer spiral inductors in a 0.8 μm CMOS with four levels of AlCu interconnects.

Burghartz, et al. (1996b) later described the use of a so-called Damascene process to realize multilayer spiral inductors using low-resistivity Cu metallization on otherwise standard CMOS, high-resistivity silicon, and sapphire substrates. The process involved the deposition of an adhesion layer, two cycles of lithography and dry etching, one Cu deposition step, removal of excess metal using chemical-mechanical polishing, and electroless plating to cap the exposed Cu surface. This process was repeated for each coil layer. Maximum Q values for the three substrate types were (all for inductors of $\approx$ 1.4 nH): Q = 18 at 3.7 GHz (low-resistivity silicon), Q = 30 at 5.2 GHz (high-resistivity silicon), and Q = 40 at 5.8 GHz (sapphire). It is interesting to note that for larger inductors (L = 80 nH), the corresponding Q and frequency values were Q = 8 at 300 MHz (low-resistivity silicon), Q = 9.4 at 600 MHz (high-resistivity silicon), and Q = 13 at 600 MHz (sapphire), indicating that substrate losses alone make far less of a difference for larger inductances (Burghartz, et al. (1996b) included a plot that shows this effect quite clearly). They concluded that performance could be enhanced by increasing the spacing of the coil from the substrate with dielectric layers, avoiding highly doped regions beneath the inductors and decreasing the coil resistance. However, the multiple layers did not yield Q values as high as equivalent single-layer coils.

5.2 OTHER RF APPLICATIONS OF MICROMACHINING

There are other possible applications of micromachining for RF systems, and this area has only begun to be explored. As discussed in the Mechanical Transducers chapter, resonators for filtering and timing signal generation have been developed that cover a range of frequencies from kHz to GHz. In addition, micromachined RF switches, with their parameters optimized for higher-frequency uses, have been fabricated using a variety of designs and actuation mechanisms. Striplines, antennas, microwave circulators, and a variety of other RF devices can also potentially be microfabricated (some already have been). It is likely that in the near future, a variety of integrated RF systems will emerge that take advantage of the best properties of micromachined and conventional devices.

Packaging may also be fabricated using micromachining methods. For example, Drayton and Katehi (1995) used bulk micromachining of silicon to form integrated packaging for microwave circuits with improved control over parasitic radiation and resonances. It is likely that advanced mixed-domain RF systems can be developed that make use of micromachined components, potentially with electro-static, thermal, or optical control interfaces; low-frequency digital and analog control circuits; and a variety of silicon and non-silicon substrates. With such an approach, the best of both micromachined and conventional RF devices could be combined for a given application.

MAGNETIC AND ELECTROMAGNETIC TRANSDUCERS REFERENCES

GENERAL REFERENCES

Baltes, H. P., and Popovic, R. S., "Integrated Semiconductor Magnetic Field Sensors," Proceedings of the IEEE, vol. 74, no. 8, Aug. 1986, pp. 1107 - 1132.

Baltes, H., and Castagnetti, R., "Magnetic Sensors," Chapter 5 in "Semiconductor Sensors," Sze, S. M. [ed], John Wiley and Sons, New York, NY, 1994.

Cullity, B. D., "Introduction to Magnetic Materials," Addison-Wesley Publishing, Reading, MA, 1972.

Halliday, D., Resnick, R., and Walker, J., "Fundamentals of Physics," Fourth Edition, John Wiley and Sons, New York, NY, 1993.

Middelhoek, S., and Audet, S. A., "Silicon Sensors," Academic Press Ltd., London, U K, 1989.

SPECIFIC REFERENCES

Ahn, C. H., and Allen, M. G., "A Fully Integrated Micromagnetic Actuator with a Multilevel Meander Magnetic Core," Technical Digest of the 1992 Solid-State Sensor and Actuator Workshop, Hilton Head Island, SC, June 22 - 25, 1992, pp. 14 - 18.

Ahn, C. H., and Allen, M. G., "A Planar Micromachined Spiral Inductor for Integrated Magnetic Microactuator Applications," Journal of Micromechanics and Microengineering, vol. 3, no. 2, June 1993, pp. 37 - 44.

Ahn, C. H., Kim, Y. J., and Allen, M. G., "A Fully-Integrated Micromachined Toroidal Inductor with a Nickel-Iron Magnetic Core (The Switched DC/DC Boost Converter Application)," Proceedings of Transducers '93, the 7th International Conference on Solid-State Sensors and Actuators, Yokohama, Japan, June 7 - 10, 1993, Institute of Electrical Engineers, Japan, pp. 70 - 73.

Arai, K. I., Sugawara, W., and Honda, T., "Magnetic Small Flying Machines," Digest of Technical Papers, Transducers '95, the 8th International Conference on Solid-State Sensors and Actuators, Stockholm, Sweden, June 25 - 29, 1995, vol. 1, pp. 316 - 319.

Araki, T., and Okabe, M., "(Nd, Tb)-Fe-B Thin Film Magnets Prepared by Magnetron Sputtering," Proceedings of IEEE International Workshop on Micro Electro Mechanical Systems, San Diego, CA, Feb. 11 - 15, 1996, pp. 244 - 249.

Autino, E., Lazzari, J. P., Pisella, C., "Compatibility of Silicon Planar Heads with Conventional Thin Film Heads in Hard Disk Drives," IEEE Transactions on Magnetics, vol. 28, no. 5, Sept. 1992, pp. 2124 - 2126.

Bahl, I. J., and Bhartia, P., "Microwave Solid-State Circuit Design, John Wiley and Sons, 1988, pp. 45 - 57.

Baibich, M. N., Broto, J. M., Fert, A., Nguyen Van Dau, F., Petroff, F., Etienne, P., Creuzet, G., Friederich, A., and Chazelas, J., "Giant Magnetoresistance of (001)Fe/(001)Cr Magnetic Superlattices," Physical Review Letters, vol. 61, no. 21, Nov. 21, 1988, pp. 2472 - 2475.

Baltes, H., and Nathan, A., "Integrated Magnetic Sensors," Chapter 7 in "Sensors: A Comprehensive Survey," Göpel, W., Hesse, J., and Zemel, J. N. [eds.], VCH Verlagsgesellschaft mbH, Weinheim, Germany, 1989, pp. 195 - 215.

Barjenbruch, U., "A Novel Highly Sensitive Magnetic Sensor," Sensors and Actuators, vols. A37 - A38, June - Aug. 1993, pp. 466 - 470.

Bate, G., "Magnetism," in Chapter 34, "Magnetism and Magnetic Fields," in "The Electrical Engineering Handbook," Dorf, R. C. [ed.], CRC Press, Inc., Boca Raton, FL, 1993, pp. 811 - 826.

Bean, C. P., "Magnetism and Life," Essay 10 in Halliday, D., Resnick, R., and Walker, J., "Fundamentals of Physics," Fourth Edition, John Wiley and Sons, New York, NY, 1993, pp. E10-1 - E10-3.

Blakemore, R. P., and Frankel, R. B., "Magnetic Navigation in Bacteria," Scientific American, vol. 245, no. 6, Dec. 1981, pp. 42 - 49.

Blakemore, R., "Magnetotactic Bacteria," Science, vol. 190, no. 4212, Oct. 24, 1975, pp. 377 - 379.

Brehm, G. E., "Multifunction MMIC History from a Process Technology Perspective," IEEE Transactions on Microwave Theory and Techniques, vol. 38, no. 9, Sept. 1990, pp. 1164 - 1170.

Burghartz, J. N., Edelstein, D. C., Jenkins, K. A., Jahnes, C., Uzoh, C., O'Sullivan, E. J., Chan, K. K., Soyuer, M., Roper, P., and Cordes, S., "Monolithic Spiral Inductors Fabricated Using a VLSI Cu-Damascene Interconnect Technology and Low-Loss Substrates," Technical Digest of the International Electron Devices Meeting, San Francisco, CA, Dec. 8 - 11, 1996, pp. 99 - 102.

Burghartz, J. N., Soyuer, M., and Jenkins, K. A., "Microwave Inductors and Capacitors in Standard Multilevel Interconnect Silicon Technology," IEEE Transactions on Microwave Theory and Technology, vol. 44, no. 1, 1996, pp. 100 - 104.

Cardot, F., Gobet, J., Bogdanski, M., and Rudolf, F., "Microfabrication of High-Density Arrays of Microelectromagnets with On-Chip Electronics," Proceedings of Transducers '93, the 7th International Conference on Solid-State Sensors and Actuators, Yokohama, Japan, June 7 - 10, 1993, Institute of Electrical Engineers, Japan, pp. 32 - 35.

Chang, H., "Magnetic-Bubble Memory Technology," Marcel Dekker, New York, NY, 1978.

Chang, J. Y.-C., Abidi, A. A., and Gaitan, M., "Large Suspended Inductors on Silicon and Their Use in a 2-µm CMOS Amplifier," IEEE Electron Device Letters, vol. 14, no. 5, May 1993, pp. 246 - 248.

Chi, C.-Y., and Rebeiz, G. M., "Planar Microwave and Millimeter-Wave Lumped Elements and Coupled-Line Filters Using Micro-Machining Techniques," IEEE Transactions on Microwave Theory and Techniques, vol. 43, no. 4, Apr. 1995, pp. 730 - 738.

Chi, C.-Y., and Rebeiz, G. M., "Planar Millimeter-Wave Microstrip Lumped Elements Using Micromachining Techniques," Proceedings of the IEEE/MTT-S International Microwave Symposium, MTT '94, San Diego, CA, May 23 - 27, 1994, vol. 2, pp. 657 - 660.

Christenson, T. R., Guckel, H., Skrobis, K. J., and Jung, T. S., "Preliminary Results for a Planar Microdynamometer," Technical Digest of the 1992 Solid-State Sensor and Actuator Workshop, Hilton Head Island, SC, June 22 - 25, 1992, pp. 6 - 9.

Christenson, T. R., Klein, J., and Guckel, H., "An Electromagnetic Micro Dynamometer," Proceedings of the IEEE Micro Electro Mechanical Systems Conference (MEMS '95), Amsterdam, Netherlands, Jan. 29 - Feb. 2, 1995, pp. 386 - 391.

Clarke, J., "Advances in SQUID Magnetometers," IEEE Transactions on Electron Devices, vol. ED-27, no. 10, Oct. 1980, pp. 1896 - 1908.

Daniel, L., Sullivan, C. R., and Sanders, S. R., "Design of Microfabricated Inductors," Proceedings of the 27th Annual IEEE Power Electronics Specialists Conference (PESC '96), Baveno, Italy, June 23 - 27, 1996, vol. 2, pp. 1447 - 1455.

Darling, R. B., University of Washington, Dec. 1997, personal communication.

Donzier, E., Lefort, O., Spirkovich, S., and Baillieu, F., "Integrated Magnetic Field Sensor," Sensors and Actuators, vol. A26, nos. 1 - 3, Mar. 1991, pp. 357 - 361.

Drayton, R. F., and Katehi, L. P. B., "Micromachined Conformal Packages for Microwave and Millimeter-Wave Applications," Proceedings of the 1995 IEEE MTT-S International Microwave Symposium, Orlando, FL, May 16 - 20, 1995, vol. 3, pp. 1387 - 1390.

Eyre, B., and Pister, K. S. J., "Micromechanical Resonant Magnetic Sensor in Standard CMOS," Proceedings of Transducers '97, the 1997 International Conference on Solid-State Sensors and Actuators, Chicago, IL, June 16 - 19, 1997, vol. 1, pp. 405 - 408.

Fraden, J., "AIP Handbook of Modern Sensors," American Institute of Physics Press, New York, NY, 1993.

Fulton, T. A., Dunkleburger, L. N., and Dynes, R. C., "Quantum Interference Properties of Double Josephson Junctions," Physics Review B, vol. 6, Aug. 1, 1972, pp. 855 - 875.

Garcia, E. J., Christenson, T. R., Polosky, M. A., and Jojola, A. A., "Design and Fabrication of a LIGA Milliengine," Proceedings of Transducers '97, the 1997 International Conference on Solid-State Sensors and Actuators, Chicago, IL, June 16 - 19, 1997, vol. 2, pp. 765 - 768.

Geen, M. W., Green, G. J., Arnold, R. G., Jenkins, J. A., and Jansen, R. H., "Miniature Multilayer Spiral Inductors for GaAs MMICs," Proceedings of the 11th Annual IEEE GaAs IC Symposium, San Diego, CA, Oct. 22 - 25, 1989, pp. 303 - 306.

Gilbert, B., "Novel Magnetic-Field Sensor Using Carrier-Domain Rotation: Proposed Device Design," Electronics Letters, vol. 12, no. 23, Nov. 11, 1976, pp. 608 - 610.

Gleason, F. R., Jr., "Thin-Film Microelectronic Inductors," Proceedings of the National Electronics Conference, Chicago, IL, Oct. 19 - 21, 1964, vol. 20, pp. 197 - 199.

Goicolea, J. I., Muller, R. S., and Smith, J. E., "Highly Sensitive Silicon Carrier Domain Magnetometer," Sensors and Actuators, vol. 5, no. 2, Feb. 1984, pp. 147 - 167.

Gottfried-Gottfried, R., Budde, W., Jähne, R., Kück, H., Sauer, B., Ulbricht, S., and Wende, U., "A Miniaturized Magnetic Field Sensor System Consisting of a Planar Fluxgate Sensor and a CMOS Readout Circuitry," Digest of Technical Papers, Transducers '95, the 8th International Conference on Solid-State Sensors and Actuators, Stockholm, Sweden, June 25 - 29, 1995, vol. 2, pp. 229 - 232.

Gottfried-Gottfried, R., Zimmer, G., and Mokwa, W., "CMOS-Compatible Magnetic Field Sensors Fabricated in Standard and in Silicon on Insulator Technologies," Sensors and Actuators, vol. A27, nos. 1 - 3, May 1991, pp. 753 - 757.

Gray, P. R., and Meyer, R. G., "Analysis and Design of Analog Integrated Circuits," Second Edition, John Wiley and Sons, New York, NY, 1984.

Greenhouse, H. M., "Design of Planar Rectangular Microelectronic Inductors," IEEE Transactions on Parts, Hybrids and Packaging, vol. PHP-10, no. 2, 1974, pp. 101 - 109.

Grochowski, E., Scranton, R. A., Croll, I., "The Evolution of Magnetic Recording Heads to Maximize Data Storage Density," Proceedings of the 3rd International Symposium on Magnetic Materials, Processes, and Devices, vol. 94, no. 6, Oct. 11 - 12, 1993, pp. 3 - 16.

Grover, F. W., "Inductance Calculations," D. Van Nostrand Co., Inc., New York, NY, 1946 (also available in a 1973 edition from Dover Publications, Inc., New York, NY).

Guckel, H., Christenson, T. R., Earles, T., Klein, J., Zook, J. D., Ohnstein, T., and Karnowski, M., "Laterally Driven Electromagnetic Acutators," Technical Digest of the 1994 Solid-State Sensor and Actuator Workshop, Hilton Head Island, SC, June 13 - 16, 1994, pp. 49 - 52.

Guckel, H., Christenson, T. R., Skrobis, K. J., Jung, T. S., Klein, J., Hartojo, K. V., and Widjaja, I., "A First Functional Current Excited Planar Rotational Magnetic Micromotor," Proceedings of the IEEE Micro Electro Mechanical Systems Conference (MEMS '93), Ft. Lauderdale, FL, Feb. 1993, pp. 7 - 11.

Guckel, H., Christenson, T. R., Skrobis, K. J., Klein, J., and Karnowsky, M., "Design and Testing of Planar Magnetic Micromotors Fabricated by Deep X-Ray Lithography and Electroplating," Proceedings of Transducers '93, the 7th International Conference on Solid-State Sensors and Actuators, Yokohama, Japan, June 7 - 10, 1993, Institute of Electrical Engineers, Japan, pp. 76 - 77.

Guckel, H., Earles, T., Klein, J., Zook, D., and Ohnstein, T., "Electromagnetic Linear Actuators with Inductive Position Sensing for Micro Relay, Micro Valve and Precision Positioning Applications," Digest of Technical Papers, Transducers '95, the 8th International Conference on Solid-State Sensors and Actuators, Stockholm, Sweden, June 25 - 29, 1995, vol. 1, pp. 324 - 327.

Guckel, H., Skrobis, K. J., Christenson, T. R., Klein, J., Han, S., Choi, B., Lovell, E. G., and Chapmann, T. W., "Fabrication and Testing of the Planar Magnetic Micromotor," Journal of Micromechanics and Microengineering, vol. 1, no. 3, Sept. 1991, pp. 135 - 138.

Guzman, J. I., Mountfield, K. R., Kryder, M. H., Bojko, R., J., and Jones, R. E., Jr., "Design and Fabrication of Unshielded Dual-Element Horizontal MR Heads," IEEE Transactions on Magnetics, vol. 30, no. 6, Nov. 1994, pp. 3864 - 3866.

Hall, E. H., "On a New Action of the Magnet on Electric Currents," American Journal of Mathematics, vol. 2, 1879, pp. 287 - 292.

Hamasaki, Y., and Ide, T., Fabrication of Multi-Layer Eddy Current Micro Sensors for Non-Destructive Inspection of Small Diameter Pipes," Proceedings of the IEEE Micro Electro Mechanical Systems Conference (MEMS '95), Amsterdam, Netherlands, Jan. 29 - Feb. 2, 1995, pp. 232 - 237.

Harker, J. M., Brede, D. W., Pattison, R. E., Santana, G. R., and Taft, L. G., "A Quarter Century of Disk File Innovation," IBM Journal of Research and Development, vol. 25, no. 5, Sept. 1981, pp. 676 - 689.

Hauptmann, P., "Sensors: Principles and Applications," Prentice-Hall International (UK), Inc., Hertfordshire, UK, 1991, pp. 182 - 184.

Hetrick, R. E., "A Vibrating Cantilever Magnetic-Field Sensor," Sensors and Actuators, vol. 16, no. 3, Mar. 1989, pp. 197 - 207.

Hirano, M., Imai, Y., and Asai, K., "1/4 Miniaturized Passive Elements for GaAs MMICs," Proceedings of the 13th Annual IEEE GaAs IC Symposium, Monterey, CA, Oct. 20 - 23, 1991, pp. 37 - 40.

Hirano, M., Imai, Y., Toyoda, I., Nishikawa, K., Tokumitsu, M., and Asai, K., "Three-Dimensional Passive Elements for Compact GaAs MMICs," IEICE Transactions on Electronics, vol. E76-C, no. 6, June 1993, pp. 961 - 967.

Hunt, R. P. "A Magnetoresistive Readout Transducer," IEEE Transactions on Magnetics, vol. MAG-7, no. 1, Mar. 1971, pp. 150 - 154.

Hunt, R. P., "Magnetoresistive Head," U.S. Patent No. 3,493,694, issued Feb. 3, 1970.

Hunter, I. W., and Lafontaine, S., "A Comparision of Muscle with Artificial Actuators," Proceedings of the 1992 Solid-State Sensor and Actuator Workshop, Hilton Head Island, SC, June 22 - 25, 1992, pp. 178 - 185.

Jones, R. E., Jr., Guzman, J. I., Mountfield, K. R., and Kryder, M. H., "An Unshielded Horizontal Dual-Element MR Sensor," IEEE Transactions on Magnetics, vol. 27, no. 6, Nov. 1991, pp. 4687 - 4689.

Judy, J. W., and Muller , R. S., "Batch-Fabricated, Addressable, Magnetically Actuated Microstructures," Technical Digest of the 1994 Solid-State Sensor and Actuator Workshop, Hilton Head Island, SC, June 3 - 6, 1996, pp. 187 - 190.

Judy, J. W., Muller, R. S., and Zappe, H. H., "Magnetic Microactuation of Polysilicon Flexure Structures," Technical Digest of the 1994 Solid-State Sensor and Actuator Workshop, Hilton Head Island, SC, June 13 -16, 1994, pp. 43 - 48.

Judy, J., and Muller, R. S., "Magnetic Microactuation of Torsional Polysilicon Structures," Digest of Technical Papers, Transducers '95, the 8th International Conference on Solid-State Sensors and Actuators, Stockholm, Sweden, June 25 - 29, 1995, vol. 1, pp. 332 - 335.

Kawahito, S., Choi, S. O., Ishida, M., and Nakamura, T., "MOS Hall Elements with Three-Dimensional Microstructure," Proceedings of Transducers '93, the 7th International Conference on Solid-State Sensors and Actuators, Yokohama, Japan, June 7 - 10, 1993, Institute of Electrical Engineers, Japan, pp. 892 - 895.

Kawahito, S., Sasaki, Y., Ashiki, M., and Nakamura, T., "Micromachined Solenoids for Highly-Sensitive Magnetic Sensors," Proceedings of Transducers '91, the 1991 International Conference on Solid-State Sensors and Actuators, San Francicso, CA, June 24 - 27, 1991, pp. 1077 - 1080.

Kawahito, S., Sasaki, Y., Ishida, M., and Nakamura, T., "A Fluxgate Magnetic Sensor with Micromachined Solenoids and Electroplated Permalloy Cores," Proceedings of Transducers '93, the 7th International Conference on Solid-State Sensors and Actuators, Yokohama, Japan, June 7 - 10, 1993, Institute of Electrical Engineers, Japan, pp. 888 - 891.

Kawahito, S., Satoh, H., Sutoh, M., and Tadokoro, Y., "Micro-Fluxgate Magnetic Sensing Elements Using Closely-Coupled Excitation and Pickup Coils," Proceedings of Transducers '95, the 8th International Conference on Solid-State Sensors and Actuators, Stockholm, Sweden, June 25 - 29, 1995, vol. 2, pp. 233 - 236.

Kim, B.-K., Ko, B.-K., Lee, K., Jeong, J.-W., Lee, K.-S., and Kim, S.-C., "Monolithic Planar RF Inductor and Waveguide Structures on Silicon with Performance Comparable to Those in GaAs MMIC," Proceedings of the International Electron Devices Meeting, Washington, DC, Dec. 10 - 13, 1995, pp. 717 - 720.

Lagorce, L. K., and Allen., M. G., "Micromachined Polymer Magnets," Proceedings of the IEEE International Workshop on Micro Electro Mechanical Systems, San Diego, CA, Feb. 11 - 15, 1996, 85 - 90.

Laker, K. R., and Sansen, W. M. C., "Design of Analog Integrated Circuits and Systems," McGraw-Hill, New York, NY, 1994.

Lazzari, J. P., "Planar Silicon Heads for Disc Drive Industry," Proceedings of Transducers '97, the 1997 International Conference on Solid-State Sensors and Actuators, Chicago, IL, June 16 - 19, 1997, vol. 2, pp. 1077 - 1080.

Lazzari, J. P., Cuchet, R., "A New Magnetic Flux Tap for Saturation Control in Planar Silicon Heads," IEEE Transactions on Magnetics, vol. 31, no. 6, Nov. 1995, pp. 2690 - 2693.

Lazzari, J. P., Deroux-Dauphin, P., "A New Thin Film Head Generation," IEEE Transactions on Magnetics, vol. 25, no. 5, Sept. 1989, pp. 3190 - 3193.

Lenssen, K.-M. H., van Kesteren, H. W., Rijks, Th. G. S. M., Kools, J. C. S., de Nooijer, M. C., Coehoorn, R., and Folkerts, W., "Giant Magnetoresistance and Its Application in Recording Heads," Sensors and Actuators, vol. A60, nos. 1 - 3, May 1997, pp. 90 - 97.

Lenz, J. E., "A Review of Magnetic Sensors," Proceedings of the IEEE, vol. 78, no. 6, June 1990, pp. 973 - 989.

Liakopoulos, T. M., Zhang, W., and Ahn, C. H., "Electroplated Thick CoNiMnP Permanent Magnet Arrays for Micromachined Magnetic Device Applications," Proceedings of the IEEE International Workshop on Micro Electro Mechanical Systems, San Diego, CA, Feb. 11 - 15, 1996, pp. 79 - 84.

Liu, C., Tsao, T., Tai, Y.-C., Leu, T.-S., Ho, C.-M., Tang, W.-L., and Miu, D., "Out-of-Plane Permalloy Magnetic Actuators for Delta Wing Control," Proceedings of the IEEE Micro Electro Mechanical Systems Conference (MEMS '95), Amsterdam, Netherlands, Jan. 29 - Feb. 2, 1995, pp. 7 - 12.

Liu, C., Tsao, T., Tai, Y.-C., Liu, W., Will, P., and Ho, C.-M., "A Micromachined Permalloy Magnetic Actuator Array for Micro Robotics Assembly Systems," Digest of Technical Papers, Transducers '95, the 8th International Conference on Solid-State Sensors and Actuators, Stockholm, Sweden, June 25 - 29, 1995, vol. 1, pp. 328 - 331.

Löchel, B., Maciossek, A., Rothe, M., and Windbracke, W., "Micro Coils Fabricated by UV Depth Lithography and Galvanoplating," Digest of Technical Papers, Transducers '95, the 8th International Conference on Solid-State Sensors and Actuators, Stockholm, Sweden, June 25 - 29, 1995, vol. 2, pp. 264 - 267.

Lutes, O. S., Nussbaum, P. S., and Aadland, O. S., "Sensitivity Limits on SOS Magnetodiodes," IEEE Transactions on Electron Devices, vol. ED-27, no. 11, Nov. 1980, pp. 2156 - 2157.

Manley, M. H., Bloodworth, G. G., and Bahnas, Y. Z., "Novel Magnetic-Field Sensor Using Carrier-Domain Rotation: Operation and Practical Performance," Electronics Letters, vol. 12, no. 23, Nov. 11, 1976, pp. 610 - 611.

Mansuripur, M., "Magneto-Optical Disk Data Storage," Chapter 74.4 in "The Electrical Engineering Handbook," Dorf, R. C. [ed.], CRC Press, Inc., Boca Raton, FL, 1993, pp. 1675 - 1694.

Mathieu, N., Chovet, A., Fauquembergue, R., Descherdeer, P., and Leroy, A., "New GaAs Integrated Magnetic Field Sensors with High Sensitivities," Sensors and Actuators, vol. A27, nos. 1 - 3, May 1991, pp. 741 - 745.

Maupin, J. T., and Geske, M. L., "The Hall Effect in Silicon Circuits," in "The Hall Effect and Its Applications, Chen, C. L., and Westgate, C. R. [eds.], Plenum Press, New York, NY, 1980, pp. 421 - 445.

Milanovic, V., Gaitan, M., Bowen, E. D., and Zaghloul, M. E., "Micromachined Microwave Transmission Lines in CMOS Technology," IEEE Transactions on Microwave Theory and Techniques, vol. 45, no. 5, part 1, May 1997, pp. 630 - 635.

Milanovic, V., Gaitan, M., Bowen, E. D., Tea, N. H., and Zaghloul, M. E., "Design and Fabrication of Micromachined Passive Microwave Filtering Elements in CMOS Technology," Proceedings of Transducers '97, the 1997 International Conference on Solid-State Sensors and Actuators, Chicago, IL, June 16 - 19, 1997, vol. 2, pp. 1007 - 1010.

Miller, L. M., Podosek, J. A., Kruglick, E., Kenny, T. W., Kovacich, J. A., and Kaiser, W. J., "A μ-Magnetometer Based on Electron Tunneling," Proceedings of the 9th Annual International Workshop on Micro Electro Mechanical Systems, San Diego, CA, Feb. 11 - 15, 1996, pp. 467 - 471.

Mondal, J. P., "Octagonal Spiral Inductor Measurements and Modelling for MMIC Applications," International Journal of Electronics, vol. 68, no. 1, Jan. 1990, pp. 113 - 125.

Moskowitz, B. M., Frankel., R. B., Flanders, P. J., Blakemore, R. P., and Schwartz, B. B., "Magnetic Properties of Magnetotactic Bacteria," Journal of Magnetism and Magnetic Materials, vol. 73, no. 3, July 1988, pp. 273 - 288.

Mosser, V., Contreras, S., Aboulhouda, S., Lorenzini, P., Kobbi, F., Robert, J. L., and Zekentes, K., "High Sensitivity Hall Sensors with Low Thermal Drift Using AlGaAs/InGaAs/GaAs Heterostructures," Proceedings of Transducers '93, the 7th International Conference on Solid-State Sensors and Actuators, Yokohama, Japan, June 7 - 10, 1993, Institute of Electrical Engineers, Japan, pp. 908 - 911.

Müller, A., Simion, S., Dragoman, M., Iordanescu, S., Petrini, I., Anton, C., Vasiliche, D., Avramescu, V., Coraci, A., and Craciunoiu, F., "Passive Devices on GaAs Substrates for MMICs Applications," Proceedings of the IEEE 1996 International Semiconductor Conference, CAS '96, Sinaia, Romania, Oct. 9 - 12, 1996, vol. 1, pp. 185 - 188.

Munter, P. J. A., "A Low-Offset Spinning Current Hall Plate," Sensors and Actuators, vol. A22, nos. 1 - 3, Mar. 1990, pp. 743 - 746.

Munter, P. J. A., "Electronic Circuitry for a Smart Spinning-Current Hall Plate with Low Offset," Sensors and Actuators, vol. A27, nos. 1 - 3, May 1991, pp. 747 - 751.

Nakamura, T., and Maenaka, K., "Integrated Magnetic Sensors," Sensors and Actuators, vol. A22, nos. 1 - 3, Mar. 1990, pp. 762 - 769.

Nakazawa, H., Watanabe, Y., Morita, O., Edo, M., and Yonezawa, E., "The Two-Dimensional Micro Conveyer," Proceedings of Transducers '97, the 1997 International Conference on Solid-State Sensors and Actuators, Chicago, IL, June 16 - 19, 1997, vol. 1, pp. 33 - 36.

Nardin, M., and Najafi, K., "A Multichannel Neuromuscular Microstimulator with Bi-Directional Telemetry," Proceedings of Transducers '95, the 8th International Conference on Solid-State Sensors and Actuators, Stockholm, Sweden, June 25 - 29, 1995, vol. 1, pp. 59 - 62.

Nathan, A., Kung, B., and Manku, T., "Silicon Nanotesla Magnetotransistors - Temperature Coefficient of Resolution," Proceedings of Transducers '91, the 1991 International Conference on Solid-State Sensors and Actuators, San Francicso, CA, June 24 - 27, 1991, pp. 1073 - 1076.

Nguyen, N. M., and Meyer, R. G., "Si IC-Compatible Inductors and LC Passive Filters," IEEE Journal of Solid-State Circuits," vol. 25, no. 4, Aug. 1990, pp. 1028 - 1031.

Ogita, M., and Yasuda, S., "Nonlinearities in the AC Measurement of Hall Effect in Liquids," Journal of the Electrochemical Society, vol. 135, no. 10, Oct. 1988, pp. 2547 - 2549.

Orlando, T. P., and Delin, K. A., "Foundations of Applied Superconductivity," Addison-Wesley Publishing, Inc., Reading, MA, 1991.

Paranjpe, M., Ristic, L., and Filanovsky, I., "A 3-D Vertical Hall Magnetic Field Sensor in CMOS Technology," Proceedings of Transducers '91, the 1991 International Conference on Solid-State Sensors and Actuators, San Francicso, CA, June 24 - 27, 1991, pp. 1081 - 1084.

Partin, D. L., Heremans, J., Thrush, C. M., and Green, L., "Magnetoresistive Sensors," Technical Digest of the 1992 Solid-State Sensor and Actuator Workshop, Hilton Head Island, SC, June 22 - 25, 1992, pp. 35 - 40.

Popovic, R. S., and Baltes, H. P., "A New Carrier-Domain Magnetometer," Sensors and Actuators, vol. 4, no. 2, Oct. 1983, pp. 229 - 236.

Quandt, E., Ludwig, A., and Seemann, K., "Giant Magnetostrictive Multilayers for Thin Film Actuators," Proceedings of Transducers '97, the 1997 International Conference on Solid-State Sensors and Actuators, Chicago, IL, June 16 - 19, 1997, vol. 2, pp. 1089 - 1092.

Reyes, A. C., El-Ghazaly, S. M., Dorn, S., Dydyk, M., and Patterson, H., "Microwave Inductors on Silicon Substrates," Proceedings of the European Microwave Conference, Cannes, France, Sept. 5 - 8, 1994, vol. 2, pp. 1042 - 1047.

Reyes, A. C., El-Ghazaly, S. M., Dorn, S., Dydyk, M., Schroder, D. K., and Patterson, H., "High Resistivity Si as a Microwave Substrate," Proceedings of the 46th IEEE Electronic Components and Technology Conference, Orlando, FL, May 28 - 31, 1996, pp. 382 - 391.

Ristic, Lj., "Magnetic Field Sensors Based on Lateral Magnetotransistors," Chapter 7 in "Sensor Technology and Devices," Ristic, Lj. [ed.], Artech House, Boston, MA, 1994.

Romankiw, L. T., "Evolution of the Plating Through Lithographic Mask Technology," Proceedings of the Fourth International Symposium on Magnetic Materials, Processes and Devices - Applications to Storage and Microelectromechanical Systems (MEMS), Chicago, IL, Oct. 9 - 12, 1995 (published in 1996), pp. 253 - 272.

Rottmann, F., and Dettmann, F., "New Magnetoresistive Sensors: Engineering and Applications," Sensors and Actuators, vol. A27, nos. 1 - 3, May 1991, pp. 763 - 766.

Roumenin, C. S., "A Cross-Sensitivity Free 2-D Sensor for Magnetic Field," Digest of Technical Papers, Transducers '95, the 8th International Conference on Solid-State Sensors and Actuators, Stockholm, Sweden, June 25 - 29, 1995, vol. 2, pp. 245 - 248.

Sauer, G., Gottfried-Gottfried, R., Haase, T., and Kuck, H., "CMOS-Compatible Integration of Thin Ferromagnetic Films," Sensors and Actuators, vol. A42, nos. 1 - 3, Apr. 1994, pp. 582 - 584.

Schneider, M., Castagnetti, R., Allen, M. G., and Baltes, H., "Integated Flux Concentrator Improves CMOS Magnetotransistors," Proceedings of the IEEE Micro Electro Mechanical Systems Conference (MEMS '95), Amsterdam, Netherlands, Jan. 29 - Feb. 2, 1995, pp. 151 - 156.

Shearwood, C., Pate, M. A., Affane, W., Whitehouse, C. R., Woodhead, J., and Gibbs, M. R. J., "Electrostatic and Magnetoelastic Microactuation of Si_3N_4 Bridges," Digest of Technical Papers, Transducers '95, the 8th International Conference on Solid-State Sensors and Actuators, Stockholm, Sweden, June 25 - 29, 1995, vol. 1, pp. 336 - 339.

Shelledy, F. B., and Nix, J. L., "Magnetoresistive Heads for Magnetic Tape and Disk Recording," IEEE Transactions on Magnetics, vol. 28, no. 5. Sept. 1992, pp. 2283 - 2288.

Shibasaki, I., Kanayama, Y., Nagase, K., Ito, T., Ichimori, F., Yoshida, T. and Harada, K., "High Sensitive Thin Film InAs Hall Element by MBE," Proceedings of Transducers '91, the 1991 International Conference on Solid-State Sensors and Actuators, San Francicso, CA, June 24 - 27, 1991, pp. 1069 - 1072.

Shijo, T., and Yomamoto, H., "Large Magnetoresistance of Field-Induced Giant Ferromagnetic Layers," Journal of the Physics Society of Japan, vol. 59, 1990, pp. 3061 - 3064.

Simon, P. L. C., de Vries, P. H. S., and Middelhoek, S., "Autocalibration of Silicon Hall Devices," Digest of Technical Papers, Transducers '95, the 8th International Conference on Solid-State Sensors and Actuators, Stockholm, Sweden, June 25 - 29, 1995, vol. 2, pp. 237 - 240.

Steiner, R., Häberli, A., Steiner, F.-P., and Baltes, H., "Offset Reduction in Hall Devices by Continuous Spinning Current Method," Proceedings of Transducers '97, the 1997 International Conference on Solid-State Sensors and Actuators, Chicago, IL, June 16 - 19, 1997, vol. 1, pp. 381 - 384.

Stevens, L. D., "The Evolution of Magnetic Storage," IBM Journal of Research and Development, vol. 25, no. 5, Sept. 1981, pp. 663 - 675.

Stoessel, Z., and Resch, M., "Flicker Noise and Offset Suppression in Symmetric Hall Plates," Sensors and Actuators, vols. A37 - A38, June - Aug. 1993, pp. 449 - 452.

Sugiyama, Y., Itoh, J., and Kanemaru, S., "Vacuum Magnetic Sensor with Comb-Shaped Field Emitter Arrays," Proceedings of Transducers '93, the 7th International Conference on Solid-State Sensors and Actuators, Yokohama, Japan, June 7 - 10, 1993, Institute of Electrical Engineers, Japan, pp. 884 - 887.

Takamiya, S., and Fujikawa, K., "Differential Amplification Magnetic Sensor," IEEE Transactions on Electron Devices, vol. ED-19, no. 10, Oct. 1972, pp. 1085 - 1090.

Tang, W. C., Nguyen, T.-C.H., Judy, M. W., and Howe, R. T., "Electrostatic-Comb Drive of Lateral Polysilicon Resonators," Sensors and Actuators, vol. A21, nos. 1 - 3, Feb. 1990, pp. 328 - 331.

Terman, F. E., Radio Engineering Handbook, McGraw-Hill, New York, NY, 1943, pp. 48 - 60.

van Duzer, T., and Turner, C. W., "Principles of Superconductive Devices and Circuits," Elsevier, New York, NY, 1981.

Vishnevski, V. G., Dubinko, S. V., Levy, S. V., Nedviga, A. S., and Prokopov, A. R., "Garnet Transducers for Magnetic Field Topology Imaging," Proceedings of Transducers '97, the 1997 International Conference on Solid-State Sensors and Actuators, Chicago, IL, June 16 - 19, 1997, vol. 1, pp. 413 - 416.

Watanabe, Y., Edo, M., Nakazawa, H., and Yonezawa, E., "A New Fabrication Process of a Planar Coil Using Photosensitive Polyimide and Electroplating," Digest of Technical Papers, Transducers '95, the 8th International Conference on Solid-State Sensors and Actuators, Stockholm, Sweden, June 25 - 29, 1995, vol. 2, pp. 268 - 271.

White, R. L., "Giant Magnetoresistance Materials and Their Potential as Read Head Sensors," IEEE Transactions on Magnetics, vol. 30, no. 2, Mar. 1994, pp. 346 - 352.

White, R. L., "Giant Magnetoresistance: A Primer," IEEE Transactions on Magnetics, vol. 28, no. 5, Sept. 1992, pp. 2482 - 2487.

Wickelgren, I., "The Strange Senses of Other Species," IEEE Spectrum, vol. 33, no. 3, Mar. 1996, pp. 32 - 37.

Wiegand, J. R., "Method of Manufacturing Bistable Magnetic Devices," U.S. Patent No. 3,892,118, issued 1975.

Williams, K. R., and Muller, R. S., "Micromachined Hot-Filament Ionization Pressure Sensor and Magnetometer," Proceedings of Transducers '97, the 1997 International Conference on Solid-State Sensors and Actuators, Chicago, IL, June 16 - 19, 1997, vol. 2, pp. 1249 - 1252.

Yariv, A., and Yeh, P., "Optical Waves in Crystals," Wiley Interscience, New York, NY, 1984.

Zhang, M., and Misra, D., "A Novel 3-D Magnetic Field Sensor in Standard CMOS Technology," Proceedings of Transducers '91, the 1991 International Conference on Solid-State Sensors and Actuators, San Francisco, CA, June 24 - 27, 1991, pp. 1085 - 1088.

CITED INDUSTRY REFERENCES AND SUPPLIERS

Semiconductor Magnetic Sensors

Allegro Microsystems, Inc. (formerly Sprague Semiconductor Group), 115 Northeast Cutoff, Worcester, MA 01615, Phone: (508) 255-3476

Siemens, AG, Bereich Halbleiter, Marketing-Kommunikation, Balanstraße 73, 81541, Munich, Germany.

Non-Micromachined Flux-Gate Magnetometers

Applied Physics Systems, Inc., 897 Independence Avenue, Mountain View, CA 94043, Phone: (415) 965-0500.

Bonded Coil Isolation Transformers

Burr-Brown Corporation, International Airport Industrial Park, P.O. Box 11400, Tucson, AZ 85734, Phone: (602) 746-1111, Fax: (602) 746-7357.

Chapter 8:

CHEMICAL AND BIOLOGICAL TRANSDUCERS

1. INTRODUCTION

The subject of chemical and biological transducers covers a huge assortment of devices that interact with solids, gases, and liquids of all types, and is therefore very broad and interdisciplinary. This type of transducer is generally quite different from thermal, optical, intertial, and magnetic sensors, since they must directly interact with external chemical species. In this sense, they require "ports" or inlets in their packaging to permit this interaction (as also required for some other sensor types, such as pressure sensors). Not only is such environmental access required, but quite often the chemical species of interest are reactive, and thus the transducer must be resistant to attack.

Another key feature is that some chemical sensors are designed to identify one or more distinct compounds. In order to do this, they must have sufficient *selectivity* to make such identifications among compounds of interest, without falsely responding to potential interfering species in samples. This requirement can make the interpretation of chemical sensor signals much more complex than those of, for example, accelerometers, yet the structures of the chemical sensors themselves are often far simpler than their mechanical counterparts. Similarly, biological transducers (generally just chemical sensors and actuators used in a biological context) need to be designed so that they can carry out the necessary transduction without irritating or being attacked by biological tissues. At this point, it should be sufficient to state that most chemical and biological transducers do not require extremely sophisticated micromachining for their fabrication, yet may require considerable interdisciplinary knowledge and sophistication for their actual use.

Potential application areas include medical diagnostics, implantable biosensors, food processing/monitoring, environmental monitoring, pharmaceutical screening, etc. Overall, this area of micromachined transducers is still in relatively early stages of development (especially in terms of commercially viable products). Key areas needing further development are stability and repeatability of such transducers (the former issue can often be mitigated by using them only once, in a disposable format), testing methods, and metrics for comparision with each other and their conventionally fabricated counterparts. Due to a number of market factors, it is likely that research in this area will continue to intensify, and volume shipments of such sensors will grow.

2. CHEMICAL SENSORS

Several types of device get lumped into the category of chemical sensors, including ion/molecule specific sensors, electrochemical sensors, reaction sensors (such as calorimeters, also discussed in the Thermal Transducers chapter), gas/vapor sensors (including humidity), separation devices (chromatography, electrophoresis, mass spectroscopy, etc.), spectroscopic instruments (discussed in the Optical Transducers chapter), biosensors (transducers using biological molecules in the front-end transduction process, with a sensor in series that generates the output signal), hybrid biosensors (combinations of living cells and solid-state sensors), combinatorial assay devices (e.g., DNA hybridization), and many others.

Despite the large amount of research done in this field for many years, there are relatively few commercial successes. Certainly, there are not as yet many true "mass market" drivers such as the automotive industry for mechanical transducers. As mentioned above, one of the reasons for this has been difficulty in transitioning chemical transducers to the commercial world due to issues of reproducibility, testability, stability (in use or storage), selectivity, sensitivity to contaminants, low-cost competing "conventional" technologies, packaging difficulties, etc.

For most of those issues, proposed solutions are being investigated. Lack of reproducibility and lack of stability are two problems that, in some cases, are inherent to the type of transducer being used. Certainly, if they are based on organic molecules, their performance may decay over time, contributing to these problems. In terms of reproducibility, some types of chemical sensors are "used up" when operating, or are obscured by adsorbing contaminants. This can be avoided by using arrays where a "fresh" sensor is uncovered or used each time. Another potential solution to this is to use them differentially, with a "protected" reference sensor used as a reference. Selectivity may be limited in some devices (such as ChemFETs, as discussed below), but other sensors, such as those relying on exquisitely specific interactions between organic molecules are amazingly selective (unfortunately, some of these selective binding/recognition events are not reversible). The issue of lack of testability can be addressed by testing at the time of use (e.g., with a calibrant solution) or by other means, but it is much more difficult to test a "wet" sensor than a conventional integrated circuit (which can be tested very rapidly at the wafer scale, using purely electrical stimulus/response methods). Packaging and cost issues may well be addressed through the use of disposable micromachined devices (despite the apparent contribution to pollution, nearly all low-to-moderate cost medical devices that contact patient body fluids are disposable at present, so this would not add a large increment).

As detailed supplements to the information provided below, the reader is referred to Madou and Morrison (1989) and Janata (1989).

2.1 PASSIVE CHEMICAL SENSORS

2.1.1 CHEMIRESISTORS

Chemiresistors are simple structures in which the resistance (or impedance) between two electrical contacts is modified depending on the quantity of an unknown in the environment. Often, to obtain high sensitivity, chemiresistors are designed in the form of interdigitated electrodes embedded in a sensitive material, as illustrated below.

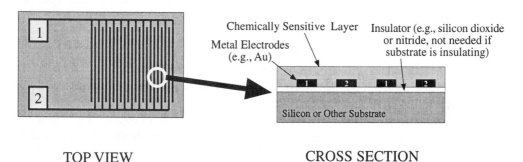

TOP VIEW   CROSS SECTION

Illustration of the basic structure of a chemiresistor, showing interdigitated electrodes coated with a chemically sensitive layer.

Many organic polymers change their conductivity if exposed to a gas. Typically, metals are substituted into the polymer, and the choice of metal often determines the primary sensitivity of the chemiresistor. An example discussed by Middelhoek and Audet (1989) is the use of copper-substituted phthalocyanine conductive polymers as a CCl_4 sensor, although the details of the transduction mechanism do not appear to be well understood. A number of gas-sensitive polymers exist, and generally exhibit relatively poor selectivity to some types of analytes. Wohltjen, et al. (1985) deposited monolayers of Cu-substituted phthalocyanine (Cu-tetracumylphenoxy phthalocyanine) on 50 interdigitated Au electrodes (on a quartz substrate) using Langmuir-Blodgett methods. The Au electrodes were used to form good ohmic contacts to the phthalocyanine layer, and were 25 μm wide, with 25 μm gaps between them, and an overlap distance of 7.25 mm. Each monolayer deposited was ≈ 2.5 nm thick, and 45 layers were deposited in total. They demonstrated responses to NH_3 (in N_2) and NO_2 at concentrations between 0.5 and 2 ppm, with response times on the order of 1 min. The chemiresistors did not respond appreciably to benzene, CO_2, and cyanogen chloride. Wohltjen, at al. (1985) noted that differential DC conductance measurements could be made (with a reference sensor kept isolated from the analyte by a paraffin wax coating but isothermal with the exposed sensor) to reduce temperature-dependent errors. It is likely that the tedious, serial Langmuir-Blodgett film preparation method could be replaced by spin-on or other methods without great compromise in performance.

Discontinuous metal films can also form the basis of chemiresistor sensors. McNerney, et al. (1972) described a mercury vapor sensor that could readily be fabricated using micromachining techniques. A discontinuous gold film (< 40 nm thick) was deposited on an insulating substrate (e.g., SiO_2 on silicon). Absorbed mercury (or other species) changed the resistance of the thin film markedly by altering the surface conductivity of the film (resistance increased due to scattering of conduction electrons as they hit the surface where Hg had adsorbed). Detector sensitivity was affected by both the film thickness (thinner films exhibited a greater fractional resistance increase per quantity of adsorbed Hg than did thicker films) and the carrier gas flow rate (optimized flow to maximize contact of the vapor with the collecting film). Detection of 50 picograms (pg) of Hg was possible, but extreme care was taken to remove H_2O vapor and acid vapors from the carrier gas (due to strong sensor sensitivity to these compounds). The sensor could be refreshed and used again since the mercury could be "desorbed" by heating the substrate to > 150°C. In a very different application, a discontinuous Pt film (≈ 2.5 nm) was used in combination with an immobilized antibody layer (as discussed in Section 2.5 below) as a sensor for biological molecules Feng, et al. (1995). The measured impedance of the film varied linearly over an analyte concentration range of 0.4 to 10 ng/ml, but as yet the underlying mechanism is not known.

An interesting analysis of the signal-to-noise limits of chemiresistors was presented by Harris, et al. (1997). These authors compared measurements using DC versus AC excitation of the chemiresistors, and demonstrated that for low-frequency excitation (on the order of 1 kHz), considerable improvements in SNR can be obtained by avoiding the excess noise (1/f spectrum) generated by these devices when DC-driven. In the cases where AC excitation is used, the distinction between chemiresistors and chemicapacitors (discussed below) becomes blurry, since they both have resistive and capacitive components.

In general, this type of sensor is very nonselective, but, as discussed below, there are some potentially viable approaches to improve upon this, particularly through combining information from arrays of different, but nonselective, sensors.

2.1.2 CHEMICAPACITORS

Often, the same interdigitated structure can be used as a sensor wherein the capacitance of the sensitive layer is measured rather than the resistance. In this case, either the dielectric constant or the physical volume of the sensitive layer varies in relation to the measurand. Such *chemicapacitor* sensors can be made sensitive to specific gases or humidity. An example is polyphenylacetylene (PPA), a polymer that can be spun on from a benzene solution (and potentially less toxic solvents) and is sensitive to CO, CO_2, N_2, and CH_4 (this is discussed in Hermans (1984)). The capacitance can be measured using many well-known techniques (AC bridge, RC time constant measurement, etc.). Problems with this approach include nonlinearity and, more importantly, lack of selectivity. More recently, Steiner, et al. (1995) demonstrated the use of polyetherurethane (PUT) as the sensing material. The PUT was deposited on interdigitated electrodes fabricated using a CMOS process, and the resulting chemicapacitor's capacitance was measured against an uncoated reference capacitor using on-chip electronics. They demonstrated sensitivity to n-octane, toluene, chloroform, and 1-propanol. The capacitance coefficients (relating capacitance to vapor concentration) of PUT were explored for a variety of analytes by Zhou, et al. (1995) and were found to vary between - 5 (for toluene) and +12 ppm/ppm (for 1-propanol), depending on the analyte in question.

Despite their relative lack of selectivity, chemicapacitors have been broadly applied in sensing humidity. Richter (1993) discusses a porous silicon-based micromachined capacitive humidity sensor where a parallel-plate electrode structure is used to measure the amount of adsorbed water in the porous silicon layer. The adsorbed water causes changes in permittivity and impedance, which are detected as measured capacitance changes. By adjusting the microstructure (pore size and pore size distribution) and doping of the material, the output characteristics and sensitivity of the humidity sensor may be controlled. Devices exhibiting good sensitivity, linearity, fast response (1 to 2 s), and very low temperature dependence have been demonstrated (unfortunately, sensor selectivity was not addressed in this paper).

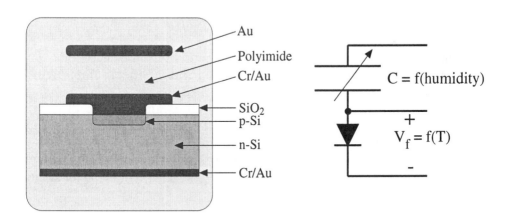

Integrated humidity and temperature sensor. After Yamamoto, et al. (1987).

Yamamoto, et al. (1987) demonstrated an integrated humidity sensor and temperature sensor (using a diode's constant-current forward voltage to sense temperature, as discussed in the Thermal Transducers chapter). A chemicapacitor humidity sensor was in series with the diode, providing access to both functions with three terminals, as illustrated below. Their process began with the growth of an 800 nm SiO$_2$ layer on an n-type silicon wafer, followed by a boron diffusion to form diode regions on the surface. Front- and back-side metallization (Cr/Au) was then deposited (patterned only on the front side), after which a polyimide layer was spun on and cured. A second Cr/Au metallization, to form the top capacitor plates, was then deposited on the polyimide and patterned. The wafers were then diced, and the individual devices were dip-coated with polyimide.

A commercial example is the Minicap™ family of capacitive humidity sensors available from Panametrics, Inc. (Waltham, MA), in which a polymer layer acts as the sensitive layer. Typical humidity sensors of this type operate over a range of 5 to 95% relative humidity, -40 to +180°C temperature and have response times on the order of minutes. Micromachined humidity sensors with integrated electronics are also available commercially from Hy-Cal Engineering (El Monte, CA). For further information on capacitive humidity sensors, the reader is referred to Bolt-shauser (1993). A review of humidity sensor principles and applications can be found in Yamazoe and Shimizu (1986) and a thorough overview of capacitive and resonant humidity sensors (see below) can be found in Boltshauser (1993).

2.1.3 CHEMOMECHANICAL SENSORS

There are some instances in which direct chemical-to-mechanical transduction is feasible. For example, a polymer or other material may be chosen that physically expands in the presence of an analyte. The resulting dimensional change can be measured using capacitive or optical sensing, or by direct mechanical means.

A use of polyimide as a chemomechanical humidity sensor was demonstrated by Kang, et al. (1997), who used a layer of it on a thin silicon diaphragm with on-board piezoresistors. As the polyimide absorbed moisture and expanded, the resulting strain of the piezoresistors was measured as an indicator of humidity. Sager, et al. (1995) studied the volume expansion of a particular polyimide useful in such applications, Pyraline® PI-2722 (DuPont Electronic Materials, Research Triangle Park, NC) and found that the expansion coefficient was 60 to 80 ppm/% RH and a time constant of expansion (normalized to the square of the layer thickness) of 2.2×10^{-3} s^{-1}.

A resonant humidity sensor was demonstrated by Boltshauser, et al. (1993) and Schroth, et al. (1995), who fabricated a polyimide layer atop a back-side-etched epitaxial silicon membrane. By thermally driving the membrane to resonance using resistors in the membrane, and sensing its deflection using piezoresistors also in the membrane, they could sense humidity via changes in the resonant frequency. The mechanisms of action in this case was that volume expansion of the polyimide caused a bending moment (and hence tensile stress) in the membrane and that the absorbed water also increased the effective mass of the resonator.

2.1.4 CALORIMETRIC SENSORS

As mentioned in the Thermal Transducers chapter, it is possible to micromachine very sensitive and fast-responding *calorimeters*, which are devices that measure excess heat generated by chemical reactions. For example, it is possibie to sense combustible gases using a calorimetric approach by oxidizing them using ambient oxygen and a suitable catalytic or enzymatic surface. A wide variety of chemical reactions can be studied with this method.

As discussed in van Herwaarden, et al. (1993), the heat of reaction, P, is given by,

$$P = Q(-\Delta H) = C \frac{dK}{dt}(-\Delta H) \quad \text{in W}$$

where
   Q = reaction rate, in mol/s
 $-\Delta H$ = energy released by the chemical reaction, in J/mol
   C = concentration of the analyte in its carrier fluid, in mol/m^3
dK/dt = catalyst or enzyme conversion efficiency, in m^3/s

Most micromachined calorimeter designs make use of a thermally isolated region, which includes a thin-film resistor with a known temperature coefficient of resistance. The thin-film resistor is used both to heat the isolated region and to sense its temperature (alternatively, a separate heater and temperature sensor can be used). If a catalyst is present and leads, for example, to the local combustion of the gas to be detected, the temperature of the membrane will rise above that expected for the input power due to the exothermic reaction. Since the thermal mass of the membrane and the TCR of the resistor can be readily determined, the temperature change due to combustion can be related back to a gas concentration.

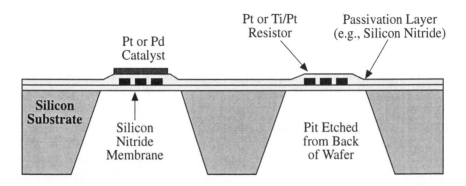

Illustration of a microcalorimetric combustible gas sensor. After Zanini, et al. (1994).

Zanini, et al. (1994, 1995) demonstrated a microcalorimetric gas sensor fabricated using KOH etching to form dual silicon nitride membranes with Pt or Ti/Pt resistors. The approximate thermal resistance of their membranes was 3,600 K/W with reported TCRs of 3,000 and 2,500 ppm for the two resistor types, respectively (assumed linear and measured using a two-point technique). A suitable catalyst (Pt or Pd) was deposited on only one of the membranes allowing differential sensing (this approach considerably improves the performance of most chemical sensors by greatly reducing drift due to thermal and adsorption phenomena). Their device operated over the concentration range of 0 to 4,000 ppm and exhibited the following sensitivities: H_2 - 2.4 ppm, CO - 1.0 ppm, C_3H_8 (propane) - 2.3 ppm, and C_3H_6 (propylene) - 3.9 ppm. Thus, while posessing high sensitivity, this sensor suffered from poor selectivity.

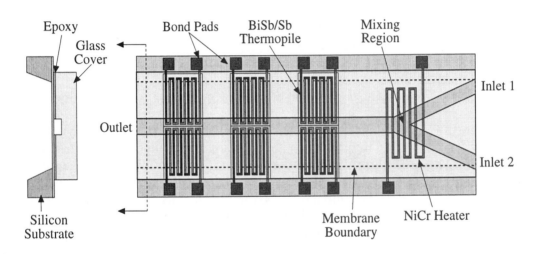

Illustration of a flow-through fluidic calorimeter, with an on-chip calibration heater. An array of thermopiles is located downstream from a mixing/reaction region where two incoming reactants converge. The cross section at the left shows the construction of the device. Adapted from Zieren and Köhler (1997).

A micromachined calorimeter with an integrated fluidic mixer/reaction region was presented by Zieren and Köhler (1997). Using a bulk-etch-released silicon nitride/silicon dioxide/silicon nitride membrane, the calorimeter consisted of an integrated thin-film NiCr heater (for calibration) and thermopiles made of $Bi_{0.87}Sb_{0.13}$/Sb thermocouples (chosen for their high thermoelectric efficiency). A glass cover was bonded over the membrane using epoxy (the glass was pre-etched to include a Y-shaped fluidic path with output dimensions of 2 mm width by 100 to 200 μm height). The two fluidic inputs converged on a mixing region (apparently no special mixing structures were used) and the resulting mixture flowed over three separate downstream thermopiles before exiting the calorimeter (the system was designed for flow rates between 10 nl/min and 100 µl/min). Detection of heat from the exothermic reaction of H_2SO_4 and NaOH was successfully demonstrated at a flow rate of 5 µl/min.

2.1.5 METAL-OXIDE GAS SENSORS

It has been known since the 1950's that adsorption of gases onto certain semiconductors can greatly modulate their resistivities. This class of sensor generally uses a semiconducting metal oxide as the transducer, typically, SnO_2 (TiO_2, In_2O_3, ZnO, WO_3, and Fe_2O_3 are also useful in some cases). The metal oxide is heated and the gas to be measured interacts with it, changing its resistance. Additives to the metal oxide can catalyze the reaction and improve sensitivity and selectivity through several proposed mechanisms (discussed below).

As explained by Morrison (1994) and Gardner (1994), in air, oxygen adsorbs on the surface and reversibly extracts an electron from the semiconductor, forming adsorbed O^- and *increasing the resistance* (for n-type semiconductors, the category of most metal oxides),

$$O_2 + 2e^- \longrightarrow 2O^-$$

In other words, the adsorption of the O_2 at the surface effectively produces an electron trap state. The trapped charge at the surface bends the energy bands of the semiconductor, depleting the surface of mobile carriers and hence increasing its resistance.

These sensors are typically used to detect combustible gases, which, when present, react irreversibly with the adsorbed forms of oxygen (O^-, O_2^-, O^{2-}) to form water (sometimes CO_2), and release electrons, thereby *reducing the resistance*,

$$H_2 + O^- \longrightarrow H_2O + e^-$$

With these two competing reactions, the resistance will increase with increasing oxygen concentration and decrease with increasing combustible gas concentration. The decrease in resistance (increase in conductivity) is due to the increase in carriers and relates to the measurand concentration through its reaction kinetics, producing a relationship where the change in resistance is inversely proportional to a fractional power of the measurand concentration,

$$\Delta R \propto \frac{1}{q\mu_n[X]^r}$$

where,
ΔR = change in resistance, in Ω

q = charge on the electron = 1.60218×10^{-19} C
μ_n = electron mobility, in $cm^2/(V \cdot s)$
[X] = measurand concentration, in mol/l
r = exponent, where $0.5 < r < 1$

In thick enough films of metal oxides, the above (surface interaction) mechanism does not explain the very large effect on resistance sometimes seen. In such cases, the metallic oxides have an open granular structure into which gases can diffuse. This gives rise to a second concentration-dependent effect, that of varying the mobility as the electrons must hop between adjacent grains over a potential barrier that is gas modulated.

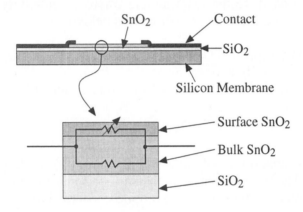

Illustration of the effect of surface resistance modulation in a thin-film metal oxide gas sensor.

In thin-film type detectors, the above case may not be very important. In terms of the physical interactions at the surface, the finite densities of electron donors (e.g., adsorbed hydrogen) or acceptors (e.g., adsorbed oxygen) bound to the surface represent surface states that can exchange electrons with the bulk semiconductor, forming a *space-charge layer* close to the surface. By changing the surface concentration of donors or acceptors, the conductivity of the space-charge region is modulated. Such a device can be considered to be a modulated conductive sheet on top of an unmodulated bulk sheet, as illustrated above (or, in some cases, a bulk sheet with a surface layer that is rendered less conductive). Since the two conductors are in parallel, the overall thickness of the film should be as small as possible to maximize the effect of the surface depletion. Maximum sensivity is obtained when the metal oxide thickness is approximately equal to the induced depletion region depth.

Generally, a catalyst such as Pt is deposited on the surface of the sensor to speed the reactions and hence the sensor's response. The catalyst reduces the activation energy for combustion as follows,

Basic Reactions	**Catalyzed Reactions**
$H_2 \longrightarrow 2H$	$H_2 + 2Pt \longrightarrow 2(Pt - H)$
$O_2 \longrightarrow 2O$	$O_2 + 2Pt \longrightarrow 2(Pt - O)$
$2H + O \longrightarrow H_2O$	$2(Pt - H) + (Pt - O) \longrightarrow 3Pt + H_2O$

Many metal oxides can be used, including the wide bandgap semiconducting oxides of tin, zinc, iron, zirconium, gallium, titanium, and others (all doped as needed), some examples of which are shown in the table below (for a detailed article listing many more, the reader is referred to Morrison (1987), Moseley (1992), and Kohl (1997)). In some cases (see Getino, et al. (1995)), arrays of differently doped metal oxides are fabricated and the overall response is calibrated for a variety of gases (as discussed for "electronic noses").

Semiconductor	Suggested Additives	Gas to Be Detected	Reference
$BaTiO_3/CuO$	La_2O_3, $CaCO_3$	CO_2	Haeusler and Meyer (1995)
SnO_2	Pt + Sb	CO	Morrison (1994)
SnO_2	Pt	alcohols	Morrison (1994)
SnO_2	Sb_2O_3 + Au	H_2, O_2, H_2S	Morrison (1994)
SnO_2	CuO	H_2S	Tamaki, et al. (1997)
ZnO	V, Mo	halogenated hydrocarbons	Morrison (1994)
WO_3	Pt	NH_3	Morrison (1994)
Fe_2O_3	Ti-doped + Au	CO	Morrison (1994)
Ga_2O_3	Au	CO	Schwebel, et al. (1997)
MoO_3	none	NO_2, CO	Guidi, et al. (1997)
In_2O_3	none	O_3 (ozone)	Wlodarski, et al. (1997)

Table of example semiconducting metal oxides suitable for use in gas sensors, additives to improve performance, and gases that can be detected.

The interaction between additives and semiconductors is not well understood, but there are several proposed mechanisms for the improved sensitivity and selectivity associated with their use. Chemical interactions whereby the additive assists the redox process of the semiconductor oxide are generally thought to be responsible for the improved performance associated with the use of catalysts such as Pt (as described above). However, an additional electronic interaction associated with the additive has also been suggested, where changes in the potential barrier at grain boundaries is considered. Vlachos, et al. (1996) modeled the depletion layer that is created as a result of the metal-semiconductor contact forced by the presence of a metallic additive. They correlated this depletion layer to an active grain size, which was smaller than their physical size (since the depletion layer decreased electron availability of the semiconducting grains). As the active grain size decreased, the conductivity of the semiconductor decreased, thereby resulting in increased sensitivity (expressed as the change in conductivity over the initial conductivity). This was related to the additive work function where larger work-function materials resulted in improved sensitivity. Experimental results verified the model with carbon monoxide, methane, and ethanol detection using a tin oxide gas sensor doped with Pt, Ni, and Pd.

For the common example of SnO_2 as a CO sensor, it can be sputtered onto a thermally isolated silicon structure (like the thermopile membranes previously discussed) and heated using an underlying, thin-film heater. SnO_2 operates well at 300°C and the excellent thermal isolation of such micromachined structures allows this temperature to be reached with relatively little input power. It should be noted that if power dissipation is not an issue, screen or transfer printing onto bulk substrates is a viable approach for the fabrication of such sensors, as demonstrated by Golovanov, et al. (1995).

A micromachined, membrane-based SnO_2 gas sensor was implemented by Chang and Hicks (1986). They formed 2 µm thick silicon membranes using a boron etch-stop layer (> 5×10^{19} cm^{-3}, ion implanted at $\approx$ 200 keV) and an EDP etch. The sensing element consisted of a lower layer of SiO_2, a polysilicon heater layer, a SiO_2 insulating layer, Al/Cr electrical interconnects, and a top SnO_2 layer. It is interesting to note that they deposited the SnO_2 from a spin-on mixture of tin (II) 2-ethylhexanoate dissolved in xylene, which was fired (presumably in an oxidizing atmosphere) to form 100 to 200 nm thick SnO_2 layers. (This is referred to as a metallo-organic deposition (MOD) technique.) Despite metal-to-SnO_2 contact problems, functional sensors were fabricated and operated at a temperature of $\approx$ 250°C using $\approx$ 150 mW of power.

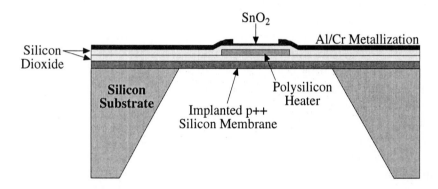

Illustration of a thin-film tin oxide gas sensor formed on a silicon membrane. Aapted from Chang and Hicks (1986). Note that contacts to the tin oxide are not shown, and were in the axis into and out of the page.

Another example of such a device is the design of Demarne and Grisel (1988) using a SiO_2 membrane to support a gold heater, capped with another SiO_2 layer and finally an upper SnO_2 layer.

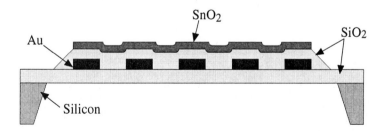

Illustration of the cross section of a micromachined tin oxide-based carbon monoxide gas sensor. After Demarne and Grisel (1988). Note that electrical contacts to the tin oxide layer are not shown.

Johnson, et al. (1988, 1990, 1994) discussed a micromachined conductivity-based gas sensor based on discontinuous, "ultra-thin" (< 10 nm) metal films (Ti-Pt) on thermally isolated dielectric membranes. They demonstrated sub-ppm detection of O_2 in CF_4 (potentially useful in plasma etchers, etc.). The dielectric window could be heated to several hundred degrees Celsius in less than one second using an integrated thin-film heater. Much like the microcalorimetric gas sensors described earlier, the temperature of the dielectric window could be monitored using integrated thermocouples and thermopiles. The thin-film resistance changes induced by the presence of a gas contains both a thermal component (varies with gas concentration and thermal conductivity) and a chemical component, both of which contribute to the measured resistance change in the constant heater current mode of operation.

However, the constant-temperature mode of operation separates these components; the heater current required to maintain the fixed temperature represents the thermal component, while the thin-film resistance change corresponds to the chemical component. Thus, by integrating both a heater and a temperature sensor they were able to add discriminating capabilities to the device, which was not previously possible.

Suehle, et al. (1993) demonstrated a microfabricated SnO_2-based gas sensor that was added on top of standard CMOS chips fabricated through the MOSIS foundry service. The structure used was a suspended thermal platform of field oxide and two upper SiO_2 layers, containing a polysilicon heater, an aluminum plate for temperature sensing, a set of aluminum contacts to the SnO_2, and the SnO_2 layer itself. Before deposition of the SnO_2, the chips (as received from MOSIS) were etched to form the undercut pit for the thermal platform by using the "open" layer to expose bare silicon. The pits were etched using EDP (at 95°C) with added $Al(OH)_3$ (aluminum hydroxide) to reduce the attack of the EDP on the exposed aluminum bond pads. The SnO_2 was then deposited by electrically heating (using the polysilicon heaters) the thermal platforms and using reactive sputtering at the same time. The devices were functional and demonstrated yet another example of post-processing prefabricated CMOS devices into functional sensors using maskless steps. Another interesting concept presented was the use of heaters on the chips to control further processing (here, the SnO_2 deposition was modified by the temperature to which the thermal platforms were electrically heated). Further work by this group (Cavicchi, et al. (1994)) yielded sensors with thermal time constants of 1 ms and thermal resistances of 8,000 K/W.

A general description of a commercial version of such a SnO_2 sensor for CO was presented by Lyle and Walters (1997). The Motorola (Phoenix, AZ) MGS1100 device consists of a bulk micromachined (using wet anisotropic etching) 2 μm thick silicon membrane with a polysilicon heater, LTO insulating layer and an SnO_2 sensing layer. The sensor was operated in a biphasic mode for each measurement cycle, using a 5 V heater drive for 5 s for a membrane temperature of 400°C and a 1 V drive for 10 s for a temperature of 80°C. The SnO_2 resistance was measured near the end of the lower temperature part of the cycle and was relatively insensitive to humidity. The reported range of sensor operation was ≈ 1 to 4,000 ppm and operation between 1 and 400 ppm was demonstrated for relative humidity values of 20 to 80%. The sensor's power dissipation was on the order of 80 mW.

2.2 WORK-FUNCTION-BASED SENSORS

This class of sensors, including the CHEMFET, utilize metal-insulator-semiconductor junctions and the fact that the work function of the materials at the interfaces can be chemically modulated.

2.2.1 ADFET GAS SENSORS

These devices utilize an extremely thin gate oxide (< 5 nm) onto which sensed gases adsorb (by both physisorption and chemisorption), hence the term *ADFET*. The electrostatic fields from the adsorbed molecules can penetrate the thin oxide and modulate the drain current of the FET. A major problem is lack of selectivity due to interference by background gases and/or contaminants and electrical noise pickup by the floating gate. The stability of these devices is often poor since degradation of the thin SiO_2 film is likely. In addition, device performance is expected to change over time due to additional growth of the native oxide as the sensor is exposed to the environment. Suspended-gate ADFETS (illustrated below) are somewhat better in performance since the suspended metal gate shields the gate from electrical noise while still allowing gases to access the gate oxide. In addition, chemically sensitive films may be deposited onto the suspended gate to increase selectivity. However, these structures will still suffer from the other limitations of standard ADFETS. For these reasons, ADFETS have not been used extensively for gas sensing and their appearance in the literature is limited (see Middelhoek and Audet (1989), Janata (1989), and Madou and Morrison (1989)).

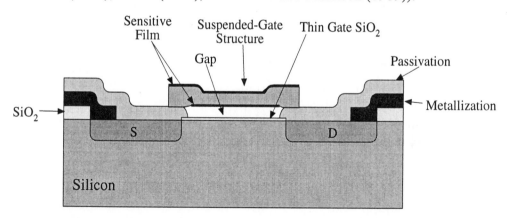

Illustration of a suspended-gate ADFET structure. After Janata (1989).

2.2.2 PLATINIDE-BASED HYDROGEN SENSORS

Lundström, et al. (1975) demonstrated a Pd gate FET structure that could be used as a hydrogen sensor (see also Armgarth and Nylander (1981) and Lundtsröm (1981)). H_2 adsorbs readily onto the Pd gate material and dissociates into H atoms (it should be noted that other platinide metals, such as Pt, Ir, etc., show similar affinities for hydrogen). The H atoms can diffuse rapidly through the Pd and adsorb at the metal/oxide interface, changing the metal work function. This shifts the drain current (or shifts the flat-band voltage of a MOS capacitor) and can be viewed as a shift in threshold voltage, V_T,

$$\Delta V_T = \frac{-\mu N \theta}{\varepsilon_o}$$

where,

μ = dipole moment of interfacial hydrogen

N = density of adsorption sites (for Pd, the average surface atom density gives N = 1.67×10^{19} m^{-2})

θ = fraction of surface sites covered, $0 < \theta < 1$

ε_o = dielectric permittivity of free space (8.85419×10^{-12} C^2/(N•m^2) or F/m^2)

They demonstrated H$_2$ sensitivity down to $\approx$ 10 ppm.

Also, other gases, such as H$_2$S and NH$_3$ can be detected if they can dissociate to release hydrogen (Lundström, et al. (1975) had observed NH$_3$ responses). Filippov, et al. (1995) described another sensor based on work-function modulation (Pd-SiO$_2$-Si MOS structure) sensitive to ammonia and carbon monoxide. The device operated using the same mechanism as the Lundström device (dissociative adsorption of hydrogen or a hydrogen-containing molecule onto the surface of the metal electrode), as well as allowing (in the SiO$_2$ layer) the CO to diffuse to the surface and to exchange oxygen atoms with the surface, changing the Fermi level of the gate material. The resulting work-function modulation changed the flat-band voltage of the gate capacitor. This voltage shift could be as much as 500 mV and was easily detectable using standard techniques. To increase sensitivity and selectivity of the device, alteration of the microstructure of the thick (60 nm) Pd electrodes was examined. Cyclic treatment of the Pd with hydrogen resulted in significant changes in the microstructure of the metal, forming a porous structure with significantly increased surface area. This expanded surface area resulted in an increase in catalytic activity of the electrode, thereby magnifying the flat-band voltage change by twofold. Dobos and Zimmer (1985) reported a similar device with a PdO-Pd gate that was lithographically "perforated" to allow CO gas to access the gate surface. They reported CO sensitivity down to $\approx$ 5 ppm but significant cross-sensitivity to butane, ethanol, and methane.

One problem with such sensors has been long-term drift, which seems to be mitigated to some extent by depositing a thin alumina (Al$_2$O$_3$) layer between the Pd gate and the underlying SiO$_2$.

Rodriguez, et al. (1992) and Hughes, et al. (1994) at Sandia National Laboratories, developed a design for a hydrogen sensor that provides a very large dynamic range (1 ppm concentration all the way to pure H$_2$) through the use of two different types of sensor: a catalytic gate transistor for low concentrations and a resistive catalytic sensor of the same Pd/Ni alloy for high concentrations.

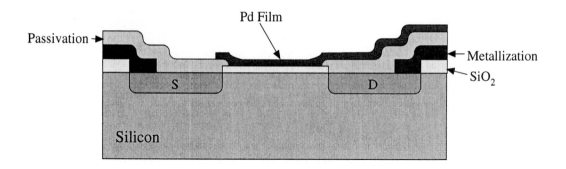

Illustration of a Pd gate FET structure. After Lundström, et al. (1975).

Tamura, et al. (1993) proposed the use of anodically oxidized tungsten films for highly sensitive hydrogen detectors (they did not build micromachined devices) as an alternative to platinide films. Since tungsten is readily available in many fabrication facilities, this may be a viable option.

2.2.3 ION-SENSITIVE FETS (ISFETS and CHEMFETS)

The fact that MOS transistors were so sensitive to surface contaminants (a serious problem in their manufacture at that time) was put to use by Bergveld in 1970 with his "ISFET" (*ion-sensitive FET*) (the original reference is Bergveld (1970)). The basic structure of the ISFET is a MOSFET, but rather than having a gate contact, the gate insulator is directly exposed to the ionic medium. The drain current is thus modulated by ion concentration rather than the applied V_{GS} (although an external electrode may be used to apply a voltage in series with the ionic modulation). Typically, an external "reference" electrode is required for stable operation of the ISFET so that the "V_{gs}" or equivalent is known (unfortunately, including such a reference is generally nontrivial). As illustrated below, the basic ISFET is an exposed gate oxide (or other dielectric material, such as silicon nitride) MOSFET and functions as a pH sensor (this is discussed in Siu and Cobbold (1979)). Commercial pH meters using such sensors are available, and an example of a manufacturer is Beckman Instruments, Inc., Fullerton, CA, who market a silicon nitride gate ISFET probe with pH accuracy of $\pm$ 0.01 pH unit.

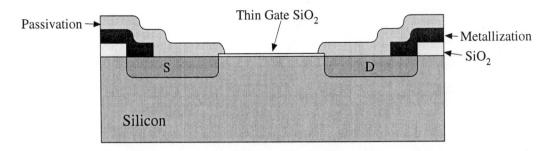

Illustration of an ion-sensitive FET, or ISFET. After Bergveld (1970).

Unlike the platinide gate hydrogen sensors, in the case of ISFETs the operational principle is not the diffusion of hydrogen (ions) through the gate oxide (the diffusion is too slow to account for the fast responses ISFETs provide), but rather reaction of the hydrogen (ions) with the SiO_2 gate material. These reactions are,

$$SiOH \Leftrightarrow SiO^- + H^+ \qquad \text{and} \qquad SiOH + H^+ \Leftrightarrow SiOH_2^+$$

and thus the surface of the gate oxide is either exposed or "blocked" by the formation of SiOH. The response of SiO_2 gate ISFETs is not very linear (particularly for low pH), but if an Al_2O_3 (alumina) gate insulator is used instead, improved linearity is obtained in the range of pH = 2.5 to 11 (in that case, the difficulty is forming a sufficiently good Al_2O_3 layer). Similarly, silicon nitride or borosilicate glass can provide good performance (see Harame, et al. (1987) and Abe, et al. (1979)). Applications of ISFET devices are discussed in Bergveld (1991).

By coating the gate region with an ion-selective membrane, one can make *CHEMFETs*, which respond specifically to certain ions. Typically, CHEMFETs are made by dip-coating ISFETS, but improved techniques, such as screen printing, exist (discussed below). Specific compounds can be sensed by using the high specificity of biological molecules such as enzymes and antibodies in the membrane. Sensors that incorporate such biological molecules are generally referred to as *biosensors*, and they are discussed below.

Good surveys of the ion sensor field can be found in Kelly and Owen (1985) and Sibbald (1986). Another interesting reference regarding MOS-based chemical sensors is Senturia, et al. (1977). In addition, a model for the response of CHEMFETs based on chemical kinetics was presented by Eddowes (1987).

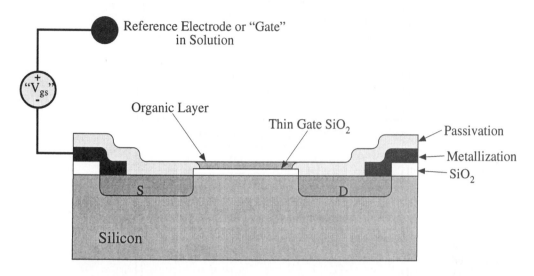

Illustration of a basic CHEMFET structure, showing the external reference electrode in the solution. After Madou and Morrison (1989).

2.3 ELECTROCHEMICAL TRANSDUCERS

There is a broad class of sensors based on a simple electrochemical electrode concept. Essentially, current is transduced from the circuit domain into the chemical domain through oxidation or reduction of chemical species at the electrode surface. These structures are among the simplest (they can be as simple as a region of bare metal in solution) and are probably the most useful chemical sensors, and also play a major role in biological interfacing (e.g., neurophysiological probes, as discussed below). While in most cases of interest, the electrode is in a liquid environment, this will not always be the case (e.g., in solid electrolyte systems, etc.). The discussion below will assume the case of liquid, aqueous solvent systems.

In a biological setting, for example, electrodes interact with a complex mixture of ions, each of which has different mobilities (e.g., roughly six orders of magnitude lower than carriers in silicon) and chemistries. In addition to the small ions shown in the table below, there are much larger charged molecules in most biological systems, with mobilities considerably lower still.

Carrier or Ion	Mobility, μ, in cm^2/(V•s)
Electron in Si	1.50×10^3
Hole in Si	4.50×10^2
H$^+$ in H$_2$O	3.63×10^{-3}
OH$^-$ in H$_2$O	2.05×10^{-3}
Cl$^-$ in H$_2$O	7.91×10^{-4}
K$^+$ in H$_2$O	7.62×10^{-4}
NH$_4^+$ in H$_2$O	7.61×10^{-4}
ClO$_4^-$ in H$_2$O	7.05×10^{-4}
Na$^+$ in H$_2$O	5.19×10^{-4}
HCO$_3^-$ in H$_2$O	4.61×10^{-4}
Li$^+$ in H$_2$O	4.01×10^{-4}

Table showing the mobilities of carriers in silicon and those of ions in a typical physiologic solution. From Sze (1981) and Bard and Faulkner (1980).

2.3.1 IONIC CAPACITANCE

When a metal electrode is placed in an ionic solution, chemical reactions result in the formation of a space charge layer (and hence capacitance) near the electrode. An example of such a reaction is a one-step electron transfer,

$$D \Leftrightarrow A^+ + e^-$$

Initially, the energy barrier for the forward reaction (*oxidation*) is lower than the reverse reaction (*reduction*). Eventually the potential that develops due to the build-up of ions in the solution (*space charge*) and electrons in the electrode reduces the barrier for the reverse reaction, equilibrium is reached, and there is zero net current across the interface unless the electrode's electrical potential is changed. The space charge layer represents a capacitance since it is a "stored" charge, Q, in proportion to the applied potential, V,

$$C = \frac{Q}{V} \quad \text{in F}$$

Looking from the electrode out into the solution, there are two groups of oriented molecules, considered to comprise three conceptual "layers" of space charge: the *hydration sheath* (one layer of oriented water molecules), the *inner Helmholtz plane*, or IHP (the location of the oxygen atoms of the hydration sheath), and the *outer Helmholtz plane* or OHP (the first layer of hydrated ions). The relationships between these three layers are illustrated below. It is interesting to note that a potential of 1 V between the electrode and the outer of these two layers (sometimes referred to as the *double layer*) (less than a 1 nm distance) is equivalent to a field strength of nearly 10^9 V/m.

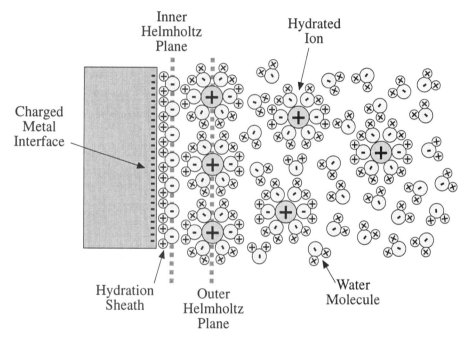

Illustration of the space charge layer in very close proximity to an electrode in solution. Adapted from Kovacs (1994).

If one assumes that the ions are formed into a "sheet" at the OHP and that the voltage drop through the dielectric layer is linear, and it is assumed to be a parallel-plate capacitor, the *Helmholtz capacitance*, C_H, can be estimated using,

$$C_H = \frac{\varepsilon_o \varepsilon_r A}{x} \quad \text{in F}$$

where,

ε_r = relative dielectric permittivity of the medium between the two plates (in this case it is water, and ε_r is 78.54 at 25°C, but can be as low as 6 at the interface due to saturated dipole orientations)

A = surface area of the electrode

x = distance to the outer Helmholtz plane (the other capacitor "plate")

Assuming an outer Helmholtz plane distance of 0.5 nm, this yields a value for C_H of ≈ 0.11 F/m². A 1 µm² plate thus has a theoretical capacitance C_H of 0.11 pF. This is the basis of electrolytic capacitors that are commonly used in electronic circuits. The actual value of the capacitance per unit area is a function of concentration, temperature, surface roughness and other factors. Also, it is known that charge in solution is not two-dimensional (i.e., a sheet) and not fixed (thus capacitance varies with DC bias, as seen in semiconductor junctions). For non-biased electrodes, the space charge voltage drop occurs within a few nanometers from the electrode surface, and shrinks even farther under bias. More complex models have been developed (discussed in Bard and Faulkner (1980) and Bockris and Reddy (1970)) that take into account the non-planar charge distribution and other effects, but for first-order estimates, the Helmholtz approximation can be adequate.

2.3.2 CHARGE TRANSFER: RESISTIVE MECHANISMS

In many situations, DC current is passed through electrodes (e.g., batteries). The movement of charge into or out of an electrode requires a shift of its potential from its equilibrium value. This shift, referred to as an *overpotential*, is defined as,

$$\eta = V - V_o$$

where V_o is the equilibrium potential and V is the applied potential. This overpotential is really (at least) four overpotentials in series, defined by Cobbold (1974) as,

$$\eta = \eta_t + \eta_d + \eta_r + \eta_c$$

where the individual overpotential components are:

η_t – due to charge transfer processes through electrode double layer

η_d – due to diffusion of reactants to and from electrode

η_r – due to (slow, rate-limiting) chemical reaction(s) preceding or following the electron transfer, occurring at or near the electrode

η_c – due to "crystallization" or incorporation of metal ions deposited on the electrode surface into the crystal lattice, adsorption/desorption, or other surface reactions

At conditions near equilibrium (small signal, as discussed below), η_t is the dominant overpotential if any steady-state current flows. At higher current densities,

η_d becomes dominant as the rate of supply of reactants from the bulk solution becomes the factor limiting the current. Returning to η_t, for an electron-transfer reaction for oxidized and reduced species O and R, respectively,

$$O + ze^- \Leftrightarrow R$$

the current density is given by the *Butler-Volmer equation,*

$$J = J_o \left(e^{\frac{(1-\beta)z\eta_t F}{RT}} - e^{\frac{-\beta z\eta_t F}{RT}} \right) \quad \text{in A/(cm}^2 \bullet \text{V)}$$

where,
J_o = exchange current density, in A/cm^2
z = charge of the ion in question
F = Faraday constant (9.64846×10^4 C/mol) = q times Avogadro's number, N_A
R = molar gas constant (8.31441 J/(mol$\bullet$K)) = Boltzmann's constant, k, times N_A
T = temperature, in K

and β is the *symmetry factor* ($0 < \beta < 1$), which reflects differences in energy barriers to electronation and de-electronation reactions (and hence determines the symmetry of the positive- and negative-current ranges of electrode J-V or I-V characteristics).

An interesting electrical circuit analog is two diodes in antiparallel, with the diode ideality factor, n, being loosely analogous (and inversely related) to the symmetry factor (if n were different for each diode, the J-V curve would be asymmetrical). The current density equation in the diode case is,

$$J = J_o \left(e^{\frac{qV}{nkT}} - e^{\frac{-qV}{nkT}} \right)$$

It is also noteworthy that there is a direct electrochemical equivalent of V_T (the *thermal voltage* commonly used in electronics), where,

$$V_T = \frac{kT}{q} = \frac{RT}{F}$$

which is useful when considering what constitutes a small electrochemical signal, for which simple linearized models can be used. For the symmetrical case (reversible chemistry is generally "non-rectifying," where $\beta = 0.5$), one obtains,

$$J = 2J_o \sinh\left(\frac{zF\eta_t}{2RT}\right)$$

which is plotted for various drive signal amplitudes below.

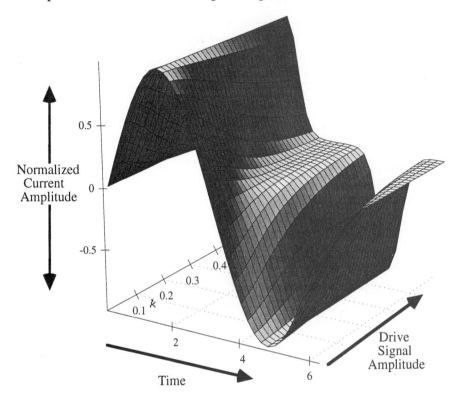

Normalized current response of an electrode to an applied sinusoidal voltage input (of peak-to-peak amplitude k, with values ranging from a few mV to 500 mV p-p), illustrating its nonlinear behavior for large signals. Adapted from Kovacs (1994).

As for semiconductor diodes, one can obtain the *incremental resistance* in the linear region,

$$R_t = \frac{\partial \eta_t}{\partial J} = \frac{RT}{J_o zF}\left[\cosh\left(\frac{zF\eta_t}{2RT}\right)\right]^{-1}$$

which, for small signals (η_t on the order of RT/F or less), yields,

$$R_t = \frac{\partial \eta_t}{\partial J} \approx \frac{RT}{J_o zF}$$

2.3.3 SPREADING RESISTANCE AND WARBURG IMPEDANCE

An additional resistive effect due to the resistance of the bulk electrolyte around the electrode is the *spreading resistance*. It is geometry specific, and can be (sometimes) derived in a closed form or estimated through a numerical integral of the resistance of symmetrical shells of electrolyte moving outward from the electrode. The spreading resistance manifests itself mainly at frequencies above $\approx$ 10 kHz, when the double-layer capacitance has essentially "shorted out" the other terms in the electrode model, shown below.

The *Warburg impedance* is another effect that is related to diffusion "waves" of ions near an electrode under sinusoidal drive. It is often represented in models as a parallel RC combination, but is actually neither resistive nor capacitive, with a nearly constant phase shift of 45°. This impedance is discussed in Cobbold (1974).

2.3.4 BASIC ELECTRODE CIRCUIT MODEL

The electrode effects described above can, to some extent, be represented as circuit elements in a simplified small-signal electrode model shown below. This model is quite useful and important because the basic electrode forms the underlying transducer in a wide variety of transducers.

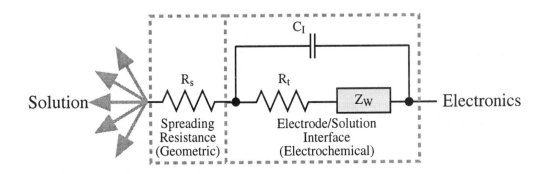

Theoretically derived small-signal equivalent-circuit representation of the electrode/solution interface showing the interfacial capacitance, C_I, the charge-transfer resistance, R_t, the parallel diffusional (Warburg) impedance, Z_W, and the spreading resistance, R_s. Adapted from Kovacs (1994).

In practice, empirical measurements are almost always necessary to characterize the operation of real electrodes. The equivalent circuit elements of electrode/electrolyte interfaces are determined by many parameters (only some of which are under the control of the researcher), so exact predictions of their behavior cannot generally be expected for realistic conditions. The circuit elements shown above, as well as many other details of using such electrodes as neural interfaces, are discussed in more detail in Robinson (1968), Cobbold (1974), and Kovacs (1994).

2.3.5 ELECTROCHEMICAL SENSING USING MICROELECTRODES

With the basic transducer being a small metal electrode insulated everywhere except at a specific location for chemical reactions, several electrochemical analytical and synthetic systems can be implemented. This type of electrode sensor is sometimes differentiated on the basis of the electrical parameter that is measured, hence called *potentiometric* or *amperometric* sensors.

Potentiometric setups are sometimes used to measure open-circuit potentials of electrochemical cells, and can be used for measurements of pH and other parameters. By actively driving an electrode, ionic species can be detected and identified. One approach to this is applying a voltage ramp to an electrode in solution and recording the current flowing as different species are oxidized or reduced. The voltages at which these reactions occur are indicative of the nature of each reaction and thus of particular species in the solution (this is called *voltammetry* since voltage is applied and current is measured). Amperometric operation of electrodes can be used to determine such things as reaction rate (current is proportional to reaction rate in electrochemical reactions), quantities of a given species (see the example of dissolved gas sensing below), etc.

Conductivity between two microelectrodes can readily be measured (with AC or DC drive signals) and can be used to estimate purity of water and other compounds (see Hoffmann, et al. (1995)) and, for example, the degradation of motor oil (Lee, et al. (1994)). The impedance spectra of microelectrodes (relative to each other or to larger counterelectrodes) can be measured and used to learn about chemical kinetics, adsorbed species on the electrodes, etc. (see Macdonald (1987)). Since numerous factors (surface roughness, microstructure, adsorbed molecules, etc.) influence the impedance of microelectrodes, electrode characterisitics are sometimes difficult to control and predict. Electrode impedance variability can have a profound impact on the interpretation of bioelectric events as well as quantification of chemical parameters. Great care must be taken to ensure repeatable electrode characteristics. This is often achieved by careful cleaning procedures, chemical modification, surface etching, and other alterations of the microelectrode surface.

Electrodes can also be deliberately driven away from equilibrium and used to locally synthesize compounds. For example, in a salt water solution, applying a sufficiently high voltage to a pair of electrodes will generate hydrogen gas at one electrode and oxygen gas at the other (this is *electrolysis* of water).

Many groups have demonstrated micromachined electrodes and have used them for electrochemical experiments and as sensors for industrial and medical applications. Due to their small size, they can offer greatly improved performance over macroscopic equivalents. For example, microelectrodes can have hemispherical diffusion gradients (as opposed to planar gradients for larger electrodes) and therefore more rapid access to ions in the solution. Arrays of microelectrodes can be placed electrically in parallel, yet be diffusionally isolated if their relative spacing is appropriately large. Lithographic fabrication of microelectrodes makes arrays quite feasible, whereas they would be quite time-consuming to fabricate via conventional means. Electrodes can also be modified for improved selectivity (or other parameters) by functionalizing the surfaces chemically, as described by Wrighton (1986). Using this approach, one electrode design can potentially be used for a variety of applications, depending on the modifications chosen. Also, electrodes in an array can be fabricated from a variety of different materials, hopefully to obtain useful information from the differences between their responses to electrochemical tests (see Glass, et al. (1992)). In addition, the required processing is often compatible with CMOS circuitry, as demonstrated by Kakerow, et al. (1994), who demonstrated an electronically addressable 400-element Pt electrode array (with two Ag reference/counter-electrodes) for potentiometric or amperometric measurements.

POTENTIOMETRIC SENSING

Potentiometric electrochemical sensing involves measuring the (generally open-circuit) oxidation/reduction potential of an electrochemical reaction as a means to measuring chemical phenomena of interest.

As described in Brett and Brett (1993), an electrode can act either as a sink or a source of electrons, to act to oxidize or reduce a species in solution (or part of the electrode itself, as in metal dissolution reactions). One can write a generic reduction reaction at one electrode as,

$$\sum_i \gamma_i [O_i] + ne^- \rightarrow \sum_i \gamma_i [R_i]$$

where,

γ_i = activity coefficient for each species, positive valued for products (reduced species) and negative valued for reactants (oxidized species)

$[O_i]$ and $[R_i]$ = concentrations of the oxidized and reduced species, respectively

$\gamma_i[O_i]$ and $\gamma_i[R_i]$ = activities, a_{Oi} and a_{Ri}, of the oxidized and reduced species, respectively

n = number of electrons transferred

The *Nernst equation* gives the potential (relative to a reference electrode potential, E°, the standard hydrogen electrode) measured for such a reaction as,

$$E = E^{\circ} + \left(\frac{RT}{nF}\right)\ln\frac{\prod\limits_{i} a_{Oi}}{\prod\limits_{i} a_{Ri}}$$

This equation can be used to predict potentials of deliberately (e.g., batteries) and accidentally (e.g., corrosion) assembled cells, and a large number of half-cell potentials (one for each of the two electrodes in a cell, tabulated with respect to the standard hydrogen electrode) are widely available in the literature (e.g., Bard and Faulkner (1980)).

There is a large variety of potential chemical measurands that can be determined using the simple approach of potentiometry, and a good general example is that of metal oxide pH sensors. Potentiometric determination of pH can readily be done using the hydrated oxides of many metals, such as Pt, Ir, and W. As discussed by Hendrikse, et al. (1997), the general form (where Me represents the metal) of the redox equilibrium reaction involved is,

$$MeO(OH) \leftrightarrow MeO_2 + H^+ + e^-$$

for which the Nernst equation becomes,

$$E = E^{\circ} + \frac{RT}{F}\ln\left(\frac{a_{MeO(OH)}[MeO(OH)]}{a_{MeO_2}[MeO_2]}\right) + \frac{RT}{F}\ln(a_{H^+})$$

where the last term can be linearly related to the definition of pH,

$$pH \equiv -\log[H^+] \text{ or, more accurately, } pH \equiv -\log(a_{H^+}[H^+])$$

Thus it can be seen that for suitable metal oxide electrodes, pH can be computed from the measured potential. Such electrodes are relatively easy to realize in micromachined forms, although non-ideal reference electrodes, such as Ag/AgCl electrodes, are often the only types that can be integrated with them (more accurate reference electrodes generally contain liquid or gaseous components). In practice, such pH sensors have some limitations, including sensitivity to elecro-chemically active substances in solution (such as O_2) and pH dependencies of the activity coefficients shown above. Despite these potential problems, the simplicity of such devices make them attractive for use in some settings, particularly when suitable metals are already present for use as electrodes for other purposes.

An example of a sensitive metal oxide for this application (sensitivity -60 to -80 mV/pH unit), iridium oxide, is discussed by Olthuis, et al. (1990, 1991) and Pásztor, et al. (1993). It is particularly noteworthy that the last reference describes a method for electrochemically depositing iridium oxide (from a solution of 0.15 g $IrCl_4$ (99.5%), 1 ml H_2O_2 (30%), and 0.5 g $(COOH)_2 \cdot 2H_2O$ in 100 ml H_2O, with the pH adjusted to 10.5 with K_2CO_3). By cycling the applied potential on thin-film Au microelectrodes 80 times between -0.2 and +0.6 V at 100 mV/s, a film of $\approx$ 50 nm thickness was deposited. This approach could be quite useful for fabricating pH sensors as a post-processing step for a variety of microstructures and CMOS devices (with a suitable plating seed layer).

At the cost of the added complexity of combining such a metal oxide electrode directly with the gate of a MOSFET (as opposed to simply fabricating an ISFET, as discussed above), Hendrikse, et al. (1997) demonstrated great improvement in drift. They subtracted the electrically measured threshold voltage for the metal-oxide-gate MOSFET from the potential of the metal oxide gate itself relative to ground (thus directly measuring the pH-dependent potential difference between the metal oxide and the electrolyte itself, and subtracting out the Nernst equation terms entirely). This unusual potentiometric approach yielded a very low-drift pH sensor.

AMPEROMETRIC SENSING

An important principle in electrochemical sensing of analyte concentrations is that of mass-transport-limited current measurement (a form of current-based, or *amperometric*, measurements). This concept applies to cases where an electro-chemical reaction can be used to deplete the analyte (reactant) at the surface of an electrode, such that diffusion of fresh reactant to the electrode from the bulk solution becomes the rate- (current-) limiting factor. As discussed in Janata (1989), if a voltage step is applied of sufficient amplitude to immediately locally deplete the reactant species of interest at the surface, the resulting limiting current is theoretically given by,

$$i_L(t) = nFAC\sqrt{\frac{D}{\pi t}} + \frac{nFACD}{r}$$

where,
 n = number of electrons transferred in the reaction
 A = area of the electrode(s), in cm^2
 C = concentration of the species of interest, in cm^{-3}
 D = diffusion çoefficient of the species of interest, in $cm^2/(V \cdot s)$
 r = radius of the electrode (assumed hemispherical), in cm

The first term of the above equation is known as the *Cottrell equation*, and predicts the transient current, while the second, geometry-dependent term (shown specifically for a hemispherical electrode) gives the steady-state current. It can be seen that relating the current to analyte concentration can be done by either factoring in the square root of time or waiting until the steady-state value is reached. For microelectrodes, which tend to be planar, the access of central regions of their surfaces to reactant may be less than that of a hemisphere, and thus the limiting current behavior is quite different. Unfortunately, a similar closed-form equation is not necessarily available for any given geometry. A series expansion solution for a circular planar electrode is,

$$i_L(t) = 4nFACD\left(1 + 0.72t^{-\frac{1}{2}} + 1.22t^{-\frac{3}{2}} + \cdots\right)$$

Vast numbers of chemical species can be sensed in this way, and they do not need to be electrochemically active. Coatings on the electrode surfaces can be engineered to convert the analyte into an electroactive species that can then be measured (indirectly) amperometrically. This approach is often employed in biosensors (discussed below) in which an enzyme (or other biomolecule) is used to perform the conversion. Some example enzymes and sensed products (all detected using a Pt electrode) are listed in the table below.

Analyte	Enzyme	Sensed Species
Glucose	Glucose Oxidase	H_2O_2 or O_2
L-Amino Acids	L-Amino Acid Oxidase	H_2O_2
Alcohols	Alcohol Oxidase	O_2
Uric Acid	Uricase	O_2
Phosphate	Phosphatase/Glucose Oxidase	O_2

Table showing some example enzymes used to convert non-electroactive analytes into species that are readily sensed amperometrically. After Janata (1989).

As an example of a direct amperometric sensor, Langereis, et al. (1997) described a thin-film Pt H_2O_2 sensor. A pair of interdigitated electrodes was used for the H_2O_2 measurements, and the temperature coefficient of resistance of the Pt wiring itself was used to provide simultaneous temperature sensing. It should be noted that they multiplied the current by the square root of time to obtain a concentration-dependent term, but apparently they did not compensate for the steady-state part of the time-dependent current.

It is often useful to determine the amount of a given gas that is in solution. A typical example is the need to know the concentration of dissolved oxygen (e.g., in blood or other physiologic solutions), although this approach can be applied to nearly any gas that can be reduced or oxidized. An amperometric approach is to use a noble metal (e.g., gold, platinum, iridium, etc.) cathode in solution (typically coated with an oxygen-permeable membrane, such as Teflon™, polyethylene, etc.) and apply a voltage between the measuring electrode and a larger counterelectrode. The reaction involving oxygen at the cathode (Cobbold (1974)) occurs at a relatively low potential (usually below those for other reactions) and is,

$$O_2 + 2H_2O + 2e^- \longrightarrow H_2O_2 + 2OH^-$$

$$H_2O_2 + 2e^- \longrightarrow 2OH^-$$

As the voltage is increased, the current that can flow eventually plateaus, and the current at which this occurs is proportional to the amount of dissolved oxygen. This amperometric technique is sometimes also referred to as *polarographic* oxygen sensing, and the electrodes themselves sometimes as *Clark electrodes* (Clark (1956)). Microelectrodes for such purposes are readily fabricated from thin-film metal regions coated with an insulator such as silicon nitride in all regions except the active electrode site and the bond pads for off-chip connections. As mentioned above, an oxygen-permeable material is applied over the active region of the electrode(s).

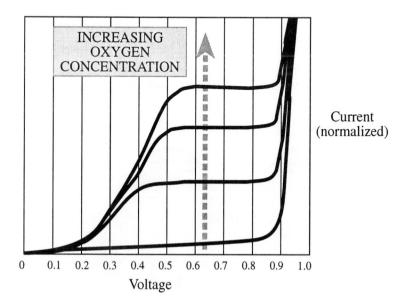

Illustration of I/V characteristics of noble metal dissolved gas sensing electrode. After Cobbold (1974).

Various non-planar electrode geometries have been discussed as a means for improving the performance of gas-sensing electrodes, generally by allowing the deposition and adhesion of thick layers of gel electrolytes and/or gas-permeable membranes (these thick layers resist the degradation phenomena that occur with thin layers and cause sensor drift). One example, presented by Conrath, et al. (1997), is the use of anisotropic wet etching of silicon to form pyramidal pits, which are metallized on their insides, or to form pyramidal hillocks, which are metallized at their tips. In both cases, relatively thick layers of the necessary coatings could be applied compared to the case with planar electrodes.

OTHER ELECTROCHEMICAL SENSING METHODS

A variety of other electrochemical sensing methods are available aside from the basic amperometric and potentiometric approaches discussed above, an example of which is *anodic stripping voltammetry*, a time-domain voltammetric technique.

Kounaves, et al. (1994) demonstrated a micromachined electrode array capable of resolving parts-per-billion concentrations of heavy metals in groundwater, drinking water, and industrial process fluids. The basic idea was to fabricate an array of microelectrodes using iridium and electroplate microscopic mercury hemispheres on them (iridium was chosen because it is insoluble in mercury, whereas most other metals suitable for electrode fabrication dissolve in it readily). When a measurement is being made, the electrodes are held at a cathodic (negative) potential so that trace heavy metal ions in solution are plated (reduced) onto the mercury and dissolve into it. The longer this is done, the more the system is allowed to integrate signal, increasing the sensitivity of the measurement. After concentrating the metal ions into the mercury, the array's potential is swept from cathodic to anodic (positive), "stripping" the metals out by reoxidizing them, each with a peak current occurring roughly at the characteristic oxidation potential of each species.

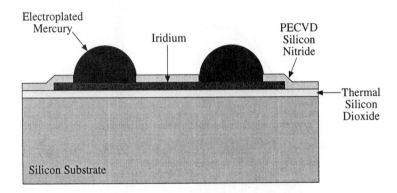

Illustration of a section of a micromachined thin-film iridium electrode array used for stripping voltammetry. After Kounaves, et al. (1994).

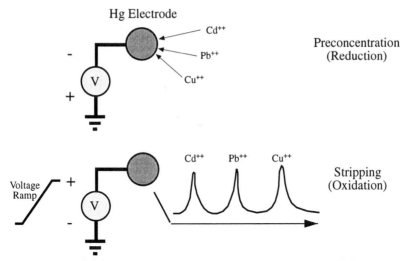

Illustration of the procedure for stripping voltammetry with a mercury electrode showing (top) the initial preconcentration of dilute ions in the solution into the mercury when the electrode is held cathodic and (bottom) the characteristic currents seen for each species when they are re-oxidized during anodic stripping.

Reay, et al. (1994) presented a complete CMOS integrated potentiostat circuit that could be used to control the operation of the stripping voltammetry probe with commands from a single-chip microcontroller or lap-top computer. The circuit provided 0.02% (of range) accuracy over 11 decades of current while consuming only 5 mW of power. The performance of the system was comparable to commercial bench-top potentiostats, but with greatly reduced, size, cost, and weight. Results for the two-chip combination were presented in Reay, et al. (1995, 1996). Such a potentiostat circuit is not limited to this particular type of analysis, and can be used in a variety of microelectrode-based sensors. Further information on IC potentiostats for sensor applications can be found in Turner, et al. (1987), Sansen, et al. (1990), and Ryan, et al. (1995).

2.4 ACOUSTIC WAVE SENSORS

Acoustic wave sensors are broadly useful, and a large number of devices have been successfully implemented and used in the detection of chemicals. The basic idea is to fabricate a sensor in which acoustic waves are propagated, and where some aspect of that propagation (e.g., velocity, amplitude, etc.) is changed by the adsorption/reaction or viscosity of the sensed species. Quite often, a chemically sensitive layer is applied to the sensor to enhance the interaction with the measurand and/or to make it more specific. Acoustic wave devices can also be used as actuators for mixing and moving fluids and particles.

Acoustic waves can be generated using piezoelectric, magnetostrictive, electrostrictive, photothermal, and other mechanisms, but piezoelectric tends to be the most popular in practice (often quartz or ZnO, with useful reviews of bulk piezoelectric resonators found in Lu and Czanderna (1984) and Benes, et al. (1995)). There are numerous types of propagated waves, transducer materials, and overall designs. The four major classes of acoustic wave transducers (illustrated below) are thickness-shear mode (TSM), surface acoustic wave (SAW), flexural plate wave (FPW), and acoustic plate mode (APM).

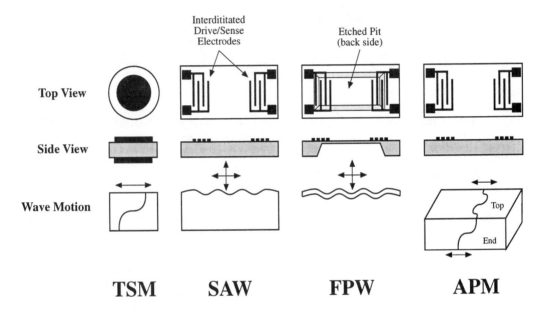

Illustration of the four major classes of acoustic wave transducers. After Grate, et al. (1993).

If the transit time of a propagated wave is altered by surface changes, the resonant frequency of an oscillator built around the transducer will thereby change. Since frequency can be measured with exquisite precision using low-cost circuitry, extremely sensitive sensors can be achieved. Alternatively, the complex impedance of the one- or two-port device can be measured relative to frequency, but this requires more complex external circuitry.

Such transducers, if operated in a self-resonant mode, can be extremely sensitive to adsorbed or otherwise bound molecules on their surfaces (see King (1964), Lu and Lewis (1972), and Vig, et al. (1996)). As an example of a biosensor application, antibodies can be coupled to the membrane and binding of antigen molecules will alter the membrane's mass, resulting in a resonant frequency shift. The frequency shift for a SAW device will be given by the *Sauerbrey equation,*

$$\Delta f = k f_o^2 \frac{\Delta m}{A}$$

where,
  k = device-dependent constant
  f_o = natural resonant frequency of the device without external mass loading, in
      Hz
Δm = mass change at surface of membrane, in g
  A = area of membrane, in cm^2

In addition to mass loading (described above), there are several other trans-duction mechanisms utilized for sensing with SAW devices, including viscoelastic, acoustoelectric and liquid density, and viscosity. Each is useful in different sensing applications and exhibits certain limitations as discussed below.

Viscoelastic sensing relies on changes in resonant frequency due to polymer mechanical modulus effects. As mentioned above, for chemical sensing, SAW devices generally employ an additional layer (often a polymer) that can recognize and bind the analyte of interest. While this bound analyte will contribute to mass loading of the device, swelling-induced modulus changes will often dominate the sensor frequency change. These devices make excellent vapor sensors but do suffer from highly temperature-dependent properties (modulus decreases with in-creasing temperature as the polymer expands).

As an acoustic wave propagates through a piezoelectric material, a layer of bound charges is formed at the surface. This bound charge generates an electric field that causes motion of charge carriers and dipoles in the adjacent medium. This process extracts energy from the propagating wave, thereby affecting its velocity and attenuation. This is known as the *acoustoelectric effect*, and has been used in gas sensors where the wave velocity changes due to this effect are significantly larger than those caused by mass loading alone. The acoustoelectric effect is independent of device frequency.

Liquid density and viscosity sensing monitors changes in wave velocity (correlated to resonant frequency) due to loading of the device by the liquid (equiv-alent to mass loading). These devices have potential for microliter viscometers and densitometers. An example of viscosity sensing using ultrasonic plate waves is discussed in Martin, et al. (1990).

An example of measuring fluid density and viscosity using flexural plate waves can be found in Wenzel and White (1990) and Costello, et al. (1993). The use of mass-loading of quartz resonators coated with organic sensing membranes to detect vapors is discussed in Wohltjen (1984). Such vapor sensors coupled with neural networks as an "odor recognizer" are discussed in Nakamoto, et al. (1991).

Further sensor applications of SAW devices are discussed in Martin, et al. (1987), Wenzel and White (1988), and Wang, et al. (1993). A recent example of the use of antibody reactions to provide extremely selective detection (in this case of a serum breast cancer antigen) is provided by Wang, et al. (1997). A useful and general overview of acoustic sensing devices can be found in D'Amico, et al. (1997).

A demonstration of ultrasonic mixing and the detection of that mixing (by enhanced local ion transport) was presented by Tsao, et al. (1991). The demonstration of acoustic mixing and transport (see Moroney, et al. 1991) implies that complete chemical synthesis/analysis systems could be integrated, especially when this approach is combined with other devices (see below and in the Microfluidic Devices chapter).

2.5 BIOSENSORS

The term *biosensor* refers to sensors wherein biologically derived molecules are used to perform an intermediate transduction between the desired measurand and some parameter readily measurable with a solid-state sensor. This approach takes advantage of the amazing selectivity of many biomolecule interactions, but unfortunately, some of the underlying binding or other chemical events are not easily reversible. Typically, an enzyme (protein), antibody (protein), polysaccharide, or nucleic acid is chosen to interact with the measurand. The underlying sensor can be nearly any of the types described above: ISFET, microelectrode, microcalorimeter, acoustic wave device, etc.

As described by Dewa and Ko (1994), four possible modes of physically coupling the biomolecules to the basic sensor are 1) membrane entrapment where a semipermeable membrane is used to physically separate the biomaterial (attached to the transducer surface) from the analyte solution; 2) physical adsorption that depends on a combination of van der Waals forces, hydrophobic forces, hydrogen bonds, and ionic forces to hold the biologically active material on the surface of the transducer; 3) matrix entrapment where a porous encapsulation matrix is formed around the biological material; and 4) covalent bonding in which the surface of the transducer is treated so it exhibits reactive groups to which the biological material can bind. These are illustrated below. In addition, the biomolecule layer may be replaced by a layer of living cells, as described in Sections 4.3 and 5.

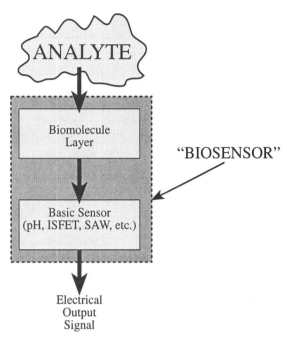

Illustration of the basic concept of a biosensor, showing the combination of biological molecules with a secondary sensor that provides an electrical signal output.

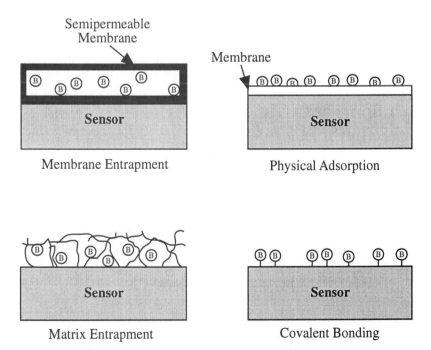

Four possible modes of physically coupling biomolecules to the basic sensor in a biosensor device. After Dewa and Ko (1994).

While each type of immobilization technique has unique application areas, all must confine the biological material such that 1) the material is constrained for the lifetime of the biosensor; 2) contact between the analyte solution and the biomaterial is possible; 3) by-products of the sensing reaction may diffuse out of the immobilization layer; and 4) the biomaterials are not inactivated by the immobilization process. The effectiveness of membrane and matrix entrapment are quite case-specific, with the relative success dependent on those items described above along with the matching of the membrane or matrix to the transducer. Physical adsorption of the biomaterial onto the transducer has the advantage of gently binding the biomaterial to the surface, thereby making distortion of the active conformation of the molecules unlikely. However, this gentle binding also makes quantification of the amount of bound biological material difficult and can facilitate desorption of the sensing layer with small changes in temperature, pH, etc. Covalent binding provides perhaps the most permanent attachment of the biomolecule to the transducer surface (thereby improving device lifetime). However, the reactive groups that bind the biomolecule may chemically modify its binding sites and may partially or entirely denature the molecules. In some cases, combinations of these methods of biosensor fabrication are employed. Several examples of different detection schemes coupled with biomolecules are given below. These examples are drawn from the large volume of available literature in this area, for a broader survey, the reader is referred to publications such as Buck, et al. (1989), Dewa and Ko (1994), and Kress-Rogers (1997).

2.5.1 RESONANT BIOSENSORS

As described in Section 2.4, acoustic wave and other resonant sensors can be used as biosensors. In general, a sensitive molecule is attached to the surface of the resonator (using one of the methods described above), and binding of analyte molecules adds mass and causes a shift in the resonant frequency. As well as references provided in Section 2.7, an interesting review of biosensor applications of acoustic wave devices is Andle and Vetelino (1994).

2.5.2 OPTICAL-DETECTION BIOSENSORS

A biosensor immunoassay based on optical diffraction was demonstrated by Tsay, et al. (1991). A silicon wafer was coated with a protein that binds a human hormone (monoclonal anti-choriogonadatropin (anti-hCG) antibody) via covalent bonds formed to aminopropyltriethoxysilane bound to the silicon. The wafer was subsequently illuminated with 254 nm UV light through a standard quartz photomask containing periodic areas of opaque and clear regions. The illumination intensity and duration was sufficient to inactivate the antibody in the exposed regions. The wafer was diced and individual chips were incubated with the analyte (serum hCG). This resulted in antigen - antibody binding in active anti-hCG regions on the chip, thereby creating a biological diffraction grating.

This grating produced a diffraction signal when illuminated with a light source such as a laser. A negative response (no anitgen - antibody binding) did not produce a diffraction signal since no biological diffraction grating was created. The hCG assay was successfully demonstrated. By including labels such as latex or colloidal gold (which increase the effective size of the bound analyte), the diffraction signal could be amplified, increasing sensitivity. Colloidal gold labeling was used for detection of β-lactam antibiotics (e.g., penicillin) in milk resulting in a detection limit of 60 ng/l. The authors note that this optical grating approach could be applied to other binding events such as DNA probes, etc.

Another optical-readout biosensor technique is *electrochemiluminescence*, or ECL. In contrast to the grating-based approach discussed above, ECL is "active" in the sense that when electrical current is passed through an appropriate mixture of compounds, light is emitted at an amplitude related to the concentration of a target analyte. Hsueh, et al. (1995, 1996) described a microfabricated electrochem-iluminescence cell for the detection of fluorescently labeled DNA. The cell was a combination of silicon and glass micromachined substrates. Two silicon chips were etched in KOH to define the solution chamber (total volume $\approx$ 85 μm) and two fluid inlet ports. A gold film electrode was evaporated on the lower silicon chamber surface (containing a pit for the chamber) and the two silicon pieces were bonded together using a low-temperature curing polyimide. The top silicon piece contained the fluid inlet ports and the top portion of the chamber, both of which were etched entirely through the wafer. A thin (1 μm) layer of electrically conductive, yet transparent indium tin oxide (ITO) was deposited on the glass substrate and then bonded to the top silicon piece using the same polyimide material. Thus one end of the chamber was coated with a gold electrode (cathode) while the other was the transparent ITO (anode). Electrogenerated chemiluminescence of tri-propylamine (TPA) and tris(2,2'-bipyridyl) ruthenium (II) (TBR) was then used to test the system for detection of DNA prepared using a selective amplification technique (the polymerase chain reaction, or PCR, as discussed in Section 2.6). Free TBR concentrations of 10^{-9} to 10^{-3} M were successfully detected and quantified using optical detection of the luminescence with an external photodetector through the glass/ITO window. Further work by this group on micromachined ECL chambers was presented in Hsueh, et al. (1997).

Also, while beyond the scope of this book, fiber-optic biosensors can also be prepared by coupling biomolecules to fibers and measuring absorbance, fluorescence, or other optical parameters. For information on such sensors, the reader is referred to Seitz (1984), Dewa and Ko (1994), Andrade, et al. (1985), and Janata (1989).

2.5.3 THERMAL-DETECTION BIOSENSORS

Thermally transduced biosenors are also feasible through binding enzymes to temperature sensors. In the presence of analyte, the heat of reaction for each enzyme is measured, and is related to its concentration. Micromachined calorimeters were described by van Herwaarden, et al. (1993), for use in both liquid and gas media. The gas calorimeter consisted of a 0.4 μm, 2.6 × 1.6 mm low-stress LPCVD silicon nitride membrane with an array of 56 aluminum and n-type LPCVD polysilicon thermocouples (with an estimated sensitivity of 10 mV/K), suspended by bulk etching from the back of a silicon support wafer. An interdigitated polysilicon electrode in the center of the membrane served as a chemiresistive sensor, on which a chemically sensitive and or catalytic layer would be deposited (neither sensitivities nor other results were presented for this design). Liquid-sample calorimeters were fabricated in a similar manner, but using a 4 or 8 μm thick n-type epitaxial layer on p-type silicon substrates to form the 3.5 × 3.5 mm membranes.

Electrochemical etching in KOH was used to define the membranes. The 120 thermocouples in this case were formed from implanted p-type silicon regions and aluminum traces (estimated sensitivity 50 to 70 mV/K). The chips were packaged such that the etched cavity contacted the test liquid thereby isolating the electronics from the liquid. Enzymes were immobilized on the underside of the membrane (in contact with the liquid), thereby allowing for selective binding of the target compound. Glucose oxidase was used with catalase for glucose detection, with an enthalpy change, $-\Delta H$ of 180 kJ/mol. Urease, for the detection of urea, yielded a $-\Delta H$ of 61 kJ/mol, and β-lactamase, for the detection of penicillin G, produced a $-\Delta H$ of 67 kJ/mol. For the glucose detection configuration, they reported a detection threshold on the order of 20 μmol, a sensitivity (in the linear region) of ≈ 45 μV/mmol and a saturation asymptote of ≈ 3 mmol. These devices are commercially available from Xensor Integration, Delft, Netherlands.

Another example of a thermal biosensor is the polysilicon-thermistor-based design of Xie, et al. (1996). They fabricated an array of four such thermistors on a quartz substrate and electrically insulated them with a layer of LTO. One enzyme was immobilized on each of the thermistors using N-hydroxysuccinimide (NHS)-activated agarose beads, 13 μm in diameter (Pharmacia Biotech, Upsalla, Sweden) and were isolated from each other using pure agarose. The analyte/enzyme combinations used were lactate/lactate oxidase, glucose/glucose oxidase, urea/urease, and penicillin/β-lactamase. Catalase was co-immobilized with the lactate and glucose oxidases to generate additional heat and to recover some of the oxygen used in the reactions. They demonstrated a limited dynamic range of sensitivity (0 to 40 mM concentrations), but this could likely be improved through better thermal isolation of the temperature sensors.

2.5.4 ISFET BIOSENSORS (CHEMFETs)

An immunoenzymatic assay utilizing covalent bonding of a molecule to which a specific receptor antibody could then be physically adsorbed was described by Colapicchioni, et al. (1991). As illustrated below, molecules of protein A were covalently bound to the gate area of an ISFET device using polyaminosiloxane (specifically, [3-(2-amionethyl) aminopropyl] trimethoxysilane and cross-linking agents such as glutaraldehyde). Protein A is a cell-wall component of the common pathogenic bacterium *Staphylococcus aureus* exhibiting high stability on exposure to high temperature, denaturing agents, and low pH. A key feature of this molecule is its ability to bind the so-called Fc region of antibodies, a section that is not involved in antigen (target) recognition.

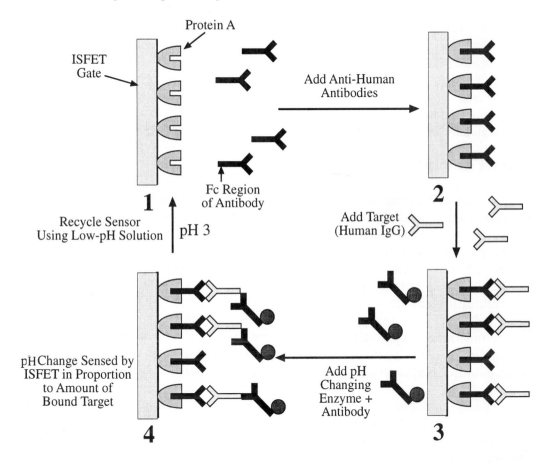

Illustration of the operating principle of an ISFET-based biosensor using antibodies to obtain analyte specificity. The target is specified by the choice of antibodies bound to a protein A layer and enzyme-labeled antibodies are used in the last step to alter the pH and allow the ISFET to sense the analyte. Adapted from Colapicchioni, et al. (1991).

The protein-A-coated ISFETs were exposed to an anti-human IgG antibody (from a rabbit), which was thus bound and specified the target antigen (note that other antibody types, e.g., against specific pharmaceuticals, could be used instead).

The target molecule was then added and allowed to be bound to the anti-human IgG. In this case, a human antibody was used as the target. The amount of bound analyte (in this case, the human antibody) was quantified by adding one more set of antibodies, this time with an enzyme that could produce a pH variation in proportion to the amount bound (and hence in proportion to the amount of analyte). The entire collection of antibodies could be released from the protein A in a low pH solution, and the sensor could be reused. Tests were performed on human immunoglobulin G (hIgG) and the herbicide atrazine with detection limits of 0.1 µg/ml and 1 ppb, respectively.

Further information on ISFET-based (pH) biosensors (or CHEMFETs), including a review of the underlying theory, and applications to glucose and penicillin sensing, can be found in a series of papers by Caras, et al. (1985a, 1985b) and Caras and Janata (1985), respectively. In addition, a useful discussion of the chemical selectivity of CHEMFETs was presented by Janata (1987).

2.5.5 OTHER pH-BASED BIOSENSORS

As described below in Section 5, optically addressable pH sensing can be accomplished by back-illuminating a silicon substrate coated on the opposite surface with a silicon dioxide/silicon nitride layer and immersed in the electrolyte to be measured. Photogenerated carriers charge the interface, and if the illumination is pulsed, an AC current signal measured from the silicon substrate to the solution can be used to determine the pH (since the liquid/dielectric interface is pH-sensitive for certain dielectrics, as discussed in Lauks (1979)). By moving the site of illumination, several regions on a single substrate can be indivdually addressed. Due to this fact, the approach is often referred to as light-addressable potentiometric sensing, or LAPS. Such pH sensors can form the basis of sensitive biosensors when coupled to appropriate biomolecules (or living cells, as discussed below).

Kung, et al. (1990) described a LAPS-based system for picogram quantitation of total DNA using DNA-binding proteins. Single-stranded DNA-binding protein (SSB) was conjugated with biotin (a linker molecule) and monoclonal anti-DNA antibody was conjugated with urease (for signal detection). These two conjugates were incubated with streptavidin and single-stranded DNA to form a complex of streptavidin-biotin-SSB-DNA-anti-DNA-urease, which was subsequently filtered through a biotin-coated nitrocellulose membrane (the biotin binds the streptavidin component of the complex). The membrane was rinsed to remove unbound reagent and then placed into a buffered urea solution in the LAPS device. The rate of conversion of urea to ammonia and carbon dioxide was monitored by the pH sensitive LAPS system. Since the amount of urease bound to the membrane is proportional to the amount of DNA, urease activity (and consequently pH) is an indicator of total DNA.

Using the system described above, Kung, et al. (1990) have successfully detected 2 to 200 pg of DNA in less than 2 hours with greater than 89% accuracy. They tested over 13 different DNA types with similar results, indicating this method detects DNA regardless of biological origin. However, the assay is sensitive to salt concentration (which affects DNA binding by SSB), pH of the incubation mixture (optimum range 6.6 to 8), filtration rate, total DNA concentration (dose effects with > 400 pg of DNA per ml), protein contaminants, and DNA fragment length (there is a minimum fragment length required for binding and small fragments may inhibit binding of larger fragments).

Therapeutic uses of DNA and monoclonal antibodies are highly sensitive to trace quantities of undesirable DNA fragments. For these high purity applications, detection of picograms of nonspecific DNA is required, rather than the amplification and identification of specific DNA fragments. Olson, et al. (1991) used a system almost identical to that of Kung, et al. (1990) for quantitation of DNA amplified by PCR, which is important for determination of the initial quantity of target DNA from an amplified sample. For this application, specific binding of a target DNA fragment was desired (rather than nonspecific DNA binding for total DNA determination). A biotinylated oligomer (DNA strand) and a fluoresceinated oligomer were hybridized to a single-stranded DNA fragment in solution. This hybridization complex is captured with a biotinylated membrane using streptavidin, as described above. A urease antifluorescein conjugate bound to the fluoresceinated probe allows for detection of the captured DNA via the urea conversions, as described above (pH monitored with a LAPS device). The assay was applied to a target DNA fragment amplified by the polymerase chain reaction (see Section 2.6 below). Both the amount of amplified target DNA and the amplification factor were determined to within a factor of two to seven depending on the starting concentration of DNA. The present sensitivity of the system is 2×10^7 starting DNA fragments. It is important to note that this dual probe format tolerates irrelevant DNA up to 10 μg, thereby providing a method for measuring a small fragment of target DNA in a large amount of irrelevant DNA.

The previous two examples illustrate the use of nucleic acids in both broad spectrum and extremely selective biosensors. All of the above examples serve to illustrate the diversity of possible biosensor configurations.

2.5.6 ELECTROCHEMICAL-DETECTION BIOSENSORS

Cyclic voltammetry has been used for detection of hybridized DNA and DNA-binding drugs. Maeda, et al. (1992) described a DNA-immobilized Au electrode, which showed an electrochemical response to a DNA-binding drug (quinacrine). In this case, the anodic peak current displayed a near linear relationship with the concentration of quinacrine in the range of 10^{-7} to 5×10^{-7} M. Millan and Mikkelsen (1993) immobilized an electroactive hybridization indicator onto glassy carbon electrodes. Target DNA hybridized on the surface affected the voltametric peak cathodic current. While the magnitude of the peak current was correlated with the amount of DNA, the technique has not yet established a quantitative means for DNA concentration determinations. Neither of these systems has been miniaturized, but micromachining techniques and IC fabrication technologies could possibly be used to do so.

An example of an amperometric biosensor is the glucose sensor presented by Hinkers, et al. (1995). They used a polyvinyl alcohol (PVA) matrix to apply the enzyme glucose oxidase (GOD) to a Pt electrode. A thick-film Ag/AgCl reference/counterelectrode was used. In the presence of oxygen, glucose present in the solution is oxidized,

$$glucose + O_2 \rightarrow gluconolactone + H_2O_2$$

The Pt electrode is held at a potential of + 700 mV with respect to the Ag/AgCl electrode, and thus any hydrogen peroxide present is oxidized, releasing hydrogen ions and causing current to flow,

$$H_2O_2 \rightarrow O_2 + 2H^+ + 2e^-$$

By measuring the current (or the pH, as discussed below), the glucose concentration can be determined. Hinkers, et al. (1995) demonstrated a flow-through glucose sensor based on this principle, and fabricated using bulk micromachining of silicon and bonding to a glass substrate (to form the flow channel). They demonstrated a linear response for concentrations up to $\approx$ 30 mmol/l, response times for 10 to 90% glucose concentration changes of 2.5 to 4.5 minutes and a drift of only a few percent over five days. The use of GOD for glucose detection is well known, and is a common biosensor application. Unfortunately, such sensors have generally not been stable long enough to make them useful with implanted insulin pumps in the treatment of diabetes. A description of a similar potentiometric sensor, but with an evaporated Ag/AgCl reference electrode can be found in Schnakenberg, et al. (1995).

2.5.7 CMOS-COMPATIBLE BIOSENSOR PROCESS

Goldberg, et al. (1994), demonstrated an approach to post-processing standard CMOS wafers into biosensors using screen-printed polymer membranes containing the desired biomolecules (in this matrix entrapment example, they used ionophores, which are compounds that provide selective responses to specific ion types). This approach allowed the entire wafer to be post-processed at once, rather than manually applying the polymer to one site at a time. Current was coupled from the polymer membranes into the aluminum conductors on the chip using a screen-printed silver epoxy intermediate stage.

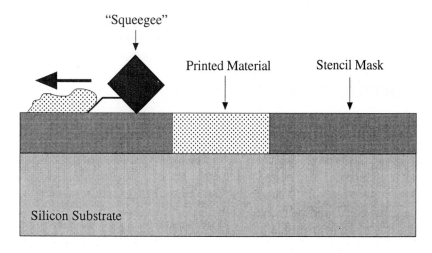

Illustration of screen printing to apply a biosensor layer above a silicon substrate. After Goldberg, et al. (1994)

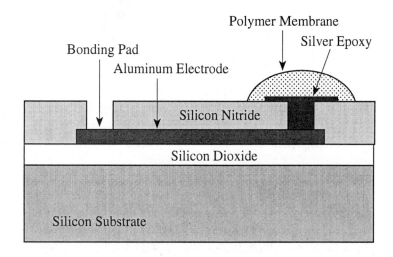

Cross-sectional view of a CMOS-compatible screen-printed polymer membrane placed atop cured silver epoxy that couples current into the underlying aluminum conductor. After Goldberg, et al. (1994).

They fabricated a complete "programmable multisensor chip" using a CMOS substrate, including a serial communication interface, A/D converter, programmable gain stage, bandgap reference, temperature sensor, and four ion sensors: potassium (using valinomycin as the ionophore), calcium (using ETH 129 as the ionophore), ammonium (using nonactin as the ionophore), and pH (using tridodecylamine as the ionophore). The fully integrated system could be self-calibrated for thermal and other drift effects using the on-chip reference and temperature sensor. This approach has great promise for future CMOS post-processed chemical sensor arrays.

2.5.8 OTHER BIOSENSOR TECHNOLOGIES

There are a variety of other biosensor approaches, including the use of whole cells as part of the sensor (see Section 5 below), and the use of surface plasmon resonance (changes in the permittivity of thin interfacial layers where the sensing molecules are located, interrogated by optical means), as discussed by Garabedian, et al. (1994), Jorgenson and Yee (1993) and many others. For further information, the reader is also referred to the general references on biosensors mentioned above.

2.6 BIOMOLECULAR GAIN MECHANISMS

As for most sensors, high sensitivity is generally desirable as long as sufficient signal-to-noise ratio is available (gain without improvement in SNR is generally useless, although SNR is seldomly reported in the sensor literature [with the marked exceptions of optical and magnetic devices]). Selective amplification of signals is therefore quite useful, and in biosensors, the specificity of biomolecular amplification mechanisms can readily be harnessed. An interesting review of such biomolecular sensors and gain via so-called *second messenger* molecules (which are generated in far larger numbers than the analyte molecules interacting with a cellular receptor), is Stieve (1983), covering cellular chemical, electrical, thermal, mechanical, and optical sensing mechanisms.

As discussed in Section 5 below, whole living cells can be used as parts of man-made sensors. This approach allows a sensor to make use of the sophisticated signal amplification mechanisms of a cells. Several other techniques exist to provide "gain" in biosensors. An interesting example is provided by methods that can amplify the amount of DNA in a sample. DNA amplification and analysis is of great importance for sequencing the human genome, genetic profiling, forensics, and unknown sample identification and classification. These techniques make use of biomolecules that normally function in cells to duplicate their genetic material during division. When properly used in vitro, tremendous amplification of trace quantities of nucleic acids is possible.

The most common, but certainly not the only such method, is the polymerase chain reaction, or PCR (for further information, see Erlich (1992). This reaction uses an enzyme, DNA polymerase, to carry out the duplication of single-stranded DNA in the presence of nucleotide triphosphates (the building blocks of DNA) and specific DNA strands (*primers*) that determine what sequence, if any, is amplified.

The specific polymerase used is obtained from a bacterium that thrives at the high temperatures of hot springs, *Thermus aquaticus* (the enzyme is referred to as taq polymerase), which was first isolated in Yellowstone National Park in 1969. Taq polymerase is thermostable, allowing an amplification scheme to operate where double-stranded sample DNA is melted apart into two single strands (at $\approx$ 90 to 95°C) and where the taq polymerase synthesizes a complementary strand for each one. Thus, the amount of DNA is doubled, and this doubling can be repeated many times simply by cycling the temperature. For example, for twenty cycles, approximately 2^{20} times the original quantity of DNA will be present in the solution (traces of DNA from amber-bound insects and even human mummies has been amplified in this way). This gain is provided without much reduction in SNR because the reaction is quite selective for specific sequences specified by DNA included in the initial mixture as primers. The key to the reaction is being able to accurately cycle the temperature in the reaction chamber to drive each round of DNA amplification. As for most thermal transduction functions, physical scaling can provide access to important properties such as reduced thermal mass, high thermal isolation, etc., and micromachined thermal cyclers for PCR illustrate this point well.

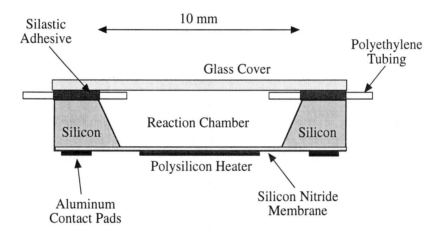

Illustration of a bulk micromachined DNA amplification chamber. After Northrup, et al. (1993).

Northrup, et al. (1993) and Northrup and White (1997a, 1997b) bulk micro-machined a reaction chamber in silicon, with a silicon nitride membrane and polysilicon heater and sealed the chamber with a glass cover (except for input and output tubes of polyethylene). They achieved PCR amplification of DNA comparable to that with commercial instruments, and newer results indicate that the PCR reaction can be carried out not only faster, but with greater "gain" using high-speed thermal cycling not readily possible with macroscopic devices (due to their larger thermal time constants) (Northrup (1995) and Woolley, et al. (1996)). As described by Northrup, et al. (1996), current versions of the device use two bonded silicon wafers with matching, anisotropically etched grooves to form a tunnel into which a disposable plastic reaction tube is inserted.

It is important to note that PCR is not the only form of DNA amplification that can be carried out. Several alternatives exist, and one that has been used in micromachined devices is an isothermal technique referred to as cycling probe technology (Tang, et al. (1997)). Due to the rapid development of new amplification and assay techniques in the molecular biology field, it is likely that many more nucleic acid amplification and analysis techniques will become available for exploitation in micromachined devices.

2.7 SELECTIVITY IMPROVEMENT USING ARRAYS

There are several potential approaches to improving chemical sensor selectivity. Naturally, it would make sense to deal with this through modifications to the sensor, but this is not always possible. In fact, many chemical sensors (particularly FET- or SAW-based) are remarkably nonselective. However, it is often feasible to fabricate arrays of nonselective sensors that, while each sensor has broad ranges of compounds it will respond to, show different response profiles to a variety of compounds. By correlating their outputs, one can often synthesize a selective sensor response.

This approach has been shown to provide enhanced selectivity, and has been particularly useful in odor recognizers, or so-called "electronic noses," as reviewed by Gardner and Bartlett (1994). They generally utilize an array of different gas sensors (semiconducting metal-oxide, conducting polymer, ADFETs, Pd gate MOS-FETS, chemoresistors, etc.). For example, Nakamoto, et al. (1991) demonstrated such a system-level implementation to allow an array of nonspecific quartz resonators, coated with different sensitive layers, to distinguish several specific odors. In order to discriminate between complex odors, the array of sensors must be combined with sophisticated pattern recognition algorithms and sensor calibration. A similar quartz resonator array system was demonstrated by Kraus, et al. (1995) using a variety of polysiloxane-based compounds as the sensitive films. In addition to general purpose arrays, the systems are often designed and implemented for a particular application, such as food characterization. Gardner and Bartlett (1995) provide a model for performance definition and standardization of electronic noses.

Fryder, et al. (1995) developed a system consisting of 15 different sensor elements (four Pd gate MOSFETS, one Pt gate MOSFET, three Ir gate MOSFETS, one Ir/Pt gate MOSFET, one Pt/Pd gate MOSFET, four chemoresistors, and one CO_2 optical absorption sensor). The array was exposed to four different alcohol vapors and water using one alcohol as a calibrant. They used a neural network for pattern recognition and were able to detect and identify each of the four types of vapors. The use of calibration significantly improved the performance. The tests were done without any contaminants in the system however, which, as the authors note, could greatly affect the results.

An electronic nose consisting of a heterogeneous array of tin-oxide gas sensors was described by Wilson and DeWeerth (1995). Three sensor types were used in the array, five of each type with each sensor operating at a different temperature. Nonlinear preprocessing (median thresholding) of the sensor signals was used to discriminate between various reducing chemicals. They found the system could discriminate between five major reducing chemicals (acetone, ammonia, carbon monoxide, hexane, and butane) across a broad range of concentrations (100 to 5,000 ppm).

There are a multitude of papers on electronic noses, many of which focus on a particular application rather than the more fundamental aspects of the technology as described above. The topics range from quality estimation of cod fillets purchased over the counter, to screening of tomatoes for determination of vegetable "stress," to wine denomination determinations. Interesting applications for alcoholic beverages are the use of shear-mode SAW sensors for the characterization of whiskey as described by Kondoh and Shiokawa (1995) and the use of an array of conducting polymers for testing beers, presented by Gardner, et al. (1994). While these applications of micromachined sensors may seem somewhat unusual, they do show how the sensors can be "trained" for a particular task (as can humans and animals), often with encouraging results. Other examples of electronic nose technologies and systems can be found in Hatfield, et al. (1994), Yea, et al. (1994), Moriizumi (1995), Cammann, et al. (1997), Di Natale, et al. (1997), and Kress-Rogers (1997).

There are several commercially available systems based on these technologies currently in use in a variety of industries (examples of vendors are Aromascan Plc, Manchester, UK, and Neotronics Scientific Ltd., Flowery Branch, GA).

2.8 COMBINATORIAL ARRAYS

Combinatorial arrays of chemicals have been synthesized to address a wide variety of analyses where a large number of possible molecule/molecule interactions must be investigated in parallel. For example, DNA sequencing can be done by allowing single-stranded DNA fragments to *hybridize* (specifically bind to) with substrate-bound, complementary sequences. In order to determine a strand's sequence, its location of binding in an ordered, combinatorial array is used. Another example of the application of combinatorial arrays is in screening potential pharmaceutical compounds from vast libraries of chemicals by looking for interactions with biomolecules (such as cell membrane receptors). Parallel chemical synthesis (potentially *in situ*) or at least ordered array fabrication are critical requirements for realization of such arrays.

Fodor, et al. (1991) described a technique for light directed, spatially addressable parallel chemical synthesis that utilized microlithographic techniques. As outlined in the figure below, a glass substrate was derivatized with linker molecules that contained amino functional groups blocked with a photochemically cleavable protecting group (nitroveratryloxycarbonyl (NVOC)). The substrate was illuminated through a photomask containing opaque and clear areas, leading to unblocking (cleavage of the protecting group) in the exposed regions. The first chemical building block (also containing a photolabile protecting group) was then washed over the entire surface, coupling to the unblocked areas. This procedure was repeated to sequentially build the desired set of chemical products.

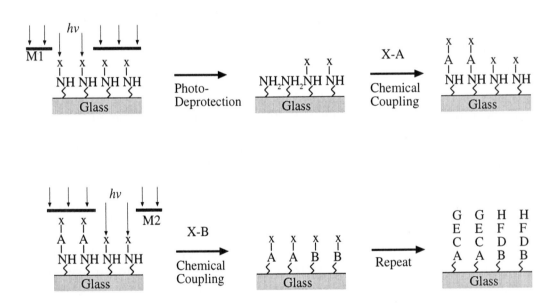

Concept of light-directed, spatially addressable parallel chemical synthesis. After Fodor, et al. (1991). A glass substrate is derivitized with amino groups (NH) that are blocked by a photolabile protecting group X. Optically driven unblocking of specific regions is accomplished by illumination through a mask M1. The amino groups in the exposed regions are now available for coupling to the first building block A (which also contains a photolabile protecting group X). A different mask M2 is used to unblock different regions of the substrate, thereby allowing chemical building block B to be coupled to the exposed amine groups (or building block A). This process is repeated until the desired set of products is obtained.

Fodor, et al. (1991) used the above technique to synthesize two pentapeptides (five amino acid proteins) in a checkerboard pattern consisting of 50×50 µm elements. A mouse antibody (3E7) with nanomolar affinity to one of the two synthesized pentapeptides was exposed to the substrate, thereby binding to the appropriate peptide site. The regions containing bound 3E7 were then detected using a fluorescein-labeled goat antibody to mouse. An intensity ratio > 21:1 was observed across the fluorescence checkerboard.

They have also demonstrated that oligonucleotides as well as peptides can be synthesized using the light directed, spatially addressable parallel chemical synthesis technique described above. This could be used to detect complimentary sequences in DNA and RNA for gene mapping, fingerprinting, diagnostics, etc. In either case, 40,000 different compounds can be synthesized in 1 cm^2 using a 50 µm checkerboard, making large-scale screening feasible in a compact array. A product based on this technology is available from Affymetrix, Inc. (Santa Clara, CA).

2.9 BIOLOGICAL CHEMICAL SENSORS

Our tongues and our noses are amazing chemical sensors, and chemical sensors are, in one form or another, ubiquitous among all living organisms. They have important roles in identifying objects, other organisms, and places as attractive or repulsive. Unfortunately, their exact operation is only moderately well understood, partly due to the relative difficulty of accessing the cells that carry out the transduction. Even so, one can learn a lot about how to build chemical sensor systems by studying the chemical sensors found throughout nature.

2.9.1 BIOLOGICAL "TASTE" SENSORS

In terms of human *gustatory* perception, there are four basic tastes: bitter, sour, salty, and sweet, which are localized to different regions of the tongue (illustrated below). Nonetheless, when combined with the signal processing power of the brain, these four basic tastes can distinguish thousands of different flavors (often in concert with the sense of smell, or *olfaction*), presumably from the superposition of the sensor signals. As discussed in Dodd and Castellucci (1991), the basic taste sensors, modified epithelial cells, are grouped together as parts of the "taste buds," as illustrated below. There are three different types of taste buds, each located primarily in certain regions of the tongue (illustrated below): *fungiform*, *foliate*, and *circumvallate* (named for their shapes). Each taste bud contains 50 to 150 receptor cells, as well as supporting cells. Chemical signals are received via an *outer taste pore* and electrical (neural) output signals are transmitted to the brain through *afferent* (output) nerve fibers at the base of each bud.

The individual receptor cells have specific responses, but also respond (to some extent) to the other basic tastes (i.e., are not perfectly selective). The basic transducers are membrane receptors that give rise to electrical potentials when their target molecule(s) are present. A wide variety of compounds can elicit the sensation of bitterness, and it is not known what chemical structures are specifically responsible. Similarly, for sweetness, it is not clear what the chemical structures are that cause the response, although the sweetest known compounds (thaumatin and monellin, 10^5 times sweeter than ordinary sugar) are proteins, and it is thought that the juxtaposition of two three-amino-acid sequences may play a role (interestingly, amino acids are not present in sugars, which are carbohydrates). Acidity (low pH) is responsible for the sensation of sourness. Saltiness is apparently perceived when ions such as sodium and potassium pass through protein-based channels in the cell membranes of the receptors.

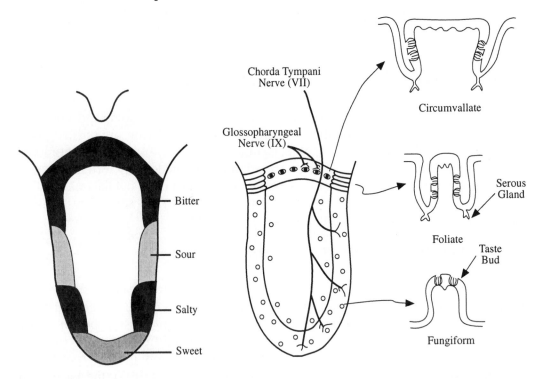

Illustration (left) of the distribution of sensors for the four basic tastes on the human tongue and (right) the neural connections carrying information from those sensors to the brain via the seventh and ninth cranial nerves. After Dodd and Castellucci (1991).

Our present understanding is that individual tastes (which can be fairly complex) are represented to the brain both by the outputs of specific receptors and by the global activity of all receptors (i.e., there are higher-level "symbols" involved in the encoding). As mentioned above, this strategy has been employed to synthesize chemical sensor selectivity from arrays of relatively nonselective sensors with somewhat different responses.

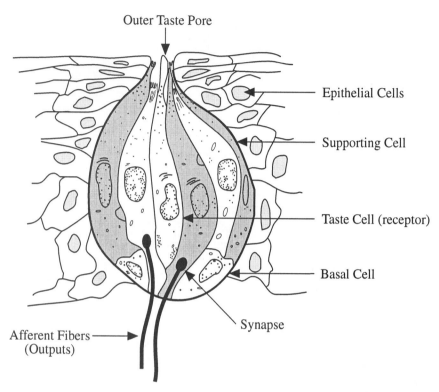

Illustration of a taste bud, showing the chemical input and neural (electrical) outputs. After Dodd and Castellucci (1991).

2.9.2 BIOLOGICAL ODOR SENSORS

Olfaction is one of the most primitive senses. Our noses, which are relatively *insensitive* when compared to those of other animals, can detect as few as 10^8 molecules in a room (parts per trillion). As for the sense of taste, the sense of smell is obtained through the combination of receptors with apparently quite non-selective responses (the exact nature of their selectivity is not yet known). Yet, it is interesting to note that the transducing cells can make fine distinctions based on molecular structure, such as being able to distinguish the isomers L-carvone (which smells like spearmint) and D-carvone, which smells like caraway seeds. As discussed in Tortora and Anagnostakos (1987), there are thought to be seven "basic" odors: "camphoraceous, musk, floral, peppermint, ethereal, pungent, and putrid." On the other hand, it has also been suggested that there may be as many as fifty distinct *primary* odors.

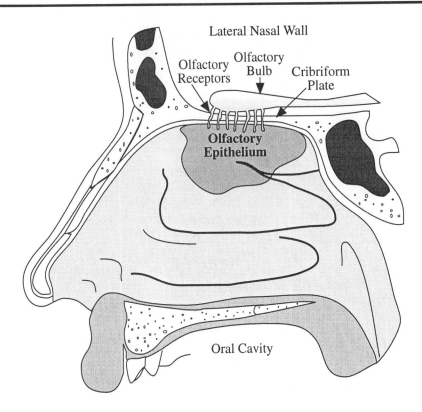

Illustration of the nasal cavity, showing the olfactory bulb (an extension of the brain) and the locations where olfactory cells protrude through to the olfactory epithelium (where odor molecules are intercepted and sensed). After Dodd and Castellucci (1991).

Molecular interactions with olfactory receptors in the nasal cavity are transduced into electrical neural signals. As illustrated below, these receptors are located in a region of the nasal cavity ≈ 5 cm^2 in area (in humans), containing the extensions of olfactory neurons. These neurons, which are unusual in that they are regenerated approximately every sixty days, are bipolar cells (meaning that the nucleus of the cells have two distinct signal-carrying protrusions). As illustrated below, the receptor cells have, at the input end, *olfactory knobs* (regions at which the transuction occurs) that are submerged in the nasal mucus. At their other end, the receptor cells have *axons*, which carry their electrical output signals to the olfactory bulb on the other side of a thin, punctuated bony plate (the *cribriform plate*). *Odorants* (molecules that give rise to sensations of smell) are absorbed in the mucus and diffuse to the olfactory knobs, where they cause biochemical cascades resulting in output signals from neurons that are sensitive to them. It is interesting to note that many odorants are not water soluble, and odorant binding proteins in the mucus serve to shuttle hydrophobic odorant molecules to receptors on the olfactory knobs (see Breer (1997)).

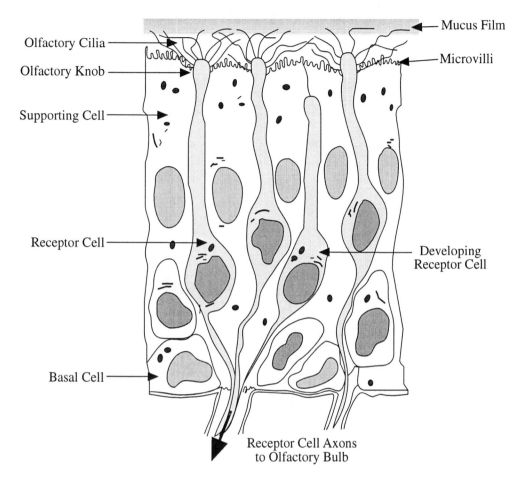

Olfactory Cilia

Olfactory Knob

Supporting Cell

Receptor Cell

Basal Cell

Mucus Film

Microvilli

Developing
Receptor Cell

Receptor Cell Axons
to Olfactory Bulb

Illustration of olfactory cells on the surface of the nasal epithelium. After Dodd and Castellucci (1991).

3. CHEMICAL ACTUATORS

There is certainly potential to fabricate chemical actuators using micromachining technologies. In this sense, "chemical actuator" can mean devices that cause chemical reactions to occur (as for electrodes, e.g., when driven to cause the oxidation or reduction of chemical species) and/or devices that result in the generation of another form of energy (light, electrical current, heat, mechanical force, etc.). As far as micromachined devices go, there has been little work in this area (with the exception of a few devices using microelectrodes, as described in Section 2.3) to date.

3.1 ELECTROCHEMICAL MECHANICAL ACTUATORS

It is relatively straightforward to use electrochemical reactions to generate mechanical force. Since electrochemistry can bring about deposition of a solid from ionic solution (e.g., electroplating) or generation of a gas from a liquid (e.g., hydrolysis of water to form H_2 and O_2), it is possible to use these change-of-state effects to generate such forces. In addition, theoretically, both types of reaction could be reversible.

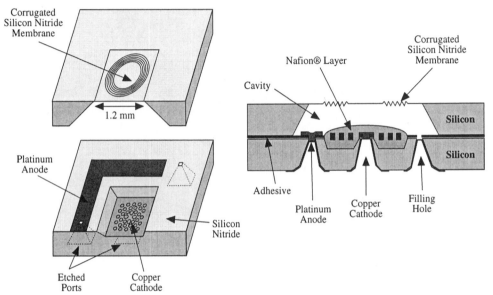

Illustration of a micromachined electrochemical actuator. After Hamberg, et al. (1995).

Hamberg, et al. (1995) described a electrolytic gas-generation-powered membrane actuator using a copper sulfate solution with a Pt and Cu electrode. If the Pt electrode is made anodic and the Cu electrode cathodic, the following reactions occur, resulting in the catalytic release of O_2 at the anode and deposition of Cu at the cathode,

$$\text{(anode)} \ 2H_2O \longrightarrow O_{2(g)} + 4H^+ + 4e^-$$

$$\text{(cathode)} \ Cu^{2+} + 2e^- \longrightarrow Cu_{(s)}$$

The net (reversible) reaction is,

$$\text{(overall reaction)} \ 2H_2O + 2Cu^{2+} \longleftrightarrow O_{2(g)} + 4H^+ + 2Cu_{(s)}$$

The reaction was carried out in a sealed chamber micromachined from two silicon dice that were glued together. The lower die, carrying sputtered Pt and Cu electrodes, was bonded to an upper die with a 1.2 mm × 1.2 mm × 1 μm silicon nitride membrane. The Cu electrode was coated with Nafion® to prevent reduction of oxygen there (this material is impermeable to anions and nonpolar species such as O_2). The two dice were attached using adhesive, which was also used to seal the filling ports after the electrolyte had been introduced (in a vacuum chamber). While current was applied, oxygen was evolved and the pressure within the micro-machined chamber increased, causing membrane deflection. The idea was to be able to hold the pressure by leaving the electrode open-circuited or decrease it back to the initial pressure by shorting or reversing the polarity of the electrodes. The ability to reverse such electrolytic reactions depends on the reversibility of the reactions, which are not always complete.

The authors demonstrated deflections of ≈ 1.5 μm (not particularly large for a membrane of that size) in times as short as 1 min, and pressure reductions back to nearly zero deflection at times as short as 5 min. The total power used was on the order of 10 μW and the pressures were on the order of 20 mbar. While the devices were not fully functional, they do demonstrate an interesting actuation method, with potential for very low operating power and no holding power requirement once actuated.

3.2 POLYMER MECHANICAL ACTUATORS

In macroscopic applications (robotics, etc.), a considerable amount of research has gone into the study of polyelectrolyte gels, some of which can contract markedly (strain > 40%) when a voltage is applied to them in a dilute salt solution. Apparently, as discussed by De Rossi, et al. (1986), electrode reactions establish pH gradients, and H^+ or OH^- ions diffuse through the gel and cause the mechanical deformation through interaction with ionized groups in it. The most common of these materials is polyacrylic acid/polyvinyl alcohol (PAA-PVA). Despite knowledge of this type of actuator dating back to the 1950's, little use of them has yet been made in micromachined devices.

There are several types of conducting polymers that have low ionization potentials or high electron affinities that permit doping with electron acceptors or donors, as discussed in Baughman, et al. (1991). Suitable polymers include poly-aniline, polyphenylene, polyacetylene, polypyrrole, and polythiophene.

Smela, et al. (1995b) demonstrated the used of doped polypyrrole actuators to generate enough force to reversibly bend four opposing flat flaps to form an open-topped box under electrical control. Polypyrroles (PPy) are conducting poly-mers that can be grown electrochemically and are capable of significant dimensional expansion (several percent) when doped with a suitable molecule such as dodecylbenzene sulfonate (anion) and cations (such as sodium) are drawn into the PPy. If the cations are removed, the material contracts back to its original length.

They showed that if suitable thin-film electrodes were coupled to a thin PPy layer, electrically movable flaps could be constructed that, when operated in an aqueous salt solution, could bend through nearly 90° within 0.5 to 10 s (depending on PPy thickness) when the voltage was raised from -1.0 to +0.35 V.

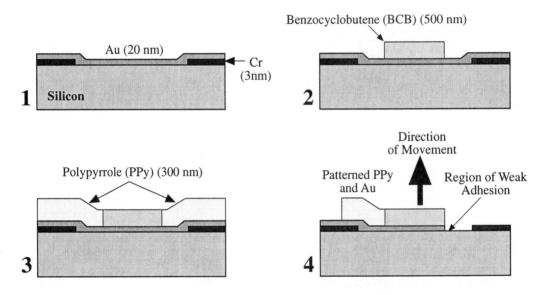

Illustration of the fabrication of polypyrrole electrochemical actuators, using selective omission of a Cr adhesion layer to allow for release of an Au layer. After Smela, et al. (1995a).

The devices were fabricated on silicon substrates by evaporating a 3 nm layer of Cr on the wafers and patterning it to remain only where the conducting/mechanical Au layer was to stick to the substrate (i.e., no Cr was deposited under the moving flaps) (see Smela, et al. (1995a). A 20 nm Au film was then evaporated, followed by a 500 nm layer of an inert, rigid polymer, benzocyclobutene (BCB), which was patterned to remain only on the moving flaps (as a stiffening layer). The final 300 nm layer of PPy was grown electrochemically onto the Au exposed by removal of the BCB and patterned to remain only on the hinge regions of the flaps. The Au was then patterned to allow release of the flaps.

Various designs were tested, and a typical flap was 100 × 100 μm, with a 10 × 100 μm active (PPy) hinge. The structures were self-releasing when operating, since they could generate enough force to pull the weakly adherent Au away from the underlying silicon in the regions where there was no Cr adhesion layer beneath it (i.e., without the use of a sacrificial layer). They could be driven to any desired angle by adjusting the applied voltage. This approach does not require high voltages or currents, nor high temperatures, but the actuators do need to operate in an electrolyte solution (not necessarily a problem for biological applications).

3.3 THIN-FILM BATTERIES

Batteries convert chemical potential energy into electrical energy by allowing chemical reactions to occur. Allesandro Volta is credited with the invention of the voltaic cell around 1800 (see Giancoli (1989)). Although developments in materials have greatly improved battery performance by increasing energy density and stabilizing the output voltage over the life of the battery, the basic principles have changed relatively little since then. It is important to note, before addressing the details, that the power density of batteries has not scaled in proportion to the miniaturization possible with modern IC technology and micromachining. Thus a significant fraction of the weight of portable computers and instruments is often due to the weight of their batteries. Another key point is that there is an entire discipline of battery research and development, and this section is not intended to serve as a comprehensive review.

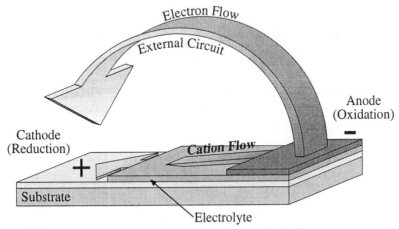

Illustration of the basic layers of a thin-film battery. Ion flow through the electrolyte and electron flow through the external circuit are driven by the redox reaction between the anode and cathode materials. Courtesy A. Flannery.

Aside from packaging, the basic components of a cell or battery are the anode, the cathode, and the electrolyte. By choosing a material for the anode that is more electropositive than the cathode, a chemical potential difference is established. To prevent the two electrodes from spontaneously reacting, they are separated by an electrolyte that allows ions to flow, but not electrons. Consequently the only path by which electron transfer can take place is through an external circuit. This oxidation-reduction (redox) reaction is designed to be thermodynamically favorable and can produce up to several volts at the terminals of a single-cell battery. Higher potentials are achieved by connecting multiple cells in series.

When an external circuit is connected between the battery terminals, electrons flow from the cathode (where oxidation takes place) to the anode (where reduction occurs). In order to prevent a build-up of charge that would oppose the reaction, the electrolyte must provide a pathway by which the ionized product from the anode can flow to the cathode and balance the negative charge resulting from electron flow. Whether or not a battery can be recharged (i.e., whether it is considered a *secondary cell*, which can be recharged, versus a *primary cell,* which cannot be recharged) depends on two primary factors. The first is the reversibility of the redox reaction that generates the electron flow. If the reaction is not thermodynamically reversible, driving the cell with an opposing potential will not restore the original reactants. The second major factor has to do with where in the cell the reaction products end up. If the oxidation-reduction products of the anode and cathode do not remain in contact with either of the electrodes but instead become dispersed throughout the non-conductive electrolyte, the electrochemical circuit will not be complete and the cell will not be rechargeable.

The classic "dry cell battery" (Leclanché cell) was made with zinc as the anode material, carbon as the cathode, and aqueous ammonium chloride (NH_4Cl) as the electrolyte. The modern alkaline battery is also comprised of zinc as the anode and manganese dioxide (MnO_2) mixed with carbon as the cathode but with KOH or NaOH as a more basic electrolyte. This increased pH reduces the corrosion rate of the zinc anode and thus increases cell lifetime. Rechargeable lead acid batteries such as those found in automobiles use lead as the anode material, lead dioxide (PbO_2) as the cathode material, and H_2SO_4 as the electrolyte. There are many other primary and secondary cell types commercially available, and a useful (but somewhat dated) review can be found in Gibilisco and Sclater (1990).

While there has been considerable interest in developing thin-film batteries, the available power decreases rapidly with their size. The amount of chemical potential energy available from a given set of reactants in a battery is a fixed physical property of the reactants and their molar quantities. An example of a pair of cathode and electrolyte materials that has been demonstrated in a thin-film battery, αV_2O_5/LIPON (amorphous vanadium pentoxide and lithium phosphorus oxynitride, discussed below), has an energy density of ≈ 700 Wh/l (Bates et al. (1993)). For comparison, high energy macroscopic lithium batteries have energy densities of ≈ 450 Wh/l, although some commercially available cells such as the Eveready CR2450 are as high as 750 Wh/l (Eveready Battery Co., Inc., St. Louis, MO). With an energy density of 700 Wh/l, a thin-film αV_2O_5/LIPON battery (with a 1 μm thick cathode) operating at ≈ 3 V would have a capacity of ≈ 120 μAh/cm^2 (Bates, et al. (1993)). This level of energy is adequate for short term back-up of CMOS memories, for on-chip power conditioning applications, and potentially to drive very low-power sensors and actuators. It should be noted that only secondary cells appear to have been investigated in thin-film formats, and the discussion below is focused thereon.

The limitations of micromachining fabrication techniques and issues of compatibility with other components of a potentially integrated system place important constraints on the selection of thin-film battery materials. Preferably the materials are stable under vacuum and at temperatures suitable for processing. Consequently, liquids (like the sulfuric acid used in the lead storage cell) are generally not a suitable choice, and solid electrolytes are used. The reaction products from the discharge of the battery must also be compatible with thin-film construction. It is also desirable that the cells (assumed to be secondary), withstand a large number of charge-discharge cycles without losing a significant percentage of their performance.

The substrate for a thin-film battery must not react with any of the other battery components. Alternatively, a potentially reactive substrate can be coated with a protective layer, as was used by Shokoohi et al. (1991) to fabricate a lithiated manganese oxide ($LiMn_2O_4$) cathode. They used 15 nm of titanium and 300 nm of gold as a buffer layer on a quartz (SiO_2) substrate to prevent it from reacting with either the lithium or the manganese in the cell. Other substrates, such as conventional glass (Jones and Akridge (1996)), alumina (Al_2O_3) (Jones and Akridge (1992)), and polyester (Bates et al. (1993)) have been used. If fabricated on a silicon substrate, suitably non-reactive barrier materials might include sputtered alumina, PECVD silicon carbide, or PECVD Teflon™-like perfluorinated hydrocarbons.

For several reasons, most thin-film batteries fabricated to date have used lithium as the anode material. As discussed in Rahner, et al. (1996), lithium is one of the most electropositive elements (its half-cell potential relative to the standard hydrogen electrode is $E_H° = -3.04$ V), giving Li-based cells the maximum practical output voltage per cell. It has a high specific capacity (3.86 Ah/g), it is very mobile in ionic form, and it can be deposited using standard IC fabrication techniques (e.g., sputtering or evaporation). Unfortunately, Li is extremely reactive and oxidizes on contact with air or water (often violently). Also, like all common alkali ions, it is not compatible with CMOS circuitry without a suitable barrier layer, such as silicon nitride. In addition, the melting point of Li is 180°C, and thus it is typically deposited near the end of the fabrication process to allow for higher temperature processing of the cathode and electrolyte. Another potential problem with Li is that during deposition it forms dendrites that can bridge across the electrolyte, shorting out the anode and cathode. To eliminate or reduce some of the problems with Li anodes, it has been combined with other materials, although generally with a significant decrease in specific capacity. For example, lithium intercalated into carbon was demonstrated by Yazami and Deschamps (1994) and Aurbach, et al. (1996), and a wide variety of lithium alloys (e.g., with Al, Sn, Pb, Mg, etc.) were discussed by Rahner, et al. (1996).

Depending on the specific battery chemistry, the electrolyte may participate in the chemical reactions or simply serve as a pathway for ions generated by them. A useful electrolyte for a thin-film battery is typically a solid and must be chemically stable in direct contact with a very reactive anode (e.g., Li). Common solid electrolytes are lithium perchlorate ($LiClO_4$) in propylene carbonate and lithium phosphorus oxynitride (LIPON) because it can readily be prepared by radio frequency magnetron sputtering of tribasic lithium phosphate (Li_3PO_4) in N_2.

The cathode material for a thin-film battery is chosen on the basis of its half-cell potential (which combined with that of the anode determines the overall cell potential) and such that it is chemically stable in contact with the electrolyte. Another key characteristic of a secondary cell cathode is the efficiency with which the cathode material reversibly interacts with the Li^+ ions. The mechanisms by which Li^+ reacts with the cathode can be divided into two groups: intercalation and polymerization. In intercalation electrodes, Li^+ moves interstitially within the lattice of the cathode material. Cathodes for thin-film batteries can include titanium disulphide (TiS_2) deposited by CVD (Takehara et al. (1991)) or PECVD (Kikkawa (1996) and Jones and Akridge (1992, 1996)), lithium-manganese-oxides (Thackeray, et al. (1992) and Shokoohi et al. (1991)), or amorphous vanadium pentoxide (αV_2O_5) (Bates, et al. (1993)). In polymerization cathodes, the electrode is depolymerized during discharge and repolymerized during charging. Liu et al. (1991) evaluated a series of candidates and found performance generally superior to that obtained with TiS_2 under identical conditions, particularly for dimercapto dithiazole electrodes. Another candidate polymer electrode is polyaniline (Yang et al. (1996)). It should be noted that the methods of preparation and the purity of cathode materials can have significant impact on the performance of thin-film batteries.

An example thin-film Li battery is illustrated below. The fabrication process used by Bates, et al. (1993) began with the DC magnetron sputter deposition of a vanadium anode and cathode (in Ar), followed by the deposition of an αV_2O_5 layer using DC magnetron sputtering of vanadium in Ar + 14% O_2. A LIPON electrolyte layer was then deposited by sputtering Li_3PO_4 in N_2, followed by an evaporated Li anode layer deposited at 10^{-6} torr. Finally, a passivation layer was deposited. The nature of the passivation layer was not specified in Bates, et al. (1993), but several are outlined in Bates, et al. (1996), including organic films such as Parylene™ (a family of polyxylylene-type polymers) and inorganic films (ceramics, such as LIPON, and metals, such as Cr, Ni, V, or Mn) in various combinations, with an outer coat of epoxy. As mentioned above, the energy density reported for this design was

700 Wh/l. The open-circuit voltages measured for this design were 1.5 to 3.6 V, and Bates, et al. (1993) noted that the batteries could be operated safely at temperatures up to the melting point of Li (180°C).

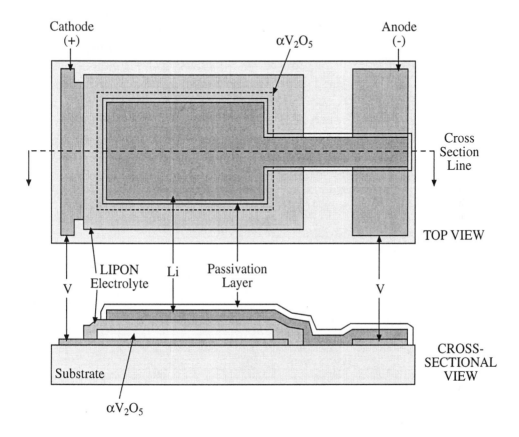

Illustration of the construction of a thin-film Li battery. Adapted from Bates, et al. (1993).

While such thin-film batteries may not be suitable for powering the more energy-hungry micromachined transducers, they will likely be quite useful where external power is periodically available. For example, they could be used in micromachined sensor systems with intermittent inductive or photovoltaic power supplies.

4. BIOELECTRIC INTERFACE DEVICES

The term *bioelectric interfaces* refers to electrode structures that transduce signals between electronics and living tissues. The structures are generally just metallic electrodes, fabricated by patterning metal traces on an insulating layer (or substrate), insulating the entire wafer, and then selectively opening regions for the electrodes (and for external interconnections). Such a device is illustrated below. Microelectrodes can be used to passively record *extracellular* signals (intercepted outside a cell) from electrically active tissues (e.g., brain neurons, peripheral nerve neurons, skeletal muscle cells, cardiac muscle cells, pancreatic islet cells, etc.), or they can be used to stimulate such cells with electrical signals. In addition, the use of microelectrodes to measure the impedance of nearby tissues can also be used to extract useful information about their position and physiologic state (see Giaever and Keese (1984, 1986) and Giaever and Keese (1993)). If properly designed, an electrode structure can serve all three functions. (For further information on the physiology and function of neurons from an engineering/physics perspective, the reader is referred to Scott (1975). A wealth of information from a neuroscience perspective can be found in Kandel, et al. (1991).)

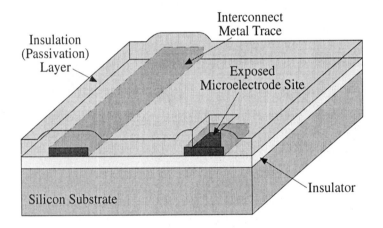

Illustration of the general structure of planar thin-film microelectrodes. Note that if the substrate is insulating, a layer of insulator is not required beneath the metallization. Also, bond-pad regions for electrical connections to the microelectrode(s) are not shown. Adapted from Kovacs (1994a).

Microelectrode devices of this nature have been used, in various forms, in the central nervous system (brain and spinal cord), the peripheral nervous system (nerves in limbs, etc.), and as platforms on which to culture cells. In all of these cases, the electrodes have the same basic structure, but the substrates are varied to suit each specific application. It should be noted that in some cases, researchers deposit platinum black (amorphous, dendritic platinum) via electroplating to reduce the electrical impedance of the electrodes, although this is seldomly done for implanted devices (see Marrese (1987) for details of platinum black deposition).

4.1 PENETRATING NEURAL PROBES

In order to record neural signals from the cortex of the brain, needle-like substrates are needed for the microelectrodes. If suitable devices are fabricated, they allow parallel access to many neurons simultaneously. Micromachining technology has been applied to such efforts since the late 1960's, as discussed in Wise, et al. (1970), Starr, et al. (1973), Wise and Angell (1975) and Pickard, et al. (1990). Active circuitry was later added to the probes themselves, as discussed in Najafi, et al. (1985) and by a separate group, Takahashi and Matsuo (1984). Since then, a great deal of work continues to be done at the University of Michigan on such probes for the cortex (see for example: Drake, et al. (1988), Ji, et al. (1991), Ji and Wise (1992), and a review by Najafi (1994)), at Stanford and Caltech (Kewley, et al. (1996, 1997)) and other sites (another interesting reference is Prohaska, et al. (1986)).

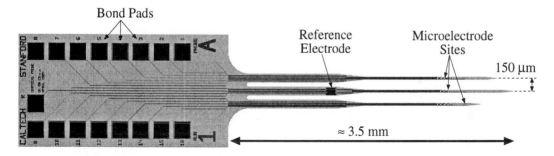

Top view of a typical penetrating neural probe. Adapted from Kewley, et al. (1996).

The original motivation for this work was to develop cortical prostheses, which would allow recording from and stimulation of regions of the brain to replace (with other hardware as well) absent or damaged bodily functions such as limb movement, sensation, vision, hearing, etc. A great number of issues have proven much more difficult than anticipated, and have thus far limited clinical use of such devices. These issues include coatings that withstand biological exposure (*bioresistant*) yet do not cause inflammation of tissues (*biocompatible*), interconnection technologies, the need for telemetry, and a lower than hoped for yield of usable neural/electronic connections (probably due to physical interactions, primarily geometric, of the neurons and electrodes). Nonetheless, the basic functionality of the probes has been demonstrated (both recording and stimulation) and progress has been made in dealing with the issues limiting their application.

The basic Michigan penetrating probe process (illustrated below) has depended on the use of deep ($\approx$15 µm) boron diffusions to define the overall shapes of the probes, which are released from the substrate using an EDP etch. Several variations on the basic process have been made over the years, but the dissolved wafer aspects remain the same. Those variations have involved changes in dielectric layers (e.g., stacked silicon dioxide/silicon nitride layers to mitigate built-in stress problems), modification of electrode structures and deposition processes, and changes in the active circuit processes included on the probes.

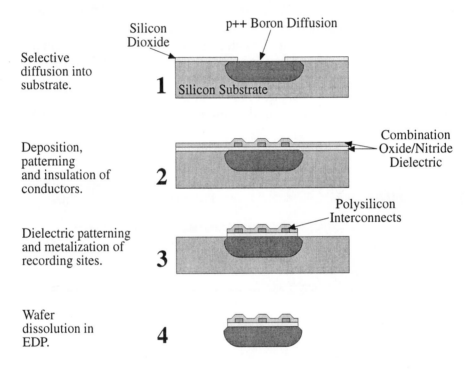

Cross-sectional illustration of the basic process used for fabrication of penetrating neural probes at the University of Michigan. After Najafi, et al. (1985). Note that electrodes, generally either gold or iridium, are deposited using a lift-off process (not shown).

Initially NMOS, and recently CMOS circuits were included on the probes for amplification, filtering, multiplexing, etc. Similarly, probes for stimulation of neural tissue have been co-integrated with circuits such as multiplexers, current sources, and digital control logic.

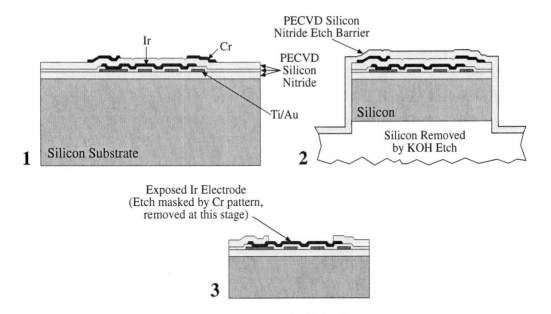

Cross-sectional illustration of a penetrating neural probe fabrication process relying on a combination of dry and wet etching for defining the probe shapes. Adapted from Kewley, et al. (1996). At the upper left, a cross section of a probe is shown after the surface features have been formed. At the upper right, a cross section is shown after a plasma etch of the silicon to define the shape of the probes, and the deposition of a front-side protective silicon nitride layer. At the bottom of the figure, a cross section of a completed probe is shown after removal of the protective silicon nitride and buried chromium etch pattern (used as an etch mask to expose the iridium electrode sites and bond pads).

An alternative method for fabricating penetrating neural probes using a combination of dry and wet etch process with multi-level metallization (for narrow probe shafts) was presented by Kewley, et al. (1996, 1997). The process flow began with deposition of a 1 μm PECVD silicon nitride insulating layer (it should be noted that some investigators feel that PECVD films are unsuitable for such applications, but experimental evidence does not support this), followed by the deposition and patterning of a 25 nm Ti, 500 nm Au bilayer using lift-off. This was followed by another silicon nitride layer (0.7 μm) and another lift-off process to deposit a 30 nm Cr, 300 nm Ir bilayer for the microelectrode sites, using magnetron sputtering. A 30 nm Cr layer was deposited and patterned to later serve as a silicon nitride etch mask, and a top silicon nitride layer (1 μm) was then deposited. The probe shapes were defined using a CHF_3/O_2 plasma etch of the silicon nitride films and a SF_6/C_2ClF_5 plasma etch into the bulk silicon. After the probe etch, a fourth layer of silicon nitride was deposited to protect the fronts of the probes using the subsequent KOH etch to thin the probes to 25 μm (the end point was optically determined). The top silicon nitride protective layer was then removed using an RIE etch, with the buried Cr layer serving to pattern the third silicon nitride layer to open the electrode sites and bond pads. The resulting probes had extremely sharp tips (< 1 μm), which are advantageous in reducing the force necessary for penetration of the brain. In combination with a separate 18-channel CMOS preamplifier/programmable filter array, cortical signals were recorded from anesthetized rats using electrodes spaced at 40 μm intervals and covering 280 μm of cortical depth.

The boron/EDP approach described above has also been used by the Michigan team to fabricate integrated silicon "ribbon cables" for connections between the probes and other circuits, as discussed in Hetke, et al. (1990, 1991). Such cables may be critical to overcome problems with separate interconnects that must be coupled to the probes after fabrication (typically, Teflon™-coated gold wires are used, but they can suffer from leakage currents, and the bonds to the probes must be manually made and overcoated with a potting compound).

This approach has also been used to make microelectrode arrays wherein the entire electrode region is flexible. Bell, et al. (1997) fabricated electrode arrays over 40 mm in length (25 mm for the electrode-bearing region itself) and only 4 μm thick where electrodes are located. This array was designed to be inserted into the helical turns of the cochlea (the transducer part of the inner ear), and was tested in cadaver cochlea and those of live guinea pigs. By directly stimulating the neurons in the cochlea, some level of hearing can be restored in individuals with damaged transducer cells (hair cells) since they are bypassed by the direct stimulation. Since sound frequencies are encoded by distance along the cochlea, an inserted electrode array offers the ability to stimulate sensations of different frequencies.

The same boron-etch-stop technology used for the Michigan probes was further extended to manufacture micromachined components to hold many of the microelectrode arrays in a "3-D" grid that could be used to "address" a larger volume of tissue as discussed in Hoogerwerf and Wise (1991).

4.2 REGENERATION NEURAL ELECTRODES

Unlike the central nervous system, peripheral nerves (e.g., those in the limbs) will regenerate when cut and if the two nerve ends are brought near to each other. If a substrate perforated by via holes is placed in the path of the regenerating nerve fibers (axons), they will be held spatially fixed with respect to microelectrodes on the surface of the substrate, hopefully forming a stable mapping between the neural signals and the hardware. Recording from peripheral nerves was attempted by Marks (1969) and Mannard, et al. (1974) using non-micromachined devices, and by Matsuo, et al. (1978) and Edell (1986) using bulk micromachined silicon devices.

An approach using thin-film iridium electrodes, with plasma etching to define via holes and wafer lapping to thin the substrates was demonstrated by Kovacs (1991) and Kovacs, et al. (1992). Akin and Najafi (1991) and Bradley, et al. (1992) demonstrated thin-membrane-based regeneration electrode arrays, fabricated using the boron-doped/EDP etch approach discussed above for penetrating probes. A combination of wet etching to form a membrane (using a etch time, rather than an etch-stop layer, to define the membrane thickness), and plasma etching to open via holes through it, was used by Kovacs, et al. (1994) to fabricate arrays that were used to record from nerves in the rat hind limb and the auditory nerve of the bullfrog.

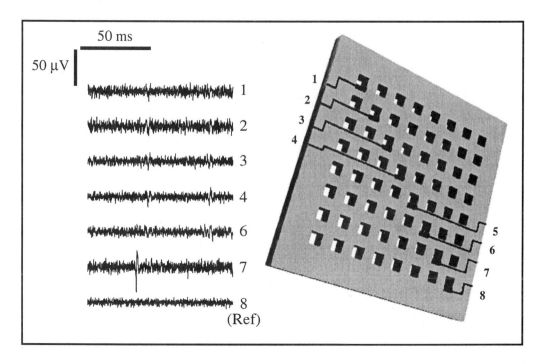

Illustration of the layout of electrodes on a regeneration interface membrane, showing the locations of eight electrodes and corresponding (actual) signals. After Kovacs, et al. (1994).

This fabrication process was felt to be compatible with CMOS circuits from a commercial vendor, although this combination was not tested. The approach taken was to use wet etching to form a membrane for the via holes and microelectrodes and then plasma etch the via holes and contacts for microelectrodes and bond pads, as illustrated below.

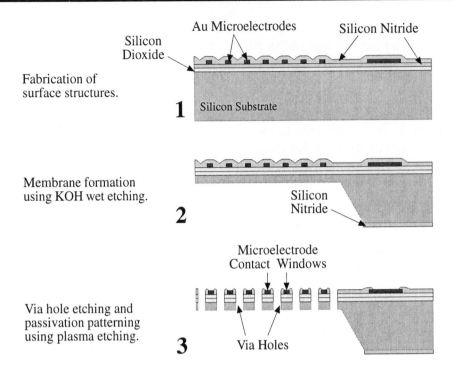

Illustration of a process used to fabricate regeneration neural interfaces. After Kovacs, et al. (1994).

As for the penetrating probes, there are many barriers to their successful clinical use. In addition to the issues of coatings and interconnect, the relatively low probability of having the most electrically active region of a peripheral nerve axon (so-called *nodes of Ranvier*) near enough to an electrode to record a signal unfortunately appears to be quite low. In addition, it has been seen that the regenerating axons can, as they expand in diameter, crowd each other against the rigid walls of the via holes, leading to damage to their structure (and presumably function).

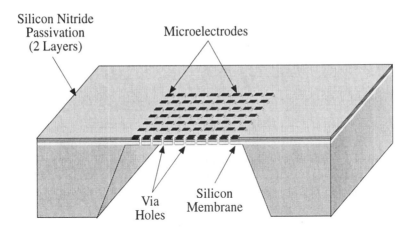

Illustration of the structure of a regeneration electrode array structure fabricated using bulk wet etching to form a silicon membrane and plasma etching to form the via holes. After Kovacs, et al. (1994).

4.3 CULTURED CELL SYSTEMS

Several groups have used metal microelectrodes like those described above to record from cells cultured on glass or silicon substrates (see Stenger and McKenna (1994) for a good overview, as well as Thomas, et al. (1972), Gross (1979), Pine (1980), Droge, et al. (1986), Novak and Wheeler (1986), Israel, et al. (1990), Martinoia, et al. (1993), and Mohr, et al. (1995)). One of the objectives of this work has been to study networks of natural neurons as they form on the substrates. With such a simplified model of the nervous system, it might be possible to investigate memory (long-term and short-term potentiation) and other neural phenomena. The use of electrodes to actively interact with cells (e.g., prevent them from adhering in certain regions, move them, etc.) in a culture system has also been investigated, and is discussed by Fuhr and Shirley (1995) and Fuhr (1996).

In addition, *hybrid biosensors*, as discussed below, can be built using neurons or other cell types that can autonomously generate action potentials. By monitoring their patterns of action potentials during exposure to various chemicals (e.g., libraries of potential pharmaceutical agents, environmental samples, etc.), they can serve to detect potential toxicity and/or beneficial effects.

4.3.1 SURFACE MODIFICATION FOR CULTURE SYSTEMS

In order to make the surfaces of the underlying electrode arrays more bio-compatible and to control cellular adhesion and patterning, several groups have done research on means to modify the surfaces using chemical means. Kleinfeld, et al. (1988) used siloxane-based chemistry to modify the surfaces of the dielectric layer (above silicon) to control the growth patterns of neurons. Soekarno, et al. (1993) used similar methods to selectively modify the surfaces of borosilicate glass substrates, and also demonstrated patterned growth of cells. This has potential to allow "custom" networks of real neurons to be grown. Hickman and Stenger (1994) describe various techniques for modifying surfaces to enhance the viability and adhesion of cultured neurons. Stenger and Hickman (1994) and Wheeler and Brewer (1994) describe approaches toward lithographic patterning of neurons, building on the work of Kleinfeld. An issue that remains to be addressed for some applications of this work is the long-term stability (more than weeks, in this case) of the surface-modified substrates in terms of their continued ability to influence cell growth patterns over time.

5. HYBRID BIOSENSORS

Micromachined electrode arrays with living cells above them can be used as sensors for a variety of purposes. The idea is to culture the cells above the electrodes and measure some aspect of their physiologic state in response to test analyte, thus forming hybrid biosensors for screening of pharmaceuticals or detection of toxins.

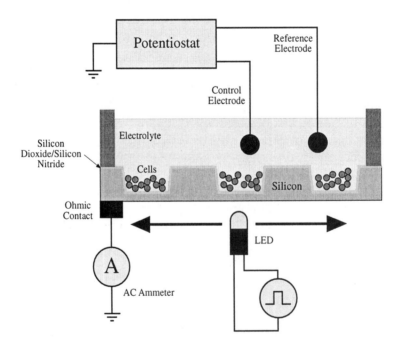

Illustration of a hybrid biosensor system based on measurement of extracellular pH. After Parce, et al. (1989)

Extracellular pH has been used as an overall indicator of cellular metabolism by Parce, et al. (1989) and Owicki, et al. (1990) using light-addressable wells containing cells instead of electrodes. Small wells were plasma etched into a silicon substrate and filled with cultured cells (covered by an appropriate culture medium). As discussed in Section 2.5, a LED was scanned beneath the wells and pulsed to generate electron-hole pairs in the silicon. The induced carriers would charge up the silicon dioxide/silicon nitride/electrolyte capacitance and the pH could be computed from the resulting AC signal (the devices are referred to as light-addressable potentiometric sensors, or LAPS). Since pH can be used to infer the metabolic state of the cells, this device can indirectly assess that important parameter for use in pharmaceutical testing/screening, toxin detection, etc. An instrument based on this principle is currently available from Molecular Devices, Inc (Sunnyvale, CA). However, in some cases, it may be desirable to measure more sensitive changes in a cell's physiologic state than pH.

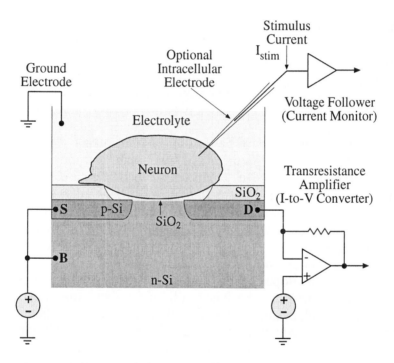

Illustration of the use of an "open-gate" FET to form a direct electrical connection to a neuron. Adapted from Fromhertz, et al. (1991).

Fromherz, et al. (1991) demonstrated coupling of neurons directly to the gate of a metal-free, "open-gate" FET device immersed in an electrolyte. An individual *Retzius cell* (a neuron taken from a leech) was attached to the end of a glass pipette and manually mounted on the gate of the device. The electrolyte was maintained at ground potential while the bulk silicon, source and drain were held at positive voltages (a p-channel FET was used). A microelectrode inserted through the cell membrane injected current to stimulate the cell in some cases, while the membrane voltage was simultaneously measured using a voltage follower. In another case, measurements were made without the intracellular electrode. A transresistance amplifier (current-to-voltage converter) was used to monitor the source-drain current. Electrical activity (action potentials), both spontaneous and stimulated, were shown to modulate the source-drain current in the FET device. This neuron-Si junction is outlined in the figure above. An important property of this approach is that a large percentage of the intracellular signal (25% was reported) is available as an electronic signal, presumably owing to the tight coupling between the cell and the FET. This could provide a higher signal-to-noise ratio than recording with conventional metal electrodes. It should also be noted that the authors reported variable signal levels, presumably due to variations in the degree of cell-to-FET coupling (a similar problem exists for the coupling of conventional microelectrodes to cells).

Interestingly, such FET structures were considered much earlier by Matsuo and Wise (1974) as a means of recording extracellular action potential signals, although this was apparently not demonstrated. The authors pointed out that their structures, incorporating a 100 nm silicon nitride layer above a 100 nm thermal silicon dioxide gate region, exhibited very linear pH response, but were light sensitive (due to photogenerated carriers, a potential problem for most such devices).

Impedance measurements of cells cultured over planar electrode arrays can also be used for monitoring cell attachment, spreading, and motility. Giaever and Keese (1984, 1986) and Lo, et al. (1995) described a system where cells were cultured in a dish containing a 250 μm diameter electrode and a larger counterelectrode. The impedance of the small electrode was monitored (with respect to the larger one) over time as cells were plated onto the surface and spread, divided, and moved. The average impedance of the entire population of cells over the electrode was thus measured. They were able to detect changes in cell morphology and movement and have developed models to extract the resistivity of the cell layer; the average distance between the basal cell surface and the substrate; and the capacitance of the apical, basal, and lateral cell membranes from the measurements. The technology has shown promise for use in monitoring cellular response to different agents and is a good indicator of general cellular health. A commercial version of their instrument is available from Applied Biophysics, Inc., Troy, NY. For additional information on impedance-based hybrid biosensors, the reader is referred to Giaever and Keese (1991, 1993), Borkholder, et al. (1996), and Ehret, et al. (1997).

CHEMICAL AND BIOLOGICAL TRANSDUCERS REFERENCES

GENERAL REFERENCES

Bard, A. J., and Faulkner, L. R., "Electrochemical Methods, Fundamentals and Applications," John Wiley and Sons, New York, NY, 1980.

Bockris, J. O'M., and Reddy, A. K. N., "Modern Electrochemistry," Plenum Publishing Corp., 1970.

Brett, C. M. A., and Brett, A. M. O., "Electrochemistry: Principles, Methods and Applications," Oxford University Press, Oxford, UK, 1993.

Cobbold, R. S. C., "Transducers for Biomedical Measurements," John Wiley and Sons, New York, NY, 1974.

Janata, J., "Principles of Chemical Sensors," Plenum Press, New York, NY, 1989.

Janata, J., Josowicz, M., and DeVaney, D. M., "Chemical Sensors," Analytical Chemistry, vol. 66, no. 12, June 15, 1994, pp. 207R - 228R.

Kandel, E. R., Schwartz, J. H., and Jessell, T. M. [eds.], "Principles of Neural Science," Third Edition, Elsevier Science Publishing Co., New York, NY, 1991.

Kress-Rogers, E. [ed.], "Handbook of Biosensors and Electronic Noses," CRC Press, Inc., Boca Raton, FL, 1997.

Madou, M. J., and Morrison, S. R., "Chemical Sensing with Solid State Devices," Academic Press, Inc., Boston, MA, 1989.

Middelhoek, S., and Audet, S. A., "Silicon Sensors," Chapter 6 in "Silicon Sensors for Chemical Signals," Academic Press, Inc., London, UK, 1989, pp. 249 - 286.

Stenger, D. A., and McKenna, T. M. [eds.], "Enabling Technologies for Cultured Neural Networks," Academic Press, Inc., San Diego, CA, 1994.

Sze, S. M. [ed.], Semiconductor Sensors, John Wiley and Sons, New York, NY, 1994.

Sze, S. M. [ed], "Physics of Semiconductor Devices," Second Edition, John Wiley and Sons, New York, NY, 1981.

SPECIFIC REFERENCES

Abe, H., Esashi, M., and Matsuo, T., "ISFETs Using Inorganic Gate Thin Films," IEEE Transactions on Electron Devics, vol. ED-26, no. 12, Dec. 1979, pp. 1939 - 1944.

Akin, T., and Najafi, K., "A Micromachined Silicon Sieve Electrode for Nerve Regeneration Applications," Proceedings of Transducers '91, the 1991 International Conference on Solid-State Sensors and Actuators, San Francisco, CA, June 24 - 27, 1991, pp. 128 - 131.

Andle, J. C., and Vetelino, J. F., "Acoustic Wave Biosensors," Sensors and Actuators, vol. A44, no. 3, Sept. 1994, pp. 167 - 176.

Andrade, J. D., Vanwagenen, R. A., Gregonis, D. E., Newby, K., and Lin, J.-N., "Remote Fiber-Optic Biosensors Based on Evanescent-Excited Fluoro-Immunoassay: Concept and Progress," IEEE Transactions on Electron Devices, vol. ED-32, no. 7, July 1985, pp. 1175 - 1179.

Armgarth, M., and Nylander, C., "A Stable Hydrogen-Sensitive Pd Gate Metal-Oxide Semiconductor Capacitor," Applied Physics Letters, vol. 39, no. 1, July 1, 1981, pp. 91 - 92.

Aurbach, D., Markovsky, B., and Ein-Eli, Y., "Investigation of Graphite-Lithium Intercalation Anodes for Li-Ion Rechargeable Batteries," Section 2.2 in "New Promising Electrochemical Systems for Rechargeable Batteries," Barsukov, V., and Beck, F. [eds.], Kluwer Academic Publishers, Dordrecht, Netherlands, 1996, pp. 63 - 75.

Bates, J. B., Dudney, N. J., and Weatherspoon, K. A., "Packaging Material for Thin Film Lithium Batteries," U.S. Patent No. 5,561,004, issued Oct. 1996.

Bates, J. B., Gruzalski, G. R., Dudney, N. J., Luck, C. F., Yu, X.-H., Jones, S. D., "Rechargeable Thin-Film Lithium Microbatteries," Solid State Technology, vol. 36, no. 7, July 1993, pp. 59 - 64.

Baughman, R. H., Shacklette, L. W., Elsenbaumer, R. L., Plichta, E. J., and Becht, C., "Micro Electromechanical Actuators Based on Conducting Polymers," in "Molecular Electronics," Lazarev, P. I. [ed.], Kluwer, Inc, Amsterdam, Netherlands, 1991, pp. 267 - 289.

Bell, T. E., Wise, K. D., and Anderson, D. J., "A Flexible Micromachined Electrode Array for a Cochlear Prosthesis," Proceedings of Transducers '97, the 1997 International Conference on Solid-State Sensors and Actuators, Chicago, IL, June 16 - 19, 1997, vol. 2, pp. 1315 - 1318.

Benes, E., Gröschl, M., Burger, W., and Schmid, M., "Sensors Based on Piezoelectric Resonators," Sensors and Actuators, vol. A48, no. 1, May 1995, pp. 1 - 21.

Bergveld, P., "Development of an Ion-Sensitive Solid-State Device for Neurophysiological Measurements," IEEE Transactions on Biomedical Engineering, vol. BME-17, no. 1, 1970, pp. 70 - 71.

Bergveld, P., "Future Applications of ISFETs," Sensors and Actuators, vol. B4, nos. 1 - 2, May 1991, pp. 125 - 133.

Boltshauser, T., "CMOS Humidity Sensors," Dissertation for Doctor of Technical Sciences, Diss. ETH Nr. 10320, Swiss Federal Institute of Technology (ETH), Zurich, Switzerland, 1993.

Boltshauser, T., Häberli, A., and Baltes, H., "Piezoresistive Membrane Hygrometers Based on IC Technology," Sensors and Materials, vol. 5, no. 3, 1993, pp. 125 - 134.

Borkholder, D. A., Maluf, N. I., and Kovacs, G. T. A., "Impedance Imaging for Hybrid Biosensor Applications," Proceedings of the Solid-State Sensor and Actuator Workshop, Hilton Head Island, SC, June 3 - 6, 1996, pp. 156 - 160.

Bradley, R. M., Smoke, R. H., Akin, T., and Najafi, K., "Functional Regeneration of Glossopharyngeal Nerve Through Micromachined Sieve Electrode Arrays," Brain Research, vol. 594, 1992, pp. 84 - 90.

Breer, H., "Sense of Smell: Signal Recognition and Transduction in Olfactory Receptor Neurons," Chapter 22 in "Handbook of Biosensors and Electronic Noses," Kress-Rogers, E. [ed.], CRC Press, Inc., Boca Raton, FL, 1997, pp. 521 - 532.

Buck, R. P., Hatfield, W. E., Umaña, M., and Bowden, E. F. [eds.], "Biosensor Technology," Marcel Dekker, Inc., New York, NY, 1989.

Cammann, K., Buhlmann, K., Schlatt, B., Müller, H., and Choulga, A., "Multicomponent Polymer Electrolytes: New Extremely Versatile Receptor Materials for Gas Sensors (VOC Monitoring) and Electronic Noses (Odour Identification/Discrimination)," Proceedings of Transducers '97, the 1997 International Conference on Solid-State Sensors and Actuators, Chicago, IL, June 16 - 19, 1997, vol. 2, pp. 1395 - 1398.

Caras, S. D., and Janata, J., "pH-Based Enzyme Potentiometric Sensors. Part 3. Penicillin-Sensitive Field Effect Transistor," Analytical Chemistry, vol. 57, no. 9, Aug. 1985, pp. 1924 - 1925.

Caras, S. D., Janata, J., Saupe, D., and Schmitt, K., "pH-Based Enzyme Potentiometric Sensors. Part 1. Theory," Analytical Chemistry, vol. 57, no. 9, Aug. 1985, pp. 1917 - 1920.

Caras, S. D., Petelenz, D., and Janata, J., "pH-Based Enzyme Potentiometric Sensors. Part 2. Glucose-Sensitive Field Effect Transistor," Analytical Chemistry, vol. 57, no. 9, Aug. 1985, pp. 1920 - 1923.

Castellucci, V. F., "The Chemical Senses: Taste and Smell," Chapter 32 in "Principles of Neural Science," Second Edition, Kandel, E. R., and Schwartz, J. H. [eds.], Elsevier, New York, NY, 1985, pp. 409 - 425.

Cavicchi, R. E., Suehle, J. S., Chaparala, P., Kreider, K. G., Gaitan, M., and Semancik, S., "Micro-Hotplate Gas Sensor," Proceedings of the 1994 Solid-State Sensor and Actuator Workshop, Hilton Head Island, SC, June 13 - 16, 1994, pp. 53 - 56.

Chang, S.-C., and Hicks, D. B., "Tin Oxide Microsensors on Thin Silicon Membranes," Record of the IEEE Solid-State Sensors Workshop, 1986 (no page numbers used).

Clark, L. C., "Monitor and Control of Blood and Tissue Oxygen Tension," Transactions of the American Society of Artificial Internal Organs, vol. 2, 1956, pp. 144 - 156.

Colapicchioni, C., Barbaro, A., Porcelli, F., and Giannini, I., "Immunoenzymatic Assay Using CHEMFET Devices," Sensors and Actuators, vol. B4, nos. 3 - 4, June 1991, pp. 245 - 250.

Conrath, N., Czupor, N., Steinkuhl, R., Sundermeier, C., Trau, D., Wittkampf, M., Hinkers, H., Meusel, M., Chemnitius, G., Knoll, M., Spener, F., and Cammann, K., "Threedimensional [sic] Structured Transducers for Chemical and Biochemical Sensors," Proceedings of Transducers '97, the 1997 International Conference on Solid-State Sensors and Actuators, Chicago, IL, June 16 - 19, 1997, vol. 1, pp. 481 - 484.

Costello, B. J., Wenzel, S. W., and White, R. M., "Density and Viscosity Sensing with Ultrasonic Flexural Plate Waves," Proceedings of Transducers '93, the 7th International Conference on Solid-State Sensors and Actuators, Yokohama, Japan, June 7 - 10, 1993, Institute of Electrical Engineers, Japan, pp. 704 - 707.

D'Amico, A., Di Natale, C., and Verona, E., "Acoustic Devices," Chapter 9 in "Handbook of Biosensors and Electronic Noses," Kress-Rogers, E. [ed.], CRC Press, Inc., Boca Raton, FL, 1997, pp. 197 - 223.

De Rossi, D., Chiarelli, P., Buzzigoli, G., Domenici, C., and Lazzeri, L., "Contractile Behavior of Electrically Activated Mechanochemical Polymer Actuators," Transactions of the American Society for Artificial Internal Organs, vol. 32, 1986, pp. 157 - 162.

Demarne, V., and Grisel A., "An Integrated Low-Power Thin-Film CO Sensor on Silicon," Sensors and Actuators, vol. 13, no. 4, Apr. 1988, pp. 301 - 314.

Dewa, A. S., and Ko, W. H., "Biosensors," Chapter 9 in "Semiconductor Sensors," Sze, S. M. [ed.], John Wiley and Sons, New York, NY, 1994, pp. 415 - 472.

Di Natale, C., Macagnano, A., Paolesse, R., Tarizzo, E, D'Amico, A., Davide, E., Boschi, T., Faccio, M., Ferri, G., Sinesio, F., Bucarelli, F. M., Moneta, E., and Quaglia, G. B., "A Comparison between an Electronic Nose and Human Olfaction in a Selected Case Study," Proceedings of Transducers '97, the 1997 International Conference on Solid-State Sensors and Actuators, Chicago, IL, June 16 - 19, 1997, vol. 2, pp. 1335 - 1338.

Dobos, K., and Zimmer, G., "Performance of CO-Sensitive MOSFETs with Metal Oxide Semiconductor Gates," IEEE Transactions on Electron Devics, vol. ED-32, no. 7, July 1985, pp. 1165 - 1169.

Dodd, J., and Castellucci, V. F., "Smell and Taste: The Chemical Senses," Chapter 34 in "Principles of Neural Science," Third Edition, Kandel, E. R., Schwartz, J. H., and Jessell, T. M. [eds.], Elsevier Science Publishing Co., Inc., New York, NY, 1991, pp. 512 - 529.

Drake, K. L., Wise, K. D., Farraye, J., Anderson, D. J., and BeMent, S. L., "Performance of Planar Multisite Microprobes in Recording Extracellular Single-Unit Intracortical Activity," IEEE Transactions on Biomedical Engineering, vol. 35, no. 9, Sept. 1988, pp. 719 - 732.

Droge, M., Gross, G., Hightower, M., and Cziny, L., "Multielectrode Analysis of Coordinated, Multisite, Rhythmic Bursting in Cultured CNS Monolayer Networks," The Journal of Neuroscience, vol. 6, June 1986, pp. 1583 - 1592.

Eddowes, M. J., "Response of an Enzyme-Modified pH Sensitive Ion-Selective Device; Analytical Solution for the Response in the Presence of pH Buffer," Sensors and Actuators, vol. 11, no. 3, Apr. 1987, pp. 265 - 274.

Edell, D. J., "A Peripheral Nerve Information Transducer for Amputees: Long-Term Multichannel Recordings from Rabbit Peripheral Nerves," IEEE Transactions on Biomedical Engineering, vol. 33, no. 2, Feb. 1986, pp. 203 - 214.

Ehret, R., Baumann, W., Brischwein, M., Schwinde, A., Stegbauer, K., and Wolf, B., "Monitoring of Cellular Behavior by Impedance Measurements on Interdigitated Electrode Structures," Biosensors and Bioelectronics, vol. 12, no. 1, 1997, pp. 29 - 41.

Erlich, H. A. [ed.], "PCR Technology: Principles and Applications for DNA Amplification," Oxford University Press, New York, NY, 1992.

Feng, C.-D., Nelson, T. E., Hardman, S., Hesketh, P. J., Maclay, G. J., Gendel, S. M., and Stetter, J. R., "Impedance Analysis of Ultrathin Platinum Film Immunosensors with Different Thickness and Macro Geometry," Proceedings of Transducers '95, the 8th International Conference on Solid-State Sensors and Actuators, Stockholm, Sweden, June 25 - 29, 1995, vol. 2, pp. 538 - 541.

Filippov, V. I., Terentjev, A. A., and Yakimov, S. S., "Electrode Structure Effects on the Selectivity of Gas Sensors," Sensors and Actuators, vol. B28, no. 1, July 1995, pp. 55 - 58.

Fodor, S. P. A., Read, J. L., Pirrung, M. C., Stryer, L., Lu, A. T., and Solas, D., "Light-Directed, Spatially Addressable Parallel Chemical Synthesis," Science, vol. 251, no. 4995, Feb. 15, 1991, pp. 767 - 773.

Fromherz, P., Offenhäusser, A., Vetter, T., and Weis, J., "A Neuron-Silicon Junction: A Retzius Cell of the Leech on an Insulated-Gate Field Effect Transistor," Science, vol. 252, no. 5010, May 31, 1991, pp. 1290 - 1293.

Fryder, M., Holmberg, M, Winquist, F., and Lundström, I., "A Calibration Technique for an Electronic Nose," Proceedings of Transducers '95, the 8th International Conference on Solid-State Sensors and Actuators, and Eurosensors IX, Stockholm, Sweden, June 25 - 29, 1995, vol. 1, pp. 683 - 686.

Fuhr, G., "Examples of Three-Dimensional Micro-Structures for Handling and Investigation of Adherently Growing Cells and Sub-Micron Particles," Proceedings of the 2nd International Symposium on Miniaturized Total Analysis Systems (μTAS '96), Basel Switzerland, Nov. 19 - 22, 1996, pp. 39 - 54.

Fuhr, G., and Shirley, S. G., "Cell Handling and Characterization Using Micron and Submicron Electrode Arrays: State of the Art and Perspectives of Semiconductor Microtools," Journal of Micromechanics and Microengineering, vol. 5, no. 2, June 1995, pp. 77 - 85.

Garabedian, R., Gonzalez, C., Richards, J., Knoesen, A., Spencer, R., Collins, S. D., and Smith, R. L., "Microfabricated Surface Plasmon Sensing System," Sensors and Actuators, vol. A43, nos. 1 - 3, May 1994, pp. 202 - 207.

Gardner, J. W., "(Bio)chemical Microsensors," Chapter 9 in "Microsensors: Principles and Applications," John Wiley and Sons, West Sussex, UK, 1994, pp. 224 - 249.

Gardner, J. W., and Bartlett, P. N., "A Brief History of Electronic Noses," Sensors and Actuators, vol. B18, nos. 1 - 3, Mar. 1994, pp. 211 - 220.

Gardner, J. W., and Bartlett, P. N., "Performance Definition and Standardization of Electronic Noses," Proceedings of Transducers '95, the 8th International Conference on Solid-State Sensors and Actuators, and Eurosensors IX, Stockholm, Sweden, June 25 - 29, 1995, vol. 1, pp. 671 - 674.

Gardner, J. W., Pearce, T. C., Friel, S., Bartlett, P. N., and Blair, N., "A Multisensor System for Beer Flavour Monitoring Using an Array of Conducting Polymers and Predictive Classifiers," Sensors and Actuators, vol. B18, nos. 1 - 3, Mar. 1994, pp. 240 - 243.

Getino, J., Gutiérrez, J., Arés, L., Robla, J. I., Sayago, I., Horrillo, M. C., and Agapito, J. A., "Integrated Sensor Array for Gas Analysis in Combustion Atmospheres," Proceedings of Transducers '95, the 8th International Conference on Solid-State Sensors and Actuators, Stockholm, Sweden, June 25 - 29, 1995, vol. 1, pp. 803 - 806.

Giaever, I., and Keese, C. R., "Cell Substrate Electrical Impedance Sensor with Multiple Electrode Array," U.S. Patent No. 5,187,096, issued Feb. 16, 1993.

Giaever, I., and Keese, C. R., "Electric Measurements Can Be Used to Monitor the Attachment and Spreading of Cells in Tissue Culture," BioTechniques, vol. 11, no. 4, Oct. 1991, pp. 504 - 510.

Giaever, I., and Keese, C. R., "Monitoring Fibroblast Behavior in Tissue Culture with an Applied Electric Field," Proceedings of the National Academy of Sciences, vol. 81, June 1984, pp. 3761 - 3764.

Giaever, I., and Keese, C. R., "Use of Electric Fields to Monitor the Dynamical Aspect of Cell Behavior in Tissue Culture," IEEE Transactions on Biomedical Engineering, vol. BME-33, no. 2, Feb. 1986, pp. 242 - 247.

Giancoli, D. C., "Physics for Scientists and Engineers with Modern Physics," Second Edition, Prentice-Hall, Inc., Englewood Cliffs, NJ, 1989, pp. 585 - 588.

Gibilisco, S., and Sclater, N., "Encyclopedia of Electronics," Second Edition, TAB Books, Inc., Blue Ridge Summit, PA, 1990, pp. 101 - 106.

Glass, R. S., Perone, S. P., Ciarlo, D. R., and Kimmons, J. F., "Electrochemical Sensor/Detector System and Method," U.S. Patent No. 5,120,421, issued Jun. 9, 1992.

Goldberg, H. D., Brown, R. B., Liu, D. P., and Meyerhoff, M. E., "Screen Printing: A Technology for the Batch Fabrication of Integrated Chemical-Sensor Arrays," Sensors and Actuators, vol. B21, no. 3, Sept. 1994, pp. 171 - 183.

Golovanov, V., Solis, J. L., Lantto, V., and Leppävuori, S., "Different Thick-Film Methods in Printing of One-Electrode Semiconductor Gas Sensors," Proceedings of Transducers '95, the 8th International Conference on Solid-State Sensors and Actuators, Stockholm, Sweden, June 25 - 29, 1995, vol. 2, pp. 874 - 877.

Grate, J. W., Martin, S. J., and White, R. M., "Acoustic Wave Microsensors," Parts 1 and 2, Analytical Chemistry, vol. 65, 1993, pp. 940A - 948A, pp. 987A - 996A.

Gross, G., "Simultaneous Single Unit Recording In Vitro with a Photoetched Laser Deinsulated Gold Multi Microelectrode Surface," IEEE Transactions on Biomedical Engineering, vol. BME-26, no. 5, May 1979, pp. 273 - 279.

Guidi, V., Cardinali, G. C., Dori, L., Faglia, G., Ferroni, M., Martinelli, G., Nelli, P., and Sberveglieri, G., "Novel Gas Sensor Based on Thin MoO_3 Film and Low Power Consumption Micromachined Si-Based Structure," Proceedings of Transducers '97, the 1997 International Conference on Solid-State Sensors and Actuators, Chicago, IL, June 16 - 19, 1997, vol. 2, pp. 943 - 946.

Haeusler, A., and Meyer, J.-U., "A Novel CO_2 Sensor Based on Changes in Conductivity of Metal Oxides," Proceedings of Transducers '95, the 8th International Conference on Solid-State Sensors and Actuators, Stockholm, Sweden, June 25 - 29, 1995, vol. 2, pp. 866 - 869.

Hamberg, M., Neagu, C., Gardeniers, J. G. E., Ijntema, D. J., and Elwenspoek, M., "An Electrochemical Micro Actuator," Proceedings of the 1995 IEEE Micro Electro Mechanical Systems Workshop (MEMS '95), Amsterdam, Netherlands, Jan. 29 - Feb. 2, 1995, pp. 106 - 110.

Harame, D. L., Bousse, L. J., Shott, J. D., and Meindl, J. D., "Ion-Sensing Devices with Silicon Nitride and Borosilicate Glass Insulators," IEEE Transactions on Electron Devices, vol. ED-34, no. 8, Aug. 1987, pp. 1700 - 1707.

Harris, P. D., Andrews, M. K., and Partridge, A. C., "Conductive Polymer Sensor Measurements," Proceedings of Transducers '97, the 1997 International Conference on Solid-State Sensors and Actuators, Chicago, IL, June 16 - 19, 1997, vol. 2, pp. 1063 - 1066.

Hatfield, J. V., Neaves, P., Hicks, P. J., Persaud, K., and Travers, P., "Towards an Integrated Electronic Nose Using Conducting Polymer Sensors," Sensors and Actuators, vol. B18, nos. 1 - 3, Mar. 1994, pp. 221 - 228.

Hendrikse, J., Olthuis, W., and Bergveld, P., "A Drift Free Nernstian Iridium Oxide pH Sensor," Proceedings of Transducers '97, the 1997 International Conference on Solid-State Sensors and Actuators, Chicago, IL, June 16 - 19, 1997, vol. 2, pp. 1367 - 1370.

Hermans, E. C. M., "CO, CO_2, CH_4 and H_2O Sensing by Polymer Covered Interdigitated Electrode Structures," Sensors and Actuators, vol. 5, no. 3, May 1984, pp. 181 - 186.

Hetke, J. F., Najafi, K., and Wise, K. D., "Flexible Silicon Interconnects for Microelectromechanical Systems," Proceedings of Transducers '91, the 1991 International Conference on Solid-State Sensors and Actuators, San Francisco, CA, June 24 - 27, 1991, pp. 764 - 767.

Hetke, J.F., Najafi, K., and Wise, K. D., "Flexible Miniature Ribbon Cables for Long-Term Connection to Implantable Sensors," Sensors and Actuators, vol. A23, nos. 1 - 3, Apr. 1990, pp. 999 - 1002.

Hickman, J. J., and Stenger, D. A., "Interactions of Neurons with Defined Surfaces," Chapter 4 in "Enabling Technologies for Cultured Neural Networks," Stenger, D. A., and McKenna, T. M. [eds.], Academic Press, Inc., San Diego, CA, 1994, pp. 51 - 76.

Hinkers, H., Dumschat, C., Steinkuhl, R., Sundermeier, C., Cammann, K., and Knoll, M., "Microdialysis System for Continuous Glucose Monitoring," Proceedings of Transducers '95, the 8th International Conference on Solid-State Sensors and Actuators, Stockholm, Sweden, June 25 - 29, 1995, vol. 1, pp. 470 - 473.

Hoffmann, B., Gadau, M., Paeschke, M., Hintsche, R., "Conductivity Measurements with Miniaturised Thin Film Metal Electrodes," Proceedings of Transducers '95, the 8th International Conference on Solid-State Sensors and Actuators, Stockholm, Sweden, June 25 - 29, 1995, vol. 2, pp. 837 - 840.

Hoogerwerf, A. C., and Wise, K. D., "A Three-Dimensional Neural Recording Array," Proceedings of Transducers '91, the 1991 International Conference on Solid-State Sensors and Actuators, San Francisco, CA, June 24 - 27, 1991, pp. 120 - 123.

Hsueh, Y. T., Smith, R. L., and Northrup, M. A., "A Microfabricated, Electrochemiluminescence Cell for the Detection of Amplified DNA," Sensors and Actuators, vol. B33, nos. 1 - 3, July 1996, pp. 110 - 114.

Hsueh, Y.-T., Collins, S. D., and Smith, R. L., "DNA Quantification with an Electrochemiluminescence Microcell," Proceedings of Transducers '97, the 1997 International Conference on Solid-State Sensors and Actuators, Chicago, IL, June 16 - 19, 1997, vol. 1, pp. 175 - 178.

Hsueh, Y.-T., Smith, R. L., and Northrup, M. A., "A Microfabricated, Electrochemiluminescence Cell for the Detection of Amplified DNA," Proceedings of Transducers '95, the 8th International Conference on Solid-State Sensors and Actuators, Stockholm, Sweden, June 25 - 29, 1995, vol. 1, pp. 768 - 771.

Hughes, R. C., Moreno, D. J., Jenkins, M. W., and Rodriguez, J. L., "The Response of the Sandia Robust Wide Range Hydrogen Sensor to H_2-O_2 Mixtures," Proceedings of the 1994 Solid-State Sensor and Actuator Workshop, Hilton Head Island, SC, June 13 - 16, 1994, pp. 57 - 60.

Israel, D. A., Edell, D. J., and Mark, R. G., "Time Delays in Propagation of Cardiac Action Potential," American Journal of Physiology, vol. 258, no. 6, part 2, June 1990, pp. 1906 - 1917.

Janata, J., "Chemical Sensitivity of Field-Effect Transistors," Sensors and Actuators, vol. 12, no. 2, Aug./Sept. 1987, pp. 121 - 128.

Ji, J., and Wise, K. D., "An Implantable CMOS Circuit Interface for Multiplexed Microelectrode Recording Arrays." IEEE Journal of Solid-state Circuits, vol. 27, no. 3, Mar. 1992, pp. 433 - 443.

Ji, J., Najafi, K., and Wise, K. D., "A Low-Noise Demultiplexing System for Active Multichannel Microelectrode Arrays." IEEE Transactions on Biomedical Engineering, vol. 38, no. 1, Jan. 1991, pp. 75 - 81.

Johnson, C. L., Najafi, N., Wise, K. D., and Schwank, J. W., "Detection of Semiconductor Process Gas Impurities Using an Integrated Ultra-Thin-Film Detector," Digest of Technical Papers, SRC TECHCON '90, San Jose, CA, Oct. 1990, p. 417.

Johnson, C. L., Schwank, J. W., and Wise, K. D., "Integrated Ultra-Thin-Film Gas Sensors," Sensors and Actuators, vol. B20, no. 1, May 1994, pp. 55 - 62.

Johnson, C. L., Wise, K. D., and Schwank, J. W., "A Thin-Film Gas Detector for Semiconductor Process Gases," Digest of the International Electron Devices Meeting (IEDM), San Francisco, CA, Dec. 1988, p. 662.

Jones, S. D., and Akridge, J. R., "A Microfabricated Solid-State Secondary Li Battery," Proceedings of the 10th International Conference on Solid State Ionics, Singapore, Dec. 3 - 8, 1995, published in Solid State Ionics, Diffusion and Reaction, vols. 86 - 88, part 2, July 1996, pp. 1291 - 1294.

Jones, S. D., and Akridge, J. R., "A Thin Film Solid State Microbattery," Proceedings of the 8th International Conference on Solid State Ionics, Lake Louise, Alta., Canada, Oct. 20 - 26, 1992, published in Solid State Ionics, Diffusion and Reaction, vols. 53 - 56, part 1, July - Aug. 1992, pp. 628 - 634.

Jorgenson, R. C., and Yee, S. S., "A Fiber-Optic Chemical Sensor Based on Surface Plasmon Resonance," Sensors and Actuators, vol. B12, no. 3, Apr. 1993, pp. 213 - 220.

Kakerow, R., Manoli, Y., Mokwa, W., Rospert, M., Meyer, H., Drewer, H., Krause, J., and Cammann, K., "A Monolithic Sensor Array of Individually Addressable Microelectrodes," Sensors and Actuators, vol. A43, nos. 1 - 3, May 1994, pp. 296 - 301.

Kang, J., Kim, Y., Kim, H., Jeong, J., and Park, S., "Comfort Sensing System for Indoor Environment," Proceedings of Transducers '97, the 1997 International Conference on Solid-State Sensors and Actuators, Chicago, IL, June 16 - 19, 1997, vol. 1, pp. 311 - 314.

Kelly, R. G., and Owen, A. E., "Microelectronic Ion Sensors: A Critical Survey," IEE Proceedings, vol. 132, part I, no. 5, Oct. 1985, pp. 227 - 236.

Kewley, D. T., Hills, M. D., Borkholder, D. A., Opris, I. E., Maluf, N. I., Storment, C. W., Bower, J. M., and Kovacs, G. T. A., "Plasma-Etched Neural Probes," Proceedings of the Solid-State Sensor and Actuator Workshop, Hilton Head Island, SC, June 3 - 6, 1996, pp. 266 - 271.

Kewley, D. T., Hills, M. D., Borkholder, D. A., Opris, I. E., Maluf, N. I., Storment, C. W., Bower, J. M., and Kovacs, G. T. A., "Plasma-Etched Neural Probes," Sensors and Actuators, vol. A58, no. 1, Jan. 1997, pp. 27 - 35.

Kikkawa, S., "Titanium Disulphide Thin Film Prepared by Plasma-CVD for Lithium Secondary Battery," Ceramics International, vol. 23, no. 1, 1996, pp. 7 - 11.

King, W. H., "Piezoelectric Sorption Detector," Analytical Chemistry, vol. 36, no. 9, Aug. 1964, pp. 1735 - 1739.

Kleinfeld, D., Kahler, K., and Hockberger, P. E., "Controlled Outgrowth of Dissociated Neurons on Patterned Substrates," Journal of Neuroscience, vol. 8, no. 11, 1988, pp. 4098 - 4120.

Kohl, D., "Semiconductor and Calorimetric Sensor Devices and Arrays," Chapter 23 in "Handbook of Biosensors and Electronic Noses," Kress-Rogers, E. [ed.], CRC Press, Inc., Boca Raton, FL, 1997, pp. 533 - 561.

Kondoh, J., and Shiokawa, S., "Liquid Identification Using SH-SAW Sensors," Proceedings of Transducers '95, the 8th International Conference on Solid-State Sensors and Actuators, Stockholm, Sweden, June 25 - 29, 1995, vol. 2, pp. 716 - 719.

Kounaves, S. P., Deng, W., Hallock, P. R., Kovacs, G. T. A., and Storment, C. W., "Iridium-Based Ultramicroelectrode Array Fabricated by Microlithography," Analytical Chemistry, vol. 66, 1994, pp. 418 - 423.

Kovacs, G. T. A., "Introduction to the Theory, Design, and Modeling of Thin-Film Microelectrodes for Neural Interfaces," Chapter 7 in "Enabling Technologies for Cultured Neural Networks," Stenger, D. A., and McKenna, T. M. [eds.], Academic Press, Inc., San Diego, CA, 1994, pp. 121 - 166.

Kovacs, G. T. A., "Regeneration Microelectrode Arrays for Direct Interface to Nerves," Proceedings of Transducers '91, the 1991 International Conference on Solid-State Sensors and Actuators, San Francisco, CA, June 24 - 27, 1991, pp. 116 - 119.

Kovacs, G. T. A., Storment, C. W., and Rosen, J. M., "Regeneration Microelectrode Array for Peripheral Nerve Recording and Stimulation," IEEE Transactions on Biomedical Engineering, vol. 39, no. 9, Sept. 1992, pp. 893 - 902.

Kovacs, G. T. A., Storment, C. W., Halks-Miller, M., Belczynski, C. R., Della Santina, C. C., Lewis, E. R., and Maluf, N. I., "Silicon-Substrate Microelectrode Arrays for Parallel Recording of Neural Activity in Peripheral and Cranial Nerves," IEEE Transactions on Biomedical Engineering, June 1994, vol. 41, no. 6, pp. 567 - 577.

Kraus, G., Hierlemann, A., Gauglitz, G., and Göpel, W., "Analysis of Complex Gas Mixtures by Pattern Recognition with Polymer Based Quartz Microbalance Sensor Arrays," Proceedings of Transducers '95, the 8th International Conference on Solid-State Sensors and Actuators, Stockholm, Sweden, June 25 - 29, 1995, vol. 1, pp. 675 - 678.

Kung, V. T., Panfili, P. R., Sheldon, E.L., King, R. S., Nagainis, P. A., Gomez, B., Ross, D. A., Briggs, J., and Zuk, R. F., "Picogram Quantitation of Total DNA Using DNA-Binding Proteins in a Silicon Sensor-Based System," Analytical Biochemistry, vol. 187, no. 2, 1990, pp. 220 - 227.

Langereis, G. R., Olthius, W., and Bergveld, P., "Measuring Conductivity, Temperature and Hydrogen Peroxide Concentration Using a Single Sensor Structure," Proceedings of Transducers '97, the 1997 International Conference on Solid-State Sensors and Actuators, Chicago, IL, June 16 - 19, 1997, vol. 1, pp. 543 - 546.

Lauks, I. R., "pH Measurements Using Polarizable Electrodes," IEEE Transactions on Electron Devices," vol. ED-26, no. 12, Dec. 1979, pp. 1952 - 1959.

Lee, H.-S., Wang, S. S., Smolenski, D. J., Viola, M. B., and Klusendorf, E. E., "In Situ Monitoring of High-Temperature Degraded Engine Oil Condition with Microsensors," Sensors and Actuators, vol. B20, no. 1, May 1994, pp. 49 - 54.

Liu, M., Visco, S. J., and De Jonghe, L. C., "Novel Solid Redox Polymerization Electrodes," Journal of the Electrochemical Society, vol. 138, no. 7, July 1991, pp. 1891 - 1895.

Lo, C., Keese, C. R., Giaever, I., "Impedance Analysis of MDCK Cells Measured by Electrical Cell-Substrate Impedance Sensing," Biophysical Journal, vol. 69, no. 6, Dec. 1995, pp. 2800 - 2807.

Lu, C., and Czanderna, A. W. [eds.], "Applications of Piezoelectric Quartz Crystal Microbalances," Volume 7 of "Methods and Phenomena: Their Application in Science and Technology," Wolsky, S. P., and Czanderna, A. W. [series eds.], Elsevier Science Publishers Co., Amsterdam, Netherlands, 1984.

Lu, C.-S., and Lewis, O., "Investigation of Film-Thickness Determination by Oscillating Quartz Resonators with Large Mass Load," Journal of Applied Physics, vol. 43, no. 11, Nov. 1972, pp. 4385 - 4390.

Lundström, I., "Hydrogen Sensitive MOS-Structures Part 1: Principles and Applications," Sensors and Actuators, vol. 1, no. 4, June 1981, pp. 403 - 426.

Lundström, I., Shivaraman, S., Svensson, C., and Lundkvist, L., "Hydrogen Sensitive MOS Field-Effect Transistor," Applied Physics Letters, vol. 26, no. 2, Jan. 15, 1975, pp. 55 - 57.

Lyle, R. P., and Walters, D., "Commercialization of Silicon-Based Gas Sensors," Proceedings of Transducers '97, the 1997 International Conference on Solid-State Sensors and Actuators, Chicago, IL, June 16 - 19, 1997, vol. 2, pp. 975 - 977.

Macdonald, J. R., "Impedance Spectroscopy," John Wiley and Sons, New York, NY, 1987.

Maeda, M., Mitsuhashi, Y., Nakano, K., and Takagi, M., "DNA-Immobilized Gold Electrode for DNA-Binding Drug Sensor," Analytical Sciences, vol. 8, Feb. 1992, pp. 83 - 84.

Mannard, A., Stein, R. B., and Charles, D., "Regeneration Electrode Units: Implants for Recording from Peripheral Nerve Fibers in Freely Moving Animals," Science, vol. 183, no. 4124, Feb. 8, 1974, pp. 547 - 549.

Marks, A. F., "Bullfrog Nerve Regeneration into Porous Implants," Anatomical Record, vol. 163, 1969, p. 226.

Marrese, C. "Preparation of Strongly Adherent Platinum Black Coatings," Analytical Chemistry, vol. 59, no. 1, Jan. 1987, pp. 217 - 218.

Martin, B. A., Wenzel, S. W., and White, R. M., "Viscosity and Density Sensing with Ultrasonic Plate Waves," Sensors and Actuators, vol. A22, nos. 1 - 3, Mar. 1990, pp. 704 - 708.

Martin, S. J., Ricco, A. J., and Hughes, R. C., "Acoustic Wave Devices for Sensing in Liquids," Proceedings of Transducers '87, the 4th International Conference on Sensors and Actuators, Tokyo, Japan, June 2 - 5, 1987, pp. 478 - 481.

Martinoia, S., Bove, M., Carlini, G., Ciccarelli, C., Grattarola, M., Storment, C., and Kovacs, G., "A General-Purpose System for Long-Term Recording from a Microelectrode Array Coupled to Excitable Cells," Journal of Neuroscience Methods, vol. 48, nos. 1 - 2, June 1993, pp. 115 - 121.

Matsuo, T., and Wise, K. D., "An Integrated Field-Effect Electrode for Biopotential Recording," IEEE Transactions on Biomedical Engineering, vol. BME-21, no. 6, Nov. 1974, pp. 485 - 487.

Matsuo, T., Yamaguchi, A., and Esashi, M., "Fabrication of Multi-Hole-Active Electrode for Nerve Bundle," Journal of Japan Society of Medical Electronics and Biological Engineering, July 1978 (in Japanese).

McNerney, J. J., Buseck, P. R., and Hanson, R. C., "Mercury Detection by Means of Thin Gold Films," Science, vol. 178, no. 4061, Nov. 10, 1972, pp. 611 - 612.

Millan, K. M., and Mikkelsen, S. R., "Sequence-Selective Biosensor for DNA Based on Electroactive Hybridization Indicators," Analytical Chemistry, vol. 65, no. 17, Sept. 1993, pp. 2317 - 2323.

Mohr, A., Finger, W., Föhr, K. J., Nisch, W., and Göpel, W., "Performance of a Thin Film Microelectrode Array for Monitoring Electrogenic Cells In Vitro," Proceedings of Transducers '95, the 8th International Conference on Solid-State Sensors and Actuators, Stockholm, Sweden, June 25 - 29, 1995, vol. 2, pp. 479 - 482.

Moriizumi, T., "Biomimetic Sensing Systems with Arrayed Nonspecific Sensors," Proceedings of Transducers '95, the 8th International Conference on Solid-State Sensors and Actuators, Stockholm, Sweden, June 25 - 29, 1995, vol. 1, pp. 39 - 42.

Moroney, R. M., White, R. M., and Howe, R. T., "Microtransport Induced by Ultrasonic Lamb Waves," Applied Physics Letters, vol. 59, no. 7, Aug. 12, 1991, pp. 774 - 776.

Morrison, S. R., "Chemical Sensors," Chapter 8 in "Semiconductor Sensors," Sze, S. M. [ed.], John Wiley and Sons, New York, NY, 1994.

Morrison, S. R., "Selectivity in Semiconductor Gas Sensors," Sensors and Actuators, vol. 12, no. 4, Nov./Dec. 1987, pp. 425 - 440.

Moseley, P. T., "Materials Selection for Semiconductor Gas Sensors," Sensors and Actuators, vol. B6, nos. 1 - 3, Jan. 1992, pp. 149 - 156.

Najafi, K., "Solid-State Microsensors for Cortical Nerve Recordings," IEEE Engineering and Biology Magazine, June/July 1994, pp. 375 - 387.

Najafi, K., Wise, K. D., and Mochizuki, T., "A High-Yield IC-Compatible Multichannel Recording Array," IEEE Transactions on Electron Devices, vol. ED-32, no. 7, July 1985, pp. 1206 - 1211.

Nakamoto, T., Fukuda, A., and Moriizumi, T., "Perfume and Flavor Identification by Odor Sensing System Using Quartz-Resonator Sensor Array and Neural-Network Pattern Recognition," Proceedings of Transducers '91, the 1991 International Conference on Solid-State Sensors and Actuators, San Francisco, CA, June 24 - 27, 1991, pp. 355 - 358.

Northrup, M. A., and White, R. M., "Microfabricated Reactor," U.S. Patent No. 5,646,039, issued July 8, 1997.

Northrup, M. A., Beeman, B., Hadley, D., Landre, P., and Lehew, S., "Integrated Miniature DNA-Based Analytical Instrumentation," Proceedings of the 2nd International Symposium on Miniaturized Total Analysis Systems (μTAS '96), Basel Switzerland, Nov. 19 - 22, 1996, pp. 153 - 157.

Northrup, M. A., Ching, M. T., White, R. M., and Watson, R. T., "DNA Amplification with a Microfabricated Reaction Chamber," Proceedings of Transducers '93, the 7th International Conference on Solid-State Sensors and Actuators, Yokohama, Japan, June 7 - 10, 1993, Institute of Electrical Engineers, Japan, pp. 924 - 926.

Northrup, M. A., Lawrence Livermore National Laboratory, Livermore, CA, personal communication, 1995.

Northup, M. A., and White, R. M., "Microfabricated Reactor," U.S. Patent No. 5,639,423, issued June 17, 1997.

Novak, J. L., and Wheeler, B. C., "Recording from the *Aplysia* Abdominal Ganglion with a Planar Microelectrode Array," IEEE Transactions on Biomedical Engineering, vol. BME-33, no. 2, Feb. 1986, pp. 196 - 202.

Olson, J. D., Panfili, P. R., Zuk, R. F., and Sheldon, E. L., "Quantitation of DNA Hybridization in a Silicon Sensor-Based System: Application to PCR," Molecular and Cellular Probes, vol. 5, 1991, pp. 351 - 358.

Olthuis, W., Robben, M. A. M., Bergveld, P., Bos, M., and van der Linden, W. E., "pH Sensor Properties of Electrochemically Grown Iridium Oxide," Sensors and Actuators, vol. B2, no. 4, Oct. 1990, pp. 247 - 256.

Olthuis, W., van Kerkhof, J. C., Bergveld, P., Bos, M., and van der Linden, W. E., "Preparation of Iridium Oxide and Its Application in Sensor-Actuator Systems," Sensors and Actuators, vol. B4, nos. 1 - 2, May 1991, pp. 151 - 156.

Owicki, J. C., Parce, J. W., Kersco, K. M., Sigal, G. B., Muir, V. C., Venter, J. C., Fraser, C. M., and McConnell, H. M., "Continuous Monitoring of Receptor-Mediated Changes in the Metabolic Rates of Living Cells," Proceedings of the National Academy of Sciences, vol. 87, 1990, pp. 4007 - 4011.

Parce, J. W., Owicki, J. C., Kersco, K. M., Sigal, G. B., Wada, H. G., Muir, V. C., Bousse, L. J., Ross, K. L., Sikic, B. I., and McConnell, H. M., "Detection of Cell-Affecting Agents with a Silicon Biosensor," Science, vol. 246, no. 4927, Oct. 13, 1989, pp. 243 - 247.

Pásztor, K., Sekiguchi, A., Shimo, N., Kitamura, N., and Masuhara, H., "Iridium Oxide-Based Microelectrochemical Transistors for pH Sensing," Sensors and Actuators, vol. B12, no. 3, Apr. 1993, pp. 225 - 230.

Pickard, R. S., Wall, P., Ubeid, M., Ensell, G., and Leong, K. H., "Recording Neural Activity in the Honeybee Brain with Micromachined Silicon Sensors," Sensors and Actuators, vol. B1, nos. 1 - 6, Jan. 1990, pp. 460 - 463.

Pine, J., "Recording Action Potentials from Cultured Neurons with Extracellular Microcircuit Electrodes," Journal of Neuroscience Methods, vol. 2, no. 1, Feb. 1980, pp. 19 - 31.

Prohaska, O. J., Olcaytug, F., Pfunder, P., and Dragaun, H., "Thin-Film Multiple Electrode Probes: Possibilities and Limitations," IEEE Transactions on Biomedical Engineering, vol. BME-33, no. 2, Feb. 1986, pp. 223 - 229.

Rahner, D., Machill, S., Siury, K., Kloß, M., and Plieth, W., "Intercalation Materials for Lithium Rechargeable Batteries," Section 2.1 in "New Promising Electrochemical Systems for Rechargeable Batteries," Barsukov, V., and Beck, F. [eds.], Kluwer Academic Publishers, Dordrecht, Netherlands, 1996, pp. 35 - 61.

Reay, R. J., Kounaves, S. P., and Kovacs, G. T. A., "Integrated CMOS Potentiostat," Digest of Technical Papers from the 1995 International Solid-State Circuits Conference, San Francisco, CA, Feb. 16 - 18, 1994, pp. 162 - 163.

Reay, R. J., Storment, C. W., Flannery, A. F., Kounaves, S. P., and Kovacs, G. T. A., "Microfabricated Electrochemical Analysis System for Heavy Metal Detection," Proceedings of Transducers '95, the 8th International Conference on Solid-State Sensors and Actuators, Stockholm, Sweden, June 25 - 29, 1995, vol. 2, pp. 932 - 935.

Reay, R. J., Storment, C. W., Flannery, A. F., Kounaves, S. P., and Kovacs, G. T. A., "Microfabricated Electrochemical Analysis System for Heavy Metal Detection," Sensors and Actuators, vol. B34, nos. 1 - 3, Aug. 1996, pp. 450 - 455.

Richter, A., "Design Considerations and Performance of Adsorptive Humidity Sensors with Capacitive Readout," Proceedings of Transducers '93, the 7th International Conference on Solid-State Sensors and Actuators, Yokohama, Japan, June 7 - 10, 1993, Institute of Electrical Engineers, Japan, pp. 310 - 313.

Robinson, D. A., "The Electrical Properties of Metal Microelectrodes," Proceedings of the IEEE, vol. 56, no. 6, June 1968, pp. 1065 - 1071.

Rodriguez, J. L., Hughes, R. C., Corbett, W. T. and McWhorter, P. J., "Robust, Wide Range Hydrogen Sensor," Digest of the International Electron Devices Meeting (IEDM), San Francisco, CA, Dec. 1992, pp. 521 - 524.

Ryan, J. G., Barry, L., Lyden, C., Alderman, J., Lane, B., Schiffner, L., Boldt, J., and Thieme, H., "A CMOS Chip-Set for Detecting 10 ppb Concentrations of Heavy Metals," Digest of Technical Papers from the 1995 International Solid-State Circuits Conference, San Francisco, CA, Feb. 15 - 17, 1995, pp. 158 - 159.

Sager, K., Schroth, A., and Gerlach, G., "Humidity-Dependent Mechanical Properties of Polyimide Films and Their Use for IC-Compatible Humidity Sensors," Proceedings of Transducers '95, the 8th International Conference on Solid-State Sensors and Actuators, Stockholm, Sweden, June 25 - 29, 1995, vol. 2, pp. 736 - 739.

Sansen, W., de Wachter, D., Callewaert, L., Lambrechts, M., and Claes, A., "A Smart Sensor for the Voltammetric Measurement of Oxygen or Glucose Concentrations," Sensors and Actuators, vol. B1, nos. 1 - 6, Jan. 1990, pp. 298 - 302.

Schnakenberg, U., Lisec, T., Hintsche, R., Kuna, I., Uhlig, A., and Wagner, B., "Novel Potentiometric Silicon Sensor for Medical Devices," Proceedings of Transducers '95, the 8th International Conference on Solid-State Sensors and Actuators, Stockholm, Sweden, June 25 - 29, 1995, vol. 2, pp. 959 - 962.

Schroth, A., Sager, K., Gerlach, G., Häberli, A., Boltshauser, T., and Baltes, H., "A Resonant Polyimide-Based Humidity Sensor," Proceedings of Transducers '95, the 8th International Conference on Solid-State Sensors and Actuators, Stockholm, Sweden, June 25 - 29, 1995, vol. 2, pp. 740 - 742.

Schwebel, T., Fleischer, M., and Meixner, H., "CO-Sensor for Domestic Use Based on High Temperature Stable Ga_2O_3 Thin Films," Proceedings of Transducers '97, the 1997 International Conference on Solid-State Sensors and Actuators, Chicago, IL, June 16 - 19, 1997, vol. 1, pp. 547 - 550.

Scott, A. C., "The Electrophysics of a Nerve Fiber," Reviews of Modern Physics, vol. 47, no. 2, Apr. 1975, pp. 487 - 533.

Seitz, W. R., "Chemical Sensors Based on Fiber Optics," Analytical Chemistry, vol. 56, no. 1, Jan. 1984, pp. 16A - 34A.

Senturia, S. D., Sechen, C. M., and Wishneusky, J. A., "The Charge Flow Transistor: A New MOS Device," Applied Physics Letters, vol. 30, no. 2, Jan. 15, 1977, pp. 106 - 108.

Shokoohi, F. K., Tarascon, J. M., and Wilkens, B. J., "Fabrication of Thin-Film $LiMn_2O_4$ Cathodes for Rechargeable Microbatteries," Applied Physics Letters, vol. 59, no. 10, Sept. 1991, pp. 1260 - 1262.

Sibbald, A., "Recent Advances in Field-Effect Chemical Microsensors," Journal of Molecular Electronics, vol. 2, no. 2, Apr. - June 1986, pp. 51 - 83.

Siu, W. M., and Cobbold, R. S. C., "Basic Properties of the Electrolyte-SiO2-Si System: Physical and Theoretical Aspects," IEEE Transactions on Electron Devices, vol. ED-26, no. 11, Nov. 1979, pp. 1805 - 1815.

Smela, E., Inganäs, O., and Lundrström, I., "Differential Adhesion Method for Microstructure Release: An Alternative to the Sacrificial Layer," Proceedings of Transducers '95, the 8th International Conference on Solid-State Sensors and Actuators, Stockholm, Sweden, June 25 - 29, 1995, vol. 1, pp. 218 - 219.

Smela, E., Inganäs, O., and Lundrström, I., "Self-Opening and Closing Boxes and Other Micromachined Folding Structures," Proceedings of Transducers '95, the 8th International Conference on Solid-State Sensors and Actuators, Stockholm, Sweden, June 25 - 29, 1995, vol. 2, pp. 350 - 351.

Soekarno, A., Lom, B., and Hockberger, P. E., "Pathfinding by Neuroblastoma Cells in Culture Is Directed by Preferential Adhesion to Positively Charged Surfaces," Neuroimage, vol. 1, 1993, pp. 129 - 144.

Starr, A., Wise, K. D., and Csongradi, J., "An Evaluation of Photoengraved MicroElectrodes for Extracellular Single-Unit Recording," IEEE Transactions on Biomedical Engineering, vol. BME-20, no. 4, July 1973, pp. 291 - 293.

Steiner, F.-P., Hierlemann, A., Cornila, C., Noetzel, G., Bächtold, M., Korvink, J. G., Göpel, W., and Baltes, H., "Polymer Coated Capacitive Microintegrated Gas Sensor," Proceedings of Transducers '95, the 8th International Conference on Solid-State Sensors and Actuators, Stockholm, Sweden, June 25 - 29, 1995, vol. 2, pp. 814 - 817.

Stenger, D. A., and Hickman, J. J., "Lithographic Definition of Neuronal Microcircuits," Chapter 5 in "Enabling Technologies for Cultured Neural Networks," Stenger, D. A., and McKenna, T. M. [eds.], Academic Press, Inc., San Diego, CA, 1994, pp. 77 - 98.

Stieve, H., "Sensors of Biological Organisms - Biological Transducers," Sensors and Actuators, vol. 4, no. 4, Dec. 1983, pp. 689 - 704.

Suehle, J. S., Cavicchi, R. E., Gaitan, M., and Semancik, S., "Tin Oxide Gas Sensor Fabricated Using CMOS Micro-Hotplates and In-Situ Processing," IEEE Electron Device Letters, vol. 14, no. 3, Mar. 1993, pp. 118 - 120.

Takahashi, K., and Matsuo, T., "Integration of Multi-Microelectrode and Interface Circuits by Silicon Planar and Three-Dimensional Fabrication Technology," Sensors and Actuators, vol. 5, no. 1, Jan. 1984, pp. 89 - 99.

Takehara, Z.-I., Ogumi, Z., Yoshiharu, U., Endo, E., and Kanamori, Y., "Thin Film Solid-State Lithium Batteries Prepared by Consecutive Vapor-Phase Processes," Journal of the Electrochemical Society, vol. 138, no. 6, June 1991, pp. 1574 - 1582.

Tamaki, J., Shimanoe, K., Yamada, Y., Yamamoto, Y., Miura, N., and Yamazoe, N., "Dilute Hydrogen Sulfide Sensing Properties of Copper Oxide-Tin Oxide Thin Film Prepared by Low-Pressure Evaporation," Proceedings of Transducers '97, the 1997 International Conference on Solid-State Sensors and Actuators, Chicago, IL, June 16 - 19, 1997, vol. 2, pp. 987 - 990.

Tamura, H., Saito, S., and Ito, K., "Hydrogen-Sensitive Film Consisting of Anodically Oxidized Tungsten," Proceedings of Transducers '93, the 7th International Conference on Solid-State Sensors and Actuators, Yokohama, Japan, June 7 - 10, 1993, Institute of Electrical Engineers, Japan, pp. 310 - 313.

Tang, T., Ocvirk, G., Harrison, D. J., "Iso-Thermal DNA Reactions and Assays in Microfabricated Capillary Electrophoresis Systems," Proceedings of Transducers '97, the 1997 International Conference on Solid-State Sensors and Actuators, Chicago, IL, June 16 - 19, 1997, vol. 1, pp. 523 - 526.

Thackeray, M. M., de Kock, A. , Rossouw, M. H., Liles, D., Bittihn, R., and Hoge, D., "Spinel Electrodes from the Li-Mn-O System for Rechargeable Lithium Battery Applications," Journal of the Electrochemical Society, vol. 139, no. 2, Feb. 1992, pp. 363 - 366.

Thomas, C., Springer, P., Loeb, G., Berwald-Netter, Y., and Okum, L., "A Miniature Microelectrode Array to Monitor the Bioelectric Activity of Cultured Cells," Experimental Cell Research, vol. 74, 1972, pp. 61 - 66.

Tortora, G. J., and Anagnostakos, N. P., "Principles of Anatomy and Physiology," Fifth Edition, Harper and Row, New York, NY, 1987, p. 371.

Tsao, R. T., Moroney, R. M., Martin, B. A., and White, R. M., "Electrochemical Detection of Localized Mixing Produced by Ultrasonic Flexural Waves," IEEE Ultrasonics Symposium, Lake Buena Vista, FL, Dec. 8 - 11, 1991, pp. 937 - 940.

Tsay, Y. G., Lin, C. I., Lee, J., Gustafson, E. K., Appelqvist, R., Magginetti, P., Norton, R., Teng, N., and Charlton, D., "Optical Biosensor Assay (OBA^{TM})," Clinical Chemistry, vol. 37, no. 9, 1991, pp. 1502 - 1505.

Turner, R. F. B., Harrison, D. J., and Baltes, H. P., "A CMOS Potentiostat for Amperometric Chemical Sensors," IEEE Journal of Solid-State Circuits, vol. 22, no. 3, June 1987, pp. 473 - 478.

van Herwaarden, A. W., Sarro, P. M., Gardner, J. W., and Bataillard, P., "Microcalorimeters for (Bio)Chemical Measurements in Gases and Liquids," Proceedings of Transducers '93, the 7th International Conference on Solid-State Sensors and Actuators, Yokohama, Japan, June 7 - 10, 1993, pp. 411 - 414.

Vig, J. R., Filler, R. L., and Kim, Y., "Chemical Sensor Based on Quartz Microresonators," Journal of Microelectromechanical Systems, vol. 5, no. 2, June 1996, pp. 138 - 140.

Vlachos, D. S., Papadopoulos, C. A., and Avaritsiotis, J. N., "On the Electronic Interaction between Additives and Semiconducting Oxide Gas Sensors," Applied Physics Letters, vol. 69, no. 5, July 1996, pp. 650 - 652.

Wang, A. W., Costello, B. J., and White, R. M., "An Ultrasonic Flexural Plate-Wave Sensor for Measurement of Diffusion in Gels," Analytical Chemistry, vol. 65, 1993, pp. 1639 - 1642.

Wang, A. W., Kiwan, R., White, R. M., and Ceriani, R. L., "A Silicon-Based Ultrasonic Immunoassay for Detection of Breast Cancer Antigens," Proceedings of Transducers '97, the 1997 International Conference on Solid-State Sensors and Actuators, Chicago, IL, June 16 - 19, 1997, vol. 1, pp. 191 - 194.

Wenzel, S. W., and White, R. M., "A Multisensor Employing an Ultrasonic Lamb-Wave Oscillator," IEEE Transactions on Electron Devices, vol. 35, no. 6, June 1988, pp. 735 - 743.

Wenzel, S. W., and White, R. M., "Flexural Plate-Wave Gravimetric Chemical Sensor," Sensors and Actuators, vol. A22, nos. 1 - 3, Feb. - Apr. 1990, pp. 700 - 703.

Wheeler, B. C., and Brewer, G. J., "Multineuron Patterning and Recording," Chapter 8 in "Enabling Technologies for Cultured Neural Networks," Stenger, D. A., and McKenna, T. M. [eds.], Academic Press, Inc., San Diego, CA, 1994, pp. 167 - 186.

Wilson, D. M., and DeWeerth, S. P., "Nonlinear Preprocessing for Smart Chemical Sensing Systems," Proceedings of Transducers '95, the 8th International Conference on Solid-State Sensors and Actuators, Stockholm, Sweden, June 25 - 29, 1995, vol. 1, pp. 814 - 817.

Wise, K. D., and Angell, J. B., "A Low-Capacitance Multielectrode Probe for Use in Extracellular Neurophysiology," IEEE Transactions on Biomedical Engineering, vol. BME-22, no. 3, May 1975, pp. 212 - 219.

Wise, K. D., Angell, J. B., and Starr, A., "An Integrated-Circuit Approach to Extracellular Microelectrodes, IEEE Transactions on Biomedical Engineering, vol. BME-17, no. 3, July 1970, pp. 238 - 246.

Wlodarski, W., Sun, H.-T., Gurlo, A., and Göpel, W., "Sol-Gel Prepared In_2O_3 Thin Films for Ozone Sensing," Proceedings of Transducers '97, the 1997 International Conference on Solid-State Sensors and Actuators, Chicago, IL, June 16 - 19, 1997, vol. 1, pp. 573 - 576.

Wohltjen, H., "Mechanism of Operation and Design Considerations for Surface Acoustic Wave Device Vapour Sensors," Sensors and Actuators, vol. 5, no. 4, July 1984, pp. 307 - 325.

Wohltjen, H., Barger, W. R., Snow, A. W., and Jarvis, N. L., "A Vapor-Sensitive Chemiresistor Fabricated with Planar Microelectrodes and a Langmuir-Blodgett Organic Semiconductor Film," IEEE Transactions on Electron Devices, vol. ED-32, no. 7, July 1985, pp. 1170 - 1174.

Woolley, A. T., Hadley, D., Landre, P., deMello, A. J., Mathies, R. A., and Northrup, M. A., "Functional Integration of PCR Amplification and Capillary Electrophoresis in a Microfabricated DNA Analysis Device," Analytical Chemistry, vol. 68, no. 23, Dec. 1996, pp. 4081 - 4086.

Wrighton, M. S., "Surface Functionalization of Electrodes with Molecular Reagents," Science, vol. 231, no. 4733, Jan. 3, 1986, pp. 32 - 37.

Xie, B., Mecklenberg, M., Dzgoev, A., and Danielsson, B., "Simultaneous Determination of Glucose, Lactate, Urea and Penicillin in Mixed Samples Using an Integrated Thermal Biosensor Array," Proceedings of the 2nd International Symposium on Miniaturized Total Analysis Systems (µTAS '96), Basel Switzerland, Nov. 19 - 22, 1996, pp. 95 - 99

Yamamoto, T., Murakami, K, Shimizu, H., and Takai, T., "An Integrated Temperature and Humidity Sensor," Proceedings of the 4th International Conference on Sensors and Actuators, Transducers '87, Tokyo, Japan, June 2 - 5, 1987, pp. 658 - 660.

Yamazoe, N., and Shimizu, Y., "Humidity Sensors: Principles and Applications," Sensors and Actuators, vol. 10, nos. 3 - 4, Nov. - Dec. 1986 pp. 379 - 398.

Yang, L., Qiu, W., Liu, Q., "Polyaniline Cathode Material for Lithium Batteries," Proceedings of the 10th International Conference on Solid State Ionics, Singapore, Dec. 3 - 8, 1995, published in Solid State Ionics, Diffusion and Reactions, vols. 86 - 88, part 2, July 1996, pp. 819 - 824.

Yazami, R., and Deschamps, M., "The Carbon-Lithium Negative Electrode for Lithium-Ion Batteries in Polymer Electrolyte," Proceedings of the Materials Research Society Symposium on Solid State Ionics (IV), Boston, MA, Nov. 28 - Dec. 1, 1994, vol. 369, pp. 165 - 176.

Yea, B., Konishi, R., Osaki, T., and Sugahara, K., "The Discrimination of Many Kinds of Odor Species Using Fuzzy Reasoning and Neural Networks," Sensors and Actuators, vol. A45, no. 2, Nov. 1994, pp. 159 - 165.

Zanini, M., Visser, J. H., Rimai, L., Soltis, R. E., Kovalchuck, A., Hoffman, D. W., Logothetis, E. M., Bonne, U., Brewer, L., Bynum, O. W., and Richard, M. A., "Fabrication and Properties of a Si-Based High Sensitivity Microcalorimetric Gas Sensor," Proceedings of the 1994 Solid-State Sensor and Actuator Workshop, Hilton Head Island, SC, June 13 - 16, 1994, pp. 176 - 178.

Zanini, M., Visser, J. H., Rimai, L., Soltis, R. E., Kovalchuck, A., Hoffman, D. W., Logothetis, E. M., Bonne, U., Brewer, L., Bynum, O. W., and Richard, M. A., "Fabrication and Properties of a Si-Based High Sensitivity Microcalorimetric Gas Sensor," Sensors and Actuators, vol. A48, no. 3, May 1995, pp. 187 - 192.

Zhou, R., Hierlman, A., Weimar, U., and Göpel, W., "Gravimetric, Dielectric and Calorimetric Methods for the Detection of Organic Solvent Vapours Using Poly(etherurethane) Coatings," Proceedings of Transducers '95, the 8th International Conference on Solid-State Sensors and Actuators, Stockholm, Sweden, June 25 - 29, 1995, vol. 2, pp. 833 - 836.

Zieren, M., and Köhler, J. M., "A Micro-Fluid Channel Calorimeter Using BiSb/Sb Thin Film Thermopiles," Proceedings of Transducers '97, the 1997 International Conference on Solid-State Sensors and Actuators, Chicago, IL, June 16 - 19, 1997, vol. 1, pp. 539 - 542.

CITED INDUSTRY REFERENCES AND SUPPLIERS

Polyimides

DuPont Electronic Materials, P.O. Box 13999, 14 T.W. Alexander Drive, Research Triangle Park, NC 27709-4425, Phone: (800) 557-4505, Fax: (302) 992-5843.

Integrated Humidity Sensors

Hy-Cal Engineering, 9650 Telstar Avenue, El Monte, CA 91731, Phone: (818) 444-4000, Fax: (818) 444-7075.

Metal Oxide Gas Sensors

Motorola, Inc., P.O. Box 20912, Phoenix, AZ 85036, Phone: (800) 441-2447, http://www.mot.com/

Microcalorimeter Biosensors

Xensor Integration, P.O. Box 3233, 2601 DE Delft, the Netherlands, Phone: +31 15 697696, Fax: +31 15 697748.

ISFET pH Sensors

Beckman Instruments, Fullerton, CA, Phone: (800) 854-8067, http://www.beckman.com

"Electronic Noses"

Aromascan, Plc, Manchester, UK, Phone: USA Office (603) 598-2922, UK 44 1270216444, http://www.aromascan.com

Neotronics Scientific, P.O. Box 2100 Flowery Branch, GA 30542, Phone: (800) 535-0606, Fax: (770) 967-1854, http://www.neotronics.com

Hybrid Biosensors

Molecular Devices, Corp., 1311 Orleans Drive, Sunnyvale, CA 94025, Phone: (408) 747-1700, Fax: (408) 747-3602.

Applied Biophysics, Inc., 1223 Peoples Avenue, Troy, NY 12180-3535, Phone: (518) 276-2165, Fax: 518-276-6380, http://www.biophysics.com

Enzyme Immobilization Materials for Biosensors

Pharmacia Biotech, S-75182, Uppsala, Sweden, Phone: #46 18 16 50 00, Fax: #46 18 16 54 58.

Chapter 9:

MICROFLUIDIC DEVICES

1. INTRODUCTION

A fluid is a material (gas or liquid) that deforms continually under shear stress. This simply means that the material can flow and has no rigid three-dimensional structure. Intermolecular interactions only involve immediately neighboring molecules. Under most circumstances (exceptions include extremes in ambient temperature and pressure), liquids and gases may be treated identically, with the exceptions that gases generally need complete containment and that gases are generally compressible, while liquids are generally incompressible.

Micromachining has numerous applications in fluidics, and its use in this area has become even more important as people strive to create complete fluidic systems in miniaturized formats. Micromachining should certainly allow for the implementation of such systems, since many of the key building blocks (flow channels, flow restrictors, mixers, pumps, valves, sensors, etc.) are either fairly mature or are already under development. A broad variety of materials are available for fabricating the systems or their components, including glasses, plastics/polymers, metals, ceramics, and semiconductors. Thus techniques currently exist for generating a wide variety of structures and functions, and system concepts are rapidly evolving. However, to take full advantage of the power of these technologies, one must deal with significant additional issues, such as packaging, interfaces between components (often made from different materials), and testing. In the early development of electronic systems, similar issues were present and were dealt with by standardized packaging, interconnects like the now-ubiquitous printed circuit board (and the integrated circuit interconnects themselves) and with modern electronic test equipment. Unfortunately, the analogy is not perfect, since electronic components do not generally require consideration of flowing materials, chemical compatibilities, high pressures, packaging with openings for fluid access, etc. It is likely, however, that as high-volume markets are identified, mass production issues will drive the development of solutions to these problems.

Applications of micromachined fluidic systems include chemical analysis; biological and chemical sensing; drug delivery; molecular separation; amplification, seqencing or synthesis of nucleic acids; environmental monitoring; and many others. Potential benefits include reduced size, improved performance, reduced power consumption, disposability, integration of control electronics, and lower cost. Naturally, however, none of these benefits is *guaranteed* simply by using micromachining, and these technologies must be applied with care.

For real-world applications, there are many differences between macroscopic and micromachined fluidic devices, summarized *(in very general terms)* in the table below. The effects of scaling down fluidic functions and systems are quite case-specific. For example, these effects are considered for chromatographic and electrophoretic separations by Manz, et al. (1990a), who noted the dramatic improvements in performance afforded by scaling. Some general conclusions can certainly be drawn that apply to a variety of common applications.

Issue	Macroscopic	Micromachined
Unwanted turbulent flow?	Y	N
Very small dead volume?	varies	Y
Problems purging bubbles?	N	Y
Efficient liquid pumps available?	Y	not yet
Efficient liquid valves available?	Y	not yet
Efficient gas pumps available?	Y	N
Efficient gas valves available?	Y	Y
Simple interconnect scheme?	Y	N
Chemical resistant materials available?	Y	varies
Low power?	N	varies
Sub-cm^2 volume?	N	Y
High surface-area-to-volume ratio?	N	Y
Batch fabricated?	N	Y (not packaging)

Table giving a general comparison of performance of macroscopic versus micromachined fluidic devices based on currently available technology. Clearly, this comparison is generalized and intended as an introduction to the issues discussed below.

As illustrated in the above table, performance considerations important to micromachined fluidic systems generally include minimal dead volume, low leakage or gas permeation, good flow/volume control accuracy, rapid mechanical/diffusion mixing times, appropriate chemically/biologically compatible surfaces, etc. The relative importance of each of these issues is very dependent on the nature of the microfluidic system in question.

While a complete review of all microfluidics research and applications is beyond the scope of this book, representative examples are given below to help the reader survey current research directions in this field. Modeling of fluidic systems is not discussed herein, and the reader is referred to Ulrich, et al. (1995) and Ilzhöfer, et al. (1995) for examples of its use with respect to micromachined fluidic structures. In addition, a general review of microfluidic devices can be found in Shoji and Esashi (1994) and of general fluidics in White (1991).

1.1 BASIC FLUID PROPERTIES AND EQUATIONS

An important fluidic parameter is *density*, ρ, and some examples for common fluids and solids are provided in the table below. If it is relatively constant over wide pressure variations, this shows that a particular fluid is incompressible (most liquids are, and gases are not). Density is defined as,

$$\rho = \frac{m}{V} \quad \text{in kg/m}^3 \text{ or g/cm}^3$$

where m is the mass of fluid, in kg or g, and V is the volume of fluid, in m^3 or cm^3.

Substance	Pressure	Density, kg/m^3
Water	1 atm	998
Water	50 atm	1,000
Seawater	1 atm	1,024
Ice	1 atm	917
Ethanol	1 atm	791
Acetone	1 atm	792
Air	1 atm	1.21
Air	50 atm	60.5
Mercury	1 atm	13,600
Iron	1 atm	7,900
Quartz	1 atm	2,650

Table of example densities of selected substances at specified pressures (Weast (1987) and Halliday, et al. (1993)).

Most gases have a pressure/volume relationship that is estimated (assuming intermolecular interactions are small) by the *ideal gas law*,

$$PV = nRT$$

where,
P = pressure, in Pa = N/m^2
V = volume, in m^3
n = number of moles (mol)
R = gas constant, 8.31451 (N•m)/(mol•K)
T = absolute temperature, in K

It should be noted that this is only useful, in general, for first-order approximations but becomes more valid as gas densities decrease. This relationship illustrates the basis for potential pressure-based actuation schemes for use in micromachined devices: 1) increasing the number of pressure generating molecules, n, (e.g., by electrolysis of water), 2) increasing the pressure by heating, or 3) changing the volume (e.g., with a pump).

Pascal's principle is a statement of the fundamental concept that pressure applied to an enclosed fluid is transmitted to every portion of the fluid (and hence the vessel it is in), and applies whether or not the fluid is incompressible.

Archimedes' principle states that the buoyant force acting on an immersed body is equal in magnitude (but acting in the opposite direction) to the force of gravity on the displaced fluid.

Viscosity, μ, is a measure of how resistant a fluid is to flow (e.g., honey is more viscous than water) and is analogous to friction between solid objects (conversion of mechanical energy into thermal energy). It is important to note that *absolute* (or dynamic) viscosity is referred to here, as opposed to the kinematic viscosity, which is μ divided by the density of the fluid. This property can be very important to micromachined devices, since (as discussed below), it is a key determining factor in flow rates. In general, the viscosity of liquids decreases rapidly with temperature (see below), the viscosity of low-pressure gases increases with temperature, and viscosity always increases with pressure (see White (1991) for a detailed explanation). Viscosity can be found to be given in many differing units in the literature, but the most common is the poise (g/(s•cm)).

Substance	Temperature (°C)	Viscosity (centipoise = 0.01 g/(s•cm)
Water	0	1.787
Water	20	1.002
Water	100	0.2818
Water Vapor	100	0.01255
Blood	37	4.5 to 5.5
Acetone	25	0.316
Ethanol	0	1.773
Ethanol	20	1.200
Isopropanol	15	2.86
Mercury	0	1.685
Mercury	20	1.554
Air	0	0.01708
Air	18	0.01827
Carbon Dioxide	0	0.01390
Carbon Dioxide	20	0.01480
Nitrogen	27.4	0.01781
Xenon	20	0.02260

Table of example viscosities at selected temperatures. From Weast (1988) and Tortora and Anagnostakos (1987).

While approximate equations to derive viscosity from first principles are available (see Cheremisinoff (1981)), in practice, highly accurate tabulated data are used. For a given substance, viscosity varies over a fairly wide range with temperature, as illustrated by the plot of μ versus T for water shown below. For gases, viscosity versus temperature can be estimated using *Sutherland's formula* (Blackburn, et al. (1960)),

$$\mu(T) = \mu_o \left(\frac{T_o - C}{T - C} \right) \left(\frac{T}{T_o} \right)^{\frac{3}{2}}$$

where,

T_o = reference temperature, in K or °C
μ_o = viscosity at reference temperature, in poise
C = gas-specific constant, in K or °C

For liquids, the viscosity versus temperature can be estimated over limited temperature ranges using (Blackburn, et al. (1960)),

$$\mu(T) = \mu_o\, e^{-\lambda(T - T_o)}$$

where λ is a liquid-specific constant, in K^{-1} or $°C^{-1}$.

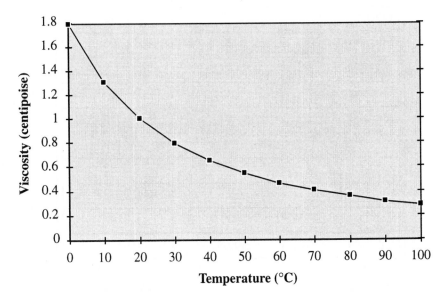

Plot of viscosity versus temperature for water at one atmosphere. Data from Weast (1988).

The *volume flow rate* of a fluid, Q, in a channel is given by,

$$Q = A\bar{v} \quad \text{in m}^3\text{/s}$$

where A is the cross-sectional area of the flow channel, in m^2, and $\bar{v}$ is the average velocity of fluid (over the channel cross section), in m/s.

The *mass flow rate* is simply the product of the volume flow rate and the density of the fluid.

In conventional-scale fluidics, for steady-state flow of an incompressible, frictionless fluid in a channel of varying dimension, the pressure and the velocity of the flow along a streamline (path of a fluid molecule or particle in steady flow) are related by *Bernoulli's equation*,

$$P_1 + \frac{1}{2}\rho v_1^2 + \rho g y_1 = P_2 + \frac{1}{2}\rho v_2^2 + \rho g y_2$$

where,

P_1 and P_2 = static pressures at two locations in the channel, in Pa
v_1 and v_2 = velocities at two locations along a streamline in the channel, in m/s
g = force of gravity ≈ 9.80 m/s^2
y_1 and y_2 = heights above a reference plane, in m

and where irrotational flow can be understood by considering that in such a flow, a test body in the fluid will not rotate about an axis through its center of mass.

For level flow (no effect of gravity due to height differences), as encountered in many fluidic systems, the equation can be rearranged to show the fact that the pressure difference is proportional to the square of the velocity difference,

$$P_2 = P_1 + \frac{1}{2}\rho \left(v_1^2 - v_2^2\right)$$

Essentially, this shows that mechanical energy is conserved and that as the cross-sectional area of the channel is increased, the velocity of the fluid decreases and the pressure increases. These equations assume *inviscid* (or zero viscosity) fluids, but realistic fluids have viscous drag effects and boundary layers resisting flow at the interfaces with flow channels or surfaces. Since microfluidic devices generally have high surface-area-to-volume ratios, these equations will generally not apply well. As a counterexample, an unusual fluid, liquid helium-4 at T < 4.2 K, flows without friction in small channels (this is referred to as *superfluidity*).

For *Newtonian* fluids, a shear stress, τ, is linearly proportional to shear rate, with the proportionality constant being the viscosity, μ,

$$\tau = \mu \frac{dv}{dy}$$

where y is a distance measured in a direction normal to the surface of a flow channel or plate wall. Thus, by definition, the coefficient of viscosity is constant over all shear rates for a Newtonian fluid. For a typical flow in a channel, v(y) is not linear in y, and a typical flow profile is illustrated below. Shear stresses are maximal at the walls and at their minimum in the center of the flow.

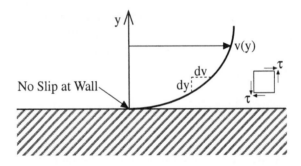

Illustration of a typical Newtonian flow in a channel. The small square at the right shows the directions of the shear forces on a particle of fluid in the flow.

The assumption of Newtonian (or near-Newtonian) behavior allows the use of several helpful approximations to the general equations of fluid flow and considerably simplifies analysis, particularly for long flow channels and nozzles. Most fluids typically encountered in microfluidics applications may be treated as Newtonian for purposes of first approximation. As described in Mazumdar (1992) non-Newtonian fluids include 1) time-independent fluids with yield stresses (minimum forces required to initiate flow) such as ketchup or mayonnaise, 2) time-dependent fluids with shear rates that are both a function of magnitude of deformation and time history (*thixotropic* fluids show a reversible decrease in shear stress with time at constant force, and *rheopectic* fluids show a reversible increase), and 3) *viscoelastic* fluids (e.g., certain polymer gels) permit some of the energy of deformation to be recovered (as opposed to dissipated in viscous fluids) as in elastic solids. Most body fluids, or other fluids containing high molecular weight solutes, are non-Newtonian. As an example, for blood, the apparent viscosity at high shear rates is smaller in tiny tubes than in large ones. This effect is referred to as *shear-thinning*. Transport in microchannels is generally characterized by very high shear rates. Thus, if the fluids exhibit any shear-thinning, non-Newtonian behavior will be accentuated in such channels.

In addition to the assumption of Newtonian behavior, the other typical assumptions when performing first-order analysis of fluidic systems are incompressible and steady (time-invariant) flow. These assumptions may be invalid in some microfluidic devices, and must be made with care (and with experimental support). It should be noted that if irrotationality is assumed, viscosity is neglected, since viscous flow is rotational.

1.2 TYPES OF FLOW

Fluid flow can be categorized in terms of the mode of flow: 1) steady or *laminar* flow is flow in which the velocity of a given point in the fluid does not change with time (there are well-defined, stable stream lines) and 2) *turbulent* flow, in which this is not true. Both laminar and turbulent flow occur in natural systems (e.g., the circulatory system), but turbulent flow is more often encountered in nature.

For any fluid flow it is possible to define the dimensionless *Reynolds number*, which is the ratio of the inertial forces to the viscous forces, and it indicates the relative turbulence (or smoothness) of the flow stream,

$$R_e = \frac{\rho \, v \, L}{\mu}$$

where L is the characteristic length, in m.

For a given geometry and a given fluid, there is a *transitional* (critical) value of the Reynolds number, R_{et}, above which the flow is said to be turbulent and below which the flow is laminar (dominated by viscous losses). For typical *macroscopic* fluidic devices, the transition from laminar to turbulent flow has been empirically determined to occur around $R_{et} \approx 2,300$ for flows in large circular ducts. For non-circular channels, the range may be between 2,000 and 3,000. The classic approximation of $R_{et} \approx 2,300$ is empirically derived for such large duct flows and may apply at microscopic scales. It must also be noted that a local R_e does not necessarily predict the nature of the flow, since upstream conditions may alter the flow conditions where R_e is defined, unless the incoming flow profile is not changing with distance, as discussed below.

In general, flow is effectively laminar in most microstructures, and a survey of microfluidic devices by Gravesen, et al. (1993) indicates that the majority fall into this category. However, considerable experimental work must still be done in the millimeter- to micrometer-scale range to verify this. Laminar flow rules out turbulent mixing (common in macroscopic systems) since it makes diffusion the dominant mixing mechanism. However, over short distances, diffusion can occur relatively quickly, as discussed in Section 3.1 below. Also, because laminar flow tends to reduce dead volumes, and other reasons discussed below, microfluidic devices can provide the designer with useful options compared to macroscopic equivalents.

Despite the above statement, it is often useful to be able to estimate whether or not a particular flow will be turbulent, particularly with channel sizes of larger dimensions (e.g., > 100 μm). As stated above, for a macroscopic circular pipe, with a *fully developed flow*, R_{et}, has been empirically determined to be ≈ 2,300. It should be noted that the term fully developed refers to the concept that the fluid has flowed in a channel beyond the so-called entry length, X_e, beyond which the flow profile no longer changes with distance. Predicting turbulence for microstructures may be complicated by the fact that the flow paths are sometimes shorter than the length required for fully developed laminar or turbulent flow.

For purposes of predicting whether or not a flow is turbulent, one can estimate the Reynolds number for a given channel by replacing the characteristic length in the basic equation with the *hydraulic diameter*, D_h,

$$R_e = \frac{\rho \, v \, D_h}{\mu}$$

where D_h is defined in terms of the cross-sectional area, A, and the wetted perimeter, P, as,

$$D_h \equiv \frac{4A}{P} \quad \text{in m}$$

and where area refers to the cross-sectional area and the hydraulic radius, $R_h = D_h/4$. A table of hydraulic diameters for various cross-sectional shapes is provided below (for cautious use, with the assumption that $2{,}000 < R_{et} < 3{,}000$ for non-circular channels). As mentioned above, it is important to consider that for microfluidic devices, channels may be shorter than X_e, so the assumed R_{et} values may not be valid. Such estimates must be used carefully.

For gas flows, it is possible to have flow rates in microstructures near the velocity of sound with low Reynolds numbers (laminar flow). This is generally not possible in macroscopic structures, where turbulence typically occurs even at subsonic flow rates. In microscopic gas flows, there can be slip at the solid interface (e.g., for extremely small channels, with dimensions approaching the mean free path of gas molecules), allowing for higher flow rates than predicted by theory used with macroscopic flows. Such flow-rate effects in ultra-small channels have been demonstrated by Arkilic, et al. (1994, 1997), and other references to similar phenomena can be found in Pfahler, et al. (1990), Jerman (1991a), and Gravesen, et al. (1993).

For liquid flows, a fundamental assumption in macroscopic structures, that viscosity is independent of channel dimensions, appears to be violated in some microstructures, manifesting as higher than expected flow rates as dimensions are reduced. It is likely that this effect may be related to the polar nature of some of the fluids tested (see Gravesen, et al. (1993)).

Cross Section	Formula	Variables
Circle	$D_h = D$	D = diameter
Annulus	$D_h = (D - d)$	d = inner diameter
Rectangle	$D_h = \dfrac{2ab}{(a+b)}$	a, b = sides
Triangle, equilateral	$D_h = \dfrac{\sqrt{3}}{3}\, a$	a = side
Triangle, general	$D_h = \sqrt{\dfrac{16s(s-a)(s-b)(s-c)}{(a+b+c)}}$ where $s = \dfrac{1}{2}(a+b+c)$	a, b, c = sides

Table of formulas for the hydraulic diameter, D_h. Adapted from Cheremisinoff (1981).

1.3 BUBBLES AND PARTICLES IN MICROSTRUCTURES

Bubbles have increased importance in microstructures versus their macroscopic counterparts. Small volumes of one fluid within another fluid take on spherical shapes ("droplets") due to surface (interfacial) tension. If the fluids have different densities, the droplets will move upward or downward since the buoyant forces acting on them is not equal to the force of gravity. If such droplets contact surfaces where the surface tension is greater than the buoyancy force, the droplets will stay in place.

As described in Fay (1994), the buoyant force, F_B, on a spherical air bubble (or droplet of other fluid type) of radius r (in m), in a liquid of density, ρ, is given by (Archimedes' principle),

$$F_B = \rho g\, \frac{4}{3}\pi\, r^3$$

(When the bubble is comprised of a material with non-negligible density, the value of r used in the above equation should be the difference in density between the bubble and the medium.)

For the example of a bubble, the force acting to hold it in place on a surface is the interfacial tension force, F_I,

$$F_I = \pi d \gamma$$

where γ is the interfacial surface tension, in N/m (7.27×10^{-2} N/m for air/water at 20°C, 1 atm, from a table covering various temperatures and pressures, in White (1991)), and d is the diameter of the contact area of the bubble, in m.

By equating the interfacial tension and gravitational forces, one can see that bubbles with diameters greater than a few millimeters will generally not adhere to surfaces, while smaller bubbles generally will. Unfortunately, smaller bubbles are more the exception than the norm for micromachined fluidic systems.

For bubbles in flow channels, the pressure drop across a liquid/gas interface, ΔP, and the pressure difference, P_d, needed to move them are given by (after Matsumo and Colgate (1990)),

$$\Delta P = \frac{2\gamma \cos\theta}{r} \qquad P_d = \frac{2\gamma_f}{r}$$

where r is the channel radius, in m, and γ_t is a frictional surface tension parameter, in N/m.

If a bubble fully covers the cross-sectional area of a capillary, there will be no net force ΔP, and P_d may be quite reasonable. However, if the bubble ends up at a region with different radii of curvature (e.g., the capillary changes radius), ΔP may be significant and relatively large P_d values may be required to purge the bubble. As discussed below in Section 5.2.3, one can fabricate bubble valves that take advantage of this to controllably form and collapse bubbles and hence control the flow of a fluid. For further information, a useful reference on bubbles, drops, and particles is Clift, et al. (1978) and an interesting theoretical analysis of bubble formation and transport in beer can be found in Shafer and Zare (1991). Another interesting reference is Jacobsen (1975), which describes the use of circulating bubbles (or regions of immiscible liquids) as a digital memory device.

A simple, and often applicable solution to the problem of bubbles in microfluidic systems was discussed by Zengerle, et al. (1995b). By priming a fluidic system with a gas that was highly solubility in water, such as CO_2, it was shown that any bubbles formed during filling with solution were quickly dissolved. Carbon

dioxide has nearly three times the solubility in water than air. Since air typically only contains 350 ppm CO_2, even aerated water will seldomly be saturated with the CO_2. It is worth noting that there are other gases, such as C_2H_2, H_2S, SO_2, and NH_3 (listed in order of increasing solubility) that could be used, although all of them less safely. This simple approach does prove quite effective in many fluidic system settings, and could likely be extended to nonaqueous fluids by the proper choice of gas.

When considering the movement of a particle in a fluid (sphere with radius r in m) in a homogeneous medium of viscosity, μ, the friction coefficient, f, (in kg/s) is given by *Stokes' law* (assuming an infinitely large region of fluid and a small Reynolds number),

$$f = 6\pi r \mu$$

The force on a moving particle is this frictional coefficient times the velocity of the particle and, as explained below, determines its velocity in an electric field if the particle is charged (e.g., in electrophoresis). Particles in microfluidic systems are of considerable importance because they may be of comparable size to flow channels and may become lodged between control surfaces of valves, pumps, etc. At present, careful filtration of fluids is generally required with microfluidic devices to prevent such problems.

1.4 CAPILLARY FORCES

For a small flow channel, or capillary, the surface tension force tending to draw liquid into the channel (assuming a round channel) is (White (1991)),

$$F_I = 2\pi r \gamma \cos(\theta)$$

where θ is the contact angle between liquid and surface (e.g., at the meniscus).

For a vertical capillary, gravitational force on the rising column of liquid of height, h, opposes the capillary force, and is given by,

$$F_g = \rho g \pi r^2 h$$

Equating these two forces gives the maximum rise in the height of a fluid in a capillary against gravity of,

$$h = \frac{2\gamma \cos(\theta)}{\rho g r}$$

Thus the height of the fluid column will greatly increase as the size of the channel is decreased. If the capillary force is not opposed by gravity (e.g., horizontal capillary), very long lengths of channel can be filled with fluid using this force alone. Such surface tension effects are significant in the priming of liquid flow structures (e.g., pumps).

1.5 FLUIDIC RESISTANCE

Analogous to electrical resistance, *fluidic resistance* is defined as the ratio of pressure drop over flow rate,

$$R = \frac{\Delta P}{Q} \quad \text{in } (N\bullet s)/m^5$$

where ΔP is the pressure difference, in N/m^2, and Q is the volume flow rate, in m^3/s.

For a pipe with a circular cross section, assuming laminar flow and a Newtonian fluid (assumptions that may not apply in all situations), flow is given by the *Hagen-Poiseuille equation*,

$$Q = \frac{\pi r^4}{8 \mu L} \Delta P$$

where r is the channel radius, in m, and L is the channel length, in m. From this equation, the resistance is simply,

$$R = \frac{8 \mu L}{\pi r^4}$$

For rectangular channels, one can approximate the fluidic resistance for channels with widths, w, much larger than depths, h, as,

$$R = \frac{12 \mu L}{w h^3}$$

(Note that the roles of w and h are interchangeable and the approximation simply assumes a relatively high aspect ratio.)

For rectangular channels with low aspect ratios, one can use,

$$R = \frac{12\,\mu\,L}{w\,h^3}\left[1 - \frac{h}{w}\left(\frac{192}{\pi^5}\sum_{n=1}^{\infty}\frac{1}{n^5}\tanh\left(\frac{n\,\pi\,w}{h}\right)\right)\right]^{-1}$$

(For more details, see Foster and Parker (1970).)

Another useful geometry in micromachining is the triangular cross section available via (111)-plane-dependent etching of silicon, producing 54.74° angles to a (100) wafer's surface (shown below assuming an anodically bonded glass cover plate, after Lammerink, et al. (1995)).

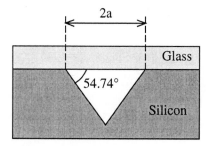

Illustration of a typical micromachined fluidic channel fabricated using anisotropic etching of (100) silicon and, as shown, sealed with an anodically bonded glass cover. After Lammerink, et al. (1995).

As given by Lammerink, et al. (1995), the fluidic resistance for such a channel is,

$$R = \frac{17.4\,L\,\mu}{a^4}$$

where, a is one-half of the width of the channel, in m.

For other non-circular cross sections, generally the case in micromachined fluidic systems, resistance values can be derived from the *Navier-Stokes equation*. For crude approximations, the equation for cylindrical channels can be modified by replacing the radius, r, with the hydraulic radius, R_H, for the cross section in question. However, fully developed flow solutions are known for a wide variety of channel cross sections, as discussed in White (1991).

1.6 FLUIDIC CAPACITANCE

Compliant elements of a fluidic system exhibit the fluidic equivalent of *capacitance* as a pressure-dependent volume change,

$$C = \frac{dV}{dP} \quad \text{in m}^5/\text{N}$$

As described in Zengerle and Richter (1994), the fluidic capacitance for a square membrane (typical in micromachined systems) is,

$$C = \frac{6 \, w^6 \left(1 - v^2\right)}{\pi^4 E \, h^3}$$

where,
w = membrane width, in m
E = Young's modulus of membrane, in N/m^2
h = membrane thickness, in m
v = Poisson's ratio of membrane (dimensionless)

For a circular membrane of radius r (in m), the capacitance is given by (Lammerink, et al. (1995),

$$C = \frac{\pi \, r^6 \left(1 - v^2\right)}{16 \, E \, h^3}$$

If the fluid itself is compressible, it may represent a capacitance that can be defined in terms of the change in the number of molecules, n, in a fixed volume, V, with respect to pressure changes (cautiously assuming an ideal gas),

$$C = \frac{\partial n}{\partial P} = \frac{\partial}{\partial P}\left(\frac{PV}{RT}\right) = \frac{V}{RT}$$

The slight compressibility of most liquids (compressibility of gases, however, is generally very large), which is related to their bulk modulus of elasticity, is important for the propagation of acoustic energy.

1.7 FLUIDIC INDUCTANCE

In a manner analogous to electrical inductance, fluidic systems are capable of storing kinetic energy in fluidic inductance, H (in kg/m^4),

$$\Delta P = H \frac{dQ}{dt}$$

For incompressible and inert fluids in tubes of constant cross section, A, the fluidic inductance is given by,

$$H = \frac{\rho L}{A} \text{ in kg/m}^4$$

For micromachined fluidic systems, the channels are typically rigid and, if liquids are used, they are incompressible, so fluidic resistance and inductance dominate.

For both macro- and microscopic fluidic systems, frequency and time-domain analyses can be carried out in a manner directly analogous to those used in electronics. For example, Lammerink, et al. (1995) implemented a micromachined hydraulic astable multivibrator using direct translation from a two-transistor RC oscillator (pressure-controlled valves replaced the transistors, round silicon membranes replaced the electrical capacitors and narrow flow channels replaced the electrical resistors). It is interesting to note that their multivibrator had an operating frequency of 0.18 Hz and an output pressure swing of 90% of the supply pressure. For distributed fluidic RLC circuits (e.g., macroscopic hydraulic systems), transmission-line analyses can often be extremely useful (see Kirshner and Katz (1966)).

2. FLOW CHANNELS

Flow channels for fluids and gases are fundamental building blocks of microfluidic systems, analogous to wires or thin-film electrical interconnects in conventional integrated circuits. As described in the Micromachining Techniques chapter, it is readily possible to fabricate flow channels in silicon using various etching techniques (anisotropic wet etching, isotropic wet etching, dry etching, laser-driven etching, etc.). Similarly, flow channels can be fabricated in glasses and other materials, but generally, only isotropic etches are readily available (one important exception is the use of dry etching of organic materials such as polyimides).

A wide variety of fluidic channels have been fabricated using micromachining approaches. These channels generally fall into one of three categories based on the methods used for their fabrication: bulk micromachined, surface micromachined (additive), and others (methods such as molding, etch pit replication, etc.). When considering different micromachined channel fabrication technologies, the most important factors are generally:

1) Available channel cross-sectional areas and other geometric constraints (note that most surface micromachined fluidic channels have significant channel height constraints, leading to surface areas that scale directly with the channel cross section).

2) Channel interior surface materials (key issues are typically whether or not all surfaces are of one material type, and if the material(s) is (are) compatible with any biological or chemical fluids they may encounter).

3) Complexity of fabrication (hence yield and cost).

Secondary factors that are typically more application-specific include whether or not the channels are optically accessible (important for applications using colorimetry, fluorimetry, etc., to assess chemical reactions), interior wall roughness, hermeticity, burst pressure, etc.

2.1 BULK MICROMACHINED CHANNELS

Sobek, et al. (1994) fabricated a micromachined flow cell for use in flow cytometry (measurement and/or counting of cells and potentially other biologically derived particles) from fused silica (quartz) wafers. They etched the flow channels in fused silica substrates (Hoya T-4040) with 50°C 7:1 buffered oxide etch (the reported etch rate was 23 μm/h), using 2.5 μm LPCVD polysilicon masks that were later removed with 20 wt% KOH at 60°C. The perpendicular through-holes necessary for the design were fabricated using electrochemical-discharge machining and two fused silica wafers were aligned and bonded at 1,000°C. The design of the flow path (shown in the right-hand view below) allows hydrodynamic focusing of the injected fluid into a very thin stream ($\approx$ 10 μm in diameter). (Further examples of micromachined sheath flow chambers for cytometry can be found in Miyake, et al. (1991), Altendorf, et al. (1997) and Larsen, et al. (1997).)

Several groups have fabricated similar channels in glass (primarily for electrophoresis applications) using fluoride-based aqueous etching, such as Harrison, et al. (1993a), Jacobson, et al. (1994c), Manz, et al. (1994b), and Woolley and Mathies (1994). In addition, photosensitive Foturan® glass has been used to fabricate a variety of fluidic channel structures, as described by Möbius, et al. (1995).

Flow channels for an electrophoresis system using micromachined Pyrex™ plates was demonstrated by Harrison, et al. (1993), who micromachined the channels in Pyrex™ using Au or Cr masks and $HF:HNO_3:H_2O$ (20:14:66) etchant. After stripping the masking layer, a cover glass plate was attached by melting the two glass parts together at 650°C.

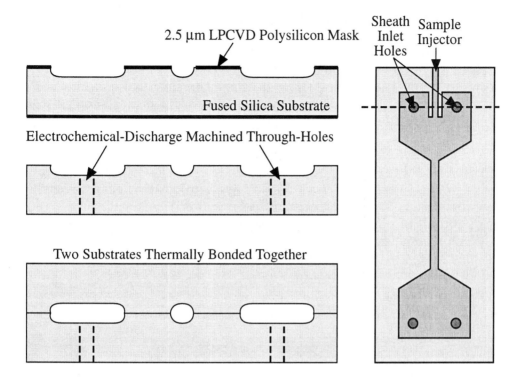

Micromachined flow cytometry cell showing process cross section (left) and top view (right). Sheath fluid is injected into the holes perpendicular to the flow path, and the sample is injected parallel to the flow path. After Sobek, et al. (1994).

Woolley and Mathies (1994) provided a detailed fabrication procedure for such channels etched into soda lime glass. After a careful H_2SO_4/H_2O_2 cleaning and methylation in HMDS, MicroPosit™ S-1400-31 photoresist (Shipley, Inc., Newton, MA) was applied and 150°C hard-baked (60 min) to form the etch mask. The etching was carried out using a 1:1 mixture of two buffered oxide etch (BOE) etchants (NH_4F/HF, mixture of BOE 5:1 and BOE 10:1) for 15 min to achieve 8 μm etch depths (30 min etches resulted in 16 μm depths, but with increased and nonuniform undercutting of the photoresist mask). They formed closed channels by thermally bonding such an etched glass plate to a second, un-etched plate through which 800 μm access holes were drilled using a diamond drill. The thermal channel sealing protocol used was: ramp 5 °C/min to 500°C, hold for 30 min, ramp 5 °C/min to 550°C, hold for 30 min, ramp 5 °C/min to 600°C, hold for 2 h, ramp -5 °C/min to 550°C, hold for 1 h, ramp -5 °C/min to 500°C, hold for 30 min, and cool to room temperature. Unfortunately, it is not clear from their paper what was the significance of the complex ramping scheme.

Bloomstein and Ehrlich (1991, 1992) demonstrated the use of laser-assisted chemical etching to realize complex flow channels in a serially fabricated manner. This approach may be necessary in cases where the desired channel geometries are difficult or impossible to produce using "conventional" micromachining. This approach to channel fabrication takes an amount of time that is proportional to channel volume, but the actual parts fabricated may serve as molds for subsequent parallel replication of the shapes.

Chen and Wise (1994) developed needle-like neural probes, using an EDP-etched dissolved wafer process, that contained integral flow channels within their shafts as well as thin-film electrical recording electrodes for neural signals. The probes were designed to allow the controlled injection of minute quantities (on the order of 100 pl) of physiologically active substances directly into tissue regions from which recordings were being made. Their flow channels were 10 μm wide and 4 mm long, and achieved the desired 100 pl/s flow rates at 11 torr drive pressure and 1.3 mm/s flow velocity.

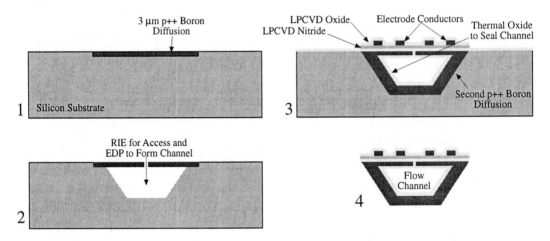

Illustration of flow channel process. After Chen and Wise (1994).

The microchannels in the probes were fabricated (as illustrated above) beginning with a shallow (3 μm) p++ boron diffusion to define the "roofs" of the channels. This was followed by an RIE etch to provide access through the p++ silicon to allow the EDP etchant to form the hollow flow channels and thermal oxidation (with LPCVD oxide) to seal them. The next step was fabrication of the electrode conductors and insulators, with subsequent EDP etching of the wafer to release the probes.

The general approach of sealed flow channels could be used within a silicon substrate without dissolving it away, and the second boron diffusion would then not be necessary. Chen and Wise applied this technique in such a manner to realize an ink-jet printer head (Chen and Wise (1995)). Alternatively, boron diffusions could be entirely avoided by using a deposited dielectric as the channel roof.

Tjerkstra, et al. (1997) demonstrated a variety of fluidic channel fabrication techniques based on combining wet and dry bulk etching with deposition of dielectric films to form the walls of the channels. One approach involved bulk-etching channel shapes into a silicon substrate, coating the channel walls with a conformal film (e.g., 50 nm of LPCVD silicon nitride and top layer of 600 nm TEOS silicon dioxide), anodically bonding the silicon to a glass wafer and etching away the silicon. The channels thus formed had two different materials exposed on their interior surfaces (glass and silicon dioxide) and were relatively thin-walled. This process, combined with a thick polymer coating over the thin-walled channels (for mechanical protection and planarization), was used by Spiering, et al. (1997) to fabricate a prototype capillary electrophoresis system.

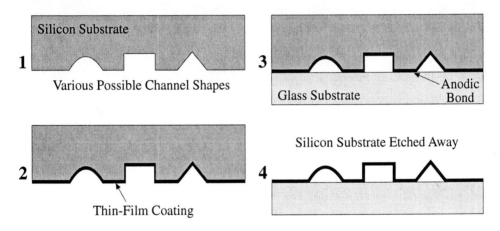

Illustration of a method to fabricate thin-film fluidic channels by first etching the channel shapes into a silicon substrate (a hypothetical trio of channel shapes are shown, representing wet isotropic [left], dry anisotropic [center], and wet anisotropic [right]), coating the channels with a thin-film layer, anodically bonding using an intermediate silicon dioxide layer (not shown), and finally etching away the silicon substrate. Adapted from Tjerkstra, et al. (1997).

Another approach presented by this group (illustrated below) was the formation of channels that were "buried" in a silicon substrate and formed through a narrow slit, which can later be sealed through the deposition of a suitable thin-film material. As the sealing material was deposited, the interior surfaces of the channels became uniformly coated, and eventually, the slit used to form the channel was coated and sealed. A key advantage of this approach is that all interior surfaces of the resulting fluidic channels were thus coated with the same material. The process used by

Tjerkstra, et al. (1997) began with a deep RIE etch of narrow slits (100 μm deep and 4 μm wide) into a silicon substrate, followed by the deposition of LPCVD silicon nitride. The silicon nitride coated all surfaces of the slits, but was selectively removed from their bottoms using RIE, leaving the slit sidewalls coated. The buried channels were then formed using wet anisotropic (KOH) or isotropic (HNA) etching, which proceeded only at the bottoms of the slits (the remaining surfaces were protected by silicon nitride) and formed the channels. After removal of the silicon nitride using a wet HF etch, the channels were sealed by depositing LPCVD silicon nitride on all surfaces.

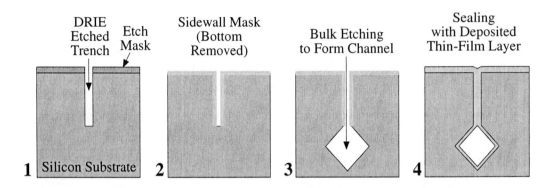

Illustration of a process for forming buried, bulk-etched fluidic channels in a silicon substrate, and subsequently sealing them with a deposited thin-film layer. Adapted from Tjerkstra, et al. (1997).

Tjerkstra, et al. (1997) also experimented with coating the insides of channels with various LPCVD films. They found that for LPCVD silicon nitride, the film thickness decreased exponentially from an opening to deeper regions of a channel. For example, an LPCVD film that was 2.0 μm thick at the opening of a 96 μm semicircular channel decreased to ≈ 60 nm at a distance of 1.5 cm down the channel. The thickness of TEOS silicon dioxide decreased even faster. However, other film deposition methods, such as vapor phase Parylene™ coating (discussed below) may offer better performance in such applications.

Flannery, et al. (1997) demonstrated a simple fabrication process for bulk micromachined fluidic channels that were also sealed using a deposited thin-film layer. The approach taken was to use a thermal SiO_2 mask and an isotropic SF_6 plasma etch to define the channel cross section through a "slit" mask (generally less than ≈ 4 μm in width) in the silicon dioxide. The channels were then sealed using a PECVD amorphous, hydrogenated silicon carbide film (a-SiC:H). It should be noted that any PECVD deposition process that provides sufficiently good conformal coverage (particularly beneath the overhanging SiO_2) can be used to fabricate such channels.

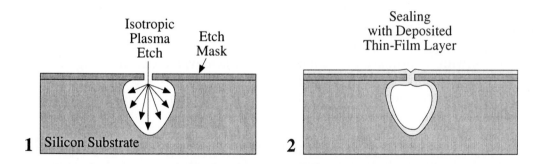

Illustration of bulk micromachined fluidic channels fabricated by etching isotropically through a "slit" mask and then sealing the channel with a deposited thin-film layer. After Flannery, et al. (1997).

2.2 SURFACE MICROMACHINED CHANNELS

Lin, et al. (1993) presented an approach to surface micromachined flow channel fabrication, also on needle-like probes similar to those of Chen and Wise (1994) (and also defined using p++ boron doping and an EDP release etch). Their technique made use of a sacrificial channel definition layer of 5 μm phosphorus-doped glass (PSG) and 3 μm LPCVD SiO_2 (LTO) that was encapsulated with a 1 μm LPCVD silicon nitride layer through which etch access holes were defined using a plasma etch. At this point, the probes were etched in HF to remove the sacrificial PSG and LTO, leaving empty flow channels that were sealed using a 1.5 μm LPCVD silicon nitride layer.

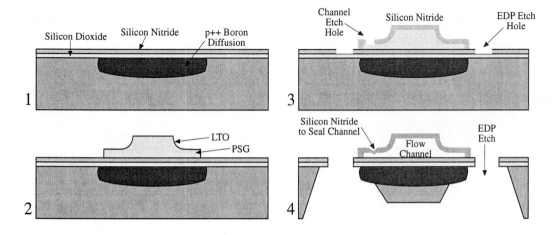

Surface micromachined flow channel process. After Lin, et al. (1993).

Since an underlying LPCVD layer had already been deposited, the flow channels were completely surrounded by nitride. This approach used only surface micromachining for channel formation, and can be generalized to other channel materials, with its primary limitation being the achievable channel height. In an interesting application of this technology, a similar process was used by Desai, et al. (1997) to fabricate microfluidic nozzles for injection of fluid droplets into a mass spectroscopy system.

Man, et al. (1997) presented a technique for forming surface micromachined organic fluidic channels at low temperatures (compatible with on-chip electronics if used on silicon substrates, for example). Their approach was to vapor-deposit a 2 μm thick poly-p-xylylene (Parylene™ C) layer on the substrate, followed by a 20 μm layer of AZ4620 photoresist (note that the thickness of this sacrificial layer determined the final inner height of the channels), which was hard-baked at 120°C for 3 min. A second 2 μm layer of poly-p-xylylene was then deposited and roughened with a short O_2 plasma treatment. A 40 μm layer of photosensitive polyimide (OCG Probimide™ 7020, Olin Microelectronic Materials, Tempe, AZ) was then spun on and baked at 90°C for 2 min and 110°C for 7 min. The polyimide was then exposed and developed, followed by the use of a patterned aluminum mask to etch the exposed top poly-p-xylylene layer using an O_2 plasma. The sacrificial photoresist thus exposed was removed using an acetone bath at 40 to 50°C over 20 to 30 min (dependent on the channel cross section).

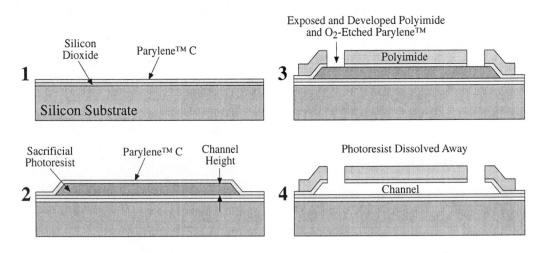

Illustration of surface micromachined organic fluidic channels fabricated using a sacrificial photoresist layer, Parylene™ C channel lining, and a polyimide structural layer. Adapted from Man, et al. (1997).

Channels with interior heights of 25 μm were demonstrated, and, subject to the constraints of obtainable photoresist thickness and polyimide conformality during the spin-on step, a wide range of heights could theoretically be obtained (e.g., 0.1 to 100 μm). A key feature of this process is that the channels are uniformly coated with Parylene™ C, which is known to be very compatible with biological compounds (and is compatible with the polymerase chain reaction for DNA amplification). However, as for all surface micromachined channel processes, limitations on channel height force the use of relatively large amounts of substrate surface area if large channel cross-sectional areas are required. Webster and Mastrangelo (1997) combined this type of channel with prefabricated photodiodes on the silicon substrate to provide on-chip fluorescence detection during gel-based DNA electrophoresis.

An alternative surface micromachined channel process using organic materials was presented by Guérin, et al. (1997). They used multi-layers of a specialized photoresist (the resist is referred to as SU-8, consisting of a mixture of resin, solvent, and a photoinitiator, as discussed by Despont, et al. (1997)) and an epoxy sacrificial layer to define the bores of the channels. They fabricated channels with cross sections as small as 25×50 μm, with lengths on the order of 1 cm. Unfortunately, the authors did not address the issue of chemical compatibility of the channels, which is of prime interest in many biological and chemical applications of such fluidic channels.

Papautsky, et al. (1997) presented a technique for fabricating electroplated metal fluidic channels above a substrate. Using photoresist template and sacrificial (to define the inner channel diameter) layers and electroplated gold, channels with heights (inside vertical dimensions) of 13 to 50 μm were fabricated with widths from 30 μm to 1.5 mm. A limitation of this approach is that the acetone used to dissolve the inner photoresist sacrificial layer takes very long times to clear lengthier channels (e.g., 7 hours to clear a 30 μm high, 1 mm wide, and 7 mm long channel, which is considerably longer than the times reported by Man, et al. (1997) for similar channels). However, such channels are fabricated at low temperatures and could theoretically be made on a wide variety of substrates. In addition, if the inside surfaces could be coated with non-metallic films, their compatibility with biological fluids could probably be increased. An alternative method for electroplating fluidic channels was reported by Joo, et al. (1995) in which the deposited metal was deliberately plated beyond photoresist template structures, forming sealed channels around them. The photoresist was then dissolved out with acetone (this can be a troublesome release mechanism, since acetone significantly swells most photoresists, potentially subjecting microstructures to large forces).

Kämper, et al. (1997) presented a variety of fluidic channel structures that were fabricated using LIGA (additive) or electric discharge machining (EDM, subtractive). While the basic fluidic structures are thereby limited to metal structural materials, very high aspect ratios are possible (> 100 for LIGA and > 15 for EDM).

2.3 OTHER CHANNEL TYPES AND SYSTEM APPROACHES

VerLee, et al. (1996) presented mechanically fabricated fluidic channels in acrylic (and other organic) substrates. This approach, while not "micromachined," is important in that it can serve as an interface between micromachined fluidic devices and macroscopic fluidic connections. After thermal "preshrinking" of the plastic blocks, they were milled on a numerically controlled machine to form the channels, which were diamond fly cut to obtain very smooth surfaces. Multiple such blocks, with or without intervening flexible membranes (to form valves, etc.), were thermally bonded together under pressure. Typical values for acrylic were 45 psi pressure and a thermal cycle of a 2 h ramp to 126°C, a 4 h hold, and a 2 h ramp to room temperature. Materials used included PolyCast™ Acrylic (Polycast Technology, Corp., Stamford, CT), polycarbonate, α-butylstyrene, etc.). Mechanical drift during bonding was stated to be on the order of ± 25 µm, which would allow quite precise alignment to micromachined devices (although these authors did not attempt that). Complex fluidic functions were demonstrated, equivalent to a bench-top blood chemistry analyzer in a 58 × 18 × 5 cm block, with 254 valves and 24 liquid/air sensors (optical).

Poplawski, et al. (1994), presented a simple approach for integrating flow channels with silicon chips containing sensors. Their approach was to etch flow channels in a glass substrate, masked using a sputter-deposited silver layer, and then to use a polymer "gasket" to seal the channels to a silicon "flip-chip." Conductive epoxy bumps were used to connect between metallization on the glass (the same silver layer used as an etch mask) and on the silicon.

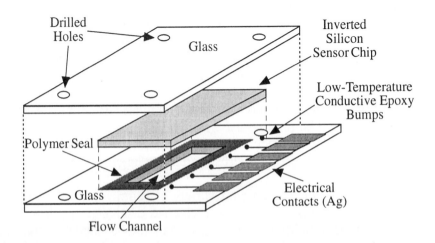

Illustration of a technique for integrating micromachined structures into assembled fluidic systems. After Poplawski, et al. (1994).

Lammerink, et al. (1996) presented the concept of combining discrete silicon-based fluidic devices on a silicon/glass (anodically bonded) or molded plastic "mixed circuit board" (MCB). The MCB would carry both fluidic and electrical connections between components. They demonstrated a complete absorption-based flow injection analyzer (mixing two chemical streams together in a controlled fashion and determining the extent of a chemical reaction via optical absorption measurements). The system consisted of a stack of a MCB with two micromachined pumps, two flow sensors, and an optical absorption cell, atop two conventional printed circuit boards with control circuitry (unfortunately, the authors did not describe the assembly methods used to couple the fluidic chips to the MCB). In many applications of micromachined fluidic systems, it is likely that such "hybrid" assembly approaches will be far more practical than attempting to build truly monolithic fluidic "system-on-a-chip" devices.

3. FLUIDIC CHANNEL APPLICATIONS

A logical extension of simple flow channels is to design channels with more complex shapes and multiple flow paths, providing the capabilities of mixing, amplification, and logic functions. According to some authors, the definition of a "pure" fluidic device is providing a desired function (e.g., amplification, switching, etc.) without moving parts; but in practice, fluidic systems typically contain moving and fixed components.

Work on macroscopic fluidic devices is well documented back to the early 1700's and, on the scale of viaducts and other large-scale structures, has been going on for millennia. In the late 1950's through the 1960's, fluidic logic was heavily researched as a means of realizing radiation hardened backup or primary control systems for aircraft and other vehicles. Today, the automatic transmissions of automobiles use "hydraulic logic," which employs most of the same concepts. They are robust and manufactured in large numbers.

3.1 MIXERS

The controlled mixing of two or more fluids is an important function in chemical analysis/synthesis, drug dosing, compound labeling, cell lysis, etc. In general, the mixing of two fluids requires that they be interspersed so that their area of contact is maximized, allowing diffusion to complete the mixing process. In macroscopic and some mesoscopic fluidic devices, turbulent flow can be used to accomplish this. However, as discussed by Evans, et al. (1997), this approach is not feasible for micromachined fluidic devices (with typical sizes no larger than 10

× 10 × 0.5 mm and volumes to be mixed in the picoliter to microliter range), since turbulent flows cannot realistically be achieved. Thus, for micromachined devices, increasing the contact area of fluids to be mixed must be accomplished via other means (generally geometric or through the use of mechanical agitation).

For larger-scale fluidic devices, turbulent flow (i.e., flow described by $R_e >$ R_{et}) improves mixing but can be susceptible to dead volume problems. On a macroscopic scale, certain junction shapes such as sharp corners encourage turbulence, improving mixing. These approaches may not work well at the scales of microfluidic devices, however.

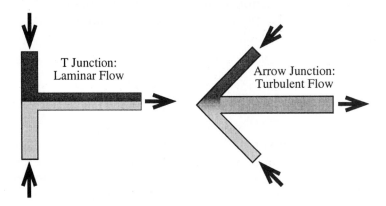

Illustration of shapes that can be used to enhance mixing at macroscopic scales. After Krulevitch (1995).

Another approach is to force flow around a turn with a small radius of curvature, which induces a secondary, rotating flow due to centripetal acceleration (although laminar flow can redevelop after the turn for small values of R).

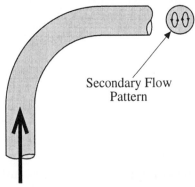

Illustration of centripetal mixing during fluid transit around a corner. After Krulevitch (1995).

In microfluidic devices, diffusion is the key to mixing, and is well modeled by *Fick's law* (for non-extreme temperatures), which gives the flux of diffusing species as a function of the change in concentration with distance, x,

$$J = -D \frac{\partial C}{\partial x}$$

where,

J = particle/molecule/ion flux, in particles/(m²•s)

D = diffusion coefficient, in m²/s (strongly temperature-dependent), examples of which are shown in the table below

C = concentration, in particles/m³

The average time for a particle to diffuse a given distance depends on the square of that distance. As discussed in Evans, et al. (1997), a diffusion mixing time scale, T_D, can be written,

$$T_D = \frac{L^2}{D}$$

where L is the relevant mixing length, in m (generally an in-plane distance for microfluidics).

Molecule	Molecular Weight, AMU	Diffusion Coefficient in Water, μm²/s
H⁺	1	9,000
Na⁺	23	2,000
O_2	32	1,000
Glycine	75	1,000
Hemoglobin	6×10^4	70
Myosin	4×10^5	10
Tobacco Mosaic Virus	4×10^7	5

Table of molecular weights and appproximate diffusion coefficients (room temperature) for several molecules relevant to biological applications of microfluidics. Note that the diffusion coefficients are given in μm²/s. Data from Brody and Yager (1997).

This equation predicts that using diffusion alone for mixing can sometimes result in very long mixing times (e.g., ten seconds for L = 100 μm, and over four minutes for L = 500 μm, for the example of NaCl in water). This fact can have considerable impact on flow-based systems, where the dwell-time of mixing fluids must then be increased, at the expense of flow rate. This also indicates that fluid-filled chambers between which no mixing is desired can be "isolated" for certain time periods simply by the fact that diffusion cannot occur fast enough to allow mixing (this was demonstrated by Albin, et al. (1996) in a multi-well chemical reactor design for DNA amplification wherein a single filling channel connected several reaction chambers together, yet mixing was prevented over several minutes simply by diffusion time constants).

Branebjerg, et al. (1994) investigated the mixing efficiencies of various channel shapes for "mini" channels (300 μm width × 600 μm depth) and "micro" channels (180 μm width × 25 μm depth). The mini channels were fabricated using conventional machining in acrylic and the micro channels were etched into a glass wafer. The mini channels were 100 mm long, while the micro channels were 5 mm long.

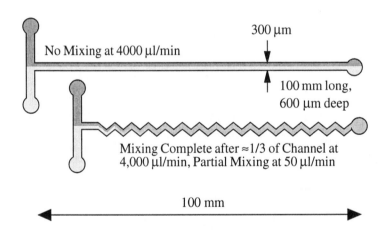

Illustration of miniature fluidic channels used to compare mixing in macroscopic and microscale fluidics. After Branebjerg, et al. (1994).

They injected the pH indicator bromothymol blue (yellowish) into one of the input ports, and NaOH into the other — mixing was indicated by the formation of a dark blue product. Their results indicated that in the mini case, turbulent flow was caused by the sharp corners, resulting in full mixing by the time the fluid had traversed one-third of the channel. However, in the micro case, turbulence did not occur, and *mixing was by diffusion only*. This confirmed the expectation (discussed above) that flows are almost entirely laminar for the smaller-scale structures.

3.1.1 LAMINATING MIXERS

Given the laminar nature of most flows at the microscopic scale, a logical strategy is to try to "fold" or "laminate" two or more fluids together to increase the contact area and enhance diffusion. Several micromachined devices have been demonstrated with this in mind.

Branebjerg, et al. (1996) and Larsen, et al. (1996) demonstrated a rapid fluidic mixer based on repeated lamination of two fluid flows. The fabricated various mixers by etching channels in silicon and in glass, anodically bonded the two substrates together. In between channels in the two materials were thin, chevron-shaped slitted plates fabricated on the silicon substrates. The lamination effect was accomplished through the action of the separation plates on the incident fluid flows. At flow rates of 0.5 to 12 µl/min (equal for the two fluids), mixing times on the order of 100 to 300 ms were demonstrated for aqueous solutions.

3.1.2 PLUME MIXERS

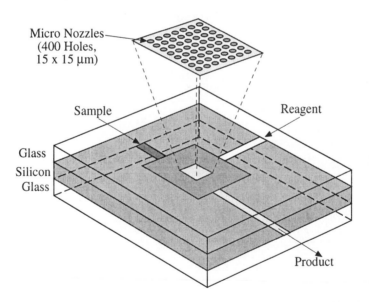

Illustration of a micromachined mixer based on the formation of an array of microscopic plumes that increase the contact area between the two liquids to be mixed. Adapted from Miyake, et al. (1993).

Miyake, et al. (1993) designed a mixer that takes advantage of the behavior of a fluid leaving a narrow nozzle. A micromachined array of 400 nozzles, each 15 µm on a side, generates an array of small plumes, which increase the contact area between the two liquids. This speeds up diffusion, and for the device described, resulted in homogeneous mixing in 1.2 seconds (in a 0.5 μl volume at a 45 μl/min flow rate). This type of mixer was later used in a absorptiometric micro flow cell for chemical analysis by Miyake, et al. (1997).

3.1.3 ACTIVE MIXERS

Other mixing approaches rely on the addition of external energy (other than hydraulic) to further agitate and intermix the fluids. One example of such an active mixer (discussed in Section 6.7 below) is the use of ultrasonic traveling wave pumps to drive mixing by moving fluid in a circulating path, as described by Moroney and White (1991).

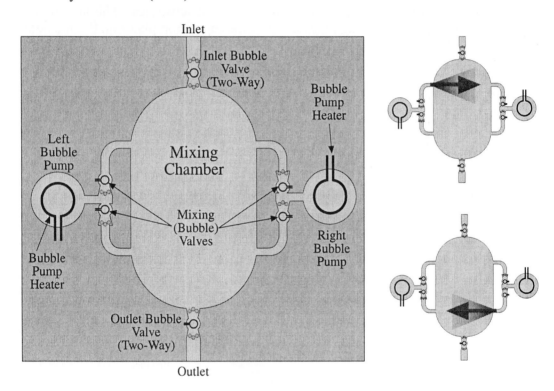

Illustration of a micromachined fluidic mixer based on chaotic advection. Mixing is accomplished through the use of vapor bubble valves and pumps in an otherwise passive structure. The layout of the design is shown at the left, and one cycle of the mixer is shown at the right (with each subsequent cycle in the opposite direction to the previous one). Adapted from Evans, et al. (1997).

Evans, et al. (1997) demonstrated the use of thermally generated vapor bubbles to pump fluid back and forth through a multi-ported mixing chamber ($\approx 1.6 \times 0.6$ mm in area, by 100 μm deep). Such bubble pumps are discussed in Section 6.1

below. In this case, the mixing principle used was *chaotic advection*, in which initially neighboring particles become widely separated in a chaotic flow field (one in which initial conditions profoundly influence the path and final position of individual particles).

As illustrated above, their design incorporated two large bubble pumps that could generate "push" or "pull" forces depending on whether a vapor bubble was expanding or collapsing, respectively. The bubble pumps were connected to the mixing chamber via four ports, each with its own bubble valve. The bubble valves, making use of the fact that a vapor bubble in a channel can support a pressure differential across it if the surface areas of its two faces in the channel differ (such valves are discussed in more detail below). By controlling the state of the bubble valves synchronously with the pumps, fluidic "dipoles" (push and pull on opposite sides of the mixing chamber could be created to move fluid alternately from left-to-right across the top of the chamber and then right-to-left across the bottom, with each plug of fluid being inverted ("first-in-last-out") in the process.

The specific devices demonstrated by Evans, et al. (1997) were fabricated by etching the fluidic channels 100 μm deep into a silicon wafer using deep reactive ion etching, creating access ports through the silicon using KOH anisotropic etching with a patterned silicon nitride mask, and bonding the silicon substrate to a quartz wafer on which aluminum interconnects and polysilicon heaters (with a silicon dioxide passivation layer) were prefabricated. It should be noted, however, that the chaotic advection approach can be implemented using a variety of other means.

3.2 DIFFUSION-BASED EXTRACTORS

As described above, for the vast majority of microfluidic devices and typical flow rates, flows will be laminar. Thus, mixing, if desired, must be accomplished using diffusion or via the application of an external energy source (e.g., acoustic energy or hydraulic pressure). The fact that on microfluidic scales, two fluid streams can be flowing next to each other with only diffusional mixing can be harnessed as a means of extracting desired or unwanted molecules (or suspended particles) from one of the streams.

If the two fluid streams of interest are forced to flow in parallel in a single channel, and the relative widths of each stream can be adjusted (hence the relative volumes into which molecules from each can diffuse), extraction of molecules can be accomplished on the basis of diffusion coefficient differences. For example, if high diffusion coefficient molecules, such as Na^+ must be removed from a fluid such as blood plasma, the plasma can be flowed next to a diluent stream (e.g., water). Over the distance that the two fluid streams flow in contact, only molecules that can diffuse fast enough to cross into the diluent will be extracted. In the example, Na⁺, with its high diffusion coefficient will move much more rapidly than larger molecules in the plasma, permitting the concentration of the former to be greatly reduced in the plasma. It should be noted that such extraction cannot necessarily be specific to a particular diffusion coefficient, but rather, selects molecules into broad ranges based on being above or below a certain diffusion coefficient window. This effect is controllable to a large extent by channel dimensions and flow rate, setting the dwell time wherein the streams are in contact.

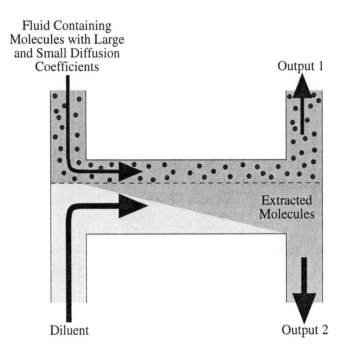

Illustration of a diffusion-based extractor. At the left, two fluid streams enter the central chanel, where they flow in parallel. The upper flow in this example contains molecules with large and small diffusion coefficients. While the fluids flow in contact with each other, high diffusion coefficient molecules diffuse into the diluent stream, while low diffusion coefficient molecules do not. Adapted from Brody and Yager (1997).

Such a device was demonstrated by Levin and Giddings (1991) and Giddings (1993), and implemented in a micromachined form by Brody and Yager (1997). The micromachined version was implemented using bulk anisotropic etching of (100) silicon with EDP. Fluidic input/output ports were formed by etching all the way through the substrates (≈ 400 μm), with 10 μm deep fluidic channels formed on one side, followed by the anodic bonding of a glass cover over the channels. They used test solutions of 0.5 μm diameter fluorescent polystyrene spheres (Duke Scientific, Palo Alto, CA) and the fluorescent dye carboxyfluorescein (a relatively small molecule) in one fluid stream and pure water in the other. It was demonstrated that a considerble amount of the fluorescent dye would diffuse across to the water, but the spheres would not. This approach is relatively simple to implement and, through appropriate design of channel dimensions and flow rates, could be used for a variety of extraction tasks. In addition, if it was desired to remove small molecules from particulate-laden flows, it could effectively serve as a particulate filtering mechanism (as opposed to conventional filters, which rely on passing fluids through porous materials that can eventually occlude). A "T"-shaped structure has also been developed that takes advantage of this extraction principle to monitor quickly diffusing molecules in potentially particulate-filled sample streams by allowing them to diffuse into an adjacent stream containing a fluorescent indicator (Weigl, et al. (1996)). The underlying principles of the "T"-shaped structure used as a sensor are discussed in Galambos, et al. (1997).

3.3 FLUIDIC AMPLIFIERS AND LOGIC

Fluidic amplifiers, logic, multivibrators, etc., have all been implemented in micromachined devices, and have a long history (see, for example, Kirshner (1966)). In 1919, famed inventor Nicola Tesla filed a patent for a "valvular conduit," which was a fluidic rectifier (Tesla (1920)). This design provides low resistance to fluid flow in the forward direction and high resistance to flow in the reverse direction. It is important to note that this differential fluidic resistance requires a relatively high fluid velocity. As the flow rate goes to zero, the differential resistance also disappears.

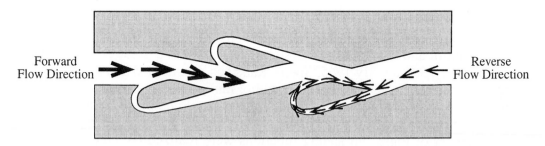

Forward Flow Direction

Reverse Flow Direction

Illustration of Tesla's fluidic rectifier patent. After Humphrey and Tarumoto (1965).

An important concept for the implementation of fluidic devices is the *Coanda effect* (Coanda (1934)), which states that flow emerging from a free jet opening will tend to follow nearby angled or curved surfaces and will attach itself to those surfaces if the angles are not too steep.

The underlying principle is that the jet stream tends to capture or "entrain" adjacent fluid molecules, and when it is near to a surface, the supply of these molecules is limited and a lower-pressure region forms, drawing the jet toward the surface. This principle was eventually applied to fabricating "multi-state" jets, which could be set to one of three or more stable states, as illustrated below. Many past and present fluidic devices make use of the Coanda effect.

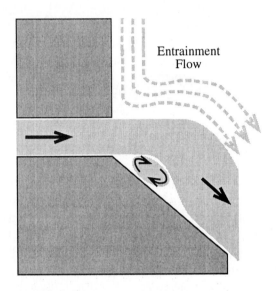

Entrainment
Flow

Illustration of the Coanda effect. After Humphrey and Tarumoto (1965).

In the late 1950's, there was great interest in temperature, shock, and radiation resistant control and computational systems, which led to a great deal of research in the following decade. By 1965, Sperry Rand had constructed a prototype fluidic computer with ≈ 250 NOR gates, each with a fan-in and fan-out of four (Jacoby (1965)). Eventually, however, electronic systems could be made suitably rugged, and the predicted large growth of fluidic circuits did not materialize. Nonetheless, a large number of logic and analog functions were realized (some examples are shown below).

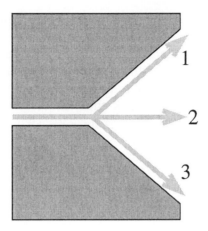

Example of the three stable states of a wide angle output ("diffuser") for a jet.

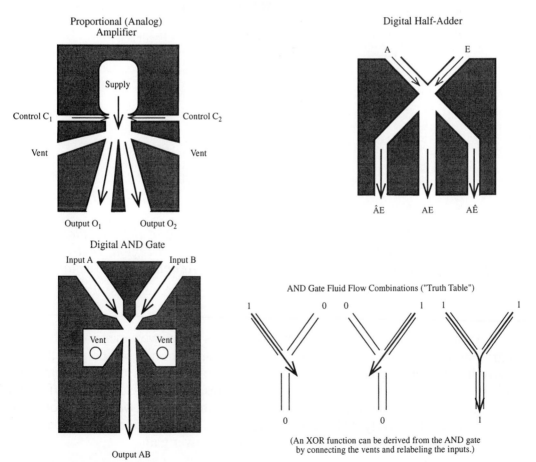

Illustration of examples of fluidic analog and digital functions. After Foster and Parker (1970).

Fluidic oscillators can be realized without any compliant regions (i.e., no fluidic "capacitances," as had been used by Lammerink, et al. (1995)) by employing fluidic feedback instead. Gebhard, et al. (1997) demonstrated such an oscillator using a V-shaped fluidic circuit with feedback channels. Fluid flowing into the structure from the inlet would preferrentially follow one of the two output channels due to the Coanda effect. The feedback channel for that output path would route part of the output flow back to the neck of the "V," switching the flow to the other output channel. The other output channel and its symmetrical feedback channel would repeat the process in the opposite direction, for oscillations at a flow-rate-dependent frequency. For their LIGA-fabricated prototype, with a height of 500 μm and supply nozzle width of 100 μm, they reported oscillation frequencies in the range of 250 to 390 Hz for deionized water pressures of ≈ 0.6 to 2.5 atm.

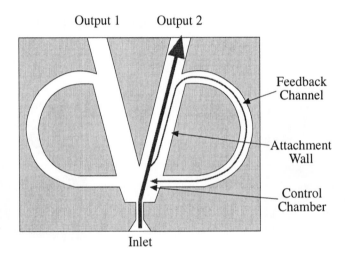

Illustration of a fluidic oscillator structure making use of the Coanda effect to cause part of the output flow to be fed back into the inlet region, redirecting the main flow to the other of the two output channels. Adapted from Gebhard, et al. (1997).

Another important class of fluidic structure is the *vortex device*, which uses a control stream to deflect a normally straight stream into a spiral course, thereby greatly increasing its path length and pressure drop (when in the spiral path, acceleration of the main fluid stream also contributes to the pressure drop). Typical flow and power gains reported for macroscopic vortex devices were on the order of 200 and 300, respectively. These relatively simple amplification devices could be applied in a variety of situations to provide fluidic power gain.

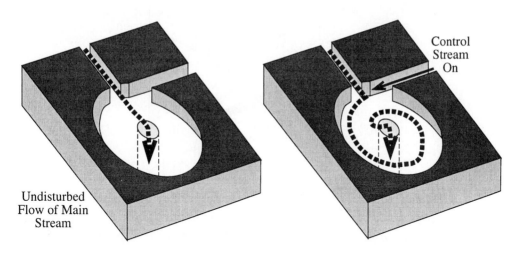

Illustration of a vortex device without (left) and with (right) control stream acting on the main flow stream. After Humphrey and Tarumoto (1965).

It is interesting to note that external forces (e.g., inertial) can distort the fluid flow, thus allowing the vortex device to be used, for example, as an acceleration sensor.

In the 1960's, injection molded and photosensitive plastics; machined, compression molded, electroplated or photolithographically etched metals; and photosensitive ceramics were used to fabricate complex fluidic devices, often stacked to form "integrated circuits."

Zdeblick and Angell (1988) demonstrated the first micromachined silicon fluidic amplifier. They used fluorine-based plasma etching and a photoresist mask to fabricate silicon channels many tens of microns deep. This work demonstrated that micromachining techniques could be used to revive and greatly scale down fluidic amplifiers and logic. Vollmer, et al. (1993) demonstrated the use of LIGA techniques to fabricate bistable (wall attachment via Coanda effect) fluidic elements with nozzle widths of 30 µm and wall heights of 500 µm. Their devices operated with gases such as nitrogen, argon, and carbon dioxide at pressures on the order of 30 kPa (≈ 0.3 atm). Blankenstein, et al. (1996) demonstrated multi-ported fluidic switches, etched into silicon and capped with glass, that used differential pressures between ports to direct a central fluidic stream to one of the outputs.

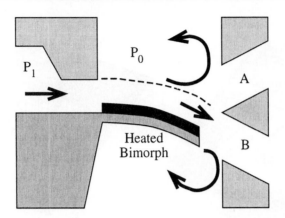

Illustration (side view) of the basic principle of a thermal bimorph Coanda effect cantilever directing fluid flow into one of two outlets. After Döring, et al. (1992).

Döring, et al. (1992) demonstrated a thermal bimorph cantilever that combined the use of the Coanda effect with forced convection, to allow a laminar flow to be steered into one of two outlet ports under electrical control. They used an electro-chemical etch-stop and plasma etch edge release to form 11 μm thick n-epitaxial cantilevers that included a p++ boron diffusion for heating resistors and an 11 μm aluminum layer to form the bimorph with silicon (the cross section of their devices is illustrated below). They measured the temperature coefficient of the thermal resistor (p++ silicon) to be 1.25×10^{-3} °C^{-1} and used real-time resistance measurements to determine the average temperature of the cantilevers (this approach can be used for real-time deflection feedback). At power levels of $\approx$ 1 W, deflections on the order of 15° were obtained from the initially upward deflected state of the cantilevers (due to internal stresses). At pressures of 2 to 7 bar and flow rates of 100 to 150 ml/min, free water jets could be deflected with time constants on the order of 1 ms. This type of actuator was later used to control a micromachined valve by using the directed flow to lift (or press down against the valve seat) a valve membrane, thus opening it (Trah, et al. (1993)).

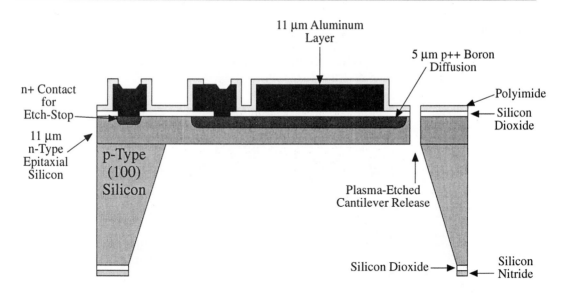

Cross-sectional illustration of the structure of a micromachined thermal bimorph fluidic device. After Döring, et al. (1992).

4. FLUIDIC SENSORS

There is a large variety of micromachined sensors that can be used in fluidic applications and considerable overlap with those presented in other chapters. In the discussion below, the reader is referred to the relevant chapters.

4.1 FLOW SENSORS

Fluid-dependent flow sensors measure thermal dilution or tracer (such as optical dyes or ionic conductors) transit time, injecting pulses (of heat and charge, respectively) and detecting them downstream (thermal flow sensors are discussed in the Thermal Transducers chapter).

Fluid-independent flow sensors measure pressure across a flow restriction or drag force exerted on an object in the flow stream. A potential advantage of this approach to flow sensing is that the fluid is not heated, a potentially important issue if flows containing biological molecules are to be measured.

An example is the bulk micromachined, cantilever drag-force flow sensor fabricated by Gass, et al. (1993a, 1993b) wherein the deflection of a cantilever in the flow stream is measured piezoelectrically and interpreted as a flow rate. This approach can provide relatively high linearity and can sense the direction as well as magnitude of flow. They demonstrated linearity of such a sensor (on the order of a few percent) over flow rates of 0.1 to 200 µl/min with a sensitivity of 0.5 mV/µl/min (directly from the piezoresistive bridge).

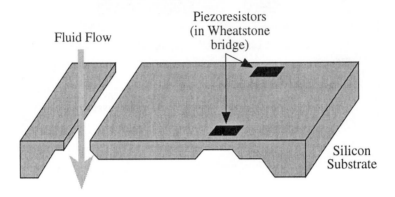

Illustration of a micromachined mechanical (drag-force) flow sensor. After Gass, et al. (1993a, 1993b).

Lift force can also be used to measure flow, as demonstrated by Svedin, et al. (1997). Their device (illustrated below) consisted of a pair of bulk micromachined torsional "airfoils," mounted at an angle in a flow channel. In operation, the airfoil is torqued on one plate (upstream) more than the other (downstream) plate, and the signals from strain gauges on both airfoils are combined to null common-mode offsets. As for drag-force flow sensors, the required structures can readily be fabricated using micromachining approaches.

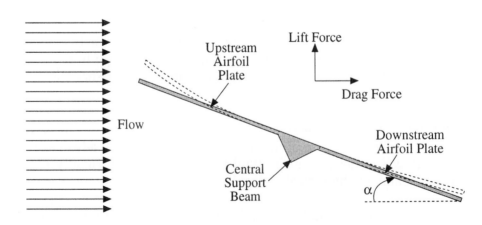

Illustration of a dual-airfoil lift-force flow sensor, showing unequal deflection of the upstream and downstream airfoils. Adapted from Svedin, et al. (1997).

As mentioned above, pressure sensors can be used to measure flow by essentially sampling the pressure drop along a flow channel with known fluidic resistance, R, and computing flow from the fluidic equivalent of Ohm's law,

$$Q = \frac{\Delta P}{R}$$

In practice, however, capacitances of the pressure sensors (typically due to membrane compliance), as well as the fluidic inductance (momentum) of the fluid in the flow channel must be taken into account if dynamic flow measurements are desired.

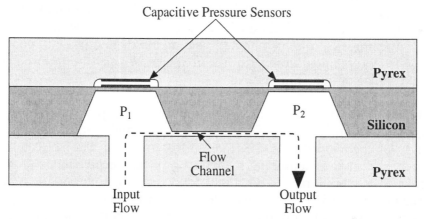

Illustration of a micromachined flow sensor using the approach of differential pressure sensing at both sides of a precise fluid flow channel. Adapted from Oosterbroek, et al. (1997).

Oosterbroek, et al. (1997) demonstrated a micromachined pressure/flow sensor based on this principle, combining two capacitive pressure sensors with a precise flow channel (illustrated above). This approach is general, and can be applied to a variety of monolithic or hybrid micromachined structures.

4.2 VISCOSITY/DENSITY SENSORS

In numerous industrial and research settings, it is important to be able to quantify the viscosity of fluids. There are a variety of methods used in macroscopic devices (see Walters and Jones (1995)), but most micromachined devices are of the vibrational type. Such viscometers make use of the fact that the viscous damping of a resonator in contact with a fluid is proportional to the product of the density and the viscosity of the fluid (hence they are, whether desired or not, sensing both

parameters). However, in typical settings, the density of a liquid should remain fairly constant.

Several micromachined viscometers have been demonstrated, including acoustic wave (Grate, et al. (1993)), shear mode (Martin, et al. (1994, 1995)) and Lamb wave devices (Martin, et al. (1990)). As an example Lamb wave (otherwise known as flexural plate wave, or FPW) device, Costello, et al. (1993) fabricated 3 × 8 mm membranes of 3.5 μm thick LPCVD silicon nitride, with ZnO acoustic transducers. Interdigitated aluminum fingers were fabricated and coated with RF-magnetron sputtered ZnO, and finally capped with a 0.3 μm aluminum ground plane. When used as a delay-line oscillator, the oscillation frequency could be used to compute fluid density, while the viscosity was measured by driving the acoustic transducers and measuring the frequencies of maximum transmission and transmission loss. A useful feature of this type of design is that the back side (no lithographic features) of the membrane is the region that comes into contact with the fluid, so the chemical compatibility of the sensor can be tailored with the appropriate choice of thin-film material.

Brand, et al. (1997) fabricated a thermally driven, piezoresistively sensed, low-frequency viscometer using KOH bulk micromachining to form a thin silicon membrane. On the membrane's center, a p-diffusion heater was fabricated to serve as a heater, and at the membrane's edge, a p-diffusion piezoresistor served to detect transverse vibrations. In air, the 2.5 × 2.5 mm membranes were found to have Q factors of ≈ 115, using 16 mW DC bias and 13 mW AC signal input to the heater, with the Q factor dropping as low as 6 in viscous liquids (the resonant frequency of the membrane, ≈ 14 kHz in the present example, is largely a function of fluid density). A noteworthy feature of this work was the use of polydimethylsiloxane (PDMS) formulations with viscosities varying over five orders of magnitude, yet with densities varying by only 20%. This allows the effect of density on resonator damping to be neglected, and measured Q factors can then be used to determine viscosity. This is also significant because polymers such as PDMS are often used for molding, and monitoring their viscosity during polymerization can be quite useful. These devices allowed viscosities in the range of 10^{-3} to 1 Pa•s (1 to 1,000 centipoise) to be measured.

Fluid density can also be measured by flowing a fluid through a structure with which the mass of a known volume of fluid can be determined. Enoksson, et al. (1995) presented such a sensor, fabricated using bulk micromachining of silicon. They flowed fluid through a hollow, resonating single-crystal silicon structure and measured the mass-loading-induced frequency shift. They achieved fluid density measurement sensitivities in the range of -230 to -320 ppm per kg/m³ with resonant frequencies on the order of 10 kHz. In addition, a thermally driven density sensor, fabricated by post-processing standard CMOS integrated circuits, was presented by Westberg, et al. (1997).

4.3 OTHER FLUIDIC SENSORS

There are a number of other sensors with relevance to fluidic systems. Mechanical pressure sensors are discussed in the Mechanical Transducers chapter, and thermal (gas) pressure sensors are discussed in the Thermal Transducers chapter. In addition to basic pressure measurement and indirect flow rate measurement, pressure sensors can be used to detect the onset or extent of turbulence by monitoring the nonstationary component of the pressure (i.e., high-pass filtering the pressure sensor output).

In many fluidic systems, chemical sensors are required, as discussed in the Chemical and Biological Transducers chapter. Examples of such transducers that are applicable to fluidic systems include temperature sensors, work-function modulated sensors, electrochemical sensors (e.g., conductivity or pH), acoustic wave sensors (e.g., adsorption), hybrid biosensors (i.e., using cells as part of the sensor), optical sensors (e.g., fluorescence, turbidity/opacity, colorimetry, etc.), adsorption sensors (e.g., resonators), surface plasmon resonance sensors, and others.

5. VALVES

Valves are critical components of microfluidic and macroscopic systems. An ideal valve would have a currently (and probably always) unachievable set of characteristics:

- zero leakage
- zero power consumption
- zero dead volume
- infinite differential pressure capability
- insensitivity to particulate contamination
- zero response time ("infinitely fast" state change)
- potential for linear operation
- ability to operate with liquids and gases of any density/viscosity/chemistry

Clearly, none of these characteristics is fully attainable in macroscopic valves, so they should not be expected of micromachined valves. In practice, only an approximation of a *subset* of the ideal characteristics are important in a given application — valves are fairly application specific.

Micromachined (and macroscopic) valves can readily be categorized in terms of whether or not they have a powered actuation mechanism (*passive* or *active*).

While this is clearly not the only way of breaking down valves into categories, employing it in the context of studying the construction of the valves is useful. Other important characteristics of valves include whether they are normally open or normally closed, are for gas or liquid, are proportional or digital, etc.

5.1 PASSIVE VALVES

Several designs have been proposed or implemented that are completely passive and require no external power or control. Such valves are often used as passive check valves for pumps, such as our hearts.

Among the structurally simplest micromachined passive valve design is that of Smith and Hök (1991).

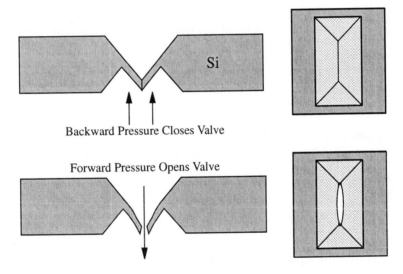

Illustration of a passive microfluidic valve fabricated using two-sided bulk etching. After Smith and Hök (1991).

The fabrication sequence involved using oxide masks to etch pyramidal pits into a silicon wafer, followed by removal of oxide and heavy boron doping (> 10^{19} cm^{-3}). This was followed by dopant-selective etching with no masking (KOH was used in some of the devices) to form the valve leaflets and the opening slit (it forms because the etch rate does not drop to zero at the p++ interface, and the bottom of the V-grooves are exposed to the etchant the longest). This process (illustrated below) is a simple approach to fabrication of unidirectional valves, but may have unacceptable leak rates for some applications. Slit widths obtained were on the order of 10 μm.

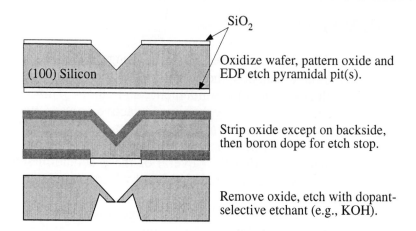

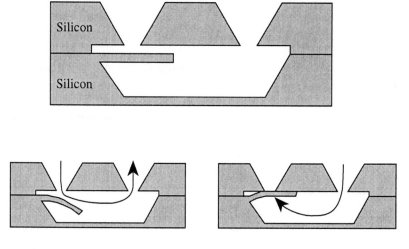

Illustration of a passive valve fabrication process. After Smith and Hök (1991).

Illustration of a two-wafer stack, bulk micromachined passive valve design. After Tiren, et al. (1989).

Another passive valve design, this time based on a two-wafer stack, was proposed by Tiren et al. (1989). In this case, a single cantilever acts as a seal when forced against the upper wafer by reverse flow, as illustrated above. In addition an interesting bulk micromachined passive valve design combined with a fuel injection nozzle (for automotive use) can be seen in Giachino and Kress (1986, 1988).

5.2 ACTIVE VALVES

Active microfluidic valves have been developed using a wide range of actuation schemes, including thermal expansion, thermopneumatic, pneumatic, piezoelectric, shape memory alloy, electrostatic, electromagnetic, and others. Unfortunately, at present, there is no "ideal" actuation mechanism for active valves. Each one has its own set of advantages and drawbacks, as discussed in the Mechanical Transducers chapter (these properties are summarized below for those valves not requiring external hydraulic or pneumatic drive). Several of these actuation schemes are covered in a review article on micromachined valves by Barth (1995).

Thermal expansion actuation can generate large forces, but it is relatively slow (especially with passive cooling) and may undesirably heat the fluid being valved, and generally consume a large amount of power (dissipated within a reduced volume). Despite the drawbacks, thermal actuation is presently the most popular for micromachined valves because as devices are scaled, thermal mass goes down and it is possible to make very thin, high thermal resistance membranes to minimize losses. Also, this approach can generate relatively large forces. Linear thermal expansion, thermal bimorph actuation, and the use of shape memory alloys have essentially the same advantages and drawbacks as direct thermal actuation.

Piezoelectric actuation can yield very high forces, but very small movements for even very large voltages.

Electrostatic actuation can provide high forces (nonlinear) and large movements, but also requires high voltages.

For macroscopic valves, electromagnetic (e.g., solenoid) actuators are generally used since they generate large forces and movements quickly and with relatively little power. Unfortunately, it has been difficult to micromachine equivalent electromagnetic actuators, although progress continues to be made in this area.

Despite such problems, micromachined valves offer greatly reduced volumes and weights, although the power consumption and volume (packaged) *may not actually be less than that of macroscopic (but "miniature") valves* (for example, a large assortment of miniature, but not micromachined, valves and pumps can be obtained from the Lee Company, Westbrook, CT). Typical miniature solenoid valves from the Lee Co., are compatible with a wide variety of chemicals, operate on less than 1 W of power, and actuate in as little as 830 µs. Some variants of the miniature valves have Teflon™ inner surfaces, allowing compatibility with nearly any chemicals. The smallest of the miniature conventional valves have volumes on the order of 0.5 cm^3, certainly comparable to packaged micromachined valves demonstrated to date.

Perhaps the first active micromachined valve was that demonstrated by Terry (1975) and Terry, et al. (1979) as a component of an integrated gas chromatography system. While external actuation was used, the basic concepts of the micromachined silicon valve seats spelled out in this work continue to be used in subsequent micromachined valves. Two designs were tested, one using a silicon diaphragm/silicon valve seat arrangement, and the other using a Teflon™-coated polyimide (Kapton™) membrane/silicon valve seat design. Both designs required hand assembly (as is presently still the case, in general) and used external actuation, but demonstrated principles that would later be applied to truly batch-fabricated micromachined valves.

5.2.1 PNEUMATIC VALVE ACTUATION

Various approaches have been taken to fabricate micromachined valves that are actuated using external pneumatic power. This technique avoids the need to locally generate the forces required to open or close a valve, but results in the need for external fluidic interconnects (as opposed to electrical connections, for example).

An organic membrane-based valve for controlling aqueous biomedical fluids was demonstrated by Vieider, et al. (1995), using external pneumatic actuation. Their design made use of a spun-on layer of a two-component silicone elastomer (type not specified) to form a membrane supporting a moving silicon "plunger" that could be pneumatically driven to close a fluid flow channel.

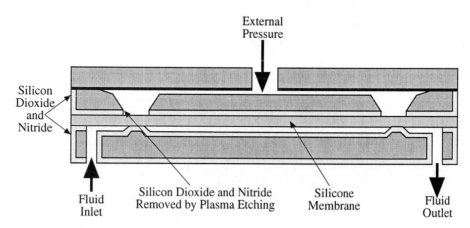

Illustration of a silicone elastomer membrane, micormachined pneumatically actuated microvalve. After Vieider, et al. (1995).

Their design consisted of three wafers in a stack, the central one containing the movable plunger and silicone membrane. To fabricate the membranes, silicone was spun onto what would become the central wafer after the anisotropic etching was completed, so that the silicone was held in place by thin silicon dioxide/nitride membranes where there was no silicon. A plasma etch was then used to remove the silicon dioxide/nitride from the silicone in the regions where it was to move. They demonstrated a 500 kPa pressure difference across the membrane, tested flows up to 275 µl/min, and were unable to detect any leakage through the structure.

Bruns (1992) also demonstrated a pneumatically actuated gas valve with a trapped polymeric (polyimide) membrane. They obtained helium-leak-test tight seals and demonstrated more than 2.5 million cycles without failures, with a 15 ms valve response time.

5.2.2 THERMOPNEUMATIC VALVE ACTUATION

One of the earliest successful self-contained micromachined active valves was the so-called "fluistor" (a term coined later for the commercially available version) developed by Zdeblick and Angell (1987) (also see Zdeblick (1988), and Zdeblick (1989)). They demonstrated a thermopneumatically actuated valve wherein liquid trapped in a sealed cavity is pressurized (and/or partly vaporized) by dissipation of power by a thin-film heater. The cavity is formed by bulk micromachining in silicon, and is sealed by anodically bonding a Pyrex™ wafer, with patterned heater resistors, onto the silicon.

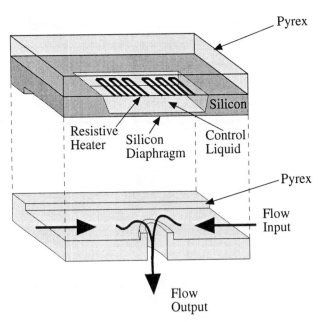

Illustration of a micromachined thermal expansion valve. After Zdeblick, et al. (1994).

The expanding working fluid trapped in the chamber forces a thin silicon membrane against a valve seat fabricated in the underlying Pyrex™ wafer, closing the valve (typical membrane excursion is $\approx$ 50 µm). The working fluid must be sealed into the device after anodic bonding. By appropriate choice of the working fluid, actuation temperature can be specified (typical valves operate at temperatures above the commercial temperature range maximum of 55°C). Similar valves in normally closed configurations (Zdeblick, et al. (1994)) and liquid handling capability (Henning, et al. (1997)) have also been developed by this group. It should be noted that these valves generally require on the order of 1 to 2 W of power to operate. Such devices are commercially available through Redwood Microsystems, Inc., Menlo Park, CA.

A version of the thermal expansion valve described above, designed for proportional control of refrigerant fluid flows (sometimes called a "metering valve"), was demonstrated by Henning, et al. (1997). Their normally open valve was capable of sustaining differential pressures of up to 140 psi ($\approx$ 1 MPa) and allowing flow rates of 2 to 10 g/s of a refrigerant fluid (R-134a). On-to-off times were on the order of 1 s, with power consumption reduced to < 0.5 W, considerably less than prior valves of this type.

Organic membranes also appear to be quite useful in such valve designs. Fahrenberg, et al. (1995) demonstrated a thermal expansion actuated valve fabricated using thermoplastic molding to form two interlocking polymethylmethacrylate elements with a captured polyimide membrane with integrated (sputtered) thin-film heating resistors. Molded cavities with ports were filled with thermally expandable liquid.

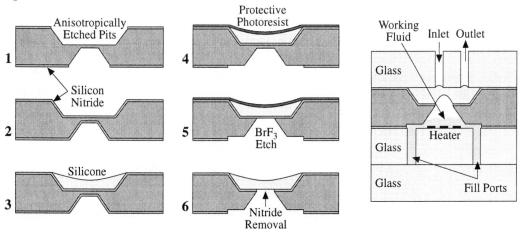

Illustration of a micromachined, silicone membrane, thermopneumatic valve. At the left, the fabrication steps are shown (numbered 1 through 6) and at the right, the valve is shown as assembled, and in the "closed" state. Adapted from Yang, et al. (1997).

Yang, et al. (1997) demonstrated the use of a silicone membrane to fabricate an integrated normally open microvalve (illustrated above). They took advantage of the low Young's modulus ($\approx$ 1 MPa) of silicone rubber to allow for very large membrane excursions for a relatively low input power. The valves were fabricated by bulk etching from both sides of a silicon wafer to form a 50 µm thick silicon membrane in between two cavities (by etching 235 µm deep into both sides of a 520 µm thick wafer). Following the deposition of a 0.5 µm silicon nitride film on all surfaces, silicone rubber (MRTV1, American Safety Technologies, Inc., Roseland, NJ) was molded into the cavities on one side of the wafer (the cured thickness of the silicone was 60 to 70 µm). The back-side silicon nitride was then removed selectively below the silicon membranes using an SF_6/O_2 plasma, the thin silicon membrane was etched away using BrF_3 vapor, followed by removal of the remaining silicon nitride above the silicon membrane's former location releasing the silicone membrane. These membrane-bearing devices were glued (using epoxy) to an underlying Pyrex™ substrate with a prefabricated thin-film gold heater. The valve's working fluid (3M PF5060 fluorocarbon, 3M, Inc., Minneapolis, MN) was sealed within the expansion cavity. Similarly, a Pyrex™ upper substrate with mechanically drilled inlet and outlet holes was glued above the silicon substrate to form the valve seat.

The authors reported 860 µm peak membrane deflection at 970 mW of input power, corresponding to a developed pressure of 7 psi. They demonstrated that a power input of 280 mW was sufficient to shut off a 20 psi airflow of 1.3 lpm. One current drawback of such valves, however, is that silicones tend to be highly permeable to most liquids that could be used as the working fluids in such devices. This problem could potentially be solved through the addition of a thin-film barrier layer onto membranes.

5.2.3 PHASE-CHANGE VALVE ACTUATION

An alternative thermally driven valve concept is that of using a phase change of the fluid to be controlled by the valve to form a "plug" that impedes fluid flow. The two useful phase changes for a liquid are to gas or solid. Both have been used to fabricate valves. A key feature of these approaches is that they essentially create zero dead volumes in the fluid channels.

Relatively simple micromachined valve structure can be designed to take advantage of thermal formation of bubbles from the fluid within the valve. Such *bubble valves*, were demonstrated by Evans, et al. (1997), who noted that a bubble can readily support a pressure differential across it if the radii of curvature of its front and rear surfaces, r_1 and r_2, respectively, are different. The magnitude of the pressure difference is given by,

$$\Delta P = \gamma \left(\frac{1}{r_1} - \frac{1}{r_2} \right)$$

where γ is the interfacial surface tension for the fluid/gas of interest (7.27×10^{-2} N/m, or 0.608 atm•μm for air/water at 20°C, 1 atm pressure). The necessary bubble(s) can be formed by locally vaporizing some of the fluid in the channel, and the process is reversible via cooling.

In their bubble valve design, Evans, et al. (1997) used deep reactive ion etching of silicon to fabricate channels of constant height (100 μm), such that only the radii in the plane were of any consequence. By varying the fluid channel radii appropriately, and including an array of silicon pillars to constrain the location of the bubbles, pressure differentials of ≈ 0.05 atm were demonstrated in uni- and bidirectional valves. With the heat for bubble formation generated using polysilicon heaters, the valves were operable at frequencies of 0.5 Hz. A similar "bubble power" approach can also be used to pump fluid, as discussed below.

The phase change between solid and liquid was demonstrated as a valve technology by Kaartinen (1996). The concept is to thermally bias (cool) valve regions of a fluidic system such that when not warmed by a local heater, the liquid in the valve freezes, closing the valve. Kaartinen demonstrated a complete fluidic chemical reactor system fabricated using electroforming (thick electroplating onto sacrificial forms, which are later removed to form the fluidic channels) and external heater resistors. Using carefully designed thermally conductive paths, the valves were biased at -20°C through connection to a -25°C heat sink. Upon powering a particular valve's heating resistor, a valve could be opened in 100 ms. Re-freezing occurred within 500 ms, with a frozen fluid volume within the valve of ≈ 50 nl. While the pressure resistance of such valves should be extremely large, Kaartinen only reported testing to tens of atmospheres. A very interesting observation was that the heat of fusion released by the freezing fluid within the valve could be monitored via valve temperature (e.g., using the heating resistor's temperature coefficient of resistance or an external sensor) and used to measure valve closure (or, in an analogous manner, valve opening). This approach could readily be applied to a variety of micromachined fluidic systems, particularly with the availability of low-cost Peltier effect thermoelectric cooling modules to provide the necessary bias cooling.

5.2.4 SOLID-EXPANSION THERMAL VALVE ACTUATION

Jerman (1991a, 1991b, 1994) demonstrated normally closed thermal bimorph (differential thermal expansion, as discussed in the Thermal Transducers chapter) actuation scheme for a microvalve for gas flow control using an aluminum layer fabricated on an approximately round silicon membrane. By varying the power dissipated in the diffused resistors underlying the aluminum annulus (and thus the temperature of the bilayer), the position of the valve boss could be controlled, opening the valve as the temperature increases.

An important feature of the design is a thin SiO$_2$ "hinge" surrounding the membrane to increase thermal isolation and, more importantly, provides a hinged boundary condition for the membrane, allowing it to move the valve boss away from the seat when the membrane is heated.

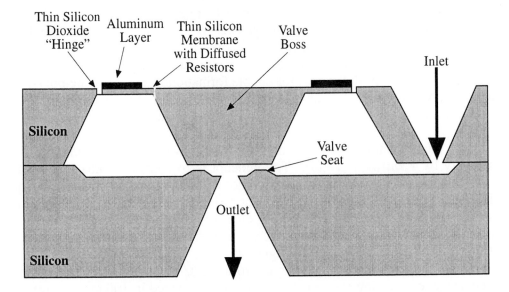

Illustration of a normally closed gas valve, showing two-wafer construction and thermal bimorph actuation scheme (note that the valve is shown partly open). After Jerman (1991a).

On/off flow ratios (defining leakage) of 5,000:1 were achieved, with gas flows of 0 to 150 cc/min possible at inlet pressures of 1 to 50 psig, with an input power of 150 mW required for any appreciable flow and more power required for higher flow rates (on the order of 500 mW for full flow). Despite using a thermal actuation scheme that might be sensitive to ambient temperature variations, these valves apparently operate in a useful manner over -20 to +85°C, greater than the commercial temperature range (0 to +55°C). This type of valve (model 4425) is commercially available from EG&G IC Sensors, Inc., Milpitas, CA.

Barth, et al. (1994) and Barth and Gordon (1994) demonstrated another bimorph actuated valve, using electroplated nickel on silicon as the bilayer. Their valve, normally closed and for controlling gas, is heated using an electroplated nickel resistor on the membrane, the resistor being electrically isolated from the nickel on the membrane.

While few details of fabrication or test results were presented, the valves were able to operate over a pressure range of 0 to 200 psi and flow rates of 0 to 600 sccm. These valves require ≈ 1 W of power for full flow. They were developed for Hewlett-Packard internal use, and are not available commercially.

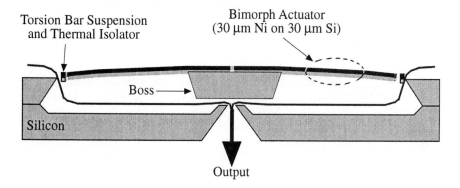

Illustration of a thermal bimorph actuated, normally closed gas valve (shown in the open [heated] state). After Barth, et al. (1994).

Busch and Johnson (1990) described a shape-memory-alloy (SMA) actuated normally closed gas valve that was similar in structure to the Jerman valve, but it used a SMA membrane that was biased with an external BeCu spring to keep the valve closed. Heating of the SMA caused it to shorten in length, overcoming the force of the spring and opening the valve. Such valves are commercially available through Microflow, Inc., Dublin, CA. For further information on SMA valves, the reader is referred to Johnson and Ray (1991) and Skrobanek, et al. (1997).

A thermally driven valve based on a buckled membrane was presented by Popescu, et al. (1995), and a thermal bimorph driven valve with an integrated flow sensor was demonstrated by Franz, et al. (1995). In addition, a thermal bimorph cantilever was used by Trah, et al. (1993) to steer a fluid flow toward or away from a valve membrane, allowing it to be opened or closed (the basic fluid-steering structure is discussed above in Section 3.3).

5.2.5 PIEZOELECTRIC VALVE ACTUATION

Shoji, et al. (1991) demonstrated the use of a separately fabricated stacked piezoelectric actuator as a means of closing a micromachined valve. An important feature of this valve in addition to the actuation scheme was that a thin polymer membrane (rather than a silicon membrane) was used to seal the inlets when deflected toward the substrate. This approach allows for considerable flexibility in choosing valve membranes with mechanical, chemical, or other properties usefully different from those of typical materials used in micromachining (e.g., deformable membranes for low-leak seals).

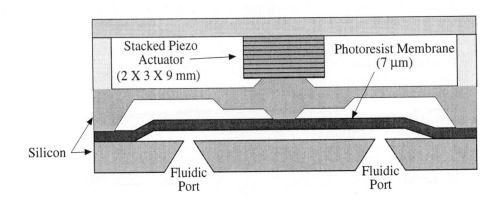

Illustration of a micromachined piezoelectric valve with an organic membrane (polymerized negative photoresist). After Shoji, et al. (1991). Actuating the piezo- electric stack would push downward on the photoresist membrane, interrupting the fluidic path between the two ports.

The silicon components of the valve consisted of two micromachined substrates mechanically held together, sandwiching the membrane between them. The lower substrate, incorporating the fluid channel, was fabricated via anisotropic wet etching and the polymer membrane was made from polymerized negative photoresist (Tokyo Ohka OMR-83) with a sacrificial positive photoresist (Hoechst AZ-4562) that was removed with acetone without dissolution of the membrane. As illustrated below, silicon dioxide and nitride membranes were formed on the lower substrate and coated with both the sacrificial positive photoresist and membrane-forming negative photoresist. Using etching from the back side of the wafer, the oxide and nitride membranes were removed and the positive resist dissolved (isopropyl alcohol was used after acetone to prevent sticking of the membrane to the substrate).

Their normally open valves operated with applied voltages up to 100 V (typical for piezoelectric actuators), had liquid (deionized water) flow rates up to 12 μl/min, and could operate at inlet pressures up to 0.5 atm. Normally closed valves were also demonstrated.

Another bulk micromachined, piezoelectrically actuated automotive fuel- injection valve and spray nozzle is described in Gardner, et al. (1990).

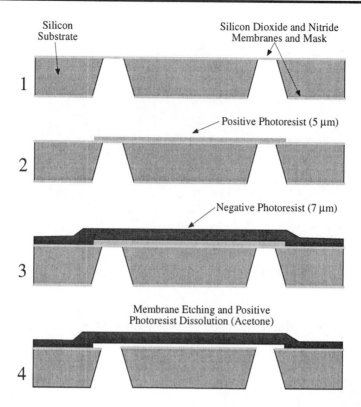

Illustration of a fabrication process for micromachined piezoelectric valve with integral organic membrane. After Shoji, et al. (1991).

5.2.6 ELECTROSTATIC VALVE ACTUATION

Huff, et al. (1993) demonstrated an electrostatically (or pneumatically) actuated valve based on a fusion bonded stack of three silicon wafers, illustrated below. They demonstrated pneumatic control of flow air rates from near zero up to ≈ 550 ml/min, and "digital" control using electrostatic actuation at ≈ 200 V. They observed that the use of electrostatic control requires extremely tight control over device tolerances (due to the highly nonlinear force generation mechanism).

There are a large number of possible electrostatically actuated valve configurations, and another interesting approach can be seen in Haji-Babaei, et al. (1997). These authors presented a gas valve in which a Cr/SiO_2 bimorph was released and allowed to curl up and away from an underlying orifice with an electrode on the substrate surface near it. When 30 V was applied between the bimorph and the substrate electrode, the valve could be closed (in the absence of gas flow). Unfortunately, detailed results were not reported.

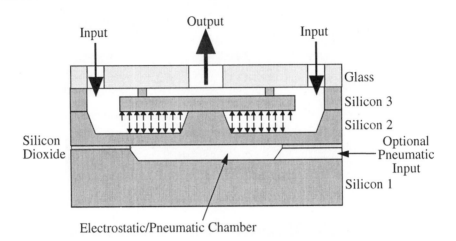

Electrostatically or pneumatically actuated valve design. After Huff, et al. (1993).

5.2.7 ELECTROMAGNETIC VALVE ACTUATION

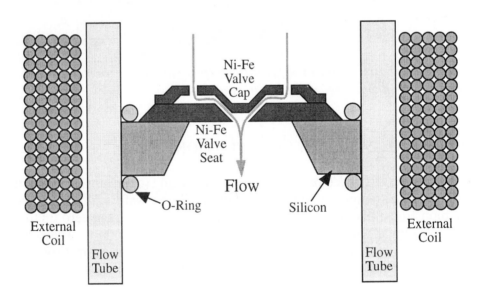

Illustration of an externally actuated micromachined electromagnetic valve. After Yanagisawa, et al. (1993).

At present, it appears that no fully micromachined electromagnetically actuated valves have been demonstrated, although Yanagisawa, et al. (1993), presented a micromachined valve that could be actuated using an external coil (illustrated above). Their valve consisted of a sputtered NiFe valve seat and a moving NiFe valve member (released using a sacrificial RF sputtered SiO_2) that could be opened or closed. Depending on built-in stress in the moving member's support springs, normally open or closed valves could be fabricated. The micromachined valve components were placed within a pipe carrying the fluid of interest, while the external magnetic field was applied via a coil placed *outside* the pipe (an interesting demonstration of electrically isolated operation).

5.2.8 BISTABLE VALVE STRUCTURES

Bistable valves have mechanical properties capable of holding them in either state (open or closed) and only require power to be applied when transitioning from one state to another. Such valves would be excellent for applications where power is at a premium, such as implantable biomedical devices, spacecraft, etc. Micromachined bistable structures had been demonstrated in the past (e.g., Huff, et al. (1991)), yet their application to valves took several years, despite the *huge* potential benefits for medical, aerospace, and instrumentation applications.

Wagner, et al. (1996) discussed the fabrication of a bistable, electrostatically actuated valve for liquids. The basic concept was to use a thin silicon membrane with a thermally grown SiO_2 layer on one side to provide built-in compressive stress to force the membrane to buckle. The membrane was bonded over two conductive electrodes (one at the bottom of each of the recessed cavities) that could be used to draw the membrane sections down, one at a time. The two electrode cavities were connected via a narrow channel. The space beneath the membrane sections was filled with air, and when one section was pulled down electrostatically, the underlying air was forced into the other chamber, causing its direction of buckling to reverse. Using such an actuator, a valve could be constructed so that one region of the membrane could close or open the valve depending on whether or not it was forced into contact with a valve seat.

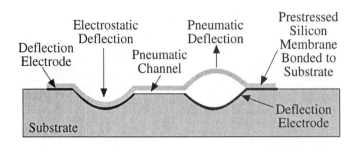

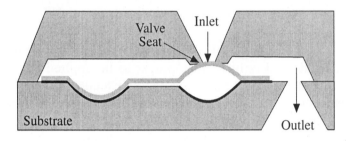

Simplified illustration of a micromachined bistable fluidic valve in which electrostatic deflection drives the inversion of one buckled membrane region, which pushes trapped gas into a pneumatically coupled second region, forcing that membrane against a valve seat. The upper illustration shows the basic operating principle and the lower illustration shows the location of the valve seat on a second silicon substrate fusion bonded to the first substrate. To open the valve, the right-hand deflection electrode would be energized. Adapted from Wagner, et al. (1996).

The bistable membrane actuator was fabricated by first etching the deflection cavities into a silicon substrate using HNA. Masked ion implantation with phosphorus was used to form the deflection electrodes. Silicon-on-insulator wafers with 7 μm upper silicon on 1 μm oxide were used to form the membranes. Before fusion bonding the two wafers together, 500 nm of thermal SiO_2 was grown on the SOI wafer to form the compressive layer responsible for buckling the 7 μm silicon membrane formed later. The two wafers were bonded together, and the SOI wafer was thinned in TMAH, using the buried 1 μm oxide layer as an etch stop. This oxide layer was then thinned using dry etching and patterned to form a mask that was used to selectively remove some membrane regions using TMAH. The membrane etch step exposed the underlying implanted electrodes, to which electrical connections were made by depositing and patterning aluminum contacts. While they demonstrated the bistable actuator (with measured deflections on the order of ± 10 μm), the authors did not present actual results for the valve.

6. PUMPS

Pumps have many applications, including transport of reagents, delivering pulsatile flows, generating pressure differences, moving cooling fluids, transporting suspended particles or cells, etc. For many proposed microfluidic systems, micromachined pumps are a critical, but not yet very practical, component. Micromachined pumps have been investigated by many groups, and several have been shown to function.

The pump principles most likely to be applicable to micromachined devices are membrane (with passive check valves), rotary, diffuser, electrohydrodynamic, electrophoretic/electroosmotic, and ultrasonic pumping. Membrane pumps commonly suffer from leakage and/or fouling of the check valves, but comprise the bulk of micromachined pumps since they can be more readily implemented than most other approaches. Valved, rotary, and ultrasonic pumps have also been implemented using micromachining techniques. Piston or "syringe" pumps have yet to be implemented in micromachined form, but can already be reduced to extremely small volumes using conventional, non-lithographic fabrication methods. It is also important to note that, to date, the vast majority of micromachined pumps have been designed with consideration of pumping liquids only (likely because they tend to be inertia-based, rather than positive-displacement designs).

6.1 BUBBLE PUMPS

One of the simplest pump design is the bubble pump, which makes use of the repetitive formation and collapse of vapor bubbles, generally formed by locally heating the fluid to be pumped. While the fluid may become quite hot locally, there appear to be several applications in which this approach is quite suitable. Evans, et al. (1997) demonstrated bubble pumps based on deep reactive ion etching of silicon to form circular chambers ($\approx 200~\mu$m in diameter and 100 μm deep) along fluidic channels, at which points thin-film polysilicon heaters were fabricated for bubble formation. They reported pump displacements of 3.1 nl and operating frequencies of 0.5 Hz.

6.2 MEMBRANE PUMPS

Membrane pumps are, by definition, reciprocating devices. One can define a stroke efficiency, η_p, which relates the volume actually pumped to the volume variation (amplitude) of the pump chamber (see Stemme and Stemme (1993)),

$$\eta_p = \frac{V_o}{2V_x}$$

where V_o is the volume of fluid pumped, and V_x is the volume variation of the pump chamber. The heart is a familiar example of a reciprocating pump, and the medical analog for stroke efficiency is *ejection fraction*, which perhaps better expresses the fact that the amount ejected is the parameter of interest.

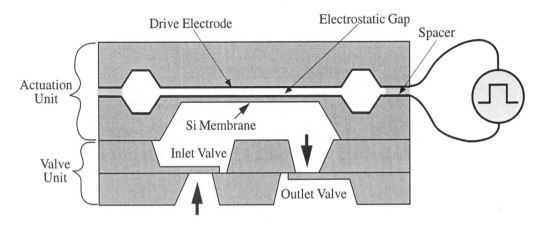

Illustration of a micromachined, electrostatically actuated, reciprocating liquid pump. After Zengerle, et al. (1995a).

Zengerle, et al. (1995a) demonstrated an electrostatically actuated bidirectional silicon membrane micropump. Their design consisted of an electrostatically driven membrane to generate pressure pulses and two passive valves to control the fluid flow, as illustrated below. A four-wafer stack of bulk micromachined wafers was combined to form the pumps. The pump was driven with voltage pulses of 150 to 200 V amplitude at frequencies from 0.1 Hz to > 10 kHz. At low drive frequencies (< 800 Hz), the pump operated in the forward mode, but at higher frequencies (2 to 8 kHz), the pump operated in the reverse direction due to phase shift between the response of the valve and the pressure pulses driving the pumping action (their paper includes a mathematical analysis).

Various designs of the pump had maximum flow rates of 250 to 850 µl/min (forward) and 200 to 350 µl/min (reverse). The maximum pump rate achieved was 850 µl/min at a maximum back-pressure of 310 cm H_2O. As expected for electrostatic actuation, the power dissipation was low, on the order of 1 mW.

In practice, these pumps are very sensitive to particulates and require the use of a micropore filter and high-voltage power supply that are generally much larger than the pump itself. Micromachined pumps of this type do, however, have tremendous potential, especially if these functions become more integrated into the pumps.

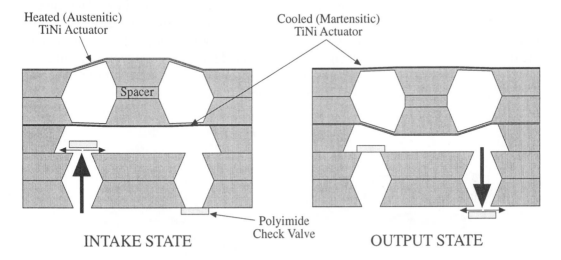

Illustration of a thermally driven, TiNi actuator membrane pump with polyimide check valves. Adapted from Benard, et al. (1997).

A thermally driven, TiNi actuator membrane pump with polyimide check valves was presented by Benard, et al. (1997). A pair of opposed TiNi membranes provided push-pull actuation for the pump assembly, consisting of two actuator wafers bonded together (with a spacer to ensure that the membranes can mechanically deflect), in turn bonded to a trio of wafers containing the check valves. Using deionized water, they reported pump rates as high as 50 µl/min, when driven at 0.9 Hz with 0.6 V, 0.9 A pulses (0.54 W).

Judy, et al. (1991) fabricated an electrostatically actuated pump that was entirely surface micromachined, using polysilicon electrodes encapsulated between 0.2 µm LPCVD silicon nitride layers, combined with sacrificial PSG layers. Each pump consisted of an input valve, pumping membrane, and an output valve. With voltages as low as 50 V, the devices were apparently actuated, but no pumping results were reported.

Surface micromachined electrostatic pumps, while limited in throughput by the volumes achievable with relatively thin sacrificial layers, appear to have considerable potential for further research, particularly for very low fluid volumes and flow rates. If surface area usage is not a great concern, many such pumps could be placed in parallel to improve the flow rates achievable.

6.3 DIFFUSER PUMPS

Diffuser pumps have inlet and outlet ports with increasing and decreasing cross-sectional areas (in the direction of flow), respectively, coupled to a chamber with an oscillating pressure (e.g., a cavity with a driven membrane). The kinetic energy (flow velocity) of the fluid is transformed into potential energy (pressure) in the pump ("pressure recovery"), but the efficiency of this process is greater in the diffuser direction than in the nozzle direction. Thus ports conduct more fluid in the diffuser direction than in the nozzle direction, resulting in a net pumping action (this effect is sometimes referred to as "flow rectification").

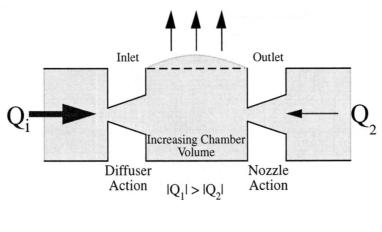

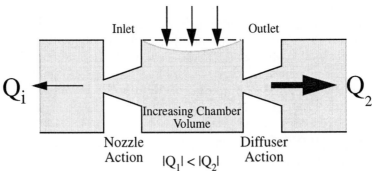

Conceptual illustration of a diffuser pump. After Stemme and Stemme (1993).

As discussed by Stemme and Stemme (1993) and Olsson, et al. (1995, 1996), the pressure differences across each port, acting as a diffuser or nozzle respectively, are given by,

$$\Delta P_d = \frac{\rho\, v_d^2}{2}\, \xi_d \qquad \text{and} \qquad \Delta P_n = \frac{\rho\, v_n^2}{2}\, \xi_n$$

where,

ΔP_d and ΔP_n = pressure difference across the diffuser and nozzle

ξ_d and ξ_n = pressure loss coefficients for the diffuser and nozzle

v_d and v_n = mean velocities at the diffuser and nozzle

If ξ_n is greater than ξ_d, pumping will occur (a thorough overview of the underlying principles is presented in Olsson, et al. (1996) and simulation approaches in Olsson, et al. (1997b)). While these pumps are inherently open flow systems and cannot generate large pressure differentials, they could potentially be quite useful in low-pressure fluidic systems.

Precision machined brass prototype pumps of this type were used to demonstrate this principle for liquids (Stemme and Stemme (1993)) and for gases (Olsson, et al. (1997c)). Olsson, et al. (1995a, 1995b, 1996), demonstrated a micromachined silicon diffuser pump fabricated using HNA as an isotropic silicon etchant in a silicon substrate that was anodically bonded to a glass cover. They bonded piezoelectric disks to both faces of the assembly and excited several pumps at resonance frequencies on the order of 1.3 kHz to reach a maximum pump rate of 230 μl/min at a pressure of 170 cm H_2O for methanol (water was also pumped, at lower achievable flow rates and pressures due to higher frictional losses). In addition, Olsson, et al. (1997a) demonstrated the use of micromachined molds for fabricating plastic diffuser valves. Deep reactive etching of silicon was used to form templates for electroplated nickel structures, which were in turn removed and used to form the molds for plastic injection molding or hot embossing. The resulting plastic pumps, driven with external piezoelectric disk actuators, achieved flow rates as high as 1.9 ml/min and a maximum pressure of 7.7 kPa ($\approx$ 0.08 atm). Gerlach (1997) presented a gas bulk micromachined gas pump based on this principle, and reported a zero-load pump rate of 7.5 ml/min and a maximum pump pressure of 2.8 kPa when pumping air. Similar diffuser pumps were also demonstrated by Forster, et al. (1995) and in addition, they demonstrated the use of Tesla's "fluidic rectifiers" (Tesla (1920)) in such pumps.

6.4 ROTARY PUMPS

Ahn and Allen (1995) demonstrated a magnetically driven (with monolithic magnetic coils), *rotary ("jet-" or impeller-type) micropump* fabricated using multi-level organic template (polyimide) electroplating. The magnetic stator and the central pin were fabricated using high-permeability electroplated permalloy (81% Ni, 19% Fe). Multi-level electroplating allowed for the magnetic coils for the stators to be monolithically fabricated. The template material used for the electroplating was 60 μm thick Probimide™ 349 polyimide (Olin Microelectronic Materials, Tempe, AZ). The 50 μm tall, 500 μm diameter rotors were separately fabricated on separate wafers and assembled by hand into the pumps, which were then sealed with a Pyrex™ top plate. Using 200 to 500 mA to power the pumps with a driving voltage of < 3 V (0.6 to 1.5 W), the flow rate demonstrated was 24 µl/min at 5,000 rpm with an achievable differential pressure of 10 kPa (≈ 0.1 atm).

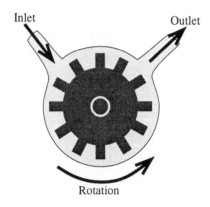

Illustration of the basic concept of a rotary "jet-type" pump. After Ahn and Allen (1995).

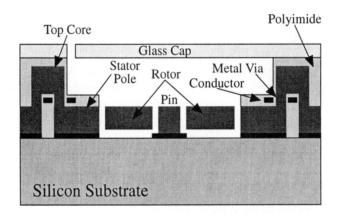

Cross-sectional llustration of a micromachined magnetic rotary pump. After Ahn and Allen (1995). Note that the magnetic coils were monolithically integrated.

A LIGA-fabricated in-line gear pump was demonstrated by Dewa, et al. (1997). In this type of pump, tight tolerances between the teeth of the two gears, and the surrounding walls, allow for the formation of low-leakage sliding seals. Fluid is pumped by the action of the turning gears, with fluid being carried along the walls of the cavity enclosing the gears, and these pumps are generally self-priming. The pump, illustrated below, consisted of two PMMA gears with an inset electroplated NiFe (78:22 ratio) bar for magnetic drive of one of the gears. Both gears were 1.4 mm in diameter, 200 µm thick, and with 24 teeth, and the overall dimensions of the pump itself were 3.175 mm in diameter by 600 μm thick. Drive power was supplied via an external rotating permanent magnet. At 5,000 rpm, the pump produced a flow rate of 350 μl/min of water, and ≈ 140 cm of water pressure head (≈ 0.014 atm, or 1.4 kPa) at the output and 29 cm of water vacuum at the intake. The pumps were tested for leak rates over four months, at which point the leak rate for 25 cm of water pressure was 10 μl/min, up from the initial value of 0.01 μl/min.

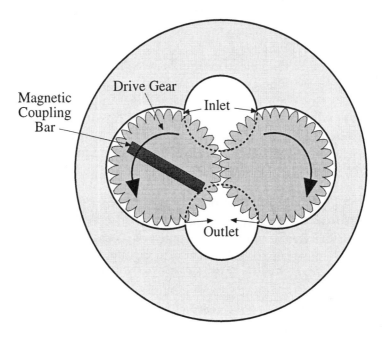

Illustration of a LIGA-fabricated, in-line, magnetically driven gear pump. Adapted from Dewa, et al. (1997).

6.5 ELECTROHYDRODYNAMIC PUMPS

Electrohydrodynamic (EHD) pumps have no moving parts and, like electroosmotic or electrophoretic pumps described below, make use of electric fields for pumping. EHD pumps make use of, and depend on, the electrical properties of the fluid being pumped (this method is not readily suitable for concentrated ionic conductors, such as most biological fluids, but EHD pumping of these fluids may be possible through scaling). Typically, the conductivity of the fluid to be pumped must lie in the range of 10^{-14} to 10^{-9} S/cm (Richter, et al. (1991a, 1991b)). An applied electric field is used to induce charge in the liquid and also to electrostatically move the induced charges (typically, hundreds of volts are required). Classic references in this field include Pickard (1965), Melcher (1966), and Steutzer (1968).

EHD pumping requires either free charges in the fluid (e.g., ions injected into the fluid by electrochemical reactions, hence the name *"injection-type"* EHD pumps) or a gradient/discontinuity in conductivity and/or permittivity in the fluid(s) to be pumped (the gradient/discontinuity may be achieved through layered fluids, suspended particles, or induced anisotropy, hence, the term *"non-injection"* EHD pumps).

6.5.1 INJECTION-TYPE EHD PUMPS

Injection-type EHD pumps rely on the fact that controllable concentrations of ions can be created from dielectric liquids at voltages above $\approx$ 100 kV/cm (not difficult to generate in micromachined structures, since this corresponds to 10 V/μm) at which point even previously insulating fluids can support nonlinearly increasing current with rising voltages. The current flowing under those conditions consists of *homocharges* (same polarity as the generating electrode) that arise through electrochemical reactions at the electrodes. The injected charges can then be acted upon by an electric field to pump the fluid.

Richter, et al. (1991a, 1991b) demonstrated a micromachined injection-type EHD pump consisting of two stacked perforated silicon dice, anodically bonded using sputtered 7740 Pyrex™ glass, and mounted on a ceramic substrate, as illustrated below. The 3 × 3 mm grids of 70 or 140 μm through holes in the 380 μm thick silicon wafers were etched using 33% KOH at 80°C and a PECVD silicon nitride masking layer. The 4 μm 7740 Pyrex™ bonding layer was reactively sputtered in 10% O_2/Ar and the anodic bond was formed at 450°C with 50 V applied, for a grid separation of 350 μm.

Ethanol, methanol, propanol, and acetone could be pumped, as could deionized water, although electrolysis (and bubble formation) interfered with pumping. At 700 V, a maximum pressure of 2.84 kPa was achieved, and in a separate experiment, a maximum flow rate of 14 ml/min was achieved at a pressure head of 420 Pa.

It should be noted that there is potential for considerably lowered operating voltages if the devices are further scaled down geometrically.

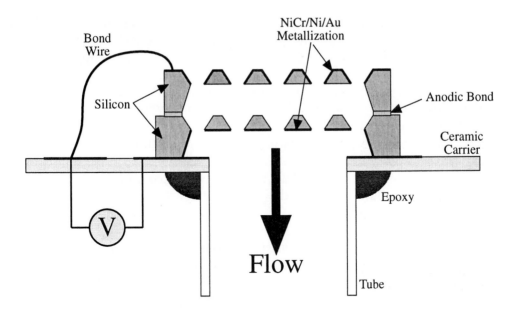

Illustration of a micromachined injection-type EHD pump (not to scale). After Richter, et al. (1991a, 1991b).

6.5.2 NON-INJECTION-TYPE EHD PUMPS

As explained by Fuhr, et al. (1994), there are several basic approaches to obtaining the required conductivity and/or permittivity gradients for EHD pumping (as an alternative to ion injection) and applying traveling wave potentials through multiple electrodes in order to pump fluid. A traveling wave of potential applied to multiple electrodes will pump the fluid, since the induced charges at the interface or through the gradient will lag the moving image charge at the electrodes, inducing a traction force on the fluid, as illustrated below.

A useful non-injection EHD pump approach is to induce a conductivity/permittivity gradient by establishing a temperature gradient across the flow channel, since typically *conductivity increases and permittivity decreases with temperature*. Thus the charge relaxation time, τ, decreases through the fluid toward the heated region, allowing for the desired charge lag,

$$\tau = \frac{\varepsilon_o \varepsilon_r(T)}{\sigma(T)}$$

(For example, for water, σ increases $\approx 2.2\%/°C$ and ε_r decreases $\approx 0.46\%/°C$.) A very interesting effect is that the direction of pumping depends on the velocity of propagation of the applied traveling wave.

Another approach to obtaining the desired charge lag is to use two different fluids (e.g., a gas layer atop a liquid layer) to create an outright discontinuity in those properties, but this may not be as practical for micromachined applications.

One can also use dielectric particles added to the fluid to enable pumping, since the particles can be polarized by the applied traveling wave of potential (referred to as traveling wave *dielectrophoresis*). If the delay between the applied potential and the particle movements is greater than the time for one-half of the traveling wave to pass, the particles will be pumped with the fluid flow. If the delay is less than the time for one half of the wave to pass, the particles will be pumped against the fluid flow. If the density of particles in the fluid is high enough, the fluid will also be pumped.

Alternatively, in some situations (e.g., conducting fluids), intense electric fields near the electrodes can give rise to local temperature gradients that can induce sufficient conductivity/permittivity gradients to allow for pumping, although this is highly geometry-dependent (stability of pumping improves with miniaturization).

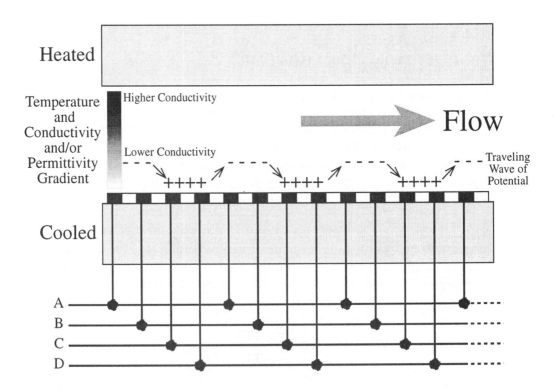

Conceptual illustration of non-injection EHD pumping using a thermal gradient, practically achievable on a microscopic scale using, for example, a thin-film heater.

An important point is that the necessary signal frequency (corresponding to traveling wave velocity) *increases in proportion to the conductivity of the solution*, and the optimal frequency is a function of the fluid to be pumped.

A thorough review of the underlying theory can be found in Fuhr, et al. (1994), who (in the same paper) demonstrated micromachined EHD pumps based on multiple 1 μm thick electroplated gold electrodes on quartz or oxidized silicon wafers, sealed from above using a Plexiglass™ cover with machined fluid transport channel(s). The streaming velocity of suspended latex particles was found to vary with the applied voltage as $V^{2.5}$, between the minimum voltage at which pumping was seen (10 V) to the maximum that could be applied (40 to 50 V, limited by bubble formation through electrolysis). Various frequencies were tested, and the predicted proportionality of drive frequency to solution conductivity was verified.

It should be noted that for non-injection EHD pumps, a likely approach to application in micromachined systems might be through the use of integrated thin-film heaters to create thermal gradients, and the electrodes can be insulated with a thin-film dielectric (hopefully with high ε_r value, optimally close to that of water (80) to prevent large decreases in field strength) to prevent electrolysis. While this type of pump cannot generate large differential pressures, it should be quite adequate for many integrated fluidic systems. For further information, another useful reference on micromachined, non-injection EHD pumps is Bart, et al. (1990).

6.6 ELECTROOSMOTIC/ELECTROPHORETIC PUMPS

As for electrohydrodynamic pumps, there are no moving parts required for *electroosmotic* or *electrophoretic pumping*. Rather than relying on injection of ions or gradients in conductivity/permittivity as in EHD pumping, electroosmotic or electrophoretic pumping relies on the continuous presence of ions in a suitable solvent. Electrophoretic/electroosmotic pumping can be used to move liquids or ions at flow rates in the range of microns up to millimeters per second for a wide range of flow channel cross-sectional areas. Electrophoresis is essentially the movement (drift) of ions relative to solvent molecules under an externally generated electric field, and it is the ions themselves that are "pumped." As discussed by Manz, et al. (1994), this approach can be used to move or switch the direction of "plugs" of ions through flow channels on a fluidic device, all under direct electrical control.

The underlying physical principle of electrophoresis is that the velocity of an ion of charge z_i, in a homogeneous medium, is proportional to the applied electric field, as given by,

$$v_i = \mu_i{}^* E$$

where μ_i^* is the mobility of ion species i, in $cm^2/(V \cdot s)$ (note that this is not viscosity, but rather, a function of viscosity), and,

$$\mu_i{}^* = \frac{z_i e}{f_i}$$

where q is the charge on the electron = 1.602177×10^{-19} C, and f_i is the friction coefficient of ion species i, in $(V^2 \cdot s)/cm^2$.

The friction coefficient is given by Stokes' law for a radius r_i, and gives a velocity equation of (noting that, potentially confusingly, μ here represents viscosity, not mobility),

$$v_i = \frac{z_i e}{6\pi \, r_i \, \mu} E$$

Under an applied electric field, ions will move at characteristic velocities determined by their charges, radii and mobilities, and this is the basis for electrophoretic separation of a sample into its constituent species for analysis.

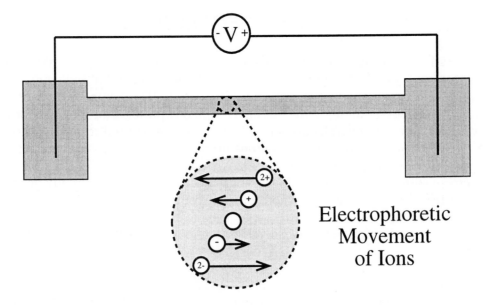

Conceptual illustration of electrophoresis. After Manz, et al. (1994).

For electrophoretic separations, the spatial resolution of the separated ions cannot exceed that of the initially injected "plug" (it is always lower in practice, due to diffusion and other effects tending to spatially spread the ions), and thus it is desirable to inject very small plugs into the separation channel.

If electrophoresis is carried out in a medium that physically interacts with the migrating ions (such as a gel or micromachined structures in the flow path), the friction coefficient is no longer given by Stokes' law and becomes dependent on the nature of the medium (e.g., the porosity of a gel) and the size/shape of the moving molecules. In such cases, the physical interactions between the medium and the ions can offer higher separation resolution than for simple liquid media.

Electroosmosis is more of a *bulk* phenomenon than electrophoresis, and results in pumping of electrically neutral fluid, but can only occur if there are immobilized charges on the walls of the flow channel (due to ionized chemical species on the wall itself or adsorbed charges). As for any fixed charges in ionic solutions, a *double-layer* (as discussed in the Chemical and Biological Transducers chapter) of opposite charges is formed in the solution near the channel walls, and the charges in this double layer can then be moved under the influence of an externally applied electric field, as in electrophoresis. The only region where there are significant (relative to the buffer) concentrations of charges is this double layer. Thus the thin layer near the channel walls will move and, in doing so, will osmotically draw solvent along with it.

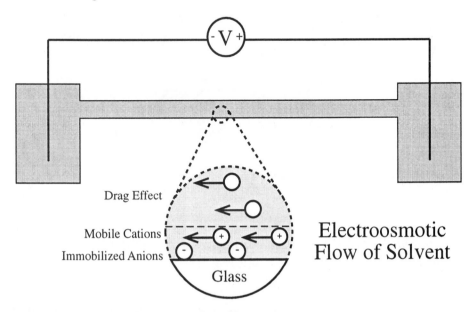

Conceptual illustration of electroosmotic pumping. After Manz, et al. (1994).

As for electrophoresis, the rate of electroosmotic pumping is also linearly dependent on the applied potential. A factor of critical importance in electroosmotic flow is that, unlike pressure difference induced flow, with a parabolic velocity profile in a channel, *electroosmotic flow yields a rectangular flow profile*, allowing "plugs" of fluid to be transported without much geometric distortion.

Another related type of pumping is the use of switched electrophoretic fields to physically manipulate particles. An interesting example is given by Moesner and Higuchi (1995) in which they use arrays of microelectrodes to move particles in the 5 to 400 μm range in controllable patterns using a technique akin to that involved in moving charge in CCDs (creating potential wells into which the particles are electrophoresed). By forming moving potential wells on the surface of the electrode array, the particles can be physically swept along well-defined paths. (Integrated electrophoresis systems are further discussed below.)

6.7 ULTRASONIC PUMPS

Ultrasonic, or *traveling wave* pumps, are useful in situations where a (large) pressure head is not necessary. Given that limitation, such pumps can be used to move or mix fluids with simple mechanical structures.

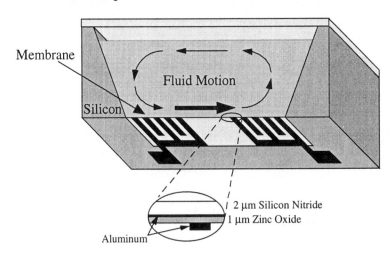

Cross-sectional illustration of an ultrasonic mixer/pump device. After Moroney and White (1991).

Moroney and White (1991) demonstrated ultrasonic mixers/pumps that consisted of a KOH-etched tub and LPCVD silicon nitride membrane with an aluminum ground plane, RF magnetron-sputtered zinc oxide piezoelectric actuator layer, and interdigitated aluminum drive/sense electrodes. With a 4 μm total membrane thickness and electrodes spaced at 100 μm, the flexural plate waves launched traveled at velocities of 100 to 500 m/s, for a drive frequency of 1 to 5 MHz.

They demonstrated (using 2.5 μm polystyrene spheres) that fluid was transported in the direction of wave propagation and at a velocity proportional to the square of the acoustic wave amplitude. The maximum speed they reported was 130 μm/s for a drive voltage of 7.1 VRMS at 3.5 MHz. They also demonstrated that ultrasonic drive can be used both to induce *mixing* and to pump red blood cells (many such ultrasonic pump designs can also be used as mixers).

Miyazaki, et al. (1991) discussed the possibilities of implementing an ultrasonic pump using progressive flexural waves applied to the walls of a thin metal pipe (this approach may be suitable for a fully micromachined ultrasonic pump).

6.8 VACUUM PUMPS

To date, it appears that no true vacuum pumps have been implemented via micromachining. Nonetheless, several groups are pursuing integrated mass spectrometers, micromachined electron microscopes, etc., and micromachined vacuum pumps would be quite useful in these applications.

Low Vacuum	760 torr (1 atm) to 25 torr
Medium Vacuum	25 to 10^{-3} torr
High Vacuum	10^{-3} to 10^{-6} torr
Very High Vacuum	10^{-6} to 10^{-9} torr
Ultrahigh Vacuum	10^{-9} torr and below

Table of typical ranges of vacuum, as generally used in vacuum engineering.

The above table outlines the commonly used definitions of the different ranges of vacuum given in torr (mm Hg), which, as a unit of measure is technically obsolete, but is still generally used in industry (pressure conversion constants are shown in the table below). In general, a "high vacuum" is said to exist if the mean free path of gas atoms is much greater than the linear dimensions of the vacuum vessel, and the mean free path, λ, can be approximated (ignoring temperature and molecular size factors) as,

$$\lambda = \frac{5.0 \times 10^{-3}}{P(torr)} \quad \text{in cm}$$

In low vacuum regimes (higher pressures), flows are viscous, transitioning to molecular flow in the high vacuum regions. In higher vacuums, molecules are primarily on the surfaces and can move freely (with little or no mutual interference).

It is interesting to note that today's vacuum technology covers a range of ≈ *19 orders of magnitude,* with the lower limit decreasing as technology improves.

Table of Pressure Conversions							
Units	Pa (N•m^{-2})	dyn•cm^{-2}	bar	atm	torr (mm Hg)	inches Hg	ft. H$_2$O (4°C)
1 Pa (N•m^{-2})	1	10	10^{-5}	9.869×10^{-6}	7.501×10^{-3}	2.953×10^{-4}	3.3456×10^{-4}
1 dyn•cm^{-2}	0.1	1	10^{-6}	9.869×10^{-7}	7.501×10^{-4}	2.953×10^{-5}	3.3456×10^{-5}
1 bar	10^5	10^6	1	0.9869	750.0617	29.530	33.456
1 atm	101,325.0	1,013,250	1.023250	1	760	29.9213	33.90
1 torr (mm Hg)	133.3224	1,333.224	1.333×10^{-3}	1.316×10^{-3}	1	0.0394	0.0446
1 inch Hg	3,386.388	33,863.88	0.03386388	0.03342105	25.4	1	1.13295
1 ft. H$_2$O (4°C)	2,988.98	29,889.80	0.0298898	0.0294989	22.42	0.882646	1

Table of pressure conversion factors. After Beni, et al. (1990) and the Lucas NovaSensor pressure converter slide rule (1994).

Vacuum pumps can be grouped into two categories:

1) *Compression* or *gas transfer* types that operate by removing gas from the pumped volume through one or more stages of compression — rotary, roots, ejector, turbomolecular and diffusion pumps are in this category.

2) Pumps that operate via *condensation* or *chemical binding* of gas molecules — sorption, ion, and cryopumps are in this category (as are "getters," which are chemicals that are heated or otherwise activated to scavenge gas molecules by reacting with them).

In theory, either of the two approaches could be applied to micromachined vacuum pumps, but for continuous operation, it seems more likely that the devices will be of the compression or gas transfer type. The operating ranges of typical vacuum pumps are illustrated in the chart below.

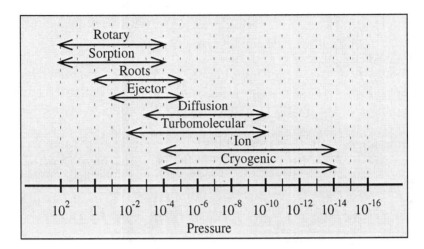

Approximate pressure ranges of operation for major types of vacuum pumps. After Beni, et al. (1990).

7. DROPLET GENERATORS

As a lower-cost alternative to laser printers, ink-jet printers have become extremely popular, and are currently available in color models, with resolution approaching that of conventional photographs. The basic principle of such printers is to use precisely generated and timed droplets of ink to form characters on paper. There are two major categories of ink-jet printing systems, based on whether or not ink droplets are continuously ejected, namely, continuous or "drop-on-demand."

Some early micromachining work focused on the development of ink-jet printers, such as the use of pyramidal anisotropically etched pits in silicon to form nozzles (Bassous (1975), Bassous, et al. (1977), and Bassous and Baran (1978)), the use of grooves through (110) silicon with doped sidewalls as controllable means to electrostatically charge ejected droplets (Bassous, et al. (1977)), and the sandwiching of isotropically etched nozzles and anisotropically etched ink cavities along the surface of (110) silicon between an underlying thick glass plate and an overlying thin glass plate with a piezoelectric actuator for ejection (Petersen (1979)).

A great deal of ink-jet printer development was also underway at Hewlett-Packard, Inc., as well as Canon, Inc., and the history of the HP efforts (including the details of turning the basic invention into a product) are described in Nielsen (1985), Allen, et al. (1985), Bhaskar and Aden (1985), and Siewell, et al. (1985).

The HP drop-on-demand ink-jet technology relies on thin-film resistors to rapidly form a bubble of vapor from the ink beneath an exit nozzle (reported bubble formation times were on the order of 1 µs, with peak pressures of 14 atm and a lifetime of $\approx$ 20 µs). The bubble, acting as a piston, displaces fluid away from itself, and if the nozzle is well designed, much of the fluid is ejected out the exit hole of an orifice. Since the vapor bubble formation is reversible, as the heater resistor is turned off, passive cooling causes the bubble to collapse rapidly, but by this point a bubble of ink would have been ejected through the orifice (at a velocity on the order of 10 m/s), and the bubble's collapse serves to draw ink into the assembly from a reservoir. For reliable ink droplet ejection, the temperature at the heater must reach $\approx$ 90% of the ink's critical temperature (beyond which the liquid phase no longer exists), $\approx$ 330°C as reported for the early ThinkJet™ inks.

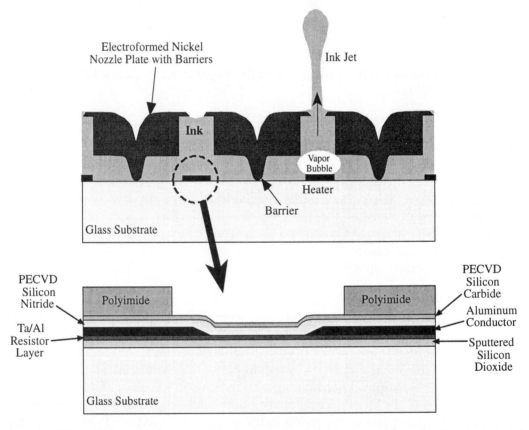

Illustration of Hewlett-Packard ink-jet devices, showing bonded electroformed nickel nozzle plate and multi-layer thin-film heaters. After Allen, et al. (1985) and Bhaskar and Aden (1985).

The 1985 HP articles describe the fabrication of the ink-jet heads by bonding an electroformed (plated onto a reusable template, or "mandrel" and then released from it) nickel orifice plate onto a glass substrate containing TaAl thin-film resistors. The electroformed orifice plates were made by using an etched stainless steel mandrel, on which dry film photoresist was laminated and patterned, followed by nickel electroplating and a final step in which the electroplated nickel was peeled away from the mandrel (it is weakly adherent since the steel has a surface oxide). Barrier regions of the orifice plate serve to hydraulically isolate individual orifice regions. In keeping with this early work, current Hewlett-Packard ink-jet print heads do not contain any circuitry, and are disposable.

Smith, et al. (1994) used a combination of dry etching, boron diffusion etch stop and EDP etching to realize conical orifice structures for a continuous ink-jet print head. For such electrostatically deflected ink jets, directional control of the droplets is critical, and by making conical shaped orifice structures, the longer channel lengths helped to improve the alignment of the ejection paths of the droplets.

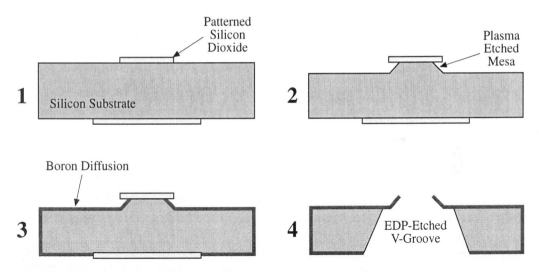

Illustration of a fabrication method for conical ink-jet orifice structures. After Smith, et al. (1994).

Krause, et al. (1995) demonstrated CMOS-compatible, micromachined drop-on-demand ink-jet printer heads that were being combined with on-chip control circuitry. They used (110) silicon in order to allow for the etching of 100:1 vertical ink channels with KOH (40 to 50 wt% at 80°C for > 90:1 aspect ratio). They fabricated thin membranes of silicate glass and thin-film metals (Ti/Cu/Ni/Au) under which were located hafnium diborane (HfB_2) thin-film heaters (while the details of this unusual choice was not explained in their paper, perhaps it allowed for optimization of the thermal properties).

In operation, < 6 μs, 6 mW electrical pulses are used to heat the ink quickly, form vapor bubbles, and eject a single droplet per pulse. While the details of the fabrication process are covered primarily in a related patent application, it is available for perusal since applications in Europe are "laid open" when filed (Berghof, et al. (1992)).

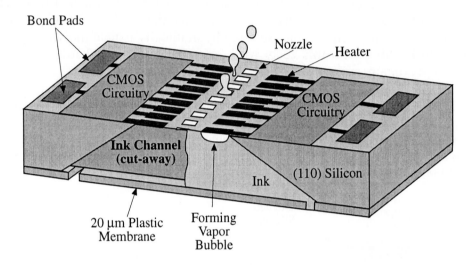

Illustration of a fully integrated CMOS ink-jet print head design. After Krause, et al. (1995).

In a very different application of controlled droplet formation, Wallman, et al. (1995) demonstrated a micromachined, piezoelectrically actuated flow-through cell for ion exchange chromatography system. The eluent from the ion exchange column was sampled to form picoliter-range droplets that were deposited on an electrophoresis gel moved under the sampler. The electrophoresis gel was then run in a conventional manner, allowing for electrophoresis of the linear array of droplets without undue use of the ion exchange eluent. Further work by this group is reported in Nilsson, et al. (1996).

In some cases, droplets may be required in large numbers but without the need for position or timing control (e.g., an atomizer). Kurosawa, et al. (1995) demonstrated a surface acoustic wave atomizer based on y-cut LiNbO$_3$ that could form a fine mist from fluid pumped onto its surface at a maximum rate of 0.1 ml/min.

8. OTHER DEVICES

A variety of fluidic devices that are difficult to classify in any of the above categories are discussed in this section.

8.1 CONTROL OF MACROSCOPIC FLOWS WITH MICRODEVICES

Coe, et al. (1994, 1995) demonstrated a micromachined array of electrostatically (or external piezoelectrically) actuated gas jets. It has been proposed to use such jets to control macroscopic flows such as those over control surfaces of aircraft wings or exiting attitude control jets. The basic idea is to exploit hydrodynamic instabilities at flow boundaries such that a small amount of input power from the microjets controls a much larger amount of energy.

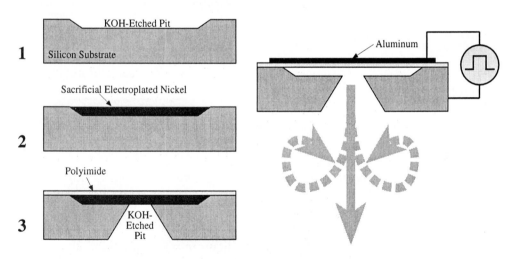

Illustration of an electrostatically actuated microjet. After Coe, et al. (1995). The process steps are shown at the left, and device operation at the right.

The microjets were fabricated via double-side-etched (using KOH with SiO_2 masking) holes through a silicon substrate with a metallized polyimide film fabricated on one side so that the film could be deflected toward the substrate electrostatically, ejecting a jet of gas out the other side. In order to allow the polyimide to be spun on the wafer after through-hole formation, the shallower region of the hole (for electrostatic actuation) was temporarily filled with an electroplated nickel sacrificial layer that was later wet etched away.

An alternative approach to macroscopic flow control was described by Liu, et al. (1995), using magnetically actuated flaps that can extend into the boundary layer of the flow stream of air surrounding a delta wing (if a suitable, asymmetrical pattern of such interaction is set up, rolling forces can theoretically be generated on a tail-less, macroscopic delta wing).

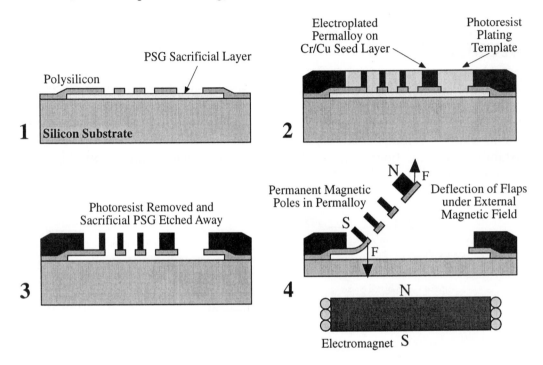

Illustration of the fabrication and operation of permanent magnet flap actuators. After Liu, et al. (1995).

Their approach was to fabricate polysilicon moving members above a 3 μm PSG sacrificial layer and a 0.5 μm phosphorous-doped PSG layer (for in turn doping the polysilicon). A seed layer (20 nmCr/180nm Cu) was evaporated on the wafer, followed by application and patterning of a 5 μm photoresist template for electroplating. Under an external magnetic field, $Ni_{80}Fe_{20}$ permalloy was electroplated on the seed layer, after which the photoresist is removed and the PSG was etched away in 50% HF (minimal or no attack of permalloy, Cu or Cr). Apparently, an in-house rinse/drying process was used to ensure release, but the authors did not discuss it. Under the external magnetic field of an electromagnet, the flaps could be deflected (over 60° deflection was achieved at an external magnetic field of 800 gauss) with generated forces on the order of 100 μN and rolling moments on a miniature delta-wing prototype were demonstrated.

8.2 PARTICLE FILTERS AND TRAPS

As mentioned above, many micromachined pumps, valves, and other structures are quite susceptible to particulate contamination, and thus the use of particulate filters will be very important for many microfluidic systems that may deal with "real-world" samples. In some cases, the utility of micromachining such filters is not clear, since a wide variety of conventional filters are available commercially at low cost and high performance. However, micromachining techniques can certainly be used to produce filters with very narrow statistical distributions of pore sizes, as compared to traditional filters.

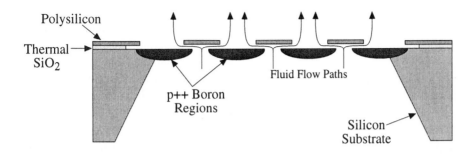

Illustration of a micromachined particle filter. Adapted from Kittilsland, et al. (1990).

Micromachined filters may be useful where not prohibited for cost reasons or where they must be integrated with another silicon structure. A wide variety of micromachining techniques could be used to fabricate filter structures. For example, Kittilsland, et al. (1990) demonstrated such a filter using a perforated, suspended 1.5 μm polysilicon membrane above a sieve structure of heavily boron doped silicon. The holes in the polysilicon and the heavily doped silicon were offset so that there was no direct path through any corresponding pairs on both layers. This forced fluid to take an indirect path between the layers, and the layer-to-layer spacing determined the maximum particle size that could pass. Their designs had 10 μm holes in both layers, and an interlayer spacing that was determined by the thickness of a sacrificial thermal SiO_2 layer between the polysilicon and the boron-doped regions (they fabricated gaps as narrow as 50 nm). In some cases, a thin ($\approx$ 10 nm) thermal SiO_2 layer was grown on the entire released structure to make it more hydrophilic (the unmodified structures were hydrophobic).

In addition, van Rijn and Elwenspoek (1995) demonstrated micromachined sieve-type filters with pore sizes between 0.5 and 10 μm in diameter through 1 μm thick LPCVD silicon nitride membranes. Further work by van Rijn, et al. (1997)

provided models and experimental results regarding the deflection and maximum pressure loads for such membrane filters.

Alternative approaches to membrane filters have also been demonstrated in micromachined formats, including the use of deep reactive ion etched structures to block particulates and narrow vertical slits between bonded wafers (see Brody, et al. (1995)) to provide size selectivity.

In some cases, particles are deliberately introduced into a fluidic stream. For example, many biological assays make use of small beads of glass or other materials (e.g., magnetic) that are coated with specific (or generic, in some cases) binding molecules. The beads are mixed into a solution being investigated, and bind to molecules of interest, for example, nucleic acids. The beads are then somehow separated from the bulk solution and the target molecules are either used in place or released for further processing.

The beads are often made of ferromagnetic materials, and are separated from the bulk fluid using localized magnetic fields, which temporarily induce magnetic dipoles in the particles. Once the particles are captured, the external magnetic fields can be turned off, releasing the particles. Ahn, et al. (1996) and Liakopoulos (1997) demonstrated planar electromagnetic coils integrated with micromachined fluidic channels to form such magnetic bead separators. Using organic templates on a glass substrate, the electromagnet coils were formed using electroplated Cu, with NiFe (Permalloy) flux concentrating cores electroplated in a subsequent step. The glass substrate was then anodically bonded to a silicon substrate into which the fluidic channels had been previously formed using KOH anisotropic wet etching. The completed magnetic separators, each with three separately controllable coils, were used to capture 0.83 µm beads (10% solid content, dispersed in water) in a few drops of fluid, using a current of 300 mA for 10 s. While external magnetic coils can certainly be fabricated at low cost for this purpose, there may be applications in which full integration of such coils may be advantageous.

9. MICROFLUIDIC SYSTEM ISSUES

Unlike several other types of micromachined transducers (e.g., mechanical or optical), where the transducers often are handled similarly to conventional integrated circuits (e.g., soldered onto printed circuit boards), the construction of microfluidic systems presents a unique set of problems. For present micromachined systems, there are no fluidic interconnect analogs of bond wires or ribbon cables, although substrate-based interconnect schemes have been researched. Overall packaging for single transducers or entire systems is still in its early stages of development.

9.1 INTERCONNECTS

Several methods have been experimentally used to join together the various elements of compound microfluidic systems. One potential interconnect scheme involves the use of external tubing to form single-point fluidic interconnections, as shown conceptually below.

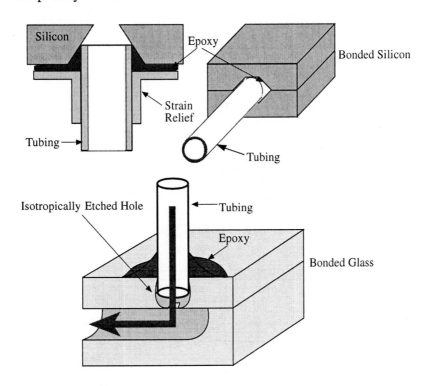

Illustration of conceptual off-chip fluidic interconnect schemes (top two figures after Krulevitch (1995)).

Off-chip fluidic interconnect schemes have been used (some examples are illustrated above) in various forms in many devices demonstrated to date, but so far, most have required precision manual fabrication and are not readily suited to mass production. Examples of hand-fitted and glued off-chip interconnects can be seen in the majority of microfluidics papers, and some (such as that described by Spiering, et al. (1997)) are designed to minimize dead volume (spaces in which fluid can be trapped or slowed from rejoining the main flow, potentially leading to unintentional contamination after a compound has moved through a region).

A potentially multi-point fluidic interconnect scheme involves the use of stacked substrates with various interconnected flow channels, analogous to a multilayer electrical printed circuit board. One example of this approach is that described by Poplawski, et al. (1994) and illustrated in Section 2 above.

Mensinger, et al. (1995) proposed a micromachined fluid reactor with access ports for optical sensing at the output. In their design, several layers of PMMA or polycarbonate were embossed to form channels and layer-to-layer interconnects and were bonded together by melting. Further system design ideas along these lines were discussed by Möbius, et al. (1995).

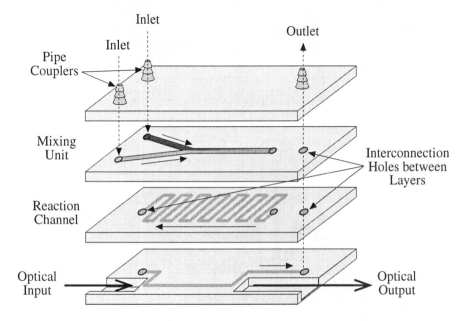

Illustration of integrated fluid reactor fabricated from stacked organic substrates with channels and through-holes. After Mensinger, et al. (1995).

Another example of stacked interconnecting layers for microfluidic systems (demonstrated on a macroscopic scale using machined Plexiglass™), including a valveless flow injection system (external piston pumps were used to control the fluid streams), was described in Fettinger, et al. (1993). A similar approach involved bonding individual fluidic components to a substrate (which may be a stack) in which interconnecting channels are prefabricated (as discussed by Bergveld (1995)).

Another approach toward minimizing interconnect complexity is the use of x-y addressing schemes (e.g., multiplexers or cross-point switches). Pan, et al. (1997) demonstrated both pneumatically and electrostatically latched, multiplexed valve manifolds. In the pneumatically latched case, the simultaneous application of a pneumatic "storage enable" signal and "data signal" to a given valve address allowed it to be programmed into an open or closed state, with the "data" stored as pressure. In the electrostatically latched case, the valves were actuated pneumatically, but the "memory" was provided by electrostatic hold-down signals. In both cases, successful demonstration of the concepts was achieved, showing that simple, local memory devices could be used to control flow in large fluidic manifolds. This approach may have utility in situations where the complexity and parallelism of fluidic systems is large.

These approaches appear to be leading toward some technologies that are analogous to the printed circuit boards and multiplexed signal routing for conventional electronics, and which will hopefully confer similar advantages in terms of simplifying system interconnects. However, even if multi-point interconnects such as these are used on such substrates, there must still be connections to the outside world.

9.2 PACKAGING/SYSTEM INTEGRATION

As mentioned above, packaging for microfluidic devices is typically case-by-case customized and generally requires ports to and from the outside world. Well-designed packaging minimizes dead volume without affecting performance. Some devices built to date have used modified transistor or other multi-pin metal packages. Snap-together plastic packages have also been applied, and recently injection-molded packages (over metal lead frames) have been used for some devices. It appears that, at present, these packaging methods still require a gluing or sealing step of some sort. For example, the thermal bimorph valve of Barth, et al. (1994) was packaged using a two-part metal package and an epoxy hermetic seal, as illustrated below.

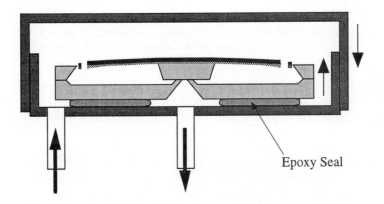

Epoxy Seal

Illustration of packaging for a thermal bimorph valve. After Barth, et al. (1994).

Schomberg, et al. (1994) presented a stacking scheme (illustrated below) that combined packaging and interconnect functions by stacking and sealing multiple fluidic modules.

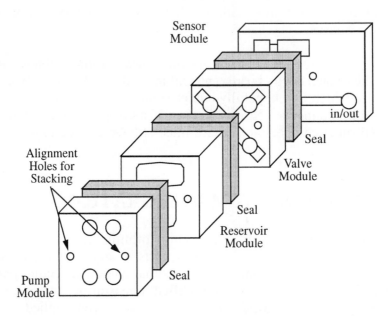

Illustration of a stacked fluidic module scheme. After Schomberg, et al. (1994).

A similar approach was used by van der Schoot, et al. (1994) to package an ISFET-based flow cell, complete with macroscopic piezoelectric micropumps and an underlying printed circuit board for electrical interconnects. The system is illustrated above. They used a photolithographically patterned polysiloxane ring to ensure a good seal between the silicon ISFET array chip and the module stacked above it. The system used one pump to wash the ISFETs with calibrant solution and another to pull a sample onto the array, with the metal inlet tube being used as an electrochemical pseudo-reference electrode for the ISFET array. Further discussion of this fluidic stack approach can be found in Verpoorte, et al. (1994). Hoffmann and Rapp (1995) also described a fluidic system integration approach based on the combination of plastic and silicon (or other) components into stacked assemblies.

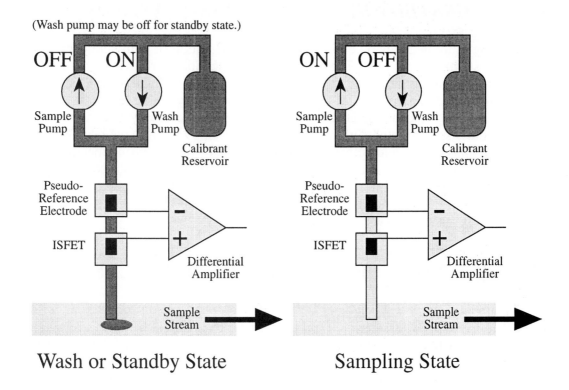

Illustration of a washing/calibration (left) and sampling (right) scheme for an integrated chemical analysis system. After van der Schoot, et al. (1994).

9.3 DESIGN FOR DISPOSAL OR REUSE

Whether a microfluidic device (or system) is to be disposed of after one or more uses, or used "permanently," has considerable impact on system design. In some applications, such as medical, where reuse may require resterilization and potentially other steps, disposable devices are often quite attractive. For reusable devices, dead space, purging, washing/carryover, repeatability, operational lifetime, filtering, and other issues may be critical. For disposable devices, these issues may not be very important, and cost considerations will probably dominate.

10. INTEGRATED CHEMICAL ANALYSIS SYSTEMS

There is tremendous potential for the implementation of complete chemical analysis systems "on a chip." There are many applications for devices that could identify a wide variety of unknowns (e.g., carry out environmental testing or DNA sequencing) or perhaps synthesize desired products from available reactants (i.e., a "chemical factory on a chip"). Gas/liquid chromatography, capillary electrophoresis, and even mass spectrograph systems (Nathanson, et al. 1995) all have been (or are being) developed, and could potentially incorporate many of the sensing mechanisms discussed above, as well as acoustic/electrophoretic transport, localized heaters, and others. There is a multitude of possible chemical analysis system applications of micromachining, a complete survey of which is beyond the scope of this book. However, a few examples are presented below.

10.1 SCALING ISSUES FOR CHEMICAL ANALYSIS

For chemical analyzers/detectors, there are some key considerations with respect to the effects of sampling and detection. If chemical samples are taken from sources with low concentrations of target molecules, an important issue is *whether or not sufficient numbers of those molecules are obtained to make a measurement.* As sensor and system sizes decrease, so do the probabilities that they will come into contact with target molecules if no active "gathering" or amplification methods are used. For example, in an environmental monitoring application, one can have a sensor passively exposed to molecules drifting by, or actively gathering samples (e.g., pumping fluid through) with some way to sense or trap molecules (i.e., preconcentrate them). In some cases, such as detection/analysis of nucleic acids, amplification methods exist (such as the polymerase chain reaction, discussed below) that can greatly increase the number of target molecules by replicating them.

It is important to consider such scaling effects on the minimum detectable concentration, as discussed by Manz, et al. (1990a). This issue is simply that with a perfect detector (able to detect a single target molecule), as the concentration of the target molecule decreases, the volume that must be sampled by the detector increases. The volume containing a single molecule (assuming uniform distribution of molecules) is given by,

$$V_{sm} = \frac{1}{CN_A}$$

where C is the concentration of molecular species, in mol/l, and N_A is Avogadro's number = 6.02214×10^{23} molecules/mol. For example, if one has a sensor that addresses a $10 \times 10 \times 10$ μm cubic region (e.g., the cross section of a square flow channel over a comparable length), this is a volume of 1 pl, giving a theoretical single-molecule detection limit concentration of 1.7×10^{-12} mol/l.

Similarly, one can determine the *spatial resolution* of a given sensor since a detector with a given response time (time to respond if molecule is at the same location as the detector) will have a spatial resolution that is limited by the diffusion rate of target molecules (i.e., if detection takes a certain amount of time, the detector is "blind" to molecules that could not reach it quickly enough to be sensed). This means that as the response time of a molecular sensor decreases, the effective volume it is addressing decreases (this may not be a problem for micromachined systems, since the volumes decrease greatly).

If amplification of the target molecules or some form of amplification within the sensor is possible (e.g., fluorescent detection, wherein a single molecule can produce a large number of fluorescent emissions, increasing the SNR), real-world sensitivities can be very impressive (however, the fundamental limitation is still having at least a single molecule to work with).

Clearly, many other scaling issues affect micromachined transducers and structures. These are generic (such as thermal scaling) or domain-specific (such as chemical detectability issues), and can typically be at least estimated from first principles. *By examining such matters, one can often make critical decisions regarding the utility of micromachining devices.*

10.2 GAS CHROMATOGRAPHY SYSTEMS

The basic idea of gas chromatography is to pass an unknown mixture of gases through a narrow column with known surface characteristics, and as the gas mixture (plus a carrier gas) moves, interactions with the column walls give rise to differing effective mobilities of the individual gases. Thus each constituent gas reaches the end of the column at a different time, allowing them to be identified on the basis of these differing mobilities. The gases reaching the end of the column can be detected by their thermal conductivities (which can differ greatly).

Terry, et al. (1979) demonstrated an entire gas chromatography system on a single two inch wafer. The column was formed by wet isotropic etching using a silicon dioxide mask. A Pyrex™ cover was anodically bonded on to seal the column. An etched silicon valve body, combined with a macroscopic solenoid attached to the wafer, provided the ability to admit a "pulse" of an unknown gas at the front of a stream of carrier gas.

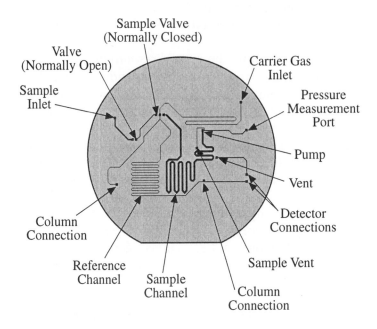

Illustration of a micromachined gas chromatography system. After Terry, et al. (1979).

The thermal conductivity detector used was a 100 nm nickel-film resistor on a sputtered Pyrex™ membrane, heated with a current source. The voltage across the detector directly indicated the local thermal conductivity (making use of the known temperature coefficient of resistance of the nickel film).

10.3 LIQUID CHROMATOGRAPHY SYSTEMS

Manz, et al. (1990b) demonstrated a micromachined open column liquid chromatography device (a mobility-based separation scheme based on interactions of a gas or liquid with the surfaces of a long separation column). Using oxide/nitride/oxide masking and an HNA etchant, isotropic column walls were etched into a silicon substrate, which was later oxidized and the column was sealed by anodically bonding a glass wafer above the oxide. A trio of thin-film platinum electrodes were used as a conductimetric detector (the cited reference does not give results for the entire system).

10.4 ELECTROPHORESIS SYSTEMS

Electrophoresis is a powerful chemical analysis technique that makes use of the different mobilities of ions to separate and identify them under electric-field drive. The basic principles of this approach, which can be used to identify nucleic acids, proteins, and a number of other types of molecules, are discussed in Landers, et al. (1993) and Kuhr and Monnig (1992).

Several micromachined electrophoretic devices have been discussed and/or demonstrated, mainly fabricated through anisotropic etching of glass (see Sethi (1987), Pace (1990), Sethi, et al. (1993), Clark, et al. (1993), Jacobson, et al. (1994a, 1994b, 1994c), Harrison, et al. (1993a, 1993b, 1994), Burggraf, et al. (1993), Manz (1994), Manz, et al. (1994, 1995)). The reader is also referred to interesting patents on the subject, issued to Sethi, et al. (1987, 1990) and Pace (1990). Electrophoretic sample injection, mixing, on-column reaction and detection have all been demonstrated (see, for example, Chiem, et al. (1997)).

Woolley and Mathies (1994, 1995) used such lithographically fabricated electrophoresis channels in glass to achieve separations of DNA molecules from 70 to 1,000 bases in as little as 120 s with 3.5 cm long channels. They used a hydroxymethyl cellulose sieving matrix in the channels to improve fractionation of the DNA fragments, which were obtained by using restriction enzymes to cleave the sample DNA into varying lengths. The separations were approximately ten times faster than comparable separations done using conventional capillaries ten times longer, yet had comparable resolution. The Woolley and Mathies (1994) paper gives extremely detailed fabrication and procedures for using the arrays, and is a key reference. They carried out extensive work to verify the reproducibility of this approach for DNA separation, to compare electrophoretic sample injection methods (as discussed below), and to examine effects of field strength on separation quality. Such separation channels could be fabricated in large arrays to enable simultaneous separation and sequencing of large numbers of DNA samples.

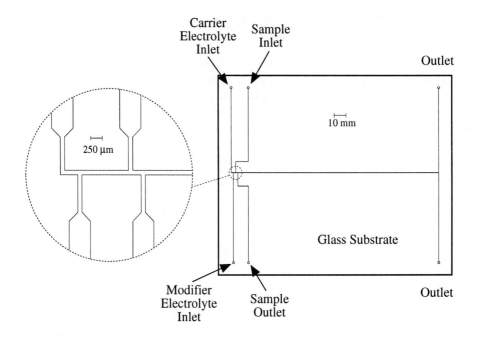

Example layout of a micromachined capillary electrophoresis device. After Manz, et al. (1995).

Harrison, et al. (1994) and Jacobson, et al. (1994b) demonstrated the use of multiple flow channels and applied potentials to obtain electrical control by gating a "plug" of sample (from an injection channel crossing the separation channel) into the separation channel. A key advantage of this approach is that, if the applied potentials are appropriately controlled during the separation, leakage flow from the injection channel can be reduced or eliminated, increasing the SNR of the separation. As mentioned above, Woolley and Mathies (1994) carried out a thorough comparison of such electrically controlled injection methods. Bousse, et al. (1997) demonstrated an extension of this approach to convert a single stream of sample into multiple parallel streams.

Detection of the electrophoretically separated plugs of each ion species is typically done using fluorescence (a dye is often coupled to the species of interest), optical absorption, or electrical conductivity measurements. Detection in most of the micromachined glass-based electrophoresis systems discussed above has been done using a fluorescence microscope (UV source and optical microscope with UV-blocking filter and imager or photomultiplier).

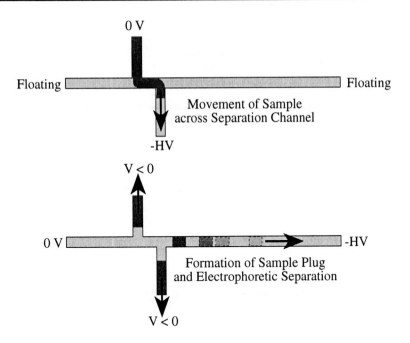

Illustration of electrically controlled sample plug formation and electrophoretic separation, as described by Harrison, et al. (1994).

A method of direct electronic detection is the use of conductivity or impedance measurement methods to detect the passage of slugs of separated ions by the change in conductivity at the detector (see Huang, et al. (1987) and Huang and Zare (1991)). A micromachined conductivity detector for capillary electrophoresis, coupling to two ends of a conventional capillary that was split into two pieces, was demonstrated by Reay, et al. (1994). The detectors were positioned along the capillary (which was used in a conventional capillary electrophoresis system), and since the electrodes were perpendicular to the high applied potential, they did not have any appreciable voltage across them and could reject common-mode noise in the high-voltage power supply.

The coupling and flow channel for the detector were fabricated using plasma etching into a silicon wafer and anodically bonding a glass cover plate with patterned Pt detector electrodes (electroplated with platinum black) and Au or Pt bond pads.

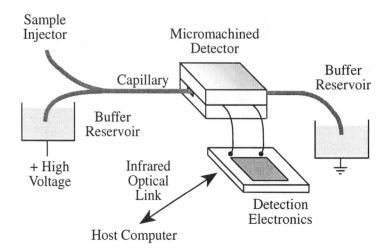

Illustration of the operation of micromachined capillary electrophoresis detector. After Reay, et al. (1994).

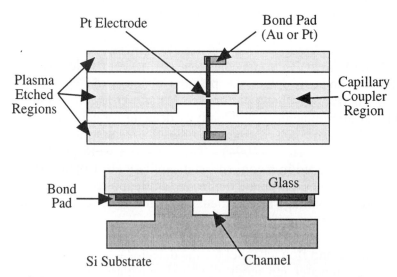

Illustration of the structure of a micromachined capillary electrophoresis detector (top view above, and cross section below). After Reay, et al. (1994).

A miniaturized sine-wave synthesizer, lock-in amplifier, and A/D converter were used to make the measurements, which were communicated to a host computer via a free-space infrared digital link, and yielded a detection limit better than 7×10^{-7} mol/l.

As mentioned above, the effective mobilities of molecules can be affected by the medium through which they are moving, which may be a gel and interact with long molecules such as DNA by physical hooking and then elongation/release. In an effort to better understand the physical interactions of molecules with gels and to investigate a more controllable alternative, Volkmuth and Austin (1992) etched 0.15 µm, 1 µm SiO_2 posts (spaced 2 µm center-to-center) on a silicon wafer and then anodically bonded a Pyrex™ cover plate to the posts. They were able to electrophorese fluorescently labeled (with ethidium bromide) 100 kbase DNA molecules (30 µm stretched length) through the post arrays and monitor the physical interactions in real time using video epifluorescence microscopy. They observed that as the DNA molecules traversed the array, they repeatedly became hooked on posts and slid off. They verified that the array of micromachined posts could separate DNA molecules on the basis of their lengths, as do gels, but without the need for an organic matrix. This approach has the potential to allow fabrication of varying-sized arrays of posts (potentially varying within a single array) to allow enhanced fractionation of DNA molecules on the basis of length. Applications of such structures in sorting cells, viruses, macromolecules, etc., can be found in the patent to Austin, et al. (1995).

Whole cells can also be transported using electrophoretic, electroosmotic, or electrohydrodynamic pumping. For example, Harrison and Li (1997) demonstrated the use of electroosmotic force to transport whole, non-adherent (i.e., not anchored to surfaces) cells, such as red blood cells, yeast, and *E. coli* bacteria. They also demonstrated electrokinetic mixing of cells with chemical agents that disrupted the cells (referred to as *lysis*) and made their contents available for analysis. An example of the use of such methods to perform physiologic assays on cells transported in such a manner was presented by Andersson, et al. (1997).

10.5 CELL FUSION DEVICES

In vitro cell fusion has historically been a valuable tool in cell membrane research, genetic mapping, and genetic engineering (DNA-transfection, etc.). Zimmermann (1986) provides a review of the classical electrical fusion techniques for joining two or more cell membranes together. The cells to be joined are aligned and brought into intimate contact with one another by a relatively low intensity inhomogenous alternating electric field applied between two electrodes in solution. Once in position, a field pulse of high intensity (kV/cm regime) and short duration (ns to µs) is applied across the electrodes causing an electrical breakdown of the cellular membranes. If the magnitude and duration of the pulse is not too large, this breakdown is reversible and "self-healing." The membrane breakdown allows transfer of intracellular material from one cell to another, and in some cases fuses the two membranes together.

Sato, et al. (1990) presented a bulk micromachined device designed to bring two cells together at the base of a pyramidal pit in silicon and then to allow a 1 kV/cm field strength pulse to be applied locally, fusing the cells. The cell fusion system consisted of 1,584 microchambers fabricated by joining two silicon dice, each with an array of pyramidal pits etched all the way through them. The two dice were aligned so that the narrow orifices formed at the tips of the pyramids were brought together. On the lower die, stripe electrodes, spaced 200 μm apart, ran between the rows of orifices, allowing the voltage pulse for fusion to be applied. Lettuce cells, with a mean diameter of 35 μm, were used for the preliminary experiments, and suspended in a 0.5 M sorbitol solution. Each microchamber was loaded with one cell of each of the two types to be fused via a similar pyramidal pit array loaded with the cells using a pressure gradient, aligned with the fusion array, and emptied into it by reversing the applied pressure. Twenty volt pulses of 150 μs duration provided the field strength for fusion, with a maximum fusion rate of 52%.

Lee, et al. (1995) also described a micromachined device for the fusion of biological cells. Two copper microelectrodes were fabricated on a silicon substrate and insulated with a 20 μm thick polyimide layer, which was patterned to form fluidic channels over the electrodes. There were two inlet ports that allow for two cells to be injected into the fusion chamber simultaneously. An AC field (8 V p-p at 1 MHz) was applied across the electrodes to align two cells (by dielectrophoresis, as discussed in Section 6.5.2 above) and to bring them into intimate contact. Attempts to electrofuse the two cells together, however, were unsuccessful.

10.6 DNA AMPLIFICATION (PCR) SYSTEMS

As described in the Chemical and Biological Transducers chapter, the polymerase chain reaction, or PCR, makes use of an enzyme, DNA polymerase, and nucleotide triphosphates (building blocks of DNA) to copy DNA using a single-stranded template. When the "melting" temperature of the DNA is reached, the two strands of the double-helix separate, exposing the bases to the polymerase, which builds new complementary strands for each one thus melted apart. Commercial instruments for PCR use relatively large chambers with a large thermal mass for this temperature cycling, thereby limiting their speed and confining their use to the laboratory (due to their bulk and power requirements). Micromachined PCR reactors, discussed in the Chemical Transducers chapter, have been realized by Northrup, et al. (1993) and Northrup (1995). The focus in this section is fluidic systems, and it is worth mentioning an extension of the basic PCR reactor into a more complete DNA amplification/analysis system.

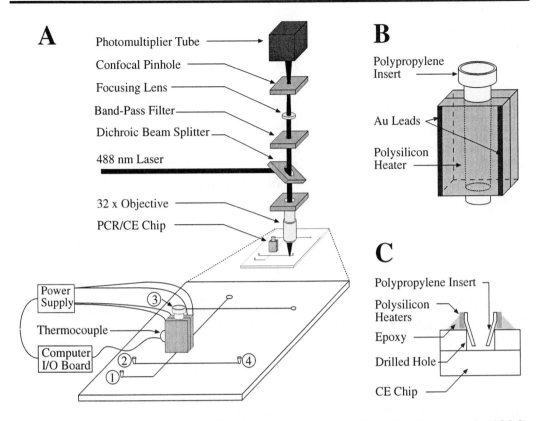

A

Photomultiplier Tube

Confocal Pinhole

Focusing Lens

Band-Pass Filter

Dichroic Beam Splitter

488 nm Laser

32 x Objective

PCR/CE Chip

Power Supply

Thermocouple

Computer I/O Board

③ ② ① ④

B

Polypropylene Insert

Au Leads

Polysilicon Heater

C

Polypropylene Insert

Polysilicon Heaters

Epoxy

Drilled Hole

CE Chip

Illustration of an integrated PCR - CE microdevice. After Woolley, et al. (1996). (A) A laser-excited confocal fluorescence detection system interrogates an integrated PCR - CE device. The injection channel connects reservoirs 1 and 3 while reservoirs 2 and 4 are connected by the separation channel. (B) Expanded view of the micro-PCR chamber. (C) Expanded cross-sectional view of the PCR and CE device junction.

Woolley, et al. (1996) described a microfabricated DNA analysis system that coupled a PCR amplification reactor with a capillary electrophoresis chip. The PCR reaction chamber was fabricated by bonding two identical silicon pieces together using polyimide as an adhesive. Each silicon piece had integrated polysilicon heaters and a groove (formed by KOH etching) extending 85% of the thickness of the wafer. As shown in the figure below, the two silicon pieces were bonded together such that the grooves aligned to form a channel designed to receive a polypropylene reaction chamber insert. This insert extended 1 mm beyond the end of the PCR reaction chamber allowing insertion directly into the capillary electrophoresis chip. The CE chip consisted of an injection channel (50 μm wide, 8 μm deep, and 12 mm long) to which the PCR chamber coupled, crossing a separation channel (100 μm wide, 8 μm deep, and 46 mm long) where the CE was performed. The separation channel was interrogated using laser-excited confocal fluorescence detection to analyze the migrating fluorescently labeled DNA strands. This system allowed for rapid DNA amplification and analysis without any sample handling being required beyond initial sample injection.

Rapid thermal cycling of the PCR chamber was accomplished at a rate of 10 °C/s heating and 2.5 °C/s cooling (actually linear approximations to exponential temperature-versus-time curves), and CE DNA separations were completed in under 120 s. Functionality of the system was demonstrated with two assays. A β-globin target was amplified and analyzed in under 20 min, while an assay for genomic *Salmonella* was performed in under 45 min. In comparison, a commercial system cycling at 1 °C/s would have taken at least six times as long for the PCR amplification (with 45 min for the separation).

10.7 MULTIFUNCTIONAL FLUIDIC SYSTEMS

As commonly used in the micromachining community, the term "lab-on-a-chip" is misleading, since reagents, power sources, and user interfaces seldomly can be scaled as aggressively as the micromachined fluidic and other structures that often make up the "chip," which is only part of any usable system. However, the number and complexity of the functions integrated into microscale systems has increased. In general, however, the more complex fluidic systems have been implemented mainly using conventional technologies such as molded plastics, with selected functions handled by micromachined components integrated therein. In some cases, potentially dictated by sample sizes, the fluidic microsystems contain no components fabricated using micromachining techniques.

An example of an integrated fluidic analytical system was presented by Anderson, et al. (1997). The device contained at its core an optically addressed, ordered combinatorial array of bound nucleotides with which amplified nucleotides from a sample are bound. The known location of a particular nucleotide sequence allows positive identification of a complementary strand from the sample (target strand) by the location at which it binds (hybridizes) with the bound probe (a single strand of DNA that is complementary to the target strand). The fluidic system was implemented in disposable cartridge form by injection molding of polycarbonate or polypropylene, and insertion of the only microfabricated component: the lithographically defined nucleotide array. On-board reaction chambers had volumes between 5 and 20 μl. Externally powered and controlled pneumatic valves were used to control fluid flow, and externally controlled pressures and vacuum were used to move fluids. The system carried out many separate sample preparation and analysis functions, including cell disruption (lysis), nucleotide extraction (e.g., RNA from HIV viruses), mixing of reagents, polymerase chain reaction (amplification of nucleotides), fragmentation of amplified nucleotides, hybridization with the combinatorial array, and washing off of unhybridized nucleotides prior to optical scanning.

This type of disposable, multi-function fluidic module is likely to be employed in a number of analytical and diagnostic settings, and as long as the assays are cost-competitive with existing methods, they may be used in large numbers.

11. BIOLOGICAL FLUIDIC SYSTEMS

There are many fluid flow channels in most living organisms. In mammals, there are fluidic systems for blood, lymphatic fluid, cerebrospinal fluid, lacrimal fluid, saliva, bile, urine, etc. In addition, the gastrointestinal tract handles the flow of fluid, fluid/solid mixtures, and solids. While detailed study of these various systems is beyond the scope of this book, a brief overview is presented. The largest and most varied biological fluidic system is the vascular (circulatory) system, and most examples will be drawn from it.

11.1 BIOLOGICAL FLOW CHANNELS

The most common biological flow channel is the blood vessel, either arterial or venous. Blood is a complex mixture of dissolved ions, organic molecules and particles (cells), functioning to: 1) transport oxygen from the lungs to the rest of the body, waste products (including carbon dioxide) from tissues to excretory organs, nutrients from the digestive systems to other regions, and hormones/enzymes throughout the body; 2) maintain continuous flow despite punctures (via clotting); and 3) help regulate pH, osmolarity, and temperature throughout the body.

Vessel	Diameter (cm)	Total Area (cm^2)	Length (cm)	Wall Thickness (cm)	Average Pressure (mm Hg)	Average Velocity (cm/s)
Aorta	2.5	5	50	0.2	100 ± 20	48
Arteries	0.4	24	50	0.1	90 ± 20	45
Arterioles	0.005	40	1	0.02	60 ± 5	5
Capillaries	0.0008	2,500	0.1	0.0001	20	0.1
Venules	0.002	250	0.2	0.0002	10	0.2
Veins	0.5	100	2.5	0.05	5	1.0
Vena Cava	3.0	7	50	0.15	1	38

Table of approximate fluidic parameters of the human circulatory system. After Mazumdar (1992) and Fay (1994). (Note that due to different geometric categories used by the two sources, approximations were made in combining the information.)

These channels are optimized for the flow of blood, and exhibit many properties currently unachievable via micromachining: self-repair (to some extent), ability to constrict and change their mechanical properties under external (neural/hormonal) control, integrated valves (in veins) with seamless transitions, complete compatibility with biological molecules and cells, etc. In humans, roughly five liters of blood (average 4 to 5 liters for females and 5 to 6 liters for males) is distributed throughout the entire circulatory system. In all but the aorta, the Reynolds number is low enough to guarantee laminar flow. At the level of capillaries, with diameters smaller than 10 μm, there are billions of flow channels, some of which are so small that red blood cells (average diameter 8 μm) must deform to fit through. The table below gives some of the relevant channel diameters and velocities for the human circulatory system, for comparison to micromachined examples presented elsewhere.

11.2 BIOLOGICAL VALVES

Both passive and active valves exist in the bodies of many living organisms. In humans, passive valves (venous and cardiac) are involved in pumping blood, while active valves (sphincters) are involved in gastrointestinal and urinary functions.

11.2.1 PASSIVE BIOLOGICAL VALVES

The most numerous valves are those in the veins, their role being the pumping of blood in the limbs against gravity to return it to the heart for oxygenation. The venous valves are passive in the sense that the energy that drives their opening and closing is not derived from the valve structures themselves, but from surrounding skeletal muscles. As illustrated below, when the muscles contract, blood is pumped toward the heart. These valves form a huge distributed pump throughout the limbs.

In some individuals the venous valves fail in the limbs, leading to a condition referred to as *varicose veins*. Due to a variety of factors, including heredity, aging, pregnancy, etc., the process can lead to grossly enlarged veins due to pressure overloading. Superficial veins are most susceptible since they do not benefit from the mechanical support of surrounding musculature that deeper veins have.

Interestingly, if veins are surgically harvested and used in grafting to replace arteries, their orientation relative to the direction of blood flow must be reversed to prevent them from valving it off and defeating the purpose of the graft(s).

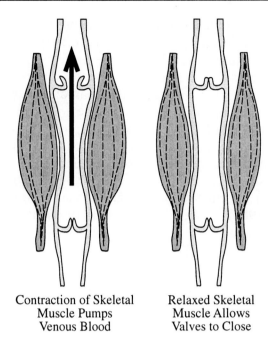

Contraction of Skeletal
Muscle Pumps
Venous Blood

Relaxed Skeletal
Muscle Allows
Valves to Close

Illustration of pumping action of the passive check valves of veins, with actuation provided by external muscles. After Tortora and Anagnostakos (1987).

11.2.2 ACTIVE BIOLOGICAL VALVES

There are several active valves in the body, and these are generally sphincters, such as those found in the digestive and urinary systems. In these valves, smooth (involuntary) and, in some cases, skeletal (consciously controlled) muscles, are arranged in circular patterns around the sphincter such that their contraction generates closing forces.

11.3 BIOLOGICAL PUMPS

The heart is probably the most well-known biological pump, and the basic design is scalable down to extremely small dimensions. In an average person, with a heart rate of $\approx$ 70 beats/min, their heart beats roughly 37 million times per year, or roughly 3 billion times during a lifetime. The heart pumps an average of 5.25 l/min, corresponding to a total pumped volume of roughly *200 billion liters* in a lifetime. While the heart is certainly "macroscopic," it is constructed from an array of microscopic structures (myocardial cells, the specialized muscle cells of the heart) working in synchrony (the microstructure of muscles is discussed in the Mechanical Transducers chapter).

Microscale biological pumps also exist, such as those used by blood-sucking insects (e.g., mosquitoes, fleas, etc.). The *cibarial pump* of such insects is a flexible cavity structure driven by external muscles (Bennet-Clark (1963)). Such pumps allow blood-sucking insects to achieve impressive flow rates through an orifice (the proboscis) that is roughly 10 μm in diameter. For example, the blood-sucking bug, *Rhodnius prolixus*, can consume a 300 mg ($\approx$ 300 μl) blood meal in 15 minutes (Bennet-Clark (1963)), for a pump rate of 20 μl/min (far less than the best micromachined pumps). There is likely to be plenty of information useful to the designer of micromachined pumps that can be gleaned from the study of such natural examples, particularly in light of their very high power efficiencies. Very interesting further information on this subject can be found in Novotny and Wilson (1997), Pappas (1988), Visagie and Schmidt (1994), and Daniel, et al. (1989).

Microfluidic Devices 883

MICROFLUIDIC DEVICES REFERENCES

GENERAL REFERENCES

Blackburn, J. F., Reethof, G., and Shearer, J. L. [eds], "Fluid Power Control," MIT Press/John Wiley and Sons, Cambridge, MA, 1960.

Chereminisof, N. P., "Fluid Flow," Ann Arbor Science (Butterworth Group), Ann Arbor, MI, 1981.

Elwenspoek, M., Lammerink, T. S. J., Miyake, R., and Fluitman, J. H. J., "Towards Integrated Microliquid Handling Systems," Journal of Micromechanics and Microengineering, vol. 4, 1994, pp. 227 - 245.

Fay, J. A., "Introduction to Fluid Mechanics," MIT Press, Cambridge, MA, 1994.

Gravesen, P., Branebjerg, J., and Søndergård Jensen, O. S., "Microfluidics - A Review," Journal of Micromechanics and Microengineering, vol. 3, no. 4, Sept. 1993, pp. 168 - 182.

Halliday, D., Resnick, R., and Walker, J., "Fundamentals of Physics," John Wiley and Sons, New York, NY, 1993.

Incropera, F. P., and DeWitt, D. P., "Fundamentals of Heat and Mass Transfer," John Wiley and Sons, 1990, Chapter 10, "Boiling and Condensation," pp. 588 - 637.

Lucas NovaSensor Pressure Conversion Slide Rule, Lucas NovaSensor, Inc., Fremont, CA, 1994.

Shogi, S., and Esashi, M., "Microflow Devices and Systems," Journal of Micromechanics and Microengineering, vol. 4, 1994, pp. 157 - 171.

Urbanek, W., Zemel, J. N., and Bau, H. H., "An Investigation of the Temperature Dependence of Poiseuille Numbers in Microchannel Flow," Journal of Micromechanics and Microengineering, vol. 3, no. 4, 1993, pp. 206 - 209.

Weast, R. C. [ed.], "CRC Handbook of Chemistry and Physics," CRC Press, Inc., Boca Raton, FL, 1988.

White, F. M., "Viscous Fluid Flow," McGraw-Hill, New York, NY, 1991.

SPECIFIC REFERENCES

Ahn, C. H., Allen, M. G., Trimmer, W., Jun, Y. J., and Erramilli, S., "A Fully Integrated Micromachined Magnetic Particle Separator," Journal of Microelectromechanical Systems, vol. 5, no. 3, Sept. 1996, pp. 151 - 158.

Ahn, C. H., and Allen, M. G., "Fluid Micropumps Based on Rotary Magnetic Actuators," Proceedings of the 1995 IEEE Micro Electro Mechanical Systems Workshop (MEMS '95), Amsterdam, Netherlands, Jan. 29 - Feb. 2, 1995, pp. 408 - 412.

Albin, M., Kowallis, R., Picozza, E., Raysberg, Y., Sloan, C., Winn-Deen, E., Woudenberg, T., and Zupfer, J., "Micromachining and Microgenetics: What Are They and Where Do They Work Together?," Technical Digest, Solid-State Sensor and Actuator Workshop, Hilton Head Island, SC, June 3 - 6, 1996, pp. 253 - 257.

Allen, R. R., Meyer, J. D., and Knight, W. R., "Thermodynamics and Hydrodynamics of Thermal Ink Jets," Hewlett-Packard Journal, vol. 36, no. 5, May 1985, pp. 21 - 27.

Altendorf, E., Zebert, D., Holl, M., and Yager, P., "Differential Blood Cell Counts Using a Micromachined Flow Cytometer," Proceedings of Transducers '97, the 1997 International Conference on Solid-State Sensors and Actuators, Chicago, IL, June 16 - 19, 1997, vol. 1, pp. 531 - 534.

Anderson, R. C., Bogdan, G. J., Barniv, Z., Dawes, T. D., Winkler, J., and Roy, K., "Microfluidic Biochemical Analysis System," Proceedings of Transducers '97, the 1997 International Conference on Solid-State Sensors and Actuators, Chicago, IL, June 16 - 19, 1997, vol. 1, pp. 477 - 480.

Andersson, P. E., Li, P. C. H., Smith, R., Szarka, R. J., and Harrison, D. J., "Biological Cell Assays on an Electrokinetic Microchip," Proceedings of Transducers '97, the 1997 International Conference on Solid-State Sensors and Actuators, Chicago, IL, June 16 - 19, 1997, vol. 2, pp. 1311 - 1314.

Arkilic, E. B., Schmidt, M. A., and Breuer, K. S., "Gaseous Flow in Microchannels," Proceedings of the Workshop on Application of Microfabrication to Fluid Mechanics, ASME Winter Annual Meeting, Chicago, IL, Nov. 1994, pp. 57 - 65.

Arkilic, E. B., Schmidt, M. A., and Breuer, K. S., "Gaseous Slip Flow in Long Microchannels," Journal of Microelectromechanical Systems, vol. 6, no. 2, June 1997, pp. 167 - 178.

Austin, R. H., Volkmuth, W. D., and Rathbun, L. C., "Microlithographic Array for Macromolecule and Cell Fractionation," U.S. Patent No. 5,427,663, issued June 27, 1995.

Bart, S. F., Tavrow, L. S., Mehregany, M., and Lang, J. H., "Microfabricated Electrohydrodynamic Pumps," Sensors and Actuators , vol. A21, nos. 1 - 3, Feb. 1990, pp. 193 - 197.

Barth, P. W., "Silicon Microvalves for Gas Flow Control," Proceedings of Transducers '95, the 8th International Conference on Solid-State Sensors and Actuators, and Eurosensors IX, Stockholm, Sweden, June 25 - 29, 1995, vol. 2, pp. 276 - 277.

Barth, P. W., and Gordon, G. B., "High Performance Micromachined Valve Orifice and Seat," U.S. Patent No. 5,333, 381, issued Aug. 2, 1994.

Barth, P. W., Beatty, C. C., Field, L. A, Baker, J. W., and Gordon, G. B., "A Robust, Normally-Closed Silicon Microvalve," Technical Digest, Solid-State Sensor and Actuator Workshop, Hilton Head Island, SC, June 13 - 16, 1994, pp. 248 - 250.

Bassous, E., "Nozzles Formed in Mono-Crystalline Silicon," U.S. Patent No. 3,921,916, issued Nov. 25, 1975.

Bassous, E., and Baran, E. F., "The Fabrication of High Precision Nozzles by the Anisotropic Etching of (100) Silicon," Journal of the Electrochemical Society, vol. 125, 1978, p. 1321.

Bassous, E., Kuhn, L., Reisman, A., and Taub, H. H., "Ink Jet Nozzle," U.S. Patent No. 4,007,464, issued Feb. 8, 1977.

Benard, W. L., Kahn, H., Heuer, A. H., and Huff, M. A., "A Titanium-Nickel Shape-Memory Alloy Actuated Micropump," Proceedings of Transducers '97, the 1997 International Conference on Solid-State Sensors and Actuators, Chicago, IL, June 16 - 19, 1997, vol. 1, pp. 361 - 364.

Beni, G., Hackwood, S., Belinski, S., Shirazi, M., Li, S., and Karrupiah, L., "Vacuum Mechatronics," Artech House, Boston, MA, 1990.

Bennet-Clark, H. C., "Negative Pressures Produced in the Pharyngeal Pump of the Blood Sucking Bug Rhodineus Prolixus," Journal of Experimental Biology, vol. 40, 1963, pp. 223 - 229.

Berghof, W., Hagemeyer, F., Neubauer, L., and Wehl, W., "Membrane Structure with Buried Heating Elements for Backshooter Inkjet Printheads," German Patent Application No. 42 14 554, 1992.

Bergveld, P., "The Challenge of Developing μTAS," van den Berg, A., and Bergveld, P. [eds.], Proceedings of Micro Total Analysis Systems Conference, Twente, Netherlands, Nov. 21 - 22, 1994, Kluwer Academic Publishers, Dordrecht, Netherlands, 1995, pp. 1 - 4.

Bhaskar, E. V., and Aden, J. S., "Development of the Thin-Film Structure for the ThinkJet Printhead," Hewlett-Packard Journal, vol. 36, no. 5, May 1985, pp. 27 - 33.

Blankenstein, G., Scampavia, L., Branebjerg, J., Larsen, U. D., and Ruzicka, J., "Flow Switch for Analyte Injection and Cell/Particle Sorting," Proceedings of the 2nd International Symposium on Miniaturized Total Analysis Systems, μTAS '96, Basel Switzerland, Nov. 19 - 22, 1996, pp. 82 - 84.

Bloomstein, T. M., and Ehrlich, D. J., "Laser-Chemical Three-Dimensional Writing for Micro-electromechanics and Application to Standard-Cell Microfluidics," Journal of Vacuum Science and Technology B, vol. 10, no. 6, Nov./Dec. 1992, pp. 2671 - 2674.

Bloomstein, T. M., and Ehrlich, D. J., "Laser-Chemical Three-Dimensional Writing of Multima-terial Structures for Microelectromechanics," Proceedings of the IEEE 1991 Workshop on Micro Electro Mechanical Systems (MEMS '91), Nara, Japan, Jan. 30 - Feb. 2, 1991, pp. 202 - 203.

Bousse, L., Kopf-Sill, A., and Parce, J. W., "An Electrophoretic Serial to Parallel Converter," Proceedings of Transducers '97, the 1997 International Conference on Solid-State Sensors and Actuators, Chicago, IL, June 16 - 19, 1997, vol. 1, pp. 499 - 502.

Brand, O., English, J. M., Bidstrup, S. A., and Allen, M. G., "Micromachined Viscosity Sensor for Real-Time Polymerization Monitoring," Proceedings of Transducers '97, the 1997 International Conference on Solid-State Sensors and Actuators, Chicago, IL, June 16 - 19, 1997, vol. 1, pp. 121 - 124.

Branebjerg, J., Fabius, B., and Gravesen, P., "Application of Miniature Analyzers from Microfluidic Components to μTAS," van den Berg, A., and Bergveld, P. [eds.], Proceedings of Micro Total Analysis Systems Conference, Twente, Netherlands, Nov. 21 - 22, 1994, pp. 141 - 151.

Brenebjerg, J., Gravesen, P., Krog, J. P., and Nielsen, C. R., "Fast Mixing by Lamination," Proceedings of the 9th Annual Workshop on Micro Electro Mechanical Systems, San Diego, CA, Feb. 11 - 15, 1996, pp. 441 - 446.

Brody, J. P., and Yager, P., "Diffusion-Based Extraction in a Microfabricated Device," Sensors and Actuators, vol. A58, no. 1, Jan. 1997, pp. 13 - 18.

Brody, J. P., Osborn, T. D., Forster, F. K., and Yager, P., "A Planar Microfabricated Fluid Filter," Proceedings of Transducers '95, the 8th International Conference on Solid-State Sensors and Actuators, and Eurosensors IX, Stockholm, Sweden, June 25 - 29, 1995, vol. 1, pp. 779 - 782.

Bruns, M. W., "Silicon Micromachining and High-Speed Gas Chromatography," Proceedings of the 1992 International Conference on Industrial Electronics, Control, Instrumentation and Automation (IECON '92), San Diego, CA, Nov. 9 - 13, 1992, vol. 3, pp. 1640 - 1644.

Burggraf, N., Manz, N., Verpoorte, E., de Rooij, N. F., and Widmer, H. M., "Synchronized Cyclic Capillary Electrophoresis - A Novel Approach to Ion Separations in Solution," Proceedings of Transducers '93, the 7th International Conference on Solid-State Sensors and Actuators, Yokohama, Japan, June 7 - 10, 1993, Institute of Electrical Engineers, Japan, pp. 399 - 401.

Busch, J. D., and Johnson, A. D., "Prototype Microvalve Actuator," IEEE Micro Electro Mechanical Systems Conference, Napa Valley, CA, Feb. 11 - 14, 1990, IEEE Press, pp. 44 - 41.

Chen, J., and Wise, K. D., "A High-Resolution Silicon Monolithic Nozzle Array for Inkjet Printing," Proceedings of Transducers '95, the 8th International Conference on Solid-State Sensors and Actuators, and Eurosensors IX, Stockholm, Sweden, June 25 - 29, 1995, vol. 2, pp. 321 - 324.

Chen, J., and Wise, K. D., "A Multichannel Neural Probe for Selective Chemical Delivery at the Cellular Level," Proceedings of the Solid-State Sensor and Actuator Workshop, Hilton Head Island, SC, June 13 - 16, 1994, pp. 256 - 259.

Chiem, N., Colyer, C., and Harrison, D. J., "Microfluidic Systems for Clinical Diagnostics," Proceedings of Transducers '97, the 1997 International Conference on Solid-State Sensors and Actuators, Chicago, IL, June 16 - 19, 1997, vol. 1, pp. 183 - 186.

Clark, M. G., Lee, R. A., Lowe, C. R., Maynard, P., Sethi, R. S., and Weir, D. J., "Sensor Devices," U.S. Patent No. 5,194,133, issued Mar. 16, 1993.

Clift, R., Grace, J., and Weber, M., "Bubbles, Drops and Particles," Academic Press, Inc., New York, NY, 1978.

Coanda, H., "Procede et Dispositif Pour Faire Devier, une Veine Fluide Penetrant Autre Fluides," French Patent No. 788,140, issued 1934.

Coe, D. J., Allen, M. G., Smith, B. L., and Glezer, A., "Addressable Micromachined Jet Arrays," Proceedings of Transducers '95, the 8th International Conference on Solid-State Sensors and Actuators, and Eurosensors IX, Stockholm, Sweden, June 25 - 29, 1995, vol. 2, pp. 329 - 332.

Coe, D. J., Allen, M. G., Trautman, M. A., and Glezer, A., "Micromachined Jets for Manipulation of Macro Flows," Proceedings of the Solid-State Sensor and Actuator Workshop, Hilton Head Island, SC, June 13 - 16, 1994, pp. 243 - 247.

Costello, B. J., Wenzel, S. W., and White, R. M., "Density and Viscosity Sensing with Ultrasonic Flexural Plate Waves," Proceedings of Transducers '93, the 7th International Conference on Solid-State Sensors and Actuators, Yokohama, Japan, June 7 - 10, 1993, Institute of Electrical Engineers, Japan, pp. 704 - 707.

Daniel, T. L., Kingsolver, J. G., and Meyhöfer, E., "Mechanical Determinants of Nectar-Feeding Energetics in Butterflies: Muscle Mechanics, Feeding Geometry, and Functional Equivalence," Oecologia, vol. 79, 1989, pp. 66 - 75.

Desai, A., Tai, Y.-C., Davis, M. T., and Lee, T. D., "A MEMS Electrospray Nozzle for Mass Spectroscopy," Proceedings of Transducers '97, the 1997 International Conference on Solid-State Sensors and Actuators, Chicago, IL, June 16 - 19, 1997, vol. 2, pp. 927 - 930.

Despont, M., Lorenz, H., Fahrni, N., Brugger, J., Renaud, P., and Vettiger, P., "High-Aspect-Ratio, Ultrathick, Negative-Tone Near-UV Photoresist for MEMS Applications," Proceedings of the IEEE 10th Annual Workshop of Micro Electro Mechanical Systems (MEMS '97), Nagoya, Japan, Jan. 26 - 30, 1997, pp. 518 - 522.

Dewa, A. S., Deng, K., Ritter, D. C., Bonham, C., Guckel, H., and Massood-Ansari, S., "Development of LIGA-Fabricated, Self-Priming, In-Line Gear Pumps," Proceedings of Transducers '97, the 1997 International Conference on Solid-State Sensors and Actuators, Chicago, IL, June 16 - 19, 1997, vol. 2, pp. 757 - 760.

Döring, C., Grauer, T., Marek, J., Mettner, M., Trah, H. P., and Willman, M., "Micromachined Thermoelectrically Driven Cantilever Beams for Fluid Deflection," Proceedings of the IEEE Micro Electro Mechanical Systems Workshop (MEMS '92), Travemünde, Germany, Feb. 4 - 7, 1992, pp. 12 - 18.

Enoksson, P., Stemme, G., and Stemme, E., "Silicon Tube Structures for a Fluid Density Sensor," Proceedings of Transducers '95, the 8th International Conference on Solid-State Sensors and Actuators, and Eurosensors IX, Stockholm, Sweden, June 25 - 29, 1995, vol. 1, pp. 540 - 543.

Evans, J., Liepmann, D., and Pisano, A. P., "Planar Laminar Mixer," Proceedings of the IEEE 10th Annual Workshop of Micro Electro Mechanical Systems (MEMS '97), Nagoya, Japan, Jan. 26 - 30, 1997, pp. 96 - 101.

Fahrenberg, J., Bier, W., Maas, D., Menz, W., Ruprecht, R., and Schomburg, W. K., "A Microvalve System Fabricated by Thermoplastic Molding," Journal of Micromechanics and Microengineering, vol. 5, no. 2, June 1995, pp. 169 - 171.

Fettinger, J. C., Manz, A., Lüdi, H., and Widmer, H. M., "Stacked Modules for Micro Flow Systems in Chemical Analysis: Concept and Studies Using an Enlarged Model," Sensors and Actuators, vol. B17, no. 1, Nov. 1993, pp. 19 - 25.

Flannery, A. F., Mourlas, N. J., Storment, C. W., Tsai, S., Tan, S. H., and Kovacs, G. T. A., "PECVD Silicon Carbide for Micromachined Transducers," Proceedings of Transducers '97, the 1997 International Conference on Solid-State Sensors and Actuators, Chicago, IL, June 16 - 19, 1997, vol. 1, pp. 217 - 220.

Fluidonics Research Laboratory Staff, "Fluidic Systems Design Guide," Fluidonics Research Laboratory, Salt Lake City, UT, 1966.

Forster, F. K., Bardell, R. L., Afromowitz, M. A., Sharma, N. R., and Blanchard, A., "Design, Fabrication and Testing of Fixed-Valve Micro-Pumps," Proceedings of the 1995 ASME International Mechanical Engineering Congress and Exposition, San Francisco, CA, Nov. 12 - 17, 1995, pp. 39 - 44.

Foster, K., and Parker, G. A., "Fluidics: Components and Circuits," Wiley Interscience, New York, NY, 1970.

Franz, J., Baumann, H., and Trah, H.-P., "A Silicon Microvalve with Integrated Flow Sensor," Proceedings of Transducers '95, the 8th International Conference on Solid-State Sensors and Actuators, and Eurosensors IX, Stockholm, Sweden, June 25 - 29, 1995, vol. 2, pp. 313 - 316.

Fuhr, G., Schnelle, T., and Wagner, B., "Travelling Wave-Driven Microfabricated Electrohydro-dynamic Pumps for Liquids," Journal of Micromechanics and Microengineering, vol. 4, no. 4, Dec. 1994, pp. 217 - 226.

Galambos, P., Forster, F. K., and Weigl, B. H., "A Method for Determination of pH Using a T-Sensor," Proceedings of Transducers '97, the 1997 International Conference on Solid-State Sensors and Actuators, Chicago, IL, June 16 - 19, 1997, vol. 1, pp. 535 - 538.

Gardner, R. C., Giachino, J. M., Horn, W. F., Rhoades, M. K., Wells, M. D., and Yockey, S. J., "Fuel Injector with Silicon Nozzle," U.S. Patent No. 4,907,748, issued Mar. 13, 1990.

Gass, V., van der Schoot, B. H., and de Rooij, N. F., "Nanofluid Handling by Micro-Flow-Sensor Based on Drag Force Measurements," Proceedings of the IEEE MEMS Workshop, Ft. Lauderdale, FL, Feb. 1993, pp. 167 - 172.

Gass, V., van der Schoot, B. H., Jeanneret, S., and de Rooij, N. F., "Integrated Flow-Regulated Silicon Micropump," Proceedings of Transducers '93, the 7th International Conference on Solid-State Sensors and Actuators, Yokohama, Japan, June 7 - 10, 1993, Institute of Electrical Engineers, Japan, pp. 1048 - 1051.

Gebhard, U., Hein, H., Just, E., and Ruther, P., "Combination of a Fluidic Micro-Oscillator and Micro-Actuator in LIGA-Technique for Medical Application," Proceedings of Transducers '97, the 1997 International Conference on Solid-State Sensors and Actuators, Chicago, IL, June 16 - 19, 1997, vol. 2, pp. 761 - 764.

Gerlach, T., "Pumping Gases by a Silicon Micro Pump with Dynamic Passive Valves," Proceedings of Transducers '97, the 1997 International Conference on Solid-State Sensors and Actuators, Chicago, IL, June 16 - 19, 1997, vol. 1, pp. 357 - 360.

Giachino, J. M., and Kress, J. W., "Method for Fabricating a Silicon Valve," U.S. Patent No. 4,628,576, issued Dec. 16, 1986.

Giachino, J. M., and Kress, J. W., "Silicon Valve," U.S. Patent No. 4,756,508, issued July 12, 1988.

Giddings, J. C., "Field Flow Fractionation: Analysis of Macromolecular, Colloidal, and Particlate Materials," Science, vol. 260, 1993, pp. 1456 - 1465.

Grate, J. W., Martin, S. J., and White, R. M., "Acoustic Wave Microsensors," (Parts I and II) Analytical Chemistry, vol. 65, no. 21, Nov. 1, 1993, pp. 940 - 948, 987 - 996.

Greenwood, J. R., "The Design and Development of a Fluid Logic Element," S.B. Thesis, Massachusetts Institute of Technology, 1960.

Guérin, L. J., Bossel, M., Demierre, M., Calmes, S., and Renaud, Ph., "Simple and Low Cost Fabrication of Embedded Microchannels by Using a New Thick-Film Photoplastic," Proceedings of Transducers '97, the 1997 International Conference on Solid-State Sensors and Actuators, Chicago, IL, June 16 - 19, 1997, vol. 2, pp. 1419 - 1421.

Haji-Babaei, J., Kwok, C. Y., and Huang, R. S., "Integrable Active Microvalve with Surface Micromachined Curled-Up Actuator," Proceedings of Transducers '97, the 1997 International Conference on Solid-State Sensors and Actuators, Chicago, IL, June 16 - 19, 1997, vol. 2, pp. 833 - 836.

Harrison, D. J., and Li, P. C. H., "Transport, Manipulation and Reaction of Biological Cells On-Chip Using Electrokinetic Effects," Analytical Chemistry, vol. 69, no. 8, Apr. 15, 1997, pp. 1564 - 1568.

Harrison, D. J., Fan, Z., and Seiler, K., "Integrated Electrophoresis Systems for Biochemical Analyses," Proceedings of the 1994 Solid-State Sensor and Actuator Workshop, Hilton Head Island, SC, June 13 - 16, 1994, pp. 21 - 24.

Harrison, D. J., Fan, Z., Seiler, K., and Flurri, K., "Miniaturized Chemical Analysis Systems Based on Electrophoretic Separations and Electroosmotic Pumping," Proceedings of Transducers '93, the 7th International Conference on Solid-State Sensors and Actuators, Yokohama, Japan, June 7 - 10, 1993, Institute of Electrical Engineers, Japan, pp. 403 - 406.

Harrison, D. J., Glavina, P. G., and Manz, A., "Toward Miniaturized Electrophoresis and Chemical Analysis Systems on Silicon: An Alternative to Chemical Sensors," Sensors and Actuators, vol. B10, no. 2, Jan. 1993, pp. 107 - 116.

Henning, A. K., Fitch, J., Hopkins, D., Lilly, L., Faeth, R., Falsken, E., and Zdeblick, M., "A Thermopneumatically Actuated Microvalve for Liquid Expansion and Proportional Control," Proceedings of Transducers '97, the 1997 International Conference on Solid-State Sensors and Actuators, Chicago, IL, June 16 - 19, 1997, vol. 2, pp. 825 - 828.

Hoffmann, W., and Rapp, R., "Integrated Microanalytical System with Electrochemical Detection," Proceedings of Transducers '95, the 8th International Conference on Solid-State Sensors and Actuators, and Eurosensors IX, Stockholm, Sweden, June 25 - 29, 1995, vol. 2, pp. 955 - 958.

Huang, X., and Zare, R. N., "Improved End-Column Conductivity Detector for Capillary Zone Electrophoresis," Analytical Chemistry, vol. 63, Oct. 1, 1991, pp. 2193 - 2196.

Huang, X., Pang, T. J., Gordon, M. J., and Zare, R. N., "On-Column Conductivity Detector for Capillary Zone Electrophoresis," Analytical Chemistry, vol. 59, no. 23, Dec. 1, 1987, pp. 2747 - 2749.

Huff, M. A., Gilbert, J., and Schmidt, M. A., "Flow Characteristics of a Pressure-Balanced Microvalve," Proceedings of Transducers '93, the 7th International Conference on Solid-State Sensors and Actuators, Yokohama, Japan, June 7 - 10, 1993, pp. 98 - 101.

Huff, M. A., Nikolich, A. D., and Schmidt, M. A., "A Threshold Pressure Switch Utilizing Plastic Deformation of Silicon," Proceedings of Transducers '91, the 1991 International Conference on Solid-State Sensors and Actuators, San Francisco, CA, June 24 - 27, 1991, pp. 177 - 180.

Humphrey, E. F., and Tarumoto, D. H. [eds.], "Fluidics," Fluid Amplifier Associates, Boston, MA, 1965.

Ilzhöfer, A., Ritter, B., and Tsakmakis, Ch., "Development of Passive Microvalves by the Finite Element Method," Journal of Micromechanics and Microengineering, vol. 5, no. 3, Sept. 1995, pp. 226 - 230.

Jacobsen, S. C., "Fluid System and Method for Coding Information," U.S. Patent No. 3,895,641, issued July 22, 1975.

Jacobson, S. C., Hergenröder, R., Koutny, L. B., and Ramsey, J. M., "High-Speed Separations on a Microchip," Analytical Chemistry, vol. 66, no. 7, Apr. 1, 1994, pp. 1114 - 1118.

Jacobson, S. C., Hergenröder, R., Koutny, L. B., Warmack, R. J., and Ramsey, J. M., "Effects of Injection Schemes and Column Geometry on the Performance of Microchip Electrophoresis Devices," Analytical Chemistry, vol. 66, no. 7, Apr. 1, 1994, pp. 1107 - 1113.

Jacobson, S. C., Hergenröder, R., Moore, A. W., and Ramsey, J. M., "Electrically Driven Separations on a Microchip," Proceedings of the 1994 Solid-State Sensor and Actuator Workshop, Hilton Head Island, SC, June 13 - 16, 1994, pp. 65 - 68.

Jacoby, M., "Digital Applications of Fluid Amplifiers," in "Fluidics," Humphrey, E. F., and Tarumoto, D. H. [eds.], Fluid Amplifier Associates, Boston, MA, 1965, pp. 240 - 249.

Jerman, H., "Electrically-Activated, Normally-Closed Diaphragm Valve," Proceedings of Transducers '91, the 1991 International Conference on Solid-State Sensors and Actuators, San Francisco, CA, June 24 - 27, 1991, pp. 1045 - 1048.

Jerman, H., "Electrically-Activated, Normally-Closed Diaphragm Valves," Journal of Micromachining and Microengineering, vol. 4, 1994, pp. 210 - 216.

Jerman, H., "Semiconductor Microactuator," U.S. Patent No. 5,271,597, issued Dec. 3, 1991.

Johnson, D. A., and Ray, C. A., "Shape Memory Alloy Film Actuated Microvalve," U.S. Patent No. 5,325,880, issued Oct. 22, 1991.

Joo, Y., Dieu, K., and Kim, C.-J., "Fabrication of Monolithic Microchannels for IC Chip Cooling," Proceedings of the IEEE 1995 Micro Electro Mechanical Systems Workshop (MEMS '95), Amsterdam, Netherlands, Jan. 29 - Feb. 2, 1995, pp. 362 - 367.

Judy, J. W., Tamagawa, T., and Polla, D. L., "Surface-Machined Micromechanical Membrane Pump," Proceedings of the IEEE 1991 Workshop on Micro Electro Mechanical Systems (MEMS '91), Nara, Japan, Jan. 30 - Feb. 2, 1991, pp. 182 - 186.

Kaartinen, N., "Micro Electro Thermo Fluidic (METF) Liquid Microprocessor," Proceedings of the 9th Annual Workshop on Micro Electro Mechanical Systems (MEMS '96), San Diego, CA, Feb. 11 - 15, 1996, pp. 395 - 399.

Kämper, K.-P., Ehrfeld, W., Döpper, J., Hessel, V., Lehr, H., Löwe, H., Richter, Th., and Wolf, A., "Microfluidic Components for Biological and Chemical Microreactors," Proceedings of the 10th Annual Workshop of Micro Electro Mechanical Systems (MEMS '97), Nagoya, Japan, Jan. 26 - 30, 1997, pp. 338 - 343.

Kirshner, J. M. [ed.], "Fluid Amplifiers," McGraw-Hill, New York, NY, 1966.

Kirshner, J. M., and Katz, S., "Fluid Circuit Theory," Chapter 11 in "Fluid Amplifiers," Kirshner, J. M. [ed], McGraw-Hill, New York, NY, 1966, pp. 146 - 186.

Kittilsland, G., Stemme, G., and Nordén, B., "A Sub-Micron Particle Filter in Silicon," Sensors and Actuators, vol. A23, nos. 1 - 3, Apr. 1990, pp. 904 - 907.

Krause, P., Obermeier, E., and Wehl, W., "Backshooter - A New Smart Micromachined Single-Chip Inkjet Printhead," Proceedings of Transducers '95, the 8th International Conference on Solid-State Sensors and Actuators, and Eurosensors IX, Stockholm, Sweden, June 25 - 29, 1995, vol. 2, pp. 325 - 328.

Krulevitch, P., Microfluidics Section in "MEMS for Medical and Biotechnological Applications," Course Notes, UCLA Extension, Mar. 1995.

Kuhr, W. G., and Monnig, C. A., "Capillary Electrophoresis," Analytical Chemistry, vol. 64, 1992, pp. 389R - 407R.

Kurosawa, M., Watanabe, T., and Higuchi, T., "Surface Acoustic Wave Atomizer with Pumping Effect," Proceedings of the IEEE 1995 Micro Electro Mechanical Systems Workshop (MEMS '95), Amsterdam, Netherlands, Jan. 29 - Feb. 2, 1995, pp. 25 - 30.

Lammerink, T. S. J., Speiring, V. L., Elwenspoek, M., Fluitman, J. H. J., and van den Berg, A., "Modular Concept for Fluid Handling Systems - A Demonstrator Micro Analysis System," Proceedings of the 9th Annual Workshop on Micro Electro Mechanical Systems (MEMS '96), San Diego, CA, Feb. 11 - 15, 1996, pp. 389 - 394.

Lammerink, T. S. J., Tas, N. R., Berenschot, J. W., Elwenspoek, M. C., and Fluitman, J. H. J., "Micromachined Hydraulic Astable Multivibrator," Proceedings of the IEEE 1995 Micro Electro Mechanical Systems Workshop (MEMS '95), Amsterdam, Netherlands, Jan. 29 - Feb. 2, 1995, pp. 13 - 18.

Landers, J. P., Oda, R. P., Spelsberg, T. C., Nolan, J. A., and Ulfelder, K. J., "Capillary Electrophoresis: A Powerful Microanalytical Technique for Biologically Active Molecules," BioTechniques, vol. 14, no. 1, 1993, pp. 98 - 110.

Larsen, U. D., Blankenstein, G., and Branebjerg, J., "Microchip Coulter Particle Counter," Proceedings of Transducers '97, the 1997 International Conference on Solid-State Sensors and Actuators, Chicago, IL, June 16 - 19, 1997, vol. 2, pp. 1319 - 1322.

Larsen, U. D., Branebjerg, J., and Blankenstein, G., "Fast Mixing by Parallel Multilayer Lamination," Proceedings of the 2nd International Symposium on Miniaturized Total Analysis Systems, µTAS '96, Basel Switzerland, Nov. 19 - 22, 1996, pp. 228 - 230.

Lee, S-W, Choi, J-H, Kim, Y-K, "Design of a Biological Cell Fusion Device," Proceedings of Transducers '95, the 8th International Conference on Solid-State Sensors and Actuators, and Eurosensors IX, Stockholm, Sweden, June 25 - 29, 1995, vol. 1, pp. 377 - 380.

Levin, S., and Giddings, J. C., "Continuous Separation of Particles from Macromolecules in Split-Flow Thin (SPLITT) Cells," Journal of Chemical Technology and Biotechnology, vol. 50, 1991, pp. 43 - 56.

Liakopoulos, T. M., Choi, J.-W., and Ahn, C. H., "A Bio-Magnetic Bead Separator on Glass Chips Using Semi-Encapsulated Spiral Electromagnets," Proceedings of Transducers '97, the 1997 International Conference on Solid-State Sensors and Actuators, Chicago, IL, June 16 - 19, 1997, vol. 1, pp. 485 - 488.

Lin, L., Pisano, A. P. and Muller, R. S., "Silicon Processed Microneedles," Proceedings of Transducers '93, the 7th International Conference on Solid-State Sensors and Actuators, Yokohama, Japan, June 7 - 10, 1993, pp. 237-240.

Liu, C., Tsao, T., Tai, Y.-C., Leu, T.-S., Ho, C.-H., Tang. W.-L., and Miu, D., "Out-of-Plane Permalloy Magnetic Actuators for Delta-Wing Control," Proceedings of the IEEE 1995 Micro Electro Mechanical Systems Workshop (MEMS '95), Amsterdam, Netherlands, Jan. 29 - Feb. 2, 1995, pp. 7 - 12.

Man, P. F., Jones, D. K., and Mastrangelo, C. H., "Microfluidic Plastic Capillaries on Silicon Substrates: A New Inexpensive Technology for Bioanalysis Chips," Proceedings of the 10th Annual Workshop of Micro Electro Mechanical Systems (MEMS '97), Nagoya, Japan, Jan. 26 - 30, 1997, pp. 311 - 316.

Manz, A., "Electrophoretic Separating Device and Electrophoretic Separating Method," U.S. Patent No. 5,296,114, issued Mar. 22, 1994.

Manz, A., Effenhauser, C. S., Burggraf, N., Harrison, D. J., Seiler, K., and Flurri, K., "Electroosmotic Pumping and Electrophoretic Separations for Miniaturized Chemical Analysis Systems," Journal of Micromechanics and Microengineering, vol. 4, 1994, pp. 257 - 265.

Manz, A., Graber, N., and Widmer, H. M., "Miniaturized Total Chemical Analysis Systems: A Novel Concept for Chemical Sensing," Sensors and Actuators, vol. B1, nos. 1 - 6, Jan. 1990, pp. 244 - 248.

Manz, A., Miyahara, Y., Miura, J., Watanabe, Y., Miyagi, H., and Sato, K., "Design of an Open-Tubular Column Liquid Chromatograph Using Silicon Chip Technology," Sensors and Actuators, vol. B1, nos. 1 - 6, Jan. 1990, pp. 249 - 255.

Manz, A., Verpoorte, E., Raymond, D. E., Effenhauser, C. S., Burggraf, N., and Widmer, H. M., "μ-TAS: Miniaturized Total Chemical Analysis Systems," in Proceedings of the Micro Total Analysis Systems Conference (μTAS '94), van den Berg, A., and Bergveld, P. [eds.], Twente, Netherlands, Nov. 21 - 22, 1994, Kluwer Academic Publishers, Dordrecht, Netherlands, 1995, pp. 181 - 190.

Martin, B. A., Wenzel, S. W., and White, R. M., "Viscosity and Density Sensing with Ultrasonic Plate Waves," Sensors and Actuators, vol. A22, nos. 1 - 3, Mar. 1990, pp. 704 - 708.

Martin, S. J., Cernosek, R. W., and Spates, J. J., "Sensing Liquid Properties with Shear-Mode Resonator Sensors," Proceedings of Transducers '95, the 8th International Conference on Solid-State Sensors and Actuators, and Eurosensors IX, Stockholm, Sweden, June 25 - 29, 1995, vol. 2, pp. 712 - 715.

Martin, S. J., Frye, G. C., and Wessendorf, K. O., "Sensing Liquid Properties with Thickness-Shear Mode Resonators," Sensors and Actuators, vol. A44, no. 3, Sept. 1994, pp. 209 - 218.

Matsumo, H. and Colgate, J. E., "Preliminary Investigation of Micropumping Based on Electrical Control of Interfacial Tension," Proceedings of the IEEE 1990 Micro Electro Mechanical Systems Workshop (MEMS '90), Napa Valley, CA, 1990, pp. 105 - 110.

Mazumdar, J. N., "Biofluid Mechanics," World Scientific Press, River Edge, NJ, 1992.

Melcher, J. R., "Traveling-Wave Induced Electroconvection," Physics of Fluids, vol. 10, 1966, pp. 1548 - 1555.

Mensinger, H., Richter, T., Hessel, V., Döpper, J., and Ehrfeld, W., "Microreactor with Integrated Static Mixer and Analysis System," van den Berg, A., and Bergveld, P. [eds.], Proceedings of Micro Total Analysis Systems Conference (μTAS '94), Twente, Netherlands, Nov. 21 - 22, 1994, Kluwer Academic Publishers, Dordrecht, Netherlands, 1995, pp. 237 - 243.

Miyake, R., Lammerink, T. S. J., Elwenspoek, M., and Fluitman, J. H. J., "Micro Mixer with Fast Diffusion," Proceedings of the IEEE 1993 Micro Electro Mechanical Systems Workshop (MEMS '93), Ft. Lauderdale, FL, Feb. 2 - 7, 1993, pp. 248 - 253.

Miyake, R., Ohki, H., Yamazaki, I., and Yabe, R., "A Development of Micro Sheath Flow Chamber," Proceedings of the IEEE 1991 Micro Electro Mechanical Systems Workshop (MEMS '91), Nara, Japan, Jan. 30 - Feb. 2, 1991, pp. 265 - 270.

Miyake, R., Tsuzuki, K., Takagi, T., and Imai, K., "A Highly Sensitive and Small Flow-Type Chemical Analysis System with Integrated Absorptiometric Micro-Flow Cell," Proceedings of the 10th Annual Workshop of Micro Electro Mechanical Systems (MEMS '97), Nagoya, Japan, Jan. 26 - 30, 1997, pp. 102 - 107.

Miyazaki, S.-I., Kawai, T., and Aragi, M., "A Piezo-Electric Pump Driven by a Flexural Progressive Wave," Proceedings of the IEEE 1991 Workshop on Micro Electro Mechanical Systems (MEMS '91), Nara, Japan, Jan. 30 - Feb. 2, 1991, pp. 283 - 288.

Möbius, H., Ehrfeld, W., Hessel, V., and Richter, Th., "Sensor Controlled Processes in Chemical Reactors," Proceedings of Transducers '95, the 8th International Conference on Solid-State Sensors and Actuators, and Eurosensors IX, Stockholm, Sweden, June 25 - 29, 1995, vol. 1, pp. 775 - 778.

Moesner, F., and Higuchi, T., "Devices for Particle Handling by an AC Electric Field," Proceedings of the IEEE 1995 Micro Electro Mechanical Systems Workshop (MEMS '95), Amsterdam, Netherlands, Jan. 29 - Feb. 2, 1995, pp. 66 - 71.

Moroney, R. M., White, R. M., and Howe, R. T., "Ultrasonically Induced Microtransport," Proceedings of the IEEE 1991 Micro Electro Mechanical Systems Workshop (MEMS '91), Nara, Japan, Jan. 30 - Feb. 2, 1991, pp. 277 - 282.

Nathanson, H., Liberman, I., and Freidhoff, C., "Novel Functionality Using Micro-Gaseous Devices," Proceedings of the IEEE 1995 Micro Electro Mechanical Systems Workshop (MEMS '95), Amsterdam, Netherlands, Jan. 29 - Feb. 2, 1995, pp. 72 - 76.

Nielsen, N. J., "History of ThinkJet Printhead Development," Hewlett-Packard Journal, vol. 36, no. 5, May 1985, pp. 4 - 20.

Nilsson, J., Laurell, T., Wallman, L., and Drott, J., "A Flow-Through Liquid Picoliter Sampling Cell," Proceedings of the 2nd International Symposium on Miniaturized Total Analysis Systems, µTAS '96, Basel, Switzerland, Nov. 19 - 22, 1996, pp. 88 - 90.

Northrup, M. A., Ching, M. T., White, R. M., and Watson, R. T., "DNA Amplification with a Microfabricated Reaction Chamber," Proceedings of Transducers '93, the 7th International Conference on Solid-State Sensors and Actuators, Yokohama, Japan, June 7 - 10, 1993, Institute of Electrical Engineers, Japan, pp. 924 - 926.

Northrup, M. A., Lawrence Livermore National Laboratory, Livermore, CA, personal communication, 1995.

Novotny, V., and Wilson, M. R., "Why Are There No Small Species among Xylem-Sucking Insects?," Evolutionary Biology, vol. 11, 1997, pp. 419 - 437.

Olsson, A., Enoksson, P., Stemme, G., and Stemme, E., "A Valve-Less Planar Pump in Silicon," Proceedings of Transducers '95, the 8th International Conference on Solid-State Sensors and Actuators, and Eurosensors IX, Stockholm, Sweden, June 25 - 29, 1995, pp. 291 - 294.

Olsson, A., Larsson, O., Holm, J., Lundbladh, L., Öhman, O., and Stemme, G., "Valve-Less Diffuser Micropumps Fabricated Using Thermoplastic Replication," Proceedings of the 10th Annual Workshop of Micro Electro Mechanical Systems (MEMS '97), Nagoya, Japan, Jan. 26 - 30, 1997, pp. 305 - 310.

Olsson, A., Stemme, G., and Stemme, E., "A Valve-Less Fluid Pump with Two Pump Chambers," Sensors and Actuators, vol. A47, nos. 1 - 3, Mar. - Apr. 1995, pp. 549 - 556.

Olsson, A., Stemme, G., and Stemme, E., "Micromachined Diffuser/Nozzle Elements for Valve-Less Pumps," Proceedings of the 9th Annual Workshop on Micro Electro Mechanical Systems (MEMS '96), San Diego, CA, Feb. 11 - 15, 1996, pp. 378 - 383.

Olsson, A., Stemme, G., and Stemme, E., "Simulation Studies of Diffuser and Nozzle Elements for Valve-less Micropumps," Proceedings of Transducers '97, the 1997 International Conference on Solid-State Sensors and Actuators, Chicago, IL, June 16 - 19, 1997, vol. 2, pp. 1039 - 1042.

Olsson, A., Stemme, G., and Stemme, E., "The First Valve-Less Diffuser Gas Pump," Proceedings of the 10th Annual Workshop of Micro Electro Mechanical Systems (MEMS '97), Nagoya, Japan, Jan. 26 - 30, 1997, pp. 108 - 113.

Oosterbroek, R. E., Lammerink, T. S. J., Berenschot, J. W., van den Berg, A., and Elwenspoek, M. C., "Designing, Realization and Characterization of a Novel Capacitive Pressure/Flow Sensor," Proceedings of Transducers '97, the 1997 International Conference on Solid-State Sensors and Actuators, Chicago, IL, June 16 - 19, 1997, vol. 1, pp. 151 - 154.

Pace, S. J., "Silicon Semiconducting Wafer for Analyzing Micronic Biological Samples," U.S. Patent No. 4,908,112, issued Mar. 13, 1990.

Pan, J., Y., VerLee, D., and Mehregany, M., "Latched Valve Manifolds for Efficient Control of Pneumatically Actuated Valve Arrays," Proceedings of Transducers '97, the 1997 International Conference on Solid-State Sensors and Actuators, Chicago, IL, June 16 - 19, 1997, vol. 2, pp. 817 - 820.

Papautsky, I., Frazier, A. B., and Swerdlow, H., "A Low Temperature IC Compatible Process for Fabricating Surface Micromachined Metallic Microchannels," Proceedings of the 10th Annual Workshop of Micro Electro Mechanical Systems (MEMS '97), Nagoya, Japan, Jan. 26 - 30, 1997, pp. 317 - 322.

Pappas, L. G., "Stimulation and Sequence Operation of Cibarial and Pharyngeal Pumps during Sugar Feeding by Mosquitoes (*Diptera: Culicidae*)," Annals of the Entomological Society of America, vol. 81, no. 2, Mar. 1988, pp. 274 - 277.

Petersen, K. E., "Fabrication of an Integrated, Planar Silicon Ink-Jet Structure," IEEE Transactions on Electron Devices, vol. ED-26, no. 12, Dec. 1979, pp. 1918 - 1920.

Pfahler, J., Harley, J., Bau, H., and Zemel, J., "Liquid Transport in Micron and Submicron Channels," Sensors and Actuators, vol. A22, nos. 1 - 3, Mar. 1990, pp. 431 - 434.

Pickard, W., "Progress in Dielectrics," vol. 6, Heywood, London, UK, 1965.

Popescu, D. S., Dascalu, D. C., Elwenspoek, M., and Lammerink, T., "Silicon Active Microvalves Using Buckled Membranes for Actuation," Proceedings of Transducers '95, the 8th International Conference on Solid-State Sensors and Actuators, and Eurosensors IX, Stockholm, Sweden, June 25 - 29, 1995, vol. 2, pp. 305 - 308.

Poplawski, M. E., Hower, R. W., and Brown, R. B., "A Simple Packaging Process for Chemical Sensors," Proceedings of the Solid-State Sensor and Actuator Workshop, Hilton Head Island, SC, June 13 - 16, 1994, pp. 25 - 28.

Reay, R. J., Dadoo, R., Storment, C. W., Zare, R. N., and Kovacs, G. T. A., "Microfabricated Electrochemical Detector for Capillary Electrophoresis," Proceedings of the Solid-State Sensor and Actuator Workshop, Hilton Head Island, SC, June 13 - 16, 1994, pp. 61 - 64.

Richter, A., Plettner, A., Hofmann, K. A., and Sandmaier, H., "A Micromachined Electrohydro-dynamic (EHD) Pump," Sensors and Actuators, vol. A29, no. 2, Nov. 1991, pp. 159 - 168.

Richter, A., Plettner, A., Hofmann, K. A., and Sandmaier, H., "Electrohydrodynamic Pumping and Flow Measurement," Proceedings of the IEEE 1991 Workshop on Micro Electro Mechanical Systems (MEMS '91), Nara, Japan, Jan. 30 - Feb. 2, 1991, pp. 271 - 276.

Sato, K., Kawamura, Y., Tanaka, S., Uchida, K., and Kohida, H., "Individual and Mass Operation of Biological Cells Using Micromechanical Silicon Devices," Sensors and Actuators, vol. A23, nos. 1 - 3, Apr. 1990, pp. 948 - 953.

Schomberg, W. K., Büstgens, B., Fahrenberg, J., and Maas, D., "Components for Microfluidic Handling Modules," van den Berg, A., and Bergveld, P. [eds.], Proceedings of Micro Total Analysis Systems Conference (μTAS '94), Twente, Netherlands, Nov. 21 - 22, 1994, pp. 255 - 258.

Sethi, R. S., Brettle, J., and Lowe, C., "Chromatographic Separation Device," UK Patent No. 2,191,110A, issued Dec. 9, 1987.

Sethi, R. S., Brettle, J., and Lowe, C., "Chromatographic Separation Device," U.S. Patent No. 4,891,120, issued Jan. 2, 1990.

Shafer, N. E., and Zare, R. N., "Through a Beer Glass Darkly," Physics Today, Oct. 1991, pp. 48 - 52.

Shoji, S., and Esashi, M., "Microflow Devices and Systems," Journal of Micromechanics and Microengineering, vol. 4, no. 4, Dec. 1994, pp. 157 - 171.

Shoji, S., van der Schoot, B., de Rooij, N., and Esashi, M., "Smallest Dead Volume Microvalves for Integrated Chemical Analyzing Systems," Proceedings of Transducers '91, the 1991 International Conference on Solid-State Sensors and Actuators, San Francisco, CA, June 24 - 27, 1991, pp. 1052 - 1055.

Siewell, G. L., Boucher, W. R., and McClelland, P. H., "The ThinkJet Orifice Plate: A Part with Many Functions," Hewlett-Packard Journal, May 1985, vol. 36, no. 5, pp. 33 - 37.

Skrobanek, K. D., Kohl, M., and Miyazaki, S., "Stress-Optimized Shape Memory Microvalves," Proceedings of the 10th Annual Workshop of Micro Electro Mechanical Systems (MEMS '97), Nagoya, Japan, Jan. 26 - 30, 1997, pp. 256 - 261.

Smith, L., and Hök, B., "A Silicon Self-Aligned Non-Reverse Valve," Proceedings of Transducers '91, the 1991 International Conference on Solid-State Sensors and Actuators, San Francisco, CA, June 24 - 27, 1991, pp. 1049 - 1051.

Smith, L., Söderbärg, A., and Björkengren, U., "Continuous Ink-Jet Print Head Utilizing Silicon Micromachined Nozzles," Sensors and Actuators, vol. A43, nos. 1 - 3, May 1994, pp. 311 - 316.

Sobek, D., Senturia, S. D., and Gray, M. L., "Microfabricated Fused Silica Flow Chambers for Flow Cytometry," Proceedings of the Solid-State Sensor and Actuator Workshop, Hilton Head Island, SC, June 13 - 16, 1994, pp. 260 - 263.

Spiering, V. L., van der Moolen, J. N., Burger, G.-J., and van den Berg, A., "Novel Microstructures and Technologies Applied in Chemical Analysis Techniques," Proceedings of Transducers '97, the 1997 International Conference on Solid-State Sensors and Actuators, Chicago, IL, June 16 - 19, 1997, vol. 1, pp. 511 - 514.

Stemme, E., and Stemme, G., "Valveless Diffuser/Nozzle-Based Fluid Pump," Sensors and Actuators, vol. A39, no. 2, Nov. 1993. pp. 159 - 167.

Steutzer, O. M., "Ion Drag Pumps," U.S. Patent No. 3,398,685, issued 1968.

Svedin, N., Stemme, E., and Stemme, G., "A New Bi-Directional Gas-Flow Sensor Based on Lift Force," Proceedings of Transducers '97, the 1997 International Conference on Solid-State Sensors and Actuators, Chicago, IL, June 16 - 19, 1997, vol. 1, pp. 145 - 148.

Tang, T., Ocvirk, G., Harrison, D. J., "Iso-Thermal DNA Reactions and Assays in Microfabricated Capillary Electrophoresis Systems," Proceedings of Transducers '97, the 1997 International Conference on Solid-State Sensors and Actuators, Chicago, IL, June 16 - 19, 1997, vol. 1, pp. 523 - 526.

Terry, S. C., "A Gas Chromatography System Fabricated on a Single Silicon Wafer Using Integrated Circuit Technology," Doctoral Dissertation, Stanford ICL Technical Report #4603-1, May 1975.

Terry, S. C., Jerman, J. H., and Angell, J. B., "A Gas Chromatographic Air Analyzer Fabricated on a Silicon Wafer," IEEE Transactions on Electron Devices, vol. ED-26, no. 12, Dec. 1979, pp. 1880 - 1886.

Tesla, N., "Valvular Conduit," U.S. Patent No. 1,329,559, issued Feb. 1920.

Tiren, J., Tenerz, L., and Hök, B., "A Batch-Fabricated Non-Reverse Valve with Cantilever Beam Manufactured by Micromachining of Silicon," Sensors and Actuators, vol. 18, nos. 3 - 4, July 1989, pp. 398 - 396.

Tjerkstra, R. W., de Boer, M., Berenschot, E., Gardeniers, J. G. E., van den Berg, A., and Elwenspoek, M., "Etching Technology for Microchannels," Proceedings of the 10th Annual Workshop of Micro Electro Mechanical Systems (MEMS '97), Nagoya, Japan, Jan. 26 - 30, 1997, pp. 147 - 151.

Tortora, G. J., and Anagnostakos, N. P., "Principles of Anatomy and Physiology," Harper and Row Publishers, New York, NY, 1987.

Trah, H.-P., Baumann, H., Döring, C., Goebel, H., Grauer, T., and Mettner, M., "Micromachined Valve with Hydraulically Actuated Membrane Subsequent to a Thermoelectrically Controlled Bimorph Cantilever," Sensors and Actutors, vol. A39, no. 2, Nov. 1993, pp. 169 - 176.

Ulrich, J., Füller, H., and Zengerle, R., "Static and Dynamic Flow Simulation of a KOH-Etched Micro Valve," Proceedings of Transducers '95, the 8th International Conference on Solid-State Sensors and Actuators, and Eurosensors IX, Stockholm, Sweden, June 25 - 29, 1995, 1995, vol. 2, pp. 17 - 20.

van der Schoot, B. H., Verpoorte, E. M. J., Jeanneret, S., Manz, A., and de Rooij, N. F., "Microsystems for Analysis in Flowing Solutions," in Proceedings of the Micro Total Analysis Systems Conference (μTAS '94), van den Berg, A., and Bergveld, P. [eds.], Twente, Netherlands, Nov. 21 - 22, 1994, Kluwer Academic Publishers, Dordrecht, Netherlands, 1995, pp. 181 - 190.

van Rijn, C. J. M., and Elwenspoek, M. C., "Micro Filtration Membrane Sieve with Silicon Micromachining for Industrial and Biomedical Applications," Proceedings of the IEEE 1995 Micro Electro Mechanical Systems Workshop (MEMS '95), Amsterdam, Netherlands, Jan. 29 - Feb. 2, 1995, pp. 83 - 87.

van Rijn, C., van der Wekken, M., Nijdam, W., and Elwenspoek, M., "Deflection and Maximum Load of Microfiltration Membrane Sieves Made with Silicon Micromachining," Journal of Microelectromechanical Systems," vol. 6, no. 1, Mar. 1997, pp. 48 - 54.

VerLee, D., Alcock, A., Clark, G., Huang, T. M., Kantor, S., Nemcek, T., Norlie, J., Pan, J., Walsworth, F., and Wong, S. T., "Fluid Circuit Technology: Integrated Interconnect Technology for Miniature Fluidic Devices," Technical Digest, Solid-State Sensor and Actuator Workshop, Hilton Head Island, SC, June 3 - 6, 1996, pp. 9 - 14.

Verpoorte, E. M. J., van der Schoot, B. H., Jeanneret, S., Manz, A., Widmer, H. M., and de Rooij, N. F., "Three-Dimensional Micro Flow Manifolds for Miniaturized Chemical Analysis Sytems," Journal of Micromechanics and Microengineering, vol. 4, no. 4, Dec. 1994, pp. 246 - 256.

Vieider, C., Öhman, O, and Elderstig, H., "A Pneumatically Actuated Micro Valve with a Silicone Rubber Membrane for Integration with Fluid-Handling Systems," Proceedings of Transducers '95, the 8th International Conference on Solid-State Sensors and Actuators, and Eurosensors IX, Stockholm, Sweden, June 25 - 29, 1995, vol. 2, pp. 284 - 286.

Visagie, E. J., and Schmidt, R., "Structure and Function of the Frontal Pit of Lipoptena spp. (*Diptera: Hippoboscidae*)," Journal of Morphology, vol. 221, 1994, pp. 133 - 138.

Volkmuth, W. D., and Austin, R. H., "DNA Electrophoresis in Microlithographic Arrays," Nature, vol. 358, Aug. 13, 1992, pp. 600 - 602.

Vollmer, J., Hein, H., Menz, W., and Walter, F., "Bistable Fluidic Elements in LIGA-Technique as Microactuators," Proceedings of Transducers '93, the 7th International Conference on Solid-State Sensors and Actuators, Yokohama, Japan, June 7 - 10, 1993, pp. 116 - 119.

Wagner, B., Quenzer, H. J., Hoerschelmann, S., Lisec, T., and Juerss, M., "Bistable Microvalve with Pneumatically Coupled Membranes," Proceedings of the 9th Annual Workshop on Micro Electro Mechanical Systems (MEMS '96), San Diego, CA, Feb. 11 - 15, 1996, pp. 384 - 388.

Wallman, L., Drott, J., Nilsson, J., and Laurell, T., "A Micromachined Flow-Through Cell for Continuous Pico-Volume Sampling in an Analytical Flow," Proceedings of Transducers '95, the 8th International Conference on Solid-State Sensors and Actuators, and Eurosensors IX, Stockholm, Sweden, June 25 - 29, 1995, vol. 2, pp. 303 - 304.

Walters, K., and Jones, W. M., "Measurement of Viscosity," Chapter 2, Part 1, in "Instrumentation Reference Book," Noltingk, B. E. [ed.], Butterworth-Heinemann Ltd., Oxford, UK, 1995, pp. 41 - 48.

Webster, J. R., and Mastrangelo, C. H., "Large-Volume Integrated Capillary Electrophoresis Stage Fabricated Using Micromachining of Plastics on Silicon Substrates," Proceedings of Transducers '97, the 1997 International Conference on Solid-State Sensors and Actuators, Chicago, IL, June 16 - 19, 1997, vol. 1, pp. 503 - 506.

Weigl, B. H., Holl, M. R., Schulte, D., Brody, J. P., and Yager, P., "Diffusion-Based Optical Chemical Detection in Silicon Flow Structures," Proceedings of the 2nd International Symposium on Miniaturized Total Analysis Systems (μTAS '96), Basel, Switzerland, Nov. 19 - 22, 1996, pp. 174 - 184.

Westberg, D., Paul, O., Andersson, G. I., and Baltes, H., "A CMOS-Compatible Devices for Fluid Density Measurements," Proceedings of the IEEE 10th Annual Workshop of Micro Electro Mechanical Systems (MEMS '97), Nagoya, Japan, Jan. 26 - 30, 1997, pp. 278 - 283.

Woolley, A. T., and Mathies, R. A., "Ultra-High-Speed DNA Fragment Separations Using Micro-fabricated Capillary Array Electrophoresis Chips," Proceedings of the National Academy of Science, vol. 91, Nov. 1994, pp. 11348 - 11352.

Woolley, A. T., and Mathies, R. A., "Ultra-High-Speed DNA Sequencing Using Capillary Elec-trophoresis Chips," Analytical Chemistry, vol. 67, 1995, pp. 3676 - 3680.

Woolley, A. T., Hadley, D., Landre, P., deMello, A. J., Mathies, R. A., and Northrup, M. A., "Functional Integration of PCR Amplification and Capillary Electrophoresis in a Microfabricated DNA Analysis Device," Analytical Chemistry, vol. 68, no. 23, Dec. 1996, pp. 4081 - 4086.

Yanagisawa, K., Kuwano, H., and Tago, A., "An Electromagnetically Driven Microvalve," Pro-ceedings of Transducers '93, the 7th International Conference on Solid-State Sensors and Actuators, Yokohama, Japan, June 7 - 10, 1993, pp. 102 - 105.

Yang, X., Grosjean, C., Tai., Y.-C., and Ho, C.-M., "A MEMS Thermopneumatic Silicone Membrane Valve," Proceedings of the IEEE 10th Annual Workshop of Micro Electro Mechanical Systems (MEMS '97), Nagoya, Japan, Jan. 26 - 30, 1997, pp. 114 - 118.

Zdeblick, M. J., and Angell, J. B., "A Microminiature Electric-to-Fluidic Valve," Proceedings of Transducers '87, the 4th International Conference on Solid-State Transducers and Actuators, Tokyo, Japan, June 2 - 6, 1987, pp. 827 - 829.

Zdeblick, M. J., "A Planar Process for an Electric-to-Fluidic Valve," Doctoral Dissertation, Stanford University, Stanford, CA, June 1988.

Zdeblick, M. J., "Integrated, Microminiature Electric to Fluidic Valve," U.S. Patent No. 4,824,073, issued Apr. 18, 1989.

Zdeblick, M. J., Anderson, R., Jankowski, J., Kline-Schoder, B., Christel, L., Miles, R., and Weber, W., "Thermpneumatically Actuated Microvalves and Integrated Electro-Fluidic Circuits," Proceedings of the Solid-State Sensor and Actuator Workshop, Hilton Head Island, SC, June 13 - 16, 1994, pp. 251 - 255.

Zdeblick, M., J., and Angell, J. B., "Microminiature Fluidic Amplifier," Sensors and Actuators, vol. 15, no. 4, Dec. 1988, pp. 427 - 433.

Zengerle, R., and Richter, M., "Simulation of Microfluid Systems," Journal of Micromechanics and Microengineering, vol. 4, no. 4, Dec. 1994, pp. 192 - 204.

Zengerle, R., Kluge, S., Richter, M., and Richter, A., "A Bidirectional Silicon Micropump," Proceedings of the IEEE 1995 Micro Electro Mechanical Systems Workshop (MEMS '95), Amsterdam, Netherlands, Jan. 29 - Feb. 2, 1995, IEEE, pp. 19 - 24.

Zengerle, R., Leitner, M., Kluge, S., and Richter, A., "Carbon Dioxide Priming of Micro Liquid Systems," Proceedings of the IEEE 1995 Micro Electro Mechanical Systems Workshop (MEMS '95), Amsterdam, Netherlands, Jan. 29 - Feb. 2, 1995, IEEE, pp. 340 - 343.

Zimmermann, U., "Electrical Breakdown, Electropermeabilization and Electrofusion," Review of Physiology, Biochemistry and Pharmacology, vol. 105, 1986, pp. 175 - 256.

CITED INDUSTRY REFERENCES AND SUPPLIERS

Non-Micromachined Microvalves

Electro-Fluidic Systems Technical Handbook, 6th Edition, The Lee Company, 2 Pettipaug Road, P.O. Box 424, Westbrook, CT, 06498-0424, Phone: (800) 533-7584

Thermally Actuated Valves

Redwood MicroSystems, Inc., 959 Hamilton Avenue, Menlo Park, CA 94025, Phone: (415) 617-1200, http://www.redwoodmicro.com

EG&G IC Sensors, 1701 McCarthy Boulevard, Milpitas, CA 95035, Phone: (408) 432-1800, Fax: (408) 434-6687

INDEX

904

908

EPILOGUE

Prof. Henry Baltes, ETH Zurich

In his *Critique of Pure Reason* Immanuel Kant (1724 - 1804) condensed his quest for wisdom as follows: All the interests in my reason, speculative as well as practical, combine in the three following questions: "What can I know? What ought I to do? What may I hope?" After reading the *Micromachined Transducer Sourcebook*, similar questions came to my mind: 1. What did we learn? 2. Where do we go from here? 3. Where will the technology go?

1. What did we learn? Before we get carried away by the amazing micromachining techniques and microtransducers, we learn in Chapter 1 that they are not a marketable end in themselves, but are a powerful enabling technology. These techniques and devices can enable and add value to systems that constitute the so-called MEMS market.

The *Sourcebook* illustrates how microtransducers, while relying on silicon technology, are quite different from silicon IC devices. Electronic microtransducers involve a wealth of physical effects beyond the closed world of electrical effects basic to IC devices. Designing microtransducers requires in-depth knowledge of the physics, chemistry, or biology of the targeted non-electrical sensor input or actuator output signal area. To be successful, a fluidic microtransducer designer must know fluid mechanics and the specific application field, in addition to silicon device design, circuit design, and fabrication technology. And the successful designer of microtransducers for biological applications might, like the author of the *Sourcebook*, need to combine backgrounds in both electrical engineering and medicine or biology. Thanks to the author's expertise in the latter, the book enriched my knowledge of biological micro-actuator action. Packaging of microtransducers, in contrast to conventional IC packaging is another area where differences are obvious.

We learn that silicon technology is very powerful and that we should stick to it as far as possible. Even as we narrow the microtransducer approach to silicon technology and its legitimate and illegitimate offspring, there is still a lot of room for creativity. These and further answers to my first question come directly from the *Sourcebook* itself, whereas further thought and analysis are required for the other two questions.

2. Where do we go from here? While the book teaches a wealth of astute techniques and devices, in future development work we have to be very selective. We must choose the few useful from the many technically appealing approaches, yielding practical devices rather than research toys.

One of my corporate project partners put it like this: "If the simple and stupid device works reliably within the cost limit, then it ain't stupid, but the smart choice." As I learned from my industrial collaboration projects, the fabrication cost limit for a silicon microtransducer chip (including circuits) is below rather than above one dollar, most of the time.

For the future research of my laboratory, the book encourages me to put more emphasis on microtransducer packaging and to push our development of microtransducer CAD tools and related measurements of process-dependent, non-electrical material properties (Baltes (1996)). We should develop test structures to characterize non-electrical materials parameters relevant for microtransducer performance and routinely place them on microtransducer production wafers (analogous to the electrical test structures found on most IC wafers). The future potential for joint university-industry microtransducer research is, in my opinion, overwhelming.

3. Where will the technology go? According to Edmund Burke (1791) "you can never plan the future by the past," whereas Lord Byron (1821) believed that "the best prophet of the future is the past" (Kipfer (1994)). While Burke may have a point when it comes to predicting world history in general, Byron's view is not out of line in the case of extrapolating the development of a defined technical area in a not too remote future. Here are some expectations (hopes?) for the future of micromachined transducers.

The extensive use of the pressure sensors in vehicle emission control systems results from two factors (Frank (1996)). Government legislation forced a change in automotive engine control. Semiconductor sensors proved to be more cost-effective and more reliable than previous mechanical devices. Continued concern about environmental protection, safety, and health on the one hand, and the proliferation of microcontrollers requiring sensor inputs and providing improved control functions on the other hand, will enhance the demand for microtransducers (Baltes (1996)).

Predicted semiconductor capabilities (0.07 μm minimum feature size by the year 2010) mean that the technology that makes the microtransducer smart is increasing in capability at a phenomenal rate (Frank (1996), Sasaki (1996)). By the year 2010 we probably will have not only the pressure sensor with integrated microcontroller, the HVAC sensor and the viscometer on a chip, but even the taste sensor, the "micronose" odor sensor, and the atomic force microscope on a chip. Besides legislation, the timing depends, in each case, on when the technology can yield a significant cost reduction edge.

Henry Baltes
Zurich, August 1997

EPILOGUE REFERENCES

Baltes, H., "Future of IC Microtransducers," Sensors and Actuators, vol. A56, nos. 1 - 2, Aug. 1996, pp. 179 - 192.

Kipfer, B. A. [ed.], "Bartlett's Book of Business Quotations," Little, Brown and Co., Boston, 1994.

Frank, R., "Understanding Smart Sensors," Artech House, Boston, MA, 1996.

Sasaki, H., "Multimedia Complex on a Chip," Technical Digest of the 1996 IEEE International Solid-State Circuit Conference, ISSCC, San Francisco, CA, Feb. 8 - 10, 1996, pp. 16 - 19.